Springer Proceedings in Physics 139

Springer Proceedings in Physics

For further volumes:
http://www.springer.com/series/361

Jiří Náprstek • Jaromír Horáček • Miloslav Okrouhlík
Bohdana Marvalová • Ferdinand Verhulst
Jerzy T. Sawicki
Editors

Vibration Problems ICOVP 2011

The 10th International Conference on Vibration Problems

Springer

Editors
Jiří Náprstek
Republic Institute of Theoretical
and Applied Mechanics
Academy of Sciences of the
Czech Republic
Prosecka 76
190 00 Prague 9
Czech Republic
naprstek@itam.cas.cz

Miloslav Okrouhlík
Institute of Thermomechanics
Academy of Sciences of the
Czech Republic
Dolejskova 1402/5
18200 Prague
Czech Republic
ok@it.cas.cz

Ferdinand Verhulst
Mathematisch Instituut
University of Utrecht
Budapestlaan 6
3584 CD Utrecht
Netherlands
f.verhulst@uu.nl

Jaromír Horáček
Institute of Thermomechanics
Academy of Sciences of the
Czech Republic
Dolejskova 1402/5
18200 Prague
Czech Republic
jaromirh@it.cas.cz

Bohdana Marvalová
Department of Engineering Mechanics
Technical University of Liberec
Studentska 2
46117 Liberec
Czech Republic
bohda.marvalova@tul.cz

Jerzy T. Sawicki
Center for Rotating Machinery
Cleveland State University
Dynamics a
Euclid Avenue 2121SH 248
44115-2214 Cleveland Ohio
USA
J.SAWICKI@csuohio.edu

ISSN 0930-8989 e-ISSN 1867-4941
ISBN 978-94-007-2068-8 e-ISBN 978-94-007-2069-5
DOI 10.1007/978-94-007-2069-5
Springer Dordrecht Heidelberg London New York

Library of Congress Control Number: 2011936970

Printed on acid-free paper

Springer is part of Springer Science+Business Media (www.springer.com)

International Federation for the Promotion of Mechanism and Machine Science

International Conference on Vibration Problems

Technical University of Liberec Czech Republic

Preface

The tenth International Conference on Vibration Problems (ICOVP 2011) in Prague represents a continuation of a tradition which started at the A. C. College, Jalpaiguri, India in 1990. We well remember the further fruitful meetings in Jalpaiguri (1993), then at the University of North Bengal (1996) and at Jadavpur University in West Bengal (1999). This successful series continued in Moscow (2001), then at the Technical University of Liberec, Czech Republic (2003) and in Istanbul (2005). The last two assemblies of experts in dynamics took place at Bengal Engineering and Science University, Shibpur (2007) and at the Indian Institute of Technology, Kharagpur (2009). The ICOVP Conferences have a long tradition, and attract a large number of highly qualified participants. This, together with the top-quality papers presented and excellent organization, have established them as a high-level forum where engineers, researchers, university teachers, students and other professionals can present recent developments and discuss the scientific, technical and experimental results and ideas in various areas of rational, experimental and applied Dynamics.

Dynamics as a scientific discipline draws inspiration from a large variety of engineering areas, such as Mechanical and Civil Engineering, Aero and Space Technology, Wind and Earthquake Engineering and Transport and Building Machinery. Moreover, the basic research in Dynamics nowadays includes many fields of theoretical physics and various interdisciplinary subject areas. It is encouraging that the ICOVP Conferences have matured into a reference platform reflecting the state of the art of Dynamics in the broadest sense of the term. Indeed, the most recent ICOVP Conference, held in Prague this year, covered all branches of Dynamics and offered the most up-to-date results of development to participants from 40 countries.

The ICOVP 2011 concentrated numerous papers of a very high scientific and technical level. It has shown that the ICOVP Conferences have become increasingly attractive for participants from the global scientific community. The International Scientific Committee has had an extremely difficult task choosing the best contributions from the 280 submitted abstracts for oral presentation and for publication in the Proceedings. Almost 200 papers were tentatively selected and their full text version submitted. Afterwards, two or more reviewers carefully assessed these individual papers and decided on their final acceptance and inclusion in the Conference Proceedings.

As the second step, the International Scientific Committee was approached to select the best papers. After very strict evaluation, the respective authors were invited to prepare revised versions for submission to the special volume *Springer Proceedings in Physics – Vibration Problems ICOVP 2011*. Following the next review process, 110 manuscripts were included in the final set of papers and sorted into nine chapters. We would like to express our deep gratitude to all authors and reviewers for their enormous effort and patience related to the preparation of this volume.

Looking through the volume, highly promising trends are noticeable. Clearly many challenging topics are being investigated using extremely sophisticated combinations of analytic and numeric procedures. Many qualitatively new phenomena have been identified theoretically and verified experimentally. Non-conventional methods based on various aspects of topology are being developed and applied directly in relevant exploration. Research in Dynamics is adopting more and more discrete models inspired by principles of digital technology itself. Uncertainty modeling is making good progress, offering various new approaches in the realm of Dynamics and Vibration Control. The formulation of problems of Stochastic Dynamics reflects better natural conditions and provides realistic results which have broad applications. Progress is being made in optimal random filtering, which provides realistic models of various physical phenomena on the basis of large data processing or data mining. Non-linear problems of every type related to flow-induced vibrations are widely discussed, providing applicable results not only in basic but also in applied research and industry.

Taking these aspects into account, we are firmly convinced that this volume, published by the highly reputed publishing house Springer Science+Business Media, represents a well-balanced overview of theoretical, numerical and experimental work on fundamental and applied studies performed in Dynamics and related branches during recent years. It is a great privilege to show recognition to all the authors for their invaluable contributions and their willingness to share their knowledge with other readers. Their contribution has become the backbone of the scientific success of the Conference and ensures the high quality of this special volume.

Concluding the introductory remarks, our warmest gratitude should be expressed primarily to the founders of the excellent ICOVP tradition, namely Professor M. M. Banerjee and Professor P. Biswas. At the same time, we cannot complete the preface to this special volume without expressing our sincerest thanks to a number of individuals whose invaluable help made possible the organization of the ICOVP 2011 Conference in Prague. These are the members of the International Scientific Committee and the highly efficient Local Organizing Committee under the chairmanship of Professor B. Marvalová. Her far-sighted management and enormous effort were essential for the success of the Conference. Our genuine thanks should be addressed to our other colleagues at the Technical University of

Liberec, who supported us during the entire period of the Conference and during the preparation of this volume.

The edition of the Proceedings was supported by the International Federation for the Promotion of Mechanism and Machine Science IFToMM.

Jiří Náprstek
On behalf of the editors

Contents

Part I
Keynote Lectures

Bifurcation and Chaos of Multi-body Dynamical Systems

Jan Awrejcewicz and G. Kudra

Abstract Triple physical pendulum in a form of three connected rods with the first link subjected to an action of constant torque and with a horizontal barrier is used as an example of plane mechanical system with rigid limiters of motion. Special transition rules for solutions of linearized equations at impact instances (Aizerman-Gantmakher theory) are used in order to apply classical tools for Lyapunov exponents computation as well as for stability analysis of periodic orbits (used in seeking for stable and unstable periodic orbits and bifurcations of periodic solutions analysis). Few examples of extremely rich bifurcational dynamics of triple pendulum are presented.

Keywords Pendulum • Impact • Bifurcation • Periodic orbit • Quasi-periodic orbit • Chaotic attractor • Lyapunov exponents • Non-smooth dynamics

1 Introduction

A single or a multiple pendulum (in their different forms) are very often studied theoretically or experimentally [1–3]. A single pendulum plays an important role in mechanics since many interesting non-linear dynamical behavior can be illustrated and analyzed using this simple system. But a single degree-of-freedom models are only the first step to understand a real behavior of either natural or engineering systems. Many physical objects are modeled by a few degrees of freedom and an attempt to investigate coupled pendulums is recently observed.

J. Awrejcewicz (✉)
LOEWE-Zentrum AdRIA, TU Darmstadt, Bartningstr. 53, 64289 Darmstadt, Germany

J. Awrejcewicz • G. Kudra
Department of Automation and Biomechanics, Technical University of Lodz, Lodz, Poland
e-mail: jan.awrejcewicz@p.lodz.pl; grzegorz.kudra@p.lodz.pl

J. Náprstek et al. (eds.), *Vibration Problems ICOVP 2011: The 10th International Conference on Vibration Problems*, Springer Proceedings in Physics 139,
DOI 10.1007/978-94-007-2069-5_1,

On the other hand, it is well known that impact and friction accompanies almost all real behavior, leading to non-smooth dynamics. The example of modeling of the piston – connecting rod – crankshaft system by the use of triple physical pendulum with rigid limiters of motion is presented in the work [4].

The non-smooth dynamical systems can be modeled as the so-called piece-wise smooth systems (PWS) and they are also interesting from a point of view of their bifurcational behavior, since they can exhibit certain non-classical phenomena of non-linear dynamics [5, 6]. One of the important tools of non-linear dynamics is the linear stability theory, useful among others in the analysis of bifurcations of periodic solutions and in the identification of attractors through Lyapunov exponents. These tools are well-developed and known in the case of smooth systems. However the same tools with small modifications [6, 7] can be also used for the PWS systems. The modifications consist in the suitable transformation of the perturbation in the point of discontinuity, accordingly to the so called Aizerman-Gantmakher theory [8, 9].

In the present paper some examples of identification of attractors in the system of triple physical pendulum with the horizontal barrier are given. The system used is a special case of the more general model of triple pendulum investigated in earlier works of the authors.

2 Event Driven Model of Mechanical System with Limiters of Motion

Let us assume firstly more general case of mechanical system of n-degrees-of-freedom with vector of generalized coordinates $\mathbf{q}(t) = [q_1(t), \ldots, q_n(t)]^T$, symmetric $n \times n$ mass matrix $\mathbf{M}(\mathbf{q}, t)$ and $n \times 1$ force vector $\mathbf{f}(\mathbf{q}, \dot{\mathbf{q}}, t)$. The system is subjected to m rigid unilateral constraints $\mathbf{h}(\mathbf{q}, t) = [h_1(\mathbf{q}, t), \ldots, h_m(\mathbf{q}, t)]^T \geq 0$. We define a set $I = \{1, 2, \ldots, m\}$ of indices of all defined unilateral constraints h_i and the set $I_{act} = \{i_1, i_2, \ldots, i_s\}$ of indices of s constraints permanently active on a certain time interval $[t_i, t_{i+1}]$. Physically it means that the system slides along obstacles with indices from the set I_{act}.

In the case of frictionless constraints, the system on time interval $[t_i, t_{i+1}]$ is governed by the following set of differential and algebraic equations (DAEs)

$$\mathbf{M}(\mathbf{q}, t)\,\ddot{\mathbf{q}} = \mathbf{f}_{\mathbf{q}}(\mathbf{q}, \dot{\mathbf{q}}, t) + \left(\frac{\partial \mathbf{h}_{act}(\mathbf{q}, t)}{\partial \mathbf{q}^T}\right)^T \boldsymbol{\lambda}_{act},$$
$$0 = \mathbf{h}_{act}(\mathbf{q}, t), \quad 0 = \dot{\mathbf{h}}_{act}(\mathbf{q}, t) = \frac{\partial \mathbf{h}_{act}(\mathbf{q}, t)}{\partial \mathbf{q}^T}\dot{\mathbf{q}} + \frac{\partial \mathbf{h}_{act}(\mathbf{q}, t)}{\partial t} \tag{1}$$

with the following event functions determining the time instances t_{i+1}

$$\boldsymbol{\lambda}_{act} = [\lambda_{i_1}, \lambda_{i_2}, \ldots, \lambda_{i_s}]^T > 0,$$

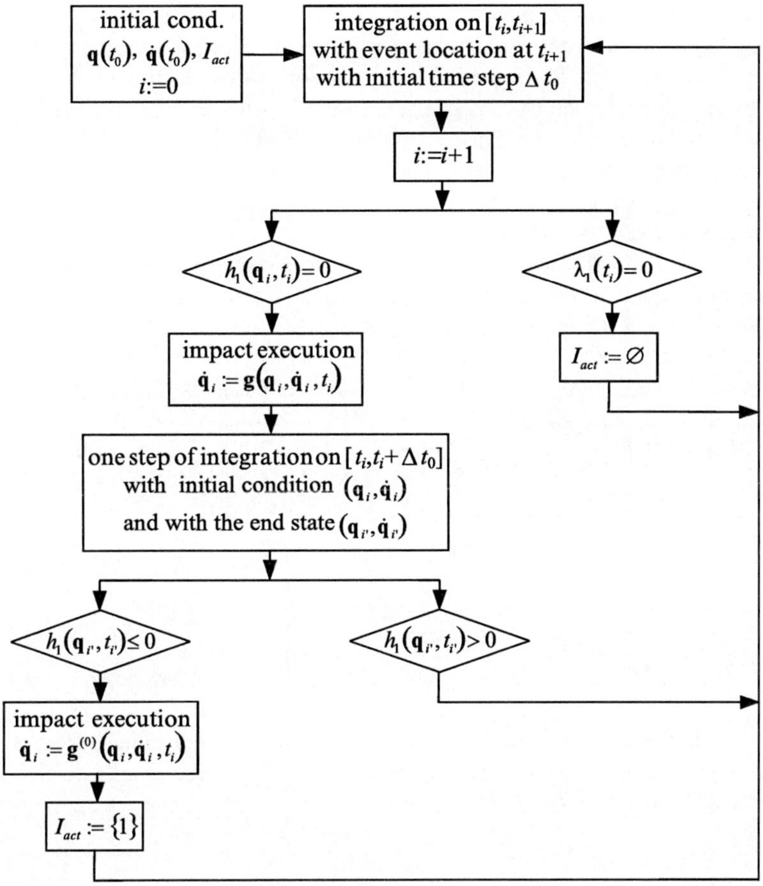

Fig. 1 Scheme for the numerical simulation of the system

$$\mathbf{h}_{inact}(\mathbf{q},t) = \left[h_{j_1}(\mathbf{q},t), h_{j_2}(\mathbf{q},t), \ldots, h_{j_{m-s}}(\mathbf{q},t)\right]^T > 0, \tag{2}$$

where $\mathbf{h}_{act}(\mathbf{q},t) = [h_{i_1}(\mathbf{q},t), h_{i_2}(\mathbf{q},t), \ldots, h_{i_s}(\mathbf{q},t)]^T$ is the vector of s constraints permanently active on $[t_i, t_{i+1}]$, $\boldsymbol{\lambda}_{act}$ is the vector of non-negative Lagrange multipliers and $\mathbf{h}_{inact}$ is the vector of $m-s$ inactive constraints, i.e. constraints which indices belong to the set $I \backslash I_{act} = \{j_1, j_2, \ldots, j_{m-s}\}$. The event t_{i+1} is determined by the use detection of zero-crossing of any component of $\boldsymbol{\lambda}_{act}$ or $\mathbf{h}_{inact}$. At time instance t_{i+1} the suitable changes in initial conditions (due to the impact) and in the set I_{act} take place and the next piece of solution $[t_{i+1}, t_{i+2}]$ is governed by the new DAEs. In this way the system has been modeled as a piece-wise smooth (PWS) DAEs.

The algorithm for the execution of changes in the system state and changes in the set I_{act} at each event time t_j, used in our numerical simulation, is presented in Fig. 1. Because of the limited space, we restrict this scheme to the simplified case, where only one constraint $h_1(\mathbf{q}, t)$ is defined ($I = \{1\}$). In the Fig. 1 the following notations are used: $\mathbf{q}_j = \mathbf{q}(t_j)$, $\dot{\mathbf{q}}_j = \dot{\mathbf{q}}(t_j)$, $t_{j'} = t_j + \Delta t_0$ and the function $\mathbf{g}(\mathbf{q}, \dot{\mathbf{q}}, t)$ represents the impact law with the restitution coefficient e while the function $\mathbf{g}^{(0)}(\mathbf{q}, \dot{\mathbf{q}}, t)$ represents impact with the restitution coefficient equal to zero independently from the system parameters.

The applied impact model is the generalized Newton's (restitution coefficient) impact law based on the reference [5], and has the following final form for the impact with the obstacle defined by $h_i(\mathbf{q}, t) = 0$:

$$\mathbf{g}(\mathbf{q}, \dot{\mathbf{q}}, t) = \begin{bmatrix} (\nabla_{\mathbf{q}} h_i(\mathbf{q}, t))^T \\ \begin{bmatrix} \mathbf{t}_1^T \\ \cdots \\ \mathbf{t}_{n-1}^T \end{bmatrix} \cdot \mathbf{M}(\mathbf{q}, t) \end{bmatrix}^{-1} \cdot \left(\begin{bmatrix} -e(\nabla_{\mathbf{q}} h_i(\mathbf{q}, t))^T \\ \begin{bmatrix} \mathbf{t}_1^T \\ \cdots \\ \mathbf{t}_{n-1}^T \end{bmatrix} \cdot \mathbf{M}(\mathbf{q}, t) \end{bmatrix} \dot{\mathbf{q}} + \begin{Bmatrix} -(e+1)\dfrac{\partial h_i(\mathbf{q}, t)}{\partial t} \\ 0 \\ \cdots \\ 0 \end{Bmatrix} \right), \tag{3}$$

where $\mathbf{t}_j$ are the base vectors of the subspace of the configuration space $\mathbf{q}$, tangent to the impact surface $h_i(\mathbf{q}, t)$ at the impact point. For more details on the impact model see works [3, 4, 7].

3 Linear Stability Model

For the dynamical system in the form

$$\dot{\mathbf{x}} = \mathbf{f}(\mathbf{x}, t), \tag{4}$$

where $\mathbf{x} = [\mathbf{q}^T, \dot{\mathbf{q}}^T]^T$, the small perturbation of the solution is governed by the following linear equations

$$\boldsymbol{\delta}\dot{\boldsymbol{x}} = \frac{\partial \mathbf{f}(\mathbf{x}, t)}{\partial \mathbf{x}^T} \boldsymbol{\delta} \boldsymbol{x}(t), \tag{5}$$

where we have assumed $\delta t = 0$ since the perturbation in time is independent from the perturbation $\boldsymbol{\delta x}$ ($\delta \dot{t} = 0$). The Eq. 5 are useful among others in the stability and bifurcation analysis of periodic solutions, as well as in the Lyapunov exponents calculation.

In the case of non-smooth dynamical system we cannot apply directly the linear stability theory since the Jacobian in (5) is not determined. But in the case of the PWS system the function $\mathbf{f}(\mathbf{x}) = \mathbf{f}_i(\mathbf{x})$ is sufficiently smooth on each time interval $[t_i, t_{i+1}]$ between two successive discontinuity points and the linear stability can applied using variational Eq. 5 on intervals $[t_i, t_{i+1}]$, and applying at each discontinuity point t_i special transformation rules accordingly to the Aizerman-Gantmakher theory (for $\delta t = 0$):

$$\delta \boldsymbol{x}_i^+ = \frac{\partial \mathbf{g}_i(\mathbf{x}_i^-, t_i)}{\partial \mathbf{x}^T} \delta \boldsymbol{x}_i^- + \left[\frac{\partial \mathbf{g}_i(\mathbf{x}_i^-, t_i)}{\partial \mathbf{x}^T} \mathbf{f}_i(\mathbf{x}_i^-, t_i) + \frac{\partial \mathbf{g}_i(\mathbf{x}_i^-, t_i)}{\partial t} - \mathbf{f}_{i+1}(\mathbf{x}_i^+, t_i) \right] \delta t_e \tag{6}$$

where

$$\delta t_e = -\frac{\dfrac{\partial event_i(\mathbf{x}_i^-, t_i)}{\partial \mathbf{x}^T} \delta \boldsymbol{x}_i^-}{\dfrac{\partial event_i(\mathbf{x}_i^-, t_i)}{\partial \mathbf{x}^T} \mathbf{f}_i(\mathbf{x}_i^-, t_i) + \dfrac{\partial event_i(\mathbf{x}_i^-, t_i)}{\partial t}},$$

and where $\mathbf{x}_i^- = \lim_{t \to t_i^-} \mathbf{x}(t)$, $\mathbf{x}_i^+ = \lim_{t \to t_i^+} \mathbf{x}(t)$, $\delta \boldsymbol{x}_i^+ = \lim_{t \to t_i^+} \delta \boldsymbol{x}(t)$, $\delta \boldsymbol{x}_i^- = \lim_{t \to t_i^-} \delta \boldsymbol{x}(t)$, $\mathbf{g}_i(\mathbf{x})$ is the function representing jump in the system state $\mathbf{x}_i^+ = \mathbf{g}_i(\mathbf{x}_i^-)$ in the discontinuity point and $event_i(\mathbf{x}, t)$ is the scalar function used for detection of the discontinuity instance at t_i $(event_i(\mathbf{x}_i^-, t_i) = 0)$.

The linearized differential-algebraic equations of the system (1) are

$$\mathbf{M}(\mathbf{q}, t)\,\delta\ddot{\mathbf{q}} = \frac{\partial \mathbf{f}(\mathbf{q}, \dot{\mathbf{q}}, t)}{\partial \mathbf{q}^T} \delta\mathbf{q} + \frac{\partial \mathbf{f}(\mathbf{q}, \dot{\mathbf{q}}, t)}{\partial \dot{\mathbf{q}}^T} \delta\dot{\mathbf{q}} + \frac{\partial}{\partial \mathbf{q}^T} \left(\left(\frac{\partial \mathbf{h}_{act}(\mathbf{q})}{\partial \mathbf{q}^T} \right)^T \boldsymbol{\lambda}_{act} \right) \delta\mathbf{q}$$

$$+ \left(\frac{\partial \mathbf{h}_{act}(\mathbf{q})}{\partial \mathbf{q}^T} \right)^T \delta\boldsymbol{\lambda}_{act} - \left(\frac{\partial \mathbf{M}(\mathbf{q}, t)}{\partial \mathbf{q}^T} \delta \boldsymbol{q} \right) \ddot{\mathbf{q}}, \; 0 = \frac{\partial \mathbf{h}_{act}(\mathbf{q}, t)}{\partial \mathbf{q}^T} \delta \boldsymbol{q}$$

$$0 = \dot{\mathbf{q}}^T \frac{\partial^2 \mathbf{h}_{act}(\mathbf{q}, t)}{\partial \mathbf{q} \partial \mathbf{q}^T} \delta \boldsymbol{q} + \frac{\partial \mathbf{h}_{act}(\mathbf{q}, t)}{\partial \mathbf{q}^T} \delta \dot{\mathbf{q}}, \tag{7}$$

where

$$\ddot{\mathbf{q}} = \mathbf{M}(\mathbf{q}, t)^{-1} \left(\mathbf{f}(\mathbf{q}, \dot{\mathbf{q}}, t) + \left(\frac{\partial \mathbf{h}_{act}(\mathbf{q}, t)}{\partial \mathbf{q}^T} \right)^T \boldsymbol{\lambda}_{act} \right)$$

and where we have also assumed $\delta t = 0$.

We have applied Eq. 7 together with the transformation rules (6) in the Lyapunov exponents calculation for the mechanical system presented in the Sect. 2. Note that

Eq. 6 with the impact law $\mathbf{g}_i(\mathbf{x},t) = \mathbf{g}_i^{(0)}(\mathbf{x},t)$ with the restitution coefficient equal to zero applied in the case where the sliding motions starts (see Fig. 1), gives the perturbation $(\boldsymbol{\delta}\mathbf{q}, \boldsymbol{\delta}\dot{\mathbf{q}})$ consistent with the algebraic equations in (7) and the perturbation vector $\boldsymbol{\delta}\mathbf{x}^{+}$ lies in the $(2n\text{-}2)$-dimensional subspace (in the case of only one constraint permanently active).

In the well-known algorithm of Lyapunov exponents computation the Gram-Schmidt reorthonormalization procedure is applied after some time of integration of variational equations. After use of this procedure to the vector of perturbations $\boldsymbol{\delta x}$ fulfilling $2s$ algebraic equations in (7) (in the case of s constraints permanently active), we obtain the new set of perturbation vectors, from which $2n\text{-}2s$ satisfy the algebraic equations and $2s$ of them do not. Then in our procedure we simply set that $2s$ vectors to zero vectors, obtaining the new "degenerated" set of orthonormal vectors, satisfying algebraic equations.

4 Triple Pendulum Model

Three joined stiff links coupled with viscous damping and moving on the plane are presented in Fig. 2. The system position is defined by three angles ψ_i $(i = 1,2,3)$, and each of the first body is under action of constant torque q_1. The set of possible

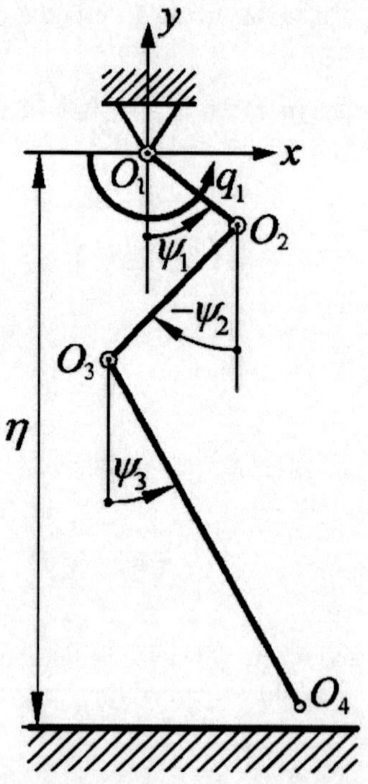

Fig. 2 Mechanical system

configurations of the system is bounded by the horizontally situated rigid and frictionless barrier.A vector of generalized coordinates is the vector of three angles $\mathbf{q} = \boldsymbol{\psi} = [\psi_1, \psi_2, \psi_3]^T$. The mass matrix, force vector and the set of algebraic equations defining rigid obstacle are as follows

$$\mathbf{M}(\mathbf{q}, t) = \mathbf{M}(\boldsymbol{\psi}) = \begin{bmatrix} 1 & \nu_{12}\cos(\psi_1 - \psi_2) & \nu_{13}\cos(\psi_1 - \psi_3) \\ \nu_{12}\cos(\psi_1 - \psi_2) & \beta_2 & \nu_{23}\cos(\psi_2 - \psi_3) \\ \nu_{13}\cos(\psi_1 - \psi_3) & \nu_{23}\cos(\psi_2 - \psi_3) & \beta_3 \end{bmatrix},$$

$$\mathbf{f}(\mathbf{q}, \dot{\mathbf{q}}, t) = \mathbf{f}(\boldsymbol{\psi}, \dot{\boldsymbol{\psi}}, t) = -\mathbf{N}(\boldsymbol{\psi})\,\dot{\boldsymbol{\psi}}^{2} - \mathbf{C}\dot{\boldsymbol{\psi}} - \mathbf{p}(\psi) + \mathbf{f}_e(\boldsymbol{\psi}, \dot{\boldsymbol{\psi}}, t), \quad (8)$$

$$h_1(\boldsymbol{\psi}) = \eta - l_1\cos\psi_1,\ h_2(\boldsymbol{\psi}) = \eta - \sum_{i=1}^{2} l_i\cos\psi_i,\ h_3(\boldsymbol{\psi}) = \eta - \sum_{i=1}^{3} l_i\cos\psi_i$$

where

$$\mathbf{N}(\boldsymbol{\psi}) = \begin{bmatrix} 0 & \nu_{12}\sin(\psi_1 - \psi_2) & \nu_{13}\sin(\psi_1 - \psi_3) \\ -\nu_{12}\sin(\psi_1 - \psi_2) & 0 & \nu_{23}\sin(\psi_2 - \psi_3) \\ -\nu_{13}\sin(\psi_1 - \psi_3) & -\nu_{23}\sin(\psi_2 - \psi_3) & 0 \end{bmatrix}$$

$$\mathbf{C} = \begin{bmatrix} c_1 + c_2 & -c_2 & 0 \\ -c_2 & c_2 + c_3 & -c_3 \\ 0 & -c_3 & c_3 \end{bmatrix}, \quad \mathbf{p}(\boldsymbol{\psi}) = \begin{Bmatrix} \sin\psi_1 \\ \mu_2\sin\psi_2 \\ \mu_3\sin\psi_3 \end{Bmatrix}, \quad \mathbf{f}_e = \begin{Bmatrix} q_1 \\ 0 \\ 0 \end{Bmatrix},$$

and where $\dot{\boldsymbol{\psi}}^{2} = [\dot{\psi}_1^2, \dot{\psi}_2^2, \dot{\psi}_3^2]^T$, where l_i is non-dimensional length of i-th link, c_i is non-dimensional damping coefficient in the i-th joint while ν_{ij} and μ_i are other non-dimensional parameters of the system.

The system response is obtained numerically by the use of the Runge-Kutta integration method of the differential equations between each two successive discontinuity points (where the activity of the obstacles changes: the impact takes place or the time interval of sliding begins or ends). These points are detected by halving integration step until obtaining assumed precision. After the simulation of the system, the next step was the stability analysis of the solution in the investigated model, which in fact is piece-wise smooth (PWS) one. The classical methods and algorithms basing on the linear perturbation equations are used with the modifications taking into account the perturbations jump in the discontinuity points [9]. The numerical software for Lyapunov exponents calculation and periodic orbit stability analysis (seeking for periodic orbits and their bifurcations analysis) was developed.

For more details on modeling, relations between real and non-dimensional parameters, numerical algorithms, etc., see works [3, 4, 7].

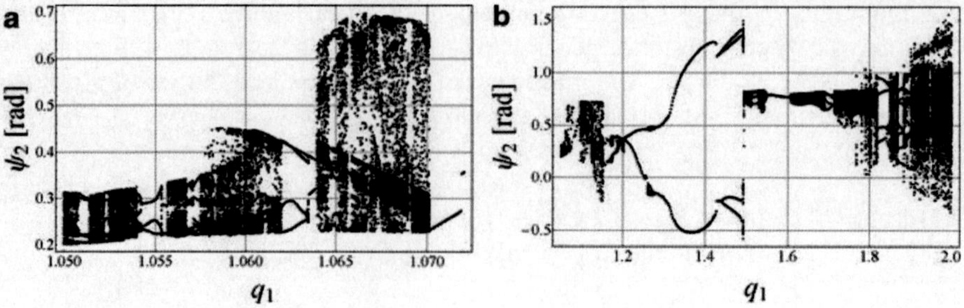

Fig. 3 Bifurcational diagrams

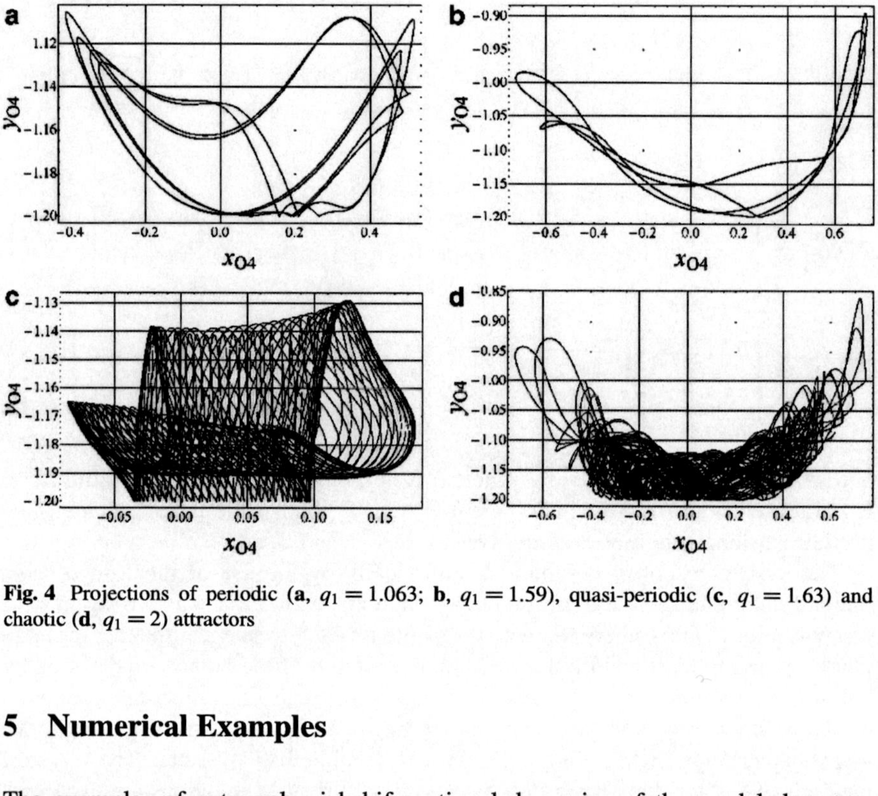

Fig. 4 Projections of periodic (**a**, $q_1 = 1.063$; **b**, $q_1 = 1.59$), quasi-periodic (**c**, $q_1 = 1.63$) and chaotic (**d**, $q_1 = 2$) attractors

5 Numerical Examples

The examples of extremely rich bifurcational dynamics of the modeled system is presented for the following non-dimensional parameters: $l_1 = O_1\ O_2 = 0.05$, $l_2 = O_2\ O_3 = 0.02$, $l_3 = O_3\ O_4 = 1$, $\eta = 1.2$ and $c_1 = c_2 = c_3 = 0.8$. The restitution coefficient is $e = 0.8$ and contact between links and obstacles is assumed to be frictionless. The externally applied torque q_1 is used as bifurcational parameter.

In Fig. 3 one can find two bifurcational diagrams where the parameter q_1 is increasing quasi-statically. In Fig. 3a the relative change of the torque is very small (about 2%) but the richness and number of bifurcational phenomena observed is

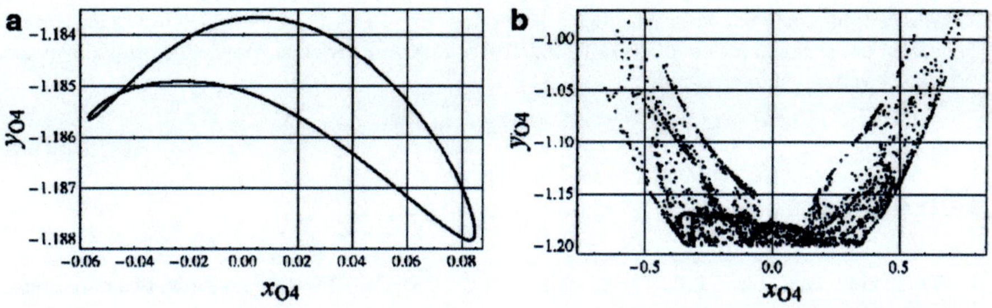

Fig. 5 Projections of Poincaré sections of quasi-periodic (**a**, $q_1 = 1.063$) and chaotic (**b**, $q_1 = 2$) attractors

Table 1 Lyapunov exponents

Figure	λ_1	λ_2	λ_3	λ_4	λ_5	λ_6	Attractor
3a	*0.00*	−0.02	−0.03	−0.24	−0.35	−5.98	Periodic
3b	*0.00*	−0.02	−0.02	−0.22	−0.47	−8.44	Periodic
3c and 4a	*0.00*	*0.00*	−0.01	−0.08	−0.54	−10.28	Quasi-periodic
3d and 4b	**0.05**	*0.00*	−0.03	−0.19	−0.52	−10.03	Chaotic

extremely large. Both bifurcational diagrams start just after disappearing of stable equilibrium position for q_1 equal about 1.01. The next Figures exhibit exemplary periodic, quasiperiodic and chaotic attractors observed on bifurcational diagrams. Fig. 4 presents trajectory projections while Fig. 5 shows corresponding Poincaré sections (performed by the use of plane $\psi_1 = 0$). The verification of a kind of each the attractor is performed by the use of Lyapunov exponents presented in Table 1.

6 Conclusions

This paper briefly reports the larger project of investigations of the flat triple physical pendulum with arbitrary situated barriers imposed on the position of the system. The Aizerman-Gantmakher theory, handling with perturbed solution in points of discontinuity, is used to extend classical method for computing Lyapunov exponents for the multi-degree of freedom mechanical system with rigid barriers imposed on its position. Some examples of identification of attractors in the system of triple pendulum with horizontal barrier are presented, including periodic, quasi-periodic and chaotic attractors. We have focused on the calculation of Lyapunov exponents, however the same methods can be used in the stability and bifurcation analysis of periodic solutions. Let us also note, that the mentioned above methods are suitable for analysis of classical bifurcations occurring in non-smooth systems. For non-classical bifurcations (like grazing bifurcation as an example) the developed methods may not be sufficient.

Acknowledgments The work has been supported by the Ministry of Science and Higher Education under the grant no. 0040/B/T02/2010/38. J. Awrejcewicz acknowledges support of the Alexander von Humboldt Award.

References

1. Baker, G.L., Blackburn, J.A.: The Pendulum. A Case Study in Physics. Oxford University Press, Oxford (2005)
2. Zhu, Q., Ishitobi, M.: Experimental study of chaos in a driven triple pendulum. J. Sound Vib. **227**(1), 230–238 (1999)
3. Awrejcewicz, J., Supeł, K.G., Wasilewski, G., Olejnik, P.: Numerical and experimental study of regular and chaotic motion of triple physical pendulum. Int. J. Bifurcation Chaos **18**(10), 2883–2915 (2008)
4. Awrejcewicz, J., Kudra, G.: The piston - connecting rod - crankshaft system as a triple physical pendulum with impacts. Int. J. Bifurcation Chaos **15**(7), 2207–2226 (2005)
5. Brogliato, B.: Non-smooth Mechanics, Springer, London (1999)
6. Leine, R.L., van Campen, D.H., van de Vrande, B.L.: Bifurcations in nonlinear discontinuous systems. Nonlinear Dyn. **23**, 105–164 (2000)
7. Awrejcewicz, J., Kudra, G.: Stability analysis and Lyapunov exponents of a multi-body mechanical system with rigid unilateral constraints. Nonlinear Anal. Theor. Methods Appl. **63**(5–7) (2005)
8. Aizerman, M.A., Gantmakher, F.R.: On the stability of periodic motions. J. Appl. Math. Mech. **22**, 1065–1078 (1958)
9. Müller, P.C.: Calculation of Lyapunov exponents for dynamic systems with discontinuities. Chaos Solitons Fractals **5**, 1671–1691 (1995)

Stochastic Model Updating: Perturbation, Interval Method and Bayesian Inference

John E. Mottershead, H. Haddad Khodaparast, R.P. Dwight, and K.J. Badcock

Abstract Stochastic model updating methods are described, including probabilistic perturbation methods, interval techniques and Bayesian inference. Particular attention is paid to aleatory uncertainty such as variability in nominally identical test structures due to manufacturing tolerances. In such cases the updating parameter distributions are meaningful physically either as PDFs or as intervals. Stochastic model updating is an inverse problem, generally requiring multiple forward solutions. The use of meta models (response-surface mapping techniques and the Kriging predictor) as surrogates for the full FE model are explained. The procedure is illustrated in each case by experimental examples, including model updating of a structure with uncertain locations of two internal beams.

Keywords Model updating • Aleatory and epistemic uncertainty • Perturbation method • Interval model updating • Kriging predictor • Bayesian inference

1 Introduction

Deterministic model updating of finite element models [1, 2] has become a mature technology. It is a classical inverse problem in the sense that a measurable output is used to correct a set of analytical parameters that are themselves inaccessible to measurement. As with most inverse problems the resulting system of equations are ill-posed and require regulatisation, discussed in full Ahmadian et al. [3].

J.E. Mottershead (✉) • H.H. Khodaparast • K.J. Badcock
Centre for Engineering Dynamics, University of Liverpool, Liverpool, UK
e-mail: j.e.mottershead@liv.ac.uk; hhadad@liv.ac.uk; k.j.badcock@liv.ac.uk

R.P. Dwight
Department of Aerodynamics, TU Delft, Delft, The Netherlands
e-mail: r.p.dwight@tudelft.nl

J. Náprstek et al. (eds.), *Vibration Problems ICOVP 2011: The 10th International Conference on Vibration Problems*, Springer Proceedings in Physics 139,
DOI 10.1007/978-94-007-2069-5_2,

Variability in performance has become one of the major challenges facing researchers and engineering scientists concerned with the robust reliability of engineering products, such as aeroplanes, helicopters and cars. This has resulted in the emergence of powerful probabilistic and non-probabilistic (interval, fuzzy etc.) methods combined with modern computer systems and codes. Research has mostly concentrated on the forward propagation of parameter uncertainty to determine output distributions of measurable quantities, such as natural frequencies, mode shapes and damping values [4–6]. An aspect of uncertainty of particular interest is the inherent variability that arises from the effect of accumulated manufacturing tolerances or from wear so that nominally identical engineering systems behave differently. This type of uncertainty is said to be irreducible, or aleatory, whereas epistemic uncertainty is due to lack of knowledge and is reducible by the processing of information or data.

The purpose of the present article is to summarise very recent results [7–9] on stochastic model updating. In this type of problem variability on a measurement is used to evaluate uncertainty (pdfs, intervals, fuzzy membership functions etc.) on inaccessible parameters. There are a variety of approaches, including the probabilistic, interval and Bayesian methods described in this article. The conventional approach requires multiple forward solutions by Monte-Carlo simulation; a process that becomes prohibitively expensive when the analysis calls for large-scale solutions by finite element code. The perturbation method often provides a low-cost solution to many engineering problems when the uncertainty is bounded within an acceptably small and well-understood region. The interval and Bayesian methods are not restricted to small uncertainty as is the perturbation method. There are numerous examples in engineering of uncertainty defined only by estimates of extreme values; the interval model updating method is entirely appropriate in such cases. Bayesian inference is the most sophisticated of the three methods considered and is capable of providing an estimate not only of the bounds of uncertainty but also of its distribution.

2 Model Updating by Perturbation

The stochastic model updating problem may be cast as,

$$\bar{\boldsymbol{\theta}}_{j+1} + \boldsymbol{\Delta\theta}_{j+1} = \bar{\boldsymbol{\theta}}_j + \boldsymbol{\Delta\theta}_j + \left(\bar{\mathbf{T}}_j + \boldsymbol{\Delta}\mathbf{T}_j\right)\left(\bar{\mathbf{z}}_m + \boldsymbol{\Delta}\mathbf{z}_m - \bar{\mathbf{z}}_j - \boldsymbol{\Delta}\mathbf{z}_j\right) \quad (1)$$

where $\bar{\mathbf{z}}_j + \boldsymbol{\Delta}\mathbf{z}_j \in \Re^{n\times 1}$ is the vector of estimated outputs (e.g. eigenvalues and eigenvectors), $\bar{\mathbf{z}}_m + \boldsymbol{\Delta}\mathbf{z}_m \in \Re^{n\times 1}$ is the vector of measured data, $\bar{\boldsymbol{\theta}}_j + \boldsymbol{\Delta\theta}_j \in \Re^{m\times 1}$ is the vector of system parameters and $\bar{\mathbf{T}}_j + \boldsymbol{\Delta}\mathbf{T}_j$; $\boldsymbol{\Delta}\mathbf{T}_j = \sum_{k=1}^{n} \frac{\partial \bar{\mathbf{T}}_j}{\partial z_{mk}} \Delta z_{mk}$ is a transformation matrix. Each term in (1) consists of a mean part and a random-variable part (denoted by the overbar and the Δ respectively).

Application of the perturbation method [7], by separating the zeroth order and first order terms from Eq. 1, leads to,

$$\mathbf{O}(\mathbf{\Delta}^0): \quad \bar{\mathbf{\theta}}_{j+1} = \bar{\mathbf{\theta}}_j + \bar{\mathbf{T}}_j \left(\bar{\mathbf{z}}_m - \bar{\mathbf{z}}_j\right) \tag{2}$$

$$\mathbf{O}(\mathbf{\Delta}^1): \quad \begin{cases} \mathrm{cov}\left(\mathbf{\Delta\theta}_{j+1},\mathbf{\Delta\theta}_{j+1}\right) = \mathrm{cov}\left(\mathbf{\Delta\theta}_j,\mathbf{\Delta\theta}_j\right) \\ -\mathrm{cov}\left(\mathbf{\Delta\theta}_j,\mathbf{\Delta z}_j\right)\bar{\mathbf{T}}_j^T + \bar{\mathbf{T}}_j\mathrm{cov}\left(\mathbf{\Delta z}_m,\mathbf{\Delta z}_m\right)\bar{\mathbf{T}}_j^T \\ -\bar{\mathbf{T}}_j\mathrm{cov}\left(\mathbf{\Delta z}_j,\mathbf{\Delta\theta}_j\right) + \bar{\mathbf{T}}_j\mathrm{cov}\left(\mathbf{\Delta z}_j,\mathbf{\Delta z}_j\right)\bar{\mathbf{T}}_j^T \end{cases} \tag{3}$$

where the measurements and parameters are assumed to be uncorrelated. This assumption is strictly not justifiable though in practice the difference obtained when the correlation is included has been found to be negligible. The computational advantage is considerable since Eqs. 2 and 3 require only the first order sensitivities (gradients) of the outputs with respect to the parameters, whereas the second-order sensitivities (Hessians) are needed when the correlation is included.

The following terms in Eq. 3 may be expressed as,

$$\mathrm{cov}\left(\Delta\mathbf{z}_j,\Delta\mathbf{z}_j\right) = \bar{\mathbf{S}}\mathrm{cov}\left(\Delta\mathbf{\theta}_j,\Delta\mathbf{\theta}_j\right)\bar{\mathbf{S}}^{\mathrm{T}} \tag{4}$$

$$\mathrm{cov}\left(\Delta\mathbf{\theta}_j,\Delta\mathbf{z}_j\right) = \mathrm{cov}\left(\Delta\mathbf{\theta}_j,\Delta\mathbf{\theta}_j\right)\bar{\mathbf{S}}^{\mathrm{T}} \tag{5}$$

$$\mathrm{cov}\left(\Delta\mathbf{z}_j,\Delta\mathbf{\theta}_j\right) = \bar{\mathbf{S}}\mathrm{cov}\left(\Delta\mathbf{\theta}_j,\Delta\mathbf{\theta}_j\right) \tag{6}$$

when the uncertainties are small and parameter distribution is Gaussian ($\bar{\mathbf{S}}$ denotes the matrix of mean sensitivities). If that is not the case, then forward propagation using the asymptotic integral is recommended.

3 Interval Model Updating

Interval model updating [8] is concerned with a mapping from the parameter space to the space of the outputs. The vertices of the parameter hypercube map onto vertices of the output space in a limited number of special cases, such as when the parameters are mass of stiffness sub-structure coefficients, typically m_j, k_j where, $\mathbf{M} = \mathbf{M}_0 + \sum_{j=1}^{p} m_j\mathbf{M}_j$, $\mathbf{K} = \mathbf{K}_0 + \sum_{j=1}^{p} m_j\mathbf{K}_j$ and the output consists only of the system natural frequencies. In that case interval model updating is straightforward and may be achieved by deterministic model updating from the output-space vertices.

In most practical applications vertex-to-vertex mapping does not occur and it is then necessary to take a large number of output samples for mapping in the reverse direction to define the vertices of the parameter hypercube. This type of inverse problem is generally solved by many forward solutions for each output data sample. Fortunately the mapping may be achieved by the use of a low-cost meta-model (or emulator), which avoids the multiple running of large-scale finite element codes. The Kriging predictor is one such meta-model. It is trained using deterministic finite element results: the rth 'true' data sample, $\mathbf{z}^{(r)} \in \Re^{n\times 1}$, is determined from the rth set of parameters, $\boldsymbol{\theta}^{(r)} \in \Re^{p\times 1}$, in a finite element run. The mean Kriging model for one output term may then be written, for any set of parameters $\boldsymbol{\theta}$, as

$$\hat{z}_i = \beta_i + \mathbf{b}_i^T\boldsymbol{\theta} + \frac{1}{2}\boldsymbol{\theta}^T\mathbf{B}_i\boldsymbol{\theta} + \varepsilon_i(\boldsymbol{\theta}) \tag{7}$$

where the terms β_i $\mathbf{b}_i$ and $\mathbf{B}_i$ consist entirely of regression coefficients fitted to the FE results. Any error, $\varepsilon_i(\boldsymbol{\theta})$, in (7) is a random function of the parameters since any lack of fit is due only to an incomplete set of regression terms. Consequently the errors at different data samples are correlated. The random function $\varepsilon_i(\boldsymbol{\theta})$ has zero mean so that the training points are predicted exactly. Its covariance takes the form,

$$\varepsilon_i(\boldsymbol{\theta}) = \text{cov}\left(\varepsilon_i\left(\boldsymbol{\theta}^{(r)}\right), \varepsilon_i(\boldsymbol{\theta})\right) = \sigma_i^2 \prod_{j=1}^{p} C_{ji}\left(\theta_j^{(r)}, \theta_j\right) \tag{8}$$

when the parameters are uncorrelated, σ_i^2 is the variance of the ith output and $C_{ji}(\theta_j^{(r)}, \theta_j)$ denotes the correlation of the jth parameter. Practical application requires the mean Kriging predictor, which may be shown to be,

$$\hat{z}_i = \beta_i + \mathbf{b}_i^T\boldsymbol{\theta} + \frac{1}{2}\boldsymbol{\theta}^T\mathbf{B}_i\boldsymbol{\theta} + \boldsymbol{\lambda}_i^T\mathbf{r}_i(\boldsymbol{\theta}) \tag{9}$$

where $\mathbf{r}_i(\boldsymbol{\theta})$ is a vector with terms $\prod_{j=1}^{p} C_{ji}(\theta_j^{(r)}, \theta_j)$ and the terms of $\boldsymbol{\lambda}_i$ are weighted functions of the differences $z_i - (\beta_i + \mathbf{b}_i^T\boldsymbol{\theta}^{(r)} + \frac{1}{2}\boldsymbol{\theta}^{(r)T}\mathbf{B}_i\boldsymbol{\theta}^{(r)}), r = 1 : n_s$.

The interval model updating method is explained in Fig. 1. Firstly, deterministic model updating is carried out using a single data point close to the mean measurement. This results in a point in the parameter space. An initial hypercube is then constructed around this point and a Kriging meta-model is created using parameters selected from the hypercube. Deterministic model updating at all measured outputs is carried out using the meta-model. If all the sets of updating parameters are located within the initial hypercube, then the bounds of the parameters are defined on the hypercube and interval model updating is complete. If some of the updating parameters lie outside the initial hypercube then the Kriging model is possibly inaccurate for those data. The parameter hypercube must be revised and a new Kriging model created. The process of deterministic model updating is repeated with adjustment to the parameter hypercube as necessary until all parameters are within the last-created hypercube.

A very significant advantage of the Kriging approach is that it provides a mapping from any system of parameters to any form of output. The parameters chosen can be very general, including parameters not readily available for conventional model updating, as in the second experimental example in Sect.5.

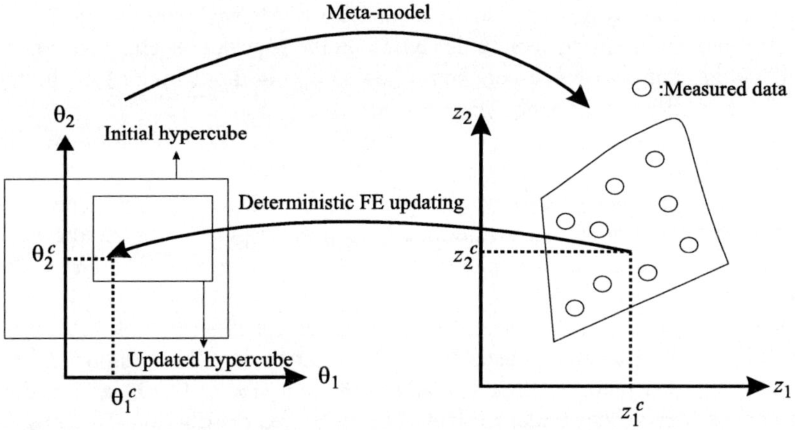

Fig. 1 Interval model updatingprocess (Reprinted from [8]. With permission from Elsevier)

4 Bayesian Inference

As a formal theorem, Bayes' rule (named after the Reverend Thomas Bayes, 1702–1761) is valid in terms of probabilities or probability densities. Its proof is given most readily in terms of the former. From the definition of joint probability,

$$\mathrm{P}(A|B)\,\mathrm{P}(B) = \mathrm{P}(A, B) \tag{10}$$

and by symmetry,

$$\mathrm{P}(B|A)\,\mathrm{P}(A) = \mathrm{P}(A, B) \tag{11}$$

so that,

$$\mathrm{P}(A|B) = \frac{\mathrm{P}(B|A)\,\mathrm{P}(A)}{\mathrm{P}(B)} \tag{12}$$

It is common to think of Bayes' rule in terms of updating a hypothesis A in the light of new data B. Specifically, the posterior probability is calculated by multiplying the prior probability by the likelihood that B will occur given A. The denominator is a normalising constant computed as,

$$\mathrm{P}(B) = \sum_i \mathrm{P}(B, A_i) = \sum_i \mathrm{P}(B|A_i)\mathrm{P}(A_i) \tag{13}$$

where i is the index on hypotheses considered. The summation is an approximation to an integral over an unbounded domain, generally a very expensive computation usually carried out by Markov-Chain Monte-Carlo (MCMC) simulation. Sampling is performed efficiently by techniques based on the Metropolis–Hastings algorithm or the Gibbs sampler, which concentrate on sampling those regions of the model space with significant probability.

In a simple model updating experiment, B might be a collection of natural frequencies and A a set of spring stiffnesses drawn from a distribution. Subscript i denotes the sample number, so that A_i is the ith set of stiffnesses. P(B) is determined entirely from the FE model. The likelihood function P($B|A$) ensures that the estimate is determined so that the FE natural frequencies converge upon their measured counterparts.

MCMC simulation is a stochastic process such that $f(\boldsymbol{\theta}^{k+1}|\boldsymbol{\theta}^{k},\ldots,\boldsymbol{\theta}^{1}) = f(\boldsymbol{\theta}^{k+1}|\boldsymbol{\theta}^{k})$. The parameter distribution, $\boldsymbol{\theta}^{k+1}$, at the $k+1^{\text{th}}$ iteration, given all the preceding distributions, depends only on $\boldsymbol{\theta}^{k}$ and $f(\boldsymbol{\theta}^{k+1}|\boldsymbol{\theta}^{k})$ is independent of k. It can be shown that under certain conditions the distribution $\boldsymbol{\theta}^{k}$ tends to an equilibrium independent of the initial estimate $\boldsymbol{\theta}^{0}$.

5 Experimental Studies

Two experimental studies are described. In the first study a distribution of plate thicknesses (mean and standard deviation) is determined from measured variation in natural frequencies by the perturbation approach. In the second study a frame structure with two randomly located internal beams is considered. The distributions of beam locations are determined by interval model updating and also by Bayesian inference.

5.1 Plate Thickness Distribution by Perturbation

Ten aluminium plates were prepared so that a distribution of thicknesses, close to Gaussian, was obtained by machining: mean value 3.975 mm and standard deviation 0.163 mm. All ten plates had the same overall dimensions, length 0.4 m and width 0.1 m. A hammer test was carried out to determine the natural frequencies of the plates using four uniaxial fixed accelerometers and free-free boundary conditions. Variability was found in the thicknesses of the individual plates when measured with a micrometer, and the plates were parameterised in four $0.1 \times 0.1\ \text{m}^2$ square regions so that $\boldsymbol{\theta} = \left(t_1\ t_2\ t_3\ t_4\right)^T$. For each of the four parameters the initial means and standard deviations were assumed to be 4 mm and 0.8 mm respectively. The first six measured natural frequencies were used in the analysis and the updated parameter standard deviations, determined after 50 iterations of Eq. 3 are given in Table 1. It is seen that the standard deviations are significantly improved by updating – the initial, values of the means were close to the measured mean thicknesses.

Table 1 Updated standard deviations of the aluminium plates

	Measured SD	Initial SD	Updated SD
SD(t_1) mm	0.159	0.8	0.129
SD(t_2) mm	0.161	0.8	0.204
SD(t_3) mm	0.164	0.8	0.166
SD(t_4) mm	0.167	0.8	0.206

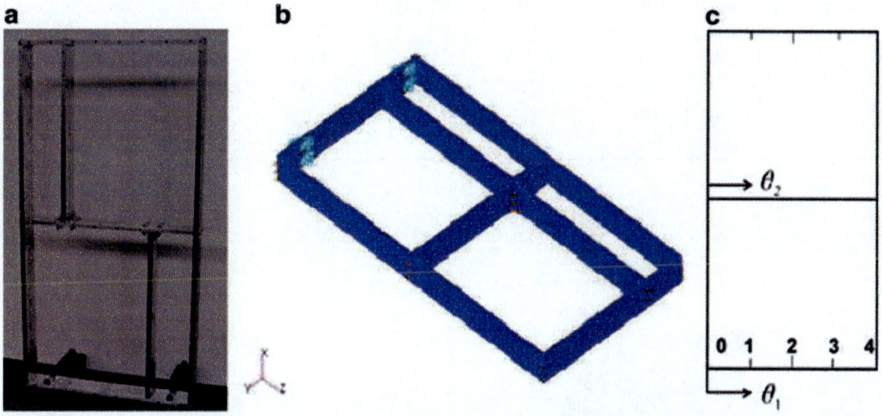

Fig. 2 Structure with internal beams: (**a**) physical structure; (**b**) FE model (**c**) beam locations (Reprinted from [8]. With permission from Elsevier)

5.2 Frame Structure with Randomly Located Internal Beams

The frame structure shown in Fig. 2 has two internal beams parameterised by $\boldsymbol{\theta} = (\theta_1, \theta_2)^T$ and independently located on the interval $(0, 4) \otimes (0, 4)$. In the experiment the beams were each fixed at one of the three locations $\theta = (1, 2, 3)$ giving nine different combinations of beam locations.

The Kriging model was obtained using the first six natural frequencies from the FE model, sampled by *central composite design* (CCD), a Design of Experiments technique with the central point as the median value. In addition to the nine physically sampled parameter locations, a further 16 numerical samples were included at $\theta = (0.5, 1.5, 2.5, 3.5)$- details can be found in [8, 9]. Interval model updating [8] was carried out by completing 9 deterministic model updates for each of the nine beam configurations and the results are summarised in Table 2. An example of how the output space converges upon the measurements is shown in Fig. 3.

A two-level approach to uncertainty modelling of the beam locations was adopted by the Bayesian method [9]. We begin with the distribution of θ and choose independent uniform distributions for each parameter, which corresponds closely to the interval representation of uncertainty described above,

$$\theta_1 \colon U\left(\Phi_1^-, \Phi_1^+\right) \tag{14}$$

Table 2 Updated parameters- interval method

True parameters		Initial parameters		Updated parameters	
θ_1	θ_2	θ_1	θ_2	θ_1	θ_2
1.0	1.0	1.6	1.6	1.04	1.02
1.0	2.0	1.6	2.4	1.0	2.15
1.0	3.0	1.6	2.4	1.0	3.08
2.0	1.0	1.6	1.6	2.04	0.90
2.0	2.0	2.4	2.4	2.13	2.0
2.0	3.0	2.4	2.4	1.95	3.09
3.0	1.0	2.4	1.6	2.98	0.89
3.0	2.0	2.4	1.6	2.99	1.83
3.0	3.0	2.4	2.4	2.93	2.98

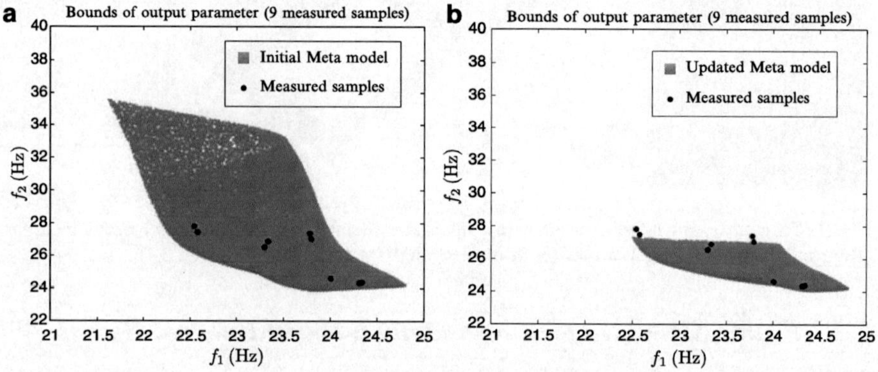

Fig. 3 Example of output bounds – interval method (Reprinted from [8]. With permission from Elsevier)

$$\theta_2: U\left(\Phi_2^-, \Phi_2^+\right) \tag{15}$$

where the *hyperparameters* $\mathbf{\Phi} = \left(\Phi_1^-, \Phi_1^+, \Phi_2^-, \Phi_2^+\right)$ define the lower and upper bounds of the distributions. The hyperparameters $\mathbf{\Phi}$ represent epistemic uncertainty whereas the aleatory uncertainty, which represents the real variability with the population, is represented by $\mathbf{\Theta} = (\theta_1, \theta_2)$.

The goal of the analysis can be formulated as identifying values for the hyperparameters. These are unknown, so under the Bayesian framework prior distributions are specified that capture any *a priori* knowledge about the structural variability,

$$\Phi_{1,2}^-: T \tag{16}$$

$$\Phi_{1,2}^+: U\left(\Phi_{1,2}^-, 4\right) \tag{17}$$

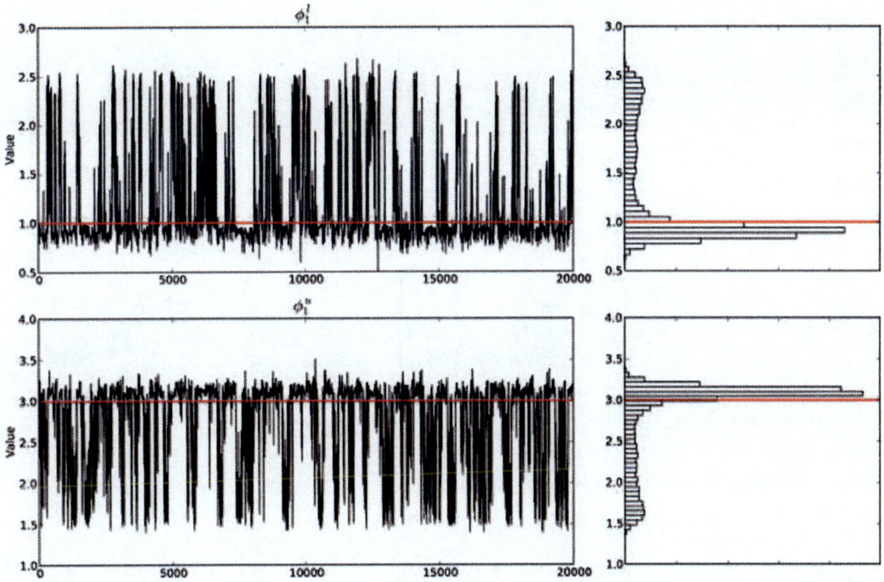

Fig. 4 Convergence of MCMC iteration (final 20,000 samples) and histogram

where U is a uniform distribution on the interval given by the arguments and T is a triangular distribution with the pdf,

$$f_T(\phi) = \begin{cases} 0 & \text{if } \phi < 0 \text{ or } \phi > 4 \\ \dfrac{4-\phi}{8} & \text{otherwise} \end{cases} \tag{18}$$

where ϕ is a deterministic sample from Φ. Finally the observed frequencies are assumed to follow a multivariate normal distribution with covariance matrix $\boldsymbol{\Sigma}_x^2$,

$$\mathbf{X} = N\left(\boldsymbol{\mu}_x,\ \boldsymbol{\Sigma}_x^2\right) \tag{19}$$

In order to determine the likelihood $f(\mathbf{x}|\phi)$- equivalently $\mathrm{P}(B|A)$ in Eq. 12 – it is necessary to calculate $\boldsymbol{\mu}_\varphi$ and $\boldsymbol{\Sigma}_\phi^2$, which are defined by integrals evaluated by probabilistic collocation [?].The posterior $f(\phi|\mathbf{x})$ – equivalently $\mathrm{P}(A|B)$in Eq. 12 – is sampled using Metropolis-Hastings MCMC, with 10^5 samples (plus an initial throw-away 'burn-in' of 10^4 samples). This process, carried out efficiently using the Kriging predictor, results in histograms, as shown in Fig. 4. The peaks indicate the most likely values of ϕ. The estimate from the Nelder-Mead simplex method, with starting points obtained from all the histograms, is $\phi_{MAP} = (0.9221,\ 3.0778,\ 0.8561,\ 3.1439)$. A full description of the method is given by Dwight et al. [9]. An example showing the output bounds with this definition of the hyperparameters is shown in Fig. 5. The ellipses denote integer numbers of standard deviations.

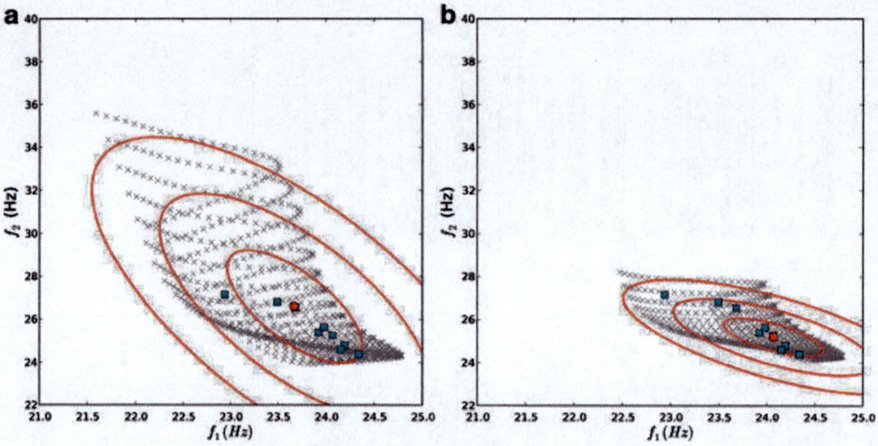

Fig. 5 Example of output bounds – Bayesian inference

It is seen that the estimate ϕ_{MAP} is very similar to the results obtained by interval updating in Table 2. Also the output-space results, in Figs. 3 and 5 are almost identical. Figure 3 shows some data points just outside the predicted bounds on the natural frequencies, probably due to other uncertainties not accounted for, such as joint stiffness which differ as a result of repeated assembly and disassembly of the structure when the beams were moved to different positions.

6 Conclusions

The theory of stochastic model updating was described briefly and three methods were demonstrated using measured natural frequencies from (a) a collection of plates with varying thickness and (b) a frame structure with randomly located internal beams. In both cases the three methods were shown to result in meaningful improvements to initial finite element models. Separating different sources of uncertainty, typically aleatory uncertainty in test samples, epistemic uncertainty in measurements and in mathematical (FE) models (parametric uncertainty and uncertainty in constitutive equations), is an open problem. Further research is needed on the choice of stochastic model updating method for different classes of problem. The application of these techniques to industrial-scale engineering problems remains as a significant challenge to researchers and industrial engineers.

Acknowledgements The research described in this paper was partly funded by the European Commission for the Marie Curie Excellence Team ECERTA under contract MEXT-CT-2006-042383.

References

1. Mottershead, J.E., Friswell, M.I.: Model updating in structural dynamics: a survey. J. Sound Vib. **167**(2), 347–375 (1993)
2. Friswell, M.I., Mottershead, J.E.: Finite Element Model Updating in Structural Dynamics. Kluwer Academic Publishers, Dordrecht (1995)
3. Ahmadian, H., Mottershead, J.E., Friswell, M.I.: Regularisation methods for finite element model updating. Mech. Syst. Signal Process. **12**(1), 47–64 (1998)
4. Moens, D., Vandepitte, D.: A survey of non-probabilistic uncertainty treatment in finite element analysis. Comput. Method Appl. Mech. Eng. **194**(12–16), 1527–1555 (2005)
5. Worden, K., Manson, G., Lord, T.M., Friswell, M.I.: Some observations on uncertainty propagationthrough a simple nonlinear system. J. Sound Vib. **288**(3), 601–621 (2005)
6. Adhikari, S., Friswell, M.I.: Random matrix eigenvalue problems in structural dynamics. Int. J. Num. Methods Eng. **69**(3), 562–591 (2007)
7. Khodaparast, H.H., Mottershead, J.E., Friswell, M.I.: Perturbation methods for the estimation of parameter variability in stochastic model updating. Mech. Syst. Signal Process. **22**(8), 1751–1773 (2008)
8. Khodaparast, H.H., Mottershead, J.E., Badcock, K.J.: Interval model updating with irreducible uncertainty using the Kriging predictor. Mech. Syst. Signal Process. **25**(4), 1204–1226 (2011)
9. Dwight, R.P., HaddadKhodaparast, H. and Mottershead, J.E.: Identifying structural variability using Bayesian inference. Mech. Syst. Signal Process. submitted

Impact Induced Stress Wave Energy Flux – Validation of Numerical and Experimental Approaches

M. Okrouhlík

Abstract The presented contribution is devoted to analysis of accuracy and validity of both experimental and numerical approaches employed when evaluating the stress wave energy. In the studied case it is shown how much of the impact wave energy, which is predominantly of axial (longitudinal) nature, is transferred into torsional (shear) energy mode as well as into other energy modes not seen by experiment.

Keywords Energy flux • Finite element method • High-speed strain gauge measurement • Validation of results

1 Introduction

Always, one has to ponder about what is a 'true' approach to the modeling of nature. Thomas Aquinas (1225–1274) claimed that the truth is an agreement of reality with perception. Today, however, the perceived reality depends on observation tools being used.

When trying to reveal the 'true' behavior of a mechanical system we are using the experiment.

When trying to predict the 'true' behavior of a mechanical system we are accepting a certain theoretical model and then solve it analytically and/or numerically.

Physical laws, however, cannot (in mathematical sense) be proved. It is often claimed that it is the experiment that is the only and ultimate judgment of the validity of the theory, model or a hypothesis being used.

M. Okrouhlík (✉)
Institute of Thermomechanics, Academy of Sciences of the Czech Republic, Dolejskova 5, 182 00 Prague, Czech Republic
e-mail: ok@it.cas.cz

J. Náprstek et al. (eds.), *Vibration Problems ICOVP 2011: The 10th International Conference on Vibration Problems*, Springer Proceedings in Physics 139,
DOI 10.1007/978-94-007-2069-5_3,

A few contradicting quotations might nicely obscure our reasoning.

Experimental science does not receive truth from superior science. She is the mistress and the other sciences are her servants. Roger Bacon: On Experimental Science (1268).

Experiment, indeed, is a necessary adjunct to a physical theory; but it is an adjunct, not the master. Truesdel, C. and Toupin. R.: The classical and field theories. In Encyclopedia of Physics, Volume III/1, p. 227, Springer, Berlin, 1960.

Many, otherwise elegant and beautiful theories were rejected because they had not agreed with experimental observation – I do not know, however, any big theory having been created as a direct generalization of an experiment. Hawkins, S.: Black Holes and Baby Universes and other Essays.

Being armed with these inspiring doubts, let's try to solve a particular stress flow problem numerically and experimentally.

2 Stress Wave Energy Flux Through Spiral Slot of a Tube Induced by Axial Impact – Preliminaries

We study the energy flux in a thin walled tube depicted in Fig. 1. The energy flux is the time rate of energy. If only mechanical energy is considered, then the energy flux is simply the sum of strain and kinetic energy rates carried by stress waves. The mechanical energy contained in a body depends on the spatial and temporal distribution of strains and velocities associated with all the body particles.

The experiment can observe and register the time distributions of surface strains at a certain location only, so the subsequent energy assessment has to be based on approximate approaches assuming that the internal strain quantities are the same as those measured at the surface and, furthermore, are uniformly distributed across the considered cross sectional area.

The transient finite element (FE) 3D analysis commonly provides for a detailed description of displacement and velocity fields within the considered body allowing thus to assess the energy flux from the transient spatial and temporal distributions of

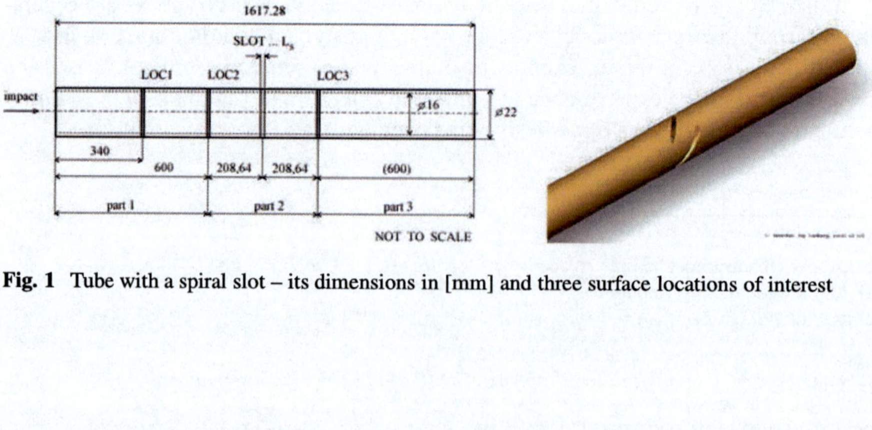

Fig. 1 Tube with a spiral slot – its dimensions in [mm] and three surface locations of interest

energy in individual parts of the body. Of course, the FE method also has a limited range of validity. Its detailed study, with the emphasis on FE technology employed in this particular case, is presented in [1, 2] and [8].

The 1D experimental approach, based on measuring and registering surface strains supply valid results provided that the stress wave pattern is of 1D or of axisymmetric nature – i.e. for long thin bodies loaded by relatively long pulses. In a case where a geometrical irregularity in the path of propagating waves is present, the level of induced stress wave 'three-dimensionality' as well as the consequent experiment reliability is difficult to predict.

That's why the stress wave energy results in the whole body obtained by the transient 3D finite element analysis are compared with those of the experiment based on evaluation of the recorded history of surface strains, measured at specified surface locations.

Elastic waves propagating through an axially impacted cylindrical tube carry mechanical energy. When the incident wave reaches the slot area a part of incoming energy is reflected, a part of it is transmitted. From that moment both reflected and transmitted waves are propagating, evoking torsional phenomena. The initial idea, based on 1D stress wave reasoning, was to study the task experimentally (by registering time history of surface strains at certain locations in front of and behind the slot area) and by subsequent data analysis to determine how the incident longitudinal (IL) energy is decomposed into transmitted longitudinal (TL), transmitted torsional (TT), reflected longitudinal (RL) and reflected torsional (RT) energy modes. However the subsequent FE analysis showed that studied stress waves in the impacted tube with spiral slots exhibit to a certain extent the 3D wave pattern and beside the energy modes 'visible' by experiment, there are additional energy modes to be taken into account.

In a purely mechanical process, when the heat flux, heat sources and the energy dissipation are not considered, the principle of conservation energy requires that the total mechanical power (the rate of change of total mechanical energy) equals the sum of stress power and the time rate of kinetic energy $W = W_{\mathrm{int}} + W_{\mathrm{kin}}$.

Using standard notation of continuum mechanics textbooks and considering the Cauchy's equations of motion and constitutive equations in the form of the Hooke's law one can express the internal stress power and the time rate of kinetic energy, see [3, 4] and [5] as follows

$$W_{\mathrm{int}} = \dot{E}_{\mathrm{s}} = \frac{\mathrm{d}}{\mathrm{d}t} \int_V \frac{1}{2} C_{ijkl}\, \varepsilon_{kl}\, \varepsilon_{ij}\, \mathrm{d}V, \quad W_{\mathrm{kin}} = \dot{E}_{\mathrm{k}} = \frac{\mathrm{d}}{\mathrm{d}t} \int_V \frac{1}{2} \rho\, \dot{u}_i\, \dot{u}_i\, \mathrm{d}V,$$

$$W_{\mathrm{kin}} = \dot{E}_{\mathrm{k}} = \frac{\mathrm{d}}{\mathrm{d}t} \int_V \frac{1}{2} \rho\, \dot{u}_i\, \dot{u}_i\, \mathrm{d}V, \quad W = \dot{E}_{\mathrm{s}} + \dot{E}_{\mathrm{k}} = \dot{T}$$

where $\dot{E}_{\mathrm{s}}$ and $\dot{E}_{\mathrm{k}}$ are the time rates of strain (potential) and kinetic energies respectively while $\dot{T}$ is the time rate of total mechanical energy in the body.

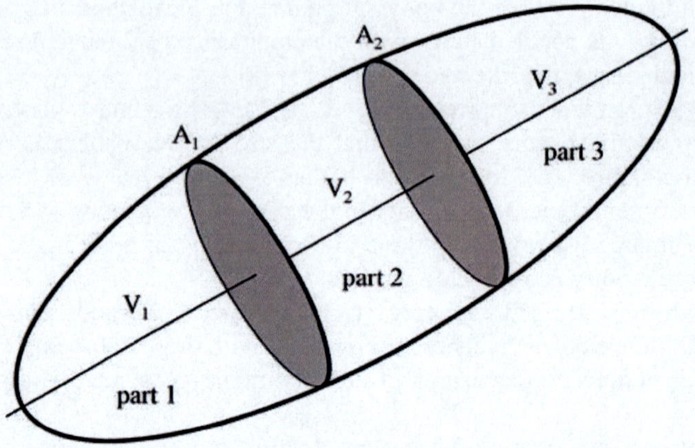

Fig. 2 Three parts of a body – notice three parts of the body indicated in Fig. 1

Similarly, the energy flux through a cross sectional area A of a body is

$$W(t) = \int_A s_i(t)\, v_i(t)\, \mathrm{d}A,$$

where $s_i(t)$ and $v_i(t)$ are components of stress $\mathbf{s}(t)$ and velocity $\mathbf{v}(t)$ vectors respectively. Furthermore, the stress vector is related to stress components by $s_i = \sigma_{ji}\, n_i$, where n_i are components of the normal vector to the surface $\mathrm{d}A$ and σ_{ij} are engineering stress components.

The energy flow through a cross sectional area A is $W = \int_A n_i\, \sigma_{ij}\, v_j\, \mathrm{d}A$.

2.1 *Finite Element Analysis*

The FE computation was carried out in the Institute of Thermomechanics. In FE analysis the energy flux computation can be assessed by analyzing the time history of energy in the whole body and in its parts.

For a body dissected by two cross sections, say A_1 and A_2, into three parts, $p = 1, 2, 3$ – as shown in Fig. 2, the total mechanical energy in one of its parts, consisting of strain E_s and kinetic E_k energies, can be expressed by

$$T^{(p)} = E_\mathrm{s}^{(p)} + E_\mathrm{k}^{(p)} = \frac{1}{2}\left(\mathbf{q}^{(p)}\right)^\mathrm{T} \mathbf{K}^{(p)} \mathbf{q}^{(p)} + \frac{1}{2}\left(\dot{\mathbf{q}}^{(p)}\right)^\mathrm{T} \mathbf{M}^{(p)} \dot{\mathbf{q}}^{(p)}, \quad p = 1, 2, 3, \ldots$$

where $\mathbf{K}, \mathbf{M}$ are global stiffness and mass matrices and $\mathbf{q}, \dot{\mathbf{q}}$ are global displacements and velocities respectively. For more details see [1, 2] and [6].

Assuming that the input energy enters the body in part 1 only, then the energy flow through a cross sectional area A_1 can be expressed by the time rates of energies in corresponding parts of the body as follows

$$W^{(A_1)} = \dot{T}^{(2)} + \dot{T}^{(3)} = \dot{T} - \dot{T}^{(1)}.$$

Similarly, for the energy flow through the cross section A_2 we get

$$W^{(A_2)} = \dot{T}^{(3)} = \dot{T} - \dot{T}^{(1)} - \dot{T}^{(2)}.$$

2.2 Experimental Analysis

Experiment was conceived and carried out at Uppsala University, Sweden. The experimental data were kindly provided by Prof. B. Lundberg and his team. In experimental analysis it is convenient to evaluate the amount of energy in the body as the cumulative energy flux through a cross sectional area, observed within a specified time interval $\Delta t = t_2 - t_1$. It is assumed that the energy enters the considered part of the body through the mentioned cross sectional area only.

So, within the specified time interval, the integral of the energy flux through the cross-sectional area A_2, i.e. $W^{(A_2)}$, can be expressed as the mechanical energy in part 3 (see Fig. 1) at two distinct time instants, say t_1 and t_2 respectively, as $\int_{t_1}^{t_2} W^{(A_2)}(t)\,\mathrm{d}t = \int_{t_1}^{t_2} \dot{T}^{(3)}(t)\,\mathrm{d}t = T^{(3)}(t_2) - T^{(3)}(t_1)$.

In the experiment the evaluation of mechanical energy is based on 1D stress wave reasoning, assuming that the longitudinal and torsional waves are independently propagating (with velocities $c_0 = \sqrt{E/\rho}$ and $c_T = \sqrt{G/\rho}$ respectively) through the circular tube with inner and outer radii r_1 and r_2 respectively. The material properties are defined by Young modulus E, Poisson ratio υ and density ρ.

Under these assumptions, see [7], the energies associated with longitudinal T_L and torsional T_T waves are obtained as integrals of energy flows through the cross sectional area within the specified time interval $\langle t_1, t_2 \rangle$.

$T_L = AEc_0 \int_{t_1}^{t_2} \varepsilon_L^2\,\mathrm{d}t$, $T_T = 4Gkc_T/b^2 \int_{t_1}^{t_2} \varepsilon_T^2\,\mathrm{d}t$, where the constants appearing in above formulas are $A = \pi(r_2^2 - r_1^2),$, $c_0 = \sqrt{E/\rho},$, $G = \frac{1}{2}E/(1+\upsilon),$, $k = \pi(r_2^4 - r_1^4)/32$, $c_T = \sqrt{G/\rho}$.

To evaluate the amount of 'longitudinal' and 'torsional' energy modes, using above formulas, one has to register time distributions of axial and shear strains measured at a specified location. Using the above formulas we tacitly assume that the measured surface quantities are attributed to strains across the cross sectional area and are uniformly distributed there. The value of each energy mode is obtained as the cumulative energy flux within the specified time interval. The evaluated energy values are constant – they do not depend on time. To extrapolate the validity of obtained results would require that outside the considered time interval the energy flux is of steady-state nature.

2.3 Energy Modes Associated with Displacement and Velocity Components

The total mechanical energy can be expressed as the sum of the individual strain and kinetic *energy contributions* associated with particular vector components of displacements and velocities appearing in expressions $(E_s)_{ij} = \frac{1}{2}\mathbf{q}_i^T\mathbf{K}\mathbf{q}_j$ and $(E_k)_{ij} = \frac{1}{2}\dot{\mathbf{q}}_i^T\mathbf{K}\dot{\mathbf{q}}_j$, where the indices $i, j = 1, 2, 3$ correspond to directions of axes of a coordinate system being employed. The body part index '$^{(p)}$' is intentionally omitted here. The individual *energy contributions* are called *energy modes* in the text.

The particular energy modes, defined in the Cartesian coordinate system, being used for FE processing, have to be expressed in the cylindrical coordinate system, in which the experimental data are available.

Using standard coordinate transformation [3], the Cartesian components of nodal displacements and velocities at each element node are recalculated into the cylindrical ones.

The total mechanical energy of a body composed of three parts – written in full by means of energy modes expressed in a cylindrical coordinate system – is

$$T^{(p)} = T_{\mathrm{tt}}^{(p)} + T_{\mathrm{rr}}^{(p)} + T_{\mathrm{aa}}^{(p)} + T_{\mathrm{ta}}^{(p)} + T_{\mathrm{ra}}^{(p)} + T_{\mathrm{tr}}^{(p)}, \quad p = 1, 2, 3, \ldots$$

Formally, there are two types of energy modes, those formed by quadratic terms, which are positive definite (i.e. $T_{\mathrm{tt}}^{(p)} + T_{\mathrm{rr}}^{(p)} + T_{\mathrm{aa}}^{(p)}$), and those formed by bilinear terms (i.e. $T_{\mathrm{ta}}^{(p)} + T_{\mathrm{ra}}^{(p)} + T_{\mathrm{tr}}^{(p)}$) that, however, are not.

The former modes – called *quadratic* in the text – represent the energy contributions **predominantly** associated with tangential (circumferential), radial and axial wave motions, while the latter – called *mixed* modes in the text – represent the rest, i.e. the modes of the tangential-axial, radial-axial and tangential-radial wave motions.

It should be reminded that in 1D simplification (the 'thin' bar theory) of wave motions we could independently think of (i) axial (longitudinal) energy mode associated with axial displacements and velocities and of (ii) tangential (torsional or shear) energy mode associated with angular motions and velocities. The experiment, based on registering axial and shears strains on the surface, 'sees' neither radial nor the mixed modes. Under accepted 1D assumptions the mentioned axial (longitudinal) and tangential (shear) waves, as well as their modes, propagate by different velocities and are independent of each other.

In contradistinction to the 1D concept of mutually independent axial (longitudinal) and tangential (shear or torsional') wave patterns, the energy modes, in a 3D formulation cannot be associated with tangential (torsional), radial and axial (longitudinal) displacements and velocities only. Always, there are the mixed modes complementing the energy balance.

3 Studied Case

The object being studied has the form of a cylindrical tube. In its the middle part there are four spiral slots going through the wall thickness as depicted in Fig. 3.

The strain gauge rosettes were located at locations 1, 2 and 3. Experimental setup is fully described in [2]. After the striker hits the tube the time distribution of axial and shear strains is registered. The registered signals have characteristic peaks, which should be properly associated with events corresponding to arrivals of incident, transmitted and reflected waves. Analyzing the experimental strain distributions in front of (at LOC2) and behind (at LOC3) the slot one can identify three types of axial strain signals associated with longitudinal energy modes, i.e. those coming to, reflected from and transmitted through the spiral slot area. In the text the following identifiers are used (i) IL ... incident longitudinal, (ii) RL ... reflected longitudinal and (iii) TL ... transmitted longitudinal. For two shear signals associated with torsional reflected and transmitted modes we use (i) RT ... reflected torsional and (ii) TT ... transmitted torsional. It should be noted, that in the experiment the energy carried by incident longitudinal (IL) pulse is taken as the input energy in contradistinction to the FE approach where the input energy equals to the kinetic energy of the striker just before the impact.

As an example the experimentally registered signal at location LOC2 (shown in Fig. 4) is attributed to two distinct parts, namely to the incident longitudinal pulse (IL) and the reflected longitudinal pulse (RL). Comparison with FE strains is shown. Differences are discussed and explained in [1].

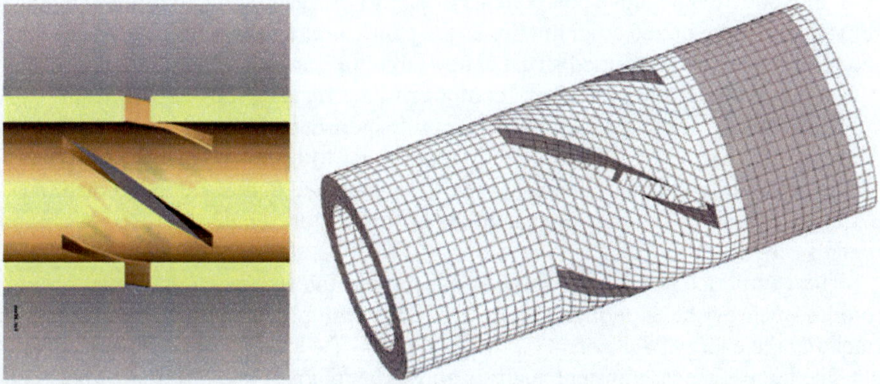

Fig. 3 The spiral slot and its FE representations in detail

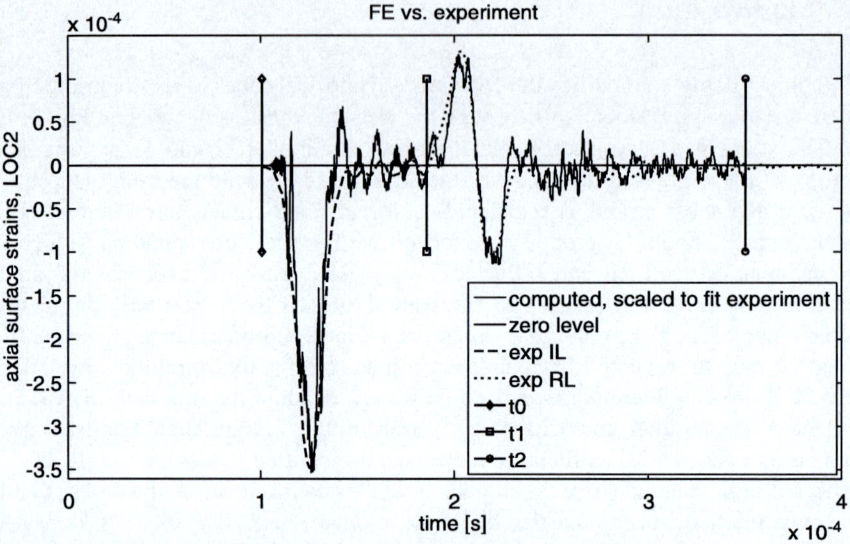

Fig. 4 Surface axial strains at LOC 2 – FE vs. experiment

4 Comparison of Experimental and FE Results and Conclusions

The FE normalized energy modes in individual parts of the tube are shown in Fig. 5. Notice the negative components of energy corresponding to bilinear modes. Physically, these mixed strain modes correspond to the strain work needed hold the body in its current deformed shape. They largely disappear when the pulse leaves the slotted area. Not '`aa`' nor '`ar`' components are registered by experiment.

The detailed comparison of values of four experimental energy modes ('RTexp', 'TTexp', 'RLexp' and 'TLexp' – they correspond to time $t_{\max}$ but for clarity are depicted as horizontal dashed lines) with those obtained numerically (denoted 'tt1', 'tt3', 'aa1' and 'aa3' – they are functions of time) is presented in Fig. 6. For more details see [6].

The comparison of results showed that beside the tangential and axial energy modes, seen by the experiment, there are radial and mixed energy modes complementing the energy balance.

The finite element transient analysis allowed explaining the differences between the experimental and numerical assessments of contributions of individual energy modes. The differences are due to the fact that the experiment, based on measuring surface strains and on 1D stress wave reasoning, can not 'notice' the radial and mixed energy modes typical for a 3D wave pattern. Furthermore, measuring the time distribution of surface strains in front of and behind the spiral slot does not allow to see the energy residing in remaining parts of the body.

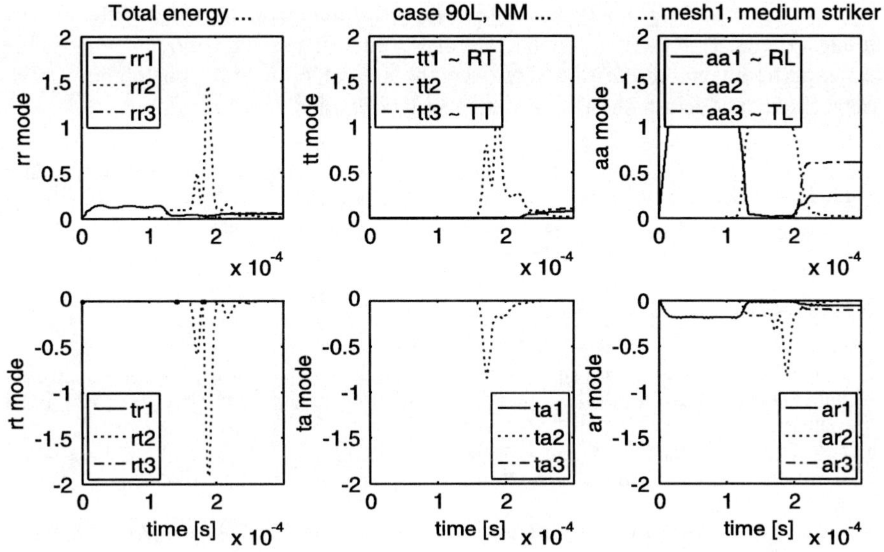

Fig. 5 Total FE energy modes. Numbers associate the energy components to body parts

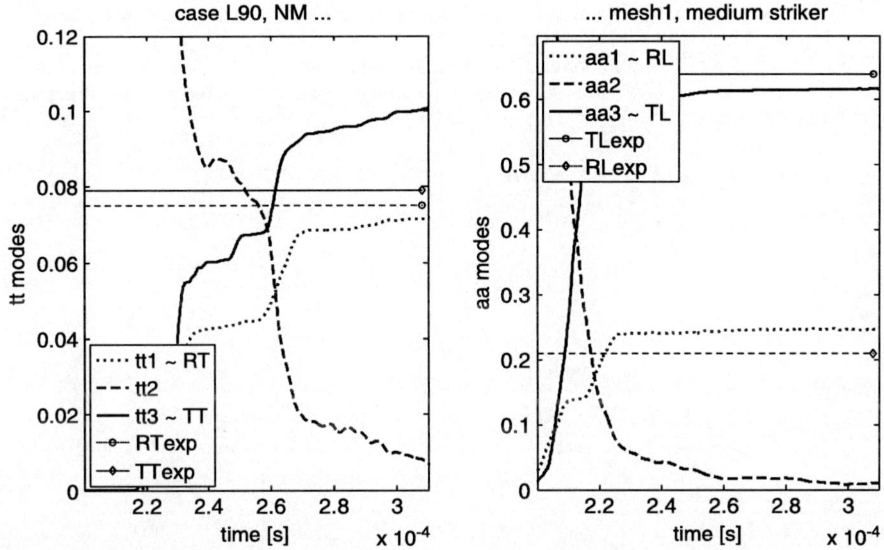

Fig. 6 Total energies – Experiment vs. FE

Of course, the notion of 'three-dimensionality' is not a question of geometry only – it also depends on the frequency spectrum of the input pulse and its relation to maximum eigenfrequencies of the considered FE model. This subject is fully treated in [1].

Nevertheless, the studied case is mainly of 1D nature with experimental energy modes (i.e. RT, RL, TT and TL) being predominant. The modes non-visible by experiment are interesting from the theoretical point but due to their small magnitude they do not compromise the results obtained experimentally.

Acknowledgments The author would like to express his thanks to B. Lundberg, S. Ptak, U. Valdek and S. Mousavi.

References

1. Okrouhlík, M., Pták, S., Valdek, U.: Self-assessment of finite element solutions applied to transient phenomena in solid continuum mechanics. Eng. Mech. **16**(2), 103–121 (2009). ISSN 1802–1484
2. Okrouhlík, M., Pták, S.: FE and experimental study of axially induced shear stress waves in a tube with spiral slots. Part 1. Research Report of the Institute of Thermomechanics, Z 1417/2008, Prague (2008)
3. Malvern, L.E.: Introduction to the Mechanics of a Continuous Medium. Prentice-Hall, Englewood Cliffs (1969)
4. Bathe, K.-J.: Finite Element Procedures. Prentice-Hall, Englewood Cliffs (1982)
5. Bathe, K.-J.: Finite Element Procedures. Prentice-Hall, Englewood Cliffs (1996)
6. Okrouhlík, M., Pták, S., Lundberg, B., Valdek, U.: FE assessment of an experiment employed for analysis of the transient stress energy flux through spiral slots in axially impacted cylindrical tube. Eng. Mech. **60**(4), 181–209 (2009)
7. Kolsky, H.: Stress Waves in Solids. Clarendon, Oxford (1953)
8. Belytschko, T., Liu, W.K., Moran, B.: Nonlinear Finite Elements for Continua and Structures. Wiley, Chichester (2000)

Nonlinear Dynamics of Pendulums System for Energy Harvesting

M. Wiercigroch, A. Najdecka, and V. Vaziri

Abstract In this paper dynamics of a parametric pendulums system operating in rotational regime has been investigated with a view of energy harvesting. The main idea is based on the conversion of the oscillatory motion of the oscillatory motion into rotation of pendulums [1]. Numerical, analytical and experimental studies have been undertaken on a parametric pendulum and a pendulum excited by a planar motion. They suggest the rotational motion is persisting and occurs for a large range of frequencies and excitation amplitudes, which are the main control parameters. These investigations reinforce the viability of this new concept of the energy conversion. A system of two pendulums has been modelled and analysed. Specifically, the dynamics of the parametric pendulums systems has been investigated numerically and experimentally focusing on synchronized rotational solutions. The target state is a synchronized counter rotation of both pendulums. A control strategy aiming to initiate and maintain the desired rotational responses, has been developed and verified numerically and experimentally.

Keywords Parametric pendulum • Coupled pendulums • Synchronization • Delayed-feedback control • Wave energy extraction

1 Introduction

The concept of using mechanical pendula systems for wave energy extraction has been given a considerable attention in recent years [1]. A parametric pendulum is a useful model of the heave excitation at the base of a suitably constrained floating structure. Most of the work in this area has been conducted in the Centre for Applied Dynamics Research at the University of Aberdeen.

M. Wiercigroch (✉) • A. Najdecka • V. Vaziri
Centre for Applied Dynamics Research, University of Aberdeen, Aberdeen, Scotland, UK
e-mail: m.wiercigroch@abdn.ac.uk

J. Náprstek et al. (eds.), *Vibration Problems ICOVP 2011: The 10th International Conference on Vibration Problems*, Springer Proceedings in Physics 139,
DOI 10.1007/978-94-007-2069-5_4, © Springer Science+Business Media B.V. 2011

Xu [2] has carried out a numerical and experimental study of a such system with energy extraction. Xu and Wiercigroch [3] investigated the existence of the rotational attractors through the parameter space and Horton et al. [4] proposed a method for parameter identification of a pendulum experimental pendulum rig for rotational motion and recorded experimental tumbling chaos. Xu et al. [5] studied the effect the interactions between the excited pendulum and electro-dynamic shaker had on pendulum dynamics. In addition, Xu and Wiercigroch [6] obtained closed form analytical expressions by first order perturbation method for the rotary solutions, which has been recently expanded to the higher order terms and extensively studies by Lenci et al. [7]. A classification of the complex responses of the parametric pendulum by the recurrence plots was proposed by Litak et al. [8].

In this paper the dynamics of a parametric pendulums system operating in rotary motion is studied with a view of energy harvesting. First we briefly review the main finding on the global dynamics of the parametric pendulum focusing on its rotary solution. In section "Pendulums System", a system of two parametric pendulum is introduced and equations of motions discussed. The focus lies here on the synchronized rotational responses. The target state is to achieve a synchronized counter rotation of both pendulums. Next, the control strategy, with the aim of initiating and maintaining the desired response, has been developed and verified numerically and experimentally.

2 Dynamics of Parametric Pendulum

The working principle of the parametric pendulum is illustrated in Fig. 1, where the pivot point is subjected to harmonic excitation in vertical direction as shown in Fig. 1. A pendulum can experiences different types of motion, which can be conveniently represented on the phase plane as depicted Fig. 1. The region of oscillatory solutions (closed loops denoted by (1)) is bounded by a critical motion described by the separatrices (2). The response outside this region is denoted by (3) and corresponds to the rotation, which is of main interest for the energy extraction purposes. For stable rotations the solution on the phase plane needs to lie sufficiently far from the separatrix to ensure that the energy dissipation will not cause the pendulum to go back to the potential well inside the region bounded by separatrices. Once the pendulum rotates its energy can be extracted directly from the rotating shaft at the pivot point, which will be the scope of the following studies.

A schematic of the parametric pendulum system is provided in Fig. 1a and the well known model for the parametrically excited pendulum is given below:

$$\theta'' + \gamma\theta' + \left(1 + y''\right) sin\theta = 0, \tag{1}$$

where γ is the coefficient of linear viscous damping, prime is the differentiation with respect to the non-dimensionalised time, $\tau = \omega_0 t$, ω_0 is the linear natural frequency.

Understanding of the global dynamics is the key in utilizing the potentials of rotary motion of a parametric pendulum and extensive numerical studies have

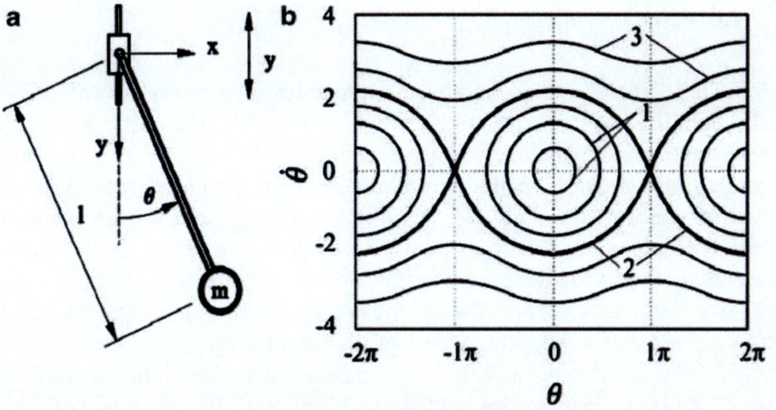

Fig. 1 Working principle of parametrically-excited pendulum. (**a**) Physical model of a parametric pendulum and (**b**) phase portraits showing different responses of the parametric pendulum

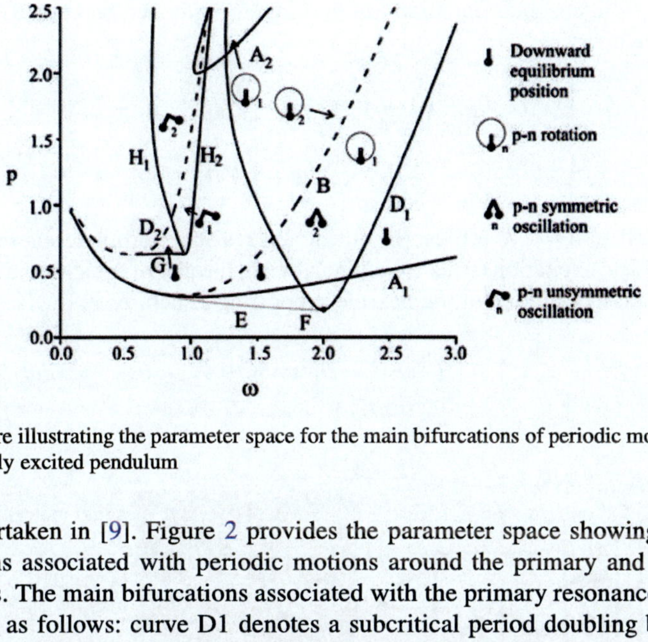

Fig. 2 Figure illustrating the parameter space for the main bifurcations of periodic motions for the parametrically excited pendulum

been undertaken in [9]. Figure 2 provides the parameter space showing the main bifurcations associated with periodic motions around the primary and secondary resonances. The main bifurcations associated with the primary resonance for oscillations are as follows: curve D1 denotes a subcritical period doubling bifurcation to period-2 oscillations, E denotes a fold bifurcation to stable period-2 oscillations, and F is a co-dimension 2 bifurcation. Similarly for rotations: A1,2 denote fold bifurcations to period-1 rotations and B denotes a period doubling flip bifurcations to period-2 rotations. It is not within the scope of this paper to describe a full description of the bifurcation scenarios throughout the illustrated parameter space, but it is sufficed to say, the motions, which co-exist at different regions are annotated for reference purposes. A1,2 and B are the important bifurcations associated with rotations.

3 Pendulums System

The first step towards understanding the dynamics of the two pendulums system includes building physical and mathematical model. The schematic representation of the system considered and its experimental rig is shown in Fig. 3. It consists of two pendulums mounted on the commonly excited flexible supporting structure. In the first stage of the study the system has been treated on the plane. It has been modelled as a four degrees-of-freedom system, where x and y denote the displacement of the structure in horizontal and vertical direction, θ_1 and θ_2 describe the angular displacement form the downward zero position. A synchronized state can be achieved due to coupling effect of the elastic base, capable of transmitting vibrations between the pendulums. The mass of the pendulums is concentrated mainly on the bob and therefore treated as a point mass. The damping effect of the base and on the shaft of the pendulum has been modelled as a viscous damping. The harmonic excitation on the base has been assumed, giving a good approximation of the wave's motion.

The non-dimensional equations of motion for the two rotational degrees of freedom are given by:

$$\theta_1'' + x'' cos\theta_1 + \left(1 + y''\right) sin\theta_1 + \gamma_\theta \theta_1' - u_1 + e_1 = 0,$$
$$\theta_2'' + x'' cos\theta_2 + \left(1 + y''\right) sin\theta_2 + \gamma_\theta \theta_2' - u_2 + e_2 = 0, \qquad (2)$$

where u_i and e_i (i = 1,2) represent control and resistive torque terms respectively. The parametric excitation terms x'', y'' included in the above equations are described by the equation of motion of the flexible supporting structure:

$$x'' + \gamma_x x' + \alpha_x x + a\left(\theta_1'' cos\theta_1 - \theta_1'^2 sin\theta_1 + \theta_2'' cos\theta_2 - \theta_2'^2 sin\theta_2\right) = 0,$$

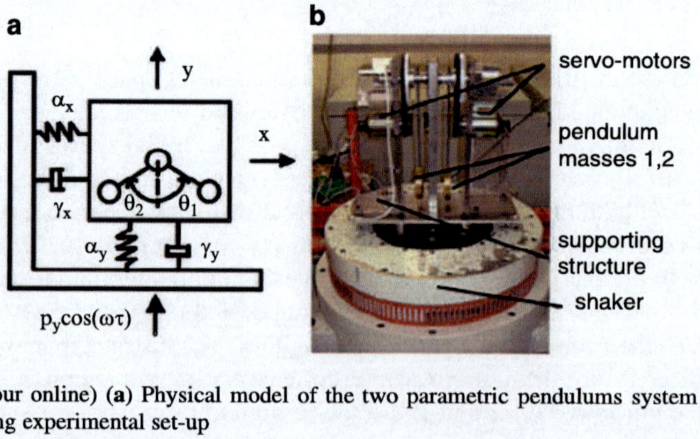

Fig. 3 (colour online) (**a**) Physical model of the two parametric pendulums system and (**b**) the corresponding experimental set-up

$$y'' + \gamma_y \left(y' + p_y\omega \sin(\omega\tau)\right) + \alpha_y \left(y - p_y \cos(\omega\tau)\right)$$
$$+ a \left(\theta_1'' sin\theta_1 + \theta_1'^2 cos\theta_1 + \theta_2'' sin\theta_2 + \theta_2'^2 cos\theta_2\right) = 0, \qquad (3)$$

where all of the system parameters are non-dimensional. γ_x and γ_y are the damping coefficients of the base in horizontal and vertical direction respectively, α_x and α_y are the stiffness coefficients, p_y is the forcing amplitude, ω is the forcing frequency, τ is the time, a is a mass ratio. All of the system parameters and variables have been rescaled with respect to the natural frequency of the pendulums. The non-dimensional form of the system equations allows comparison between different models. The result form the small scale experiment can easily be transferred on the real size installation. By changing the length of the pendulum and its natural frequency in this way, the dynamic response of the system can be adjusted to the different forcing conditions (different wavelengths). The height or frequency of the sea waves cannot be adjusted. However, once the optimal forcing parameters range for the non-dimensional system is determined, the system parameters can be adjusted to preserve the desired dynamical properties for the given sea waves conditions.

4 Dynamics and Control of Pendulums System

The dynamics of the coupled pendulums system under harmonic excitation has been studied numerically and experimentally for different levels of forcing. First the system without any control algorithm has been examined. Secondly the controlling algorithm for maintaining synchronized rotation has been applied. The experimental studies have been carried out in the Dynamics Laboratory at the University of Aberdeen. The harmonic excitation of the system has been provided by the electro-magnetic shaker. In the first stage of the study excitation acting only in the vertical direction has been considered as shown Fig. 3. The two pendulums have been fixed on the shaker as shown in the Fig. 3b and the response of the system has been observed for different initial conditions and varying frequency and amplitude.

Without control the pendulums can experience several types of dynamic behaviour. For constant amplitude and frequency of forcing different synchronized solutions have been found including equilibrium points, oscillations, rotations, and rotations of one pendulum synchronized in phase with the oscillations of the second one as shown in Fig. 4. Similar results have been obtained in the experimental studies as can be seen in Fig. 5.

After investigating several control techniques for parametric pendulum the delayed- feedback method has proved to be most efficient and robust. In this method the system can be stabilized by a feedback perturbation proportional to the difference between the present and a delayed state of the system [10], like the difference between velocities or angular displacements. During this study it

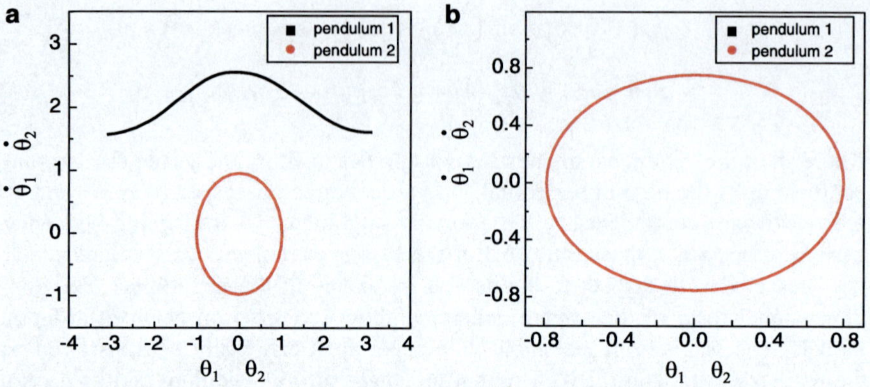

Fig. 4 (colour online) Numerical results showing (**a**) rotational and oscillatory orbit for pendulum 1 and 2 respectively and (**b**) synchronized oscillation of both pendulums

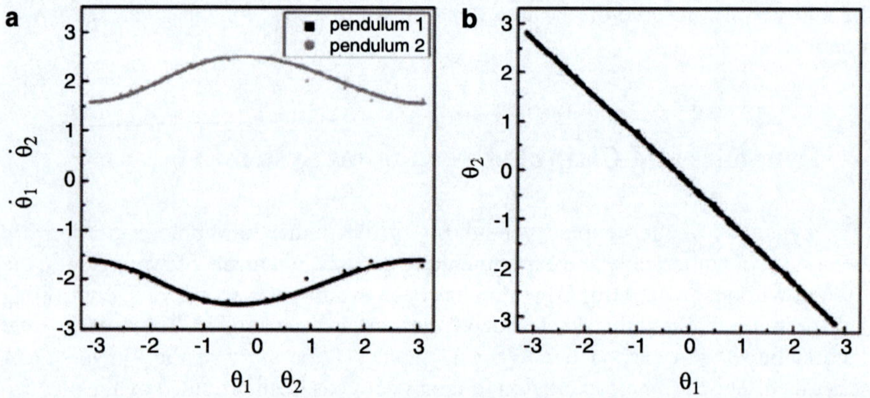

Fig. 5 Experimentally observed synchronized counter rotation of the pendulums for $\omega = 2$, $p = 0.07$. (**a**) Phase plane showing the velocity of the pendulum against its displacement for each revolution and (**b**) displacement of the pendulum 2 as a function of pendulum 1 displacement. A perfect correlation indicates complete synchronization

has been observed that when applying two independent rotation controllers to two pendulums, their responses would naturally synchronize. Therefore the separate controlling signals supplied to pendulum one and two are given by:

$$u_i = ksgn\left(\dot{\theta}_i(t)\right)(\theta_i\,(t-\tau) - \theta_i\,(t) + 2\pi) \tag{4}$$

where k is the proportionality constant, $\dot{\theta}_i$ (t), θ_i (t) and $\dot{\theta}_i(t-\tau)$ are the current velocity, current and delayed angular position of the i-th pendulum. The delay time

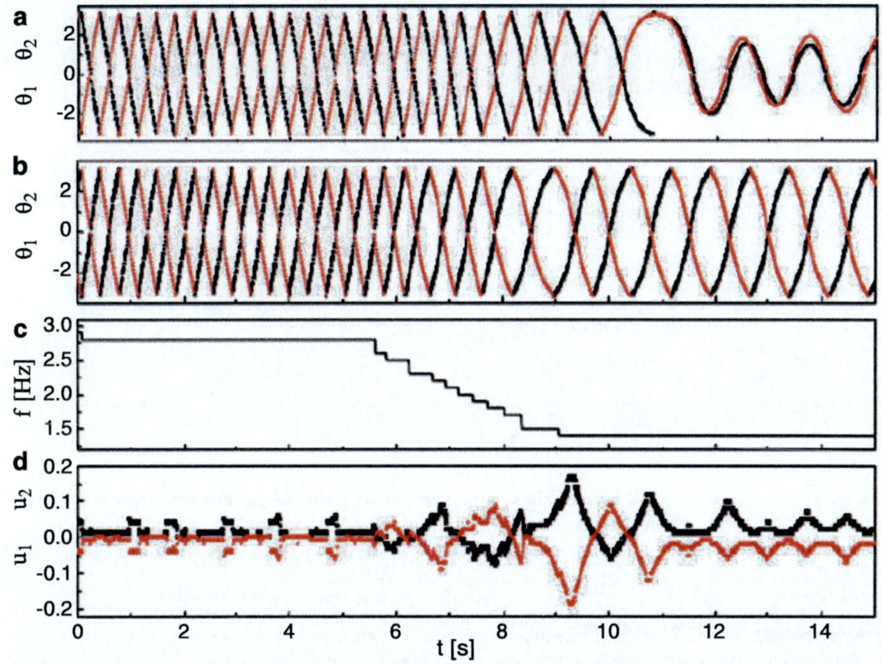

Fig. 6 (colour online) Experimental results showing the difference in the response of the system with and without control, while the forcing frequency is being changed from 3 Hz down to 1 Hz. (**a**) displacements with no control, (**b**) displacements with control, (**c**) forcing frequency, (**d**) control signals

(τ) is equal to the period of the desired periodic orbit, which for the period one rotations, being the dominant rotational response for the system considered, it can be determined from the period of the excitation. This can be extracted from the readings of the accelerometer measuring the oscillations of the pendulum base. If the pendulum rotates exactly with the period of excitation the angular displacement will increase 2π for each revolution and at the same time the control variable u_i will go to zero. The experimental and numerical result of applying two separate control signals, given by Eq. (4) to initiate and maintain rotational motion are shown in Figs. 6 and 7. Figure 6 depicts the experimental results of applying a control algorithm to maintain the synchronized rotation, while the excitation frequency is being varied. The synchronized rotation has been successfully maintained. The numerical times histories shown in Fig. 7 correlate well with the experimental results. Specifically, panels 6(b) and 7(a) depicting the controlled rotational angles, θ_1 and θ_2.

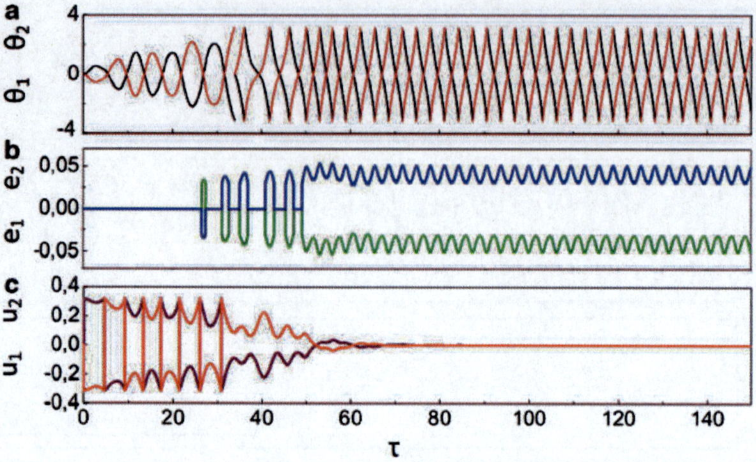

Fig. 7 (colour online) Numerical time histories showing the displacements, resistive torques corresponding to the energy extraction and controlling terms versus time, while initiating and maintaining synchronized rotational motion of two pendulums with two identical controllers

Acknowledgments We would like to acknowledge the current and former members of the Centre for Applied Dynamics Research and our collaborators who have contributed to this research. AN would like to thank the Northern Research Partnership for financial support.

References

1. Wiercigroch, M.: A new concept of energy extraction from oscillations via pendulum systems. UK Patent Application (2009, pending)
2. Xu, X.: Nonlinear dynamics of parametric pendulum for wave energy extraction. PhD thesis, University of Aberdeen, Aberdeen (2005)
3. Xu, X., Wiercigroch, M., Cartmell, M.P.: Rotating orbits of a parametrically excited pendulum. Chaos Soliton. Fract. **23**(5), 1537–1548 (2005)
4. Horton, B.W., Xu, X., Wiercigroch, M.: Transient tumbling chaos and damping identification for parametric pendulum. Philos. Tran. R. Soc. A **366**, 767–784 (2008)
5. Xu, X., Pavlovskaia, E., Wiercigroch, M., Romeo, F., Lenci, S.: Dynamic interactions between parametric pendulum and electro-dynamical shaker. Z. Angew. Math. Mech. **87**, 172–186 (2007)
6. Xu, X., Wiercigroch, M.: Approximate analytical solutions for oscillatory and rotational motion of a parametric pendulum. Nonlinear Dyn. **47**, 311–320 (2007)
7. Lenci, S., Rega, G., Pavlovskaia, E., Wiercigroch, M.: Rotating solutions and stability of parametric pendulum by perturbation method. J. Sound Vib. **310**, 243–259 (2008)
8. Litak, G., Wiercigroch, M., Horton, B.W., Xu, X.: Transient chaotic behaviour versus periodic motion of a parametric pendulum by recurrence plots. Z. Angew. Math. Mech. **90**, 33–41 (2010)
9. Horton, B.W., Sieber, J., Thompson, J.M.T., Wiercigroch, M.: Dynamics of the nearly parametric pendulum. Int. J. Non-linear Mech. **46**, 436–442 (2011)
10. Pyragas, K.: Continuous control of chaos by self-controlling feedback. Phys. Lett. A **170**, 421–428 (1992)

Part II
Vibration Problems in Structural Dynamics, Damage Mechanics and Composite Materials

Combination Resonance Characteristics of a Composite Plate Having Damage Subjected to in-Plane Pulsating Load

P.K. Datta and Sauvik Biswas

Abstract The current work predicts the occurrence of combination resonance in parametrically excited, simply supported composite plates with damage throughout the thickness. The damage is anisotropic in nature and parametrically incorporated into the composite using the concept of reduction in stiffness. Finite element is used to obtain the governing differential equation of the plate to which a modal transformation is applied in order to obtain the Matthew–Hill equation. The boundaries of instability due to combination resonance is obtained using a second-order Method of Multiple Scales (MMS) on the Matthew–Hill equation. The present work explores the effect of damage intensity and damage location on the regions of dynamic instability due to combination resonance when a narrow pulsating load having a static and a time-variant sinusoidal component is applied at the edge. Changes in the onset of instability and width of the instability region is observed due to increase in damage intensity and variation in the damage location. Further, the instability regions due to combination resonance effects are comparable to those of simple parametric regions which are usually observed in the literature. This shows that combination resonance effects are important to study the dynamic instability behavior of plate elements.

Keywords Anisotropic damage • Combination resonance • Dynamic instability

1 Introduction

Structural elements are often subjected to in-plane periodic forces during their operational lifetime. Under certain loading conditions, the structure may experience instability due to parametric resonance when the excitation frequency of the load

P.K. Datta (✉) • S. Biswas
Department of Aerospace Engineering, Indian Institute of Technology, Kharagpur 721302, India
e-mail: prosundatta@gmail.com

J. Náprstek et al. (eds.), *Vibration Problems ICOVP 2011: The 10th International Conference on Vibration Problems*, Springer Proceedings in Physics 139,
DOI 10.1007/978-94-007-2069-5_5,

becomes a function of the natural frequencies of the structure. When the excitation frequency is related to two or more natural frequencies of the structure, combination resonance phenomenon is observed. This phenomenon has been studied by many researchers [1, 2, 7] and its importance in the field of structural dynamics has been well asserted.

Damages or flaws in structures are often caused by growth of micro-cracks, accidents, projectile impacts, chemical corrosion, fatigue or due to faulty manufacturing process. While specific problems can be solved by 'hard-wiring' the damage into the model [3], a parametric model is more versatile. In situations where the exact nature of damage is unknown, yet it is necessary to identify the extent of damage, the response behavior of the structure may be monitored to do the same. However, it must be noted that the choice of a damage model depends entirely on the problem itself and parametric modeling of damage will be inadequate for large, visible discontinuities. Various researchers [4, 6, 8] have modeled damages for anisotropic materials using different internal variables.

The present work investigates the combination resonance characteristics of an anisotropically damaged composite plate when subjected to an in-plane pulsating load.

2 Mathematical Formulation

2.1 Anisotropic Damage

Anisotropic damage can be parametrically incorporated into the formulation by considering the parameter $\Gamma_i = \frac{A_i - A_i^*}{A_i}$, a representation of reduction in effective area, A_i^* [8]. Using this, the relationship between a damaged stress tensor and an undamaged stress tensor can be obtained as

$$\begin{Bmatrix} \sigma_{11}^* \\ \sigma_{22}^* \\ \sigma_{12}^* \\ \sigma_{21}^* \end{Bmatrix} = \begin{bmatrix} \frac{1}{1-\Gamma_1} & 0 & 0 \\ 0 & \frac{1}{1-\Gamma_2} & 0 \\ 0 & 0 & \frac{1}{1-\Gamma_1} \\ 0 & 0 & \frac{1}{1-\Gamma_2} \end{bmatrix} \begin{Bmatrix} \sigma_{11} \\ \sigma_{22} \\ \sigma_{12} \end{Bmatrix} \tag{1}$$

The above equation can be abbreviated as $\{\sigma^*\} = [\Psi]\{\sigma\}$, where $[\Psi]$ is a transformation matrix and can be used to relate a damaged stress-strain matrix with an undamaged one, $[D^*]^{-1} = [\Psi]^T[D]^{-1}[\Psi]$. The damaged stress-strain matrix for a two-dimensional laminate can be written as

$$[D^*] = \begin{bmatrix} f_1 E_1 & f_{12} E_2 \nu_{12} & 0 \\ f_{21} E_1 \nu_{21} & f_2 E_2 & 0 \\ 0 & 0 & 2\frac{f_1 f_2}{f_1 + f_2} G \end{bmatrix} \tag{2}$$

where the factors, $f_1 = \frac{(1-\Gamma_1)^2}{1-\nu_{12}\nu_{21}}$, $f_2 = \frac{(1-\Gamma_2)^2}{1-\nu_{12}\nu_{21}}$ and $f_{12} = \frac{(1-\Gamma_1)(1-\Gamma_2)}{1-\nu_{12}\nu_{21}}$ are defined as stiffness reduction factors.

For a damaged region, the stress-strain relation can be written as $\{\sigma^*\} = [D^*]\{\epsilon\}$ This relation can then be transformed into the general coordinate system as used in general undamaged cases.

2.2 *Method of Multiple Scales*

When a pulsating, harmonic load, $P(t) = \alpha P_{cr} + \beta P_{cr}\cos(\Omega t)$, expressed in terms of the buckling load, P_{cr}, is applied to an undamped system and a modal transformation is performed on the system, the transformed governing differential equation, can be written as

$$\{\ddot{\xi}\} + [\Lambda]\{\xi\} + 2\epsilon\cos(\Omega t)[\bar{K}_G]\{\xi\} = 0 \tag{3}$$

where $[\Lambda] = [\Phi]^T[[K] - \alpha P_{cr}[K_G]][\Phi]$, $\epsilon = \frac{\beta}{2}$ and $[\bar{K}_G] = -P_{cr}[\Phi]^T[K_G][\Phi]$. $[K_G]$ is the geometric stiffness matrix, $[\Phi]$ is the modal matrix containing the first M normal modes of vibration under the static component of the load, $P_s = \alpha P_{cr}$. $\{\xi\}$ is the vector containing the orthogonal modal coordinates. $[\Lambda]$ is the diagonal matrix containing the squares of first M frequencies of vibration under the static component of loading as the diagonal elements. Equation 3 is a form of the Matthew–Hill equation, whose component form can be written as

$$\ddot{\xi}_m + \omega^2\xi_m + 2\epsilon\cos(\Omega t)\sum_{n=1}^{M}\bar{K}_{G\,mn}\xi_n = 0 \tag{4}$$

where $m, n = 1, 2, 3, \ldots, M$. The term containing ε and Ω couples the dynamic load to the normal modes of the system, due to which a perturbation based method called Method of Multiple Scales [5] is used to solve the system of equations. The 'nearness' of Ω to the sum of ω_m and ω_n for sum type of combination resonance can be represented by introducing a small detuning parameter, σ, such that $\Omega = \omega_m + \omega_n + \epsilon\sigma$. For a second order expansion, the values of σ are obtained from the roots of a second order polynomial [1, 7] and are given by

$$\Omega = \omega_m + \omega_n \pm \frac{\beta}{2}\sqrt{\Lambda_{mn}} - \frac{\beta^2}{4}\left[\frac{\Lambda_{mn}}{8}\left(\frac{1}{\omega m} - \frac{1}{\omega n}\right) - (\hat{\chi}_m + \hat{\chi}_n)\right] \tag{5}$$

where $\Lambda_{mn} = \frac{P_{cr}^2\bar{K}_{G\,mn}\bar{K}_{G\,nm}}{\omega_m\omega_n}$ and $\hat{\chi}_i = \frac{P_{cr}^2}{2\omega_i}\left[\sum_k \frac{\bar{K}_{G\,ik}\bar{K}_{G\,ki}}{(\omega_i+\Omega)^2-\omega_k^2} + \frac{\bar{K}_{G\,ik}\bar{K}_{G\,ki}}{(\omega_i-\Omega)^2-\omega_k^2}\right]$.

When $m = n$, the Eq. 5 predicts simple resonance zones, while for $m \neq n$, combination resonance zones are obtained.

2.3 Problem Definition

For the present problem a (0/90/90/0) and a (45/−45/45/−45) simply-supported square plate having thickness ratio, $b/h = 100$ is considered under an uniaxial, partial edge loading of width $c/b = 0.2$. The static load factor, $\alpha = 0.2$, while the dynamic load factor, β, is varied from 0.0 to 1.0. The width of the instability region at $\beta = 0.8$ is taken to be a representation of the width of the instability region itself. The damage parameter representing the direction of fiber, Γ_1 is parametrically varied between 0.1 to 0.9 while Γ_2 has been set at 0.1. The implementation of the damage itself is done element-wise. A square damage patch spanning 4% of the total area is considered for all cases whose center is given by the coordinates (a^*, b^*).

3 Results and Discussion

Figure 1 validates the current formulation with Udar and Datta [7] and establishes that under a narrow edge load, regions of dynamic instability due to combination resonance is comparable to those due to simple resonance.

The change in the intensity of damage can sometimes cause a certain combination resonance to exist which may not have existed under a different damage intensity. It can be observed from Fig. 2 that when the damage ratio Γ_1/Γ_2 becomes greater than three, the instability zones due to combination resonance of $\omega_1 + \omega_4$ and $\omega_3 + \omega_4$ becomes significant while that due to combination resonance of $\omega_2 + \omega_4$ ceases to exist. Such phenomenon is not observed in the case of symmetric cross-ply.

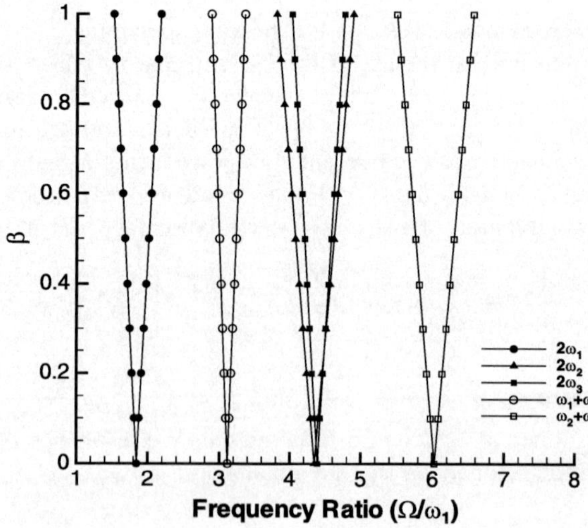

Fig. 1 Simple and combination resonance instability regions of an undamaged, edge loaded angle-ply. $c/b = 0.2$

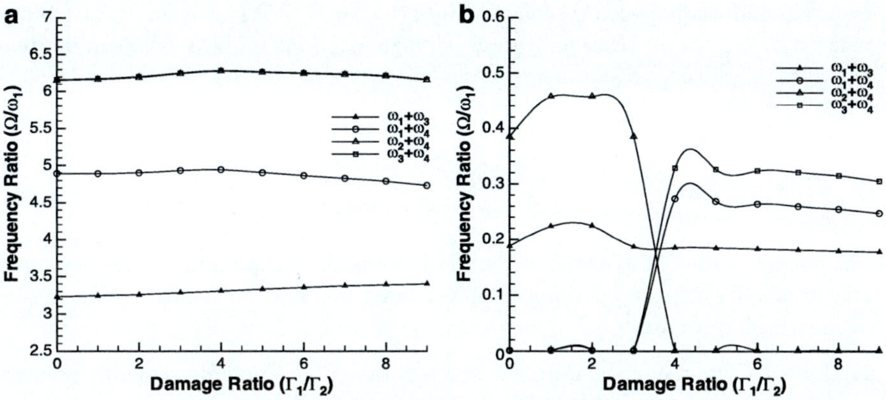

Fig. 2 Variation of (**a**) onset and (**b**) width at $\beta = 0.8$ of instability region of combination resonance with damage intensity of an edge loaded angle-ply. $c/b = 0.2$. Damage patch is located at the center of the panel

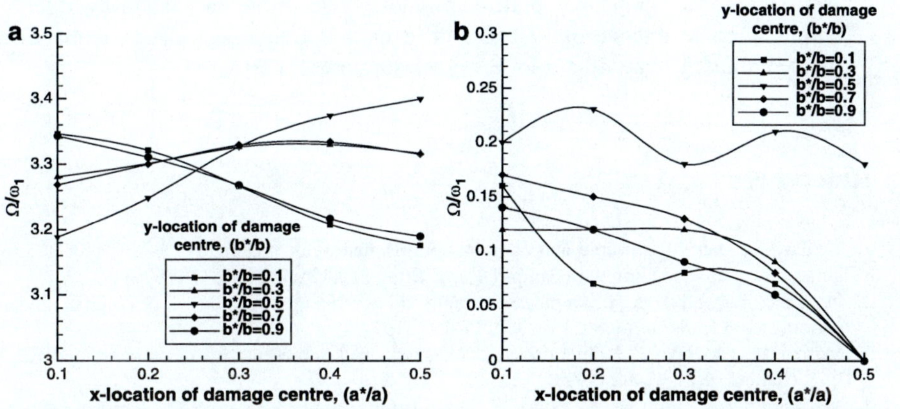

Fig. 3 Variation of (**a**) onset and (**b**) width at $\beta = 0.8$ of instability region of combination resonance ($\omega_1 + \omega_3$) with location of damage for an edge loaded angle-ply. $c/b = 0.2$. Damage ratio, $\Gamma_1/\Gamma_2 = 9$

Figure 3a plots the variation of the onset of instability regions ($\beta = 0.0$) due to damage location for a heavily damaged ($\Gamma_1/\Gamma_2 = 9.0$) angle-ply. It can be observed that for this particular combination resonance, $\Omega \approx \omega_1 + \omega_3$, the onset is advanced on the frequency ratio axis when the the damage patch is located midway near either of the simply supported sides and is delayed when the patch is located at the center of the plate. An inward shift of the onset indicates that the structure is more susceptible to instability. The width of the dynamic instability region is also dependent on the location of the damage patch. Figure 3b shows that the combination resonance due to $\Omega \approx \omega_1 + \omega_3$ ceases to exist as the damage location is moved midway between the loaded edges ($a*/a \approx 0.5$) However, the instability width remains almost invariant for all damage locations midway between the two unloaded edges ($b*/b \approx 0.5$).

4 Conclusion

The characteristics of dynamic instability regions due to combination resonance of anisotropically damaged composite plates under in-plane harmonic loading can be summarized as follows:

1. Under a narrow, edge loading, the width of the instability regions of combination resonance are comparable to those of simple resonance
2. In case of anti-symmetric angle ply, the variation in damage intensity can cause regions of dynamic instability due to combination resonance to 'open up' or 'close down'
3. The onset and width of instability regions is dependent on damage location. Anti-symmetric, angle-ply plates are more susceptible to instability due to combination resonance of first and third natural frequencies when damage is located midway near either of the simply-supported sides.

References

1. Deolasi, P., Datta, P.: Simple and combination resonances of rectangular plates subjected to non-uniform edge loading with damping. Eng. Struct. **19**(12), 1011–1017 (1997)
2. Iwatsubo, T., Sugiyama, Y., Ogino, S.: Simple and combination resonances of columns under periodic axial loads. J. Sound. Vib. **33**, 211–221 (1974)
3. Martha, L., Wawrzynek, P., Ingraffea, A.: Arbitrary crack representation using solid modeling. Eng. Comput. **9**(2), 63–82 (1993)
4. Matzenmiller, A., Lubliner, J., Taylor, R.: A constitutive model for anisotropic damage in fiber-composites. Mech. Mater. **20**(2), 125–152 (1995)
5. Nayfeh, A., Mook, D.: Nonlinear Oscillations. Wiley-Interscience, New York (1979)
6. Talreja, R.: A continuum mechanics characterization of damage in composite materials. In: Proceedings of the Royal Society of London. Series A, Mathematical and Physical Sciences, pp. 195–216. JSTOR, London (1985)
7. Udar, R., Datta, P.: Combination resonance characteristics of laminated composite plates subjected to non-uniform harmonic edge loading. Aircr. Eng. Aerosp. Tec. **78**(2), 109–119 (2006)
8. Valliappan, S., Murti, V., Wohua, Z.: Finite element analysis of anisotropic damage mechanics problems. Eng. Fract. Mech. **35**(6), 1061–1071 (1990)

On Nonlinear Vibration Analysis of Shallow Shells – A New Approach

M.M. Banerjee and J. Mazumdar

Abstract A new method for the analysis of nonlinear, large amplitude vibration of shallow shells of arbitrary shape is presented. The basic equations are derived using the concept of iso-amplitude contour lines approach. The coupled system of governing equations thus obtained, is solved using the Galerkin approach. As an illustration of the procedure, the case for a shallow dome upon an elliptical base is discussed. Some comparison is also made with previously obtained results, as available in the literature. The paper deals with first fundamental mode of vibrations only. The other mode of vibrations are not interested here as it requires to obtain the equations of iso-amplitude contour lines.

Keywords Iso-amplitude contour lines • Constant deflection contours • Shallow Shells • Nonlinear shell vibrations • CDC-Method

1 Introduction

It is well-known that the nonlinear dynamic behavior of thin shallow shell structures is of much technical importance to designers due to its wide range of applications in many fields of engineering. Containers, tanks, domes etc. are common examples of practical importance of such structures.

The problems of nonlinear vibration of shallow shells have attracted the attention of relatively few investigators in the past. Due to the very complicated nature of the

M.M. Banerjee (✉)
202 Nandan Apartment, Hill view (N), Asansol 713304, West Bengal, India
e-mail: muralimoha_banerjee@yahoo.com

J. Mazumdar
School of Electrical and Electronic Engineering and School of Mathematical Sciences, The University of Adelaide, Adelaide, Australia
e-mail: Jagan.Mazumdar@adelaide.edu.au

J. Náprstek et al. (eds.), *Vibration Problems ICOVP 2011: The 10th International Conference on Vibration Problems*, Springer Proceedings in Physics 139,
DOI 10.1007/978-94-007-2069-5_6,

basic equations governing the motion of a structure exhibiting large deflection, it has always been a difficult task for investigators to obtain even an approximate solution. Mazumdar in 1970 proposed a new approach which appeared to be quite suitable for bending analysis of elastic plates of arbitrary shapes based on the concept of iso-deflection contour lines on the bent surface of the plate [1]. This simple but efficient method is best known as Constant Deflection Contour Method or CDC-Method. Subsequently, the same method has been extended to the vibration analysis of plates and shallow shells [2, 3].

The CDC method has so far been restricted to linear analysis until an attempt has been made recently to extend it to nonlinear analysis of plates [4, 5]. In the present paper a similar approach as in [5] is undertaken for extension of the study to shallow shell analysis.

2 Derivation of Basic Equations

Consider an elastic, isotropic shallow shell of uniform thickness h subject to a continuously distributed normal load q. Let the equation of the middle surface of the shell referred to a system of orthogonal coordinates xyz, be given by [3]

$$z = \frac{x^2}{2R_x} + \frac{xy}{R_{xy}} + \frac{y^2}{2R_y} \tag{1}$$

where $r = \sqrt{x^2 + y^2}$ is small compared to the least of the radii of curvature, R_x, R_y and R_{xy} (supposed to be constants). If the shell is assumed to be comparatively thin and the displacements (u, v, w) are predominantly flexural, the strain components can be written as

$$\varepsilon_x = \frac{\partial u}{\partial x} + \frac{w}{R_x} + \frac{1}{2}\left(\frac{\partial w}{\partial x}\right)^2 - z\frac{\partial^2 w}{\partial x^2} = \frac{\sigma_x - \nu\sigma_y}{E},$$

$$\varepsilon_y = \frac{\partial v}{\partial y} + \frac{w}{R_y} + \frac{1}{2}\left(\frac{\partial w}{\partial y}\right)^2 - z\frac{\partial^2 w}{\partial y^2} = \frac{\sigma_y - \nu\sigma_x}{E}$$

$$\varepsilon_{xy} = \frac{\partial v}{\partial x} + \frac{\partial u}{\partial y} + \frac{\partial w}{\partial x}\frac{\partial w}{\partial y} + \frac{2w}{R_{xy}} - 2z\frac{\partial^2 w}{\partial x\,\partial y} = \frac{2(1+\nu)}{E}(\sigma_{xy}) \tag{2}$$

Formulating the Lagrangian with the expressions for strain energy, kinetic energy and the work done and then applying Hamilton's principle, a straightforward application of the variational calculus will yield the following equations of motion

$$D\nabla^4 w = hS(F,w) - h\left(\frac{F,_{yy}}{R_x} + \frac{F,_{xx}}{R_y} - 2\frac{F,_{xy}}{R_{xy}}\right) + q - \rho\, h\, w,_{tt} \tag{3}$$

and

$$D\nabla^4 F = -\frac{E}{2} hS\,(w, w) + E\left(\frac{w_{,yy}}{R_x} + \frac{w_{,xx}}{R_y} - 2\frac{w_{,xy}}{R_{xy}}\right) \tag{4}$$

where the operator *S(w, F)* stands for $S(w, F) \equiv \frac{\partial^2 w}{\partial x^2}\frac{\partial^2 F}{\partial y^2} - 2\frac{\partial^2 w}{\partial x \partial y}\frac{\partial^2 F}{\partial x \partial y} + \frac{\partial^2 w}{\partial y^2}\frac{\partial^2 F}{\partial x^2}$. Here 'F' denotes the well-known Airy-Stress function.

3 A New Approach

Mazumdar in [1] put forward a simple method, the so-called CDC-Method to solve the static and dynamic problems of elastic plates of arbitrary shapes. Mazumdar et al. [2, 3] applied this method for solving various problems of elastic plates and shells of arbitrary shapes, restricted to linear cases only. Following Mazumdar [1], a new idea has been put forward by Banerjee [4] to study the dynamic response of structures of arbitrary shapes based on the CDC method. While Mazumdar utilized the concept of Deflection Contour method to deduce the basic dynamical equations using elementary theory of plates and shells [1], the authors in [5] found it easy to arrive at the final equations by straightforward utilization of von Kármán field equations and then utilizing the required transformations to *u-variables.* In most practical cases, it is found that von Kármán field equations in conjunction with the CDC-Method make it easy to apply for nonlinear analyses of plates and shells.

4 Application of the CDC-Method

When the plate or the shallow shell vibrates in a normal mode, then at any instant t_θ, the intersections between the deflected surface and the parallels $z = constant$ yield contours which after projection onto the base plane $z = 0$ are a set of level curves, *u(x, y) = constant*, called the "Lines of Equal Deflections" [3], which are, in fact, iso-amplitude contour lines (Fig. 1). The boundary of the plate or the shell irrespective of any combination of support, is also a simple curve belonging to the family of lines of equal deflections. As defined by Mazumdar [1] this family of nonintersecting curves may be denoted by C_u, where $0 \leq u \leq u^*$, so that C_0 $(u = 0)$ is the boundary and C_u^* coincides with the point(s) at which the maximum $u = u^*$ is attained.

Let $u = u(x, y) = constant$ be a member of the family of iso-deflection or iso-amplitude contour lines. Using the following transformations

$$\frac{\partial w}{\partial x} = w_x = \frac{dw}{du}u_x,\ \frac{\partial w}{\partial y} = w_y = \frac{dw}{du}u_y,\ \frac{\partial^2 w}{\partial x \partial y} = w_{xy} = \frac{d^2 w}{du^2}u_x u_y + \frac{dw}{du}u_{xy},\ etc. \tag{5}$$

Fig. 1 Iso-deflection curves

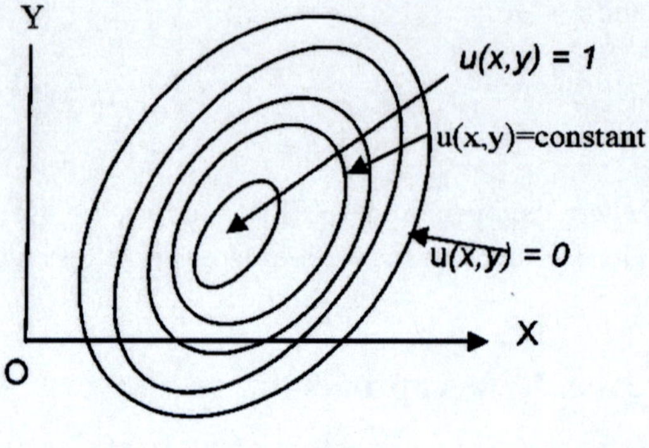

Equations 6 and 7 can be written as

$$D\sum_{i=1}^{4}\lambda_i\frac{d^{5-i}w}{du^{5-i}} = h\left[\lambda_5\frac{d}{du}\left(\frac{dw}{du}\frac{dF}{du}\right)+\lambda_6\frac{dw}{du}\frac{dF}{du}\right] - h\left[\lambda_7\frac{d^2F}{du^2}+\lambda_8\frac{dF}{du}\right]+q-\rho\, h\, w_{,tt} \tag{6}$$

$$\sum_{i=1}^{4}\lambda_i\frac{d^{5-i}F}{du^{5-i}} = -\frac{E}{2}\left[\lambda_9\frac{d}{du}\left(\frac{dw}{du}\right)^2+\lambda_{10}\left[\frac{dw}{du}\right]^2+E\left[\lambda_{11}\frac{d^2w}{du^2}+\lambda_{12}\frac{dw}{du}\right]\right. \tag{7}$$

where $\lambda_i s$ are known functions of $u_x, u_y, u_{xy}, u_{xx}, \ldots, etc.$

Since Eqs. 6 and 7 are valid for all points on the surface of the shell,

$$\iint_{\Omega}\left\{D\sum_{i=1}^{4}\lambda_i\frac{d^{5-i}w}{du^{5-i}} - h\left[\lambda_5\frac{d}{du}\left(\frac{dw}{du}\frac{dF}{du}\right)+\lambda_6\frac{dw}{du}\frac{dF}{du}\right] + h\left[\lambda_7\frac{d^2F}{du^2}+\lambda_8\frac{dF}{du}\right]-q+\rho\, h\, w_{,tt}\right\}d\Omega = 0 \tag{8}$$

and

$$\iint_{\Omega}\left\{\sum_{i=1}^{4}\lambda_i\frac{d^{5-i}F}{du^{5-i}} + \frac{E}{2}\left(\left[\lambda_9\frac{d}{du}\left(\frac{dw}{du}\right)^2+\lambda_{10}\left[\frac{dw}{du}\right]^2\right)+ - E\left[\lambda_{11}\frac{d^2w}{du^2}+\lambda_{12}\frac{dw}{du}\right]\right\}d\Omega = 0 \tag{9}$$

where the integration is over the region bounded by any contour C_u.

5 Method of Solution

It should be noted here that the above analysis is valid for any shallow shell structure. Evaluation of the integrals (8) and (9) with known values of the u-variables, will yield two ordinary differential equations; the basic fourth order partial differential equations reduce to two ordinary differential equations which make it rather easy for further study. From the resulting equation from (9), we first solve for the stress function with assumed expression of the deflected function compatible with the required boundary conditions and satisfying Galerkin's orthogonality condition. With this value of the stress function we solve the equation analogous to Eq. 8 applying Galerkin Method again, when the following "Time Differential Equation" with known constants will be obtained in the form.

$$\ddot{F}(t) + \alpha_1 F(t) + \alpha_2 F^2(t) + \alpha_3 f^3(t) = q^*, \tag{10}$$

the solution of which can be obtained and from which the subsequent analysis can be performed.

6 Specific Illustration

A specific illustration has been cited below.

6.1 *Large Vibration of a Shallow Dome upon an Elliptical Base*

Consider the vibration of a shallow dome of nonzero Gaussian curvature upon an elliptic base. Figure 2 depicts the geometry of the shell. The edges are clamped and immovable. When the shell vibrates in a normal mode, the lines of equal deflections, as described in Sect. 5, may reasonably be taken as

$$u(x, y) = 1 - \frac{x^2}{a^2} - \frac{y^2}{b^2} \tag{11}$$

Clearly, in this case $u = 0$ on the boundary and $u = u^* = 1$ at the centre of the shell. In order to obtain an approximate solutions, let us assume [5]

$$w(u, t) = W(u)\ f(t) = \sum_{j=2}^{\propto} Au^j f(t) \approx Au^2 f(t) \tag{12}$$

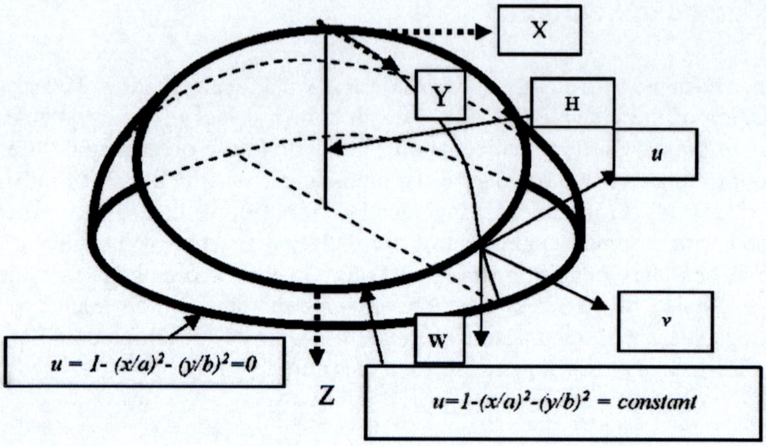

Fig. 2 Shallow dome on an elliptical dome

where $f(t)$ is an unknown function of time to be determined. Equation 10 for the present case reduces to the following equation with known coefficients

$$\rho h^2 \ddot{f} + \alpha_1 f + \alpha_2 f^2 \alpha_3 f^3 = Q^* = 5q/3 \tag{13}$$

An indirect verification of the correctness of the time differential equation may be made by considering the case for a flat plate problem. When $R_x, R_y \to \propto \alpha_2 = 0$, and further if $a = b$ the problem reduces to that of a circular plate for which Eq. 13 takes the form (for $\nu = 0.3$)

$$\rho h^2 \ddot{f} + \frac{Eh^4}{a^4}\left[9.756\ f + 4.762\ f^3\right] = \frac{5}{3}q, \quad (\textit{present study})$$

$$\rho h^2 \ddot{f} + \frac{Eh^4}{a^4}\left[9.768\ f + 4.602\ f^3\right] = \frac{5}{3}q \quad \text{(Ref. [7])} \tag{14}$$

which are in excellent agreement, considering the fact that only a single term approximation for the deflection function has been made for the present study.

6.2 *Free Linear Vibration*

Set, α_2, α_3 and Q^* each equals to zero in Eq. 13, when the linear frequency is given by

$$\omega_L^2 = \frac{Eh^2 P}{8\rho}\left[\frac{(320/3)}{12(1-\nu^2)} + 4\left(\frac{2\gamma_1}{h}\right)^2\right], \quad P = \frac{3a^4 + 2a^2b^2 + 3b^4}{a^4b^4} \tag{15}$$

where $\gamma_1 = \kappa/P$ and $\frac{2\gamma_1}{h} = \gamma^*$ represents the measure of shallowness of the shell. Equation 15, on simplification and with a little rearrangement of the parameters, is in excellent agreement with those cited in Refs. [3, 6].

If the second term in the expression for ω_L dominates the first then $\omega_{L2} = \left(\frac{2E}{P\rho}\kappa^2\right)^{1/2}$, which is exactly the same as that of Ref. [3]. It may also be noted here following Reissner [6], that the first term is predominant when γ^* or $H/h <$ or $= 25$, and the second term is predominant when $H/h \geq 25$ in order that the theory of shallow shells is applicable.

6.3 *Nonlinear Free Vibration*

Substituting $Q^* = 0$ in Eq. 13, one obtains

$$\rho h^2 \ddot{f} + \alpha_1 f + \alpha_2 f^2 \alpha_3 f^3 = 0 \tag{16}$$

This is a familiar form of time differential equation and from which the frequency ratio (Nonlinear to Linear) can be analyzed [5].

7 Results and Discussions

Frequency Analysis: Table 1 shows the values of linear frequency for a circular plates obtained using different approaches. It justifies the present approach (CDC method). Further discussion on the linear frequency is considered to be irrelevant as Eq. 15 is exactly the same as those obtained in Refs. [3, 6] and the authors have already made detailed discussion on it. Numerical results reveal that the dependence of nonlinear to linear frequency ratio on γ^* (where $\gamma^* = 2\gamma/h$ represents the measure of shallowness of the shell) and aspect ratio a/b, is significant. From the results for a spherical shell ($\gamma^* = 0$), it has been observed that the dependence on the Poisson's ratio is not so much significant though the nonlinear effect is comparatively a little lower for higher values of ν. The nonlinear effect is significant when value of a/b decreases. For a comparative study of dependence of the relative frequency ratio γ^* on the aspect ratio (a/b) of the axes of the elliptic base of the dome it confirms that the nonlinear effect is not so much dependent on aspect ratio for $\gamma^* \geq 1.5-2$. Considering all aspects as relevant from results, it appears that the values of γ^* in the range of 1–2 affect the nonlinear behavior of the vibrating shell.

Table 1 Values of $\frac{h}{a^2}\sqrt{\frac{E}{\rho}}$ in the expression for ω_L for a circular plate

ν	Ref. [6]	Present study	Ref. [7]
0	2.948	2.948	–
0.3	3.091	3.125	3.125

8 Conclusions

The application of polynomial expressions for the deflection and the stress functions in conjunction with the Galerkin procedure appears to produce highly accurate results. The comparison of results shows that using a moderately approximated expression for the deflection function yields results which are comparable to the previously obtained results using other approximate methods. It can therefore be concluded that the CDC method appears to be a simple tool to deal with the problems of nonlinear vibration of plates and shallow shells of arbitrary shapes.

In a nut shell the analysis and the results are presented here. The authors wish to publish the complete work in near future.

References

1. Mazumdar, J.: A method for solving problems of elastic plates of arbitrary shapes. J. Aust. Math. Soc. **11**, 95–112 (1970)
2. Mazumdar, J.: Transverse vibration of elastic plates by the method of constant deflection lines. J. Sound Vib. **18**, 147–155 (1971)
3. Jones, R., Mazumdar, J.: Transverse vibrations of shallow shells by the method of constant-deflection contours. J. Acoust. Soc. Am. **56**(5), 1487–1492 (1974)
4. Banerjee, M.M.: A new approach to the nonlinear vibration analysis of plates and shells. In: Transactions of the 14th International Conference on Structural Mechanics in Reactor Technology, (SMIRT-13), Division B, Paper No. 247, Lyon, France (1997)
5. Banerjee, M.M., Rogerson, G.A.: On the application of the constant deflection-contour method to non-linear vibrations of elastic plates. Arch. Appl. Mech. **72**, 279–292 (2002)
6. Reissner, E.: On axi-symmetrical vibrations of shallow spherical shells. Q. Appl. Math. **13**(3), 279–290 (1955)
7. Yamaki, N.: Influence of large amplitude on flexural vibration of elastic plates. ZAMM **41**, 501–510 (1961)

Study on the Characteristics of Noise Reduction Effects of a Sound Reduction Structure

Mitsuru Endo and Yong Su Kim

Abstract The concept of a sound reduction structure is proposed as a new method for reducing noises generated by vibrating solids. The proposed concept involves directly attaching hard sound absorbing materials to the vibrating surfaces of noise sources so as to provide a back air space. The effectiveness of the proposed concept is verified theoretically and experimentally using a one-dimensional acoustic field model. Then, the sound reduction characteristics in the three-dimensional acoustic field are clarified in detail through experiments using vibrating piston plates and a flexible flat plate with vibration modes.

Keywords Soundproofing • Characteristics of sound reduction structure • 1D and 3D acoustic fields • Hard sound absorbing material • Back air space

1 Introduction

Although there exist many developments of the sound absorbing materials or structures [1, 2], conventional noise-proofing countermeasures, which involve the installation of large amounts of sound absorbing materials on the walls surrounding noise sources, are disadvantageous with respect to cost and adversely affecting the surrounding scenery. The present paper proposes the concept of a sound reduction structure, which is directly attached to the vibrating surfaces of noise sources and involves a small amount of hard sound absorbing materials and a moderate amount of back air space. A schematic diagram of the proposed concept is shown in Fig. 1. It should be noted that there have been no works of such direct type for soundproofing until now.

M. Endo (✉) • Y.S. Kim
Tokyo Institute of Technology, Tokyo, Japan
e-mail: mendo2nd@lake.ocn.ne.jp

J. Náprstek et al. (eds.), *Vibration Problems ICOVP 2011: The 10th International Conference on Vibration Problems*, Springer Proceedings in Physics 139,
DOI 10.1007/978-94-007-2069-5_7, © Springer Science+Business Media B.V. 2011

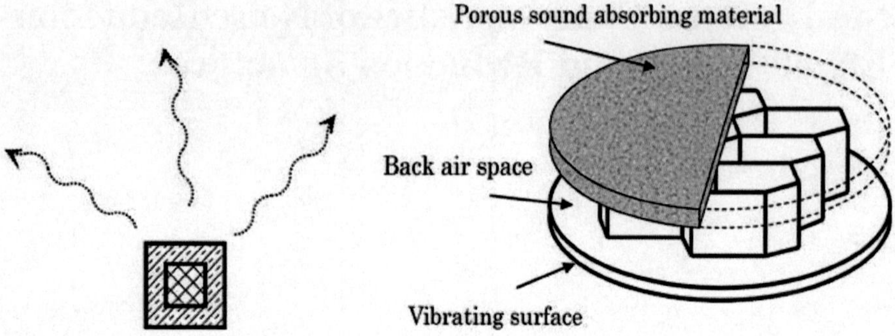

Fig. 1 Concept of proposed sound reduction structure

First, the validity of the proposed concept, i.e., whether the sound reduction effects are actually achieved for both stationary and non-stationary sounds, is verified experimentally using a one-dimensional (1D) acoustic tube. In addition, subsequent 1D theoretical analysis reveals a fairly good agreement with the experimental results. Then, experimental investigations of the noise reduction effects of the proposed sound reduction structure in a three-dimensional (3D) acoustic field are carried out using vibrating circular piston plates and a flexible flat plate with vibration modes surrounded by baffles.

In the present paper, the sound reduction effect is defined as the ratio of sound pressure generation for the case in which the specimen structure is installed on the vibrating solid surfaces to that for the case of a bare vibrating plate when the ratio is less than unity. The sound reduction effect is referred to as the normalized sound pressure in the figures.

2 Examinations by the 1D Acoustic Field Model

In this section, the effectiveness of the concept of the sound reduction structure is verified experimentally and theoretically.

2.1 Experimental Verification

Figure 2 shows the experimental apparatus for the case of stationary excitation.

One end of the tube is excited sinusoidally by an aluminum piston plate with or without the attached specimen and the other end is filled with appropriate soft sound absorbing materials so as to realize a condition that is approximately equivalent to a half-infinite length acoustic tube. For the non-stationary case, the axis of the piston

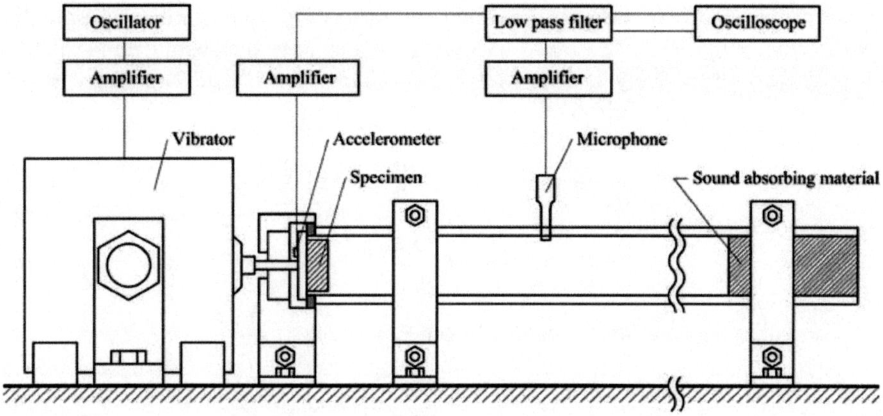

Fig. 2 Experimental setup of the 1D acoustic tube

plate is impacted by a steel ball and the generated impulsive sound is measured. The sound decreasing effects are evaluated in the frequency spectrum domain by taking the discrete Fourier Transform of the sound pressure waves.

As a result, the experimental results revealed that a maximum sound pressure reduction of approximately 70% is observed in the frequency range of $0.8 \sim 3$ kHz if we use the specimen structure constructed from the porous aluminum sound absorbing materials of thickness $h = 5$, 7, and 9 mm and a back air space of thickness $l = 20$ mm made of honeycomb plates. These results are presented in the following section.

2.2 Theoretical Examination

For a 1D acoustic field, X, the equation of motion of the enclosed air in the porous material with rigid frame is

$$\frac{\partial P}{\partial X} = -\rho_0 \frac{S}{H} \frac{\partial V}{\partial t} - \sigma V \tag{1}$$

and the equation of continuity is

$$\frac{\partial P}{\partial t} = -\frac{K_0}{H} \frac{\partial V}{\partial X} \tag{2}$$

where P, V, ρ_0, K_0, S, H, and σ are the sound pressure, the particle velocity, the air density, the bulk modulus, the structure constant, the porosity, and the resistance constant, respectively [3]. By setting S and H to be unity and σ to be zero in Eqs. 1 and 2, the governing equations for the acoustic field of air are obtained.

Next, we consider the case in which the material frame vibrates sinusoidally with velocity $U = U_0\, e^{j\omega t}$ $(\mathrm{j} = \sqrt{-1})$, where X and x are the absolute and relative coordinates, respectively, i.e., $X = x + z$, $\dot{z} = U$, *and* $V = U + u$. Therefore, Eqs. 1 and 2 are expressed as follows:

$$\frac{S}{H}\rho_0 \frac{\partial u}{\partial x} + \sigma u = -\frac{\partial P}{\partial x} - \rho_0 \frac{S}{H}\frac{\partial U}{\partial t} \tag{3}$$

$$\frac{\partial P}{\partial t} = -\frac{K_0}{H}\frac{\partial u}{\partial x} \tag{4}$$

The following equation is obtained from Eqs. 3 and 4:

$$\frac{S}{H}\rho_0 \frac{\partial^2 u}{\partial t^2} + \sigma \frac{\partial u}{\partial t} = \frac{K_0}{H}\frac{\partial^2 u}{\partial x^2} - \frac{S}{H}\rho_0 \frac{\partial^2 U}{\partial t^2} \tag{5}$$

Based on Eq. 5 by introducing the appropriate boundary conditions [4], we can evaluate the sound reduction rates for each frequency after some calculations. Then, we need the values of the parameters of the porous aluminum sound absorbing materials, i.e., *H, S,* and $\sigma(= \sigma_r + \mathrm{j}\sigma_i)$. These values were obtained from the propagation constant γ and the specific acoustic impedance W, which are measured using a two-microphone acoustic impedance measuring instrument (Brüel & Kjær Type 4206). The identified parameters are as follows:

$$H = 0.8,\ S = 9.282,\ \sigma_r = -13.923\omega + 29{,}665\ [\mathrm{kg/m^3 s}],\ \sigma_i = 29{,}665\ [\mathrm{kg/m^3 s}]$$

Figure 3 shows comparisons of the theoretical and experimental results, where h and l are the material and air space thicknesses, respectively. Both sets of results exhibit similar tendencies, whereas the deviations of the experimental results from the theoretical results are assumed to be caused by the fact that, in the experiment, a semi-infinite length acoustic tube was not realized completely.

Figure 4 shows the change in the inclinations corresponding to the thickness of the back air space. Based on Fig. 4, the sound reduction rates are observed to have a quasi-periodic inclination with respect to the sound frequency.

3 Experimental Investigation in the 3D Acoustic Field

Figure 5 shows the experimental apparatus for the measurement of the characteristics in the 3D acoustic field of the sounds generated by the specimens made of aluminum piston plates with a diameter of 50 mm and a thickness of 25 mm as base vibrating solids.

Figure 6 shows the influence of the thickness of the back air space, which was measured at a location of 90° as measured from the horizontal direction and 50 cm from the plate. These sound pressures in 3D field are supposed to show

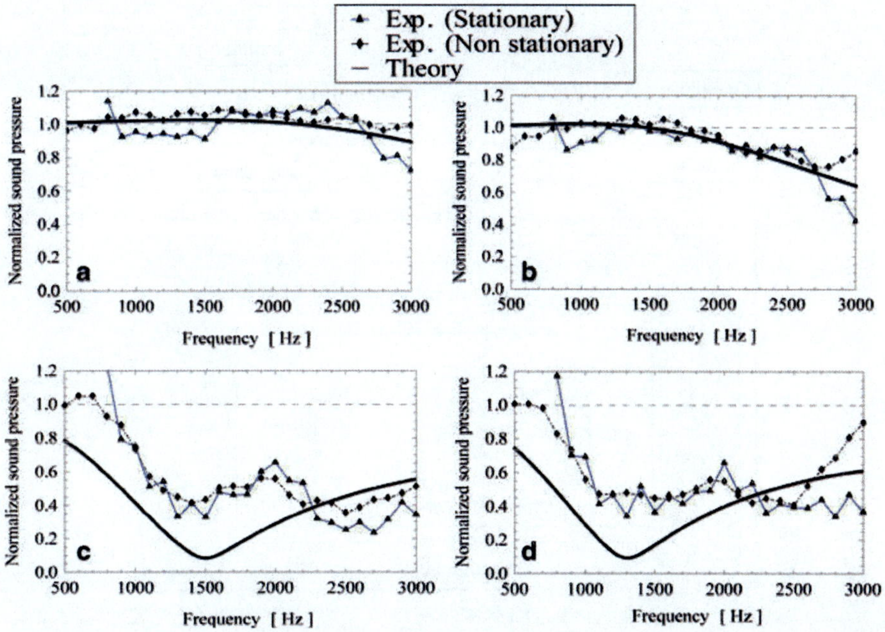

Fig. 3 Comparisons of experimental and theoretical results (**a**) $h = 7$ mm, without back air space, (**b**) $h = 9$ mm, without back air space, (**c**) $h = 7$ mm with a back air space of $l = 20$ mm, (**d**) $h = 9$ mm, with a back air space of $l = 20$ mm

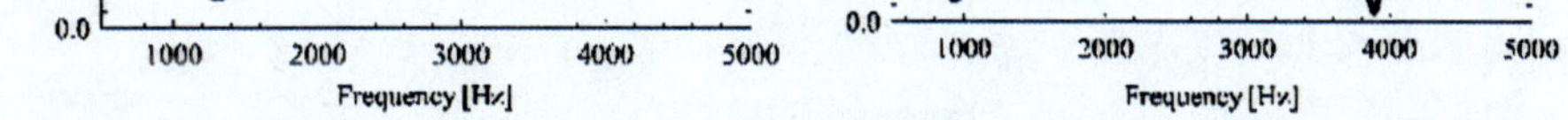

Fig. 4 Theory: sound reduction rates for $h = 9$ mm, (**a**) $l = 20$ mm, (**b**) $l = 40$ mm

the similar inclinations as in 1D field excluding the distance decay effects. Thus, as shown in Fig. 4b for 1D case, a quasi-periodic inclination with respect to the sound frequency is observed for $l = 40$ mm in Fig. 6. Further, for other observation points, i.e., $\theta = 30°$, $45°$, $60°$ (See Fig. 5), almost the same reduction tendencies were obtained. By considering those characteristics of the sound reduction rate in connection with the thickness of the back air spaces and the sound frequency, one can design, so-called, the nearly optimum sound reduction structure if the target frequency is assigned.

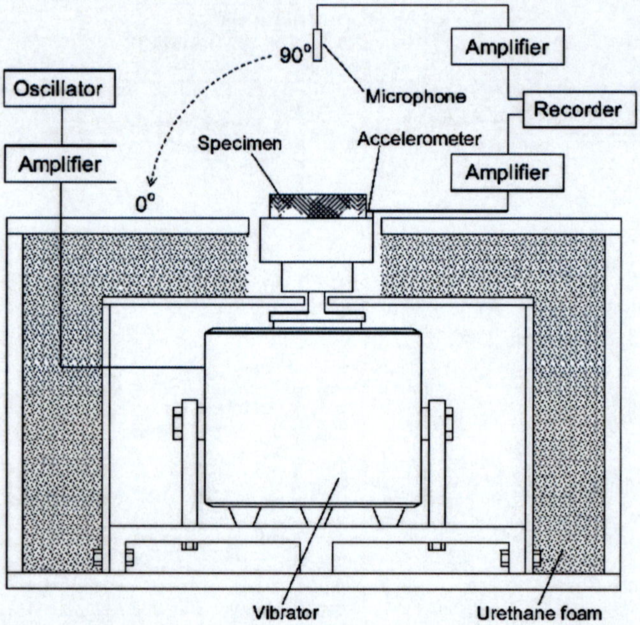

Fig. 5 Experimental apparatus for 3D measurement in a simple anechoic room

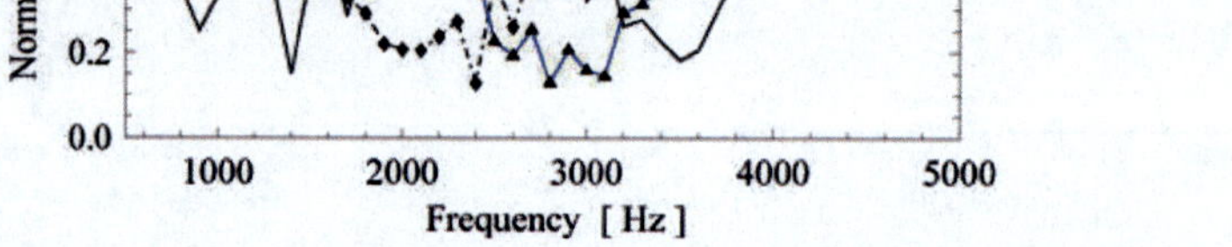

Fig. 6 Influence of thickness of back air space; *l*

Finally, an experiment was conducted in which a flexible rectangular aluminum plate having dimensions of 400 mm × 350 mm and a thickness of 5 mm, to which a specimen of $h = 5$ mm and $l = 20$ mm is attached, is excited at one point using five microphones arrayed at a height of 30 cm above the surface of the plate. The noise reduction effects evaluated in terms of the O.A. noise level are (i) 86.4 → 83.3 dB (Mic. 1), (ii) 88.7 → 86.3 dB (Mic. 2), (iii) 91.5 → 88.2 dB (Mic. 3), (iv) 91.2 → 89.3 dB (Mic. 4), (v) 86.4 → 83.7 dB (Mic. 5). Thus, a decrease in sound pressure level of approximately 3 dB O.A. was obtained at all observation points.

4 Conclusions

The conclusions of the present study are summarized as follows:

1. The effectiveness of the proposed concept of the sound reduction structure was verified experimentally and theoretically.
2. The sound reduction rate has a quasi-periodic inclination in connection with the thickness of the back air spaces and the sound frequency, which is one of the indexes for designing the sound reduction structure.
3. From a practical viewpoint, the proposed structures are expected to be applicable, for example, to flange surfaces of railway rails that generate rolling noise and to vibrating cylindrical blocks or the head surfaces of automobile engines.

References

1. Koyasu, M.: Development of sound absorbing materials. Noise Control **24**(1), 3 (2000)
2. Iida, K., et al. (ed.): Practical Noise/Vibration Control Handbook (in Japanese). NTS Inc., Tokyo (2000)
3. Zwikker, C., Kosten, C.W.: Sound Absorbing Materials. Elsevier, New York (1949)
4. Oie, S., Takeuchi, R.: Sound radiation from a surface of porous material. J. Acoust. Soc. Japan **34**(12), 704 (1978)

Delamination Detections of Laminated, Nonlinear Vibrating and Thermally Loaded Beams

Emil Manoach, Sylwester Samborski, and Jerzy Warminski

Abstract In this work, the geometrically nonlinear vibrations of fully clamped composite beams subjected to thermal changes are used to study the sensitivity of selected vibration response parameters to the presence of delamination (damage) and elevated temperature. The damage detection criterion formulated earlier for non-heated plates, based on analyzing the points in the Poincaré sections of the damaged and healthy plate, is modified and tested for the case of beams additionally subjected to elevated temperatures. The importance of the actual temperature in the process of damage detection is shown.

Keywords Beams • Nonlinear vibrations • Delamination detections • Numerical simulations • Poincaré maps

1 Introduction

The problem with the delamination in laminated structures and its early detection is very important from practical point of view. Vibration based structural health monitoring (VSHM) methods are widely used for a delamination detection. They are based on the fact that the delamination will alter the stiffness, mass or energy dissipation properties of a structure which in turn will alter its measured vibration response. The most used VSHM methods are the modal based methods [1]. According to many authors, however, there are many limitations of this group of methods (see for example [1–3]). Another problem concerning a number of VSHM

E. Manoach (✉) • S. Samborski • J. Warminski
Department of Applied Mechanics, Lublin University of Technology, Lublin, Poland

E. Manoach
Institute of Mechanics, Bulgarian Academy of Sciences, Sofia, Bulgaria
e-mail: e.manoach@imbm.bas.bg; e.manoach@pollub.pl

J. Náprstek et al. (eds.), *Vibration Problems ICOVP 2011: The 10th International Conference on Vibration Problems*, Springer Proceedings in Physics 139,
DOI 10.1007/978-94-007-2069-5_8, © Springer Science+Business Media B.V. 2011

methods is that they rely on a linear model of the structure. As the linear model itself can only approximate the actual behaviour of the vibrating structure, it will introduce computational errors [2]. These errors will be greater if the non-linearities of the system are substantial.

Temperature changes can and do affect substantially the vibration response of a structure. Thermal loads introduce stresses due to thermal expansion, which lead to changes in the modal properties. Thus, on a lot of occasions the presence of a temperature field can either mask the effect of damage or increase it, which will render a VSHM method ineffective – it might give no alarm when a fault is present or give a false alarm. This is why it is vital to be able to take into account the temperature changes when developing VSHM procedures.

The main objectives of this study are twofold: (i) to study the influence of defects, elevated temperatures and their combination on the dynamic characteristics of the laminated beams and on their geometrically nonlinear dynamic response; (ii) to test the criteria for identification of delamination in beams proposed in [3, 4] with taking into account the elevated temperature by analyzing the Poincaré map of the vibration response.

The application of the proposed approach is demonstrated on composite beams with delamination at elevated temperatures.

2 Theoretical Model

The object of investigation is a laminated beam with length l, width b having N_l number of layers, symmetrically disposed around a mid-axis (Fig. 1).

The beam is subjected to transverse load $p(x, t)$ leading to a large amplitude vibration and to temperature variation ΔT (in general it can be assumed as non-uniform along the beam thickness) with respect to a reference temperature. The geometrically nonlinear version of the Timoshenko beam theory which allow to consider the transverse shear and rotary inertia is used to describe the beam motion. The equations of motions used are written by using the effective properties of the composite beam. The analytical expressions of the studied model can be found, for example in [5, 6].

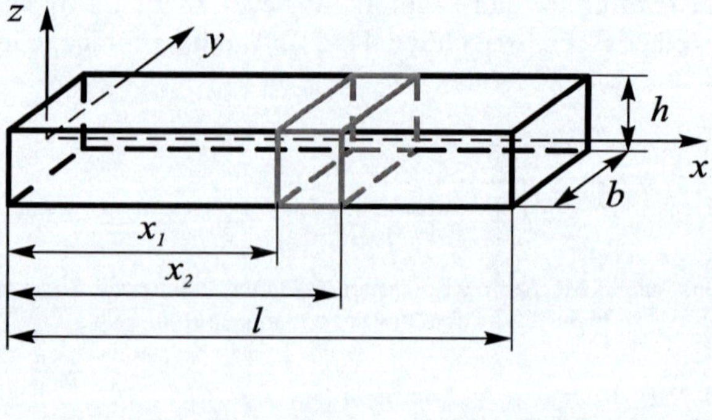

Fig. 1 Geometry of the beam. By x_1 and x_2 are denoted the beginning and the end of the delaminated area

The delamination is considered to be also symmetric with regard x axis. It results in reduction of the effective properties of the beam at the part of the cross-section of the beam having delamination. The idea how the delamination could be taken into account in the determining of the effective properties of the beam could be found for example in [7, 8]. In the theoretical model here, the variable rigidity of the beam due to the delamination is considered.

The beam is considered to be clamped, in-axis fixed and the initial conditions for all variables are set to zero. The solution of the problem is based on the numerical approach which is very similar to the one developed in [3, 9].

In most cases small damage in structures doesn't influence essentially the static response or eigen frequency of the structure. However, its influence can be observed when the structure is subjected to large dynamic loads, leading to large amplitude vibrations [4]. The identification of damage in this work is based on the criterion suggested in [3, 4].

Based on these works the following damage index is suggested:

$$I_i^d = \frac{\left|S_i^u - S_i^d\right|}{S_i^u + \varepsilon}, \quad S_i^u = \sum_{j=1}^{N_p-1} \sqrt{\left(w_{i,j+1}^u - w_{i,j}^u\right)^2 + \left(\dot{w}_{i,j+1}^u - \dot{w}_{i,j}^u\right)^2},$$

$$S_i^d = \sum_{j=1}^{N_p-1} \sqrt{\left(w_{i,j+1}^d - w_{i,j}^d\right)^2 + \left(\dot{w}_{i,j+1}^d - \dot{w}_{i,j}^d\right)^2}$$

In the above equations $i = 1, 2 \ldots N$, where N is the number of nodes, N_p is the number of points in the Poincaré map for each node and $(w_{ij}^u, \dot{w}_{ij}^u)$ and $(w_{ij}^d, \dot{w}_{ij}^d)$ denote the jth point on the Poincaré maps of the undamaged and the damaged states, respectively. It is easy to notice that S^u i and S^d represent the lengths of the lines formed by connecting the dots on the Poincaré maps for the damaged and the non-damaged beam for ith node. Therefore, the damage index is defined as the relative difference between these two lengths (see [3] for details). Parameter ε denotes a nonnegative number which depends on the geometry of the problem and on the loading. Its purpose is to avoid dividing of the denominator by very small numbers (for example at the areas close to the boundary or in a case of a curve formed from the dots in the Poincaré map with a measure close to zero). In the most cases it could be zero.

The above damage index depends on the location of the point on the beam axis, and consequently it is a function of the beam coordinate x. One can expect that the maxima of the curve $I^d(x)$ will represent the locations of the damage, i.e. $I_{\max}^d(x_d) = \max_i \left\{I_i^d\right\}$. The damage criterion based on this index presumes setting a threshold value T^d for the damage index. Thus, if the criterion $I^d(x, \Delta T) > T^d$ is fulfilled one can conclude that the beam is damaged. Moreover the areas of points (x) for which the criterion is fulfilled, form the damaged area (areas). It is important to note that the temperature changes should be taken into account. This suggestion presumes that the damage index defined by the above formulas is calculated for equal values of ΔT for the healthy and damaged beam.

3 Results and Discussions

The study focuses on delamination of a symmetric cross-ply laminated beam composed of 10 orthotropic layers, each 0.25 mm thick. The two layers located above and under the beam's neutral layer have 0° orientation with respect to x axis. The whole beam sequence of layers (plies) is $[(90/0)_2\ 0]_s$. The length of the beam is $l = 80$ mm and the width $b = 5$ mm. The material characteristics of the laminates are $E_1 = 56$ GPa, $E_2 = 16$ GPa, $\upsilon_{12} = 0.269$, $\alpha_T = 13.2 \times 10^{-6}\ K^{-1}$ and the material density $\rho = 2{,}052$ kg/m^3. These coefficients concerning rigidity properties of the beam correspond to a homogenous beam with effective properties $E_{ef} = 41.92$ GPa, $\nu_{ef} = 0.32991$. The delamination was modelled by prescribing to a small part of the beam located at $x \in (56\,\text{mm},\ 64\,\text{mm})$ (10% of the beam length, see Fig. 1) reduced rigidity $E^d_{ef} = 0.5E_{ef} = 20.96$ GPa. The beam was discretized by 40 linear beam finite elements. The damping coefficient was chosen $c_1 = 5.10^{-5}\ Nsm^{-3}$

The aim of the following example is to test the procedure for damage detection (DD) at the presence of temperature changes, to detect and localize damage (delamination) and to estimate the temperature influence on the process of DD.

First of all, the sensitivity of the first seven natural frequencies of the beams are calculated by FE method. The results show that the defects introduced due to the delamination cause very small changes in the natural frequencies (0.45% decreasing of the first natural frequency). Obviously such small changes cannot be used as an indicator for damage.

Then the forced response of the beam subjected to a harmonic loading is tested. The beam is subjected to two kind of loadings: (a) excitation with frequency of excitation equal (or very close) to the first natural frequency of the beams $\omega_e = \omega_1$ and (b) excitation equal to the half of the first natural frequency, i.e. $\omega_e = \omega_1/2$. Beams are subjected additionally to temperature changes $\Delta T = 5°K$ and $\Delta T = 10°K$.

Let us first have a look at the time histories of beam responses. In the cases when the excitation frequency is close to the first natural frequency of the beam a beating phenomenon occurs (see [3, 9]). The elevated temperature leads to an increasing the amplitude of vibrations and changes the period of beating. From the time-history diagrams (Figs. 2 and 3) it can be observed that the considered damage leads to some changes in the period of beating and small changes in the amplitude of responses (see the small figures inserted in the main figures where the time history is shown for a very short period of time).

The Poincaré maps shown in Figs. 4 and 5 confirm these observations. Nevertheless, in both cases the damage presence and its location is very well predicted by damage criterion based on the damage index as can be seen from Fig. 6. However, if one try to construct a damage index by using data for the unheated healthy beam and for the heated damaged beam the damage location cannot be predicted as can be seen from Fig. 7.

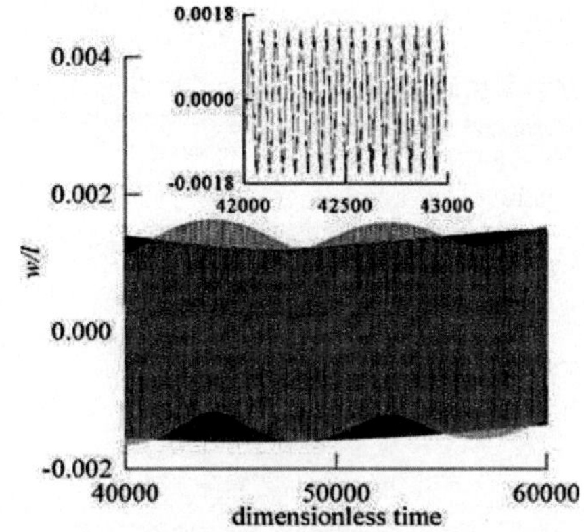

Fig. 2 Time-history diagram of the response of the beam centre of unheated healthy (*black line*) and damaged (*grey dashed line*) beam. $\omega_e = 5{,}665$ rad/s

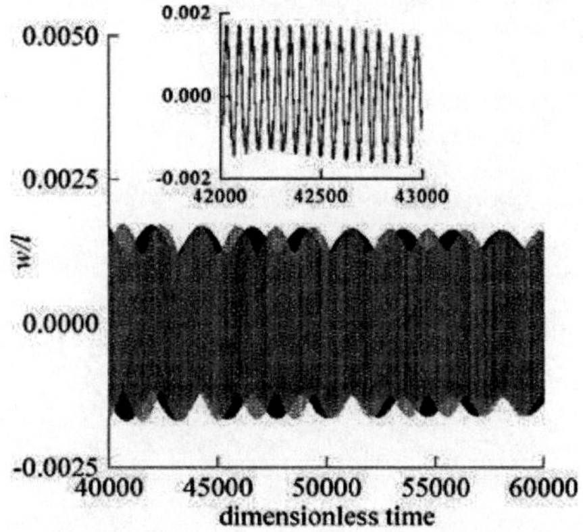

Fig. 3 Time-history diagram of the response of the beam centre of heated at $\Delta T = 10$ K healthy (*black line*) and damaged (*grey dashed line*) beam. $\omega_e = 5{,}665$ rad/s

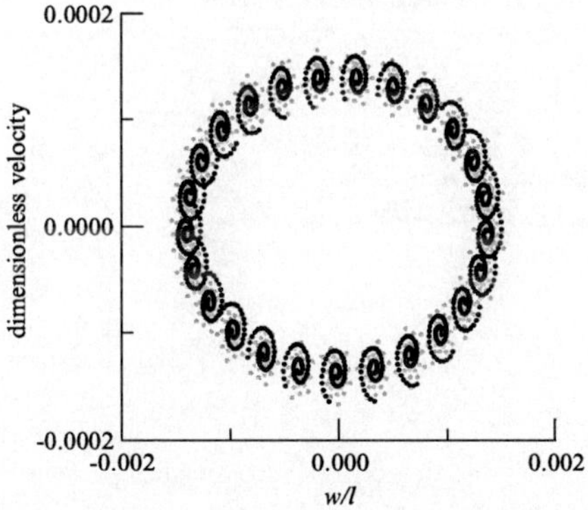

Fig. 4 Poincaré map of the response of unheated healthy (*black dots*) and damaged (*grey dots*) beam at the beam centre. $\omega_e = 5{,}665$ rad/s

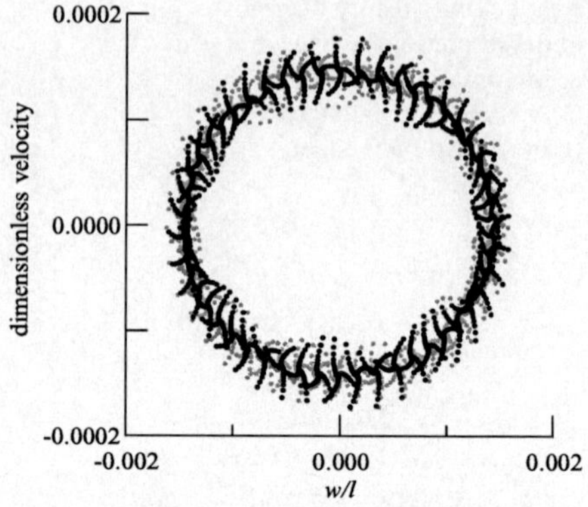

Fig. 5 Poincaré map of the response of heated at $\Delta T = 10$ K healthy (*black dots*) and damaged (*grey dots*) beam at the beam centre. $\omega_e = 5{,}665$ rad/s

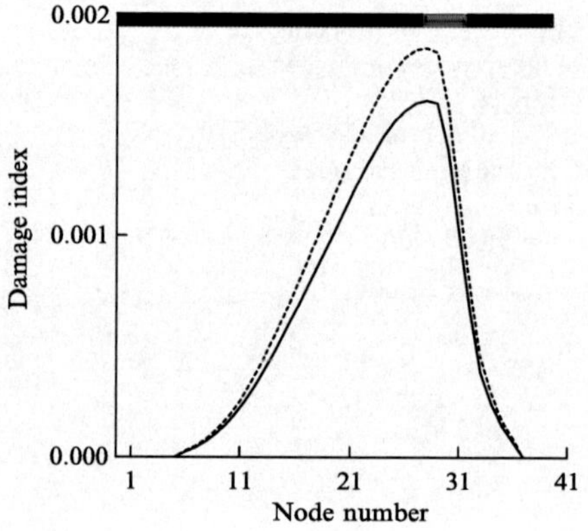

Fig. 6 Damage index for the heated and unheated beam subjected to harmonic mechanical loading with an amplitude $p = 50$ N, excitation frequency $\omega_e = 5{,}665$ rad/s. *Solid line* – $\Delta T = 0$. *Dashed line* – $\Delta T = 10$ K

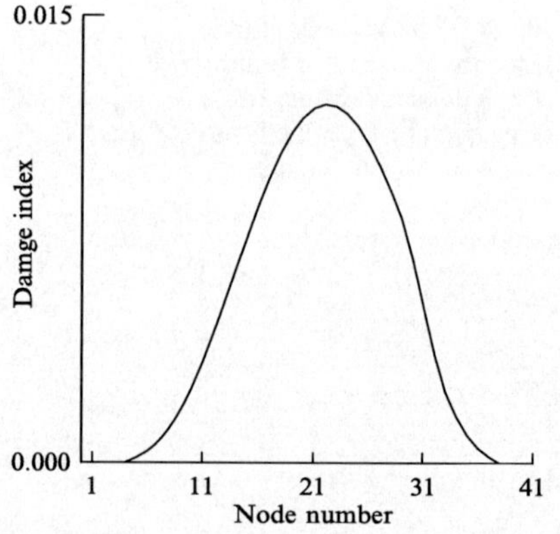

Fig. 7 Damage index computed when the healthy beam is at $\Delta T = 0$ and the damaged beam is at $\Delta T = 10$ K. $p = 50$ N, excitation frequency $\omega_e = 5{,}665$ rad/s

4 Conclusions

A numerical approach is applied to study the geometrically nonlinear vibration of thermally loaded composite beams with and without damage. Computed time domain responses are used to analyse either intact or damaged beams. Based on these analyses the damage index developed previously has been adapted and applied for a damage detection and damage location. It was demonstrated that the damage can influence substantially the time domain response of the beam despite its very small influence on the beam natural frequencies. The influence of the temperature changes is essential and it can change substantially the nonlinear dynamic response of the beam, especially for the case of fully clamped beams. Therefore temperature changes should be taken into account when developing damage assessment procedures. The potential, the sensitivity and the applicability of the developed method is to be tested for real measurements and for different defects, loading and temperature conditions.

Acknowledgments The research leading to these results has received funding from the European Union Seventh Framework Programme (FP7/2007–2013), FP7 – REGPOT – 2009 – 1, under grant agreement No: 245479. The first author would like to acknowledge the partial support received through Bulgarian NSF Grant DCVP 02/1.

References

1. Zou, Y., Tong, L., Steven, G.P.: Vibration based model-dependent damage (delamination) identification and health monitoring for composite structures – a review. J. Sound Vib. **230**, 357–378 (2000)
2. Banks, H.T., Inman, D.J., Leo, D.J., Wang, Y.: An experimentally validated damage detection theory in smart structures. J. Sound Vib. **191**, 859–880 (1996)
3. Manoach, E., Trendafilova, I.: Large amplitude vibrations and damage detection of rectangular plates. J. Sound Vib. **315**, 591–606 (2008)
4. Trendafilova, I., Manoach, E.: Vibration based damage detection in plates by using time series analysis. Mech. Syst. Signal Process. **22**, 1092–1106 (2008)
5. Altenbach, H., Altenbach, J., Kissing, W.: Structural Analysis of Laminate and Sandwich Beams and Plates. An Introduction into the Mechanics of Composites. Lubelskie Towarzystwo Naukowe, Lublin (2001)
6. Reddy, J.N.: Mechanics of Laminated Composite Plates and Shells: Theory and Analysis, 2nd edn. CRC Press, London (2003)
7. Joffe, R., Varna, J.: Analytical modeling of stiffness reduction in symmetric and balanced laminates due to cracks in 90° layers. Compos. Sci. Technol. **59**, 1641–1652 (1999)
8. Wee, Y.C., Boay, C.G.: Analytical and numerical studies on the buckling of delaminated composite beams. Compos. Struct. **80**, 307–319 (2007)
9. Manoach, E., Ribeiro, P.: Coupled, thermoelastic, large amplitude vibrations of Timoshenko beams. Int. J. Mech. Sci. **46**, 1589–1606 (2004)

Nonlinear Lateral Oscillations of Flexible Elements in Machines: Characteristic Features and Practical Applications

Vitaly Beresnevich, S. Tsyfansky, and A. Klokov

Abstract Lateral oscillations of flexible elements (belts, cables, strings, etc.) in machines under simultaneous action of kinematical and parametric excitations are studied. Mathematically the problem is presented as a partial differential equation describing interaction of forced and parametric oscillations of flexible element with due account of its geometrical, static and physical nonlinearities. By the mathematical simulation the influence of additional pulsation of axial tension force of flexible element on parameters of its forced resonant oscillations has been analyzed. New approach to the suppression of unfavorable nonlinear resonant oscillations of flexible element based on application to the system of additional parametric excitation is proposed. This method makes it possible to prevent nonlinear pulling of resonant oscillations and thanks to this extends the allowable operating frequency range of machine. The results of theoretical study are confirmed by experiments.

Keywords Flexible element • Nonlinear oscillations • Parametric excitation • Kinematical excitation • Resonant regime

1 Introduction

Flexible elements (belts, cables, guy ropes, strings, etc.) are widely used in machines and devices for various practical purposes (belt and chain transmissions, vibrating belts of vibromixers, guy ropes for stabilization of motion of machine tool, etc.) [1]. Lateral vibrations of flexible elements, which can occur during the operation of machine, are extremely detrimental. They give rise to additional dynamic loading, which encourages wearing and failure of flexible elements.

V. Beresnevich (✉) • S. Tsyfansky • A. Klokov
Riga Technical University, Riga, Latvia
e-mail: Vitalijs.Beresnevics@rtu.lv; Semjons.Cifanskis@rtu.lv; alex_klokov@inbox.lv

J. Náprstek et al. (eds.), *Vibration Problems ICOVP 2011: The 10th International Conference on Vibration Problems*, Springer Proceedings in Physics 139,
DOI 10.1007/978-94-007-2069-5_9, © Springer Science+Business Media B.V. 2011

Most of known works in the field of nonlinear dynamics of flexible elements are devoted to the analysis of free vibrations (e.g., [2, 3]). Characteristic properties of nonlinear parametric oscillations of flexible elements are analyzed in [4]. But cases of interaction between forced and parametric oscillations are usually considered only in application to the simple mechanical systems with one degree of freedom [5]. Problem of self-excited vibration suppression by parametric excitation in application to systems with finite numbers degrees of freedom is solved by A. Tondl [6]. This paper deals with the analysis of lateral oscillations of flexible elements under simultaneous action of kinematical and parametric excitations. On the base of this study new approach to the suppression of unfavorable nonlinear resonant oscillations by the additional vibration loading of the system is proposed.

2 Dynamic Model

Transverse oscillations of taut flexible element (thread) under kinematical and parametric excitations are considered (Fig. 1). Kinematical excitation is caused by forced transverse displacement of one end of the flexible element, but parametric excitation is due to additional periodic variation in time of axial tension force.

In forming of differential equation of oscillations it is supposed, that stiffness in bending of flexible element is negligible in comparison with its stiffness in tension. Weight of flexible element is ignorable in comparison with axial prestressing force T_0. Besides, it is considered that oscillations are performed in one plane, which runs along the centre line of a non-deformed flexible element. Taking the direction of the co-ordinate axis z along this centre line, the differential equation for transverse vibrations of flexible element can be stated as follows [1, 3]:

$$T_0(1+\mu \sin \Omega t)[1 + f(\varepsilon)]\left(1+b_1\frac{\partial}{\partial t}\right)\frac{\partial^2 y}{\partial z^2} - b_2\frac{\partial y}{\partial t} - \rho\left[1+\frac{1}{2}\left(\frac{\partial y}{\partial z}\right)^2\right]\frac{\partial^2 y}{\partial t^2} = 0, \tag{1}$$

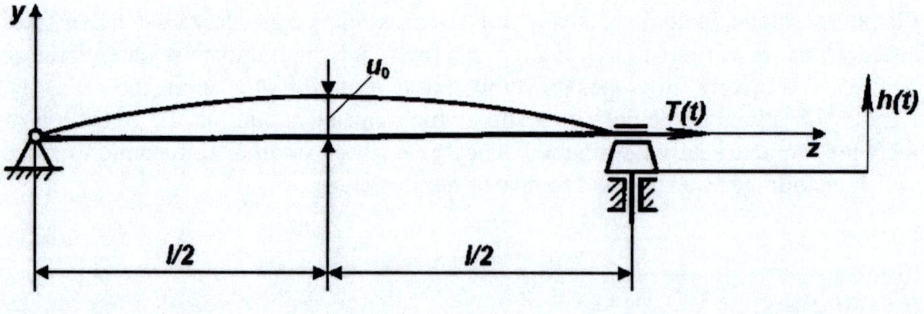

Fig. 1 Model considered in dynamic analysis

where T_0 is the prestressing force; μ and Ω are the non-dimensional amplitude and the frequency of parametric excitation; b_1 and b_2 are the coefficients of internal and external friction; y is the lateral displacement of the flexible element.

The functional $f(\varepsilon)$ in Eq. 1 takes into account additional tension caused by elastic deformation of flexible element during its oscillations (physical non-linearity). The elongation ε of flexible element can be determined by formula $\varepsilon = \frac{1}{2l}\int_0^l \left(\frac{\partial y}{\partial z}\right)^2 dz$ [3], where l is the length of flexible element.

The relationship between axial stress σ in flexible element and its elongation ε can be approximately described by the expression $\sigma = E\varepsilon - \beta\varepsilon^3$, where E is the elasticity modulus of material; β is the coefficient of non-linearity. In this case the functional $f(\varepsilon)$ can be expressed in the following form

$$f(\varepsilon) = \frac{EA}{2T_0 l}\int_0^l \left(\frac{\partial y}{\partial z}\right)^2 dz - \frac{\beta A}{8T_0 l^3}\left[\int_0^l \left(\frac{\partial y}{\partial z}\right)^2 dz\right]^3, \qquad (2)$$

where A is the cross-section area of flexible element.

Therefore an increment in tension is caused by integral elongation of flexible element and is independent of co-ordinate z. Non-linear term $[1 + \frac{1}{2}(\frac{\partial y}{\partial z})^2]$ of Eq. 1 takes into account geometrical non-linearity of flexible element [3]. In the case of forced transverse displacement of the right end of the flexible element the end boundary conditions are as follows:

$$y(z = 0, t) = 0; \; y(z = l, t) = h \sin \omega t, \qquad (3)$$

where h and ω are the amplitude and frequency of external kinematical excitation.

Equation 1, subject to the conditions (2) and (3), was solved in MATLAB environment using *pdepe* solver [7]. For this purpose initial differential Eq. 1 was transformed into equivalent set of first-order partial differential equations, and special-purpose MATLAB functions describing nonlinear characteristics have been created. Dynamics of the system also has been simulated on the specialized analogue-digital computer system developed in Riga Technical University [1, 8].

3 Characteristic Features of Oscillations and Their Application

A body of research is limited with resonance regime corresponding to the first natural frequency ω_1 of lateral oscillations of flexible element. Influence of additional pulsation of axial tension force $T(t)$ of flexible element on characteristic features of forced oscillations caused by kinematical excitation $h(t)$ have been investigated. Figure 2 shows a typical amplitude-frequency characteristic (AFC) for lateral oscillations of flexible element, which have been plotted assuming $T_0/EA = 3.5{\cdot}10^{-4}$, $b_1\omega_1 = 0.002$, $\beta/E = 0.36$, $\rho lg/EA = 5.7{\cdot}10^{-6}$ and $h/l = 0.005$. Dimensionless

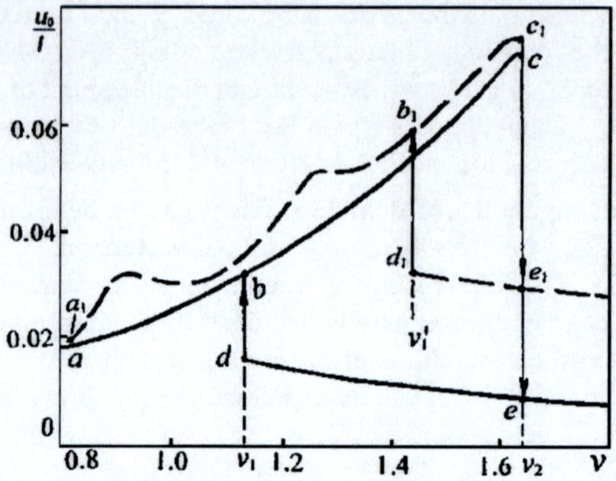

Fig. 2 Amplitude-frequency characteristics: *full lines* – purely forced oscillations ($\mu = 0$); *dotted lines* – regime resulted from interaction of forced and parametric oscillations ($\eta = 2$; $\mu = 0.2$)

displacement u_0/l (in antinodal point of the first resonant mode) is projected as amplitude on this AFC. Frequency of kinematical excitation is also laid off in dimensionless form $\nu = \omega/\omega_1$.

In the case of purely forced oscillations ($\mu = 0$, full lines on the AFC) the most intensive resonant regimes are realized within zone *bc* (zone of nonlinear pulling of vibrations). These resonant oscillations cause the appearance of high dynamic stresses in flexible element, and therefore they can be extremely detrimental for normal operation of machine. But, from the other hand, oscillations in zone *bc* have a limited reserve of stability [1]. This can result in non-stationary transfer of the system from the resonant curve *bc* to the non-resonant curve *de*. For example, such non-stationary transfer may be the result of some random perturbations, which can occur in operation regime of machine.

In accordance with the mathematical simulation, additional pulsation of axial tension force $T(t)$ of flexible element in some conditions may be used as an instrument for special-purpose control of bifurcation frequencies ν_1 and ν_2. Depending on parameters of additional parametric excitation (μ, $\eta = \Omega/\omega_1$), frequency range of dangerous resonant oscillations may be contracted or extended.

Dangerous variant of dynamic loading of flexible element is realized under the additional pulsation of axial tension force $T(t)$ with frequency $\Omega = (2 \div 2.08)\omega_1$, which is near to the top limit of the main zone of parametric resonance [4]. In such conditions a frequency range of stable forced resonant oscillations becomes sufficiently wider. As an example, Fig. 2 shows dotted resonant lines $a_1 b_1 c_1$ and $d_1 e_1$ which correspond to the case of additional pulsation of tension force $T(t)$ with frequency $\eta = 2$ and amplitude $\mu = 0.2$. This periodic pulsation of force $T(t)$ favors a shift of the opposite jump frequency ν_1 into the high-frequency range (shift of point d into the position d_1). But value of quench frequency ν_2 practically remains unaffected (compare the position of points c and c_1 on the AFC). Therefore resonant oscillations with high amplitudes become the only possible in the frequency range d–d_1. The revealed property of extension of frequency range of stable resonant forced oscillations has a negative influence on machine operation, facilitating an

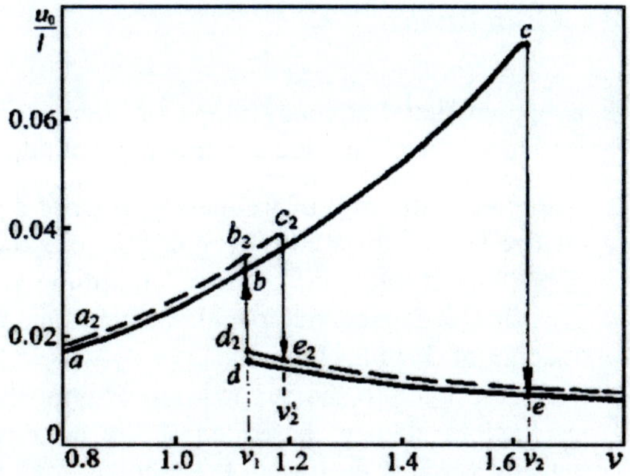

Fig. 3 Amplitude-frequency characteristics: *full lines* – purely forced oscillations ($\mu = 0$); *dotted lines* – regime resulted from interaction of forced and parametric oscillations ($\eta = 3$; $\mu = 0.25$)

increased dynamic loading of flexible elements. Therefore in designing stage of machine it is necessary to exclude the possibility of pulsation of axial tension force $T(t)$ with frequency $\Omega = (2 \div 2.08)\omega_1$.

Variant of suppression of dangerous resonant oscillations is realized in the high frequency range ($\eta = \Omega/\omega_1 = 2.8 \div 3$) of additional parametric excitation $\mu \sin\Omega t$. In this case suppression of non-linear forced resonant oscillations is achieved under the least possible amplitude $\mu = 0.20 \div 0.25$ of parametric excitation. Effect of suppression is illustrated with typical AFC shown in Fig. 3.

As it is seen, in the case of $\mu = 0$ maximal amplitudes of lateral oscillations of flexible element are realized near the resonant point c (here $u_0/l \approx 0.07$). Additional pulsation of axial tension force $T(t)$ favors a shift of the quench frequency ν_2 into the low-frequency range (shift of point c into the position c_2), but doesn't affect a value of opposite jump frequency ν_1. Therefore oscillations with small amplitudes become the only possible in the wide frequency range lain between points e_2 and e of the AFC, but maximal value of resonant amplitude is reduced in twice ($u_0/l \approx 0.035$ in the point c_2) in comparison with the case of $\mu = 0$.

On the base of this effect a new method for suppression of forced lateral oscillations of flexible element is proposed. The essence of the method may be explained using device shown in Fig. 1. During operation the right support makes vertical vibrations by the law $h(t) = h\sin\omega t$, that causes the kinematical excitation of resonant lateral oscillations of flexible element. For suppression of these oscillations an additional vibration excitation is applied to the system (e.g., by attachment the end of flexible element to a piezoelectric exciter), initiating a pulsation of axial tension force $T(t)$ with relative amplitude $\mu = 0.20 \div 0.25$ and with frequency Ω exceeding in $2.8 \div 3$ times the lowest natural frequency ω_1.

Main results of theoretical study are confirmed by experiments with a uniform rubber cord having length $l = 1.5$ m and linear density $\rho = 0.0415$ kg/m (in unloaded condition). Excitation and analysis of vibrations were carried out using standard electrodynamic and piezoelectric instrumentation [9].

4 Conclusions

It is shown that frequency range of stable forced resonant oscillations may be contracted or extended due to parameters of additional parametric excitation.

1. Sufficient extension of frequency range of stable resonant forced oscillations is realized under the additional pulsation of axial tension force $T(t)$ with frequency $\Omega = (2 \div 2.08)\omega_1$ and relative amplitude $\mu = 0.20 \div 0.25$. This effect has a negative influence on machine operation, facilitating an increased dynamic loading of flexible elements.
2. New approach to the suppression of unfavourable forced resonant oscillations of flexible element based on action on it with additional pulsation of axial tension force is proposed. It is shown, that effect of suppression is realized in the frequency range $\Omega = (2.8 \div 3)\omega_1$ under the relative amplitude of pulsation lain within the interval $\mu = 0.20 \div 0.25$.

References

1. Tsyfansky, S.L., Beresnevich, V.I., Oks, A.B.: Non-linear and Parametric Oscillations of Technological Vibromachines. Zinatne, Riga (1991) (in Russian)
2. Han, S.M., Grosenbaugh, M.A.: Non-linear free vibration of a cable against a straight obstacle. J. Sound Vib. **273**, 337–361 (2004)
3. Bondar, N.: Nonlinear Autonomous Problems in Mechanics of Elastic Systems. Budivelnik, Kiev (1971) (in Russian)
4. Beresnevich, V.: Characteristic properties of nonlinear parametric oscillations of flexible elements, scientific proceedings of Riga Technical University, Series 6: Transport and engineering. Mechanics **28**, 39–47 (2008)
5. Oks, A., Yano, S., Tsyfansky, S., Iwatsubo, T.: Suppression phenomena of resonant oscillations in strongly nonlinear systems due to additional asynchronous excitations. JSME Int. J. Ser. C **36**(1), 45–51 (1993)
6. Tondl, A., Ecker, H.: On the problem of self-excited vibration quenching by means of parametric excitation. Arch. Appl. Mech. **72**, 923–932 (2003)
7. Dukkipati, R.V.: MATLAB for Mechanical Engineers. New Age Science Limited, Tunbridge Wells (2009)
8. Cifanskis, S., Beresnevich, V.: Specialized analogue-digital computer system. In: High Tech in Latvia 2004, p. 30. AGB Publishing House, Riga (2004)
9. Tsyfansky, S., Beresnevich, V., Lushnikov, B.: Nonlinear Vibrodiagnostics of Machines and Mechanisms. RTU Publishing House, Riga (2008) (in Russian)

Appropriate Modeling of Dynamic Behavior of Quayside Container Cranes Boom Under a Moving Trolley

N. Zrnić, V. Gašić, A. Obradović, and S. Bošnjak

Abstract The paper deals with the analysis of moving trolley effects on the dynamic behavior of flexible structure of a mega high-performance quayside container crane (QCC) boom. The boom is modeled as a system with distributed parameters, comprising reduced stiffnesses and lumped masses from other parts of the upper structure. This paper looks both at the "moving force" and "moving mass" trolley modeling approaches to achieve the required performance of the QCC boom structure. Deterministic simulation for both considered approaches gives dynamic structural response of the boom for container transfer from quay-to-ship. The obtained results for "moving force" and "moving mass" models are compared in the scope of real values of parameters and future expectations in design of QCC. The conclusions lead to an appropriate way of model selection that can be used by engineers in practice.

Keywords Quayside container crane • Moving mass • Moving force • Modeling

1 Introduction

The moving load problem is one of the fundamental problems in structural dynamics. A lot of work has been reported dealing with the dynamic response of structures in the field of transportation and various constructions such as cranes, under the influence of moving loads. Extensive references to the literature on the subject can be found in the monograph by Fryba [1] with many analytical solution methods for simple cases.

N. Zrnić • V. Gašić (✉) • A. Obradović • S. Bošnjak
Faculty of Mechanical Engineering, University of Belgrade, Belgrade, Serbia
e-mail: vgasic@mas.bg.ac.rs

J. Náprstek et al. (eds.), *Vibration Problems ICOVP 2011: The 10th International Conference on Vibration Problems*, Springer Proceedings in Physics 139,
DOI 10.1007/978-94-007-2069-5_10,

The basic approaches in trolley modeling are: the “moving force” model; the “moving mass” model and the trolley “suspension model”. The simplest dynamic trolley models are the “moving force” models. The consequences of neglecting the structure-vehicle interaction in these models may sometimes be minor. In most moving force models the magnitudes of the contact forces are constant. “Moving force” models are very simple to use and yield reasonable structural results in some cases. The review paper concerning with the fundamental load problem of a uniform simply supported Euler-Bernoulli beam subjected to a constant vertical force moving at a constant speed is given in [2]. The “moving mass” model is an interactive model and includes transverse inertia effects between the beam and the mass. The interaction force between the moving mass and the structure during the time the mass travels along the structure takes into account the contribution from the inertia of the moving mass, the centrifugal force, the Coriolis force and the time-varying velocity-dependent forces. The trolley speed is assumed to be known in advance and thus not dependent on structural deformations. Generally, the dynamic structure-trolley interaction predicted by such models is very strong [3]. The study of a moving mass problem and related effects are given in [4–6]. The maximum dynamic amplification factors of beams, with different basic boundary under the action of moving forces that travel at constant speed are given in [7].

It should be mentioned that the problem of the forced motion of a beam, subjected to a moving load, is associated with serious difficulties when the effect of the mass of the load is accounted for [6]. But, on the other side it was concluded that the approximate solution for the “moving-force” model was not always an “upper bound” solution in terms of the deflection under the moving mass for the related “moving mass” problem. In other words HP Lee [5] was concluded that the inertial effect of the moving mass cannot be neglected in comparison with the gravitational effect even if the velocity of the moving mass is relatively small.

Modern high-performance mega QCCs have already tripled in outreach and load capacity compared to the first QCC built in 1959 [8]. This is not easily accomplished given the cantilevered nature of QCC. A cantilever (waterside boom) identified as the more important structural part [9] is structurally inefficient because almost all of the structural strength and weight is needed to support its own weight. In practice it is very complicated and expensive to do an experimental research on a real size mega QCC. This reason makes the investigations on mathematical models of moving load necessary, especially during the design stage and particularly having in mind the large dimensions of the boom and trolley mass.

In recent years, considerable efforts have been made in order to better understand the dynamic behavior and vibration of larges QCC under a moving trolley, particularly because of a construction of faster and heavier trolleys and the design of slender support structures without strict deflections limits. This fact resulted in several papers discussing the application of moving load problem in analysis of structural behavior of QCCs [3, 9–11]. Also, because of its great practical importance in design process engineers - practitioners dealing with constructions of QCC have studied the response of continuous beams to moving loadings [12, 13].

Therefore, the goal of this paper is to discuss both approaches ("moving force" and "moving mass") in analysis of dynamic interaction between QCC and a traveling trolley in order to suggest engineers the appropriate model to be used during the design process with respect to the level of complexity and accuracy. Particularly, because it is very difficult and expensive in practice to do an experimental research on a real-size mega QCC or even on a scale-model.

2 Mathematical Model of QCC Boom

The modeling process (consisting of several intermediate stages) of QCC boom structure is presented in details in [9]. The adopted model of the QCC boom, comprising the outline of the whole QCC FEM model, trapezoidal cross section of boom, as well as the picture of the hinge that enables understanding boundary conditions of the structural system, is shown in Fig. 1, and for a real QCC the structural parameters are taken from [3, 9–11] and are shown in Table 1. It is worthwhile to say that numerical examination of a model that is not a prototype of some real system is of little interest except for some general conclusions which can be applied to other, related configurations [14].

Discretization of the mathematical model was necessary, due to the model complexity, by five admissible functions adopted to generate an accurate estimate of the system response, as already done and verified in [9]:

$$\phi_1(x)=\frac{x}{L},\ \phi_2(x)=\sin\frac{\pi x}{L},\ \phi_3(x)=\sin\frac{2\pi x}{L},\ \phi_4(x)=\sin\frac{3\pi x}{L},\ \phi_5(x)=\sin\frac{4\pi x}{L}.$$

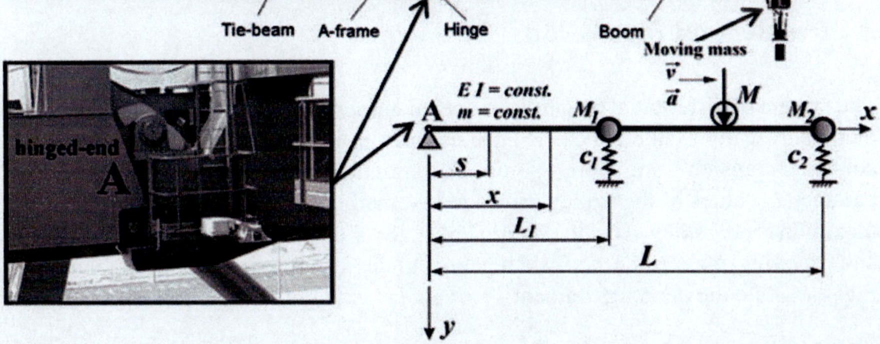

Fig. 1 Model of the QCC boom acted upon by the concentrated moving mass M

Table 1 Structural parameters of the QCC boom

L	L_l	E	I	A	m	M	M_l	M_2	c_l	c_2
[m]	[m]	[N/m^2]	[m^4]	[m^2]	[kg/m]	[t]	[t]	[t]	[N/m]	[N/m]
65.8	30	$2.1 \cdot 10^{11}$	0.25	0.193	1,849	175	3.69	21.17	$112 \cdot 10^5$	$32 \cdot 10^5$

The differential equations of motion for the "full" moving mass model are obtained by Lagrange's equations as explained in details in [9]. Deflection of the boom is assumed in the shape: $y(x,t) = \sum_{i=1}^{5} \phi_i(x) \cdot q_i(t)$. The difference between "moving force" and "moving mass" model is in contact force (1). Finally, after transformations, the contact force between the moving mass and the structure can be written as ("contribution" of the "moving mass" model is on the right side of the vertical slash):

$$F_C = M\left[g| - \ddot{y}(x,t) - 2v\dot{y}'(x,t) - v^2 y''(x,t) - ay'(x,t)\right] \tag{1}$$

In the above expression "g" is gravity acceleration, "v" is the velocity of the trolley as a moving mass, while "a" is the acceleration or deceleration (braking). In the expression for contact force the term $Mv^2y''(x,t) = -M\left(\frac{v^2}{R}\right)$ presents centripetal force, and the term $2Mv\dot{y}'(x,t)$ Coriolis force.

By applying Lagrange's equations and after transformations we obtain the final matrix form (2) for differential equations for the considered model of the boom for the mega QCC:

$$\begin{aligned}
&\sum_{j=1}^{5}\left[m_{ij} + M\phi_i(s)\phi_j(s)\right]\ddot{q}_j(t) + \sum_{j=1}^{5}\left[2Mv\phi_i(s)\phi_j{}'(s)\right]\dot{q}_j(t) \\
&\quad + \sum_{j=1}^{5}\left[c_{ij} + Mv^2\phi_i(s)\phi_j{}''(s) + Ma\phi_i(s)\phi_j{}'(s)\right]q_j(t) \\
&\quad = Mg\phi_i(s),\ i = 1,2,3,4,5.
\end{aligned} \tag{2}$$

3 Results and Discussion

The system of differential equations is solved numerically by using the Runge-Kutta method by using Mathematica [15]. Deterministic simulation is used to simulate the container transfer from shore to ship. For simulation of the trolley motion cycle, parameters values of the trolley drive are assumed in accordance with the currently maximum ones, i.e. $v = 6$ m/s, $a = 1.2$ m/s^2 [8]. For both models, "moving force" and "moving mass", the boom deflection under the moving trolley is shown in Fig. 2a, while the bending moment $M = -EIy''(x,t)$ under the trolley is shown in

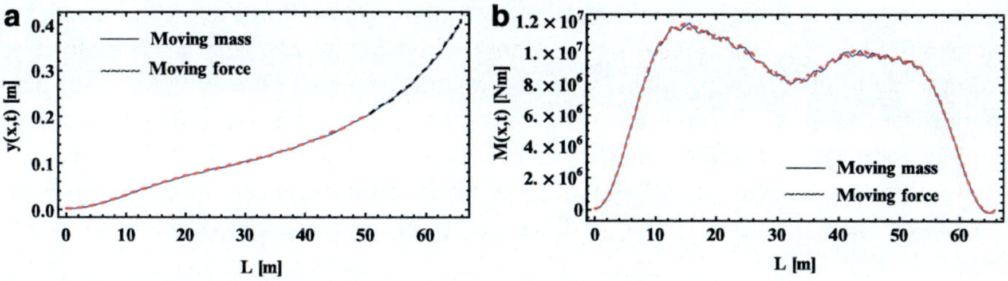

Fig. 2 (**a**) Boom deflection under the trolley, (**b**) Bending moment under the trolley

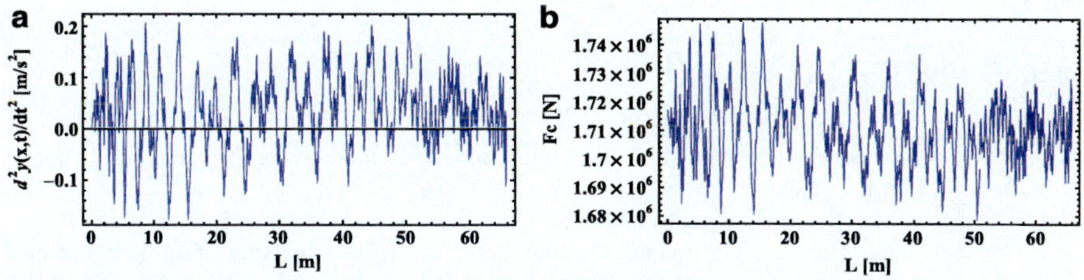

Fig. 3 (**a**) Vertical moving mass acceleration, (**b**) Contact force between the beam and the mass

Table 2 Comparison of the results for the given models

	y_{max} [m]	DAF_y	M_{max} [Nm]	DAF_M	F_c [kN]	$\ddot{y}(t)$ [m/s^2]
Mov. force	0.407	1.136	$1.174 \cdot 10^7$	1.098	1716.75	/
Mov. mass	0.409	1.142	$1.189 \cdot 10^7$	1.114	1749.60	0.21
Aps. Dev [%]	0.49	0.52	1.2	1.45	1.9	/

Fig. 2b. The moving mass acceleration in the vertical direction is shown in Fig. 3a. The contact force between the moving mass and the structure during trolley travel is shown in Fig. 3b.

For the sake of clarity comparison between the results obtained both for "moving mass" and moving force" models is shown in Table 2. The values for dynamic amplification factors (DAFs) are obtained here in relation to the maximum static deflection 0.358 m and maximum static bending moment $1.067 \cdot 10^9$ Nm for the considered structure.

4 Conclusions

This paper is an attempt to increase our understanding of the dynamics of mega QCC boom and to facilitate the decision process for model adopting. By analyzing the numerical results (Figs. 2 and 3 and Table 2) it should be evident that the dynamic structural response due to moving trolley differs from the quasistatic one.

It can be seen in Table 2 that the basic structural response "moving force" model is appropriate for use in engineering problems because it gives a very slight difference comparing to the "moving mass" model. This fact applies to the extreme up-to-date parameters of QCC. Although it very difficult to predict the future development of QCC it seems unlikely to reach in the next decades such an increase of performances that will favor "moving mass" model to be suitable for engineers in design process. However, "moving mass" model will be viable for scientific approach.

Acknowledgments A part of this work is a contribution to the Ministry of Science and Technological Development of Serbia funded project TR 35006.

References

1. Fryba, L.: Vibration of Solids and Structures under Moving Loads, 3rd rev. edn. Thomas Telford, London (1999)
2. Olsson, M.: On the fundamental moving load problem. J. Sound Vib. **145**, 299–307 (1991)
3. Zrnić, N., Bošnjak, S., Hoffmann, K.: Parameter sensitivity analysis of non-dimensional models of quayside container cranes. Math. Comput. Model. Dyn. Syst. **16**(2), 145–160 (2010)
4. Lee, H.P.: On the dynamic behaviour of a beam with an accelerating mass. Arch. Appl. Mech. **65**, 564–571 (1995)
5. Lee, H.P.: Dynamic response of a beam with a moving mass. J. Sound Vib. **191**(2), 289–294 (1996)
6. Michaltsos, G., Sophianopoulos, D., Kounadis, A.N.: The effect of a moving mass and other parameters on the dynamic response of a simply supported beam. J. Sound Vib. **191**(3), 357–362 (1996)
7. Urrutia-Galicia, J.L.: Amplification factors of beams under the action of moving point forces. Trans. Can. Soc. Mech. Eng. **18**(2), 165–172 (1994)
8. Zrnić, N., Hoffmann, K.: Ship-to-shore container cranes: from the beginning up to now (1959–2009). Österreichische Ingenieur- und Architekten-Zeitschrift **154**(7–12), S267–S276 (2009)
9. Zrnić, N., Hoffmann, K., Bošnjak, S.: Modelling of dynamic interaction between structure and trolley for mega container cranes. Math. Comput. Model. Dyn. Syst. **15**, 295–311 (2009)
10. Zrnić, N., Oguamanam, D., Bošnjak, S.: Dynamics and modelling of mega quayside container cranes. FME Trans. **34**(4), 193–198 (2006)
11. Zrnić, N., Bošnjak, S., Hoffmann, K.: Application of non-dimensional models in dynamic structural analysis of cranes under moving concentrated load. In: Troch I., Breitenecker, F. (eds.) Proceedings of 6th MATHMOD, Argesim Report No. 35, pp. 327–336. Argesim Verlag/ASIM, Vienna (2009)
12. Soderberg, E.: Megacranes: holding sway – boom motions. Port Technol. Int. 103–106 (June 2003)
13. Bhimani, A.K., Jordan, M.A.: A few facts about jumbo cranes. Presented at conference TOC Americas, Panama City, Panama, 3 Dec 2003
14. Zrnić, N., Bošnjak, S.: Comments on "modeling of system dynamics of a slewing flexible beam with moving payload pendulum". Mech. Res. Commun. **35**, 622–624 (2008)
15. Wolfram Mathematica 7 Documentation. http://www.wolfram.com

Dynamics of the Anti-backlash Designed Gearings with Elastic Elements

Vojtìch Kloucek

Abstract The paper contains a description of one of principles of the backlash-free transmissions. Specially is resolved design of the zero-backlash gearing with countershafts. The backlash elimination in gears and bearings is obtained by the preloaded torsion-bar spring. There is determined dependence of reaction forces, acting on gears and bearings, to magnitude of the torsion-bar spring preload. Using the computer program, created in the Maple environment, is simulated running of loaded gear with reversation of rotating direction. The outcomes are used for control of strength, fatigue of material and service life of gearing.

Keywords Anti-backlash transmissions • Gear trains • Precision position control • Preload • Torsion-bar spring

1 Introduction

In engineering practice there is often a need for precise positioning of physical objects such as workpieces, tools, assembled parts, transported materials, finished products etc. Positioning device can be a variable-axis of machine tool, rotary positioning table, a robotic manipulator and many other cases. In such cases it is necessary to accelerate and brake considerable weight objects. It follows that drives of positioning devices operates with a relatively low speeds and relatively large forces and torques.

In terms of type of motion drives can be categorized into (a) rotary and (b) linear. The dominant rotary drive are electric motors of various design. However, they have an optimum efficiency and loading characteristics at higher speeds and

V. Kloucek (✉)
VÚTS Liberec, a.s., Liberec, Czech Republic
e-mail: vojtech.kloucek@vuts.cz

J. Náprstek et al. (eds.), *Vibration Problems ICOVP 2011: The 10th International Conference on Vibration Problems*, Springer Proceedings in Physics 139,
DOI 10.1007/978-94-007-2069-5_11, © Springer Science+Business Media B.V. 2011

lower torques than the positioning drives require. Therefore, between the electric motor and positioning device is necessary to insert the appropriate transmission or gearbox. Gearbox, as well as any kinematic mechanism, is necessarily made with clearances and dimensional tolerances. Still current problem is the suppressing of these backlashes in reversing the shaft rotate direction. This article describes one way of suppressing the backlash in gears meshes and bearings of gearbox with spur gears and the load simulation of the gearbox.

2 Design

2.1 Principle of the Backlash-Free Gearbox

Figure 1a schematically illustrates the principle of backlash-free gearing with countershafts. Torsion bar spring is mounted at a determined preload torque, and thus act on meshing pinion wheels in opposite directions. When transmitting a rotary motion, then under the direction of rotation of input shaft performance by either one or the other path of gearing [1]. Similar methods of backslash elimination are described by [2] and [3].

2.2 Mounting Conditions

The gear by the Fig. 1a is closed loop of gears. Therefore, it is necessary in design to comply with geometric and kinematic constraints, which relate to axis distances

Fig. 1 (**a**) Anti-backlash gear assembly, (**b**) prototype design of the backlash-free gearbox

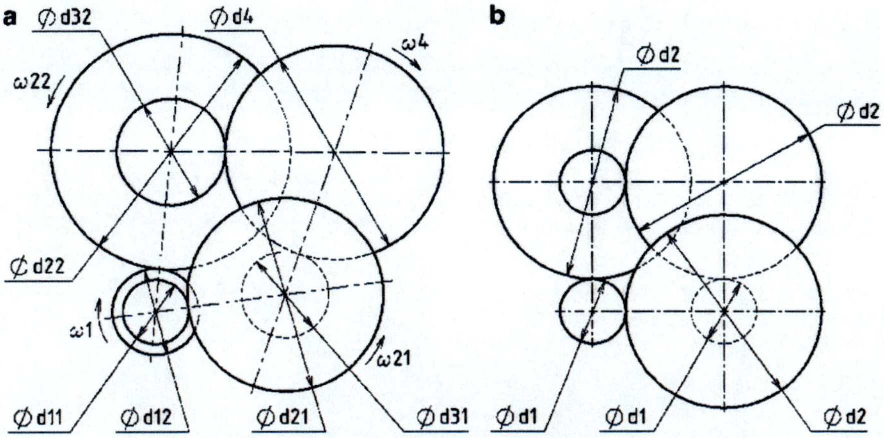

Fig. 2 Mounting conditions (**a**) general case, (**b**) prototype design

Table 1 Technical parameters of gearbox

Rated power	P	2.9	[kW]
Input torque	$M1$	18.6	[Nm]
Input speed	$n1$	1,500	[rpm]
Gear module	m	1.5	[mm]
Tooth number	$z1$	31	[–]
Tooth number	$z2$	108	[–]
Reduction ratio	i	12.1	[–]

and gear circumferential speeds. In general case (Fig. 2) must be satisfied condition for pitch diameters

$$\frac{d32}{d31} = \frac{d22}{d12} \cdot \frac{d11}{d21}. \tag{1}$$

Assuming the same module of all the gears, pitch diameters can be replaced by the number of teeth

$$\frac{z32}{z31} = \frac{z22}{z12} \cdot \frac{z11}{z21}. \tag{2}$$

The advantage of this type of gearing is high variability in dimensions and therefore large area for optimization of specific applications. For the prototype design and subsequent calculations was chosen case, when $d11 = d12 = d31 = d32 = d1$ and $d21 = d22 = d4 = d2$ (Fig. 2).

Table 1 lists technical parameters of the gearbox, which are used for subsequent calculations.

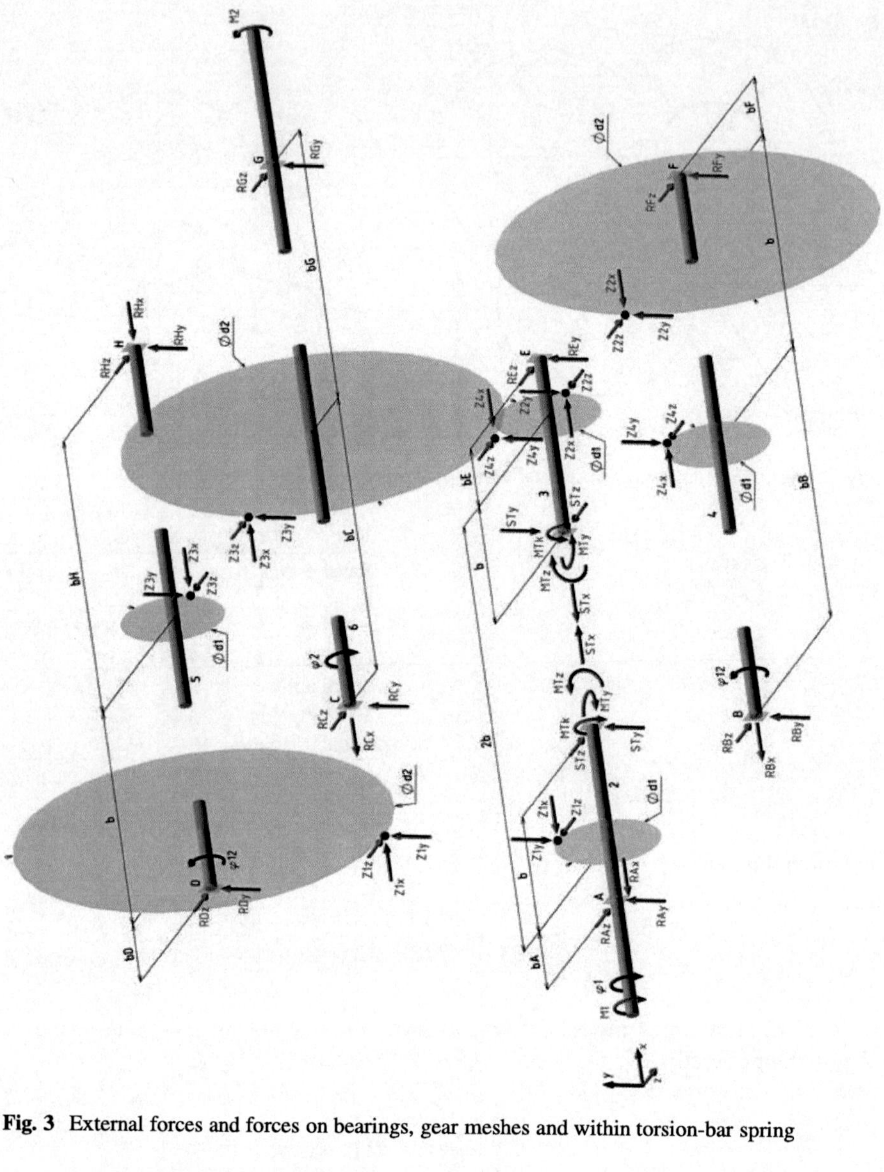

Fig. 3 External forces and forces on bearings, gear meshes and within torsion-bar spring

3 External and Internal Forces

3.1 Static Equilibrium

The gearing on Fig. 1a is a combined six-member mechanism (including frame). On Fig. 3 is release of each member. It should be determine 38 unknown forces and torques, torques Mtk (torsion-bar spring preload) a $M2$ (load on output shaft). are parameters of equation system. There are 30 equilibrium equations, the remaining eight equations resulting from the geometry of toothing (four meshes, for each one two equations).

3.2 *Dynamic Analysis*

For the dynamic analysis was created model by Fig. 4. The gear unit is connected by a ball screw with a weight, which is constrained by a ball linear guide. The input of model is known desired kinematics of weight. Subsequently, the reaction forces within the gearbox and input torque course are calculated. Mainly are controlled forces on meshes of pinions. To correct function of gearbox, they must be positive at all events.

For dynamic solution is necessary to know geometric characteristics as well as characteristics of mass. Weight mass was chosen of 100 kg. Inertia of gears, shafts and ball screw were determined from the 3D model in SolidWorks environment. For computing was created universal program in the Maple environment. All parameters are a variable, so the algorithm is applicable to any gear unit of the same design (Figs. 5–7).

4 Conclusions

Presented procedure is applicable for optimizing the dimensions of the gearboxes with same design. Created algorithm is universal, all parameters are implemented

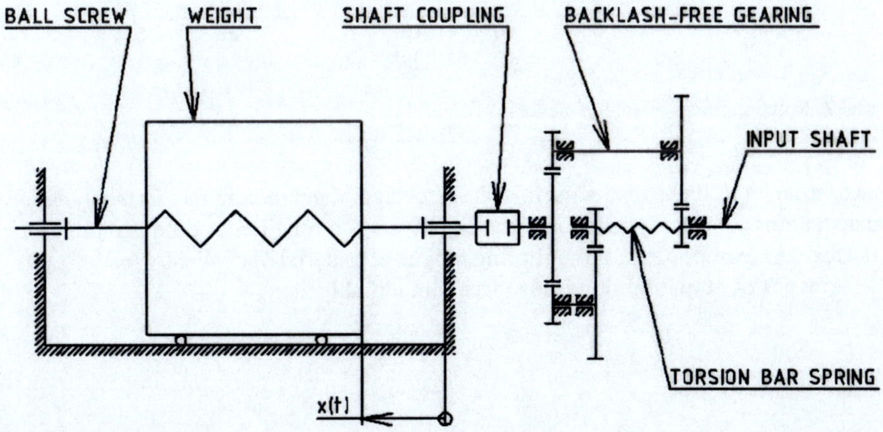

Fig. 4 Dynamic model

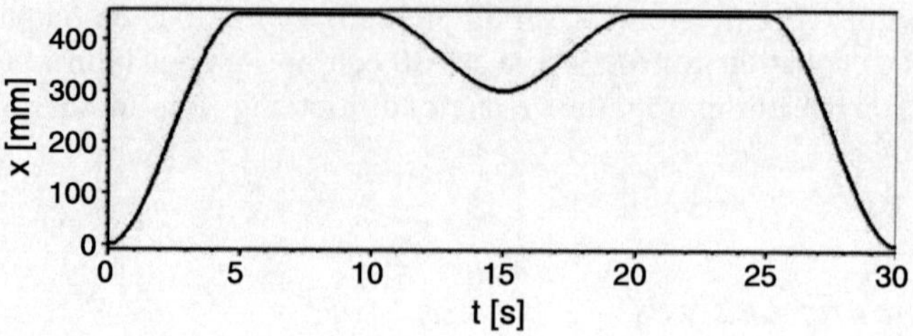

Fig. 5 Weight position vs. time

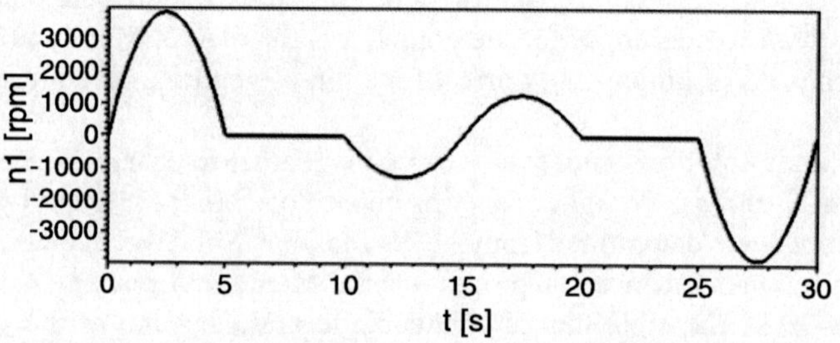

Fig. 6 Gearbox input shaft speed vs. time

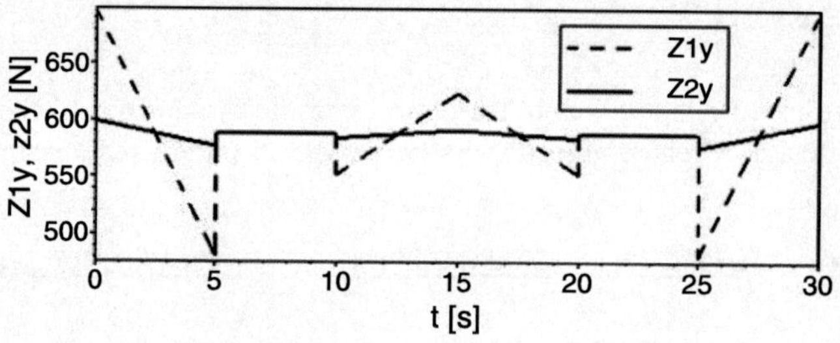

Fig. 7 Reaction forces on pinion meshes vs. time

as a variable. It, therefore, is useful for e.g. design of gearbox range. In the dynamic calculations can be any reaction force expressed as a function of time and therefore the results are important for monitoring fatigue of material of cyclically loaded parts. Kinematics of output motion is also freely modifiable.

Acknowledgments This work was supported by MPO ČR, research project TIP, no. FR–TI1/594, Research of sophisticated methods of design and development of special-purpose machines, components and peripherals of industry machines.

References

1. Hale, L.C., Slocum, A.H.: Design of anti-backlash transmissions for precision position control systems. Precis. Eng. **16**(4), 244–258 (1994)
2. Baumgarten, K., Schloeglmann, K.: Mechanical gear drive. US Patent 4,953,417, 4 Sept 1990
3. Hannel, C.L.: Anti-backlash gear assembly. US Patent 4,805,475, 21 Feb 1989

Industrial Applications of Extended Output-Only Blind Source Separation Techniques

Christophe Rutten, V.H. Nguyen, and J.C. Golinval

Abstract In the field of structural health monitoring or machine condition monitoring, most vibration based methods reported in the literature require to measure responses at several locations on the structure. In machine condition monitoring, the number of available vibration sensors is often small and it is not unusual that only one single sensor is used to monitor a machine. This paper presents industrial applications of two possible extensions of output-only Blind Source Separation (BSS) techniques, namely Principal Component Analysis (PCA) and Second Order Blind Identification (SOBI). Through the use of block Hankel matrices, these methods may be used when a reduced set of sensors or even one single sensor is available. The objective is to address the problem of fault detection in mechanical systems using subspace-based methods. The detection is achieved by comparing the subspace features between the reference and a current state using the concept of angular coherence between subspaces.

Keywords Health monitoring • Fault detection • Subspace • PCA • SOBI • Hankel matrices

1 Introduction

Blind source separation (BSS) techniques allow to recover a set of underlying sources from observations without any knowledge of the mixing process or the sources. BSS techniques were shown useful for modal identification [1], damage

C. Rutten (✉) • V.H. Nguyen • J.C. Golinval
Aerospace & Mechanical Engineering Department, Structural Dynamic Research Group, University of Liège, Liège, Belgium
e-mail: christophe.rutten@ulg.ac.be

J. Náprstek et al. (eds.), *Vibration Problems ICOVP 2011: The 10th International Conference on Vibration Problems*, Springer Proceedings in Physics 139,
DOI 10.1007/978-94-007-2069-5_12,

detection and condition monitoring [2] from output-only data. In the BSS family, on can cite for example the Principal Component Analysis (PCA) and the Second Order Blind Identification (SOBI) [3].

A drawback of many BSS techniques is the need for several sensors. However, through the use of Hankel matrices these methods can be extended to damage detection problem where only one sensor signal is available [4].

2 Principal Component Analysis

Let us assume that a dynamical system is characterized by a set of vibration features collected in the matrix $\mathbf{X} \in \Re^{(m \times N)}$, where m is the number of sensors and N is the number of samples.

PCA provides a linear mapping of data from the original dimension m to a lower dimension p. In practice, PCA is often computed by a Singular Value Decomposition (SVD) of matrix, i.e.

$$\mathbf{X} = \mathbf{U\Sigma V}^{\mathrm{T}} \tag{1}$$

Where $\mathbf{U}$ and $\mathbf{V}$ are orthonormal matrices, the columns of $\mathbf{U}$ define the principal components (PCs). The order p of the system is determined by selecting the first p non-zero singular values in Σ which have a significant magnitude ("energy") as depicted in [2].

3 Second Order Blind Identification

As described in [3], SOBI considers the observed signals as a noisy instantaneous linear mixture of source signals. In many situations, multidimensional observations are represented according to the following equation:

$$\mathbf{X}(t) = \mathbf{Y}(t) + \boldsymbol{\sigma}(t) = \mathbf{AS}(t) + \boldsymbol{\sigma}(t) \tag{2}$$

Where:

- $\mathbf{X}(t) = [x_1(t), \ldots, x_m(t)]^{\mathrm{T}}$ is an instantaneous mixture of source signals and noise.
- $\mathbf{S}(t) = \left[s_1(t), \ldots, s_p(t)\right]^T$ contains the signal from p narrow band sources $(p < m)$.
- $\mathbf{Y}(t) = [y_1(t), \ldots, y_m(t)]^T$ contains the sources assembly at a time t.

$\mathbf{A}$ is the transfer matrix between the sources and the sensor, called the mixing matrix. Under certain conditions, the mixing matrix identifies to the modal matrix of the structure and the sources correspond to normal coordinates [1]. $\boldsymbol{\sigma}(t)$ is the noise vector, modeled as a stationary white, zero-mean random process. Furthermore it is assumed to be independent of the sources.

The SOBI method aims at recovering the mixing matrix and the sources from their mixture. More details about this technique can be found in [3].

4 Hankel Matrices

Block Hankel matrices play an important role in system subspace identification [5]. Those matrices characterize the dynamics of the analyzed signals and have been used for modal identification and damage detection [4, 6, 7].

The covariance-driven block Hankel matrix is defined as follows:

$$\mathbf{H}_{r,c} = \begin{bmatrix} \Delta_1 & \Delta_2 & \cdots & \Delta_c \\ \Delta_2 & \Delta_1 & \cdots & \Delta_{c+1} \\ \vdots & \vdots & \ddots & \vdots \\ \Delta_r & \Delta_{r+1} & \cdots & \Delta_{r+c+1} \end{bmatrix}, (c \geq r) \tag{3}$$

where r, c are user-defined parameters ($r = c$, in this paper) and Δ_i represents the output covariance matrix.

The data-driven Hankel matrix is defined as:

$$\mathbf{H}_{1,2i} = \begin{bmatrix} x_1 & x_2 & \cdots & x_j \\ x_2 & x_3 & \cdots & x_{j+1} \\ \vdots & \vdots & \ddots & \vdots \\ x_j & x_{j+1} & \cdots & x_{i+j+1} \\ \hline x_{j+1} & x_{j+2} & \cdots & x_{i+j} \\ x_{j+2} & x_{j+3} & \cdots & x_{i+j+1} \\ \vdots & \vdots & \ddots & \vdots \\ x_{2i} & x_{2i+1} & \cdots & x_{2i+j-1} \end{bmatrix} \equiv \frac{\mathbf{X}_p}{\mathbf{X}_f} \equiv \frac{\text{"past"}}{\text{"future"}} \tag{4}$$

where $2i$ is a user-defined number of row blocks, each block contains m rows (number of measurement sensors), j is the number of columns (practically $j = N - 2i + 1$, N is the number of sampling points).

Enhanced-PCA (EPCA) and Enhanced-SOBI (ESOBI) techniques, as proposed in [4], consider the Hankel matrices, instead of the observation matrix $\mathbf{X}$, as input matrix. It can be shown that this procedure not only allows to enhance identification and detection but also permits the application of both methods to problems where only one sensor is available.

5 Damage Detection Problem

Damage detection is based on the concept of subspace angle introduced by Golub and Van Loan [8]. Subspace angles can be used as an indicator to quantify the spatial coherence between two data sets resulting from observation of a vibration system [2, 4].

The change in the system dynamics may then be detected by monitoring the angular coherence between subspaces estimated from a reference observation set and from the observation set of a current state of the system. A state is considered as reference if the system operates in normal conditions (i.e. damage does not exist).

In the case of EPCA, the considered subspaces are the active subspaces built by the first p columns of $\mathbf{U}$ while for ESOBI, the subspaces are built by the first columns of the mixing matrix $\mathbf{A}$.

6 Industrial Applications

6.1 *Quality Control of Electro-Mechanical Devices*

This industrial application concerns the case of electro-mechanical devices for which the overall quality at the end of the assembly line has to be assessed. A set of nine rotating devices was instrumented with two accelerometers: one triaxial accelerometer was located on the flank of the component, and one monoaxial on the top. Among this set of nine devices, five of them are known to be healthy (referenced OK-0 → OK-4) and the other four are faulty (NOK-1 → NOK-4). As it was shown in [7] that the detection is the best in the Y direction, the data in this same direction are used here to test the methods.

Detection results are presented in Fig. 1. The first healthy device (OK-0) is considered as the reference state. Both methods were able to make a clear distinction between the faulty devices and the healthy ones.

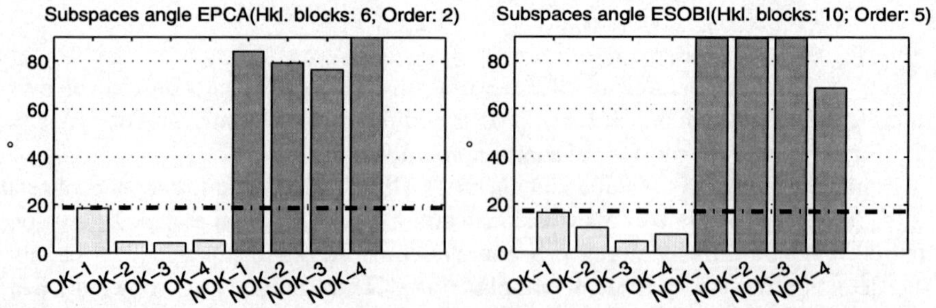

Fig. 1 Diagnosis of rotating devices with EPCA method on the *right* and ESOBI on the *left*

6.2 Quality Control of Weldings

The second example involves an industrial welding machine from a steel processing plan. The machine was instrumented with a monoaxial accelerometer on the upper forging wheel. The purpose of this wheel is to flatten the welded joint during the welding process.

The quality of the welded joints depends on several parameters. In this example, four distinct parameters were altered and multiple alteration levels were considered, leading to a batch of 27 welded joints with out-of-range parameters. Six welded joints were also realized using nominal parameters for false-positive testing (Table 1). A microscopic quality control of each welded joint was realized at the end of the campaign to assess their actual quality.

The detection results of EPCA and ESOBI are presented in Fig. 2.

The detection results are good. The subspace angles by ESOBI seem to be more consistent than EPCA.

Table 1 Welding parameters during the measurement campaign

Welding Name	Parameter	Nbr. of samples	Weld quality
OK	Nominal level	6	Good
A	−33% covering	3	Acceptable
B	−66% covering	3	Bad
C	−33% compensation	3	Good
D	−66% compensation	3	Acceptable
E	−10% current	3	Acceptable
F	−20% current	3	Bad
G	−10% forging pressure	3	Good
H	+5% forging pressure	3	Acceptable
I	−66% covering and compensation	3	Bad

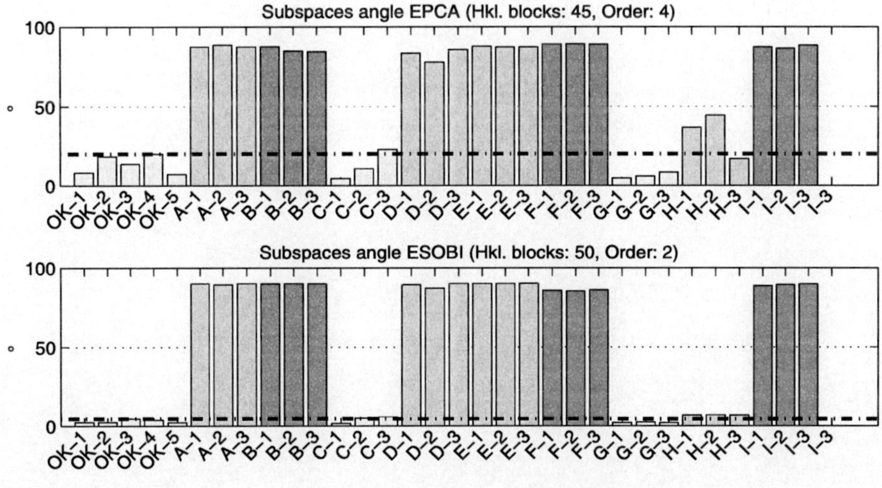

Fig. 2 Faulty welded joint detection results by EPCA (*top*) and ESOBI (*bottom*)

7 Conclusion

The presented methods where successfully applied to damage detection problems in industrial environment. By making use of the Hankel matrices, only one sensor is needed for the diagnosis which is an appreciable advantage.

Acknowledgments This work is supported by the "Mint Project" funded by the "Region Wallone" of Belgium which is greatly acknowledged.

References

1. Poncelet, F., Kerschen, G., Golinval, J.C., Verhelst, D.: Output-only modal analysis using blind source separation techniques. Mech. Syst. Signal Process. **21**, 2335–2358 (2007)
2. De Boe, P., Golinval, J.C.: Principal component analysis of a piezosensor array for damage localization. Struct. Health Monit. **2**, 137–144 (2003)
3. Belouchrani, A., Abed-Meraim, K., Cardoso, J.F., Molines, E.: A blind source separation technique using second order statistics. IEEE Trans. Signal Process. **45**, 434–444 (1997)
4. Nguyen V.H.: Damage detection and fault diagnosis in mechanical systems using vibration signals. Ph.D. Dissertation in applied science, University of Liège (2010)
5. Van Overschee, P., De Moor, B.: Subspace Identification for Linear Systems-Theory-Implementation-Applications. Kluwer Academic, Dordrecht (1997)
6. Yan, A.M., Golinval, J.C.: Null subspace based detection of structures using vibration measurements. Mech. Syst. Signal Process. **20**, 611–626 (2006)
7. Rutten, C., Loffet, C., Golinval, J.C.: Damage detection of mechanical components using null subspace analysis. Eur. J. Mech. Environ. Eng. Special Issue. **4**, 40 (2009)
8. Golub, G.H., Van Loan, C.F.: Matrix Computations, 3rd edn. The John Hopkins University press, Baltimore (1996)

The Influence of Magnetic Accumulators Equipped with Permanent Magnets on the Drive of a Mechanical System with the Periodical Reciprocating Movement and Its Behavior

Frantisek Foune, Pavel Kloucek, Pavel Sidlof, Petr Skop, Martin Pustka, and Petr Kacor

Abstract In this paper a measurement and an analysis of a common distribution mechanism for the winding of material on bobbins is described. Its machine-long rod is equipped with two energy accumulators with permanent magnets to suppress its longitudinal oscillations. On the basis of measurement data, a FEM model of the rod was proposed and used for an analysis of the causes of the increased longitudinal oscillations. Finally, a new layout of one of the accumulators was proposed. It could lead to better behavior of the whole mechanism.

Keywords Winding of bobbins • Transversal distribution mechanism • Longitudinal oscillations • Energy accumulators • Permanent magnets • Halbach array

1 Introduction

Machines on which many bobbins are wound have a common distribution mechanism for the generation of transversal movement. This mechanism has a machine-long rod carrying winding guides. The rod can be up to 50 m long and it is driven from one side by a cam or servomotor. Due to its length, the rod longitudinally oscillates and its free end moves in a different way to its beginning. Such an

F. Foune (✉) • P. Kloucek • P. Sidlof • P. Skop • M. Pustka
VUTS, a.s., Liberec, The Czech Republic
e-mail: frantisek.foune@vuts.cz; pavel.kloucek@vuts.cz; pavel.sidlof@vuts.cz; petr.skop@vuts.cz; martin.pustka@vuts.cz

P. Kacor
VSB – Technical University Ostrava, Ostrava, The Czech Republic
e-mail: petr.kacor@vsb.cz

J. Náprstek et al. (eds.), *Vibration Problems ICOVP 2011: The 10th International Conference on Vibration Problems*, Springer Proceedings in Physics 139,
DOI 10.1007/978-94-007-2069-5_13,

oscillation causes a worse quality of bobbins at the free end, increased wear of parts and noise, as well as the increased danger of rod buckling. The use of magnetic energy accumulators was suggested to remedy these shortcomings. In the following text a measurement of such a mechanism with and without magnetic accumulators is described and analyzed, and a new layout of the magnetic accumulator at the free end of the rod is proposed.

2 Settlement of Experiment

The proposed energy accumulators compose of a pair of permanent magnets which repel each other. The first magnet is firmly joined to a machine frame, the second one to a moving distribution rod. Both magnets are multipoles created from 12 segments magnetized in different directions and arranged into a ring in such a way that they form an axial Halbach array. This way, magnetic poles are transferred into one plane only. There is just one different way how to do it – to use pole pieces. However the pole pieces are passive and very heavy elements. Table 1 shows that just the setup of magnets into a Halbach array provides the highest ratio of their total repulsive force to their total mass.

Scheme of the distribution mechanism formed by a driving cam, the distribution rod and two energy accumulators is depicted in Fig. 1. The rod is situated into its right reversal point (right RP). This arrangement of the magnets allows suppressing or completely removing the buckling of the rod.

A location of sensors during the measurement on a customer testing stand is shown in Fig. 1.

Table 1 Comparison of magnets and their structures – results of FEM calculations

#	Material of magnets	Mass of magnets m_mg [g]	Mass of Fe m_Fe [g]	Total mass m [g]	Repulsive force F [N]	**Ratio of F/m [N/g]**
1	N35	298	1186	1484	902	**0.61**
2	N35	298	–	298	564	**1.9**
3	N42	182	–	182	600	**3.3**
4	N42	182	48	230	892	**3.9**
5	N42	225	–	225	1290	**5.7**

#	FeNdB magnets and structures
1	3× (Ø72× Ø25 × 4) + Pole Pieces
2	3× (Ø72× Ø25 × 4)
3	12× (1/2″ × 1/2″ × 1/2″)
4	12× (1/2″ × 1/2″ × 1/2″) + Fe wedges
5	Halbach array Ø74× Ø49 × 1/2″

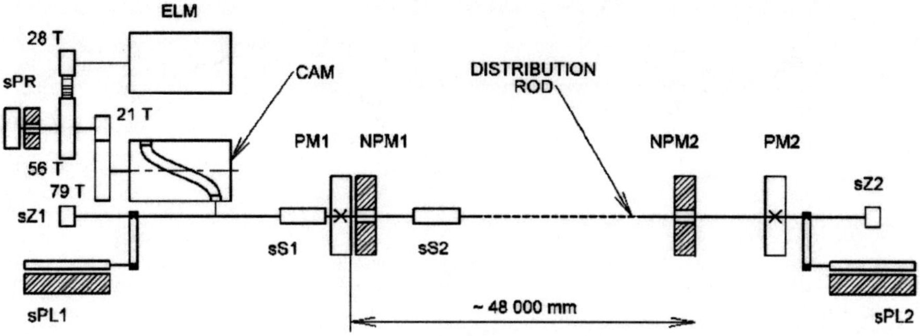

Fig. 1 Components of the stand and layout of sensors for measurement

List of sensors

sS1, sS2	A force sensor in front and behind the magnets, respectively
sPR	A rotational incremental sensor of angular position
sPL1, sPL2	A linear incremental sensor near the cam and at the free end, res.
sZ1, sZ2	An acceleration sensor near the cam and at the free end, resp.
56/28, 79/21	A timing belt and gear transmission ratios, respectively
PM1, NPM1	A movable and a stationary magnet near the cam, respectively
PM2, NPM2	A movable and a stationary magnet at the free end, respectively

The measurement was carried out without and with magnets at cam frequencies of 1.17, 1.83, 2.50, 3.17 and 3.25 Hz.

3 Results of Measurement

In Fig. 2 the results of measurement performed without and with magnets at the frequency of 3.17 Hz are presented as an example. From the comparison of these results it is clear that the longitudinal oscillation of the rod was significantly suppressed during the movement from the left to the right RP, i.e. when the rod was subjected to buckling. Simultaneously, the maximum magnitude of the force in the rod at the left RP, and the cam angular velocity variation during the movement from the left RP to the right RP were reduced substantially.

The dominant frequency of the superimposed longitudinal oscillation of the rod was determined by comparing Fourier decompositions of the cam generated motion and the rod motion at its free end. This is the seventh harmonic of the rod movement, i.e. 22.17 Hz. The cause of such high increase of the seventh harmonic could be a resonance with some natural frequency of the rod.

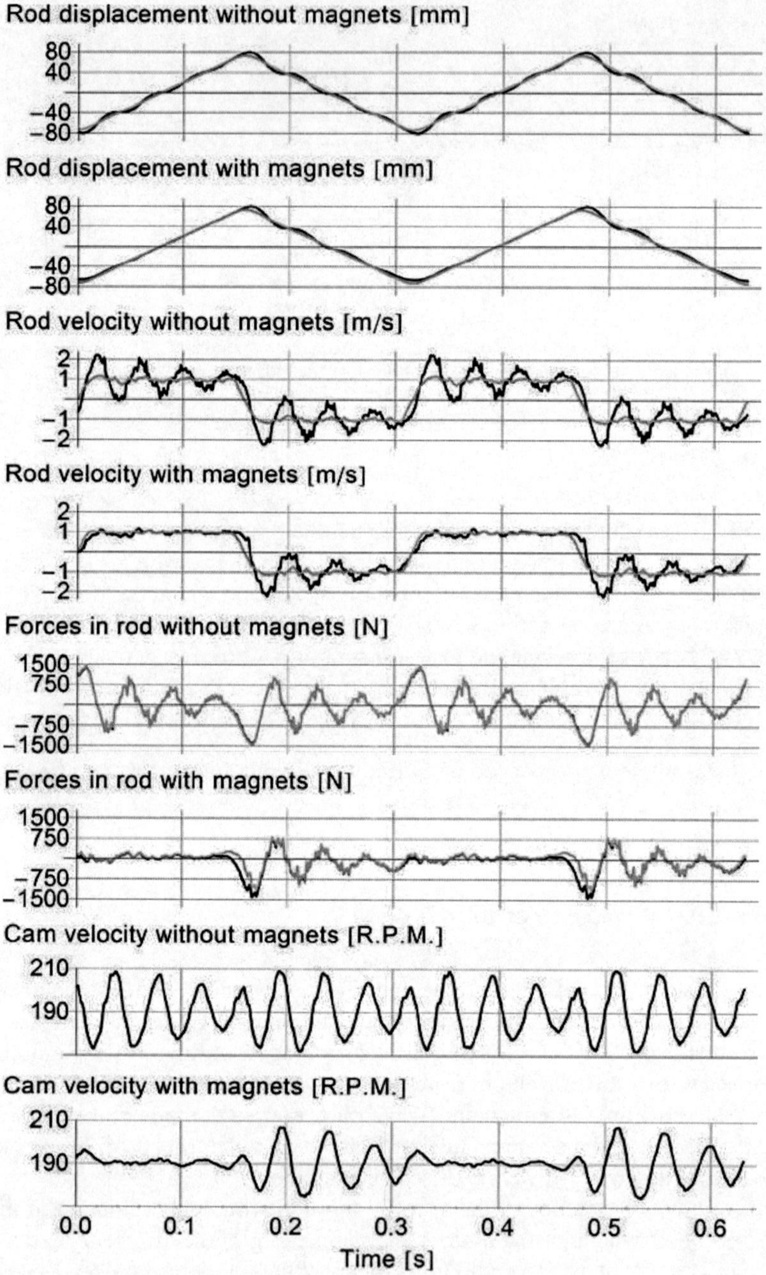

Fig. 2 Results with and without magnets. In the diagrams of displacement and velocity the *gray lines* signify the measurement near the cam, the *black lines* at the free end. In the diagrams of forces, the *black* and *gray lines* signify the force measured by the sensor sS1 and sS2, respectively

4 Experimental Analysis of the Longitudinal Oscillation of the Rod

During the measurement on the machine producer stand, the average speed of longitudinal waves in the rod was calculated from repeated impact excitations. From the speed of 4756 m/s the lowest natural frequency of the rod with one end clamped was determined. Its value is 26.1 Hz. The seventh harmonic of the rod movement, mentioned in the previous section, is very close to it.

5 Computational Analysis of the Longitudinal Oscillation

The compound distribution rod was modeled by FEM [1]. The real distribution of sections of the rod made of composite material and dural, their cross-sections, lengths, and masses were preserved. Experimentally determined stress moduli were used. Next, the discrete mass and rigidity of the distribution box were estimated and used to tune the model to match the measured data.

Subsequently this model was verified by calculations of the distribution rod behavior during kinematic excitation by the cam. The results showed a very good agreement with the measurement.

6 Design of a New Layout

The measurement of the originally designed mechanism according to Fig. 1 showed a slight drawback consisted in the diminution of the rod longitudinal oscillations only when the rod is pushed. Although the danger of buckling was suppressed to a minimum, a new solution, active also when the rod is pulled has been sought. One of the new suggestions is presented in Fig. 3. The second pair of repelling magnets was added to the energy accumulator at the free end of the rod.

The results of the new calculations are in Fig. 4. It is clear that the oscillation generated at the free end of the rod was suppressed and that the given objective was achieved.

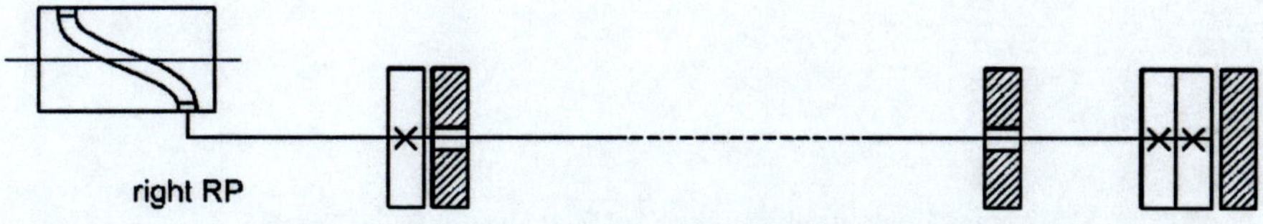

Fig. 3 Two additional repelling magnets placed into the energy accumulator at the free end

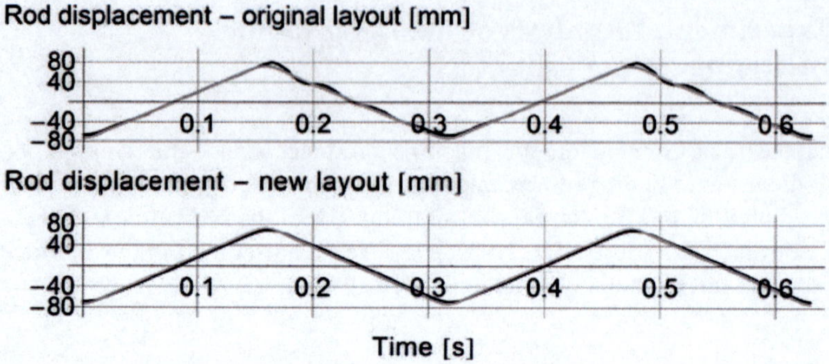

Fig. 4 Results of FEM calculations with original and new layouts of magnets, respectively. The *gray lines* signify the rod displacement near the cam and the *black lines* at the free end

7 Conclusions

With respect to the danger of buckling, the frequency of the rod motion has been limited by the machine producer to 2.67 Hz. The setup of magnet accumulators according to Fig. 1 itself will permit increasing the distribution rod frequency at least up to 3.25 Hz. It can be expected that this 20% improvement will lead to the same increase in the machine production.

The setup of permanent magnets according to Fig. 3 could suppress the rod oscillation in both directions of movement. As a result the bobbin packages could be superior in quality and the energy consumption could be lower due to a lower variation in the angular velocity of the cam. The vibrations, noise and wear of the parts of the mechanism will diminish as well.

The distribution mechanism in Figs. 1 or 3 can also be driven by a rotational or linear servomotor. Since the motion is initiated by a selectable ramp, the magnetic accumulators allow dimensioning the servo drive to a peak torque or peak force substantially lower than without them.

The design of the energy accumulators and their layout in the distribution mechanism according to Figs. 1 and 3 and according to similar setups of permanent magnets are protected by the Czech and other national patent applications.

Acknowledgement The presented work was done with the financial support of Ministry of Industry and Trade of the Czech Republic in the frame of the program TANDEM FT-TA3/017.

Reference

1. Milacek, S.: Vybrane kapitoly z dynamiky. CVUT, Praha (2003). in Czech

Theoretical Considerations and Practical Trends for Aseismic Buildings

Ferenc Kolonits and B. Csák

Abstract Two distinct aspects of Reinforced Concrete (RC) structures under seismic load are studied. Localized deterioration of multi-storey objects at intermediate levels is analyzed and the effect of non-smooth transition in stiffness on the first vibration mode is pointed at. Further, that a general collapse of a building can be avoided by more ductile joints. The application of a special plastic (3P-RESIN®) is shown.

Keywords RC structure • Multi-storey • Intermediate crash • Ductility • Resin

1 Introduction

Seismic loads may effect deterioration in rather different manners. In some cases, crashes of internal character remain highly localized in the structure, but, more often, a lot of structural joints got finished off and a collapse occurs. A possible background of the first defect will be investigated on a well-known example. In many cases, the process does not stop at that, the joints give way progressively. A collapse were escaped if the joints could bear (some, but enough) load beyond their elasticity limit. Ductility can be improved by more dense reinforcement keeping concrete in position as cracking begins, or by applying plastic additives.

F. Kolonits (✉) • B. Csák
Budapest University of Technology and Economics, Budapest, Hungary
e-mail: kolonits@vjt.bme.hu; cskbla43@gmail.com

J. Náprstek et al. (eds.), *Vibration Problems ICOVP 2011: The 10th International Conference on Vibration Problems*, Springer Proceedings in Physics 139,
DOI 10.1007/978-94-007-2069-5_14,

2 The Intermediate Storey Damage

The most notorious example for this kind of crash is that of the eight-storey Old Kobe City Hall. The Hyogo-ken Nambu earthquake (01.17.95) razed down the sixth storey, while the above ones shifted aside as a whole, otherwise there were moderate damages [1]. In the neighborhood, there occurred detriments of similar character, however, less pronounced. A series of explanation had been emerging in the literature and in discussions at courses on earthquake (EQ) safety held in Japan that time [2]. Let us see some of them.

Above the sixth floor, a later heightening took place and this made a "weak link". – The whole building had been erected in 1957, although the frame changed from Steel-RC to RC at that level. *The shear waves reflected from the top lead to the disaster.* – Well, this were similar as a plastic armor-piercing projectile worked, however, its reflected wave got tensile, more detrimental to armor, while a reflected shear remained shear and no more. *Behold soil slacking due the war bombings.* – A slacked ground would act as amplifier but uniformly and the bombs were mostly incendiaries. *Here and there, a rationale appeared that internal maxima might be inflicted by higher vibration modes* [3]. – This is the very aspect we are going to investigate now.

In a simple model, the floors are rigid and are moving in parallel, respectively. The pillars get bending moment and shear. The masses are concentrated at floor levels. The stresses are considered to be elastic – the results are not able to follow the crash-down process with, however, they identify the locations of incipient overload. For a column of height L and a × a cross section, base and top shifted in parallel, the force needed for unit deformation i.e. stiffness, due bending and shear are shown by Eq. 1, respectively

$$k_b = a^4E/L^3, k_s = a^2E/[2(1+\nu)\alpha_s L]. \tag{1}$$

The shear constant $\alpha_s = 1.5$ [4]. The M masses of floors, beams etc. are evenly distributed among the 72 columns of inter-storey support. The final model will be a vertical beam of elastic columns with point masses, clamped at the bottom, free at roof level. Details are shown in Table 1.

Table 1 Storey data of Kobe City Hall and eigenfrequencies/times

Nr	M [kg]	L [m]	a [m]	k_{sum}[N/m]	ω [1/s]	T [s]
1	27,479.2	4.15	1	243,017,935.6	13.789	0.4557
2	31,598.6	4.2	1	235,404,896.4	32.544	0.1931
3	31,598.6	3.65	0.85	188,609,516.1	52.581	0.1195
4	31,598.6	3.65	0.85	188,609,516.1	71.104	0.0884
5	31,598.6	3.65	0.85	188,609,516.1	83.356	0.0754
6	31,598.6	3.65	0.65	69,190,033.37	109.80	0.0572
7	31,598.6	3.65	0.65	69,190,033.37	139.02	0.0452
8	26,645.8	3.8	0.6	69,190,033.37	158.69	0.0396

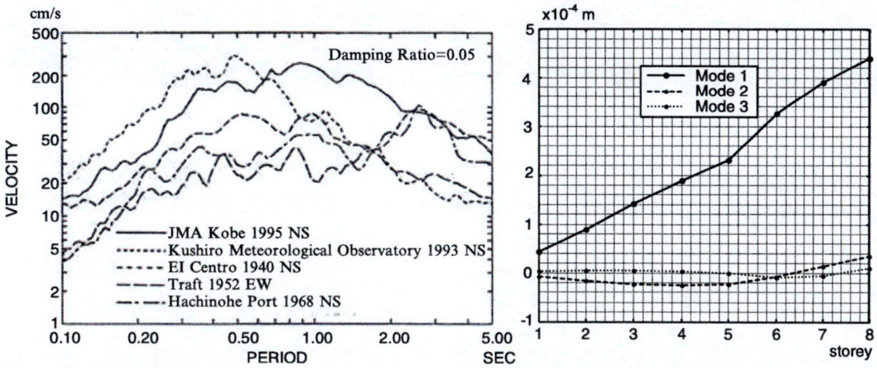

Fig. 1 Velocity response and maximal displacements at storey levels

The fundamental period is of the same order of magnitude as that given by the $0.1 \times$ (storey nr) rule-of-thumb [5]. The participation of various normal modes can be calculated on the base of response spectra. From [6], the velocity response S_v is available – the displacement response is $S_d = S_v/\omega$. The ones pertaining to the series of eigenvalues may be arranged into a diagonal matrix $<S_{v,i}/\omega_i>$. The column vector of maximal displacements $\mathbf{V}_{max}$ can be expressed with the $[\boldsymbol{\phi}_i] = \Phi$ matrix of normalized eigenvectors, by Eq. 2, **e** being an all-one load distribution vector.

$$\mathbf{V}_{max} = \Phi\langle S_{v,i}/\omega_i\rangle\ \Phi^T\langle M_i\rangle\mathbf{e}. \tag{2}$$

The responses for the first three frequencies are $S_v = 1.80$, 0.33, 0.16 m/s. The displacement responses are shown on Fig. 1. The higher-order modes do not appear intensely, the inter-storey damages can not be explained by their multi-bent form. The relations are as usual: the fundamental mode is the decisive. However, there is a local change in displacement at the sixth storey. This has a pronounced effect on both bending and shear stresses as shown in Fig. 2. (Calculated by elementary formulae, representing rather an index of the stress state).

Under given relations of masses and stiffness, by virtue of the fundamental mode, the sixth storey is prone to overload, regardless to further impairing by local changes in reinforcement etc. Sure, the stiffness had been overestimated. The floors must not be completely rigid. Suppose an implausible, one-fourth stiffness. Then, the fundamental period would be 0.9 s (just overshooting the quoted rule-of-thumb). With proportional higher periods, the responses are 2.5, 1.5 and 0.5 m/s, respectively. The relations would remain essentially the same – the influence of the first mode would decease far over the peak response at 0.9 s, but such a flaw in modeling is not real.

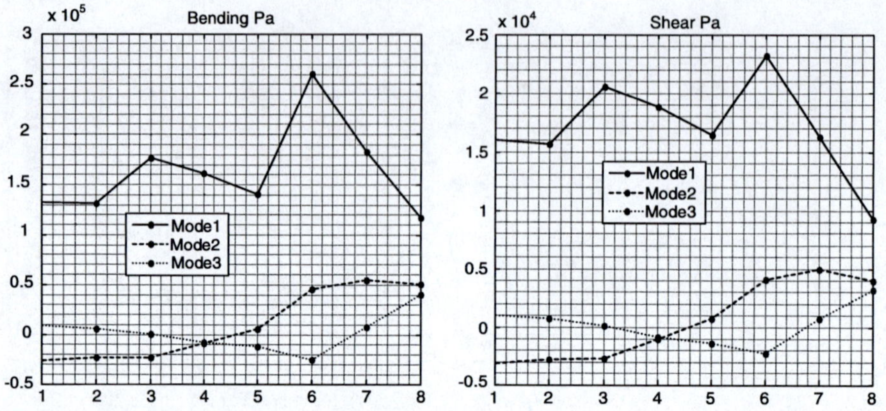

Fig. 2 *Bending* and *shear* stresses in diverse modes at storey levels

3 Joints of Enhanced Ductility

The most sensitive spots of an RC frame with respect to EQ impulses are the structural joints of pillars, floors, beams etc. They are essential in maintaining stability. The energy released by an EQ mostly rearranges the landscape. However, there remains enough to do the same to man-made structures. The result depends on whether this rearranging meant distortion (and of what a level) or raveled rebars, concrete torn away from. In order to avoid the very bad second option, the joints must be ductile, i.e. able to carry load beyond limits of their elasticity. Under reciprocating motion, that results hysteretic energy dissipation. The classic solution was to put more steel in critical areas, securing concrete from cracks as far as possible or keeping that together if crumbled [7].

The version recently studied at Budapest University of Technology (Dpt of Mechanics, Materials and Structures) changes a basic component of the concrete, the cement in a part of the critical zone. The range of experimental work had been restricted by fund reasons, authors are aware of only probing steps having been made. The additive here is a polymer compound with which an elasto-plastic ductile composite concrete is prepared: the 3P-RESIN®. The layout of the test specimen is shown on Fig. 3. As for "base line", a specimen of normal RC had been tested. It became cracked extensively after ten cycles and crumbling at the 14th. A specimen with ductile concrete joint had been tested under ±50 mm constrained deformation. Having performed 22 rounds, the load upheld had been reduced to 70% (Fig. 4). Having applied further 21 cycles, the load went down again to 52% (making a total of 36%, Fig. 5), with one rebar cracked. However, the hinge area remained flawless. Another specimen had been cycled, having got 22 rounds, further 50 times (Fig. 6). The degradation went to 23%, two rebars were cracked but the joint remained integral.

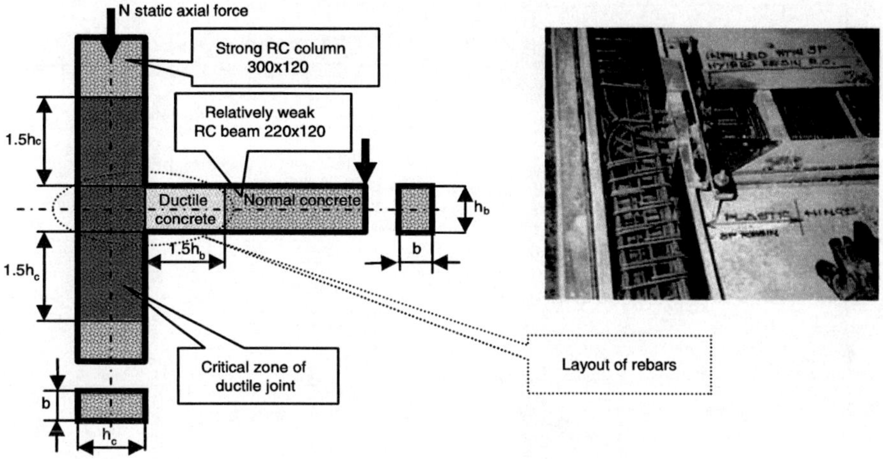

Fig. 3 Test specimen, concrete grade C16, steel B60.40

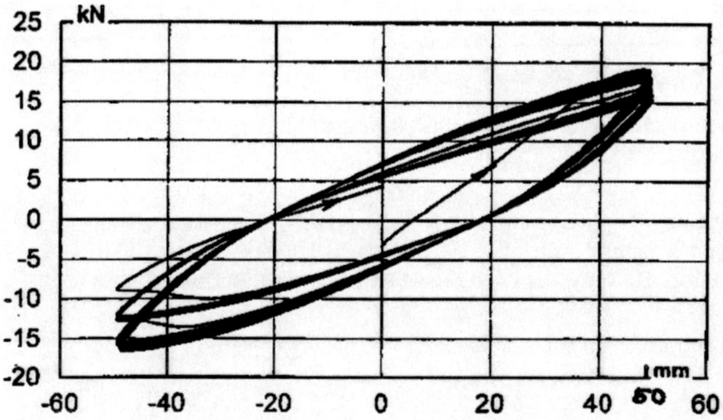

Fig. 4 Twenty-two cycles, ±50 mm, degradation to 70% (gadget weights ∼250 kN)

4 Conclusions

The storeys of the Kobe building have about the same mass, while stiffness distribution is $100 - 97 - 3 \times 78 - 3 \times 28$ [%] of. The abrupt change at the sixth storey results a local maximum of load even in the fundamental mode. Ductile joint is a very effective tool in maintaining integrity under reciprocating loads. Applying there a concrete with elasto-plastic additive, a manifold endurance is experienced. This may be very beneficial in averting structural collapse.

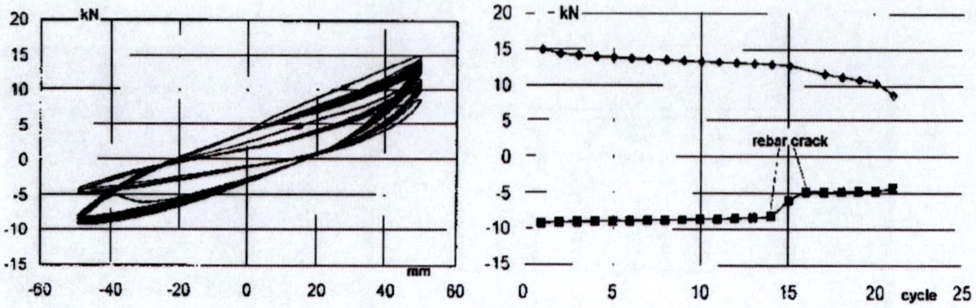

Fig. 5 Additional 21 cycles, supported force drops to 52%

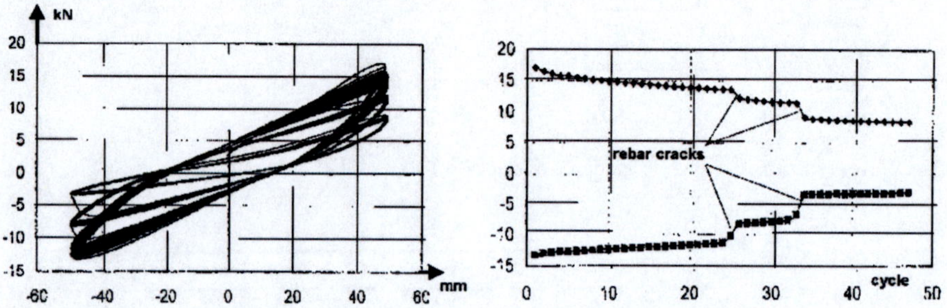

Fig. 6 Another specimen, further 50 cycles, degradation to 23%

Acknowledgments The kind help of Mr M. Ishizaki (Takenaka Co. Japan) giving access to literature on and structural data of Kobe City Hall, respectively, had been substantial. The co-operation of Messrs G. Nagy and L. Meiszel (Polinvent Ltd. Budapest) must be also appreciated.

References

1. The January 17, 1995 Kobe earthquake, EQE Summary Report (1995)
2. INCEDE: Newsl., Special Issue, Jan 1995
3. Dulácska, E.: Effect of the EQ 09.19.85 on the buildings of Mexico city, in Hungarian. Mélyépítéstudományi Szemle **36**(9), 376–386 (1986)
4. Przemiemiecki, J.S.: Effect of shear deformations. www.bth.se/tek/amt/mtd009.nsf/.../przemieniecki_p_201_208.pdf
5. Housner, G.W.: Strong ground motion, Chap. 4. In: Wiegel, R.L. (ed.) Earthquake Engineering, pp. 75–91. Prentice-Hall Inc., Englewood Cliffs (1970)
6. Kanda, J.: Building damages due to the 1995 Hyogoken-Nambu Earthquake (Paper S1-5-1). International Symposium on Seismic Safety Relating to NPP, Kobe, 3–5 Mar 1997
7. Banon, H.: Prediction of seismic damage in reinforced concrete frames, Doctoral work, Department of Civil Engineering, MIT (1980)

The Effect of Initial Stress on Nonlinear Vibrations of an Articulated Beam

Jacek Przybylski

Abstract The aim of this study is to consider the problem of nonlinear transverse vibrations of a stepped beam with immovable ends and a crack. The crack is modelled by a hinge with a massless rotational spring. The initial stress is a result of the residual force generated by a pair of piezoelectric actuators bonded to the top and bottom surfaces of the beam. A version of the Lindstedt-Poincare method is used to acquire approximate solutions to the problem. The natural frequencies are computed from the solution of three consecutive equations, which are distinguished by growing powers of the small amplitude parameter. It is shown that the residual force depends on the voltage applied to the piezoactuators, the axial stiffness ratio of the piezosegment and the beam and on the length of the piezosegment. The influence of crack locations and spring stiffness parameter on the frequency, mode shape and the amplitude-frequency relation is analysed for different values of the applied piezoelectric force.

Keywords Transverse vibrations • Amplitude-frequency relationship • Crack • Piezoactuator

1 Introduction

An internal crack appearing even in one member of a complex engineering structure can lead to the failure or catastrophic breakdown of the whole system. Special monitoring devices are engaged to control the behaviour of dynamically loaded machines, mechanisms and constructions in the case of initiation and eventual fatigue crack propagation. One of techniques leads to collecting information

---

J. Przybylski (✉)
Czestochowa University of Technology, Czestochowa, Poland
e-mail: jacek.pr@imipkm.pcz.pl

J. Náprstek et al. (eds.), *Vibration Problems ICOVP 2011: The 10th International Conference on Vibration Problems*, Springer Proceedings in Physics 139,
DOI 10.1007/978-94-007-2069-5_15,

concerning the dynamic response of structures to control a possible discrepancy between standard parameters and those demonstrating an unusual behaviour. The natural frequency and mode shape are the dynamic parameters which changes can be useful in the detection of the failure information such as the presence and severity of structure damage. Therefore free vibrations of beams with cracks have been studied extensively for three decades. The most common crack model is the one in which the rotational massless spring has been introduced at the location of a crack [1]. This rotational spring can also model a semi-rigid joint in regard to segmented beams or columns made of different materials. In this study the influence of the crack, existing at the interface of a stepped beam with immovable ends, on the natural vibration of this system is considered. The change in the cross section area of the beam results from the application of a pair of collocated piezoceramic actuators bonded to the host structure. An additional aim of the paper is to demonstrate the applicability and efficiency of the electric field applied to the piezoactuators on the stated objectives. Reassuming, the analysis of this nonlinear problem is provided to show the effects of the location of a crack and the piezo-force on the natural frequency, mode shape and the amplitude-frequency relation of the investigated system.

2 Formulation

Figure 1a presents a deflected axis of an articulated simply supported stepped beam with a rotational spring of stiffness C. Two identical piezoelectric actuators, of equal dimensions and made of the same material, are bonded collocally to the upper and lower surfaces of a beam along its second segment. The cross-section of this segment is shown in Fig. 1b.

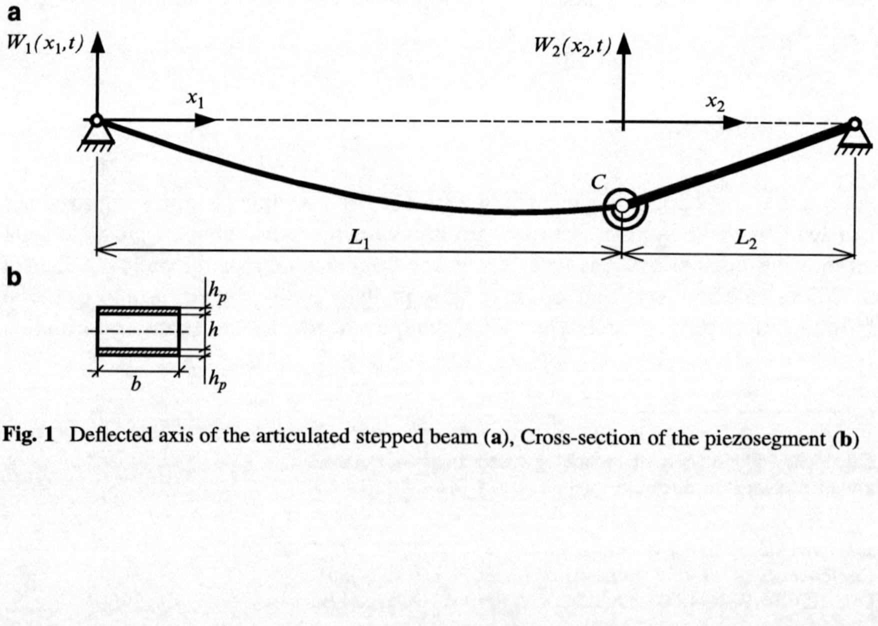

Fig. 1 Deflected axis of the articulated stepped beam (**a**), Cross-section of the piezosegment (**b**)

It has been derived in [2] that two piezoactuators of the same cross-sectional area $(b \times h_p)$, being supplied with an equal homogeneous electric field characterised by the voltage V, generate the residual piezo-force equal to:

$$F_r = F\left[1 + a\left(\frac{L}{L_2} - 1\right)\right]^{-1} \tag{1}$$

where: $a = 1 + \frac{E_p A_p}{E_b A_b}$ denotes the axial stiffness ratio of the piezosegment to that of the beam, $F = \pm 2be_{31}V$ is the piezoelectric force resulting from the voltage V applied to the piezoactuators of width b, L is the length of the beam, L_2 is the length of the piezosegment, E_b is the Young's modulus of the beam, and A_p, A_b are the cross section areas of the piezoactuators and the beam, respectively. The residual force F_r being independent of piezosegment's position, depends not only on the piezoelectric force F, but also on the axial stiffness ratio a and the ratio between the length of piezosegment L_2 to whole length of the beam L.

Hence the governing equations, describing the lateral vibrations of the i–th segment of the beam and the axial stretching force, written in the nondimensional form are

$$\frac{\partial^4 w_i(\xi_i,\tau)}{\partial \xi_i{}^4} + c_i\left[f_r - s(\tau)\right]\frac{\partial^2 w_i(\xi_i,\tau)}{\partial \xi_i{}^2} + d_i\omega^2\frac{\partial^2 w_i(\xi_i,\tau)}{\partial \tau^2} = 0, \quad (i = 1,2) \tag{2}$$

$$s(\tau) = \frac{\lambda}{2}\left(l_1 + \frac{l_2}{a}\right)^{-1}\sum_{i=1}^{2}\int_0^{l_i}\left[\frac{\partial w_i(\xi_i,\tau)}{\partial \xi_i}\right]^2 d\xi_i \tag{3}$$

where: $\xi_i = \frac{x_i}{L}, l_i = \frac{L_i}{L}, \tau = \Omega t, w_i(\xi_i,\tau) = \frac{W_i(x_i,t)}{L}$,

$$s(\tau) = \frac{S(t)L^2}{E_b I_b}, f_r = \frac{F_r L^2}{E_b I_b}, \omega^2 = \Omega^2 L^4\frac{\rho_b A_b}{E_b I_b}, \lambda = \frac{A_b L^2}{I_b}.$$

Parameters c_i and d_i, which are equal to unity for the first segment ($c_1 = d_1 = 1$), for the second segment are: $c_2 = (1 + r_m)^{-1}$, $d_2 = [\alpha_1 + (a-1)\alpha_2]\,[\alpha_1\,(1+r_m)]^{-1}$. In the applied formulas: $\alpha_1 = E_p/E_b$, $\alpha_2 = \rho_p/\rho_b$, $r_m = \alpha_1 I_p/I_b$, where I_b, I_p stand for the moments of inertia of the cross-section areas of the beam and actuators, respectively, ρ_b, ρ_p are the material densities of the beam and actuators, respectively, and Ω is a natural frequency of the system.

For the pin-ended beam from Fig. 1a, the boundary conditions are:

$$w_1(\xi_1,\tau)\big|_{\xi_1=0} = w_1^{II}(\xi_1,\tau)\big|_{\xi_1=0} = w_2(\xi_2,\tau)\big|_{\xi_2=l_2} = w_2^{II}(\xi_2,\tau)\big|_{\xi_2=l_2} = 0,$$

$$w_1(\xi_1,\tau)\big|_{\xi_1=l_1} = w_2(\xi_2,\tau)\big|_{\xi_2=0},$$

$$w_1^{II}(\xi_1,\tau)\big|_{\xi_1=l_1} = c_m\left[w_2^{I}(\xi_2,\tau)\big|_{\xi_2=0} - w_1^{I}(\xi_1,\tau)\big|_{\xi_1=l_1}\right],$$

$$w_1^{II}(\xi_1,\tau)\left|_{\xi_1=l_1} = c_2^{-1} w_2^{II}(\xi_2,\tau)\right|_{\xi_2=0},$$
$$w_1^{III}(\xi_1,\tau)\left|_{\xi_1=l_1} + [f_r - s(\tau)]\, w_1^{I}(\xi_1,\tau)\right|_{\xi_1=l_1}$$
$$- c_2^{-1} w_2^{III}(\xi_2,\tau)\left|_{\xi_2=0} - [f_r - s(\tau)]\, w_2^{I}(\xi_2,\tau)\right|_{\xi_2=0}, \tag{4}$$

where the nondimensional spring stiffness $c_m = CL(E_b I_b)^{-1}$.

3 Approximate Analytical Solution

A version of the Lindstedt-Poincare method is proposed to solve the stated nonlinear boundary value problem. One assumes the following expansions of the lateral displacement, the axial stretching force and the natural frequency into exponential series with respect to the amplitude parameter ε ($\varepsilon \ll 1$):

$$w_i(\xi_i,\tau) = \sum_{n=1}^{N} \varepsilon^{2n-1} w_{i2n-1}(\xi_i,\tau) + \mathrm{O}\left(\varepsilon^{2N+1}\right), \tag{5}$$

$$s(\tau) = \sum_{n=1}^{N} \varepsilon^{2n} s_{2n}(\tau) + \mathrm{O}\left(\varepsilon^{2(N+1)}\right), \tag{6}$$

$$\omega^2 = \omega_0^2 + \sum_{n=1}^{N} \varepsilon^{2n}\, \omega_{2n}^2 + \mathrm{O}\left(\varepsilon^{2(N+1)}\right) \tag{7}$$

Inserting Eqs. 5–7 into Eqs. 2–4 and equating coefficients of the same powers of ε one obtains an infinite set of equations of motion and axial forces, as well as associated boundary conditions. The obtained equations are subjected to consecutive analytical solutions after separating time and space variables. The solution of the first three equations gives the opportunity to numerical calculations of the two first components of the natural frequency from Eq. 7 as a function of the nondimensional piezoelectric force f, where $f = FL^2(E_b\, I_b)^{-1}$. The first component of the frequency (ω_0) symbolises its linear part. The nonlinear frequency (ω) changes with amplitude, whose maximum is expressed as

$$w_{\max} = w_{i1}\left(\xi_{\mathrm{A}}\right) = \varepsilon\sqrt{\lambda}, \quad w_{\max} \in \langle 0, 3\rangle, \tag{8}$$

where: ξ_{A} is the co-ordinate of maximal displacement, $\lambda = A_b L^2 (I_b)^{-1}$ is the nondimensional slenderness parameter of the uniform beam.

The solution method of the considered boundary value problem has been fully described in [3].

4 Exemplary Numerical Results

An aluminium alloy beam of a cross section 20×0.5 mm and a length equal to 0.25 m with two ceramic piezolayers of the thickness equal to 0.125 mm each has been taken into account. Hence, for the beam and piezocermiac Young's moduli $E_b = 70$[GPa] and $E_p = 80$[GPa], respectively, the axial stiffness parameter $a = 1.571$.

In Fig. 2 the amplitude-nonlinear frequency relationship is presented for a different spring stiffness c_m, expressing local flexibility at the crack location (here $l_1 = 0.75$). The shown curves characterise the nonlinearity of the hardening type.

Analysing the demonstrated results with those depicted in Table 1, one can conclude that the stiffer the spring, the greater the linear frequency value and the lower the influence of the amplitude on the nonlinear frequency. Two icons, additionally added in Fig. 2, illustrate the change of the first mode shape as a result of spring stiffening from $c_m = 0.5$ to $c_m = 10^6$. The effect of the crack on the stated objectives can be moderate by the piezoelectric force f, which is shown in Fig. 3

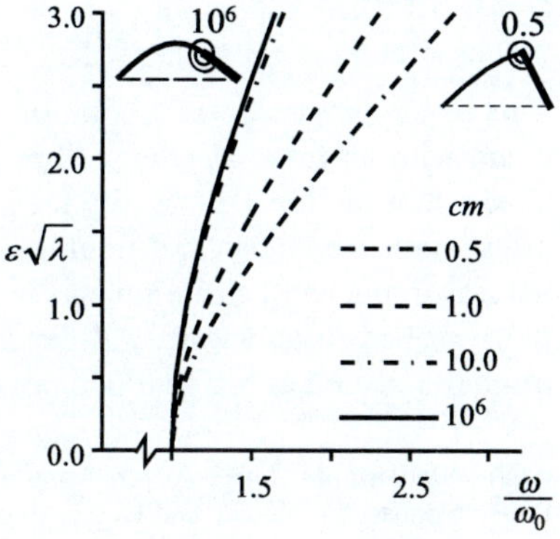

Fig. 2 Amplitude-frequency relation for different values of the rotational spring parameter c_m, $l_1 = 0.75$, $f = 0$. *Icons* show the first vibration modes for $c_m = 0.5$ and $c_m = 10^6$

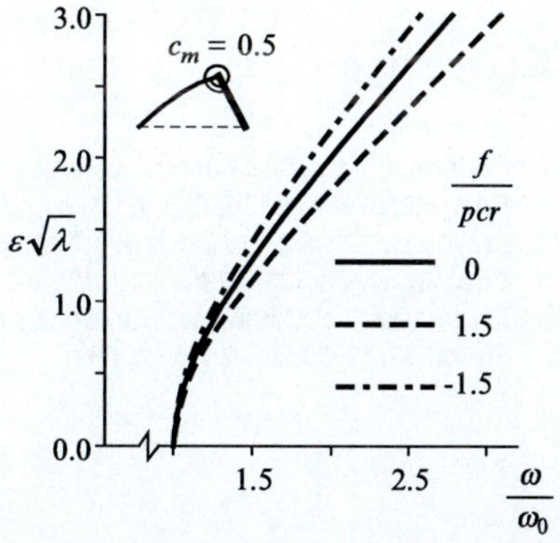

Fig. 3 Amplitude-frequency relation for different values of the piezoelectric force. Other parameters: $c_m = 0.5$, $l_1 = 0.75$

Table 1 Linear frequency ω_0 as a function of the crack location and the piezoelectric force f for different rotational spring stiffness c_m

	$l_1 = 0.75$			$l_1 = 0.25$		
c_m	$\frac{f}{p_{cr}} = -1.5$	$\frac{f}{p_{cr}} = 0$	$\frac{f}{p_{cr}} = 1.5$	$\frac{f}{p_{cr}} = -1.5$	$\frac{f}{p_{cr}} = 0$	$\frac{f}{p_{cr}} = 1.5$
0.5	5.4372	4.9646	4.4018	5.0388	4.3092	3.4428
1.0	6.9344	6.3008	5.5548	6.6945	5.7269	4.5779
10.0	10.2018	9.1582	7.9682	11.8206	10.1627	8.7564
10^6	10.8515	9.7222	8.4415	13.4045	11.5716	9.3567

and in Table 1. The stretching of the beam realised by the piezoelectric force of the value $f = -1.5p_{cr}$, where p_{cr} is the non-dimensional buckling critical force for the uniform beam ($p_{cr} = 9.86$), decreases the nonlinear effect, whereas the compressing piezoelectric force $f = 1.5p_{cr}$ increases this effect. Although the linear frequency decreases with the growing values of the piezoelectric force, this force does not affect the mode shape.

5 Conclusions

This paper has presented a solution to the nonlinear vibration problem of a stepped beam with an internal crack. After performing numerical simulations, it has been shown that the local flexibility introduced by a crack and the location of the crack affect both the linear and nonlinear frequencies as well as the mode shape. The natural frequency can be tuned by the residual force induced by the piezoelectric field applied to actuators. All deviations in the dynamic behaviour of the system might be treated as a signal of a possible crack initiation in the structure.

Acknowledgments The research work reported here has obtained the financial support of the Polish Ministry of Science and Higher Education under the grant N501 117.

References

1. Chondros, T.G.: The continuous crack flexibility model for crack identification. Fatigue Fract. Eng. Mater. Struct. **26**, 643–650 (2001)
2. Przybylski, J.: Stability of an articulated column with two collocated piezoelectric actuators. Eng. Struct. **30**(12), 3739–3750 (2008)
3. Przybylski, J.: Nonlinear vibrations of a beam with a pair of piezoceramic actuators. Eng. Struct. **31**(11), 2687–2695 (2009)

Natural Frequencies of Composite Cylindrical Helical Springs Under Compression

İlyas Kacar and Vebil Yildirim

Abstract The governing equations of cylindrical isotropic helical springs loaded axially are extended to the study of free vibration of such helices made of composite materials. Stiffness method is used based on the first order shear deformation theory. A helical element has six degrees of freedom at each node, three translations and three rotations. The element stiffness matrix is obtained exactly based on the both complementary functions and the transfer matrix methods.

Keywords Vibration • Composite springs • Axial load • Stiffness matrix

1 Introduction

Although there are a few study on the vibration of isotropic helical springs subjected to an axial load [1–6], to the authors' knowledge there is no study in the vibration analysis of composite helical springs subjected to an axial force in the related literature.

In the present work, the free vibration frequencies of unidirectional composite cylindrical helical springs subjected to the axial static load are computed numerically with the help of the stiffness matrix method. The exact numerical element stiffness matrix is obtained based on the both the complementary functions and transfer matrix methods [7, 8]. To determine static axial tip deflections of helical springs with large pitch angles due to the static axial force, the authors used analytical expression, which takes into account for the whole effect of the stress resultants such as axial and shearing forces, bending and torsional moments.

İ. Kacar (✉) • V. Yildirim
Çukurova University, Adana, Turkey
e-mail: ikacar@cuksurova.edu.tr; vebil@cukurova.edu.tr

J. Náprstek et al. (eds.), *Vibration Problems ICOVP 2011: The 10th International Conference on Vibration Problems*, Springer Proceedings in Physics 139,
DOI 10.1007/978-94-007-2069-5_16,

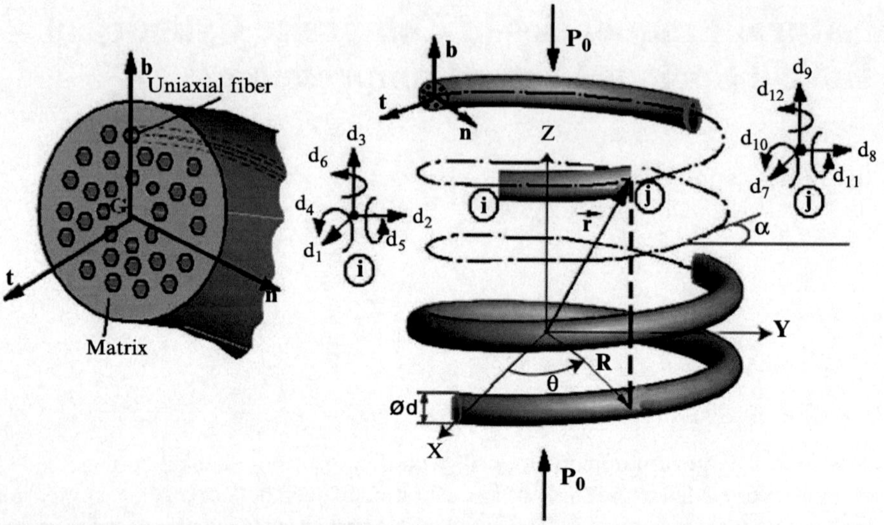

t, n, b: Local unit vectors (Frenet trihedral)
i, j, k: Global unit vectors (Cartesian Coordinates)
$d_1 = U_X, d_2 = U_Y, d_3 = U_Z, d_4 = \Omega_X, d_5 = \Omega_Y, d_6 = \Omega_Z$
$d_7 = U_X, d_8 = U_Y, d_9 = U_Z, d_{10} = \Omega_X, d_{11} = \Omega_Y, d_{12} = \Omega_Z$

Fig. 1 Degrees of freedom of a helical element in global coordinates

The global stiffness matrix, **K,** incorporates the pre-loading effects (Fig. 1). The global mass matrix, **M,** is in the diagonal form. In the solution of a large-scale eigen-value problem, the subspace-iteration method is employed. The numerical fundamental natural frequencies of such springs, ω(rad/s), are presented in this work at first time. Apart from those, the critical buckling loads are obtained based on the dynamic approach.

$$(\mathbf{K} - \omega^2\mathbf{M})\mathbf{D} = 0 \tag{1}$$

2 Determination of the Element Stiffness Matrix of a Composite Helical Bar Subjected to an Axial Static Force

For a helical bar, the Frenet unit vectors associated with the bar axes are

$$\begin{aligned}
\mathbf{t} &= d\mathbf{r}/ds = (-\cos\alpha\sin\theta)\mathbf{i} + (\cos\alpha\cos\theta)\mathbf{j} + \sin\alpha\,\mathbf{k} \\
\mathbf{n} &= (d\mathbf{t}/ds)/(dt/ds) = -\cos\theta\mathbf{i} - \sin\theta\,\mathbf{j} \\
\mathbf{b} &= \mathbf{t}x\mathbf{n} = (\sin\alpha\sin\theta)\mathbf{i} + (-\sin\alpha\cos\theta)\mathbf{j} + \cos\alpha\,\mathbf{k}
\end{aligned} \tag{2}$$

where θ is the horizontal angular displacement, α is the helix pitch angle, and the infinitesimal length of the bar is $ds = (R/\cos\alpha)d\theta$. Frenet unit vectors are related to each other with the following relations

$$d\mathbf{t}/ds = \chi\mathbf{n}, \quad d\mathbf{n}/ds = \tau\mathbf{b} - \chi\mathbf{t}, \quad d\mathbf{b}/ds = -\tau\mathbf{n} \tag{3}$$

For a cylindrical helical bar, denoting the centerline radius of the helix by R (=D/2), the curvature is $\chi = \cos^2\alpha/R$, and the tortuosity is $\tau = \sin\alpha\cos\alpha/R$. The governing equations of helical rods made of an anisotropic material and subjected to the initial force $\mathbf{T^o} = (-P_o\sin\alpha, 0, -P_o\cos\alpha)$ and initial moment $\mathbf{M^o} = (-P_oR\cos\alpha, 0, P_oR\sin\alpha)$ are obtained by Yildirim as (Fig. 1)

$$\begin{aligned}
&\frac{d\mathbf{U}}{ds} - \mathbf{A}'\mathbf{T} + \mathbf{t}\,x\,\boldsymbol{\Omega} = 0\\
&\frac{d\boldsymbol{\Omega}}{ds} - \mathbf{D}'\mathbf{M} = 0\\
&\frac{d\mathbf{T}}{ds} + (\mathbf{D}'\mathbf{M})x\mathbf{T^o} + \mathbf{p} = 0\\
&\frac{d\mathbf{M}}{ds} + (\mathbf{D}'\mathbf{M})x\mathbf{M^o} + \mathbf{t}x\mathbf{T} + (\mathbf{A}'\mathbf{T})x\mathbf{T^o} + \mathbf{m} = 0
\end{aligned} \tag{4}$$

where ***T*** and ***M*** are the internal force and internal moment vectors, ***U*** and Ω are the displacement and rotation vectors, **p** and **m** are the distributed force and moment vectors, respectively. ***A'*** and ***D'*** matrices comprise the cross-sectional rigidities for composite bars [9]. Equation 4 may be put in the form of

$$S(\theta) = \mathbf{F}(\theta)\mathbf{S}(0) \tag{5}$$

where ***S (0)*** is the state vector at $\theta = 0$. To be able to consider the variable radius of cylinder and sections, the element transfer matrix **F** is obtained by solving the following differential equation 12 times for 12 different initial conditions based on the complementary functions method [7, 8].

$$d\mathbf{F}^{*(m)}/d\theta = \mathbf{DF}^{*(m)} \ (m = 1, 2, ..., 12) \tag{6}$$

where $\boldsymbol{F}^{*(m)}$ denotes the solution when mth element of the unknown vector equals unit as its other elements are all zero. These solutions compose the exact transfer matrix for cylindrical helical springs. The elements of the state vector at both ends for an element can be expressed by the element end displacements, $\boldsymbol{d}_i$ *and* $\boldsymbol{d}_j$, and the element end forces, $\boldsymbol{p}_i$ *and* $\boldsymbol{p}_j$, as follows (Fig. 1):

$$\mathbf{S}_i = \{d_1, d_2, d_3, d_4, d_5, d_6, p_1, p_2, p_3, p_4, p_5, p_6\}^T = \{\mathbf{d}_i \mathbf{p}_i\}^T$$
$$\mathbf{S}_j = \{d_7, d_8, d_9, d_{10}, d_{11}, d_{12}, p_7, p_8, p_9, p_{10}, p_{11}, p_{12}\}^T = \{\mathbf{d}_j \mathbf{p}_j\}^T \tag{7}$$

Using the above definitions, Eq. 5 can be rearranged for an element as

$$\mathbf{S}(\theta_j) = \mathbf{F}(\theta_j - \theta_i)\mathbf{S}(\theta_i) \tag{8}$$

The element transfer matrix can be expressed in global coordinates as

$$\mathbf{F}_{ijk}(\theta_j - \theta_i) = \mathbf{T}^{-1}(\theta_j)\mathbf{F}_{tnb}(\theta_j - \theta_i)\mathbf{T}(\theta_i) \tag{9}$$

where

$$\mathbf{T} = \begin{bmatrix} \mathbf{B} & 0 & 0 & 0 \\ 0 & \mathbf{B} & 0 & 0 \\ 0 & 0 & \mathbf{B} & 0 \\ 0 & 0 & 0 & \mathbf{B} \end{bmatrix} \quad \text{and} \quad \mathbf{B} = \begin{bmatrix} -\cos\alpha\sin\theta & \cos\alpha\cos\theta & \sin\alpha \\ -\cos\theta & -\sin\theta & 0 \\ \sin\alpha\sin\theta & -\sin\alpha\cos\theta & \cos\alpha \end{bmatrix} \tag{10}$$

In the element equation, $\mathbf{p} = \mathbf{kd}$ or $\{\mathbf{p}_i \mathbf{p}_j\}^T = \mathbf{k}\{\mathbf{d}_i \mathbf{d}_j\}^T$, elements of the element stiffness matrix, ***k,*** is obtained by solving Eq. 8 12 times for 12 boundary conditions.

3 Numerical Examples

In the first example an isotropic cylindrical helical spring under axial static load compressed to half of its free axial length is considered. The results are presented in Table 1 (f_{axial} = 118.8288853 Hz).

Table 1 The present first seven natural frequencies for an isotropic cylindrical helical spring ($L_o/D = 5$, $n = 30$, $D = 10$ mm, $d = 1$ mm, $\rho = 7{,}900$ kg/m^3, $\upsilon = 0.3$, $G = 79.269$ *GPa, Fixed-Fixed*)

	f (Hz)		f/f_{axial}	
	Present Study		Becker et al. [6]	Haringx [1]
Modes	$\delta/L_o = 0$	$\delta/L_o = 0.5$	$\delta/L_o = 0.5$	
1	79.7091	55.6628	0.4727	0.4742
2	79.7386	55.6831	0.4732	0.4742
3	118.391	118.468	0.9955	1.000
4	135.043	135.147	1.140	1.140
5	191.627	187.234	1.574	1.582
6	191.775	187.486	1.574	1.582
7	236.593	236.477	1.989	2.000

Table 2 Material properties of some composites

	Graphite-epoxy AS4/3501-6	Carbon-epoxy (T300/N5208) (CFRP)	Aramid-epoxy (KEVLAR 49-EPOXY) (KFRP)
E_1 *(GPa)*	144.8	181.0	76.0
E_2 *(GPa)*	9.65	10.3	5.56
G_{12} *(GPa)*	4.14	7.17	2.30
G_{23} *(GPa)*	3.45	3.43	1.62
ρ *(kg/m³)*	1389.23	1600.0	1460.0
υ_{12}	0.3	0.28	0.34

Table 3 The first six natural frequencies for $P_0 = 0.15$ N

Modes	AS4/3501-6	T300/N5208	KFRP
1	40.5	55.1	17.2
2	41.3	55.5	18.6
3	129	160	88.8
4	130	160	89.8
5	250	304	181
6	259	317	183

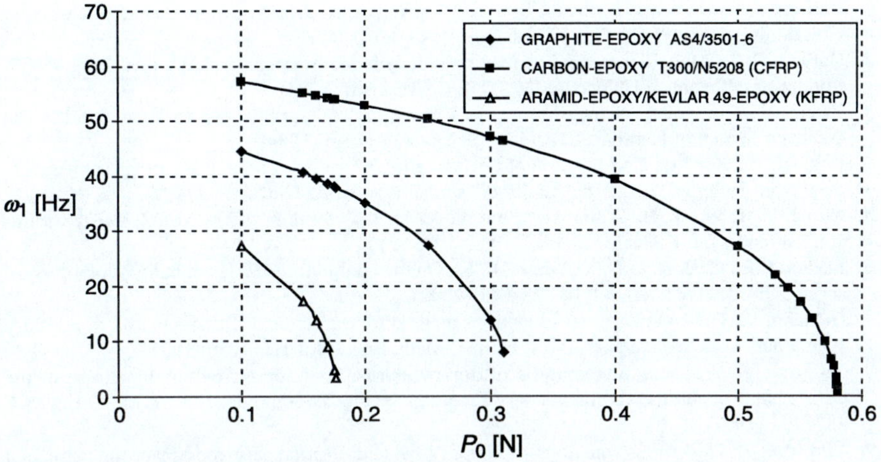

Fig. 2 Variation of the fundamental natural frequencies for various initial static forces

In the second example the geometrical properties of the springs are: D = 5 mm, d = 1 mm, n = 30, α = 15°, Fixed-Fixed. Composite material properties are presented in Table 2. The first six natural frequencies for $P_0 = 0.15$N are listed in Table 3. Variation of the fundamental natural frequencies with the numerical value of the initial axial force is illustrated in Fig. 2. As seen from Fig. 2 that the critical buckling loads are the axial loads at which the fundamental natural frequency becomes zero. Those critical buckling loads are presented in Table 4.

Table 4 Critical buckling loads obtained based on the dynamic approach

	AS4/3501-6	T300/N5208	KFRP
P_{cr} (N)	0.32	0.58	0.18

4 Conclusions

In this work, the free vibration analysis of composite cylindrical helical springs is studied by the stiffness matrix method based on the first order shear deformation theory. The element stiffness matrix obtained with the help of the complementary functions method comprises the static axial force effects. Numerical results are verified by the results for isotropic helical springs subjected to an axial force. This work is handled in the literature at first time.

References

1. Haringx, J.A.: On highly compressible helical springs and rubber rods, and their application for vibration-free mountings. Philips Res. Rep. **4**, 49–80 (1949)
2. Ünlüsoy, Y.S.: Modelling helical springs for the dynamic behavior. J. Appl. Res. **3**(9), 45–57. Middle East Technical University (METU), (In Turkish) (1983)
3. Yalçın, V.: Finite element mobility analysis of helical coil springs, MS thesis in METU The Graduate School of Natural and Applied Sciences, Ankara (1984)
4. Haktanır, V., Kıral E.: Computation of free vibration frequencies of helical springs subjected to axial load. J. Eng. **6**, 201–218. Faculty of Isparta of Akdeniz University (1992)
5. Xiong, Y., Tabarrok, B.: A finite element model for the vibration of spatial rods under various applied loads. Int. J. Mech. Sci. **34**(1), 41–51 (1992)
6. Becker, L.E., Chassie, G.G., Cleghorn, W.L.: On the natural frequencies of helical compression springs. Int. J. Mech. Sci. **44**(4), 825–841 (2002)
7. Haktanır, V.: The complementary functions method for the element stiffness matrix of arbitrary spatial bars of helicoidal axes. Int. J. Numer. Meth. Eng. **38**, 1031–1056 (1995)
8. Yıldırım, V.: Free vibration analysis of non-cylindrical coil springs by combined use of the transfer matrix and the complementary functions methods. Commun. Numer. Meth. Eng. **13**, 487–494 (1997)
9. Yildirim, V.: Governing equations of initially twisted elastic space rods made of laminated composite materials. J. Eng. Mech. **37**, 1007–1035 (1999)

Part III
Analysis of the Non-linear Deterministic/Stochastic Vibrations Phenomena

Gas Bubbles Motion in an Oscillating Fluid

V.S. Sorokin, I.I. Blekhman, L.I. Blekhman, V.B. Vasilkov, and K.S. Yakimova

Abstract Motion of a gas bubble in an oscillating fluid is of fundamental interest for the theory of various widely used technological processes, in particular, for the flotation process. Therefore, many studies have been concerned with this problem, some of those being undertaken by eminent scientists. The main remarkable effects are gas bubbles sinking in vibrating fluid's volume and asynchronous self-induced vibration of emerging air cushion. In the authors' recent papers, the problem has been solved by means of the concept of vibrational mechanics and the method of direct separation of motions; experimental studies have been also conducted. The present paper generalizes those studies. It is shown that the condition of gas bubbles sinking is strongly dependent on its own compressibility as well as on the compressibility of the surrounding medium. A formula for the average velocity of gas bubble motion, which significantly depends on the depth of its submergence and on vibration parameters, is derived. A simple physical explanation of the experimentally observed and analytically studied effects is given.

Keywords Oscillating fluid volume • Compressibility effects • Method of direct separation of motions • Sinking condition

1 Introduction

In the literature, the effect of gas bubbles sinking in a vertically oscillating fluid-filled volume has been attributed to two different mechanisms, which may be called a "wave-induced motion" [1–3] and a "vibration-induced motion" [4, 5]. The key

V.S. Sorokin (✉) • I.I. Blekhman • L.I. Blekhman • V.B. Vasilkov • K.S. Yakimova
Institute for the Problems in Mechanical Engineering RAS, Saint Petersburg, Russia

Mekhanobr-Tekhnika Research & Engineering Corp, Saint Petersburg, Russia
e-mail: slavos87@mail.ru

J. Náprstek et al. (eds.), *Vibration Problems ICOVP 2011: The 10th International Conference on Vibration Problems*, Springer Proceedings in Physics 139,
DOI 10.1007/978-94-007-2069-5_17,

feature of the first one is the gradient of the wave amplitude, while the key feature of the second mechanism is the compressibility of the bubble. In the present paper, these mechanisms are considered as being coupled. A gas bubble motion in vibrating fluid-filled volume is studied accounting for the compressibility of both the bubble and the gas-saturated layer, which is generated due to the turbulent fluid motion near its free surface [6]. In paper [1], to simplify the analysis, bubble's velocity relative to the fluid was considered to be equal to its velocity relative to the volume, whereas in the present study this assumption is omitted. A nonlinear differential equation, which describes bubble's motion relative to the volume, is derived, and, for the solution of this equation, the method of direct separation of motions is applied. The conditions of bubble sinking in gas-saturated fluid layer are determined. An expression for the critical thickness of this layer, starting from which it moves into the fluid volume, is derived, i.e. the condition of vibrational instability of the separate state of the gas-fluid system is obtained.

An approximate expression for the average velocity of bubble's motion in gas saturated layer, which strongly depends on the depth of its location and on vibration parameters, and an expression for the average velocity of penetration of this layer into the volume are derived.

Based on the obtained theoretical results, a simple physical explanation of the effects, experimentally observed in the fluid under the action of vibration, is provided. The results of recently conducted dedicated experiments are reported.

2 Governing Equations

Motion of a bubble in vertically oscillating with amplitude A and frequency ω in harmonic law $\alpha = A \sin \omega t$ fluid-filled cylindrical volume is analyzed (Fig. 1).

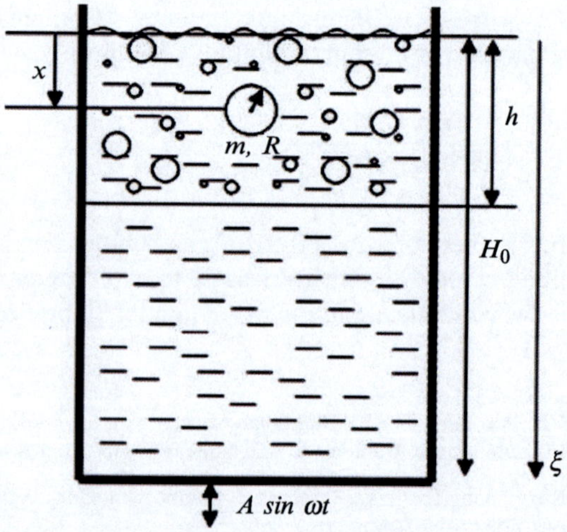

Fig. 1 Model of a bubble in gas saturated fluid

It is assumed that the fluid is saturated with gas till the certain depth $h \leq H_0$ (H_0 – the height of the fluid column) and that this fluid column may be considered as an elastic rod. The rest of the fluid is assumed to follow the oscillations of the vessel. The absolute displacement of the rod's cross-section, which in undisturbed state is situated at distance x from its free (upper) edge, is designated as $\xi(x, t)$. With the boundary conditions set as $\xi'|_{x=0} = 0,\ \xi|_{x=h} = A \sin \omega t$, this displacement is determined by the expression

$$\xi = \frac{g}{2c^2}(h^2 - x^2) + A\left(\cos\frac{\omega x}{c} \Big/ \cos\frac{\omega h}{c}\right)\sin\omega t \tag{1}$$

here c – sound speed in a gas-saturated fluid.

A pressure in an oscillating fluid at the distance x ($x \leq h$) from its free surface is defined by the formula [7]

$$P = P_e + \rho x g - \rho \int_0^x \frac{\partial^2 \xi}{\partial t^2}(x_1, t)\, dx_1, \tag{2}$$

here P_e is an external (say, atmospheric) pressure, ρ is the density of gas-saturated fluid, g – gravity acceleration.

Assuming the bubble to be compact (the condition $\frac{\omega R}{c} \ll 1$ is held, R is the radius of the bubble) and taking into account the expression (2), the equation of its motion in gas-saturated fluid can be written as

$$m\ddot{x} + \frac{d\,(m_0\dot{x})}{dt} = -F(v_{rel}) + (m - \rho V_b)\,g + \rho V_b \frac{\partial^2 \xi}{\partial t^2}(x, t)$$
$$- m\ddot{\alpha} - \frac{d\left(m_0\left(\dot{\alpha} - \frac{\partial \xi}{\partial t}(x, t)\right)\right)}{dt} \tag{3}$$

Here $v_{rel} = \dot{x} + \dot{\alpha} - \frac{\partial \xi}{\partial t}(x, t)$ is bubble's velocity relative to the surrounding medium (gas saturated fluid), m is the mass of the bubble (it accounts for the mass of a particle attached to the bubble), m_0 is the fluid added mass, defined by the formula $m_0 = \chi V_b \rho$, with χ being an added mass coefficient, V_b – instantaneous volume of the bubble (dot designates the full time derivative). $F(v_{rel}) = k_1 R^2 v_{rel}$ is the resistance force to gas bubble motion.

Assuming that gas bubble volume pulsations are small and isothermal, and taking into account expression (2), the following equation is used to describe the evolution of the volume of the bubble

$$\rho R_0 \frac{d^2 \Delta R}{dt^2} + 3\frac{\Delta R}{R_0} P_e = -\rho x g - \rho A \omega^2 f(x) \sin \omega t \tag{4}$$

where $f(x) = \frac{c}{\omega}\left(\sin\frac{\omega x}{c} / \cos\frac{\omega h}{c}\right)$, R_0 is the radius of the bubble near free surface of the fluid, $\Delta R = R - R_0$ (the surface tension is neglected, since it is assumed that the bubble has the radius, larger than 2 μm [1]).

3 Solution by the Method of Direct Separation of Motions

Inasmuch as the condition $\frac{c}{A\omega} \gg 1$ is held, the variable x can be considered as parameter while solving the Eq. 4. Thus, we obtain the following expression for the instantaneous volume of the bubble

$$V_b = V_{b0}\left(1 - \frac{\rho}{P_e}\left(xg + A\omega^2 f(x)\frac{\lambda^2}{\lambda^2 - \omega^2}\sin\omega t\right)\right) \tag{5}$$

where $\lambda = \frac{1}{R_0}\sqrt{\frac{3P_e}{\rho}}$ is the eigenfrequency of bubble's radial oscillations.

For the solution of the problem we use the concept of vibrational mechanics and the method of direct separation of motions [8]. The solutions to Eq. 3 are sought in the form

$$x = X(t) + \psi(t, \tau) \tag{6}$$

where X – "slow", and ψ – "fast", 2π – periodic in dimensionless ("fast") time $\tau = \omega t$ variable, with the period-τ average being equal zero.

As a result, considering the mass of the bubble to be negligibly small in comparison with the mass of the medium in its volume $m \ll \rho V_{b0}$, we obtain the following equation of its "slow" motions:

$$\ddot{X} + \eta\dot{X} = \frac{A^2\omega^4}{2\chi}\left[\frac{1}{c^2}\left(\frac{\omega^2}{\eta^2 + \omega^2}\left(1 + \frac{1}{\chi}\right)f'(X) - 1\right)\right.$$
$$\left. + \frac{\rho}{P_e}\frac{\lambda^2}{\lambda^2 - \omega^2}\frac{\omega^2 + \frac{1}{3}\eta^2}{\eta^2 + \omega^2}f'(X)\right]f(X) - \frac{1}{\chi}g \tag{7}$$

Here $\eta = \frac{k_1 R_0^2}{\chi\rho V_{b0}}$ (V_{b0} is the volume of the bubble near free surface of the fluid), prime designates a spatial derivative.

Vibrational force, induced on the bubble in gas-saturated fluid layer, has two components. The first one is controlled by the bubble's compressibility, while the second one is controlled by the compressibility of the surrounding medium. The ranges of the parameters, when one of these force components dominates the other, are determined.

The conditions of gas bubble sinking (or rising) in gas-saturated fluid layer, can be obtained from the Eq. 7. If the expression written in the right hand side of this equation is positive, then bubble sinks, and if it is negative, then bubble rises. Thus, the condition of bubble sinking in gas saturated fluid layer has the form

$$\frac{A^2\omega^4}{2g}\left[\frac{1}{c^2}\left(\frac{\omega^2}{\eta^2+\omega^2}\left(1+\frac{1}{\chi}\right)f'(X)-1\right)+\frac{\rho}{P_e}\frac{\lambda^2}{\lambda^2-\omega^2}\frac{\omega^2+\frac{1}{3}\eta^2}{\eta^2+\omega^2}f'(X)\right]f(X)>1 \tag{8}$$

If the inequality (8) is held at the front of the expanding area $X = h$ of the fluid-gas phase, then bubbles, generated in the layer with thickness h, will move deeper into the fluid, and the gas-saturated fluid layer will continue to expand. Thereby, by employing the obtained condition of bubble sinking (8), the condition of vibrational instability of the separate state of the gas-fluid system is determined.

An approximate expression for the velocity of bubble's "slow" motion in gas saturated fluid layer is defined from the Eq. 7 with its "slow" acceleration $\ddot{X}$ being considered as small. The velocity of this layer penetration into the volume is equal to the sinking velocity of the bubble, situated at its boundary$X = h$.

To verify analytically obtained results, a numerical experiment was conducted. Original equations of the gas bubble motion were integrated directly by means of Mathematica 7, and the obtained results are in good agreement with analytical solution.

Series of experiments has been conducted on the multi-functional vibration test rig of the Joint laboratory of Vibrational Mechanics IPME RAS and REC "Mekhanobr-Tekhnika". As a result, the assumption that the vibrational force, induced on the bubble in gas-saturated fluid, has the component controlled by the bubble's compressibility and the component controlled by the compressibility of the surrounding medium, has been entirely confirmed. It has been shown that, at some ranges of external excitation parameters, either of those can lead to gas bubble sinking.

4 Conclusions

Motion of a bubble in vibrating volume of a fluid, saturated with gas on a certain depth, is studied, with compressibility of both the bubble and the gas saturated layer being taken into account. The conditions of gas bubble sinking in the layer are determined. The expression for the critical thickness of gas-saturated fluid layer, starting from which it penetrates into the volume, is derived, i.e. the condition of vibrational instability of the separate state of the gas-fluid system is obtained. An approximate expression for the average velocity of bubble's motion in gas saturated layer, which strongly depends on the depth of its location and on vibration parameters, and an expression for the average velocity of penetration of this layer into the volume are derived. The reported results are applicable for control and optimization of relevant technological processes.

It is shown that vibrational force, induced on the bubble in oscillating gas saturated fluid, has two components controlled by the bubble's compressibility and by the compressibility of the surrounding medium, respectively.

Acknowledgments The financial support from the Russian Foundation of Fundamental Research (РФФИ), grant 10-08-00201 is gratefully acknowledged.

References

1. Blekhman, I.I., Vaisberg, L.A., Blekhman, L.I., Vasilkov, V.B., Yakimova, K.S.: "Anomalous" phenomena in fluid under action of vibration. Dokl. Phys. **53**(10), 520–524 (2008)
2. Crum, L.A., Eller, A.I.: Motion of bubbles in a stationary sound field. J. Acoust. Soc. Am. **48**(1(2)), 181–189 (1970)
3. Ganiev, R.F., Ukrainskii, L.E.: Nonlinear Wave Mechanics and Technologies (in Russian). Scientific publishing center "Regular and chaotic dynamics", Moscow (2008)
4. Bleich, H.H.: Effect of vibrations on the motion of small gas bubbles in a liquid, Jet Propulsion. J. Am. Rocket Soc. **26**(11), 958–964 (1956)
5. Sorokin, V.S., Blekhman, I.I., Vasilkov, V.B.: Motion of a gas bubble in fluid under vibration. Nonlinear Dyn. (2011). doi:10.1007/s11071-011-9966-9
6. Tatevosyan, R.A.: Analysis of the regularities of water-air system vibroturbulisation (in Russian). Theor. Fundam. Chem. Technol. **11**(1), 153–155 (1977)
7. Loytsyanskiy, L.G.: Fluid and Gas Mechanics (in Russian). Nauka, Moscow (1973)
8. Blekhman, I.I.: Vibrational Mechanics. World Scientific, Singapore/River Edge (2000)

Theoretical and Experimental Basis of Advanced Vibrational Technologies

I.I. Blekhman, L.A. Vaisberg, and D.A. Indeitsev

Abstract The paper deals with the overview of the new results obtained in the joint Vibration mechanical laboratory of two companies mentioned above.

Keywords Vibrational technologies • Theory • Experiment • Applications

1 Introduction

Vibrational technologies and machines are increasingly employed in various applications. At present, development of such technologies has advanced to the qualitatively new level. It is facilitated by the recent discoveries in this field and by the emergence of new theoretical and experimental methods. The brief overview of the new results obtained in the joint laboratory of Vibrational mechanics of the IPME RAS and REC Mekhanobr-Tekhnika (Saint-Petersburg) is given in the report. These results cover new theoretical and experimental approaches, dynamics of vibration excitation, dynamics of vibrational processes.

I.I. Blekhman (✉) • L.A. Vaisberg • D.A. Indeitsev
Institute for the Problems in Mechanical Engineering RAS, Saint Petersburg, Russia

I.I. Blekhman • L.A. Vaisberg
Mekhanobr-Tekhnika Research & Engineering Corp, Saint Petersburg, Russia
e-mail: iliya.i.blekhman@gmail.com

J. Náprstek et al. (eds.), *Vibration Problems ICOVP 2011: The 10th International Conference on Vibration Problems*, Springer Proceedings in Physics 139,
DOI 10.1007/978-94-007-2069-5_18,

2 Development of General Theoretical and Experimental Approaches

The approach of vibration mechanics and the method of direct separation of motions are the main theoretical tools for solution of the problems of the action of vibration on various media and system. They were in detail presented in books [1, 2]; its advancement is given in articles in book [3] and in presentation [4].

Vibrational stand 486 – US of the Mekhanobr – Tekhnika Corporation was used as a main tool for the experimental studies. Owing to the use of self-synchronization phenomenon this equipment provided possibility for obtaining vibrations of various forms and directions. This makes the stand a unique tool useful for investigation of various processes and devices [5].

Computer simulation methods were widely used including EDEM software employing the method of particle dynamics (molecular dynamics method [6]).

3 Problems of Vibration Excitation. Motion of Vibration Exciter Rotor in Passing Start-Stop Resonance Zone, Zommerfeld Effect

The problem of passing through a resonance zone in devices with rotating imbalanced rotors that operate within the supercritical frequency range is of significant interest for a number of engineering fields. An important characteristic of this process is the phenomenon of "sticking" rotor rotation frequency near the eigenfrequency, which was described for the first time by Zommerfeld (1902), and was named after him. A number of studies are devoted to this. Such studies began in Russia as far back as in 1939 and continued in 1953 [1, 7–11], appear to have been disregarded by the researchers of other countries. The review of early papers on this problem can be found in [8–13]. In paper [12] the problem is solved by the method of the direct separation of motions.

The setup of the system is shown in Fig. 1. An unbalanced rotor 2, driven by an electric motor, is installed on a flexible platform 1. It may move in direction x. The platform is supported on a fixed base 3 by elastic elements with rigidity c and damping element with viscous friction factor β.

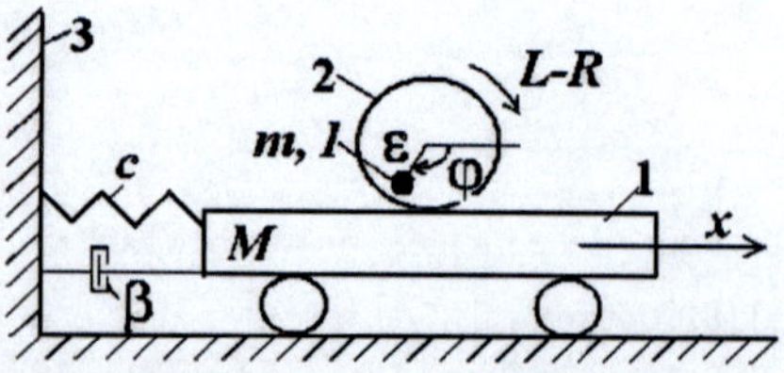

Fig. 1 The setup of system

The equations of motion of the system can be found in the publications [1–3, 8–12]. As a result in [1, 12] it is obtained two following equations

$$I\dot{\omega} = L(\omega) - R(\omega) + V(\omega), \tag{1}$$

$$\ddot{\psi} + 2n_1\dot{\psi} + b\sin\psi - \rho^2\sin^2\frac{\psi}{2} = 0 \tag{2}$$

Here ω is the rotation frequency, ψ – is a semi-slow vibrational component of the rotor rotation the angle in relation to uniform ωt, L and R – the torque of motor and resisting torque respectively, n_1 and ρ – positive parameters.

$$V(\omega) = -\frac{(m\varepsilon\omega)^2}{\mathrm{M}}\frac{n\omega^3}{(\omega^2 - p^2)^2 + 4n^2\omega^2} = -n\omega\mathrm{M}A^2 \tag{3}$$

$$\left(p^2 = c/\mathrm{M},\quad 2n = \beta/\mathrm{M},\quad A = m\varepsilon\omega^2/\mathrm{M}\sqrt{(p^2 - \omega^2)^2 + 4n^2\omega^2}\,\right)$$

is a torque refereed as vibrational torque which is an additional load on the rotor imparted by vibrations of the platform on which it is installed. The Zommerfeld effect is explained by the presence of this torque.

Equation 1 is valid, if $\dot{\omega}/\omega_0 << \omega_0$ where ω_0 is stationary value of ω, M is the mass of the systems. The value

$$q = \sqrt{b} = A\sqrt{\mathrm{M}\,|p^2 - \omega^2|\,/2I} \tag{4}$$

is the frequency of small free vibration of the "internal pendulum" (the resistance force being neglected). This frequency vanishes for $p = \omega$. To validity Eq. 2, it is necessary, that $q \ll \omega$; usually it is sufficient to have $q/\omega < 1/3$.

Analysis of Eq. 2 revealed that if in subresonance condition $\omega < p$ the stable position of the internal pendulum is $\psi_0 = 0$ and unstable position is $\psi_0 = \pi$, the post-resonance positions are quite the opposite (in absence of damping), i.e. in passing the resonance zone the internal pendulum turns backward.

These results were generalized in [13] to the case of vibrator rotor being installed on a movable platform with three degrees of freedom. The paper [14] deals with semi-slow relative vibrations of two rotors arising during disturbance of self-synchronization mode. It was found that the frequency of the internal pendulum oscillations in this case is equal to $\sqrt{2}q$ where q is determined by formula (4).

Practical value of the results obtained is that they can be used in the design of control start-stop systems of the machines under consideration [15]. These systems provide to lower the power the drive motor and the resonance amplitude.

4 Dynamics of Vibration Processes

4.1 Travel of Gas Bubble in Vibrating Vessel Filled with Liquid. Vibrational Injection. Vibro-Jet Effect

This problem is important for development of a lot of treatment processes. Most full review and own results in this problem are given in [16]. A series of theoretical and experimental studies in this field has been carried out by the authors of this report together with collaborators [17–20].

Submersion of gas bubbles into oscillating liquid observed in the experiments is of great significance. Analytical investigation involving by the method of direct separation of motions proved that a decisive cause of this phenomenon is the yieldability of the bubble and gas saturated liquid.

Bubble submersion conditions were ascertained and an expression for average velocity of immersion was obtained. The paper by V.S. Sorokin and al. summarizing these studies is presented at this conference. The theory of vibrational gas injection into liquid and vibro-jet effect are described in [20].

4.2 Behavior of Granular Media in Vibrating Vessels. Separation of Loose Particles Mixtures and Optimization of Screening Process

By contrast to many published statements the behavior of granular media under vibration is distinct from that of a viscous liquid because in the medium there arises certain vibration forces along with medium fluidization. This circumstance explains the processes of vibration transportation and specifics of granular medium behavior in vibrating communicating vessels [1, 18].

A serious challenge in some industries is separation of particles by their shape These problems were considered in [21] and solved in the patent [22]. The required result is achieved there by the use of a sloping vibrating surface of which the properties and vibration parameters have been selected in such way that flat shaped and spelled particles tend to travel upwards while cube and spherical shaped ones will roll downwards.

A series of works have been accomplished on creation of devices for a segregational ("no sieve") classification of particles by their size and shape [23, 24].

4.3 Some Other Problems

The problem of manufacturing vibratory dynamic materials and composites and vibrational control of their properties is discussed in [25] and [1–3].

Papers by K. S. Ivanov and by A. M. Guskov, E. V. Myalo and G. Ya. Panovko in the collection [3] are devoted to Chelomey's effect observed in a washer traveling along a vibrating rod.

Some investigations were aimed at enhancing grinding efficiency of the drum mills. This was achieved by periodic agitation of the grinding media. This was made possible by changing conventional shape of the operating chamber.

5 Conclusions

The paper deals with results of two research teams representing two St. Petersburg institutions and engaged in development of theoretical and experimental fundamentals, which are aimed to provide means for creation of new vibrating machines and technologies. Many of these findings have been commercialized but still greater part of them requires effort of designers and inventors.

Acknowledgements The research work reported here was made possible by grants RFBR 09-08-00620, 10-08-00201, 11-08-00278.

References

1. Blekhman, I.I.: Vibrational Mechanics, 400 p. Fizmatlit, Moscow (1994) (In Russian, Engl. trans.: Singapore et al., World Scientific Publishing Co. (2000) 510 p.)
2. Blekhman, I.I. (ed.): Selected Topics in Vibrational Mechanics, p. 409. WorldScientific Publishing Co., Singapore/River Edge (2004)
3. Beletsky, V.V., Indeitsev, D.A., Fradkov, A.L. (eds.): Nonlinear Problems in the Theory of Oscillation and Theory of Control. Vibrational Mechanics, p. 528. Nauka Press, Moscow (2009) (in Russian and in English)
4. Blekhman, I.I.: On the oscillatory strobodynamics. Accepted as a presentation on ENOC 2011, Rome, 24–29 July 2011
5. Blekhman, I.I., Vaisberg, L.A., Vasilkov, V.B., Lavrov, B.P., Yakimova, K.S.: Universal vibrational stand: experience in investigations and some results. Nauchno-technicheskie vedomosti St. Petersburg State Technological University (3), 224–227 (2003) (in Russian)
6. Arsentyev, V.A., Blekhman, I.I., Blekhman, L.I., Vaisberg, L.A., Ivanov, K.S., Krivtsov, A.M.: Methods of particles and discrete elements dynamics as a tool for studies and optimization of natural and waste mineral processing. Obogashcheniye rud (1), 30–35 (2010) (in Russian)
7. Blekhman, I.I.: Self-Synchronization of vibrators in some types of vibration machines. Inzhenerny Sbornik **16**, 49–72 (1953)
8. Blekhman, I.I.: Synchronization of Dynamic Systems, p. 894. Nauka Press, Moscow (1971) (in Russian)
9. Blekhman, I.I.: Synchronization in Science and Technology, p. 351. Nauka Press, Moscow (1981) (in Russian, English trans.: ASME Press, New York (1988) 255 p)
10. Vibration in technology. Handbook in 6 volumes, vol. 2, p. 351 (1979); vol. 4, p. 509 (1981). Mashinostroenie, Moscow (in Russian)
11. Kononenko, V.O.: Oscillatory Systems with Limited Excitation, p. 254. Nauka Press, Moscow (1964)

12. Blekhman, I.I., Indeitsev, D.A., Fradkov, A.L.: Slow motions in systems with inertial excitation. J. Mach. Manuf. Reliab. **37**(1), 21–27 (2008)
13. Blekhman, I.I., Yaroshevich, N.P.: Transition regimes in inertional excited trans-resonant vibrational devices with several degrees of freedom of the carrier system. In ref. [3], pp. 215–238
14. Potapenko, M.A.: Slow vibrations of unbalanced rotor vibration exciters in self-synchronization disturbance conditions. J. Mach. Manuf. Reliab. **37**(3), 27–29 (2008)
15. Tomchin, D.A., Fradkov, A.L.: Control of the passage of a rotor through a Resonance Zone on the basis of the Speed-Gradient Method. J. Mach. Manuf. Reliab, **34**(5), 66–71 (2005)
16. Ganiev, R.F., Ukrainsky, L.E.: Nonlinear Wave Mechanics and Technologies, 712 p. Publishing Center "Regular and Chaotic Dynamics", Moscow (2008)
17. Blekhman, I.I., Vaisberg, L.A., Blekhman, L.I., Vasilkov, V.B., Yakimova, K.S.: Anomalous phenomena in liquids under vibration action. Proc. Russ. Acad. Sci. **422**(4), 470–474 (2008) (Engl. trans.: Dokl. Phys. **53**(10), 520–524 (2008))
18. Blekhman, I.I., Blekhman, L.I., Vaisberg, L.A., Vasilkov, V.B., Yakimova, K.S.: Vibration effects in liquid, granular and mixed media: experiment, theory and applications. Obogashcheniye rud (4), 42–47 (2009) (in Russian)
19. Blekhman, I.I.,Vakulenko, S.A., Indeitsev, D.A., Mochalova, Yu.A.: Generation and movement of gas-liquid mixture in a vibrating vessel filled with free surface liquid. Transactions of XVI symposium "Dynamics of impact-vibration (non-linear) systems" "DYVIS-2009". M., 2009, pp. 61–71 (in Russian)
20. Blekhman, I.I., Blekhman, L.I., Vaisberg, L.A., Vasilkov, V.B., Yakimova, K.S.: Nonlinear effects of liquid flowing out of vibrating vessels. Trans. Russ. Acad. Sci. **391**(2), 185–188 (2003) (Engl. trans.: Dokl. Phys. **48**(7), 355–358 (2003))
21. Blekhman, I.I., Blekhman, L.I., Vaisberg, L.A., Vasilkov, V.B., Yakimova, K.S.: On separation of solids by shape on vibrating surfaces. Obogashcheniye rud (2), 23–25 (2007) (in Russian)
22. Blekhman, L.I., Vaisberg, L.A., et al.: Method of separating spherical particles from spalled ones. Patents Bull. (34), Russian Patent 2,405,634, 10 Dec 2010
23. Blekhman, L.I., Vaisberg, L.A., et al.: Vibratory classifier. Patents Bull. (36), Russian Patent 2,407,600, 27 Dec 2010
24. Arsentiev, V.A., Blekhman, I.I., Blekhman, L.I., Vasilkov, V.B., Feoktistov, A.Yu., Yakimova, K.S.: Classification of bulk flowing material by vibrational segregation. Set up, simulation, experiment, Obogashcheniye rud (5), 13–16 (2010) (in Russian)
25. Blekhman, I.I.: Vibrational dynamic materials and composites. J. Sound Vib. **317**(3–5), 657–663 (2008)

Regular and Irregular Vibrations in Nonlinear Discrete-Continuous Systems Torsionally Deformed

A. Pielorz and Danuta Sado

Abstract In the paper discrete-continuous systems torsionally deformed with a local nonlinearity are considered. The systems consist of an arbitrary number of rigid bodies connected by shafts. The nonlinearity is described by the polynomial of the third degree having a hardening characteristic. In the study a wave method is applied leading to solving equations with a retarded argument. The numerical analysis focuses on regular and irregular vibrations. Detailed considerations are presented for a two-mass system. The possibility of occurring of irregular vibrations is discussed on the basis of the Poincaré maps and on bifurcation diagrams.

Keywords Regular and irregular nonlinear vibrations • Discrete-continuous systems • Wave approach

1 Introduction

The paper deals with nonlinear vibrations of discrete-continuous mechanical systems torsionally deformed with a local nonlinearity having the characteristic of a hard type, taking into account regular and irregular vibrations. The considered systems consist of shafts with circular cross-sections connected by rigid bodies. It should be pointed out that local nonlinearities in torsional systems are justified by many engineering solutions. Local nonlinearities in the paper are described by the polynomial of the third degree.

A. Pielorz
Kielce University of Technology, Kielce, Poland
e-mail: apielorz@tu.kielce.pl

D. Sado (✉)
Warsaw University of Technology, Warsaw, Poland
e-mail: dsa@simr.pw.edu.pl

J. Náprstek et al. (eds.), *Vibration Problems ICOVP 2011: The 10th International Conference on Vibration Problems*, Springer Proceedings in Physics 139,
DOI 10.1007/978-94-007-2069-5_19,

Governing equations for multi-mass discrete-continuous systems torsionally deformed are derived in [1, 2]. In the studies a wave approach leading to solving equations with retarded argument is used, [1–3]. Irregular nonlinear vibrations including chaos are mainly considered in discrete systems, [4–7]. In [8] some numerical results concerning irregular nonlinear vibrations in a two-mass system torsionally deformed are given for local nonlinearities with a soft type characteristic. Here similar considerations are presented for systems having the hardening characteristics of the local nonlinearities. The possibility of occurring of irregular vibrations is discussed on the basis of the Poincaré maps and bifurcation diagrams. Exemplary numerical results are also given for a two-mass system. In the discussion of irregular vibrations, the approach used earlier in [4–6] for discrete systems, is generalized here to the case of discrete-continuous systems.

2 Equations

The i-th shaft in a multi-mass system torsionally deformed, $i = 1,2,\ldots,N$, is characterized by length l_i, density ρ, shear modulus G and polar moment of inertia I_{0i}, [1, 2]. The mass moment of inertia of rigid bodies, $i = 1,2,\ldots N+1$, are J_i. The first rigid body J_1 is loaded by the harmonic moment $M(t)$, and a local nonlinear discrete element with a hardening characteristic is attached to it. Equivalent external and internal damping, having coefficients d_i and D_i, are taking into account in appropriate cross-sections. It is assumed that displacements and velocities of the shaft cross-sections are equal to zero at time instant $t = 0$.

In appropriate nondimensionless quantities, the determination of angular displacements θ_i of shaft cross-sections is reduced to solving N equations, [1, 2],

$$\theta_{i,tt} - \theta_{i,xx} = 0, \; i = 1, 2, ..., N \tag{1}$$

with the following nonlinear boundary conditions

$$M_0 \sin pt - \theta_{1,tt} + K_r(D_1\theta_{1,xt} + \theta_{1,x}) - d_1\theta_{1,t} - k_1\theta_1 - k_3\theta_1^3 = 0 \text{ for } x = 0,$$

$$\theta_i(x,t) = \theta_{i+1}(x,t) \text{ for } x = \sum_{k=1} l_k, i = 1, 2, ..., N-1,$$

$$- \theta_{i,tt} - K_r B_i E_{i+1}(D_i\theta_{i,xt} + \theta_{i,x}) + K_r B_{i+1} E_{i+1}$$
$$\cdot (D_{i+1}\theta_{i+1,xt} + \theta_{i+1,x}) - E_{i+1} d_{i+1}\theta_{i,t} = 0$$

$$\text{for } x = \sum_{k=1}^{i} l_k, i = 1, 2, ..., N-1,$$

$$-\theta_{N,tt} - K_r B_N E_{N+1}(D_N\theta_{N,xt} + \theta_{N,x}) - E_{N+1}d_{N+1}\theta_{N,t} = 0$$

$$\text{for } x = \sum_{k=1}^{N} l_k, \tag{2}$$

and with zero initial conditions. Parameters M_0 and p are the amplitude and the frequency of an external loading, correspondingly. Comma denotes partial differentiation.

The solutions of Eq. 1 are sought in the form

$$\theta_i(x,t) = f_i(t-x) + g_i\left(t + x - 2\sum_{k=1}^{i-1} l_k\right), \quad i = 1, 2, ..., N. \tag{3}$$

Substituting (3) into the boundary conditions (2) we obtain the set of ordinary nonlinear differential equations with a retarded argument for unknown functions f_i and g_i which can found in [1, 2]. Such equations can be solved numerically with zero or nonzero initial conditions.

3 Numerical Results

The aim of the numerical analysis is to study the possibility of occurrence of irregular vibrations in systems considered. This is done on the basis of the Poincaré maps for the two-mass torsional system, presented in Fig. 1, characterized by the following basic parameters: $N = 1$, $E_2 = 0.8$, $K_r = 0.05$, $k_1 = 0.05$, $k_3 = 0.005$. The first two natural frequencies for the system are then $\omega_1 = 0.126$, $\omega_2 = 0.351$.

In [1, 2] it is shown that the effect of the nonlinearity is observed only in the first two resonant regions. In the case of the nonlinearity with a hardening characteristic, amplitude jumps are observed. Amplitude-frequency curves for angular displacement, presented in Fig. 2a, show such jumps for four values of the amplitude $M_0 = 0.1, 0.25, 0.5, 1.0$ and with large damping expressed by appropriate coefficients equal to $d_0 = d_1 = d_2 = D_1 = 0.1$. Jumps occur in the second resonant region. Two amplitude jumps can be obtained, with zero initial and nonzero initial

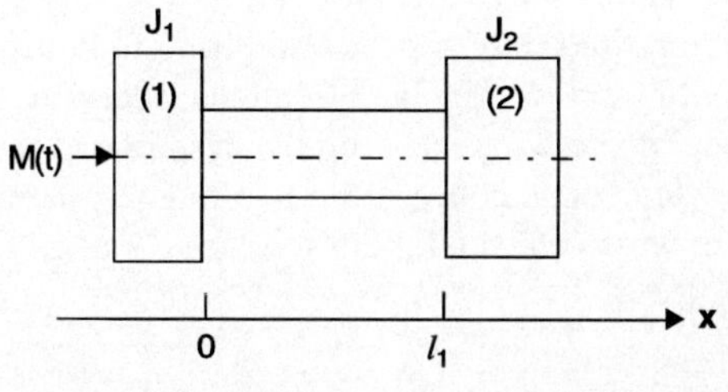

Fig. 1 Two-mass discrete-continuous system torsionally deformed

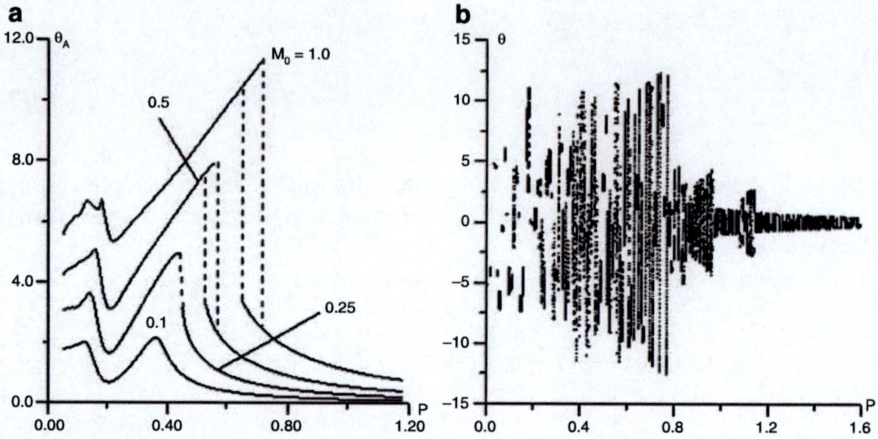

Fig. 2 (**a**) Amplitude-frequency curves for angular displacements in $x = 0$, (**b**) Bifurcation diagram for $x = 0$

conditions, [1]. In the nonzero initial case jumps are observed for larger values of the frequency p of the external moment $M(t)$. From Fig. 2a it follows that distances between jumps increase with the increase of M_0. It appears that distances of jumps increase also with the decrease of damping.

Irregular vibrations are looked for with $M_0 = 1$ and for small damping, i.e., for $d_0 = 0.001$ in the cross-section $x = 0$. The bifurcation diagram is shown in Fig. 2b, exemplary maps of Poincaré are presented in Fig. 3, and maximal Lyapunov exponents λ as functions of number of iterations n for two values of frequency p are given in Fig. 4.

In the Fig. 3d, e we can observe the strange attractors, and values of maximal Lyapunov exponents presented in Fig. 4 are positive, so in these cases (for $p = 0.37$ and $p = 0.38$) motions of angular displacements are chaotic.

4 Conclusions

From considerations in the paper it follows that in torsional discrete-continuous systems, with a local nonlinearity having a hardening characteristic, except periodic vibrations may appear also different kinds of irregular vibrations including chaotic vibrations. Irregular vibrations occur in the limited range of the change of the parameters representing the system and the external moment. Presented numerical calculations concern the two-mass system, however governing equations allow us to widen considerations to more complex systems.

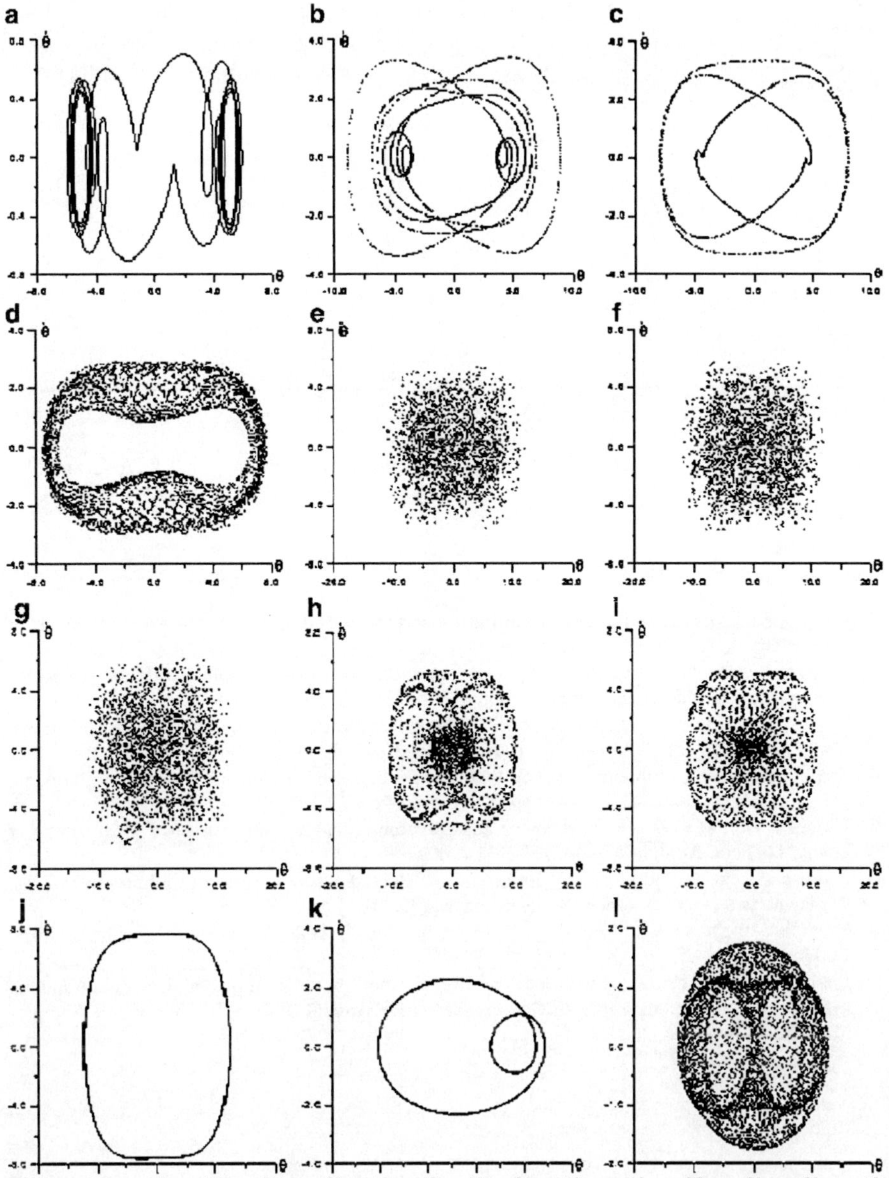

Fig. 3 Poincaré maps for **(a)** $p = 0.04$, **(b)** $p = 0.3$, **(c)** $p = 0.34$, **(d)** $p = 0.37$, **(e)** $p = 0.38$, **(f)** $p = 0.39$, **(g)** $p = 0.4$, **(h)** $p = 0.41$, **(i)** $p = 0.42$, **(j)** $p = 0.74$, **(k)** $p = 0.79$, **(l)** $p = 0.87$

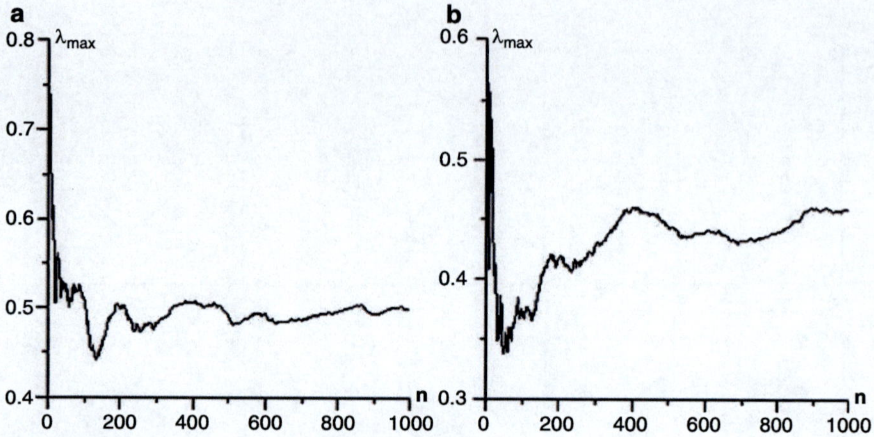

Fig. 4 Maximal Lyapunov exponents for angular displacements with (**a**) $p = 0.37$, (**b**) $p = 0.38$

References

1. Pielorz, A.: Dynamic analysis of a nonlinear discrete-continuous torsional system by means of wave method. ZAMM **75**, 691–698 (1995)
2. Pielorz, A.: Non-linear vibrations of a discrete-continuous torsional system with non-linearities having characteristic of a soft type. J. Sound Vib. **225**(2), 375–389 (1999)
3. Pielorz, A.: Nonlinear equations with a retarded argument in discrete-continuous systems. Math. Probl. Eng. **2007**(28430), 1–11 (2007). doi:10.1155/2007/28430
4. Sado, D., Gajos, K.: Note on chaos in three degree of freedom dynamical system with double pendulum. Meccanica **38**, 719–729 (2003)
5. Sado, D., Kot, M.: Chaotic vibration of an autoparametrical system with a non-ideal source of power. J. Theor. Appl. Mech. **45**, 119–131 (2007)
6. Sado, D.: Regular and Chaotic Vibrations in Selected Systems with Pendulum (in Polish). Wydawnictwa Naukowo-Techniczne, Warszawa (2010)
7. Szemplińska-Stupnicka, W.: Chaos, Bifurcations and Fractals Around Us. World Scientific, London (2003)
8. Pielorz, A., Sado, D.: Note on regular and irregular nonlinear vibrations in discrete-continuous systems. In: Proceedings of 10th Conference on Dynamical Systems – Theory and Applications, Łódź, pp. 325–330 (2009)

Stability of the Vertical Excitation of Slender Structures

J. Náprstek and C. Fischer

Abstract Vertical slender structures exposed to a strong vertical component of an earthquake excitation are endangered by auto-parametric resonance effect. In sub-critical regime the vertical and horizontal response components are independent. In post-critical regime, the non-trivial horizontal response occurs and dynamic non-linear interaction of both vertical and horizontal response components takes place.

It can be shown using a simplified analytical model, that there exist several theoretical stability limits either for the parameters of excitation or for parameters of the structure and subsoil. One corresponds to appearance of the non-trivial horizontal response. Other describes the case, when the horizontal response exhibits steadily increasing amplitude, which fact leads sooner or later to collapse of the structure.

Even if presence of the horizontal component in the system response does not automatically mean inevitable collapse of the structure, this type of behaviour should be avoided. The presented paper studies the stability conditions of the stationary response on a harmonic excitation and formulates the critical values of excitation amplitude.

Keywords Dynamic stability • Non-linear resonance • Non-lineardynamics • Auto-parametric system

1 Introduction

Many studies devoted to dynamics of slender structures (towers, masts, chimneys, bridges, etc.) related with earthquake attack have been published. They are dealing predominantly with an influence of horizontal excitation components. On the other

J. Náprstek (✉) • C. Fischer
ITAM AS CR, v.v.i., Prague, Czech Republic
e-mail: Naprstek@itam.cas.cz; FischerC@itam.cas.cz

J. Náprstek et al. (eds.), *Vibration Problems ICOVP 2011: The 10th International Conference on Vibration Problems*, Springer Proceedings in Physics 139,
DOI 10.1007/978-94-007-2069-5_20, © Springer Science+Business Media B.V. 2011

hand a strong vertical excitation component especially in the earthquake epicentre area can be decisive. The vertical component of the subsoil movement caused heavy damages or collapses of many high rise structures due to earthquake attack. Very widely used linear approach, however, usually doesn't provide any interesting knowledge in such a case. It reveals that the origin of these problems consists in auto-parametric resonance effects.

Auto-parametric systems have been recently studied intensively. These investigations are motivated by various technical branches and by basic theoretical research in rational mechanics. A theoretical outline dealing with these systems has been presented probably for the first in [1]. During this time many papers contributing to analytical, numerical as well as experimental aspects of auto-parametric systems have been published mostly by Tondl, Nabergoj and co-authors, e.g. [2–5]. Many other references can be found. Several monographs, e.g. [6] or [7], presenting a comprehensive overview of partial results and methods have appeared. A couple of papers dealing with auto-parametric systems under deterministic and random excitation have been recently published by authors of this study [8–10].

The mechanical system under study resembles inverse pendulum based systems. It shares major resonant properties with the inverse pendulum, like the main types of resonant states: chaotic and cyclic motion. Influence of the bending of the console, which the studied system comprises, is not significant but noticeable in the results. The presented contribution extends the previous paper [10] and studies in detailed analysis of the stability conditions of the stationary response on a harmonic excitation. It also identifies the critical values of excitation amplitude and formulates some design criteria.

2 Model Formulation

Let us consider the three DOF theoretical model outlined in the Fig. 1. It describes roughly a slender structure and its immediate subsoil. Influence of bending of the structure is taken into account. The system is Hamiltonian, see [11].

Non-dimensional response and excitation components are useful to be introduced (see Fig. 1 for interpretation) :

$$\zeta_0 = \frac{y_0}{l},\ \zeta = \frac{y}{l},\ \varphi,\ \psi = \frac{\varphi + \xi}{l} \tag{1}$$

The excitation process $\zeta_0(t)$ will be considered as harmonic:

$$y_0 = A_0 \sin \omega t \quad \Rightarrow \quad \zeta_0 = a_0 \sin \omega t, \quad A_0 = a_0 \cdot l \tag{2}$$

The governing 3DOF differential system in the form of Lagrange equations emerge from the kinetic and potential energies of the moving system. Finally it can be formulated as follows:

a

$y = y(t)$ - vertical displacement of the B point;
$y_0 = y_0(t)$ - kinematic excitation (seismic random process);
$\varphi = \varphi(t)$ - angular rotation of the system in the B point;
$\xi = \xi(t)$ - bending deformation of the vertical console;
M, m - foundation and structure effective masses;
C, EJ - subsoil stiffness, console bending stiffness;
η_c, η_e - viscous components of the C, EJ stiffness (Kelvin);
ρ, l - geometric parameters.

b

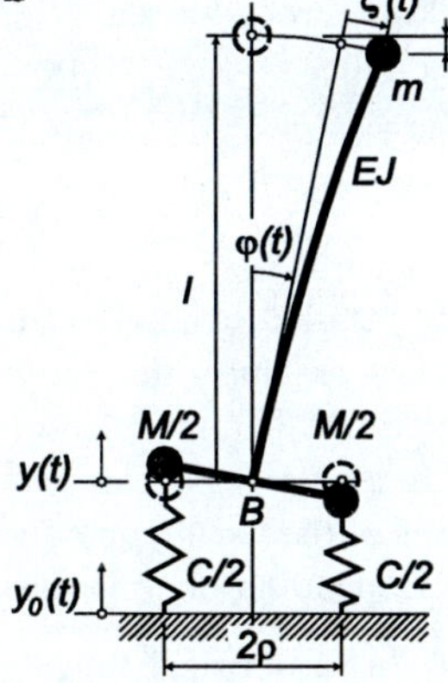

Fig. 1 Outline of the 3-DOF auto-parametric system and used symbols

$$
\begin{aligned}
\ddot{\zeta} - \kappa_0(\dot{\psi}\varphi)^{\bullet} + \omega_0^2(\zeta - \zeta_0 + \eta_c(\dot{\zeta} - \dot{\zeta}_0)) &= 0, \quad (a)\\
\ddot{\varphi} - \kappa_1(\dot{\zeta}\varphi)^{\bullet} + \kappa_1\ddot{\psi} + \kappa_1\dot{\zeta}\dot{\psi} - \kappa_1\omega_2^2\psi + \omega_1(\varphi + \eta_c\dot{\varphi}) &= 0, \quad (b)\\
\ddot{\psi} - (\dot{\zeta}\varphi)^{\bullet} + \omega_3^2(\psi - \varphi + \eta_e(\dot{\psi} - \dot{\varphi})) &= 0, \quad (c) \qquad (3)
\end{aligned}
$$

where the following notation has been used:

$$
\kappa_0 = \frac{m}{M+m}, \quad \kappa_1 = \frac{m \cdot l^2}{M \cdot \rho^2}, \quad \omega_0^2 = \frac{C}{M+m}, \quad \omega_1^2 = \frac{C}{M}, \quad \omega_2^2 = \frac{g}{l}, \omega_3^2 = \frac{6EJ}{m \cdot l^3}
$$

3 Analysis

In the post-critical state all the response components are non-trivial. Therefore expecting a single mode response, following approximate expressions can be written:

$$
\begin{aligned}
\zeta(t) &= r_c(t)\cos\omega t + r_s(t)\sin\omega t, & R^2(t) &= r_c^2(t) + r_s^2(t) \quad (c)\\
\varphi(t) &= p_c(t)\cos\frac{1}{2}\omega t + p_s(t)\sin\frac{1}{2}\omega t, & P^2(t) &= p_c^2(t) + p_s^2(t) \quad (b)\\
\psi(t) &= s_c(t)\cos\frac{1}{2}\omega t + s_s(t)\sin\frac{1}{2}\omega t, & S^2(t) &= s_c^2(t) + s_s^2(t) \quad (c) \quad (4)
\end{aligned}
$$

To make a decision between single- and multi-harmonic approximation is a delicate question. For some interesting experiences, see [12]. Another opinion can be found in [13], when internal resonance or its proximity should be analysed.

Let us use the approximations (4) in the system (3). Going through the harmonic balance procedures a differential system for unknown amplitudes **X** can be obtained:

$$\mathbf{H}(\mathbf{X})\,\frac{\mathrm{d}\mathbf{X}}{\mathrm{d}t} = \mathbf{K}(\mathbf{X}); \qquad \mathbf{H}(\mathbf{X}) \in \mathbb{R}^{6\times 6},\ \ \mathbf{K}(\mathbf{X}) \in \mathbb{R}^{6}$$
$$\mathbf{X}(t) = [r_c(t), r_s(t), p_c(t), p_s(t), s_c(t), s_s(t)]^T \tag{5}$$

The detailed structure of the matrix **H**(**X**) and the vector **K**(**X**) is described in [10].

Let us study the properties of the proposed solution (4). The system (5) for amplitudes $\mathbf{X}(t)$ is meaningful if they are functions of a "slow time," in other words if their changes within one period $T = 2\pi/\omega$ are small or vanishing and individual steps of the harmonic balance operation are acceptable.

Depending on the types of response, the two main cases can be identified:

(a) In sub-critical linear regime the vertical and horizontal response components are independent. If no horizontal excitation is taken into account, no horizontal response component is observed. In the components defined by (4), $P(t) = S(t) = 0$ and $R(t)$ is so called semi-trivial solution.

(b) If the auto-parametric resonance occurs, horizontal response components arise even if no horizontal excitement is present. All the components $R(t), P(t), S(t)$ are non-trivial.

Studying properties of the system (5), one should be able to distinguish individual cases (a) and (b).

Let us look at first at the matrix **H**(**X**) of the differential system (5). Its determinant has the following form:

$$\det(\mathbf{H}(\mathbf{X})) = \frac{1}{256}\omega^6\left(\kappa_0\left(p_c(t)^2 + p_s(t)^2\right) - 4\right)^2 \tag{6}$$

It is clear from Eq. 6 that regularity of **H** (for $\omega > 0$) depends only on $P^2(t)$:

$$P^2(t) \neq 4\frac{M}{m}\left(1 + \frac{m}{M}\right), \quad \frac{m}{M} \ll 1$$

The inverse matrix can be formulated easily:

$$\mathbf{H}(\mathbf{X})^{-1} = 2\begin{pmatrix} 0 & 0 & 0 & -\omega^{-1} & 0 & \kappa_1\omega^{-1} \\ 0 & 0 & \omega^{-1} & 0 & -\kappa_1\omega^{-1} & 0 \\ 0 & 2\delta & 0 & 0 & \delta\kappa_0 p_s(t) & \delta\kappa_0 p_c(t) \\ -2\delta & 0 & 0 & 0 & -\delta\kappa_0 p_c(t) & \delta\kappa_0 p_s(t) \\ -2\delta p_s(t) & 2\delta p_c(t) & 0 & 0 & 0 & 4\delta \\ -2\delta p_c(t) & -2\delta p_s(t) & 0 & 0 & -4\delta & 0 \end{pmatrix}$$

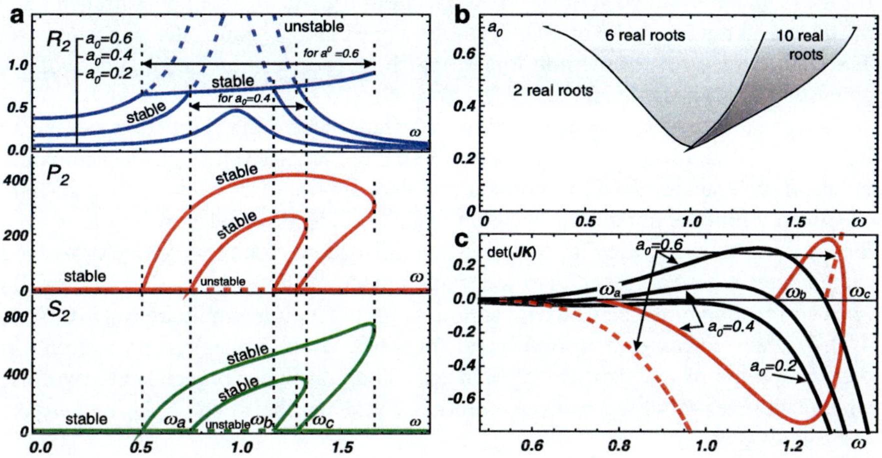

Fig. 2 *Left*: (**a**) stable and unstable branches of the resonance plots for generalised amplitudes R^2, P^2, S^2. *Right top*: (**b**) position of the bifurcation points with respect to excitation frequency ω and amplitude a_0. *Right bottom*: (**c**) values of the Jacobi determinant $\det(\mathbf{JK}(\mathbf{X}))$ along the zero solutions $\mathbf{K}(\mathbf{X}) = 0$. Model data: $M = 1990$, $m = 10$, $l = 20$, $C = 2000$, $\rho = 1$, $\eta_c = \eta_e = 0.32$, $EJ = 10^4$, $g = 9.81$

where $\delta = \omega^{-1}\left(\kappa_0\left(p_c(t)^2 + p_s(t)^2\right) - 4\right)^{-1}$. As for the real structures $m \ll M$, nonexistence of the inverse matrix does not cause problems for $\omega > 0$.

Knowing the exact form of $(\mathbf{H}(\mathbf{X}))^{-1}$, the normal form of the differential Eq. 5 can be easily established. However, as long as the matrix $\mathbf{H}(\mathbf{X})$ is regular, the original right hand side $\mathbf{K}(\mathbf{X})$ can be studied equivalently.

Let us consider stationary response of the system. In this case, the derivatives $\mathrm{d}\mathbf{X}/\mathrm{d}t$ vanish and the right hand side $\mathbf{K}(\mathbf{X})$ has to vanish too. Thus, to identify the stationary solutions, the zero solution points of $\mathbf{K}(\mathbf{X})$ should be traced. In the same time, the signum and the zero points of the Jacobi determinant of $\mathbf{K}(\mathbf{X})$ have to be checked, as in these points bifurcations could occur.

For the particular data, the dependence of $R^2(t), S^2(t)$ and $P^2(t)$ on excitation frequency ω is shown in the left hand part of Fig. 2. Three traces are presented for each variable, they correspond to excitation amplitudes $a_0 = 0.2$, 0, 4, 0.6. Regarding the topmost part of Fig. 2a, it shows the resonance curve of the semi-trivial solution. For amplitudes $a_0 > 0.2$ are the curves enriched by the lover stable branches in the resonance region. Each of this non-linear branches begins in the intersection point with the ascending part of the semi-trivial resonance curve (let us denote the point as ω_a for the curve for to $a_0 = 0.4$) and extends beyond the intersection point with the descending part of the linear resonance curve (ω_b) up to certain point ω_c. Another interesting point is the fact, that position and shape of the non-linear stable branch coincides for all different amplitudes a_0. The individual curves differ by their length, see the dotted lines connecting corresponding points for R^2, P^2 and S^2. Between the both intersections ω_a and ω_b, the upper branch R^2 is unstable and the lower R^2 branch is stable. For P^2 and S^2 holds the opposite case. The cantilevered part of the non-linear branch R^2 (between ω_b and ω_c) is twofold, consisting of both stable and unstable parts.

Figure 2b shows the areas in the $(\omega,\ a_0)$ plane, where the right hand side $\mathbf{K}(\mathbf{X})$ has 2, 6 or 10 real roots. Boundaries between these areas are formed by bifurcation points of the equation $\mathbf{K}(\mathbf{X}) = 0$ or, alternatively, by singular points of the Jacobian.

Course of the Jacobi determinant det $(\mathbf{JK}(\mathbf{X}))$ for individual zero solutions from the Fig. 2a is shown in the Fig. 2c. Let us remind the fact, that the negative sign of the Jacobi determinant correspond to the stable branch of the solution. The black solid lines correspond to the semi-trivial solutions for excitation amplitudes $a_0 = 0.2,\ 0,$ $4,\ 0.6$. Their corresponding non-linear branches (dashed lines) begin and end in the zero points of the semi-trivial branches. The non-linear branches are twofold, which fact gives six or ten real zero solutions in the part between ω_a, ω_b and ω_b, ω_c respectively.

Let us attend now in more detail the case, when semi-trivial resonance curve loses its stability and the non-linear branch arises. The amplitude corresponding to this particular case forms the upper limit of the design value for any real structure. The critical amplitude would be characterised by the following three conditions:

$$\mathbf{K}(\mathbf{X}) = 0; \quad \det(\mathbf{JK}(\mathbf{X})) = 0; \quad \frac{\mathrm{d}}{\mathrm{d}\omega}\det(\mathbf{JK}(\mathbf{X})) = 0 \tag{7}$$

Solving (7) numerically for data from the example, the following values can be obtained: $\omega = 1.0242, a_0 = 0.261$.

4 Conclusions

Non-linear auto-parametric system with three degrees of freedom has been investigated. The aim of this study was to compose a simple mathematical model which enables to assess the non-linear post-critical dynamic response of a slender vertical structure on an elastic subsoil exposed to a strong vertical excitation. It has been shown that under certain condition the semi-trivial solution loses its dynamic stability and strong horizontal response components become decisive from the point of view of various standards. Post-critical states have been thoroughly analysed with respect to excitation frequency and excitation amplitude. Stability of the individual solution branches has been assessed. The critical value of the excitation amplitude has been determined. A similar procedure can be applied also for determining dangerous values of the other structural parameters, like damping ratio etc.

Acknowledgments The support of the Czech Scientific Foundation No. 103/09/0094, Grant Agency of the AS CR No. IAA200710902, IAA200710805 and AV0Z20710524 research plan are gratefully acknowledged.

References

1. Haxton, R.S., Barr, A.D.S.: The autoparametric vibration absorber. ASME J. Appl. Mech. **94**, 119–125 (1974)
2. Nabergoj, R., Tondl, A.: A simulation of parametric ship rolling: effects of hull bending and torsional elasticity. Nonlinear Dynam. **6**, 265–284 (1994)
3. Tondl, A.: Quenching of Self-Excited Vibrations. Academia, Prague (1991)
4. Tondl, A.: To the analysis of autoparametric systems. Z. Angew. Math. Mech. **77**(6), 407–418 (1997)
5. Tondl, A., Ruijgrok, T., Verhulst, F., Nabergoj, R.: Autoparametric Resonance in Mechanical Systems. Cambridge University Press, Cambridge (2000)
6. Guckenheimer, J., Holmes, P.: Nonlinear Oscillations, Dynamical Systems and Bifurcations of Vector Fields. Springer, New York (1983)
7. Hatwal, H., Mallik, A.K., Ghosh, A.: Forced nonlinear oscillations of an autoparametric system. ASME J. Appl. Mech. **50**(3), 657–662 (1983)
8. Náprstek, J.: Non-linear self-excited random vibrations and stability of an SDOF system with parametric noises. Meccanica. **33**, 267–277 (1998)
9. Náprstek, J., Fischer, C.: Auto-parametric semi-trivial and post-critical response of a spherical pendulum damper. In: Topping B.H.V. (ed.) Proceeding of the 11th International Conference on Civil Structural & Environmental Engineering Computing. Civil-Comp Press, Malta (2007). CD paper #CC2007/2006/0020
10. Náprstek, J., Fischer, C.: Auto-parametric stability loss and post-critical behaviour of a three degrees of freedom system. In: Papadrakakis M., Stefanou G., Papadopoulos V. (eds.) Computational Methods in Stochastic Dynamics. Springer, Dordrecht, Heidelberg, London, New York (see ISBN 978-90-481-9986-0) (2011)
11. Arnold, V.I.: Mathematical Methods of Classical Mechanics. Springer, New York (1978)
12. Ren Y., Beards, C.F.: A new receptance-based perturbative multi-harmonic balance method for the calculation of the steady state response of non-linear systems. J. Sound Vib. **172**(5), 593–604 (1995)
13. Lee, W.K., Hsu, C.S.: A global analysis of an harmonically excited spring-pendulum system with internal resonance. J. Sound Vib. **171**(3), 335–359 (1994)

Bifurcation Diagram of an Impact Oscillator with Uncertain Impact Model Parameters

R.R. Aguiar, T.G. Ritto, and H.I. Weber

Abstract Understanding vibro-impact systems is basic for impact dynamics used in resonance drilling. In order to model a vibro-impact system, a contact model must be chosen. The choice of the best contact model for a certain application is not an easy task. The numerous models available in the literature are a proof that research regarding this subject has not come to an end. When considering the force-based method (or continuous analysis), one challenge is to define the values of the impact coefficients. The uncertainties related to these parameters are not aleatory, but rather epistemic. In the proposed approach these variables are going to be modeled by random variables and the probabilistic model is going to be constructed using the maximum entropy principle. The purpose of this work is the numerical/ experimental investigation of the impact force behavior in a vibro-impact system, taking into account uncertainties in the impact model parameters. The experimental data will be used to validate the mathematical model, using the bifurcation diagram. An improvement of the Peterka map is proposed to better visualize each impact force behavior, plotting one stable periodic behavior at a time and adding colors to the third coordinate, the impact force.

R.R. Aguiar (✉)
COPPE - Department of Mechanical Engineering, Universidade Federal do Rio de Janeiro, PO Box 68.503, Rio de Janeiro, RJ, Brazil
e-mail: aguiar.rr@gmail.com

T.G. Ritto
Department of Mechanical Engineering, Universidade Federal Fluminense, Rua Passo da Pátria 156, Niterói, 24210-240 Rio de Janeiro, RJ, Brazil
e-mail: thiagoritto@gmail.com

H.I. Weber
Department of Mechanical Engineering, Pontifícia Universidade Católica do Rio de Janeiro, PO Box 38097, Rio de Janeiro, RJ, Brazil
e-mail: hans@puc-rio.br

J. Náprstek et al. (eds.), *Vibration Problems ICOVP 2011: The 10th International Conference on Vibration Problems*, Springer Proceedings in Physics 139,
DOI 10.1007/978-94-007-2069-5_21,

Keywords Bifurcation diagram • Vibro-impact system • Uncertain impact coefficients • Stochastic dynamics

1 Introduction

The study of vibro-impact systems has been the aim of several researches, from the application of a percussive action in rotary drilling for improved performance [3], to ultrasonic drilling [6], impact dampers [5] and vibro-safe percussion machines. Experimental investigation of impact problems reveals typical difficulties of this subject. Each situation is different, and even each impact may present variations. Analytical models try to compensate that by making use of more elaborate models. However, when working experimentally, simple impact force models may give better results when the complexity of the system increases.

When modeling a vibro-impact system some simplifications must be done, hence, as a consequence: (1) the model chosen for the system possibly does not describe it correctly (epistemic uncertainty related to the model), (2) the impact model chosen possibly does not describe the impact correctly (epistemic uncertainty related to the impact model) and the values of the parameters used in the model possibly are not correct (epistemic uncertainty related to the parameters).

The objective of this work is to investigate the behavior of an impact pendulum (hammer) inside a vibrating structure taking into account uncertainties in the impact model parameters. In the proposed approach these variables are going to be modeled by random variables and the probabilistic model is going to be constructed using the maximum entropy principle. The idea is to use the information we have about the random variables, so that the distribution used for the random variables chosen is compatible with the physics of the problem. Experimental data will be used to validate the mathematical model, with help of the bifurcation diagram.

2 Experimental Apparatus and Mathematical Modeling (Deterministic)

The system analyzed in this work has been developed and fully described in [1]. It consists of am impact pendulum mounted inside a moving cart, excited by an AC motor, see Fig. 1a. The test rig considers different values for the impact gap and the excitation frequency. The methodology is to observe the impact force behavior as the gap is varied. In this work, results are shown for the 0.0 mm gap. In all results shown the data was acquired after the transient response.

The system is modeled as a single-degree-of-freedom pendulum embedded in a cart with prescribed movement, with θ as the generalized coordinate, see Fig. 1b. Equations of motion are:

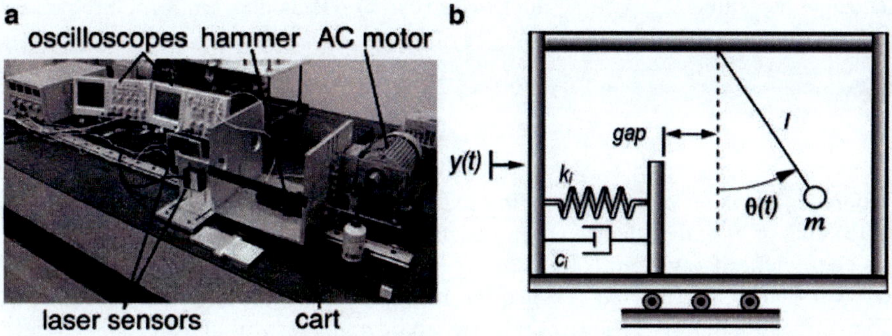

Fig. 1 (**a**) Test rig; (**b**) Physical representation

$$
\begin{aligned}
&ml^2\ddot{\theta} - mlA_0\Omega^2\cos\theta\sin\Omega t + mgl\sin\theta = 0; \;\; \text{if } l\sin\theta - \text{ gap} > 0\\
&ml^2\ddot{\theta} - mlA_0\Omega^2\cos\theta\sin\Omega t + mgl\sin\theta = F_i l; \;\; \text{if } l\sin\theta - \text{ gap} \leq 0\\
&F_i = k_i\delta + c_i\dot{\delta}
\end{aligned}
\tag{1}
$$

The impact is described using the spring-dashpot model. Impact parameters identification procedure and comparison between experimental data and deterministic numerical simulation can be found in [1]. Although the spring-dashpot model is not capable of reproducing the real impact force profile over time due to the jump caused by the damping force, this model generated satisfactory results.

3 Uncertainty Modeling

A challenge in the problem analyzed is to define the values of the impact coefficients k_i and c_i. The uncertainty related to these parameters are not aleatory, but rather epistemic [2]. In the proposed approach these variables are going to be modeled by random variables and the probabilistic model is going to be constructed using the maximum entropy principle [4]. The resulting stochastic system is computed using the Monte Carlo method. Let A represent the random variables K_i and C_i (related to the impact parameters k_i and c_i). Since there is no explicit information about their dependence, we are assuming that variables K_i and C_i are independent. The only available information is the support of the random variable: $supp = [a_1, a_2]$. The identified value $\underline{a}$ (the value identified for k_i or c_i using the deterministic model) is not completely reliable to be used as information for the mean value of A, nevertheless we believe that the true value of the parameter is in the vicinity of $\underline{a}$. Note that the support is not fixed a priori, it has to be calibrated using the experimental results. Therefore, the maximum entropy principle (using Shannon entropy measure, $S = -\int_{a_1}^{a_2} \ln(p(a))p(a)da)$ yields the Uniform probability density function [4] given by:

$$p_A(a) = \mathbb{1}_{[a_1,a_2]}(a)\frac{1}{(a_2 - a_1)}, \tag{2}$$

where $\mathbb{1}_B(a)$ is an indicator function that is equal to 1 for $a \in B$ and 0 otherwise. The value of a_1 and a_2 can be computed using the mean value of A, $\mathbb{E}\{A\} = \underline{a}$, and its coefficient of variation $CV_A = \sigma_A/\underline{a}$, where σ_A is the standard deviation of A: $a_1 = \underline{a} - \sqrt{3}\underline{a}CV_A$, $a_2 = \underline{a} + \sqrt{3}\underline{a}CV_A$. It should not be forgotten that the values of the stiffness/damping are always positive (to be coherent with the physical model), hence the coefficient of variation of A should be controlled to guarantee this fact.

4 Numerical Results

First, it is presented a comparison between numerical simulation and experimental data [1]. Figure 2a shows the bifurcation chart of the non-dimensional displacement $(l\sin(\theta)/A_0)$ in the non-dimensional frequency domain $(\Omega\,/\,\omega)$, for the condition of one impact per excitation cycle, using deterministic parameters. Comparison shows reasonable agreement.

Due to the difficulty in modeling the impact and identifying the model parameters, it might be useful to analyze the system response in the case where the impact model parameters are treated as stochastic variables. Four different coefficients of variation are chosen for the pair (K_i, C_i), $CV = \{5\%; 10\%; 15\%; 20\%\}$. No specific metric was established in order to choose which coefficient of variation best fits experimental data. By simply comparing the simulation results for each coefficient

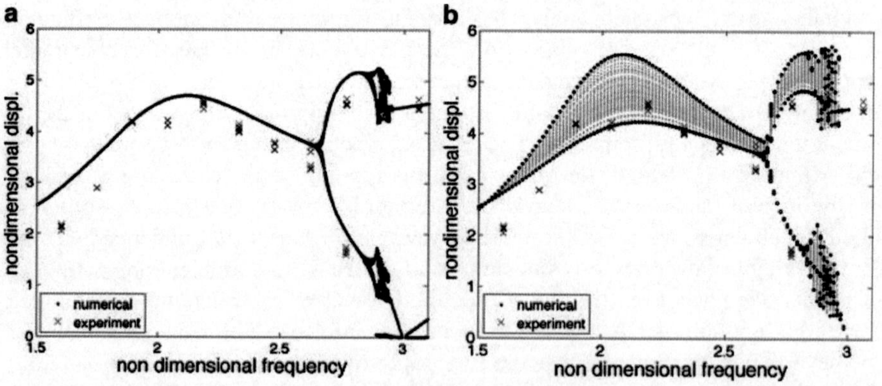

Fig. 2 Bifurcation diagram; numerical simulation/experimental data comparison, gap 0.0 mm: **(a)** deterministic impact model parameters; **(b)** stochastic model parameters, $CV = 10\%$

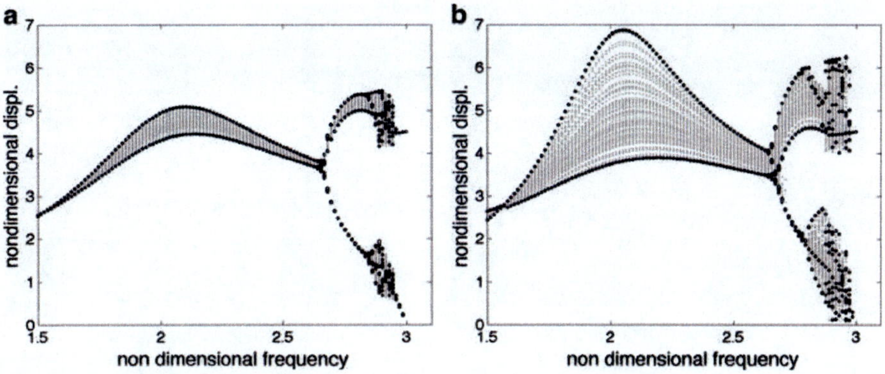

Fig. 3 Stochastic bifurcation diagram: pendulum displacement ($l\sin(\theta)/A_0$); gap 0; influence of the coefficient of variation: (**a**) $CV = 5\%$; (**b**) $CV = 20\%$

of variation with the experimental data, it was chosen that data for $CV = 10\%$ best fits the experimental results. Figure 2b shows the stochastic bifurcation diagram for the mentioned *CV*. In Fig. 3 it is shown the bifurcation diagram from simulation results for different coefficients of variation.

From charts shown in Figs. 2b and 3, it is noted that the frequency where the maximum impact force (maximum displacement) occurs in this periodic behavior (one impact per excitation cycle) does change as the impact model parameters are varied, so the condition of maximum impact force (frequency and magnitude) are directly related to the impact model parameters, and consequently the impact model used to characterize the impact force. Regarding the period-doubling bifurcation, the frequency where the bifurcation occurs does not change considerably as the coefficient of variation is changed. After the period-doubling bifurcation, it is noted that the dispersion of the nonlinear displacement in the upper branch ($2.67 < l\Omega / \omega < 2.86$) increases with the coefficient of variation *CV*, see Fig. 3. However, for the lower branch it is observed no dispersion of the points, having the same result as the deterministic bifurcation diagram, see Fig. 2a. Another relevant information is that even with a higher value of *CV*, a full agreement between numerical simulation and experimental data is not verified. Such fact might be an indication that some experimental phenomenon is not being considered in the mathematical modeling (for instance the relative velocity before the impact might have a tangential component).

To better visualize the behavior of this dynamical system, the map of regions of stable periodic behavior, the Peterka map [5], is plotted. See Fig. 4. The variable z determines the number of impacts per excitation cycle. From this chart one can see the areas where the two main periodic behaviors occur ($z = 1/1$, one impact per excitation cycle and $z = 1/2$, one impact every two excitation cycles). The maps are produced from the pairs (K_i, C_i) that generated the lower and upper curves in the bifurcation diagram, for $CV = 10\%$, see Fig. 2b. Chart shows that main regions of

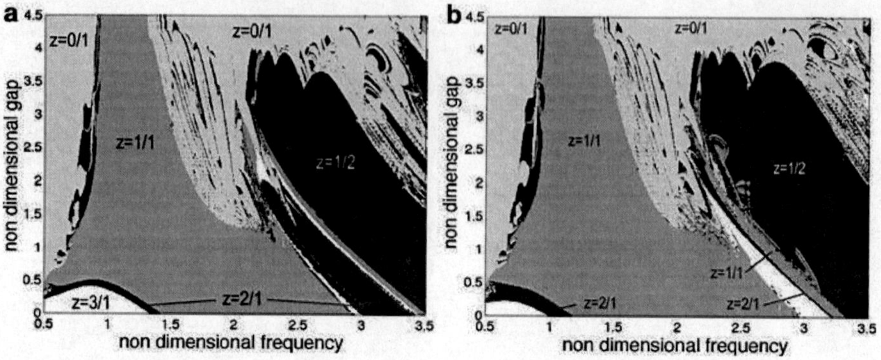

Fig. 4 Regions of stable periodic behavior considering different impact model parameters: (**a**) lower (K_i, C_i); (**b**) upper (K_i, C_i)

stable periodic behavior ($z = 1/1$ and $z = 1/2$) are not substantially affected by the difference in the impact model parameters.

5 Concluding Remarks

This paper has investigated the numerical/experimental impact force behavior of a vibro-impact system, taking into account uncertainties in the impact model parameters by means of a probabilistic approach. The value of the coefficient of variation (a measure of the dispersion) of the random variables (stiffness and damping impact parameters) were calibrated using the experimental response. It has been shown that there are some regions that are robust to a variation of the impact model parameters and others that are sensitive to them. The propagation of the uncertainty related to the impact model parameters through the system has been useful to a better understanding of the system response.

Acknowledgements The authors gratefully acknowledge the support of CNPq (*Conselho Nacional de Desenvolvimento Científico e Tecnológico*) and FAPERJ (*Fundação de Amparo à Pesquisa do Estado do Rio de Janeiro*).

References

1. Aguiar, R. R., Weber, H.I.: Mathematical modeling and experimental investigation of an embedded vibro-impact system. Nonlinear Dyn. (2010). doi:10.1007/s11071-010-9894-0
2. Ferson, S., Joslyn, C.A., Helton, J.C., Oberkampf, W.L., Sentz, K.: Summary from the epistemic uncertainty workshop: consensus amid diversity. Reliab. Eng. Syst. Safe **85**(1–3), 355–369 (2004)

3. Franca, L.F.P., Weber, H.I.: Experimental and numerical study of a new percussive drilling model with a drift and internal impacts. in: Proceedings of the 6th International Conference on Vibration Problems. Technical University of Liberec, Czech. Republic (2003)
4. Kapur, J.N., Kesavan, H.K.: Entropy Optimization Principles with Applications. Academic Press, Boston USA (1992)
5. Peterka, F.: More detail view on the dynamics of the impact damper. Facta Universitatis Series: Mechanics, Automatic Control and Robotics **3**(14), 907–920 (2003)
6. Wiercigroch, M., Neilson, R.D., Player, M.A.: Material removal rate prediction for ultrasonic drilling of hard materials using an impact oscillator approach. Phys. Lett. A **259**, 91–96 (1999)

Dynamics of Self-Excited Oscillators with Neutral Delay Coupling

K. Edelman and O.V. Gendelman

Abstract The work is devoted to analytic and numeric investigation of dynamical behavior in a system of two Van–der–Pol (VdP) oscillators coupled by non–dispersive elastic rod. The model is rigorously described by the system of nonlinear neutral differential delay equations. For the case of relatively small coupling and moderate delay, an approximate analytic investigation is performed by means of averaging procedure. If the effective coupling remains small if the system is far from the internal resonance (sine in the denominator is not small). Region of synchronization in the space of parameter is established and characteristic bifurcations are revealed. Numeric study confirms the validity of the analytic approach in the synchronization region. Beyond this region the analytic approach is no more valid. Multitude of quasiperiodic and chaotic – like orbits has been revealed. Especially interesting behavior corresponds to sequential quenching and excitation of the VdP oscillators. This regime is also explored analytically, by means of large – delay approximation, which reduces the system to perturbed discrete map.

Keywords Synchronization • Van der Pol oscillator • Time delay • Neutral delay differential equations (NDDE)

1 Introduction

The idea of reducing the hyperbolic PDE to a system of neutral differential delay equations involving the boundary conditions of the problem goes back to early 1960s [1–4]. Such systems are a subject of considerable interest and thorough study now, primarily from the viewpoint of existence of periodic solutions [5].

K. Edelman • O.V. Gendelman (✉)
Faculty of Mechanical Engineering, Technion – Israel Institute of Technology, Haifa 32000, Israel
e-mail: kobied1@techunix.technion.ac.il; ovgend@tx.technion.ac.il

J. Náprstek et al. (eds.), *Vibration Problems ICOVP 2011: The 10th International Conference on Vibration Problems*, Springer Proceedings in Physics 139,
DOI 10.1007/978-94-007-2069-5_22,

Stability of solutions in such delay differential equations also has been studied quite thoroughly [6, 7]. However, more general properties of such systems from the viewpoint of nonlinear dynamics, such as synchronization, quasiperiodicity or existence of nontrivial attractors, received less attention.

In this work, we are going to modify classical system of Van der Pol oscillators with weak elastic coupling [8], by including a nondispersive elastic rod instead of the coupling spring. In this way, we are going to obtain a hyperbolic problem with nonlinear boundary conditions. This system may be rigorously reduced to a system of two nonlinear neutral differential delay equations. By means of averaging, we are going to study the regimes of synchronization and quasiperiodicity. Further numeric simulations will reveal limitations of the standard approaches and will shed light on behavior of the system in the case of relatively large delays.

2 Description of the Model

Let us consider two oscillators coupled by elastic rod with Young modulus E, cross–section S, length L and density ρ (consequently, the sound velocity in the rod is $c = (E/\rho)^{1/2}$). The oscillators with masses m_1 and m_2 are placed in points with coordinates 0 and L, as it is shown at Fig. 1:

The system is described by the following equations:

$$\begin{aligned} m_1\ddot{u}_1 + f_1(u_1, \dot{u}_1) &= ES\left.\frac{\partial u}{\partial x}\right|_{x=0} \\ m_2\ddot{u}_2 + f_2(u_2, \dot{u}_2) &= -ES\left.\frac{\partial u}{\partial x}\right|_{x=L} \\ \frac{\partial^2 u}{\partial t^2} - c^2\frac{\partial^2 u}{\partial x^2} &= 0, 0 < x < L \end{aligned} \tag{1}$$

where u(x, t) is the displacement field of the rod, m_i and f_i, i = 1,2 are masses and spring/damper characteristics of the oscillators. Continuity of displacements and

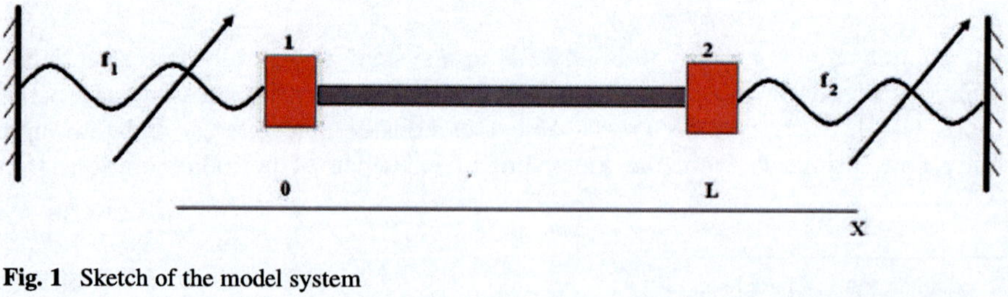

Fig. 1 Sketch of the model system

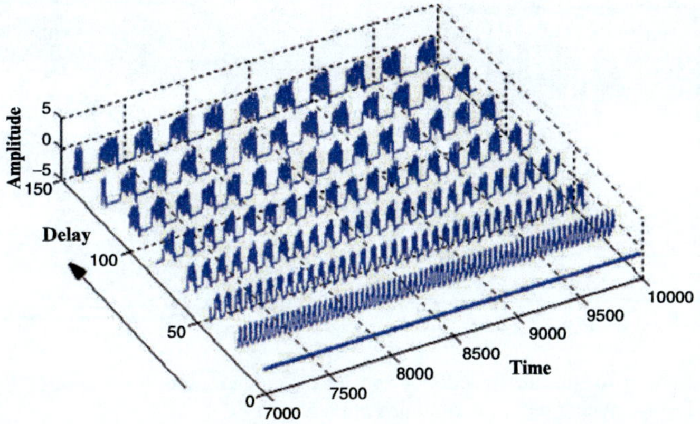

Fig. 2 Response of the right oscillator, for growing delays

velocities implies the following relationships between the displacements and the velocities of the rod and the discrete oscillators

$$u_1 = u(0), \dot{u}_1 = \left.\frac{\partial u}{\partial t}\right|_{x=0}, u_2 = u(L), \dot{u}_2 = \left.\frac{\partial u}{\partial t}\right|_{x=L} \tag{2}$$

By using the well – known general solution of the wave equation in (1) in D'Alembert form we reduce System (1 and 2) to a system of two neutral differentials – delay equations (NDDE) of the form:

$$\begin{cases} \ddot{u}_1 + \varepsilon \dot{u}_1 \left(u_1^2 - 1\right) + u_1 + \varepsilon \kappa \dot{u}_1 = -\left[\ddot{u}_2 + \varepsilon \dot{u}_2 \left(u_2^2 - 1\right) + (1 + \varepsilon\sigma)\, u_2 - \kappa \dot{u}_2\right]_{(t-\tau)} \\ \ddot{u}_2 + \varepsilon \dot{u}_2 \left(u_2^2 - 1\right) + (1 + \varepsilon\sigma)\, u_2 + \varepsilon \kappa \dot{u}_2 = -\left[\ddot{u}_1 + \varepsilon \dot{u}_1 \left(u_1^2 - 1\right) + u_1 - \kappa \dot{u}_1\right]_{(t-\tau)} \end{cases} \tag{3}$$

The system of Eq. 3 was studied numerically using a continuous explicit Runge-Kutta method. The solver was based on the **ddverk** [9] algorithm while the Runge-Kutta coefficients were taken from [10]. We have chosen the eight stage CERK in order to achieve the maximum accuracy, the numerical results were supported by analytical studies of the system of Eq. 3.

System (3) was studied with changing parameters τ (the time delay), σ (the second oscillator detuning) and ε (the small parameter related to coupling and nonlinearity). Well-known response regimes for linearly coupled VdP oscillators, such as stable synchronization, relaxation oscillations and quasi-periodic behavior, were encountered. Interesting regime, related to sequential excitation and quenching of the VdP oscillators, was revealed for relatively large delay times (Fig. 2).

Typical behavior of each oscillator in the course of the sequential quenching is presented at Fig. 3 and propagation of the wave over the rod – at Fig. 4.

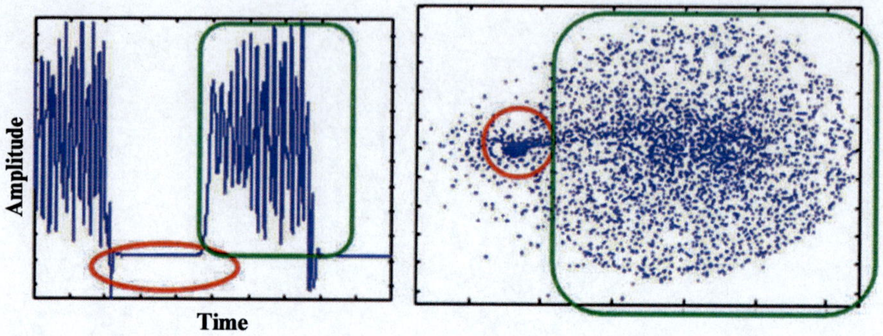

Fig. 3 Time series for terminal oscillator and local Poincare section the case of the sequential quenching. The quenching region is encircled at both plots

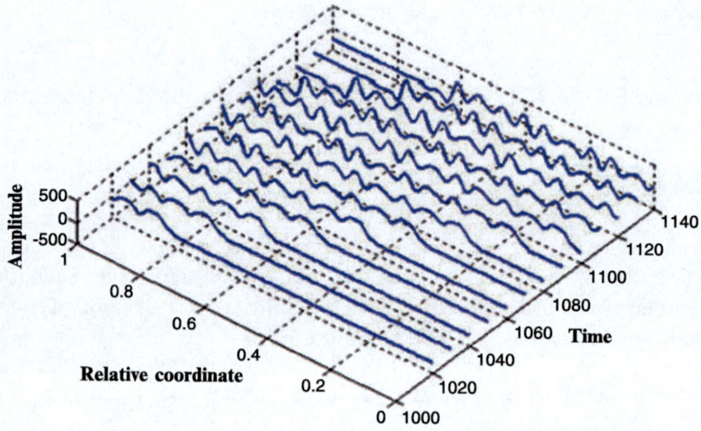

Fig. 4 Propagation of the elastic wave over the rod in the case of the sequential quenching. Relative coordinate is defined as x/L

To analyze this phenomenon, let us consider the combination of constant displacement and oscillatory responses. The following ansatz should be used:

$$u_j = u_{j0} - \frac{i}{2}\left(\phi_j e^{it} - \phi_j^* e^{-it}\right), \dot{u}_j = \dot{u}_{j0} + \frac{1}{2}\left(\phi_j e^{it} + \phi_j^* e^{-it}\right), j = 1,2 \quad (4)$$

Where u_{j0} are slow functions of time,

$$\left|\frac{1}{u_{j0}}\frac{du_{j0}}{dt}\right| \ll 1$$

Substituting (4) into (3) and considering the system in a limit of very large delay times (just in this case we really see the sequential quenching!), adopting

$\tau = T/\varepsilon^2, T = O(1)$ and making additional formal change of variables $\xi = \varepsilon^2 t$, we finally obtain:

$$
\begin{aligned}
&\varepsilon^4 u''_{10} + \varepsilon^3 u'_{10}\left(u_{10}^2 + \frac{1}{2}|\phi_1|^2 - 1\right) + u_{10} + \varepsilon^3 k u'_{10} + \\
&\quad + \left[\varepsilon^4 u''_{20} + \varepsilon^3 u'_{20}\left(u_{20}^2 + \frac{1}{2}|\phi_2|^2 - 1\right) + (1+\varepsilon\sigma)u_{20} - \varepsilon^3 k u'_{20}\right]_{\frac{\xi - T}{\varepsilon^2}} = 0 \\
&\varepsilon^2 \phi'_1 + \frac{\varepsilon}{2}\left(\frac{1}{4}|\phi_1|^2 + u_{10}^2 - 1 - i\varepsilon^2 u'_{10}u_{10}\right)\phi_1 + \frac{\varepsilon k}{2}\phi_1 + \\
&\quad + e^{-iT/\varepsilon^2}\left[\varepsilon^2 \phi'_2 + \frac{\varepsilon}{2}\left(\frac{1}{4}|\phi_2|^2 + u_{20}^2 - i\varepsilon^2 u'_{20}u_{20} - (1+i\sigma)\right)\right. \\
&\qquad\qquad \left.\phi_2 - \frac{\varepsilon k}{2}\phi_2\right]_{\frac{\xi - T}{\varepsilon^2}} = 0 \\
&\varepsilon^4 u''_{20} + \varepsilon^3 u'_{20}\left(u_{20}^2 + \frac{1}{2}|\phi_2|^2 - 1\right) + (1+\varepsilon\sigma)u_{20} + \varepsilon^3 k u'_{20} + \\
&\quad + \left[\varepsilon^4 u''_{10} + \varepsilon^3 u'_{10}\left(u_{10}^2 + \frac{1}{2}|\phi_1|^2 - 1\right) + u_{10} - \varepsilon^3 k u'_{10}\right]_{\frac{\xi - T}{\varepsilon^2}} = 0 \\
&\varepsilon^2 \phi_2 + \frac{\varepsilon}{2}\left(\frac{1}{4}|\phi_2|^2 + u_{20}^2 - i\varepsilon^2 u'_{20}u_{20} - (1+i\sigma)\right)\phi_2 + \frac{\varepsilon k}{2}\phi_2 + \\
&\quad + e^{-iT/\varepsilon^2}\left[\varepsilon^2 \phi_1 + \frac{\varepsilon}{2}\left(\frac{1}{4}|\phi_1|^2 + u_{10}^2 - 1 - i\varepsilon^2 u'_{10}u_{10}\right)\phi_1 - \frac{\varepsilon k}{2}\phi_1\right]_{\frac{\xi - T}{\varepsilon^2}} = 0
\end{aligned}
\tag{5}
$$

System (5) is a typical system of singularly perturbed delay differential equations. At this stage, we are interested only in "outer" behavior of the system, which corresponds to the slow evolution of all variables. So, in this approximation, we can omit all terms which contain ε in the orders above the main one, after some mathematical manipulations we obtain System (6) which describes a discrete map.

$$
\begin{aligned}
&\left(\frac{1}{4}V + C_1^2 - 1 + k\right)^2 V = \left(\left(\frac{1}{4}W + C_1^2 - 1 - k\right)^2 + \sigma^2\right) W \\
&\left(\frac{1}{4}V + C_1^2 - 1 - k\right)^2 V = \left(\left(\frac{1}{4}W + C_1^2 - 1 + k\right)^2 + \sigma^2\right) W
\end{aligned}
\tag{6}
$$

where V and W describe the squared amplitudes of the oscillations of the left and the right VdP oscillators respectively, C_1 and C_2 are their DC displacements.

Since obviously V, W ≥ 0, then for $|C_1| \geq 1$ the only fixed point will correspond to a trivial solution; this case corresponds to the situation of quenching. For $|C_1| < 1$, nontrivial fixed point always exists. Therefore, the sequential quenching is realized for $|C_j| > 1, j = 1, 2$

Only one extremely important question remains unresolved – whether this sequential quenching is true attractor of the system. In order to check that, let us make a certain trick – we'll include very small damping in the rod. Real attractors are robust: addition of extremely small damping will not destroy them completely.

It is easy to prove that no matter how small the damping will be, the equilibrium solutions of appropriately modified Eqs. 5 and 6 will be zero. So, the quenching phenomenon disappears. It means that the regime of sequential quenching is not robust and does not correspond to a stable attractor.

3 Conclusions

To conclude, the neutral delay differential equations of the two coupled Van der Pol oscillators demonstrate the well-known responses known also for other types of coupling. A new response regime – the sequential quenching- was observed. It was studied analytically and appeared to be a transient behavior of the system

References

1. Abolinia, V.E., Myshkis, A.D.: Mixed problem for an almost linear hyperbolic system in the plane (in Russian). Matem. Sbornik **12**, 423 (1960)
2. Bellman, R.E., Cooke, K.L.: Differential-Difference Equations. Academic, New York (1963)
3. Brayton, R.K., Miranker, W.L.: Oscillations in a distributed network. Arch. Rat. Mech. Anal. **17**, 358 (1964)
4. Cooke, K.L.: A linear mixed problem with derivative boundary conditions. Seminar on differential equations and dynamical systems (III), Lecture series 51, University of Maryland, College Park, p. 11 (1970)
5. Hale, J.K., Verduyn Lunel, S.M.: Introduction to Functional Differential Equations. Springer, New York (1993)
6. Azbelev, N.V., Berezansky, L., Simonov, P.M., Chistyakov, A.V.: The stability of linear systems with aftereffect. I. Diff. Equ. **23**, 493 (1987)
7. Berezansky, L., Braverman, E.: Global linearized stability theory for delay differential equations. Nonlinear Anal. **71**, 2614 (2009)
8. Landa, P.S.: Nonlinear Oscillations and Waves in Dynamical Systems. Kluwer Academic, Dordrecht (1996)
9. Owren, B., Zennaro, M.: Derivation of efficient continuous explicit Runge-Kutta methods. SIAM J. Sci. Stat. Comput. **13**, 1488 (1992)
10. Thompson, S., Shampine, L.: A friendly fortran 90 DDE solver. Appl. Numer. Math. **56**, 503 (2006)

High Order Nonlinearities and Mixed Behavior in Micromechanical Resonators

N. Kacem, S. Hentz, S. Baguet, and R. Dufour

Abstract This paper investigates the sensitivity of the third order nonlinearity cancelation to the mixed (hardening and softening) behavior in electrostatically actuated micromechanical resonators under primary resonance at large amplitudes compared to the gap. We demonstrate the dominance of the mixed behavior due to the quintic nonlinearities, beyond the critical amplitude when the third order mechanical and electrostatic nonlinearities are balanced. We also report the experimental observation of a strange attraction which can lead to a chaotic resonator.

Keywords Nonlinear dynamics • Mixed behavior • Strange attraction

1 Introduction

The large potential of microelectromechanical systems (MEMS) and nanoelectromechanical systems (NEMS) has been widely demonstrated for ultrasensitive force and mass sensing applications [1, 2]. However, it is a challenge to enhance their dynamic range (DR) by achieving large amplitudes comparable to the gap in the case of electrostatically driven resonators without altering their frequency stability. Combined with the noise mixing issue [3] when the resonator is driven in the nonlinear regime, this leads to a drastic limitation on the sensor resolution.

An alternative to overcome this physical limitation is the hysteresis suppression by nonlinearity cancelation which has been the aim of extensive research work

N. Kacem (✉) • S. Baguet • R. Dufour
Université de Lyon, CNRS INSA-Lyon, LaMCoS UMR525, F-69621 Villeurbanne, France
e-mail: najib.kacem@insa-lyon.fr; sebastien.baguet@insa-lyon.fr; regis.dufour@insa-lyon.fr

S. Hentz
CEA/Leti - MINATEC, Microsystems Components Laboratory, F-38054 Grenoble, France
e-mail: sebastien.hentz@cea.fr

J. Náprstek et al. (eds.), *Vibration Problems ICOVP 2011: The 10th International Conference on Vibration Problems*, Springer Proceedings in Physics 139,
DOI 10.1007/978-94-007-2069-5_23,

lately and requests the study of nonlinear dynamics of MEMS and NEMS. Most nonlinear models dealing with the nonlinearity cancelation in capacitive micro and nanoresonators includes nonlinear terms up to the third order [4, 5]. So even experimentally observed [5], the nonlinearity cancelation is limited by the high order nonlinearities which occur proportionally sooner with respect to the resonator size. Precisely, the onset of the mixed behavior is close to the critical amplitude for resonant NEMS sensors.

Indeed, accurate nonlinear models [6, 7] for large amplitudes includes all main sources of nonlinearities without Taylor series expansion of the nonlinear electrostatic forces. Kacem et al. developed complete multiphysics models including high order nonlinear terms and demonstrated analytically the ability to cancel out the nonlinearities in MEMS and NEMS capacitive clamped–clamped beam resonators [6] as well as cantilevers [8]. Moreover, Kacem and Hentz [9] identified experimentally the mixed behavior predicted in [6] i.e. they demonstrated the physical significance of the quintic nonlinear terms.

In this paper, we report experimentally the high sensitivity of the third order nonlinearity cancelation to the mixed behavior on a MEMS resonator under primary resonance at large amplitudes with respect to the gap. Based on a high order nonlinear model, the fabricated resonator was designed with a large ratio between its width and the sensing gap in order to maximize its global critical amplitude. We demonstrate the ability to reach very large amplitudes of order 90% of the sensing gap without pull-in occurrence. Moreover, strange attraction on a mixed behavior is experimentally identified which can lead in some configurations to chaos.

2 Experimental Investigations

The fabricated resonator consists in a silicon doubly clamped beam electrostatically in plane actuated and detected using two electrodes (electrode 1 for actuation and 2 for sensing) which allows for two port electric measurements.

It has been fabricated on 200 mm SOI (Silicon on insulator) wafers and microelectronic silicon processes. It is 200 μm long, 2 μm thick, 5 μm wide, the actuation gap is 1 μm, and the detection gap 300 nm. It has been designed to enable the compensation of the nonlinearities thanks to a compact analytical model [6] including all main sources of nonlinearities (electrostatic and mechanical) presenting non linear terms up to the seventh order as shown in Eq. 1.

$$\left(1+\sum_{i=1}^{4}\mu_i a_1^i\right)(\ddot{a}_1+c\dot{a}_1)+\sum_{i=0}^{2}\zeta_i a_1^i\cos(\Omega t)$$
$$+\sum_{i=0}^{7}\chi_i {a_1}^i+\sum_{i=3}^{5}\zeta_i a_1^{i-3}\cos(2\Omega t)=0 \quad (1)$$

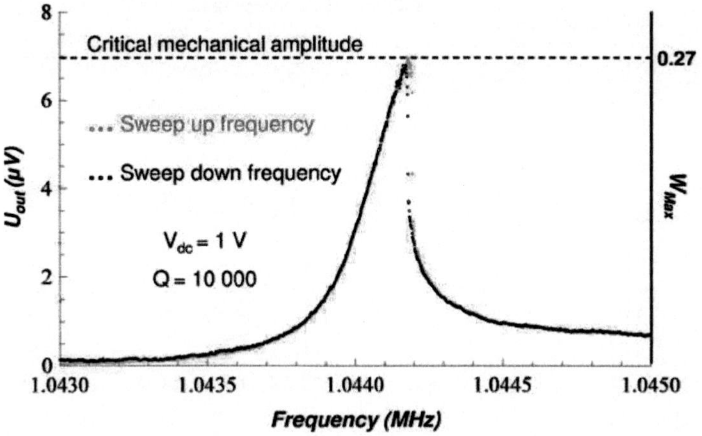

Fig. 1 Measured critical hardening resonance peak for $V_{ac} = 0.2V$. W_{Max} is the displacement of the beam at its middle point normalized by the gap g_d

The device was placed in a vacuum chamber (down to 1 mTorr), and the two-port electrical measurements were performed at room temperature using a low noise lock-in amplifier (Signal Recovery 7,280).

As a first step, the critical behavior is approached in order to slowly access the nonlinear regime. Figure 1 displays a hardening behavior for which the mechanical nonlinearities are much higher than the electrostatic nonlinearities. Indeed, for a *DC* voltage of $1V$, the nonlinear negative Duffing nonlinearity is very low which implies a high quality factor ($Q \approx 10^4$). Hence, in this configuration, the mechanical critical amplitude is smaller than the electrostatic one ($A_c \approx 80$ nm). The hardening resonance peak of Fig. 1 has been obtained in sweep up and down frequency which displays a critical behavior at an output voltage of $7\,\mu V$. The fabricated device has the characteristic of large width to gap ratio and consequently can potentially display a linear compensated behavior when both electrostatic and mechanical nonlinearities are balanced by increasing the nonlinear spring softening effect.

In order to approach this operating point (hysteresis suppression), the *DC* voltage has been increased to $3V$ and $5V$ as shown in Fig. 2 where the resonance curves have been electrically characterized in sweep up and down frequency.

Since the negative nonlinear stiffness is proportional to the square of V_{dc} (for $V_{dc} >> V_{ac}$), unlike the first peak obtained at a *DC* voltage of $3V$, the second one (left curve) measured for $V_{dc} = 5V$ displays nonlinear branches in the softening domain characterized by a complete modification of the dynamic bifurcation topology. The frequency shift between both curves due to the negative stiffness is about 2 KHz.

Moreover, the first peak measured at $V_{dc} = 3V$ displays a quality factor $Q = 6{,}000$ which has been measured on a linear curve at a low *AC* voltage and the same

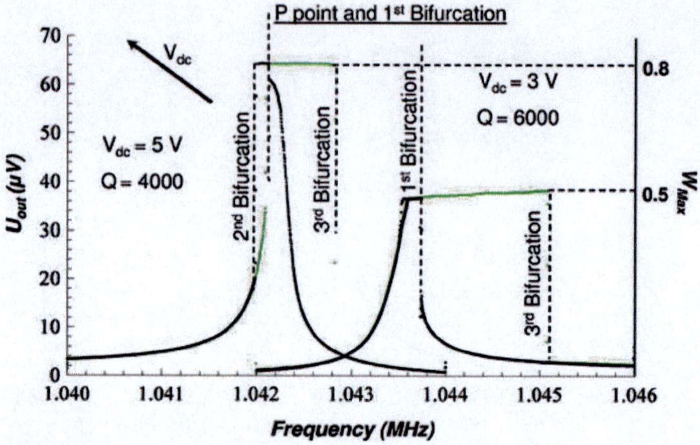

Fig. 2 Measured strange hardening and mixed resonance peaks for $V_{ac} = 0.2V$ showing the changes on the bifurcation topologies. P is the mixed behavior initiation point and the third bifurcation is the highest point in the hardening domain

DC voltage. The resonance curve, in this case, is strangely nonlinear (hardening behavior). The third bifurcation is the highest one in the hardening domain obtained in sweep up frequency where the corresponding curve displays two regimes: a first fast in amplitude variation and a second slow with a slope approaching zero. The first bifurcation is obtained in sweep down frequency and intercepts the sweep up frequency curve in a small part of the slow regime and the entire fast regime of amplitude variation. The strange nonlinear hardening behavior obtained experimentally can be explained by a strong dynamic perturbation due to the increase of the softening nonlinearitie.

In Fig. 2, the left resonance curve measured at $V_{dc} = 5V$ displays a quality factor $Q = 4,000$. Using the analytical model of a two ports nonlinear resonator developed in [6], for these parameters, the resonator should display a linear resonance peak obtained by the compensation of the mechanical and the electrostatic nonlinearities. Unlike the mixed behavior (hardening-softening) demonstrated in [9], the experimental peak measured at $V_{dc} = 5V$ displays clearly a mixed behavior starting by a softening branch and ending by a hardening one where the peak amplitude is around three times the critical amplitude displayed in Fig. 1.

Particularly, in this mixed behavior, the P point and the first bifurcation have the same frequency and the hardening domain is reduced in comparison with the first resonance curve of Fig. 2. Actually, for the fabricated resonator, the fifth order nonlinear terms are no more negligible when it is used to operate close to the hysteresis suppression point. Indeed, the compensation of the nonlinearities is sensitive to the highly unstable mixed behavior.

Once the mixed behavior was reached, we continue the increase of the DC voltage in order to track the transition from a mixed to a softening behavior as

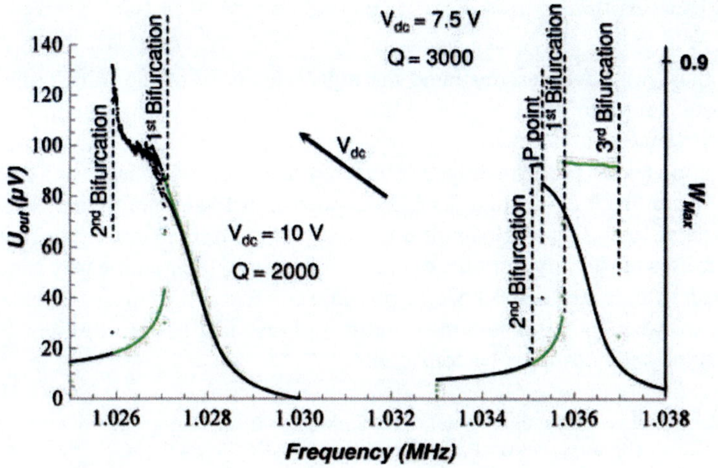

Fig. 3 Measured mixed and softening resonance peaks for $V_{ac} = 0.2V$

shown in Fig. 3. The first peak has been measured at a DC voltage of $7.5V$ where the branches obtained in sweep down frequency enable the capture of two bifurcation points: the first one is the P point which corresponds to the initiation of the mixed behavior domain and the second one is the highest bifurcation point in the hardening domain.

Then, in sweep up frequency, the output voltage V_{out} follows the curve characterized by the bifurcation points 1 and 3. When the frequency of the first bifurcation point is reached, a strange attractor brings the resonator oscillation to the upper branch in the hardening domain instead of following the sweep down frequency curve (jump to the downer branch) which should be physically easier where the basins of attraction are quite larger. This is another illustration of the physical importance of the quintic nonlinearities. Hence, it is difficult to achieve large amplitudes for MEMS and NEMS capacitive resonators without altering their frequency stability.

In order to suppress the mixed behavior, the effect of the hardening nonlinearities is reduced in the second peak of Fig. 3 for $V_{dc} = 10V$ and an estimated quality factor $Q = 2,000$. The electrostatic nonlinearities are amplified with respect to the mechanical nonlinearities which brings the third bifurcation point at the same frequency as the first one. Remarkably, the resonator has reached very large oscillations up to 90% of the sensing gap without pull-in occurrence. Nevertheless, the highest branch in the softening domain displays an inflection point (potentially due to the high order hardening nonlinearities) followed by a dynamic relaxation at the jump down bifurcation point.

3 Conclusions

In summary, we have demonstrated the high sensitivity of the third order nonlinearity cancelation to the mixed behavior in nonlinear capacitive clamped–clamped MEMS resonators under primary resonance at large oscillations. The experimental investigations were performed on a fabricated resonator designed using a complete analytical nonlinear model in order to balance the third order nonlinearities and drive the resonator linearly at high amplitudes. Beyond the critical amplitude, the quintic nonlinearities drastically limit the operating domain of the nonlinearity cancelation. Moreover, strange attractions become possible which can lead to undesirable chaotic sensors. Consequently, for nonlinear optimized resonant MEMS designs, the onset of the mixed behavior must be considered.

Acknowledgements The authors gratefully acknowledge financial support from the CEA LETI and I@L Carnot institutes (NEMS Project) and from the MNTEurop Project.

References

1. Rugar, D., Budakian, R., Mamin, H.J., Chui, B.W.: Single spin detection by magnetic resonance force microscopy. Nature **430**, 329 (2004)
2. Jensen, K., Kim, K., Zettl, A.: An atomic-resolution nanomechanical mass sensor. Nat. Nano. **3**, 533 (2008)
3. Kaajakari, V., Koskinen, J.K., and Mattila, T.: Phase noise in capacitively coupled micromechanical oscillators. IEEE Trans. Ultrason. Ferroelectr. Freq. Control **52**, 2322 (2005)
4. Kozinsky, I., Postma, H.W.Ch., Bargatin, I., Roukes, M.L: Tuning nonlinearity, dynamic range, and frequency of nanomechanical resonators. Appl. Phys. Lett. **88**, 253101 (2006)
5. Shao, L.C., Palaniapan, M., Tan, W.W.: The nonlinearity cancellation phenomenon in micromechanical resonators. J. Micromech. Microeng. **18**, 065014 (2008)
6. Kacem, N., Hentz, S., Pinto, D., Reig, B., Nguyen, V.: Nonlinear dynamics of nanomechanical beam resonators: improving the performance of NEMS-based sensors. Nanotechnology **20**, 275501 (2009)
7. Kacem, N., Baguet, S., Hentz, S., Dufour, R.: Computational and quasi-analytical models for non-linear vibrations of resonant MEMS and NEMS sensors. Int. J. Non Linear Mech. **46**, 532 (2011)
8. Kacem, N., Arcamone, J., Perez-Murano, F., Hentz S.: Dynamic range enhancement of nonlinear nanomechanical resonant cantilevers for highly sensitive NEMS gas/mass sensor applications. J. Micromech. Microeng. **20**, 045023 (2010)
9. Kacem, N., Hentz S.: Bifurcation topology tuning of a mixed behavior in nonlinear micromechanical resonators. Appl. Phys. Lett. **95**, 183104 (2009)

Part IV
Vibrations of Transport Systems

The Importance of a Detailed Vehicle Modelling in the Numerical Prediction of Railway Ground Vibrations

G. Kouroussis, E. Bergeret, C. Conti, and O. Verlinden

Abstract With the development of continuous technological innovation, the railway transport is presented as an interesting alternative to the road traffic. Some drawbacks exist, one of the most problematic being certainly the vibrations induced by the railway traffic. The presented research wants to establish a reliable methodology in order to evaluate, from the design stage of a vehicle or of a track, the efforts transmitted by the vehicle to the track/soil system and consequently the level of vibrations in the surroundings. An analysis of the interaction between the track and the soil has been performed in order to show when the track/soil uncoupling can be assumed, with the aim of working in two stages. The first step is based on the vertical dynamic behaviour of the vehicle/track subsystem, taking into account any irregularity in the rail surface. For the soil subsystem (second step), recent publications showed that the finite/infinite element method can be an interesting alternative to boundary element method. The objective of this paper is to demonstrate the real benefit of the vehicle modelling in this kind of problem. Typical railway applications (Brussels tram, Thalys HST) are proposed, showing among others that significant reduction of ground vibration level can be obtained by modifying the dynamic characteristics of the vehicle.

Keywords Ground vibrations • Multibody systems • Vehicle–infrastructure interaction

---

G. Kouroussis (✉) • E. Bergeret • C. Conti • O. Verlinden
University of Mons, Belgium
e-mail: georges.kouroussis@umons.ac.be

J. Náprstek et al. (eds.), *Vibration Problems ICOVP 2011: The 10th International Conference on Vibration Problems*, Springer Proceedings in Physics 139,
DOI 10.1007/978-94-007-2069-5_24,

1 Introduction

The development of train lines has considerably grown during the last decade. Although it has been the subject of continuous technological innovation, railway transport (high-speed train, freight, tramways, ...) is still perceived as an important source of environmental nuisances on the neighbourhood. Among these nuisances, the vibrations induced by railway traffic get the same concern as noise, passenger discomfort or visual impact. Numerical procedures allow to analyse the dynamic behaviour of a mechanical system and to propose solutions to alleviate the vibrations before its manufacturing (virtual prototyping).

If the vehicle dynamics simulation packages are now commonly used in railway industry, it is not yet the case for the track/soil vibrations which are rarely treated from the very beginning of the design. Although multibody system simulation tools like ADAMS, SIMPACK or Virtual.Lab are used to assess the dynamic performances of a vehicle, the simulation of a complete process taking into account the track and the soil is not yet usually performed: the track, and therefore the soil, are supposed to be rigid. Nevertheless some authors consider a flexible track in their simulations. For example, to study the effect of the vehicle running on bridges, Dietz et al. [1] propose to analyse by co-simulation the interaction between the vehicle and the track, considering a finite element model for the latter. Knothe and Grassie have presented an excellent survey of vehicle/track models [2], classifying the dynamic modelling of railway track in function of their application and of the purpose of the study. Unfortunately the ground vibration propagation is scarcely taken into account.

On the other hand, the developed track/soil models rarely consider a detailed modelling of the vehicle and often reduce the vehicle to a sequence of axle loads [3] or rigid wheelsets [4]. The main parts of the train are decoupled, neglecting the inertial forces of the carbodies and the bogie frames. These models are useful to understand the ground waves propagation and their refraction but it is impossible for the train constructors to analyse the effect of the vehicle parameters (suspension, unsprung mass, ...) on the environment.

To consider a complete approach of the problem, Kouroussis et al. have proposed a numerical model, based on the assumption that the vehicle/track interaction and its effect on the soil are decoupled [5]. An analysis of the interaction between the track and the soil has been performed in order to show that the track/soil uncoupling is valid, with the aim of working into two stages. This hypothesis is valid when the ballast stiffness is small compared to the foundation one. The presented research wants to establish a reliable methodology in order to predict, from the design stage of a vehicle or of a track, the efforts transmitted by the vehicle to the track/soil system and consequently the level of vibrations in the surroundings.

The objective of this paper is to illustrate the proposed approach through two important case studies from the tram to the high-speed train, covering the present commercial speeds. Various simulations are performed for verifying if the integration of vehicle dynamics into the prediction model influences the railway-induced ground vibrations.

2 Multibody/FIEM Approach

In this section, the modelling methodology selected for the ground-borne vibration prediction is briefly described. The vehicle is modelled with the help of a multibody approach: the model is formed by the combination of bodies (rigid or flexible bodies, rotating bodies like wheelsets or independent wheels) and interconnection elements, related to the suspensions. The track is defined as a finite element beam discretely supported by the sleepers through the railpad elements, considering the connection to the ground by the ballast. The equations of motion of the vehicle/track subsystem can be written as

$$\begin{pmatrix} \mathbf{M}_v & 0 \\ 0 & \mathbf{M}_t \end{pmatrix} \begin{Bmatrix} \ddot{\mathbf{q}}_v \\ \ddot{\mathbf{q}}_t \end{Bmatrix} + \begin{pmatrix} \mathbf{C}_v & 0 \\ 0 & \mathbf{C}_t \end{pmatrix} \begin{Bmatrix} \dot{\mathbf{q}}_v \\ \dot{\mathbf{q}}_t \end{Bmatrix} + \begin{pmatrix} \mathbf{K}_v & 0 \\ 0 & \mathbf{K}_t \end{pmatrix} \begin{Bmatrix} \mathbf{q}_v \\ \mathbf{q}_t \end{Bmatrix} = \begin{Bmatrix} \mathbf{f}_v \\ \mathbf{f}_t \end{Bmatrix} \tag{1}$$

where $\mathbf{q}$ represents the configuration parameters and mass $\mathbf{M}$, stiffness $\mathbf{K}$ and damping $\mathbf{C}$ matrices are defined for the vehicle (subscript v) and the track (subscript t). The two subsystems are not decoupled since vectors $\mathbf{f}_v$ and $\mathbf{f}_t$ contain the contribution of the wheel/rail contact forces, explicitly considered as non-linear

$$F_{\text{wheel/rail},i} = K_{Hz}\, d^{3/2}, \tag{2}$$

and taking into account the relative penetration d of each wheel with respect to the rail. Vector $\mathbf{f}_v$ also concerns the external forces (essentially gravity) acting on the vehicle. The first step simulates the vertical dynamic response of the vehicle/track subsystem to the rail irregularity. The roughness is reconstructed in the time domain from its power spectral density function. The simulation of the vehicle/track system is made in the time domain, with the help of a home-made C++ library.

For the soil subsystem (second step), the input forces $\mathbf{f}_{soil}$ are derived from the sleepers reaction and are directly applied on the soil surface. A recent paper [6] has shown that the finite/infinite element method (FIEM) can be an interesting alternative to boundary element method. The soil is modelled with the help of the finite element software where infinite elements are added at the border of the region of interest. A mapping formulation is preferred, allowing an elegant means to include decay functions in the ordinary finite element shape function. Viscous boundary, represented by a damping matrix $\mathbf{C}_d$ is added on the junction between finite and infinite element, for preventing any residual reflected waves in the region of interest. The governing equations of motion of the soil subsystem can be written

$$[\mathbf{M}_s]\{\ddot{\mathbf{q}}\} + [\mathbf{C}_s + \mathbf{C}_d]\{\dot{\mathbf{q}}_s\} + [\mathbf{K}_s]\{\mathbf{q}_s\} = \{\mathbf{f}_{soil}\} \tag{3}$$

where $\mathbf{q}_s$ is the vector related to the *dof* of the soil subsystem. $\mathbf{K}_s$ and $\mathbf{M}_s$ are the global mass and stiffness matrices, respectively, and $\mathbf{C}_s$ the intrinsic damping matrix, representing the viscous damping contribution. As demonstrated in [6], time domain simulation is more appropriate to simulate the soil dynamic behaviour (ground wave propagation) and to take into account possible non-linearities. The calculation time remains reasonable thanks to the explicit integration scheme coupled with parallel computer programming. With an uncoupled approach, each subsystem can be modelled in detail and we can focus on the dynamic train/track interaction which is the principal source of ground vibration. The hypothesis of decoupling is valid if the ballast stiffness is small compared to the stiffness of the soil [5]. In this case, the coupling between sleepers through the soil is limited.

3 Case Study 1: The TRAM T2000

The T2000 light railway vehicle circulating in Brussels is defined as a multi-car tramway that uses advanced technology like independent rotating wheels and low floor design. Motors are directly mounted inside the wheels and, to reduce the effect of the unsprung masses, resilient material equips the motor wheels (rubber layer between the web and the tread). The vehicle model considers the bounce and the pitch motions of each component (wheelset, bogie, carbody). The tram case is a perfect example of dynamic track interaction where the quasi-static deflection has a small effect on the ground vibrations.

Figure 1 shows the free field vertical velocity at 2 m from the track for a tram speed v_0 of 20 kmh. The results are based on the passing of the vehicle on a local discontinuity. It can be seen that the model correctly predicts the levels of the ground velocity. The signals shapes are lightly different but a good correspondence can

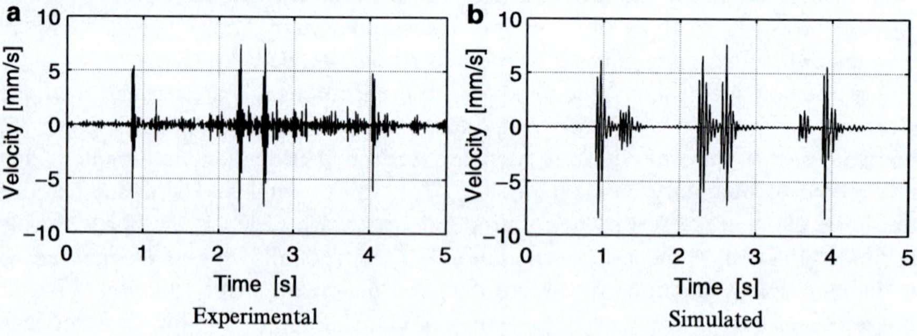

Fig. 1 Vertical ground velocity at 2 m from the track, during the passing of the tram T2000 at speed $v_0 = 20\,\text{km/h}$ (before modification)

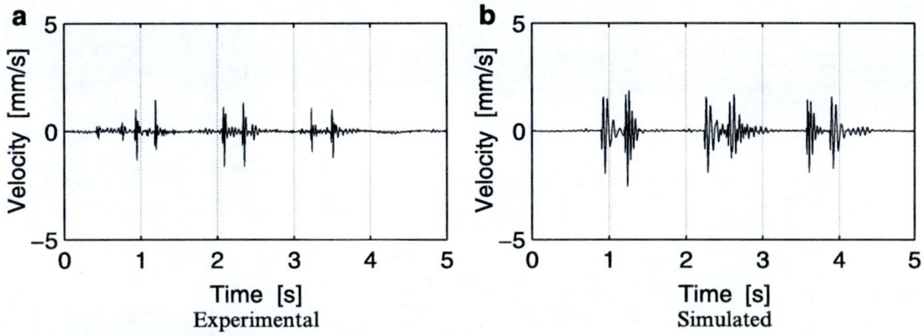

Fig. 2 Vertical ground velocity at 2 m from the track, during the passing of the tram T2000 at speed $v_0 = 20\,\mathrm{km/h}$ (after modification)

be observed on the peak particle velocity *PPV* (maximum of velocity signal). The influence of the vehicle dynamics is also emphasized through Fig. 2, related to the same vehicle but with a local modification of the resilient material (more flexible). Experimental value corroborates the numerical prediction and both results show a significant reduction of the soil vibratory level without modifying the train weight. In addition supplementary results can be found in [5] to demonstrate the real benefit of a complete model of the vehicle, with respect to a sequence of axle loads.

4 Case Study 2: The Thalys HST

When the vehicle speed is large, the quasi-static deflection becomes predominant. If the track irregularity is almost nonexistent, the pressure mechanism is the major source of vibrations. The proposed example is related to the Thalys high-speed train (HST) at Mévergnies (Belgium) along the French border. The proposed model for the vehicle considers only the vertical motion of the vehicle through a 3 *dof*-system for each axle.

Figure 3 gives an example of comparison between experiment and modelling in the case of a Thalys HST moving at a speed $v_0 = 275\,\mathrm{km/h}$ at 12 m from the track. Both time evolution and spectrum show a good agreement. The frequency content shows that the vibrations are dominated by a discrete spectrum where the fundamental bogie passage frequency at 3. 8 Hz (and particularly its harmonics) play an important role, amplified at the soil resonance frequency, at around 30 Hz [7]. Due to the complexity of the spectrum, it is difficult to show the influence of the vehicle, with mode shapes at low frequency (0. 8, 2. 5, 4. 1 and 6. 5 Hz). However this approach can be useful if integrated by the constructor in his pilot study.

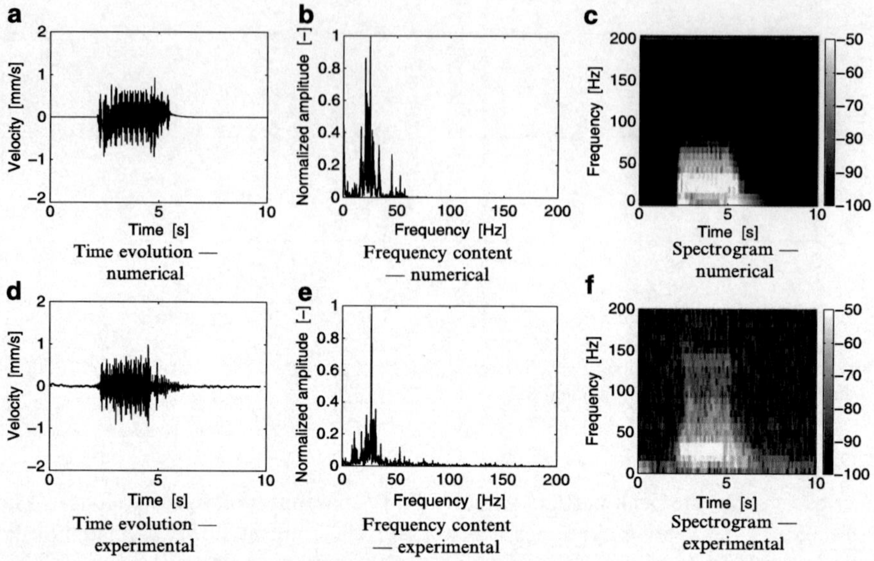

Fig. 3 Numerical (*top*) and experimental (*bottom*) vertical ground velocity at 12 m from the track, in the case of Thalys HST ($v_0 = 275\,\text{km/h}$)

5 Conclusion

In this paper, the implementation of a vehicle/track model in time domain simulation was presented and applied to the assessment of the railway ground vibrations. This approach can be easily coupled to the multibody modelling of the vehicle, fully integrating the dynamic wheel/rail interaction and also performed in time domain. The soil simulation is performed at next step of the analysis, with as input the forces acting on the ground surface. Some results are presented for a tramway case and a high-speed train. Induced vibrations are compared to experimental values in order to validate the proposed approach. These results show the important influence of the vehicle/track dynamics and the real benefit of a complete model. They demonstrate the model applicability to the low- and high-speed cases, including the vehicle/track interaction. By using this approach, solutions to alleviate the ground vibrations induced by a vehicle can be studied from beginning of the track or vehicle design.

References

1. Dietz, S., Hippmann, G., Schupp, G.: Interaction of vehicles and flexible tracks by co-simulation of multibody vehicle systems and finite element track models. Vehicle Syst. Dyn. Supp. (Dyn. Vehicle Road. Track.) **37**, 372 (2002)

2. Knothe, K., Grassie, S. L.: Modelling of railway track and vehicle/track interaction at high frequencies. Vehicle Syst. Dyn. **22**, 209 (1993)
3. Sheng, X., Jones, C. J. C., Thompson D. J.: Prediction of ground vibration from trains using the wavenumber finite and boundary element methods. J. Sound Vib. **293**(3–5), 575 (2006)
4. Auersch, L: Dynamic axle loads on tracks with and without ballast mats: numerical results of three-dimensional vehicle-track-soil models. P. I. Mech. Eng. F.-J. Rai. **220**(2), 169 (2006)
5. Kouroussis, G., Verlinden, O., Conti, C.: On the interest of integrating vehicle dynamics for the ground propagation of vibrations: the case of urban railway traffic. Vehicle Syst. Dyn. **48**(12), 1553 (2010)
6. Kouroussis, G., Verlinden, O., Conti, C.: Ground propagation of vibrations from railway vehicles using a finite/infinite-element model of the soil. P. I. Mech. Eng. F.-J. Rai. **223**(F4), 405 (2009)
7. Kouroussis, G., Verlinden, O., Conti, C.: Free field vibrations caused by high-speed lines: measurement and time domain simulation. Soil Dyn. Earthq. Eng. **31**(4), 692 (2011)

Nonlinear Sway and Active Stiffness Control of Long Moving Ropes in High-Rise Vertical Transportation Systems

Stefan Kaczmarczyk

Abstract In this paper a model to describe the lateral dynamic behaviour of long moving ropes employed in high-rise vertical transportation is developed. The model takes into account the fact that the longitudinal elastic stretching of the ropes is coupled with their transverse motions (sway) which results in cubic nonlinear terms. The governing non-stationary nonlinear equations are solved numerically to investigate the passage through resonance conditions arising during the system operation. The active stiffness control of transverse vibrations of the ropes is discussed. This involves the application of a longitudinal action at the rope end. The results of numerical simulation tests demonstrate the ability of multimodal active control to reduce non-linear low frequency sway of the ropes during and after the passage through resonance.

Keywords Nonliner sway • Vertical transportation • Cubic nonlinear terms • Active stiffness control • Longitudinal action

1 Introduction

Sustainable development practice of modern cities worldwide involves the design and construction of high-rise office and residential buildings. Currently, the world's tallest building is the Burj Khalifa in Dubai which has over 160 stories and is over 828 m high. The CN Tower in Toronto, Canada, is the second tallest (over 553 m) followed by the Taipei 101 office tower in Taiwan (over 509.2 m high).

S. Kaczmarczyk (✉)
The University of Northampton, Northampton, United Kingdom
e-mail: stefan.kaczmarczyk@northampton.ac.uk

J. Náprstek et al. (eds.), *Vibration Problems ICOVP 2011: The 10th International Conference on Vibration Problems*, Springer Proceedings in Physics 139,
DOI 10.1007/978-94-007-2069-5_25,

Vertical transportation systems are integral parts of tall office and residential buildings. In modern high-rise built environment high-speed traction-type elevator systems are employed for vertical transportation of personnel and residents. In such systems steel wire ropes are used as a means of car and counterweight suspension and for compensation of tensile forces over the traction sheave. They are driven to move at speed with their length varying with time during the system operation, resulting in time-varying dynamic characteristics and rendering the system non-stationary [1]. The dynamic response of these long slender continua exhibit large amplitudes, low-frequency non-stationary modes and non-linear dynamic phenomena which often affect the performance of the system.

In this paper a model to describe the lateral dynamic behaviour of long moving compensating ropes with time-varying length is developed. The model takes into account the fact that the longitudinal elastic stretching of the system is coupled with their transverse motions (sway) which results in cubic nonlinear terms. The governing non-stationary nonlinear equations are solved numerically to investigate the passage through resonance conditions arising during the system operation. The *active stiffness* control [2] of transverse vibrations of the ropes is then applied. This involves the application of a longitudinal action at the rope support end which causes time variation of tension and therefore resulting in variation of the transverse stiffness of the ropes [3]. The results of numerical simulation tests demonstrate the ability of multimodal active control to reduce non-linear low frequency sway of the rope during and after the passage through resonance.

2 Equations of Motion

A simplified model of an elevator system is depicted in Fig. 1. An elevator car (m_{car}) with compensating ropes of length $L(t)$ and mass per unit length m, is moving at speed v with the acceleration rate a. The system is equipped with a compensator sheave assembly of mass M. The system is excited by the building deformations that are represented by the motion $w_0(t)$ applied above the machine room level. At the compensator sheave end an axial control motion $u_0(t)$ is implemented by an actuator in order to extract energy from the system.

In the modern elevators the drive control systems allow an accurately prescribed velocity and acceleration time profiles of the car/counterweight to be realized. Thus, it is assumed that vertical transient oscillations of the car/counterweight are negligible. The strain measure due to stretching of the rope is given as $\varepsilon = u_x + \frac{1}{2}w_x^2$, where $(\bullet)_x \equiv \partial(\bullet)/\partial x$, $w(x,t)$ is the lateral response and $u(x,t)$ represents axial (longitudinal) motions of the rope.

In this study only the lower-order lateral modes are considered. Thus, bearing in mind that no interaction will take place between these lateral modes and the axial (longitudinal) modes, the equation of motion describing the lateral displacements over the spatial domain $0 < x < L(t)$ is as follows

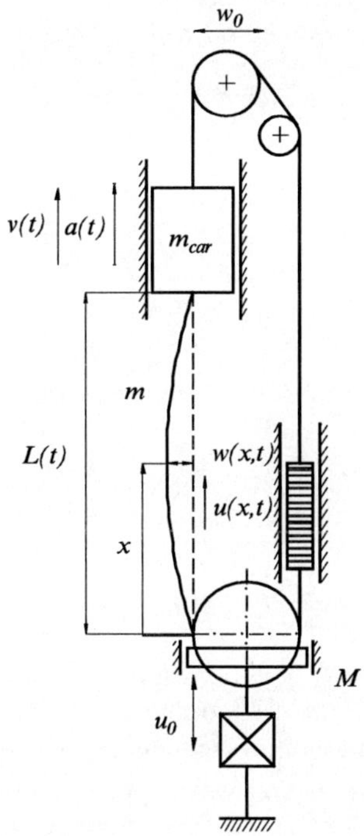

Fig. 1 A simplified model of an elevator system

$$mw_{tt} - \left\{ T_0 + m\left[\frac{1}{2}a(t)L(t) - v(t)^2 + gx\right] + \frac{EA}{L(t)}\left[\frac{1}{2}\int_0^L w_x^2 dx + u_0(t)\right]\right\} w_{xx}$$
$$+ m\left[a(t) - g\right] w_x + 2mv(t)w_{xt} = 0 \tag{1}$$

where $()_t \equiv \partial () / \partial t$, $T_0 = Mg/2$, g denotes the acceleration of gravity, E is the modulus of elasticity of the rope and A represents the effective (steel) cross-sectional area of the rope. The lateral displacements at the boundaries are defined as $w(0,t) = w(L,t) = 0$. In order to accommodate the excitation due to the building motions in the equation of motion (1) the overall lateral displacements of the rope are expressed as

$$w(x,t) = \bar{w}(x,t) + \frac{x}{L(t)} w_L(t), \quad 0 \le x \le L(t) \tag{2}$$

where $\bar{w}(x,t)$ are the displacements of the rope relative to its stretched configuration and $w_L(t)$ represent the lateral displacements of the building corresponding to the upper end of the rope (at the car end termination). Using transformation (2) in

Eq. 1 an approximate solution to the nonlinear partial differential equation of motion is sought by using the *Galerkin* method with the following finite series

$$\bar{w}(x,t) = \sum_{n=1}^{N} \Phi_n(x; L(t)) q_n(t) \tag{3}$$

where $\Phi_n(x; L(t)) = \sin \frac{n\pi}{L(t)} x,\ n = 1, 2, \ldots, N$ are the natural vibration modes of the corresponding taut string of length $L = L(t)$ with the time-variant tension given as $T(t) = T_0 + \frac{1}{2} m a(t) L(t)$, and $q_n(t)$ represent the modal coordinates.

The *Galerkin* procedure results in the following ordinary differential equation with cubic nonlinearities in terms of the modal coordinates

$$\ddot{q}_r + 2\zeta_r \omega_r(t) \dot{q}_r + \lambda_r^2 \left\{ \bar{c}^2 - v^2 + c^2 \left[\frac{1}{2} \left(\frac{w_L(t)}{L(t)} \right)^2 + \frac{u_0(t)}{L(t)} \right] \right\} q_r$$
$$+ \sum_{n=1}^{N} K_{rn}(t) q_n + \sum_{n=1}^{N} C_{rn}(t) \dot{q}_n + \left(\frac{\lambda_r}{2} c \right)^2 q_r \sum_{n=1}^{N} \lambda_n^2 q_n^2 = Q_r(t) \tag{4}$$

where $r = 1, 2, \ldots, N$ and modal damping represented by the ratios ζ_r has been added, ω_r are the undamped natural frequencies of the rope, $\bar{c} = \sqrt{T(t)/m}$ and $c = \sqrt{EA/m}$ represent the lateral wave speed and the longitudinal wave speed, respectively, $\lambda_r = r\pi / L(t)$, K_{rn}, C_{rn} are time-variant coefficients and the modal excitation is represented by function $Q_r(t)$.

3 Results

An approximate solution to Eq. 4 can be obtained by numerical simulation with an explicit Runge-Kutta (4, 5) formula applied. A case of the system with eight ropes of mass per unit length $m = 2.11$ kg/m each tensioned by a compensator assembly of mass $M = 3{,}500$ kg is considered. In this case the elevator is moving upwards at a speed of 8 m/s. The numerical tests are conducted for the system ascending with the acceleration/deceleration rate of $a = 1.2$ m/s^2 from the lowest landing upwards to the highest level. The travel height is $H = 390$ m and the building structure is subjected to a harmonic sway of frequency 0.1 Hz and amplitude 0.53 m measured at 402.8 m above the ground floor level. A multimode feedback control axial motion expressed as

$$u_0(t) = a_u \frac{\sum_{n=1}^{N} q_n \dot{q}_n}{\sum_{n=1}^{N} \omega_n^2 q_n^2} \tag{5}$$

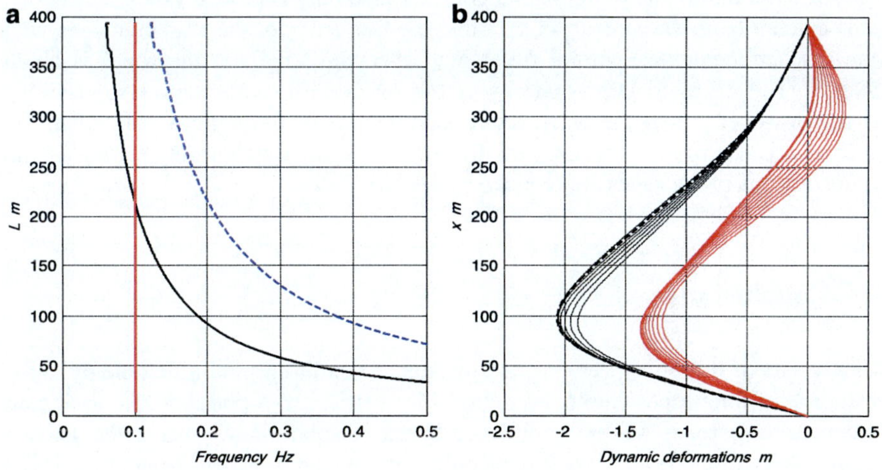

Fig. 2 The dynamic response of the compensating ropes; (**a**) the variation of the natural resonances with the length of the ropes: the first mode (*solid line curve*) and the second mode (*dashed line curve*), together with the excitation frequency (*vertical solid line*); (**b**) the dynamic shapes of the ropes during the elevator travel: without feedback control (*dashed lines*), with feedback control (*solid lines*)

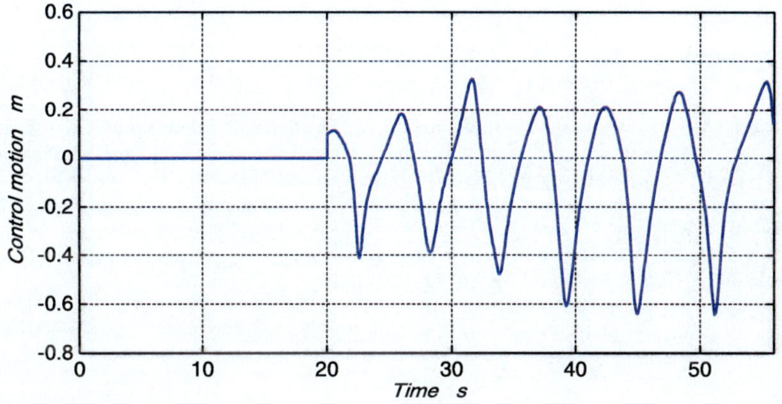

Fig. 3 Axial control motion

where a_u is the control gain. The results are shown in Fig. 2. The first two natural frequencies are determined from the eigenvalues of the stiffness matrix which accommodates the influence of speed and acceleration of the ropes. They are plotted vs. the length of the ropes in Fig. 2a The dynamic deformations of the ropes when the elevator approaches the top landing are presented in Fig. 2b to illustrate the maximum amplitudes of the ropes at various stages of the elevator travel.

The axial motion $u_0(t)$ of the compensator assembly is plotted vs. time in Fig. 3. It is evident from the plots in Fig. 2a that the frequency of the load coincides with the first (fundamental) natural frequency when the length of the ropes is about 215 m. This leads to large motions of the ropes with the amplitudes exceeding 2 m. However, when the active control system with the control gain of $a_u = 0.25$ is activated (after approx. 20 s from the elevator start instant) the maximum deformations of the ropes are reduced by approx. 33%.

4 Conclusions

The results of this study demonstrate resonance conditions that affect the dynamic response of long ropes employed in high-rise vertical transportation. In particular, compensating ropes suffer from large lateral displacements that often exceed allowable limits. The active stiffness control of the lateral response can be achieved by axial motion applied at the compensator sheave end using a modal control strategy.

Acknowledgments Support received from ThyssenKrupp Elevator Corporation for the research work reported in this paper is gratefully acknowledged.

References

1. Kaczmarczyk, S., Iwankiewicz, R., Terumichi, Y.: The dynamic behaviour of a non-stationary elevator compensating rope system under harmonic and stochastic excitations. J. Phys. Conf. Ser. **181**, 012047 (2009). doi:10.1088/1742-6596/181/1/012047
2. Chen, J.C.: Response of large space structures with stiffness control. J. Spacecr. Rockets **21**(5), 463–467 (1984)
3. Wang, L.Y., Xu, Y.L.: Active stiffness control of wind-rain induced vibration of prototype stay cable. Int. J. Numer. Methods Eng. **74**, 80–100 (2008)

The Influence of Semi-active Dampers on the Vibration Behaviour of Passenger Cars

Sebastian Schneider, Daniel Brechter, Andreas Janßen, and Heiko Mauch

Abstract The number of mechatronic components in modern car suspensions is increasing continuously to solve conflicts concerning design goals. Thus, changes in the vibration behaviour of the vehicle are caused. It needs to be ascertained whether this influence has to be taken into account when determining the fatigue life of a car and its components. Therefore, changes of the loads are studied in measurements and multi-body simulations of a passenger car with semi-active dampers. The evaluation of the forces at the wheel centre and at the shock absorber tower shows that different settings of semi-active dampers have an influence on fatigue life of the chassis and the car body. It is concluded that these effects need to be taken into account when determining fatigue life. Furthermore, multi-body simulations have been successfully applied to study the influence of semi-active dampers on the loads.

Keywords Vibration behaviour • Mechatronic suspension component • Semi-active damper • Fatigue life • Multi-body simulation

1 Introduction

Traditionally, the suspension travel response behaviour of a car is defined by adapting the force-displacement characteristics of the springs and the force-velocity characteristics of the dampers [1]. When installing passive springs and dampers, constant curves are specified. However, the limitations of these components result in a design conflict [2].

S. Schneider (✉) • D. Brechter • A. Janßen • H. Mauch
Volkswagen AG, Wolfsburg, Germany
e-mail: sebastian.schneider2@volkswagen.de; daniel.brechter@volkswagen.de; andreas.janssen2@volkswagen.de; heiko.mauch@volkswagen.de

J. Náprstek et al. (eds.), *Vibration Problems ICOVP 2011: The 10th International Conference on Vibration Problems*, Springer Proceedings in Physics 139,
DOI 10.1007/978-94-007-2069-5_26,

On the one hand, car suspensions have to fulfil various requirements regarding optimum road holding as well as good handling performance. Therefore, the dynamic tyre forces need to be reduced. On the other hand, maximum comfort of the passengers is an important design goal which is achieved by reduction of the vibrations of the car body. Furthermore, the latter is taken into account for medicinal reasons in order to avoid damage of human organs. [3]

By means of intelligent components, such as controllable shock absorbers, the mentioned design conflict can be resolved [4]. Therefore, the number of mechatronic parts in modern car suspensions is increasing continuously. The introduction of intelligent subsystems influences the vibration behaviour of the overall vehicle.

Changes in vibration behaviour result in different magnitudes of the forces acting on the vehicle chassis as well as the body during service. It needs to be ascertained whether this influence has to be taken into account when determining the fatigue life of a car and its components. Especially, it is necessary to establish whether the customer behaviour is appropriately represented on fatigue test tracks. In addition, extrapolation coefficients which are applied in calculations of fatigue life need to be verified.

In this paper, the vibration behaviour of a vehicle with semi-active dampers is studied using experimental and mathematical methods. The aim is to identify the influence of different shock absorber settings on the loads of a passenger car. Thereby, it can be determined whether it is necessary that testing procedures involve the influence of semi-active dampers on the fatigue life of a vehicle.

2 Semi-active Damper

The term 'semi-active damper' refers to the classification of vertical dynamic systems according to [1].

The main construction of the semi-active damper which is considered in this article corresponds to a conventional hydraulic shock absorber used in passenger cars. Moreover, it is equipped with a controllable valve. Thereby, the size of the flow cross-section can be adjusted during service at low energy consumption. As a result, the energy dissipation of the damper unit can be adapted. In contrast to conventional dampers, the velocity-displacement characteristics of semi-active dampers are described by diagrams instead of constant curves, Fig. 1.

Furthermore, the driver can choose from three different settings of the controllable shock absorbers (comfortable, normal and sporty performance). Thereby, the force-velocity diagram is limited to a certain baseline level. If the driver selects the sporty setting, the minimum value of the absolute damping forces is higher compared to the comfortable setting.

When adjusting the damping, the controller unit analyses the driving situation, the road condition, as well as the driver's choice of setting, within milliseconds.

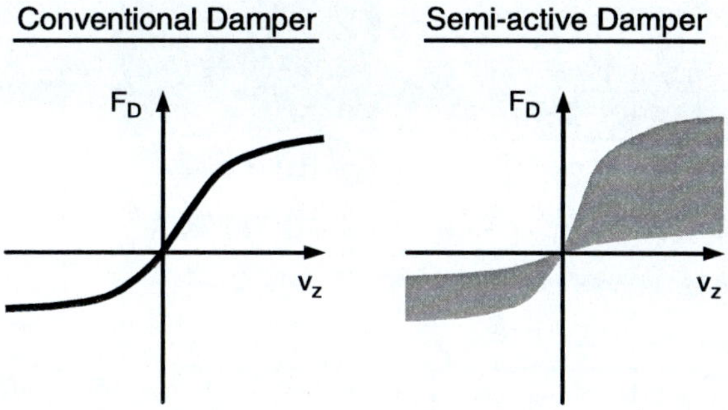

Fig. 1 Force-velocity curve of a conventional damper and force-velocity diagram of a semi-active damper

3 Method

As mentioned in Sect. 1, the influence of different damper settings on the loads of a passenger car was studied in vehicle measurements and numerical simulations. Thereby, it was possible to determine whether it is necessary to adapt and expand existing procedures for fatigue life prediction so that they take into account changes of the loads caused by intelligent components like semi-active dampers.

When identifying the vibration behaviour, first calculations were performed using a quarter car model. Thereafter, vehicle measurements were conducted. The results were used to set up and calibrate a model for multi-body simulations, Fig. 2.

3.1 *Estimation*

In preparation for vehicle measurements, a quarter car model was studied to estimate the magnitude of the influence of the semi-active damper on the vertical vibrations of the car, Fig. 3. The system representing the effective direction of the damper was described by the following equations of motion [5]

$$m_B \cdot \ddot{z}_B = -c_B \cdot (\dot{z}_B - \dot{z}_W) - k_B \cdot (z_B - z_W)\,, \tag{1}$$

$$m_W \cdot \ddot{z}_W = -c_B \cdot (\dot{z}_W - \dot{z}_B) - k_B \cdot (z_W - z_B) - c_W \cdot (\dot{z}_W - \dot{z}_E) - k_W \cdot (z_W - z_E)\,. \tag{2}$$

In contrast to the equation of motion of a vehicle with conventional dampers, the behaviour of semi-active shock absorbers was represented by a variable damping factor c_B. However, calculations were only performed using constant values for the minimum and maximum possible damping. Moreover, the road was represented by sinusoidal excitations with different frequencies and amplitudes as well as measurement signal input. These assumptions were sufficient for basic estimations.

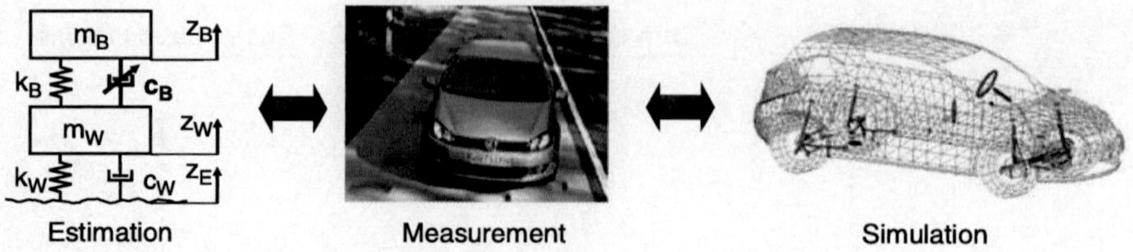

Fig. 2 Procedure for determining the influence of semi-active dampers on the loads of a car

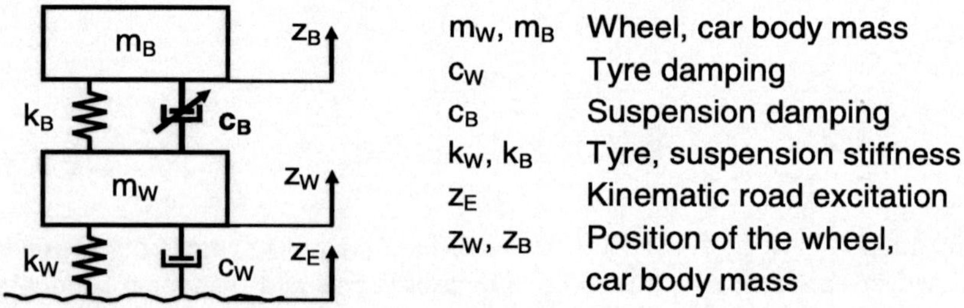

Fig. 3 Quarter car model with semi-active damper

The results of the above-mentioned calculations were used to estimate acceleration and force magnitudes during vehicle measurements beforehand.

3.2 Measurement

A passenger car was equipped with sensors on the car body and on chassis components to analyse the influence of semi-active dampers on the vibration behaviour, as well as the loads, during service. Thus, accelerations, strains and forces were measured on public roads and test tracks. The choice of measuring positions made it possible to analyse the distribution of forces. In addition, measuring wheels were used to observe the forces and torques at the wheel centre. Suspension travel was measured to determine the relative motion between the car body and the chassis. CAN-bus messages were recorded for plausibility checks of the measurement signals. Furthermore, a development ECU was used to record input, output and internal variables of the damper controller unit.

3.3 Simulation

In addition, a vehicle model for multi-body simulations was set up. The semi-active dampers and their control unit needed to be represented to thoroughly model the vibration behaviour of the passenger car. The software code of the controller unit in the multi-body model corresponded to the software on the hardware ECU of the test car. Furthermore, the model was calibrated using the results of the measurements. By means of numerical simulations using the road profile of the test track differences in vibration behaviour of the car were analysed. Hence, it was possible to eliminate the influence of varying conditions like different driver behaviour which occur during measurements.

3.4 Evaluation

The aim of the study was to determine the overall change in the vertical vibration behaviour due to different settings of the semi-active dampers. In addition, the difference between semi-active and conventional shock absorbers was analysed. Furthermore, the evaluation of the measurement and simulation signals was conducted according to a procedure applied in fatigue life prediction as follows:

1. Range pairs were counted by classification of the measurement time signals.
2. A Wöhler curve was set up using values according to a standard procedure.
3. A linear damage calculation according to Miner elementary was performed.
4. Relative damage values were calculated.
5. The power spectral density was analysed to make sure that changes in vibration behaviour were detected.

The study focused on the vertical movements of the car, because this direction is mainly affected by the damper characteristics. Furthermore, in vehicle measurements the driver strongly influences the longitudinal and lateral forces whereas the vertical direction is nearly unaffected. In addition, changes in vertical forces due to the vehicle's velocity and the road profile can be minimized by measuring on a test track applying previously defined velocities in each measurement.

4 Results

During vehicle measurements, the biggest vertical forces at the wheel centres occurred in the comfortable setting, Fig. 4. Higher damping (normal and sporty setting) caused a decrease of the vertical wheel forces. On the contrary, the highest forces at the shock absorber towers (interface of the vertical forces between the car chassis and the body) were observed in the sporty setting. Decreasing damping resulted in lower forces on the car chassis.

In the studies of the quarter car model, as well as in the multi-body simulations, the same observations were made. Furthermore, the influence of the different damper settings which was determined in first calculations with a multi-body car model corresponded very well with the results of the vehicle measurements.

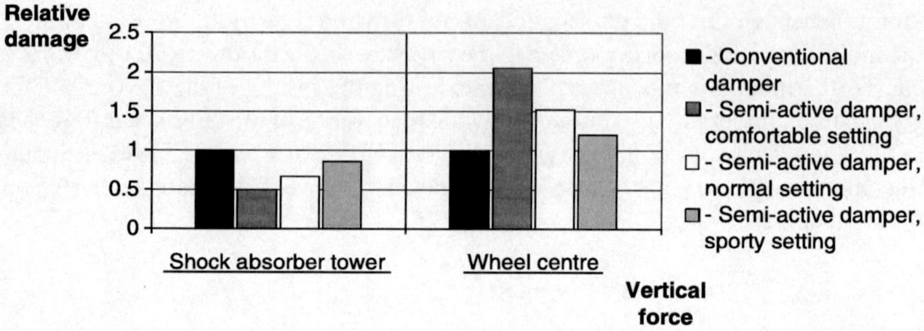

Fig. 4 Relative damage of vertical forces measured at the *shock absorber tower* and the *wheel centre* for different damper settings on a test track

5 Conclusions

The measurement results show that the influence of semi-active dampers needs to be considered when determining the fatigue life of a passenger car and its components. Furthermore, multi-body simulations can be used to reveal differences in vibration behaviour due to different damper settings. Thereby, load assumptions, the design of components as well as the following calculations of fatigue life can be improved in an early state of the design process by application of numerical simulations. In the future, multi-body simulations will be conducted to involve the influence of different damper settings during laboratory fatigue life tests.

References

1. Heißing, B., Ersoy, M. (eds.): Chassis Handbook. Vieweg and Teubner, Wiesbaden (2011)
2. Karnopp, D., Margolis, D.: Adaptive suspension concepts for road vehicles. Veh. Syst. Dyn. **13**, 145–160 (1984)
3. Guglielmino, E., et al.: Semi-active Suspension Control. Springer, Berlin (2008)
4. Elbeheiry, E.M., et al.: Advanced ground vehicle suspension systems – a classified bibliography. Veh. Syst. Dyn. **24**, 231–258 (1995)
5. Wallentowitz, H.: Vertikal – und Querdynamik von Kraftfahrzeugen. Schriftenreihe Automobiltechnik. Forschungsgesellschaft Kraftfahrwesen mbH, Aachen (2006)

Detailed Theoretical Preparation of the Drop Test of an Electron Microscope

Petr Paščenko and V. Kanický

Abstract The article deals with the preparation and execution of the drop test of the electron microscope assembly. Based on the finite element method (FEM), the simplified numerical model of the assembly is created for this purpose. The aim of the nonlinear dynamic analysis has been to obtain the size and direction of maximum inertia forces in the system during the simulated impact of the model on basis. Calculated values are verified by actual drop tests of the assembly, where the microscope is replaced by a dummy body with similar inertial characteristics. During the real tests, accelerations and strains have been measured by accelerometers and strain gauges placed in selected locations. Theoretical results are re-adjusted according to the test results. The conditions of the drop tests are governed by the internal regulations of the manufacturer. Based on the knowledge of the actual load, the load carrying structural parts of the assembly (supporting frame, horizontal frame) may be properly designed. Sufficient strength and rigidity must be guaranteed especially with regard to the transportation where rough treatment can be expected.

Keywords Drop test • Electron microscope • Dummy body • Supporting frame • Horizontal frame • Inertial forces

P. Paščenko (✉)
University of Pardubice, Pardubice, Czech Republic
e-mail: petr.pascenko@upce.cz

V. Kanický
KVD, Brno, Czech Republic
e-mail: pahavlik@volny.cz

J. Náprstek et al. (eds.), *Vibration Problems ICOVP 2011: The 10th International Conference on Vibration Problems*, Springer Proceedings in Physics 139,
DOI 10.1007/978-94-007-2069-5_27,

1 Introduction

The detail computer analysis of the transported assembly of the electron microscope is provided in this paper. The analysis is supplemented by results of a series of drop tests performed at a test-room. The drop test is prescribed by the producer as a load simulation corresponding to the improper treatment during the transport. The assembly consists of a basic supporting frame, horizontal frame, microscope and resilient pallet. The wooden box covering the whole assembly is neglected in the model due to its low weight. The nonlinear dynamic analysis of the simplified FEM beam model describes the behavior of the assembly during the simulated drop test. It results in theoretical inertial forces induced at the moment of the impact of the assembly on the ground. The inertial forces, verified by experiments, can be used for the quasi-static strength analysis of the detail shell model of both supporting and horizontal frame. Numerical analyses are then supplemented by series of real drop tests. The accelerometer is located close to the center of gravity of the microscope (COG). The exact location of the strain gages is estimated by means of the preliminary linear analysis of the detailed FEM shell model. The computational models are then corrected based on the evaluated test results. Consequently, the reliable stress distribution over both the supporting and horizontal frame can be determined. Finally, this procedure results in the proper design of the assembly intended for the transport. The theoretical basis of the drop tests is taken from [1–3].

2 Brief Description of the Problem

In the operating state, the massive body of the electron microscope produced by FEI Company is supported by four pneumatic springs on an internal horizontal frame, which is also resiliently mounted on the main supporting frame serving as the work table. The frame legs are provided with rubber pads.

For the transport, the internal resilient elements are set out of function using auxiliary fixing bolts. The main frame is resiliently mounted on a horizontal flat wooden pallet. Both longitudinal beams of the pallet are designed as a pair of parallel beams mutually resiliently constrained over their whole length in order to decrease the stiffness of the pallet with respect to the vertical shock response and to damp the rebounds (see Fig. 1).

During the transport, the supporting frame structure of the microscope may be accidentally subjected to shock loads, which could result in unacceptable residual deformations of the frame parts or impairing the device. Hence, it is necessary to assess the reliability of the transport arrangement of both the supporting frame with the microscope and the structural system of the pallet with respect to probable loads during the transport.

Both the assessment of the reliability by computation and the check of the computation results require the experimental determination of the probable load

Fig. 1 Assembly with dummy body at testing room

intensity during the transport. The necessary work needed for the proper deign of both the supporting and horizontal frame may be defined in the next few steps briefly described below (except the last one):

- Preliminary nonlinear dynamic analysis of highly simplified FEM model.
- Preliminary linear static analysis of detailed FEM model [4].
- Test preparation and procedure [5].
- Test result evaluation – comparison with the theoretical results.
- The quasi-static strength analysis of the detailed FEM model.

3 Preliminary Nonlinear Dynamic Analysis

The aim of the computational analysis is to find a suitable placement of all needed sensors and the adaptation of the measuring chain. Finally, the assessment of size and direction of maximum inertia forces in the system is needed. The simplified computational model of the assembly is created by means of the computer program COSMOS/M [6]. The model consists of the finite elements BEAM, MASS, SPRING and GAP (see scheme in Fig. 2). The real spatial frame is replaced with a planar model. The response to a simulated drop test is computed by direct integration of the equations of motion. Large displacements are considered. Mass of the microscope is $m = 559$ kg, mass of the assembly is $m = 792$ kg, total

Fig. 2 Scheme of simplified nonlinear model

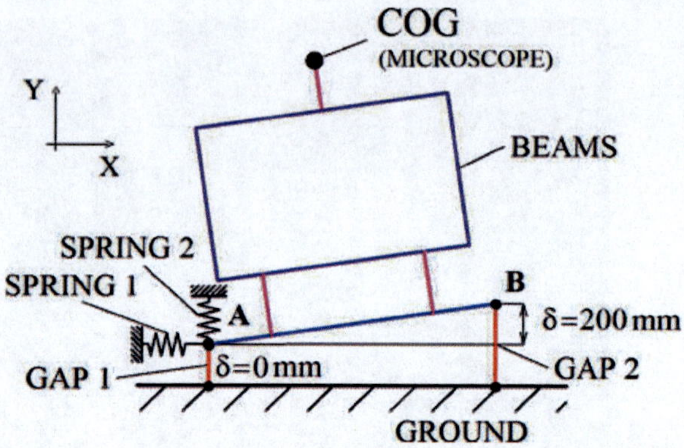

Fig. 3 Curve displacements vs. time

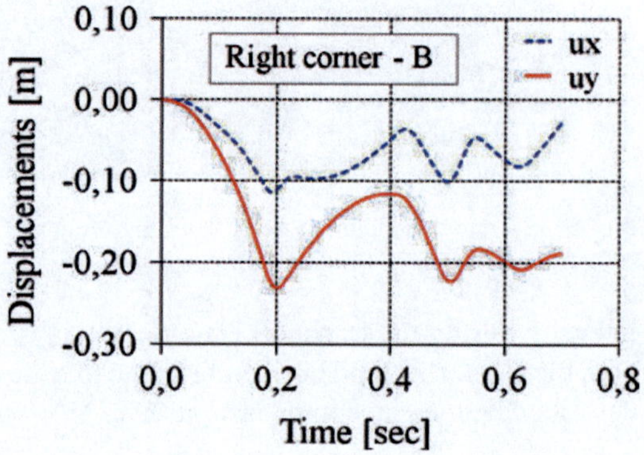

stiffness of rubber springs in both normal and transversal direction (in GAP1, GAP2) is $k = 2.4\text{E}+6\ \text{N}\,\text{m}^{-1}$, total damping coefficient of rubber springs is $b = 7{,}000\ \text{kg s}^{-1}$.

In the following text, the analysis in the dominant plane (XY, front-back direction) is described. The global coordinate system is shown in Fig. 1. The diagram in Fig. 3 shows the movement of the structure at point B dependent on time. Fall from a height of 200 mm including the compression of elastic elements in the impact of assembly on the base is apparent. Maximum vertical and horizontal movements are $u_y = -231$ mm *and* $u_x = -114$ mm, respectively.

The diagram in Fig. 4 represents the dependence of the acceleration of COG of the microscope on time. Maximum values $a_x = -9.6\ \text{m sec}^2$ and $a_y = -60.5\ \text{m}{\cdot}\text{sec}^{-2}$ are reached at the first impact of the structure at point B on the base. Knowing the mass of the microscope, the inertial force in the center of gravity can be easily determined, i.e. $F = 34.2$ kN with components $F_x = -5.37$ kN and $F_y = -33.8$ kN.

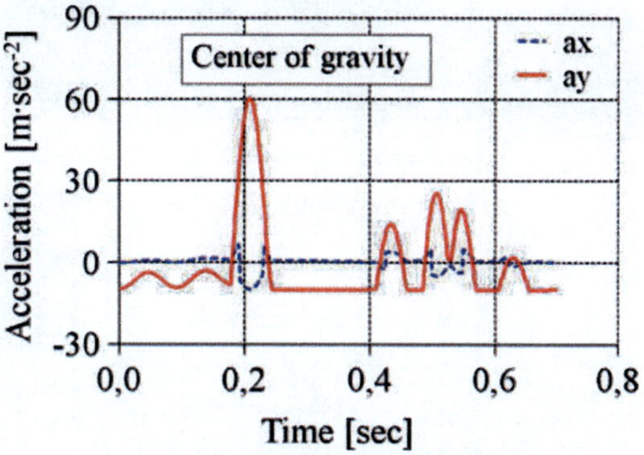

Fig. 4 Curve acceleration vs. time

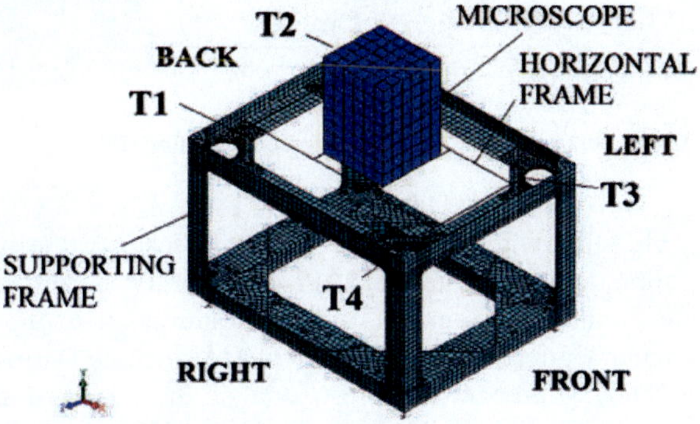

Fig. 5 Detailed models for stress analysis

4 Preliminary Linear Static Analysis

The linear static model (see Fig. 5) is loaded in the COG of the microscope by inertial force $F = 34.2$ kN. The analysis shows the most convenient places for the strain gauges location on the lower edge of the rib in each upper corner of the frame (see Fig. 6). The figure shows maximum theoretical principal stress $\sigma_I = 179.9$ Nmm^{-2} with a clearly distinct position. Finally, the comparison of the measured and theoretical computed values is intended.

5 Test Preparation and Procedure

For the preparation of the tests, the results of the preliminary numerical analyses were considered in order to obtain features of the response with respect to the choice of the measuring devices. For the measurements of acceleration components a_x, a_y, a_z in the microscope center of gravity, the tri-axial accelerometer of the type KB 103, serial No. 00 8115, (manufacturer: MMF Radebeul), mounted on a special clamp, was used (see detail in Fig. 2). The accelerometer was calibrated on the vibration

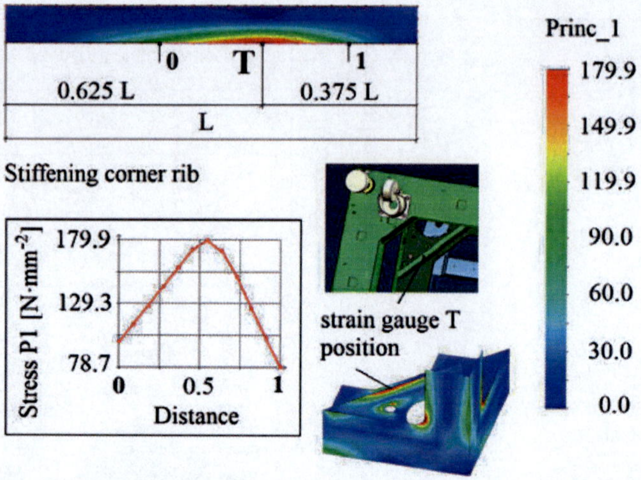

Fig. 6 Strain gauge T1÷T4 location on *stiffening rib*

table fitted with the precise standard sensor of the type SV 100, ser. No. 10004, calibration certificate No. 8012-KL-2045-04 (manufacturer: AURA Milevsko). For the strain measurements, the foil strain gages of the type 6/350 LY 11, k = 2.01 (manufacturer: Hottinger Baldwin Messtechnik Darmstadt) were used.

Having prepared the test, one of the edges of the pallet with the mounted support frame was lifted to a height of 200 mm relative to the opposite parallel edge simply supported by the steel plate (XY plane, front-back direction). The test started with the sudden removal of the calibrated rod supporting the elevated edge. Consequently, the rotational motion of the structure (canting) was initiated. Due to relatively low value of the coefficient of sliding friction, there was a large horizontal displacement. After the contact of the pallet with the ground, a multiple rebounding accompanied by sliding was observed. The test was repeated. In the second series of the tests, the drop test in perpendicular direction was performed (ZY plane, right-left direction).

6 Results Evaluation

Seven attempts in both planes (XY, YZ) were performed. The results of maximum vertical accelerations near the COG of microscope a_y are shown in Tables 1 and 2. The selected set of evaluated records of the measured quantities is presented in Figs. 7 and 8. These displayed graphs of functions correspond to the attempt No. 2.7 with the maximum acceleration. The stress-time relations are derived from measured strain courses.

Table 1 Measured vertical accelerations a_y (ms^{-2}) in front-back direction tests

Test no.	2.1	2.2	2.3	2.4	2.5	2.6	2.7
Plane XY	59.0	50.0	62.5	61.0	54.5	56.0	62.8

Table 2 Measured vertical accelerations a_y (ms^{-2}) in right-left direction

Test No.	1.1	1.2	1.3	1.4	1.5	1.6	1.7
Plane YZ	21.7	22.0	17.7	18.2	17.4	19.9	23.4

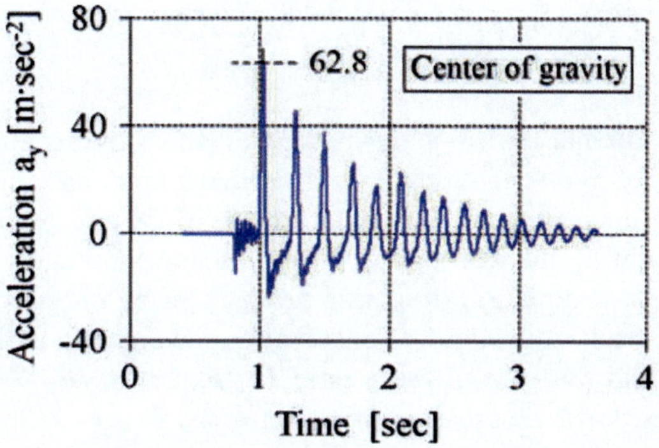

Fig. 7 Curve acceleration vs. time (measured)

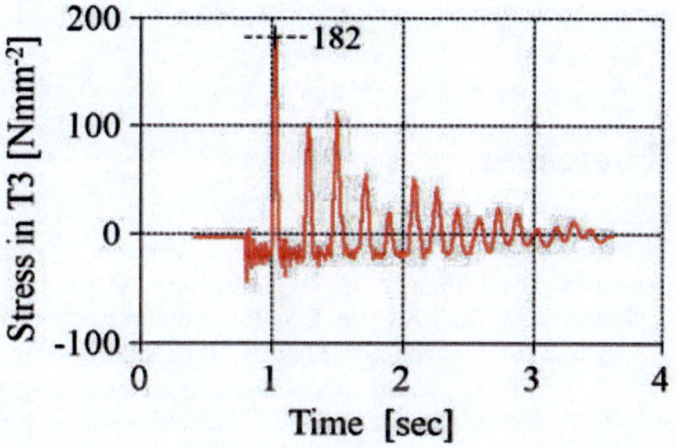

Fig. 8 Curve stress vs. time (measured)

The maximum measured acceleration and dynamic stress component in XY plane are $a_y = (57.97 + 4.83/-7.97)$ m·sec^{-2} *and* $\sigma_{T3} = 182$ Nmm^{-2} (strain gauge T3), resp.

The maximum measured acceleration and dynamic stress component in YZ plane are $a_y = (20.4 + 3.36/-2.64)$ ms^{-2} *and* $\sigma_{T3} = 87$ Nmm^{-2} (strain gauge T1), resp.

The measured values for the dominant front-back direction can be compared with the theoretical values in order to calibrate the computational model for final detail strength analysis. The satisfactory accordance of the results is evident:

- Maximum measured results: $a_y = 62.8\ \text{ms}^{-2} = 6.4\ \text{g}$, $\sigma_{T3} = 182\ \text{Nmm}^{-2}$,
- Theoretical results: $a_y = 60.5\ \text{ms}^{-2} = 6.2\ \text{g}$, $\sigma_3 = 179.9\ \text{Nmm}^{-2}$.

The measured ratio of inertial force components is $F_x / F_y = a_x / a_y = 0.23$ while the computed ratio is 0.16. The difference can be explained by the assumption of a lower coefficient of friction between the pallet and the floor than it actually was.

7 Conclusions

The numerical analyses supplemented by experiments were performed in order to receive reliable data for the final numerical analysis of the behavior of both the supporting and horizontal frames of the microscope when it is accidentally dropped during the transport. The significant dominancy of the inertial forces in the front-back direction drop was proved. Based on the comparison of theoretical and test results, the slightly conservative acceleration in the COG of the microscope $\boldsymbol{a} = \mathbf{7\ g}$ and the inertial force ratio $\boldsymbol{F_x/F_y} = \boldsymbol{a_x/a_y} = \mathbf{0.2}$ can be adopted for the final strength analysis of the both frames.

Acknowledgments The work has been supported by companies FEI Czech Republic, s.r.o. in Brno, Czech Republic and AK-Mechanika, s.r.o. in Pardubice, Czech Republic.

References

1. Yuan, C., Xianlong, J.: Dynamic response of flexible container during the impact with the ground. Int. J. Impact Eng. **37**, 999–1007 (2010)
2. Zhou, C.Y., Yu, T.X., Lee, S.W.R.: Drop/impact test and analysis of typical portable electronic devices. Int. J. Mech. Sci. **50**, 905–917 (2008)
3. Shi, B.J., Wang, S., Shu, D.W., Luo, J., Meng, H., Ng, Q.Y., Zambri, R.: Excitation pulse shape effects in drop test simulation of the actuator arm of a hard disk drive. *Microsyst. Technol.* **12**, 299–305 (2006)
4. Paščenko, P: Drop test of the new frame of the electron microscope – strength verification. *Research report of AK-Mechanika, s.r.o.* (2010)
5. Kanický, V: Drop test of the new frame of the electron microscope. *Research report of AK-Mechanika, s.r.o.* (2010)
6. SRAC (Structural Research and Analysis Corporation): FEM Computer Program COSMOS/M. SRAC (Structural Research and Analysis Corporation), Los Angeles (2010)

Substitution of Gyroscopic Stabilizer Correction Motor by Active Control of Pneumatic Spring

Michal Sivčák and J. Škoda

Abstract Ordinary gyroscopic stabilizer consists of appropriate gyroscope mounted in precession (inner gimbal) and stabilizer (outer gimbal) frames. The whole stabilizer is pivoted in base frame. In standard configuration, the air spring, damper and compensation motor are mounted between base frame and frame of stabilizer. This paper aims to analyze the possibility of simplification this standard configuration. The simplification consists of replacement of the compensation motor by active controlled air spring. Leaving out compensation motor brings minimal change to the whole system, because air spring control circuit and valve are already mounted and are used to adjust the initial position of the stabilizer frame. This solution does not increase the costs. Successful application of proposed modification could reduce the prize of the system. The issue is approached by analysis of nonlinear dynamical model of gyroscopic stabilizer with built-in air spring model, electro-pneumatic valve and correction and compensation system. Precise characteristics of applied air spring and control elements of analyzed system were measured in university laboratories.

Keywords Vibration-isolation • Gyroscopic system • Active control • Air spring • Nonlinear dynamic system

1 Introduction

Mechanical system of the gyroscopic stabilizer, which is the topic of this paper, was closely described in [1]. Kinematic scheme is depicted in Fig. 1. Gyroscope with vertical rotation axis is mounted in precession frame. The precession frame is rotary mounted on stabilizer frame with horizontal axis. The stabilizer frame

M. Sivčák (✉) • J. Škoda
Technical University of Liberec, Liberec, Czech Republic
e-mail: michal.sivcak@tul.cz; jan.skoda@gmail.com

J. Náprstek et al. (eds.), *Vibration Problems ICOVP 2011: The 10th International Conference on Vibration Problems*, Springer Proceedings in Physics 139,
DOI 10.1007/978-94-007-2069-5_28,

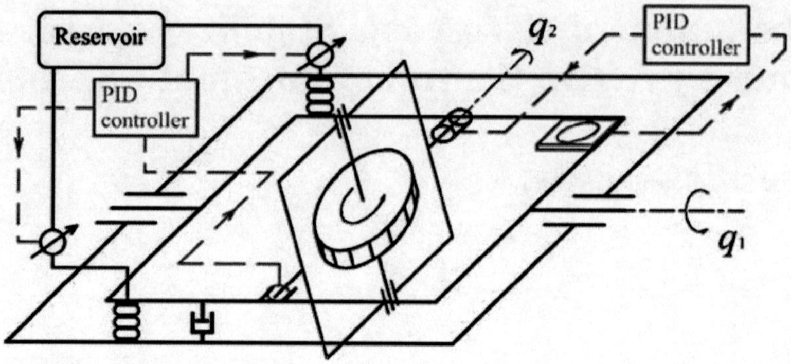

Fig. 1 Kinematic scheme

performs rotation with respect to base frame in horizontal axis perpendicular to precession frame axis. There are two air springs and damper mounted between the base frame and the stabilizer frame. The torque motor of radial correction is mounted on precession frame axis between the stabilizer and precession frame. In previous works, the compensation torque motor was mounted between the base and stabilizer frame. These torque motors are actuating devices of correction and compensation system.

Correction and compensation systems are very important parts of the gyroscopic stabilizer. They consists of two proportional (or PID) feedback loops. Correction torque motor (on precession frame axis) is driven by feedback from sensor of stabilizer frame position, which indicates direction of an apparent vertical direction. Compensation system is a feedback loop, which applies the torque on the stabilizer frame and is driven with respect to magnitude of angular displacement of the precession frame.

Use of air springs and some torque motor combination is characterized by several disadvantages like increase of stiffness, increase stabilizer frame moment of inertia and complication of construction and price increase. To correct function of vibration-isolation platform with gyroscopic stabilizer, the compensation torque of order 10^2 Nm in magnitude is required (see required system behavior in [1]). Pneumatic swivel torque motors often use the speed-reducing gearbox which increases the reduced moment of inertia and stiffness. In many cases the gearbox is self-locking. Using an active control of the already mounted air springs seems to be a logical step due to the above mentioned reasons.

2 Mathematical Model

In previous works, the model of air springs was simplified to linear model of spring with constant preload. The compensation torque dependent on angle displacement of gyroscope precession frame was applied on stabilizer frame rotation axis.

We have built modified mathematical model; also more accurate model of air springs and model of electro-pneumatic valves and their control system were applied. Changes only affect the right hand side of original motion equations of mechanical system. Original set of motion equations was derived (see [2]) using Lagrange equations of the second kind, with external torques on their right hand side. Set of motion equations:

$$\frac{d}{dt}\frac{\partial T}{\partial \dot{q}_i} - \frac{\partial T}{\partial q_i} + \frac{\partial U}{\partial q_i} = M_{s_i} + M_{d_i} + M_{pas_i} + M_{cor_i} \quad i = 1..3, \tag{1}$$

where index $i = 1$ is for stabilizer frame (q_1 is its angle displacement), $i = 2$ for precession frame (q_2 is its angle displacement) and $i = 3$ is for sensor of apparent vertical post (q_3 is angle between apparent vertical and z-axis of stabilizer frame). M_{Si} are torques of air springs, M_{ti} are torques of dampers (linear characteristics), M_{pasi} are torques of passive resistances (considered only in pivot of apparent vertical sensor – pendulum) and M_{cor2} is torque of correction motor (only on the precession frame axis). For torque of air springs we obtain

$$M_{s_1} = -r_{p1}\left(p_1 - p_a\right) S_{ef}\left(l_1(q_1)\right) + r_{p2}\left(p_2 - p_a\right) S_{ef}\left(l_2(q_1)\right). \tag{2}$$

Where on the right hand side of the equation index 1 indicates quantities of left spring and 2 indicates right spring. p_1 and p_2 are air pressures in springs, S_{ef} is function of effective area dependent on deflection of spring $l_i(q_1)$ and p_a is the atmospheric pressure.

For the air pressures inside the air springs, another two equations are added to the set of motion equations. Ideal gas law and the isothermal process are supposed. The pressure and volume of air inside the air spring are functions of time. Equation for the pressure inside the spring is

$$p_i \cdot \dot{V}\left(l_i(q_1)\right) + \dot{p}_i \cdot V\left(l_i(q_1)\right) = RTG_i, \tag{3}$$

where $i = 1..2$, G_i means mass flow of the air into or out the air spring. Using the relation between the volume of air in the spring and its effective area in case of bellow type air spring is possible to modify Eq. 3.

$$p_i \cdot S_{ef}\left(l_i(q_1)\right) \frac{dl_i}{dq_1} q_1 + p_i \cdot V\left(l_i(q_1)\right) = RTG_i \tag{4}$$

Equations 1 and 3 are complete set of equations for simulations of motion of described mechanical system. Air mass flows on the right hand sides of Eq. 4 are defined by mass flow through the control electro-pneumatic valves.

To control the pressures inside the air springs we chose the electro-pneumatic valves VY1A01 manufactured by SMC. Actually we are talking about pressure regulator. This type of valve is equipped by control circuit. It was necessary to derive the model of this pressure regulator for our simulations. Both valves, viewed in

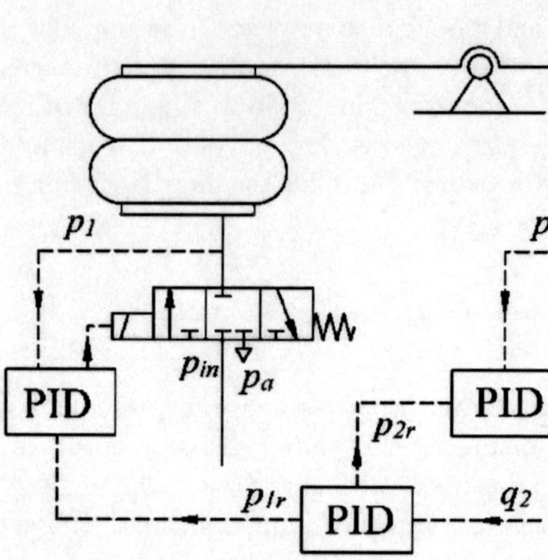

Fig. 2 Scheme of air springs control

Fig. 2, are connected to PID controller. Required pressures p_{1r} and p_{2r} are the input into the PID controllers of valves. These pressures are dependent on precession frame displacement and they are determined by PID controller of compensation system. Air mass flow through the valve from the source of pressured air to the i-th air spring can be expressed (according to ISO6358 standard):

$$G_{PA_i} = \begin{cases} p_{in} \cdot C \cdot \rho_a \sqrt{1 - \left(\dfrac{\dfrac{p_i}{p_{in}} - b_{PA}}{1 - b_{PA}} \right)^2} & \dfrac{p_i}{p_{in}} > b_{PA} \\ p_{in} \cdot C \cdot \rho_a & \dfrac{p_i}{p_{in}} \leq b_{PA} \end{cases}, \tag{5}$$

similarly the air mass flow from the i-th air spring to the atmosphere can be expressed:

$$G_{AR_i} = \begin{cases} -p_i \cdot C \cdot \rho_a \sqrt{1 - \left(\dfrac{\dfrac{p_a}{p_i} - b_{AR}}{1 - b_{AR}} \right)^2} & \dfrac{p_a}{p_i} > b_{AR} \\ -p_i \cdot C \cdot \rho_a & \dfrac{p_a}{p_i} \leq b_{AR} \end{cases}, \tag{6}$$

where p_{in} is the pressure of pressurized air, C stands for pneumatic conductivity (must be measured – ISO6358) dependent on valve opening, ρ_a signs the normal air density, b_{PA} and b_{AR} are critical pressure ratios for appropriate flow direction (must be measured – ISO6358).

Pressures p_{10} and p_{20} which establish the initial position of the system are calculated by solving Eq. 1 $i = 1$, where $q_1 = q_2 = q_3 = 0$ and $p_1 = p_{10}$, $p_2 = p_{20}$. We have to add another equation to solve initial pressures p_{10} and p_{20}. Suitable equation seems to be the one which balances the forces on pivot of stabilizer frame:

$$F_{S1}(p_{10}, q_1 = 0) + F_{S2}(p_{20}, q_1 = 0) = (m_4 + m_5 + m_6)\, g, \tag{7}$$

where F_{S1} and F_{S2} are forces of air springs and m_4, m_5 and m_6 sign weights of stabilizer frame, precession frame and gyroscope.

In previous work (1), the compensation motor torque was derived as

$$M_{com} = k \cdot q_2(t) + d \cdot q_2(t) + i \cdot \int_0^t q_2(\tau)\, d\tau, \tag{8}$$

where k, d and i are coefficients of compensation system PID feedback. It will be useful to use this expression because then we can use the same setup of PID feedbacks as in previous work [1]. Required air pressures in the springs can be obtained by solving equations:

$$M_{S_1}(p_1 = p_{1r}, p_2 = p_{2r}) - M_{S_1}(p_1 = p_{10}, p_2 = p_{20}) = M_{com}, \tag{9}$$

$$F_{S1}(p_{1r}, q_1(t)) + F_{S2}(p_{2r}, q_1(t)) = (m_4 + m_5 + m_6)\, g. \tag{10}$$

3 Simulations

Parameters of PID controllers of correction and compensation feedbacks were determined in work [1]. The same settings of these parameters were used to verify the correctness of this more realistic system. Simulations were performed using the present mathematical model with step change of transversal acceleration applied (similar as acceleration during steering maneuver) and with non zero initial displacement of stabilizer frame. Time responses of stabilizer frame and precession frame displacement are compared with the same responses of previous model in Fig. 3.

Responses on step change of transversal acceleration are shown in left column of Fig. 3. Free motion time responses with initial displacement of stabilizer frame (magnitude 0.1 rad) are shown in right column. The time responses of stabilizer frame displacement are very similar in both cases and the differences are very small within range of 10^{-5} rad. Time responses of precession frame displacement show bigger differences. Time responses of precession frame displacement of previous and present model are quite similar in first case of transversal acceleration excitation. These time responses are different in magnitude of the highest peak in second case of free motion, but behavior is similar. Steeper increase and higher magnitude of precession frame displacement are caused by slower increase

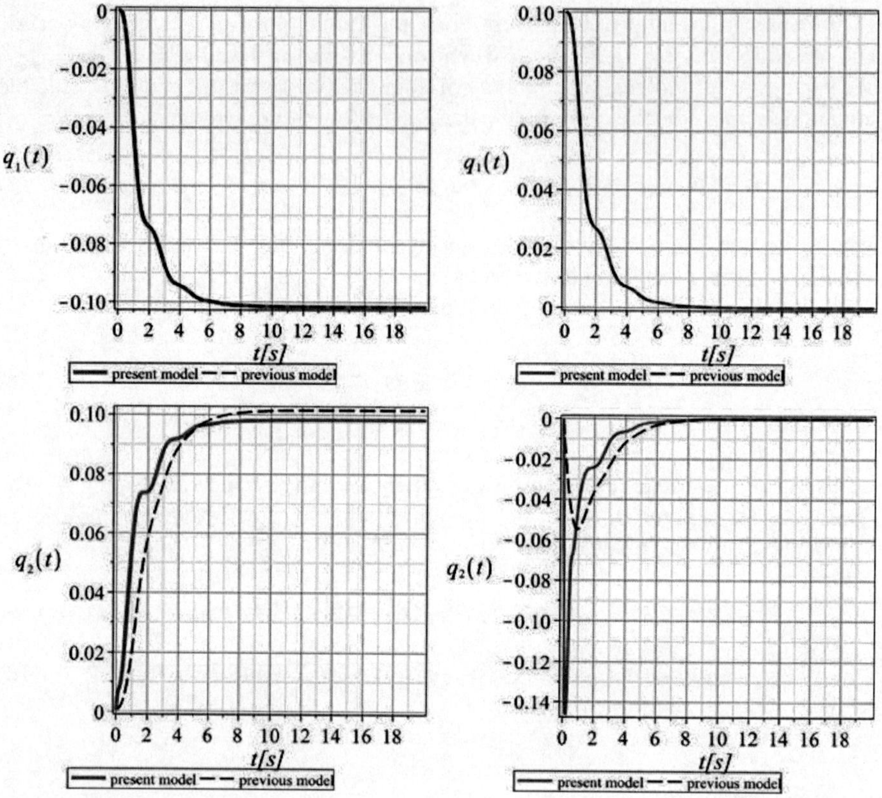

Fig. 3 Comparison of present more accurate model and previous model

and lower magnitude of compensation torque. Compensation torque acts in same direction as gyroscopic torque in these cases. These differences are caused by the time delay of air spring reaction and by nonlinearities of air springs and air mass flows through the valves.

4 Conclusions

We obtained more accurate mathematic model of one-axis vibration-isolation platform with gyroscopic stabilizer. Tubing between the air springs and their control valves are not considered in the mathematical model, but they cause another time delay of compensation system feedback loop. It is necessary to reduce their length for this reason.

The same excitation was applied on both models and time responses were compared. Performed simulations proved that it is possible to simplify the stabilizer using active control of the air springs instead of installation of the torque motor. Suggested design of the compensation system seems to be the correct direction in the development of one axis vibration-isolation platform with the gyroscopic stabilizer.

Acknowledgments The research has been supported by project GAČR 101/09/1481 "The gyroscopic stabilization of the vibration-isolation system" and by project MSM 4674788501 "Optimization of Machine Properties in Interaction with Working Processes and Human Aspects".

References

1. Sivčák, M., Škoda, J.: Radial correction controllers of gyroscopic stabilizer. J. Vibroeng. **12**(3), 300 (2010)
2. Šklíba, J., Prokop, J., Sivčák, M.: Vibroisolation system of an ambulance couch with three degrees of freedom. Eng. Mech. **13**(1), 3 (2006)

Comparison of Two Conceptions of the Vibration Isolation Systems

Jan Šklíba, M. Sivčák, and J. Čižmár

Abstract The sprung stretcher of a ground ambulance litter as the space conducting mechanism with three degrees of freedom. The first degree is determined to compensate the vertical translations of a carriage, the second and third to compensate both horizontal rotations (so called pitching and rolling). The first degree is realized with scissor or with parallelogram, on the upper base on which the double Cardane suspension is placed (as the second and third degree). The second Cardane frame is connected with an own stretcher. The vibration isolation is realized with controlled pneumatic springs. Their control has two sensing units: sensor of the relative position of the upper and lower base and sensor of the absolute angle deflection of the second Cardane frame from an horizontal plane (double electrolytic level). This level is modeled as a spherical pendulum (on the base of its identified characteristics). There was analyzed this dynamic system with five degrees of freedom. The analyze of two conceptions demonstrates that the scissor mechanism is for the complete space mechanism more useful than the parallelogram.

Keywords Parallelogram • Scissor mechanism • Level • Sensor of the absolute position

---

J. Šklíba (✉) • M. Sivčák
Technical Univerzity of Liberec, Liberec, Czech Republic
e-mail: jan.skliba@tul.cz; michal.sivcak@tul.cz

J. Čižmár
Univerzity of Defence, Brno, Czech Republic
e-mail: jan.cizmar@unob.cz

J. Náprstek et al. (eds.), *Vibration Problems ICOVP 2011: The 10th International Conference on Vibration Problems*, Springer Proceedings in Physics 139,
DOI 10.1007/978-94-007-2069-5_29,

1 Introduction

As a conducting mechanism of a driver's seat it is usually used either a scissors mechanism or a parallelogram. The advantages, resp. disadvantages of these conceptions are known. The vibration-isolation system is controlled by a relative lift, resp. by a deflection from an horizontal plane.

If we face a problem in solving a vertical vibration-isolation of an ambulance stretcher, it is the parallelogram, that proves some construction preferences.

If we want to protect the patient's body by pitching or rolling of the carriage either, it is necessary to analyze the effects of vertical vibrations on the absolute position sensors.

2 Preliminary Considerations

For a proposed motion of an ambulance carriage (3 degrees of freedom): vertical translation and two rotations around horizontal axes, the following conducting mechanism (see Fig. 1) with 3 degrees of freedom was chosen:

The first degree – parallelogram (Fig. 2-left) or scissors mechanism (Fig. 2-right).

The second and the third degree – Cardane suspension of the stretcher placed on the upper base of the first degree.

It is evident, that kinematics of both conceptions is different: The upper base of parallelogram makes a circular translation motion according to lower base, while the upper base of scissors makes a straight-lined motion.

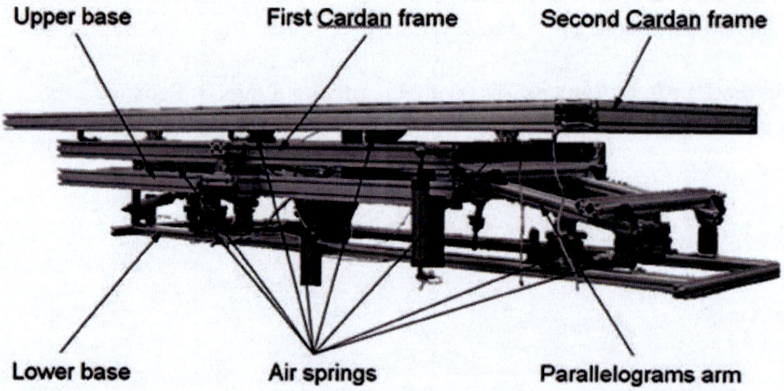

Fig. 1 The validation sample of the vibration-isolation system

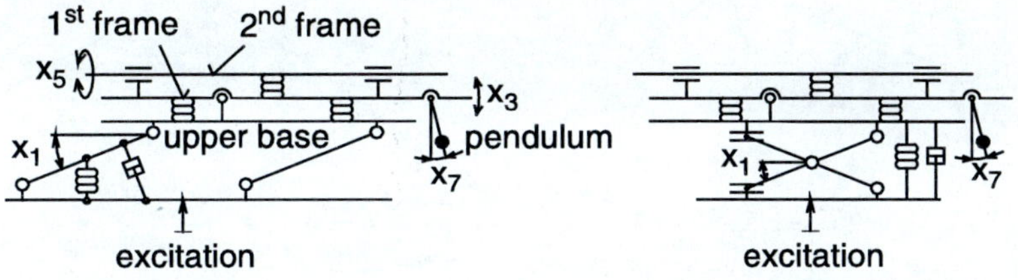

Fig. 2 Two conceptions of the vibration-isolation system

3 Dynamic Analysis

Mathematic models of the both version (see Fig. 2) are deduced (on the base of the Lagrange equations [1]) for the full nonlinear system (T – is kinetic energy, U – potential energy of the gravity forces, M_S and M_D moments of the non-conservative forces and $\vec{q} = \left[x_1 \ x_3 \ x_5 \right]$ vector of general coordinates (there are the angle deflections of the parallelogram and frames from equilibrium state).

$$\frac{d}{dt}\frac{\partial T}{\partial \dot{q}_i} - \frac{\partial T}{\partial q_i} + \frac{\partial U}{\partial q_i} = M_{Si} + M_{Di}, \quad q_i = x_{2.i-1}, \quad \dot{q}_i = x_{2.i}, \quad i = 1..3 \tag{1}$$

Small deflections are deduced from (1) a linearized system (see. [2]).

$$\mathbf{A}\ddot{\vec{q}} + (\mathbf{B}_1(t))\,\dot{\vec{q}} + (\mathbf{C}_0 + \mathbf{C}_1(t))\,\vec{q} = \vec{E}_0 + \vec{E}_1(t) + \vec{M}_D \tag{2}$$

In [2] there are deduced the explicit expressions for members of matrixes and vectors in Eq. 2 (**A** – mass matrix, $\mathbf{B}_0$ – damping matrix, $\mathbf{C}_0$ – stiffness matrix, $\mathbf{B}_1(t)$ resp. $\mathbf{C}_1(t)$ – matrix of parametric excitation, $\vec{E}_0$ – vector of the static moment (in the equilibrium state must be $\vec{E}_0 = 0$), $\vec{E}_1(t)$ vector of external excitation).

We will analyze the difference between two versions. The base of this analyze will be one-way level which is placed on first frame (see Fig. 2). We suppose that our dynamic system is uncontrolled and is excited only by vertical translation. The level is modeled as the mathematical pendulum with same natural frequency and relative damping (see [3]):

$$\ddot{x}_7 + \kappa\dot{x}_7 + \left(\frac{g - \ddot{z}}{l}\right) x_7 = -\ddot{x}_3 - \frac{g}{l}x_3 + \frac{\ddot{y}}{l} \tag{3}$$

$$z = \xi(t) + R\sin x_1, \quad y = \begin{cases} R\cos x_1 & \text{parallelogram} \\ 0 & \text{scissor} \end{cases} \tag{4}$$

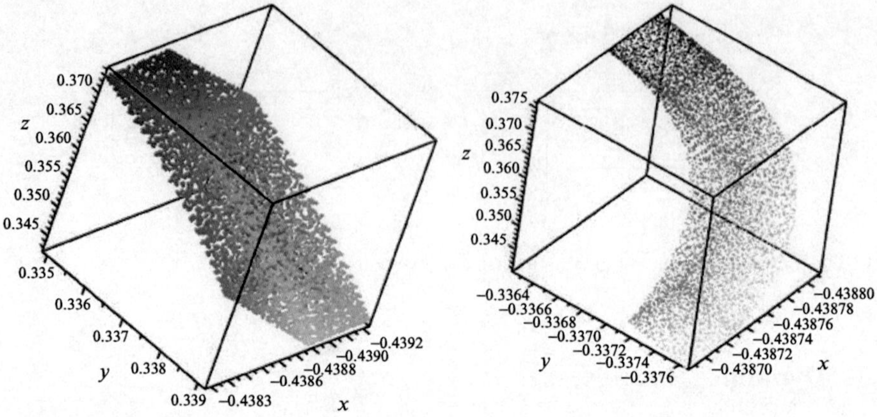

Fig. 3 Space of possible trajectories of the point (mass centre of the patient's body) on the second frame. The scissor mechanism is on the *left* and parallelogram on the *right side*

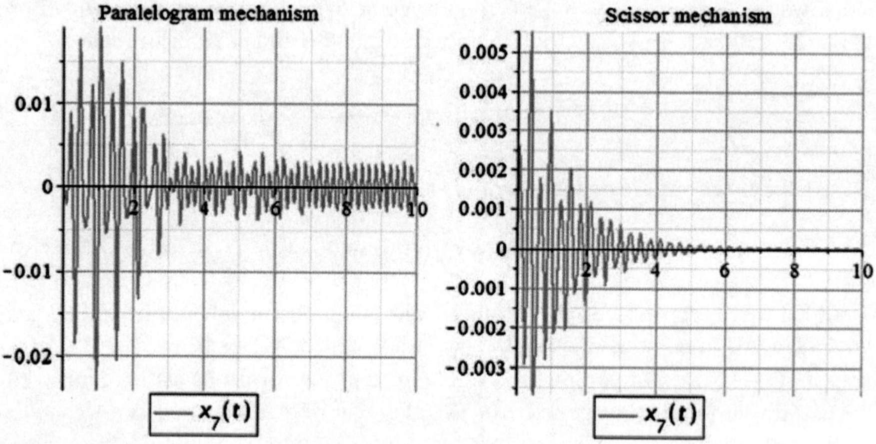

Fig. 4 The pendulum deflection [rad]; excitation by sinus function with frequency 5 Hz and amplitude 0.01 m. Angle between parallelogram arms and base is 22° in equilibrium state

The analyzed system has three natural frequencies: 2.1 Hz – parallelogram (scissor), 3.8 Hz – first frame and 4.1 Hz second frame of the Cardan suspension. The natural frequency of the mathematical pendulum is 3.5 Hz and relative damping 0.1.

If we depict the space trajectories of the mass center of patient's body, the difference between the both conceptions is evident (see Fig. 3).

The natural frequencies depend on the choice of pneumatic springs and their placement. In the over-resonance area the vertical kinematic excitation is compensated. The upper base of scissors mechanism does not move in absolute space, while the upper base of the parallelogram makes a periodic horizontal

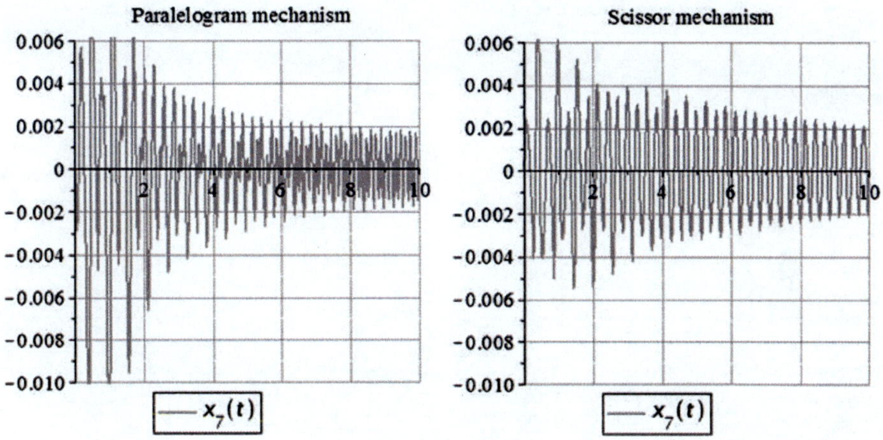

Fig. 5 The pendulum deflection [rad]; excitation by sinus function with frequency 7 Hz and amplitude 0.01 m. Angle between parallelogram arms and base is 22° in equilibrium state. It is a case of the first parametric resonance

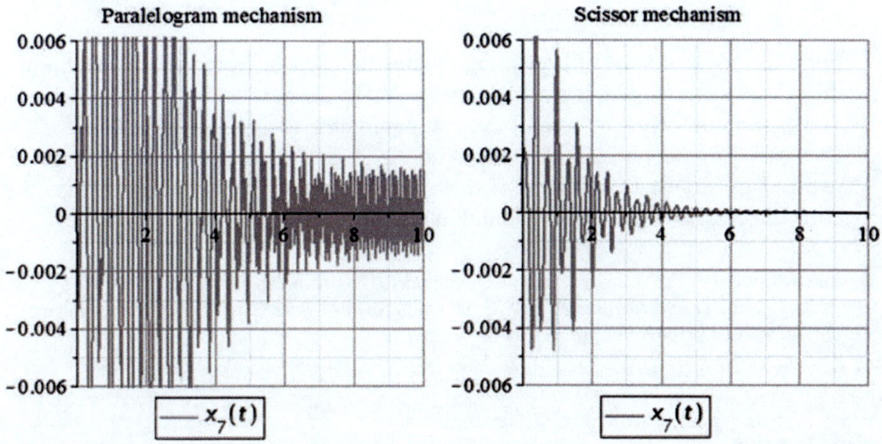

Fig. 6 The pendulum deflection [rad]; excitation by sinus function with frequency 9 Hz and amplitude 0.01 m. Angle between parallelogram arms and base is 22° in equilibrium state

motion. The sensor of absolute position, placed on the first Cardane frame, is excited by this horizontal motion.

The amplitudes of the excited oscillations are order of magnitude smaller in non-resonance area (Figs. 4 and 6) for the scissor mechanism. In the first parametric resonance are the amplitudes of oscillation in same order for both conceptions (Figs. 5 and 7).

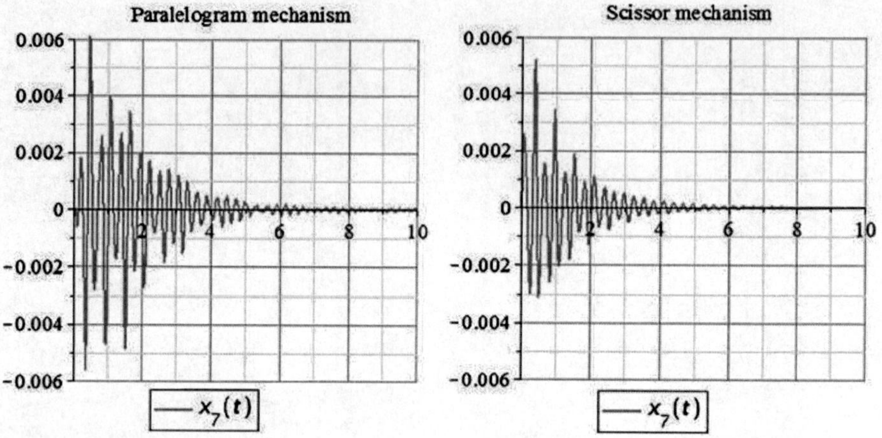

Fig. 7 The pendulum deflection [rad]; excitation by sinus function with frequency 6 Hz and amplitude 0.01 m, arms of the parallelogram are horizontal in equilibrium state

4 Conclusion

1. From the point of view of protecting the position sensor from vertical vibrations, the scissors mechanism is evidently preferred to the parallelogram.
2. The exception forms the area of the first parametric resonance.
3. The transformed horizontal motion of the upper base of the parallelogram, realized by external excitation of the position sensor is minimized in case the arms of parallelogram are horizontal in equilibrium state.

Acknowledgments The research has been supported by project GAČR 101/09/1481 "Gyroskopická stabilizace vibroizolačního systému" and MSMT 4674788501 "Optimalizace vlastností strojů v interakci s pracovními procesy a člověkem".

References

1. Šklíba, J., Prokop, J., Sivčák, M.: Vibroisolation system of the ambulance couch with three degrees of freedom. In: Engineering Mechanics 2006 – Engineering Mechanics, Brno, vol. **13**(1), 3–18 (2006) ISSN: 1210–2717 1
2. Šklíba, J., Sivčák, M., Skarolek, A.: On the stability of a vibroisolation system with more degrees of freedom. In: Vibration Problems – ICOVP 2005. Springer, Dordrecht (2006). ISBN 978-1-4020-5400-6
3. Šklíba, J., Čižmár, J., Skarolek, A.: Positions sensors on the gyroscopic stabilizer correction system. In: Colloquium Dynamics of Machines. Institute of Thermomechanics, Prague (2006). ISBN 978-80-87012-24-6

Dynamic Model of Aircraft Passenger Seats for Vibration Comfort Evaluation and Control

Z. Šika, Michael Valášek, T. Vampola, U. Füllekrug, and T. Klimmek

Abstract The paper deals with the development of the seat dynamical model for vibration comfort evaluation and control. The aircraft seats have been tested extensively by vibrations on the 6 DOF vibrating platform. The importance of the careful comfort control together with the flight mechanics control is namely stressed for the blended wing body (BWB) aircrafts. They have a very large fuselage, where the mechanical properties (accelerations, angular accelerations) vary considerably for different seat places. The model have been improved by adding of dynamical models of the aircraft passenger seats identified by the measurements on the 6 DOF vibrating platform. The experiments, their results and the identification of the dynamical seat model are described. The model is further modified by adding of the comfort evaluation norms represented by dynamical filters. The structure and identification of the seat model is briefly described and discussed.

Keywords Blended wing body aircraft • Control of aircraft comfort • Aircraft seat dynamical measurement • 6 DOF vibrating platform • Comfort standards • Aircraft seat dynamical model

---

Z. Šika • M. Valášek (✉) • T. Vampola
FME, Department of Mechanics, Biomechanics and Mechatronics, Technická 4, CTU in Prague, Praha 6, Czech republic
e-mail: Zbynek.Sika@fs.cvut.cz; Michael.Valasek@fs.cvut.cz

U. Füllekrug • T. Klimmek
Deutsches Zentrum für Luft- und Raumfahrt, Institute of Aeroelasticity, Bunsenstr. 10, 37073 Göttingen, Germany

J. Náprstek et al. (eds.), *Vibration Problems ICOVP 2011: The 10th International Conference on Vibration Problems*, Springer Proceedings in Physics 139,
DOI 10.1007/978-94-007-2069-5_30,

1 Introduction

Vibrations are present in nearly all operating ranges of all types of aircrafts. They occur in a wide range of frequencies and amplitudes. They are caused by different sources like take-off and landing, flight maneuvers in the air, aerodynamic excitations like turbulences and gusts, the engines and other on-board systems. In certain frequency and amplitude ranges the vibrations are regarded as more or less uncomfortable. At higher amplitudes, the vibrations are clearly reducing the comfort and can affect the health or even the flight safety. The improvement of comfort on vibration is an important topic when investigating overall passenger comfort. High passenger comfort might be an important reason for the decision of airline companies to procure certain types of new aircraft. The research of design and control of the new type of blended wing body (BWB) aircraft is the topic of seventh framework EC project ACFA 2020 "Active Control for Flexible 2020 Aircraft" [1]. The importance of the careful comfort control as a pendant of the flight mechanics control is namely stressed for the BWB aircrafts. They have a very large flexible fuselage [2], where the mechanical properties (accelerations, angular accelerations) vary considerably for different seat places. Consequently the ride comfort might be much more critical than on conventional aircrafts. The appropriate comfort control design needs the adequate system model. The aerodynamic model of the aircraft has to be extended by adding of seats dynamical models and the comfort evaluation dynamical models. The paper presents representation of the human comfort sensitivity by the dynamical comfort norms described by transfer functions and representation of the seat by the simplified 3D mechanical model identified from experiments.

2 Model Extension with Respect to Comfort Evaluation

To take into account the comfort sensitivity the model is completed based on the comfort standards. The current way of vibration comfort evaluation according to valid and used standards ISO 2631 and BS 6841 [3, 4] seems to be adequate for the control design purposes, nevertheless the evaluation of vibration comfort generally is not completed area [5]. Two variants are considered, firstly the vibration comfort evaluation using filtered acceleration components from all three translational and three rotational directions (filter W_b for z translational direction, filter W_d for x, y translational direction and filter W_e for φ_x, φ_y, φ_z rotational directions), secondly the low frequency filter W_f for vertical direction acceleration and sea(motion) sickness. The BS 6841 norm filters (Fig. 1) for the seat-body contact has been used, namely the W_d filter for the x, y directions, W_e filter for the φ_x, φ_y, φ_z directions (rotations) and W_b filter for the z direction. These filters in fact reflect the mechanical properties of the human bodies and the final sensitivity of human vibration perception.

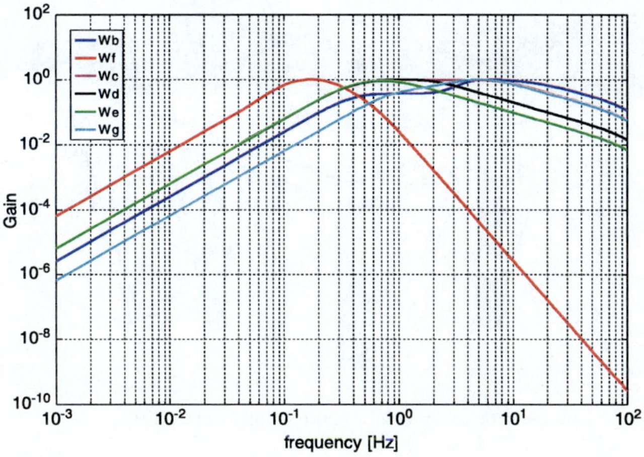

Fig. 1 Transfer functions of comfort filters based on BS 6841 norm

The complete formula of the comfort evaluation criterion has the following form

$$a = (k_x W_d a_x^2 + k_y W_d a_y^2 + k_z W_b a_z^2 + k_r W_e r_x^2 + k_r W_e r_y^2 + k_r W_e r_z^2 + k_x W_c a_{bx}^2 + k_y W_d a_{by}^2 + k_z W_d a_{bz}^2 + k_x W_d a_{fx}^2 + k_y W_d a_{fy}^2 + k_z W_b a_{fz}^2)^{1/2}, \quad (1)$$

where W_d, W_b, ... are the transfer functions of filters for separate acceleration components, k_x, k_y, ... are the weights defined by the norm and a_x, a_y, ... the corresponding acceleration components [4]. The filters for separate acceleration components have been transformed to the state-space form and appended to the original model.

3 Experimental Identification of Seat Transmissibility Functions

The experimental investigations of the dynamic characteristics of the seats are required for performing the correct comfort modeling [6]. Therefore the goal is the experimental identification of the transmissibility functions from the floor of an aircraft cabin through a typical passenger seat to the sitting person. For this purpose, a multi-axial vibration simulator MAVIS has been utilized (Fig. 2). A typical passenger seat with the test dummies is mounted on the vibrating table. All experiments were performed at the test laboratory of the Institute of Aeroelasticity of DLR in Göttingen [7].

The measurement of transmissibility functions was performed with two different tests signals. First, a sine sweep signal was used. The sine sweep signal was applied at three different levels. In addition to the translational axes x, y, z, also the rotational

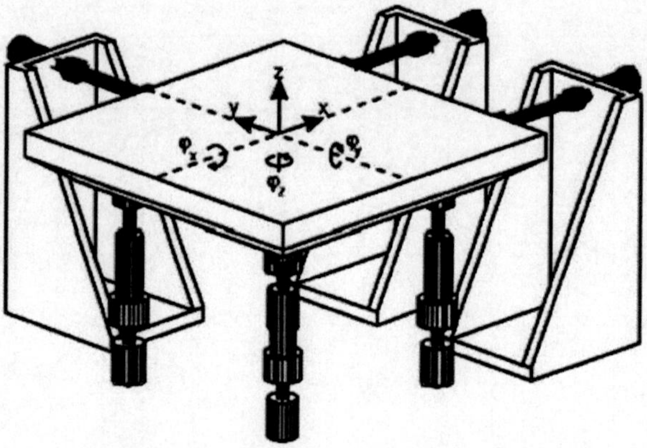

Fig. 2 Drawing of structure of multi-axial vibration simulator MAVIS

Table 1 Characteristics of applied sine sweep excitations

Level	Axis	Frequency 1–2 Hz	Frequency 10–40 Hz	Velocity
Low	x, y, z	2.5 mm	0.125 g	2 oct/min
Medium	x, y, z	5.0 mm	0.250 g	2 oct/min
High	x, y, z	10.0 mm	0.500 g	2 oct/min
Low	$\varphi_x, \varphi_y, \varphi_z$	0.025 rad	1.25 rad/s^2	2 oct/min
Medium	$\varphi_x, \varphi_y, \varphi_z$	0.050 rad	2.50 rad/s^2	2 oct/min
High	$\varphi_x, \varphi_y, \varphi_z$	0.100 rad	5.00 rad/s^2	2 oct/min

motions φ_x, φ_y, φ_z were excited. The level for the rotational axes was selected in that way that the vibration severity was comparable to the translational axes. Table 1 lists the characteristics of the sine sweep excitation for the three levels and the different excitation axes. Second, a broadband random signal was applied at one level.

Once the response at the cabin floor is available, the transmissibility functions identified here can be used for evaluation of vibration levels that the passengers are exposed to. To investigate the influence of possible nonlinearities in more detail the transmissibility functions of the different levels of sine sweep excitation were compared to each other. The structural behavior at the low level is reasonably linear. The verification of computational models should therefore be mainly accomplished with data of the low level, if the models are assumed to be linear.

4 Identification of Seat Model Based on Experimental Results

The measured frequency response data (transmissibility functions) between six components (x, y, z, φ_x, φ_y, φ_z) of acceleration of the vibration simulator table

(Fig. 2) and accelerations on the several points on the dummy (Fig. 3) are the base for the seat dynamic model creation and identification. Even though within the Matlab environment the frequency response data can be partly used directly for the control design [8], the complete incorporation of the seat dynamics needs creation and identification of the state space model. The aircraft model intended for the control design [1] has been formulated as linear, therefore the requested seat model should be also linear.

There are several possible ways how to propose the dynamic model structure, principally on the scale from fully black box models to the detailed physical models. The several different types of models have been tested for the identification target. The best one seems to be relatively simple 3D mechanical model of 12 degrees of freedom with clear physical interpretation of the model parameters (Fig. 4). Such type of model can naturally include mechanical relations between acceleration outputs in several points (Fig. 3) of the dummy. Consequently also the total number of identification parameters is relatively low with respect to other model forms. It has totally 23 mass, stiffness, damping and length parameters.

Different optimization methods can be used for the finding of model parameters in order to find solution close to the experimental frequency response for different input/output channels. The global optimization methods [9] have been used for the finding of the initial guess of the local parameter search [8]. The example of identification results is in Fig. 5 for transmissibility from y acceleration of table to y acceleration in dummy point P1. The accuracy of identification is always trade-off between fulfilments for different I/O channels and model complexity.

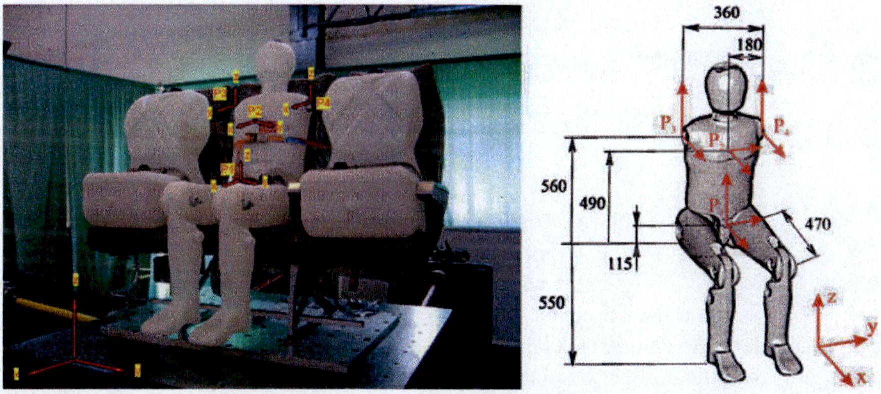

Fig. 3 Test dummy with dimensions, sensor positions and measurement directions

Fig. 4 Structure of considered 12 DOF mechanical model of seat for identification

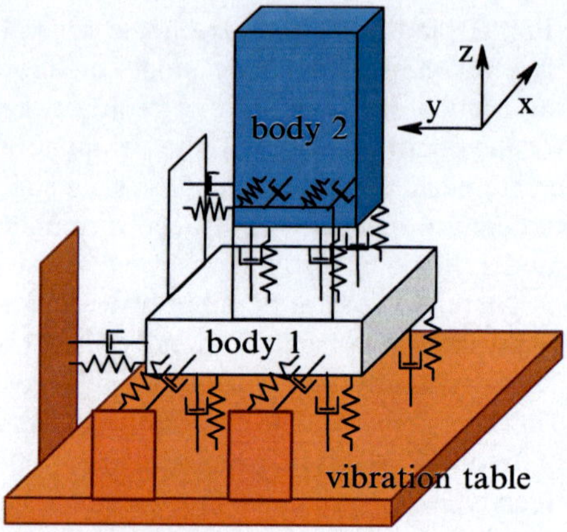

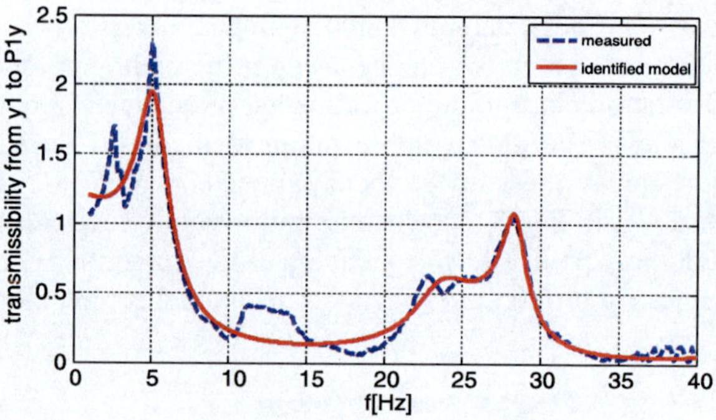

Fig. 5 Transmissibility from y acceleration of table to y acceleration in dummy point P_1

5 Conclusions

Within the context of control strategies of blended wing body aircraft, the model has been extended for the target of comfort control. The simplified mechanical model of standard aircraft seat with the passenger dummy has been proposed and identified based on the vibration experiments, taken into account many input and output acceleration channels. The human comfort sensitivity has been modeled by adding of dynamical filters from the verified comfort standards.

Acknowledgments The authors would like to thank the European Commission for funding the ACFA2020 (Active Control for Flexible 2020 Aircraft) project within the seventh Research Framework Program.

References

1. Wildschek, A., et al.: Gust load alleviation on a large blended wing body airliner. In: Proceedings of ICAS 2010, 27th International Congress of the Aeronautical Sciences, Cancun (2010)
2. Frota, J.: Novel aircraft concepts. In: Proceedings of 2nd NACRE Conference, Greenwich (2008)
3. ISO standard 2631
4. BS standard 6841
5. Valasek, M.: Derivation of objective-subjective correlation of vibration comfort. In Proceedings of Abstracts of IAVSD 2009, Lisboa: International Association for Vehicle System Dynamics, pp. 1–4 (2009)
6. Griffin, M.J.: Handbook of Human Vibration. Elsevier Academic Press, London (2004)
7. Füllekrug, U.: Experimental identification of transmissibility functions of passenger seats. Internal Research Report DLR IB 232-2010 J 09, Göttingen (2010)
8. MATLAB: Control System Toolbox, Optimization Toolbox – User's Guide. The MathWorks Inc., Natick (2001)
9. Houck, C., Joines, J., Kay, M.: A genetic algorithm for function optimization: a MATLAB implementation. Technical Report NCSU-IE TR 95-09, North California State University (1995)

Part V
Computational Methods and Mathematical Modeling in Vibration Problems

Vibration of the Package of Rods Linked by Spacer Grids

V. Zeman and Z. Hlaváč

Abstract This paper deals with modelling and vibration analysis of the large package of identical parallel rods which are linked by transverse springs (spacer grids) placed on several level spacings. The vibration of rods is caused by the support plate motion. The rod discretization by FEM is based on Rayleigh beam theory. With respect to cyclic and central package rod symmetry, the system is decomposed to identical revolved rod segments. The modal synthesis method with condensation of the rod segments is used for modelling and determination of steady forced vibration of the whole system. The presented method is the first step to modelling of the nuclear fuel assembly vibration caused by kinematical excitation determined by motion of the support plates which are part of the reactor core.

Keywords Vibration package of rods • Modal synthesis method • DOF number reduction

1 Introduction

Nuclear fuel assemblies are in term of mechanics very complicated systems of beamed type, whose basic structure is formed from large number of parallel rods fully restrained in horizontal support plates of the core shroud. The rods are linked by transverse spacer grids placed on several level spacings between plates. Vibrations of the fuel assemblies are excited by a support plate motion of the reactor core caused by seismic ground motion [1] or by pressure pulsations generated by main circulation pumps [2].

V. Zeman (✉) • Z. Hlaváč
University of West Bohemia, Czech Republic
e-mail: zemanv@kme.zcu.cz; hlavac@kme.zcu.cz

J. Náprstek et al. (eds.), *Vibration Problems ICOVP 2011: The 10th International Conference on Vibration Problems*, Springer Proceedings in Physics 139,
DOI 10.1007/978-94-007-2069-5_31,

The goal of the paper is a development of analytical method for modelling of vibrations of large package of parallel rods linked by spacer grids caused by kinematic excitation determined by the support plate motion. The developed methodology and software can be used for vibration analysis of the different large parallel beam systems.

2 Mathematical Model of the System

For the cyclic and central symmetric package of rods with respect to centre C the system decomposition on the identical revolved rod segments $s=1,\ldots,S$ (on the Fig. 1 for $S=6$) can be applied. Mathematical model of the rod segment s, in which the rods are linked by transverse springs (spacer grids) placed on several level spacings $g=1,\ldots,G$, was derived on conditions fully restrained rods in fixed lower (subscript L) and upper (subscript U) support plates. The generalized coordinate vector of one rod r in segment s is $q_{r,s}=[(q_{r,L}^{(s)})^T,(q_r^{(s)})^T,(q_{r,U}^{(s)})^T]^T$, where

$$q_{r,X}^{(s)}=\left[\xi_{r,X}^{(s)},\eta_{r,X}^{(s)},\vartheta_{r,X}^{(s)},\psi_{r,X}^{(s)}\right]^T, X=L,U, q_r^{(s)}=\left[\ldots,\xi_{r,g}^{(s)},\eta_{r,g}^{(s)},\vartheta_{r,g}^{(s)},\psi_{r,g}^{(s)},\ldots\right]^T.$$

General coordinates express displacements of the rod nodal points on the level of plate L, grids g ($g=1,\ldots,G$) and plate U. The deformations of the springs with identical stiffness k_g between two arbitrary rods on the level of grid g are expressed by lateral displacements $\xi_{r,g}^{(s)},\eta_{r,g}^{(s)}$ and bending angles $\vartheta_{r,g}^{(s)},\psi_{r,g}^{(s)}$ of the rod cross-section in contact nodal points.

The conservative mathematical model of an arbitrary isolated rod segment s was derived on the basis of beam theory in the form [3]

$$M_S\ddot{q}_s+\left(K_S+\sum_{q=1}^{Q}\sum_{g=1}^{G}K_{qg}\right)q_s=f_s(t),\; s=1,\ldots,S, \tag{1}$$

where Q designates the number of the transverse springs inside one segment and K_{qg} is stiffness matrix corresponding to the couplings q by means of the spring k_g between two rods. The mass M_S and stiffness K_S matrices of all identical parallel uncoupled rods marked by subscript R in the segment are block diagonal$X_S=\text{diag}[X_R,\ldots,X_R]$, $X=M,K$, with identical mass and stiffness rod matrices M_R, K_R derived on the basis of Rayleigh beam theory [4]. The force vector $f_s(t)$ expresses the contact forces acting in end-nodes of the rods $L_r^{(s)}$ and $U_r^{(s)}$ (see Fig. 1b). The vector of generalized coordinates of the rod segment can be partitioned in the form $q_s=[(q_L^{(s)})^T,(q_s^{(F)})^T,(q_U^{(s)})^T]^T\in\Re^{4GR+8R}$, where $q_X^{(s)}=[(q_{1,X}^{(s)})^T,\ldots,(q_{r,X}^{(s)})^T,\ldots,(q_{R,X}^{(s)})^T]^T\in\Re^{4R}$, $X=L,U$ are displacement vectors

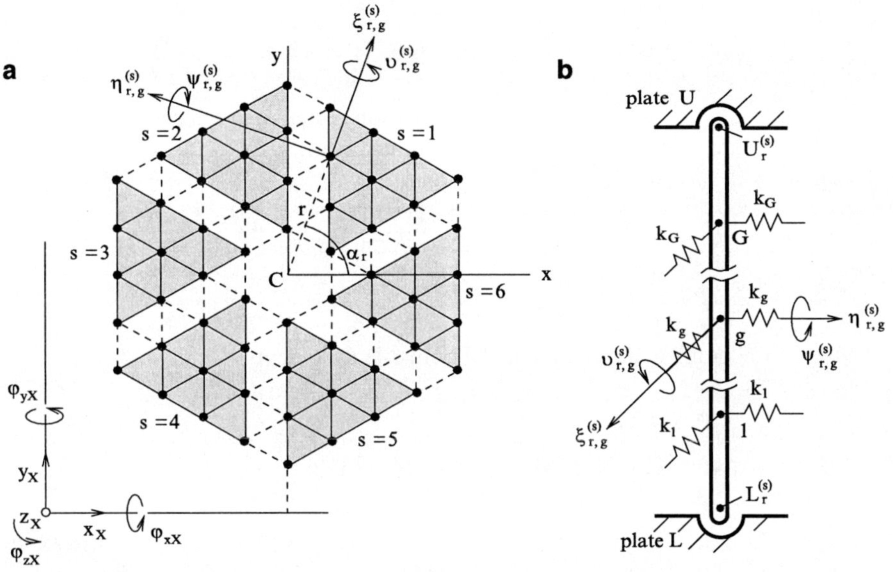

Fig. 1 The rod package cross-section (**a**) and displacements of rod r in segment s (**b**)

of end-nodes of rods $1,\dots,r,\dots,R$ coupled with moving rigid plates and $q_s^{(F)} = [(q_1^{(s)})^T,\dots,(q_r^{(s)})^T,\dots,(q_R^{(s)})^T]^T \in \Re^{4GR}$ is displacement vector of free rod nodes in contact points with the springs.

The mathematical model of the rod segment s in the decomposed block form corresponding to partitioned vector q_s can be written as

$$\begin{bmatrix} M_L & M_{L,S} & 0 \\ M_{S,L} & M_S^{(F)} & M_{S,U} \\ 0 & M_{U,S} & M_U \end{bmatrix} \begin{bmatrix} \ddot{q}_L^{(s)} \\ \ddot{q}_s^{(F)} \\ \ddot{q}_U^{(s)} \end{bmatrix} + \begin{bmatrix} K_L & K_{L,S} & 0 \\ K_{S,L} & K_S^{(F)} & K_{S,U} \\ 0 & K_{U,S} & K_U \end{bmatrix} \begin{bmatrix} q_L^{(s)} \\ q_s^{(F)} \\ q_U^{(s)} \end{bmatrix} = \begin{bmatrix} f_L(t) \\ f_s^C \\ f_U(t) \end{bmatrix}, \quad (2)$$

where stiffness submatrix $K_S^{(F)}$ includes all coupling matrices K_{qg} inside the segment. The force subvector f_s^C of dimension $4GR$ expresses the coupling forces in contact nodal points of the outer rods of segment s with adjacent segments $s-1$ and $s+1$. The displacements of the end-nodes of the rod r in the segment s coupled with plates can be expressed by the displacements of the lower and upper rigid plates in the form

$$q_{r,X}^{(s)} = T_{r,X}^{(s)} q_X, \; T_{r,X}^{(s)} \in \Re^{4,6}, \; r = 1,\dots,R, s = 1,\dots,S, X = L, U. \quad (3)$$

Relations (3) can be expressed for all rods in the segment as

$$q_X^{(s)} = T_X^{(s)} q_X,\; X = L, U,\; s = 1, \ldots, S, \tag{4}$$

where $T_X{}^{(s)} \in \Re^{4R,6}$ are the global transformation rectangular matrices. The second set of equations extracted from (2) and modified using transformation (4) for each segment $s = 1, \ldots, S$

$$M_S^{(F)} \ddot{q}_S^{(F)} + K_S^{(F)} q_S^{(F)} = -M_{S,L} T_L^{(s)} \ddot{q}_L - M_{S,U} T_U^{(s)} \ddot{q}_U - K_{S,L} T_L^{(s)} q_L$$
$$- K_{S,U} T_U^{(s)} q_U + f_s^C \tag{5}$$

allows us to complete the fee rod nodal point displacements.

3 Condensed Mathematical Model of the System

The global model of the package of rods has to large DOF number $n = 4RGS$ for calculation of the dynamic response excited by support plate motion. Therefore we assemble the condensed model of the system using the modal synthesis method [5].

Let the modal properties of the conservative model of the mutually uncoupled arbitrary rod segment with fully restrained rods in immovable support plates be characterized by segment spectral Λ_S and modal V_S matrices of order $4GR$. The vector $q_x^{(F)}$, corresponding to rod contact nodes of segment s with springs, can be approximately transformed in the form

$$q_s^{(F)}(t) = {}^m V_S x_s(t),\; s = 1, \ldots, S\,, \tag{6}$$

where ${}^m V_S \in \Re^{4GR, m_S}$ is segment modal submatrix of chosen m_s master modes included into ${}^m V_S$. The model (5) can be rewritten for each segment using (6) in the form

$$\ddot{x}_s(t) + {}^m\Lambda_S x_s(t) = -{}^m V_S^T (M_{S,L} T_L^{(s)} \ddot{q}_L + M_{S,U} T_U^{(s)} \ddot{q}_U + K_{S,L} T_L^{(s)} q_L$$
$$+ K_{S,U} T_U^{(s)} q_U + f_s^C),$$
$$s = 1, \ldots, S. \tag{7}$$

The models (7) of all segments can be then written in configuration space $x(t) = [x_s(t)]$ of the dimension $m = Sm_S$ as

$$\ddot{x}(t) + {}^m\Lambda x(t) = -{}^m V^T [M_L \ddot{Q}_L(t) + M_U \ddot{Q}_U(t) + K_L Q_L(t)$$
$$+ K_U Q_U(t) + f_C], \tag{8}$$

where $f_C = [(f_1^C)^T, \ldots, (f_S^C)^T]^T$ is global vector of coupling forces between segments and matrices

$$^m\Lambda = \mathrm{diag}[^m\Lambda_S] \in \Re^{m,m}; \ ^mV = \mathrm{diag}[^mV_S] \in \Re^{n,m};$$

$$Y_X = \mathrm{diag}[Y_{S,X} T_X^{(s)}] \in \Re^{n,6S};$$

$$Y = M, K; X = L, U$$

are block diagonal, composed from corresponding matrices of the identical rod segments and $Q_X(t) = [q_X^T(t), \ldots, q_X^T(t)]^T \in \Re^{6S}$, $X = L, U$ are extended vectors of plate displacements.

The global vector of coupling forces between segments can be calculated from identity

$$f_C = -\frac{\partial E_p}{\partial q_F} = -K_C q_F, \ q_F = \left[(q_1^{(F)})^T, \ldots, (q_S^{(F)})^T\right]^T, \tag{9}$$

where E_p is potential (deformation) energy of the spacer grids (springs) between segments (in Fig. 1 are marked by dashed lines). The expressions (9) and $q_F(t) = {}^mVx(t)$ can be substituted in Eq. 8 and then we get the condensed model of the rod package of order $m = Sm_S$

$$\ddot{x}(t) + (^m\Lambda + {}^mV^T K_C {}^mV)x(t) = -{}^mV^T[M_L\ddot{Q}_L(t) + M_U\ddot{Q}_U(t) + K_L Q_L(t) + K_U Q_U(t)]. \tag{10}$$

4 Steady Harmonic Response Excited by Support Plate Motion

Let us consider the rod package in Fig. 1, for which the only excitation source is the harmonic motion of the support plates. In consequence of lightly damped rods and spacer grids, we consider modal damping of the rod segment characterized in the space of modal coordinates x_s by diagonal matrix $D_S = \mathrm{diag}[2D_\nu \Omega_\nu]$ where D_ν are damping factors of natural modes and Ω_ν are eigenfrequencies of the segment with fully restrained rods in immovable support plates. The damping of springs is considered to be proportional to stiffnesses by coefficient β. That being simplifying supposed the conservative condensed model (10) will be completed in the form

$$\ddot{x}(t) + (D + \beta\, {}^mV^T K_C\, {}^mV)\dot{x}(t) + (^m\Lambda + {}^mV^T K_C\, {}^mV)x(t) = -{}^mV^T[M_L\ddot{Q}_L(t) + M_U\ddot{Q}_U(t) + K_L Q_L(t) + K_U Q_U(t)], \tag{11}$$

where $D = \text{diag}[D_S, \ldots, D_S] \in \Re^{m,m}$. Let us consider the harmonic motion of the plates in the complex space $q_L(t) = q_L e^{i\omega t}$ and $q_U(t) = q_U e^{i\omega t}$. Steady response can be expressed as $x(t) = x e^{i\omega t}$, where vector of complex amplitudes is

$$x = -[{}^m\Lambda + i\omega D + (1 + i\omega\beta)^m V^T K_C {}^m V - E\omega^2]^{-1} \cdot$$
$$\cdot {}^m V^T [(K_L - M_L\omega^2) Q_L + (K_U - M_U\omega^2) Q_U], \tag{12}$$

where $Q_X = [q_X^T, \ldots, q_X^T]^T$, $X = L, U$. The complex amplitudes of the free rod node displacements are $q_F = {}^m V x$ or $q_s^{(F)} = {}^m V_S x_s$, $s = 1, \ldots, S$, where $x = [x_1^T, \ldots, x_S^T]^T$.

5 Example

The methodology was tested for steady harmonic response of the package of rods with six rod segments ($S = 6$) linked by three identical spacer grids ($G = 3$) uniformly located between fully restrained ends of rods with length $l = 4$ m. Each segment has 55 identical rods ($R = 55$) in the form of steel tube ($\rho = 7,800$ kgm^{-3}, $E = 2 \cdot 10^{11}$ Pa) with outer radius 4.55 mm and inner radius 4.25 mm. The rod spacing is 13 mm and identical spacer grids are characterized by spring stiffness $k_g = 200$ N/m between adjacent rods. The complex package of rods under consideration has $n = 3{,}960$ DOF. The spectrum of eigenfrequencies is distributed between values $f_1 = 3.51$ Hz and $f_{3960} = 98.33$ Hz and is very crowded.

As an illustration, the frequency response curves of contact points lateral displacements of the random selected rod $r = 10$ in the first segment ($s = 1$) on the level grids $g = 1, 2, 3$ are shown in Fig. 2. The excitation is given by plate

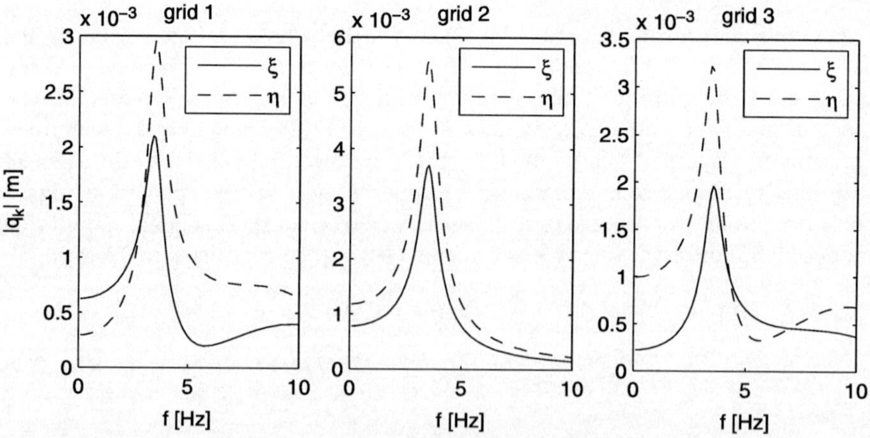

Fig. 2 Frequency response curves of rod lateral displacements in contact points

motions wired in rigid body. We assume the harmonic body motion defined in frequency area $f \in <1; 10>$ Hz by amplitude $x_L = 10^{-3}$ m in horizontal axis on the level of the lower plate and rotation amplitude $\varphi_{xL} = 0.25 \cdot 10^{-3}$ rad around this axis. The considered model (11) with 600 DOF ($m_S = 100$) was used for the calculation of the response curves. The accuracy of condensed model was tested in terms of relative errors of 80 lowest rod package eigenfrequencies defined in the form $\varepsilon_\nu = |f_\nu(m_S) - f_\nu|/f_\nu$, $\nu = 1, \ldots, 80$ where f_ν are eigenfrequencies of the full (noncondensed) model with 3,960 DOF. The relative errors ε_ν for $m_S = 100$ are less than 1%. The highest peaks correspond to the first resonance state and middle contact point displacements.

6 Conclusion

The described method enables to investigate effectively the flexural vibration of the large package of parallel rods fixed between two horizontal plates and linked by spacer grids. The special coordinate system of radial and orthogonal lateral axes for each rod makes possible to separate the complex package of rods into several identical revolved rod segments characterized by identical mass, damping and stiffness matrices.

This new approach, based on the system decomposition and modal synthesis method with reduction of DOF number, was applied to the rod package which is structurally conformable to nuclear fuel assembly. The developed methodology and software in MATLAB will be further used for modelling the nuclear fuel assembly vibration caused by kinematic excitation determined by the support plate motions of the reactor core.

Acknowledgements This work was supported by the research project MSM 4977751303 of the Ministry of Education, Youth and Sports of the Czech Republic.

References

1. Hlaváč, Z., Zeman, V.: The seismic response affection of the nuclear reactor WWER 1000 by nuclear fuel assemblies. Eng. Mech. **17**, 147 (2010)
2. Zeman, V., Hlaváč, Z.: Dynamic response of VVER 1000 type reactor excited by pressure pulsations. Eng. Mech. **15**, 435 (2008)
3. Hlaváč, Z., Zeman, V.: Transverse vibrations of the rod package linked by grids. In: Proceedings of Dynamics of Rigid and Deformable Bodies, p. 1. University J.E. Purkyně, Ústí nad Labem (2010)
4. Géradin, M., Rixen, D.: Mechanical Vibrations. Wiley, New York (1998)
5. Zeman, V.: Vibration of mechanical systems by the modal synthesis method. ZAMM. **75**, 99 (1994)

Combination Resonances and Their Bifurcations in the Nonlinear Vibromachines with a Polynomial Characteristic of Restoring Force and Periodic Excitation

V.N. Belovodskiy and M.Y. Sukhorukov

Abstract In this article main attention is paid to special cases of combination resonances, namely, low order sub- and superharmonic ones.

Mathematical models of the nonlinear machines of one mass schemes (screening, concentration tables, transporting machines) with polynomial characteristics of restoring and dissipative forces and periodic excitation are under consideration. Their dynamics in the zones of sub- and superharmonic resonances are studied. By numerical simulation and with the usage of original software the analysis of the spectral and phase composition of the oscillations is performed, the main types of bifurcations of the stationary regimes are allocated, the role of the biharmonic part of the excitation is discovered.

Keywords Vibromachine • Subharmonic resonance • Superharmonic resonance • Polyharmonic vibration • Numerical analysis

1 Introduction

In the 1970s and 1980s of the previous century, a number of devices implemented in vibromachines for compaction and transporting, peculiarities of the behavior of nonlinear systems were developed. However, inexhaustible interest in the use of polyharmonic vibration in various manufacturing processes, as it is noticed at the exhibitions of technological equipment, stimulates further research in the field of combination resonances. This work was done in this direction.

V.N. Belovodskiy (✉) • M.Y. Sukhorukov
Donetsk National Technical University, Donetsk, Ukraine
e-mail: belovodskiy@cs.dgtu.donetsk.ua; max.sukhorukov@gmail.com

J. Náprstek et al. (eds.), *Vibration Problems ICOVP 2011: The 10th International Conference on Vibration Problems*, Springer Proceedings in Physics 139,
DOI 10.1007/978-94-007-2069-5_32,

2 The Model Under Consideration

Dynamics of vibromachines of one mass scheme with kinematic or force excitation, with accuracy sufficient for practice, is described with a second order differential equation. Assuming a cubic restoring force and taking into account only resistance forces in the elastic ties we shall represent it in a dimensionless form in the following view

$$\frac{d^2\xi}{d\tau^2} + \mu\omega_0\left(1 + \beta\xi + \gamma\xi^2\right)\frac{d\xi}{d\tau} + \left(1 + \beta\xi + \gamma\xi^2\right)\xi = P\cos\eta\tau. \tag{1}$$

3 Research Tools

The research tool is numerical simulation using the developed applications of MATLAB software for constructing bifurcation (amplitude- and phase-frequency) characteristics, spectral analysis and phase composition of the stationary oscillations, identifying basins of attraction of periodic regimes [1].

The program for constructing bifurcation diagrams, for example, amplitude- and phase-frequency characteristics (AFC and PFC), is based on the implementation of the harmonic balance method to the analysis of Eq. 1, whose solution is sought in the complex form. The program provides the creation of a system of nonlinear algebraic equations and its solution with successive changes of one of the parameters of the system. Getting all the solutions in the search area is provided by a certain amount of initial test points, the formation of which is made using a Quasi-Random LpTau sequence. The bifurcation points are defined by monitoring the change of sign of the Jacobian of the system of the equations. The program of the spectral and phase analysis is based on the numerical solution of the Cauchy problem for Eq. 1, finding its steady solutions and of the subsequent numerical expansion in finite Fourier series of the form

$$\xi(\tau) = \sum_{k=0}^{N} A_k \cos\left(k\eta\tau - \varphi_k\right).$$

The program of constructing basins of attraction of periodic regimes, or more precisely their sections $\tau = \tau_0$, is based on brute force of the initial points in the given area and further constructing their orbits using the Poincare map.

4 Research Methods

Initially, by construction of amplitude- and phase-frequency characteristics frequency ranges are allocated, in which the stationary regimes of motions have pronounced polyharmonic structure. Then, by changing other parameters of the system we construct bifurcation curves. A subsequent construction of the basins of attraction gives information about stability margin. The values chosen for the simulation parameters correspond to their real range in the vibromachines of technological purposes.

5 Research Results

In Figs. 1–4 some results are presented. The harmonic components whose amplitudes were greater than 1% of the maximum one were taken into account. We note some features and formulate the following conclusions.

1. For symmetric characteristics ($\beta = 0$) from the number of the lower ones we managed to find only the resonances of the order of 1:3 (Fig. 1a, b). The presence of three pairs of subharmonic regimes is characteristic. The profiles of each group (a stable one is marked by a solid line, an unstable one – by a dashed line) differ in phase shift of harmonic components and coincide in parallel transfer, i.e. $\xi(\tau) = \xi(\tau - s)$. The elementary analysis gives grounds for generalization and allows us to formulate a hypothetical statement: In nonlinear system (1) the set of subharmonic motions of order $1 : n$ may be separated into one or more disjoint classes. Each class contains $2n$ regimes and only n of them are stable. Regimes of stable and unstable groups are identical relatively to the shear and coincide under the change of variable $\tau := \tau + 2\pi m/\omega$, where $m = 0, 1, \ldots, (n - 1)$. Or, what is the same, the phases of their harmonics differ by $2\pi km/n$, where k is the number of a harmonic, m is the number of a regime.
2. Basins of attraction (BOA, Fig. 1c) show a significant margin of stability motions of the order of 1:3. This is particularly evident in the fact that the size of the core of sections $\tau = 0$ of their basins of attraction are in the range 0.5 .. 0.6 and are quite comparable to the core of the principal regime. Judging from our statement mentioned in previous conclusion they can be regarded as sections $\tau = \tau_0 + 2\pi/n$ of the basin of attraction of any of the regimes of the stable group.
3. Figure 2 shows the bifurcation diagrams for the 1:3 resonance. Among the features we note its escalation with the introduction of asymmetry β characteristics (Fig. 2a), as well as the extreme nature of the dependence when P changes (Fig. 2b).
4. After decreasing of the resistance in the system the subharmonic regimes of 1:2 were discovered (Fig. 3). Their number and character confirms the above formulated hypothesis.
5. Asymmetry of the restoring force (Fig. 4a) and asymmetry of excitation (Fig. 4b) may be considered as factors which facilitate the appearance of the resonance 1:2. The last figure demonstrates that even small additional force at the subharmonic frequency can be sufficient for the formation of quite intense motions of the order of 1:2.

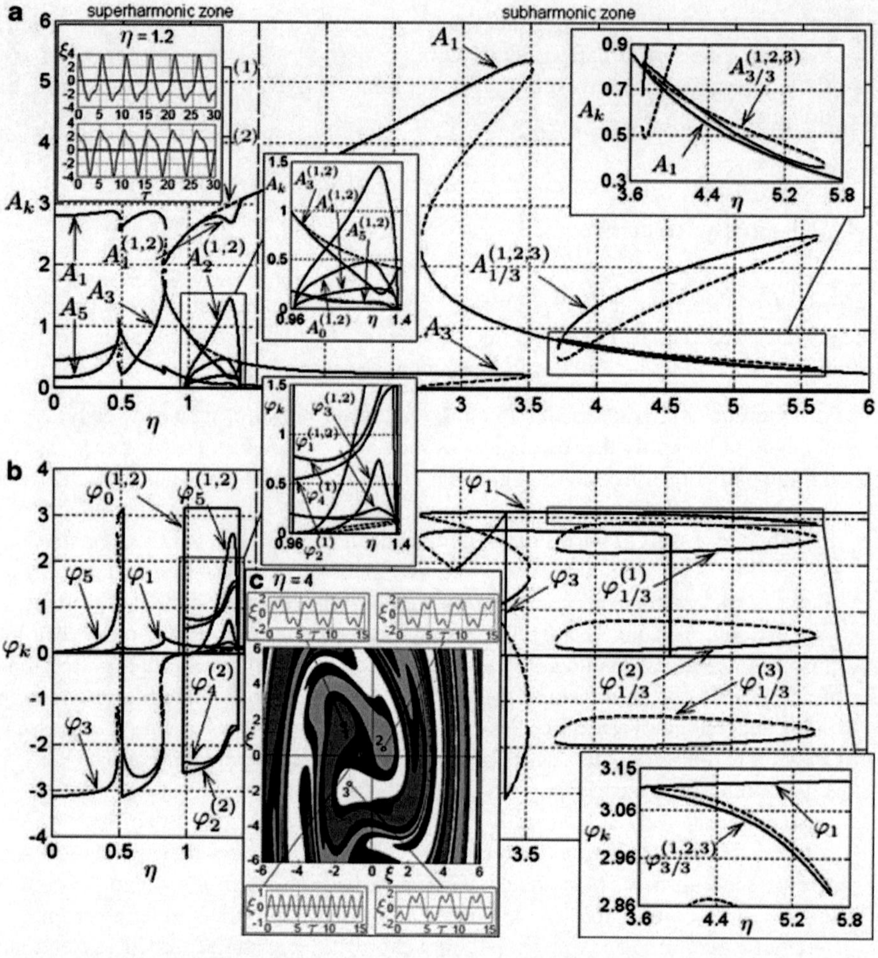

Fig. 1 AFC (**a**), PFC (**b**) and BOA (**c**) of (1) for $\mu\omega_0 = 0.1$, $\beta = 0$, $\gamma = 0.5$, $P = 10$, where $A_k^{(m)}$, $\varphi_k^{(m)}$ – are the amplitude and initial phase of k-th harmonic of m-th regime

6. Analyzing the superharmonic zone (Fig. 1a, b) one can note that these motions have diverse spectral structure and appear quite "naturally" under smooth changing of the frequency of the excitement.
7. Note "power" features of subharmonic regimes. Let us make an experiment. Consider the value of $\eta = 3.65$ (Fig. 3). It corresponds to $A_{1/2} \approx 2.16$, $A_{2/2} \approx 1.08$. Putting in (1) $\beta = \gamma = 0$, we obtain a linear vibromachine. We use the estimated ratio of $A \approx P/(\eta^2 - 1)$, which is valid for $\eta \gg 1$. We find that for the formation of polyharmonic oscillations in a linear vibromachine, shaped in such a way, it must be biharmonic excitation with the following values of its constituent power: $P_{1/2} \approx 5.03$, $P_1 \approx 13.31$. Thus, their total value amounts

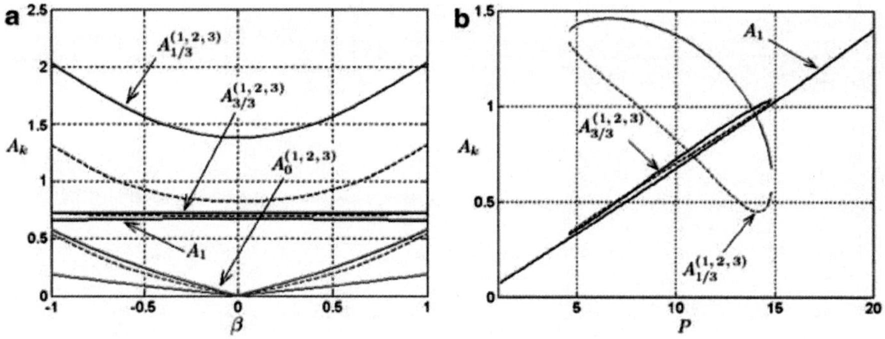

Fig. 2 Bifurcation curves for 1:3 under the varying of: (**a**) β; (**b**) P

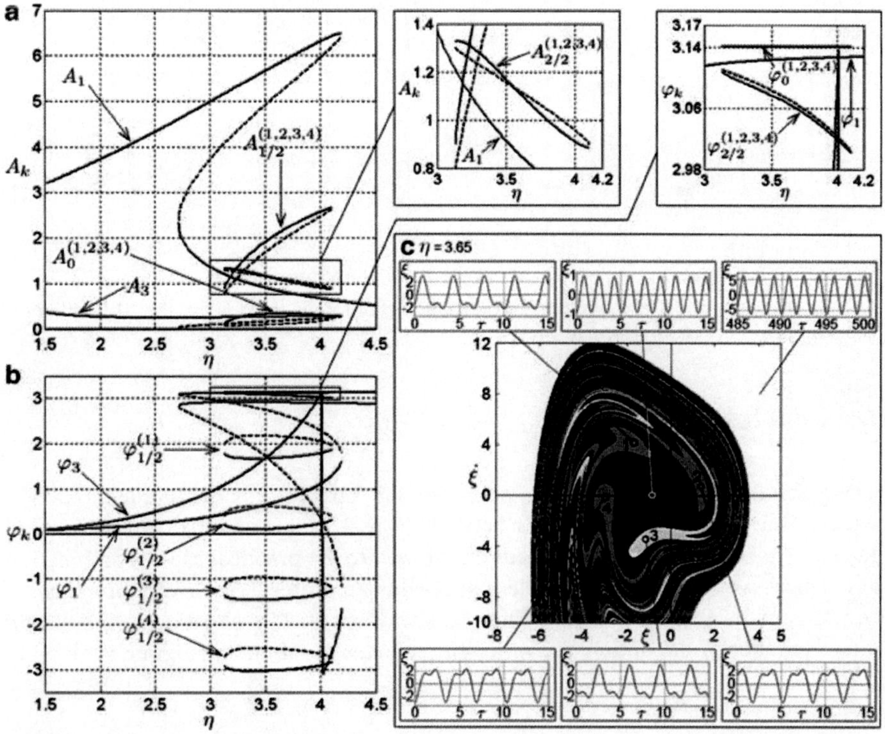

Fig. 3 AFC (**a**), PFC (**b**) and BOA (**c**) of (1) for 1:2 and $\mu\omega_0 = 0.05$

to 18.34 units versus 10 units in the non-linear one. This circumstance allows us to hope for a perceived decline in energy consumption vibromachines which operate in the regime of subresonances. Additional simplification of design, which accompanies the use of one vibration exciter instead of two, it seems obvious.

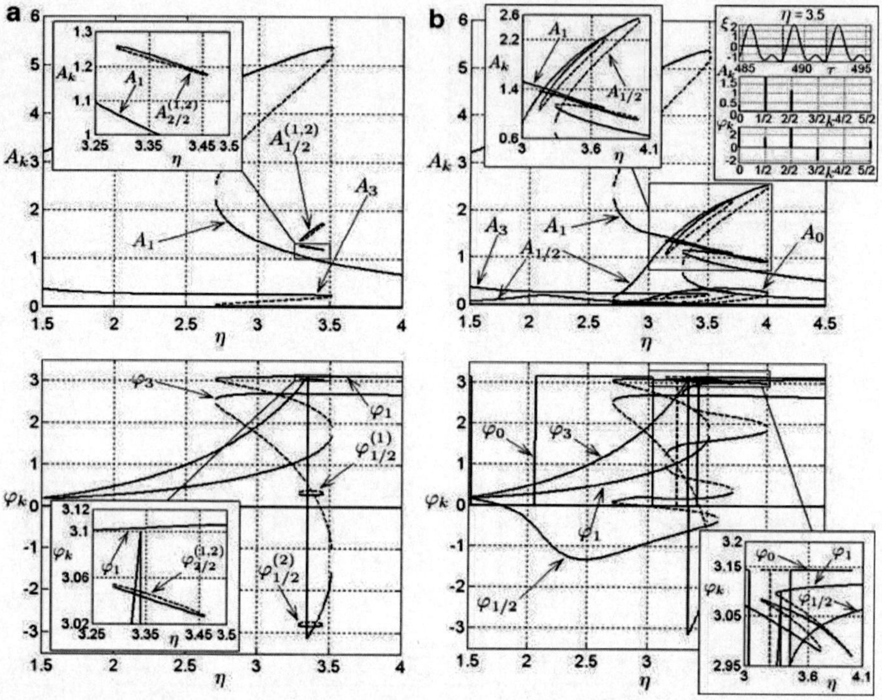

Fig. 4 AFC and PFC for 1:2 under: (**a**) asymmetric restoring force $\beta = 0.1$; (**b**) additional excitation $P_{1/2}\cos(0.5\eta\tau)$, where $P_{1/2} = 0.5$

6 Conclusions

The given results demonstrate the polyharmonic possibilities of combination resonances and some reserves of their control for real values of force and dissipation factors. Among the regular questions on the way to the practical use of combination resonances we may refer the problem of reaching a project regime of vibromachine in terms of its limited stability. The ideal way out of this situation is the elastic characteristics of the option, for which the selected combination regime is globally stable.

References

1. Belovodskiy, V.N., Sukhorukov, M.Y.: The Nonlinear Vibromachines, Superharmonic Resonances and Polyharmonic Vibrations. Advanced Technologies and Systems Engineering: Scientific Works, vol. 40, pp. 20–26. DonNTU, Donetsk (2010) (in Russian)

A Hollow Cylinder Problem in Microstretch Theory

Esin Inan and Ahmet Kiris

Abstract A hollow cylinder micro-damaged tube under pressure problem is discussed. Micro damage is modeled with Eringen's microstretch theory and the displacement and its comparison with classical result are given.

Keywords Microstretch • Damage • Vibration • Lamé Problem

1 Introduction

In the present work, an example is given for a microstructural model of damage. This model was introduced by Markov [1]. He considered the damage process as a generation and growth of microdefects as microvoids. The starting point of this though is the observation that the microcracking introduces micro-distortion tensor $\boldsymbol{\alpha}$ as an additional degree of freedom in the elementary volume of the solid, kinematically independent of the displacement field $\boldsymbol{u}\,(\boldsymbol{x})$. This tensor seems at the first sight as a geometrical characteristic of the damaged solid. The tensor α has an equivalent kinematical definition as the averaged value of distortion generalized by the set of microcracks in the small volume element. With this definition, the damaged solid becomes a micromorphic continuum in the sense of Eringen and Suhubi [2]. Thus, the micro-distorsion tensor may be defined as the damage tensor. Then the basic results concerning the kinematics and the dynamics of the

E. Inan (✉)
Arts and Science Faculty, IŞIK University, Şile, Istanbul, Turkey
e-mail: inan@isikun.edu.tr

A. Kiris
Faculty of Science and Letters, İTÜ, Maslak, Istanbul, Turkey
e-mail: kiris@itu.edu.tr

J. Náprstek et al. (eds.), *Vibration Problems ICOVP 2011: The 10th International Conference on Vibration Problems*, Springer Proceedings in Physics 139,
DOI 10.1007/978-94-007-2069-5_33,

micromorphic continua may be employed to damaged solid. More details of this model are given in [1] and developed in [3].

When switching from a macroscopic to a microscopic approach, a difficulty occurs due to the lack of sufficient knowledge of materials. This seriously limits the development of the microscaled models. To cope this unattractive feature, some simplifications are considered such as micropolar, microstretch or microisotropic theories. But, it is better to use the microstretch theory than others in damage problems since taking into account microstretching of the voids is more meaningful than the microrotation of them. In literature, some damage problems are solved by microstretch theory [4, 5], but the results are all given in terms of unknown material coefficients of microstretch medium.

To determine above mentioned unknown material coefficients, we investigated the vibration of rectangular plates modeled by microstretch theory and a method is presented for the identification of the upper bounds of microstretch elastic moduli in [5]. The frequencies of the plates are obtained by extending the Ritz method to the microstretch theory [5].

Here, we considered a hollow circular tube unbounded along its axis which occupies the region $R_1 \leq r \leq R_2$ as an example. The tube is subjected to pressure at inside and outside surfaces. General formulation of this problem is given in [3]. In this paper, we used microstretch material properties obtained from [5] for damaged material and radial displacements are obtained and then comparison with classical elasticity is given.

2 The Analysis of the Plate Vibration

2.1 Fundamental Equations

In the absence of body forces and body moments, the equations of motion in a linear homogeneous isotropic microstretch elastic solid are as follows

$$\begin{aligned}
&\left(c_1^2 + c_3^2\right) \nabla\nabla \cdot \boldsymbol{u} - \left(c_2^2 + c_3^2\right) \nabla \times \nabla \times \boldsymbol{u} + c_3^2 \nabla \times \boldsymbol{\phi} + \bar{\lambda}_0 \nabla\theta = \ddot{\boldsymbol{u}} \\
&\left(c_4^2 + c_5^2\right) \nabla\nabla \cdot \boldsymbol{\phi} - c_4^2 \nabla \times \nabla \times \boldsymbol{\phi} + \omega_0^2 \nabla \times \boldsymbol{u} - 2\omega_0^2 \boldsymbol{\phi} = \ddot{\boldsymbol{\phi}} \\
&c_6^2 \Delta\theta - c_7^2 \theta - c_8^2 \nabla \cdot \boldsymbol{u} = \ddot{\theta}
\end{aligned} \tag{1}$$

where, λ, μ are Lamé constant and shear modulus, $\kappa, \alpha, \beta, \gamma$ are the micropolar constants, λ_0, λ_1 and a_0 are the microstretch constants, j is the micro-inertia, and $\boldsymbol{u}$, $\boldsymbol{\phi}$ and θ are the displacement and the micro rotation vectors and the microstretch scalar, respectively. And,

$$c_1^2 = \frac{\lambda + 2\mu}{\rho},\ c_2^2 = \frac{\mu}{\rho},\ c_3^2 = \frac{\kappa}{\rho},\ c_4^2 = \frac{\gamma}{\rho j},\ c_5^2 = \frac{\alpha + \beta}{\rho},\ c_6^2 = \frac{2a_0}{\rho j},$$
$$c_7^2 = \frac{2\lambda_1}{3\rho j},\ c_8^2 = \frac{2\lambda_0}{3\rho j},\ \omega_0^2 = \frac{c_3^2}{j},\ \bar{\lambda}_0 = \frac{\lambda_0}{\rho}, \tag{2}$$

Now we consider a homogeneous isotropic rectangular plate with length a, width b and thickness h and the origin of a Cartesian coordinate system $(x_1,\ x_2,\ x_3)$ is located at the middle point of the plate whose normal is in the positive direction of the axes x_3. In general the maximum energy functional of a plate is given as

$$\Pi = V_{max} - T_{max} \tag{3}$$

where V_{max} and T_{max} are the linear elastic strain energy and the kinetic energy of the plate, respectively. Assuming harmonic-time dependence, the periodic displacement, micro rotation and microstretch components of the microstretch plate undergoing free vibration may be written in terms of the amplitude functions as

$$\{\boldsymbol{u}(x, y, z, t), \boldsymbol{\phi}(x, y, z, t), \theta(x, y, z, t)\} = \{(x, y, z),$$
$$\boldsymbol{u}\Phi(x, y, z), \Theta(x, y, z)\}\, e^{i\omega t}. \tag{4}$$

For brevity, the clear forms of the energy function are not given here and can be found in [5] with more details. Here, we used triplicate series of Chebyshev polynomials multiplied by a boundary function, for each amplitude functions of (3),

$$U_1(\xi, \eta, \zeta) = F_{U_1}(\xi, \eta) \sum_{i,j,k=1}^{\infty} A_{ijk} P_i(\xi) P_j(\mu) P_k(\varsigma) \tag{5}$$

and similar expressions are valid for $U_1,\ U_2,\ \Phi_1,\ \Phi_2$ and Θ with the corresponding coefficients. Now writing the extreme conditions as

$$\frac{\partial \Pi}{\partial A_{ijk}} = 0,\ \frac{\partial \Pi}{\partial B_{lmn}} = 0,\ \frac{\partial \Pi}{\partial C_{pqr}} = 0,\ \frac{\partial \Pi}{\partial \hat{A}_{\hat{i}\hat{j}\hat{k}}} = 0,\ \frac{\partial \Pi}{\partial \hat{B}_{\hat{l}\hat{m}\hat{n}}} = 0,\ \frac{\partial \Pi}{\partial \hat{C}_{\hat{p}\hat{q}\hat{r}}} = 0,$$
$$\frac{\partial \Pi}{\partial \hat{\hat{A}}_{\hat{\hat{i}}\hat{\hat{j}}\hat{\hat{k}}}} = 0, \tag{6}$$

we obtain the following eigenvalue problem,

$$(\boldsymbol{K} - \Omega^2 \boldsymbol{M})\,\boldsymbol{Z} = \boldsymbol{0}. \tag{7}$$

Here the column vector Z may be written with its sub-column vectors

$$Z = \left\{A, B, C, \hat{A}, \hat{B}, \hat{C}, \hat{\hat{A}}\right\}. \tag{8}$$

And each sub-column vector is in the following form [5]

$$A = \{A_{111,...,}A_{11N,...,}A_{1k1,...,}A_{1kN,...,}A_{I11,...,}A_{IJK}\}. \tag{9}$$

The elements of the stiffness and mass matrix are given in [5].

2.2 *Analysis of the Spectrum of the Frequencies*

The analysis of the vibration problem has shown that some additional frequencies exist among the frequencies obtained by the classical theory that arise due to the microstructure [5]. Besides, it is shown that these additional frequencies disappear when the microstretch constants are taken as zero. The inverse problem is established for the identification of the upper bounds of the microstretch elastic constants as an optimization problem where an error function is minimized. The final solution of the constants $\{\kappa, \alpha, \beta, \gamma, a_0, \lambda_0, \lambda_1\}$ are obtained by using Downhill Simplex Method and one group of results for Gauthier material is obtained as

$$\begin{aligned}
&\kappa = 132.34\ kPa, \quad \alpha = 8.3255 \times 10^{-2}\ kN, \quad \beta = 0.10282\ kN, \\
&\gamma = 3.3349\ kN, \quad a_0 = 15.947\ kN, \quad \lambda_0 = 0.57702\ kPa, \quad \lambda_1 = 34.650\ kPa.
\end{aligned} \tag{10}$$

3 Application on a Lamé's Problem

We consider a hollow circular tube unbounded along its axis, which occupies the region $R_1 \leq r \leq R_2$. The tube is subjected radial compression at its internal and external surfaces;

$$\sigma_r|_{r=R_1} = q_i \qquad i = 1, 2. \tag{11}$$

Field equations and the damage evaluation form are considered as follows

$$\begin{aligned}
&\mu \Delta \boldsymbol{u} + (\lambda + \mu)\, \nabla\nabla \cdot \boldsymbol{u} - \xi_1 \nabla \boldsymbol{\alpha} = 0 \\
&\eta_1 \Delta \boldsymbol{\alpha} + \eta_2 \left[\nabla (\nabla \cdot \boldsymbol{\alpha}) + (\nabla \cdot \boldsymbol{\alpha}) \nabla\right] - \xi_2 \boldsymbol{\alpha} + \xi_1 \boldsymbol{\varepsilon} = 0
\end{aligned} \tag{12}$$

Here $\boldsymbol{u} = \boldsymbol{u}(r)$ is the displacement field and $\boldsymbol{\alpha} = \boldsymbol{\alpha}(r)$ is the damage function. Then we may write [3]

$$\boldsymbol{u} = \boldsymbol{u}(r) = u(r)\boldsymbol{e}_r, \quad \boldsymbol{\varepsilon} = \nabla u = \left(u' - \frac{u}{r}\right)\boldsymbol{e}_r\boldsymbol{e}_r + \frac{u}{r}\boldsymbol{I}, \tag{13}$$

and

$$\boldsymbol{\alpha} = \boldsymbol{\alpha}(r) = a(r)\boldsymbol{e}_r\boldsymbol{e}_r + b(r)\boldsymbol{I}. \tag{14}$$

Now substituting these expressions into the above fundamental equations, after some mathematical manipulations, we get

$$\begin{aligned}
&(\lambda + 2\mu)\left(u' + \frac{u}{r}\right)' - \xi_1\chi = 0\\
&\eta_1\left(a'' + \frac{a'}{r} - \frac{4a}{r^2}\right) + 2\eta_2 r\left(\frac{\chi}{r}\right)' + \xi_1 r\left(\frac{u}{r}\right)' - \xi_2 a = 0\\
&\eta_1 r\left(b'' + \frac{b'}{r} + \frac{2a}{r^2}\right)' + 2\eta_2 r\left(\frac{\chi}{r}\right)' + \xi_1 r\left(\frac{u}{r}\right)' - \xi_2 r b' = 0
\end{aligned} \tag{15}$$

and extracting the later from previous one gives

$$\eta_1\left[\left(a'' + \frac{a'}{r} - \frac{4a}{r^2}\right) - r\left(b'' + \frac{b'}{r} + \frac{2a}{r^2}\right)'\right] - \xi_2\left(a - rb'\right) = 0. \tag{16}$$

The structure of this result suggests us to introduce new function F as

$$F = F(r) = a(r) - rb'(r). \tag{17}$$

Then we can write

$$F'' - \frac{1}{r}F' - \frac{\xi_2}{\eta_1}F = 0. \tag{18}$$

The solution of (18) is

$$F(r) = r\left[C_1 I_1(\kappa_1 r) + C_2 K_1(\kappa_1 r)\right] \tag{19}$$

where I_1 and K_1 are modified Bessel functions and $\kappa_1^2 = \xi_2/\eta_1$. Now after some algebra on Eq. 15b, c we arrive

$$\eta_1\left(\psi'' + \frac{\psi'}{r}\right) + 2\eta_2\left(\chi' + \frac{\chi}{r}\right) + \xi_1\left(u' + \frac{u}{r}\right) - \xi_2\psi = 0 \tag{20}$$

Here $\psi(r) = a(r) + 2b(r)$ and $\chi = \psi' + \frac{1}{r}F$. Now, defining $\Psi = \psi'$, we find

$$\left(\Psi' + \frac{\Psi}{r}\right)' - \kappa_2^2\Psi + \kappa_3^2\frac{1}{r}F = 0 \tag{21}$$

where

$$\kappa_2^2 = \frac{\xi_2(\lambda + 2\mu) - \xi_1^2}{(\eta_1 + 2\eta_2)(\lambda + 2\mu)} > 0, \quad \kappa_3^2 = \frac{2\eta_2(\lambda + 2\mu)\kappa_1^2 + \xi_1^2}{(\eta_1 + 2\eta_2)(\lambda + 2\mu)} > 0 \tag{22}$$

and $\kappa_3^2 = \kappa_1^2 - \kappa_2^2$. Thus, the general solution can be written as

$$\Psi = D_3 I_1(\kappa_2 r) + D_4 K_1(\kappa_2 r) + K_0(\kappa_1 r). \tag{23}$$

Since $\Psi = a'(r) + 2b'(r)$, then we get

$$a(r) + 2b(r) = \frac{D_3}{\kappa_2} I_0(\kappa_2 r) - \frac{D_4}{\kappa_2} K_0(\kappa_2 r) + \frac{1}{\kappa_1} K_0(\kappa_1 r) + C_5 \tag{24}$$

where

$$C_5 = -\frac{D_3}{\kappa_2} I_0(\kappa_2 R_1) + \frac{D_4}{\kappa_2} K_0(\kappa_2 R_1) - \frac{1}{\kappa_1} K_0(\kappa_1 R_1). \tag{25}$$

And after some calculations, we find

$$\begin{aligned} a(r) &= \frac{D_3}{\kappa_2} I_2(\kappa_2 r) - \frac{D_4}{\kappa_2} K_2(\kappa_2 r) + \frac{1}{\kappa_1} K_2(\kappa_1 r) + \frac{2}{\kappa_1}\left[C_1 I_2(\kappa_1 r) - C_2 K_2(\kappa_1 r)\right], \\ b(r) &= \frac{1}{r\kappa_2^2}\left[D_3 I_1(\kappa_2 r) + D_4 K_1(\kappa_2 r)\right] - \frac{1}{r}\frac{1}{\kappa_1^2} K_1(\kappa_1 r) - \frac{1}{\kappa_1}\left[C_1 I_2(\kappa_1 r)\right. \\ &\qquad \left. - C_2 K_2(\kappa_1 r)\right] + \frac{C_5}{2} \end{aligned} \tag{26}$$

Then, finally we get the displacement as

$$\begin{aligned} u(r) = C_6 r + \frac{C_7}{r} + \frac{\xi_1}{\lambda + 2\mu}&\left[\frac{D_3}{\kappa_2^2} I_1(\kappa_2 r) + \frac{D_4}{\kappa_2^2} K_1(\kappa_2 r) + \frac{C_1}{\kappa_1^2} I_1(\kappa_1 r)\right. \\ &\left. + \frac{C_2}{\kappa_1^2} I_1(\kappa_1 r) - \frac{1}{\kappa_1^2} K_1(\kappa_1 r)\right] \end{aligned} \tag{27}$$

The constants $C_1,\ C_2,\ D_3,\ D_4,\ C_6,\ C_7$ can be found from stress boundary conditions.

As an example, the expression of $u(r)$ is shown on the Fig. 1a for a Gauthier material with the following initial conditions:

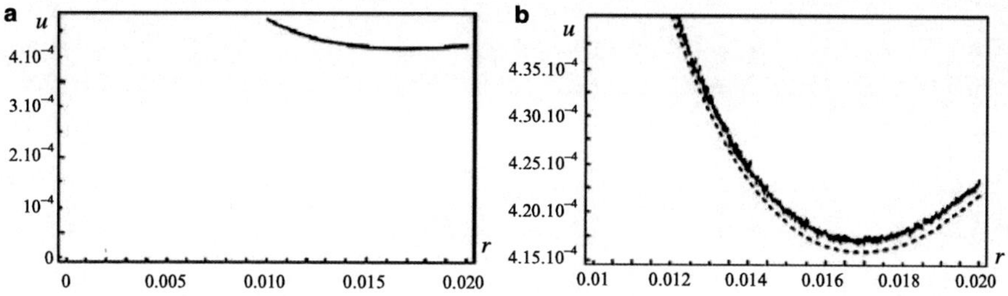

Fig. 1 The displacement in (**a**) and it's comparison with classical elasticity solution (*dashed*) in (**b**)

$$
\begin{aligned}
&R_1 = 0.01\ m,\ R_2 = 0.02\ m,\ q_1 = 0.01\ GPa,\ q_2 = 0.02\ GPa,\ \nu = 0.4,\\
&E = 5.31\ GPa,\ \xi_1 = -\lambda_0 = -5.7702 \times 10^{-7}\ GPa,\\
&\xi_2 = \lambda_1 = 3.465 \times 10^{-5}\ GPa.
\end{aligned}
\tag{28}
$$

Figure 1b shows the difference in functions $u(r)$ found for damaged and undamaged material.

As it is seen from figures, the maximum difference occurs in the mid-part of the hollow cylinder as it is expected.

4 Conclusions

The effect of micro-damages to the displacement is very small as the microstretch material properties is the order of 10^{-5} with corresponding to the classical material constants.

References

1. Markov, K.Z.: Micromorphic models for damaged solids. In: Boehler, J.P. (ed.) Yielding, Damage and Failure of Anisotropic Solids, EGF5, pp. 667–679. Mechanical Engineering Publications, London (1989)
2. Eringen, C.A., Suhubi, E.S: Nonlinear theory of micro-elastic solids-I-II. Int. J. Engng. Sci. **2**, 189–203, 389–404 (1964)
3. Inan, E., Markov, K.Z.: On a Lame's problem in the micromorphic theory of anizotropic damage. Bull. Tech. Univ. İstanbul **47**, 365–377 (1994)
4. Inan, E: Thermo-microstretch model of damage mechanics. In: Inan, E., Markov, K.Z. (eds.) Proceeding of 9th International Symposium on Continuum Models and Discrete Systems 9, pp. 719–726. World Scientific Publication, Singapore (1998)
5. Inan, E., Kiris, A.: On the identification of microstretch elastic moduli of materials by using vibrational data of plates. Int. Eng. Sci. **46**, 585–597 (2008)

Wave Approach in Discrete-Continuous Systems Torsionally Deformed Taking into Account Position-Dependent Inertia

Amalia Pielorz

Abstract The paper deals with discrete-continuous systems having position-depended inertia. The systems consist of an arbitrary number of elastic elements torsionally deformed connected by rigid bodies. The first rigid body represents an electrical motor. Constant and variable velocities of the motor are considered. The studied problems are nonlinear and they are linearized after appropriate transformations. It is shown that such problems can be investigated by means of the wave approach using the solution of the d'Alembert type. Numerical considerations are done for a three-mass model.

Keywords Discrete-continuous systems • Variable inertia • Wave approach

1 Introduction

In the paper discrete-continuous models of systems torsionally deformed having position-dependent inertia are studied. From papers [1–3] it follows that mechanism with variable inertia play an important role in systems torsionally deformed. Discrete models of mechanisms with variable inertia are considered in [3], two-mass discrete-continuous models are studied in [1, 2] and in [4, 5] multi-mass systems are discussed. In [2, 4, 5] it is assumed that the rigid body representing a motor has a constant velocity.

The aim of the paper is to represent governing equations for multi-mass systems torsionally deformed with variable inertia taking into account the assumption that the rigid body representing a motor can move with a constant or with a variable

A. Pielorz (✉)
Kielce University of Technology, Kielce, Poland
e-mail: apielorz@tu.kielce.pl

J. Náprstek et al. (eds.), *Vibration Problems ICOVP 2011: The 10th International Conference on Vibration Problems*, Springer Proceedings in Physics 139,
DOI 10.1007/978-94-007-2069-5_34, © Springer Science+Business Media B.V. 2011

velocity. In the discussion a wave approach is applied, similarly to studies of discrete-continuous systems in [6–8]. Exemplary numerical calculations are given for a three-mass system.

2 Governing Equations

Let us consider a system consisting of $N+1$ rigid bodies connected by means of shafts torsionally deformed. The x-axis is parallel to the main axis of the system, and its origin coincides with the position of the left-hand end of the first shaft in an undisturbed state at time instant $t=0$. Shaft displacements $\theta_i(x,t)$ and velocities are equal to zero at $t=0$.

The first rigid body represents an electrical motor and it is generally loaded by the motor torque $M_1 = M_0 + K(\Omega_0 - \partial\theta_1/\partial t)$, where M_0 is the nominal torque, Ω_0 is the nominal angular velocity and K is the slope of the motor characteristic. The last rigid body is loaded by an equivalent resistance torque equal to M_0. The rigid bodies in the model are characterized by the mass moment of inertia J_i, $i=1, 2, \ldots, N+1$. The mass moment of inertia of the last rigid body depends on the angular displacement, i.e., $J_{N+1} = J_{N+1}(\theta_N)$. The i-th shaft, $i=1,2,\ldots,N$, is characterized by length l_i, density ρ, shear modulus G and polar moment of inertia I_{0i}.

The determination of angular displacements θ_i of the shafts is reduced to solving N equations

$$\frac{\partial^2\theta_i}{\partial t^2} - c^2\frac{\partial^2\theta_i}{\partial x^2} = 0, \quad i = 1, 2, ..., N \tag{1}$$

with zero initial conditions and with the following boundary conditions in cross-sections where rigid bodies with $i=2, 3, \ldots, N+1$ are located

$$\theta_i(x,t) = \theta_{i+1}(x,t) \text{ for } x = \sum_{k=1}^{i} l_k, i = 1, 2, ..., N-1,$$

$$-J_{i+1}\frac{\partial^2\theta_i}{\partial t^2} - GI_{0i}\frac{\partial\theta_i}{\partial x} + GI_{0,i+1}\frac{\partial\theta_{i+1}}{\partial x} = 0 \quad \text{for } x = \sum_{k=1}^{i} l_k, i = 1, 2, ..., N-1,$$

$$J_{N+1}\frac{\partial^2\theta_N}{\partial t^2} + \frac{1}{2}\frac{\mathrm{d}J_{N+1}}{\mathrm{d}\theta_N}\left(\frac{\partial\theta_N}{\partial t}\right)^2 + GI_{0N}\frac{\partial\theta_N}{\partial x} = -M_0 \text{ for } x = \sum_{k=1}^{N} l_k. \tag{2}$$

The condition in the cross-section $x = l_1 + l_2 + \cdots + l_N$ can have various forms. Here it is assumed in the form proposed in [1, 2].

In the cross-section $x=0$ two cases are discussed

(a) the first rigid body moves with a constant velocity

$$\frac{\partial\theta_1}{\partial t} = \Omega_0 \quad \text{for} \quad x = 0, \tag{3}$$

(b) the first rigid body moves with a variable velocity

$$J_1 \frac{\partial^2 \theta_1}{\partial t^2} - G I_{01} \frac{\partial \theta_1}{\partial x} = M_0 + K \left(\Omega_0 - \frac{\partial \theta_1}{\partial t} \right) \quad \text{for} \quad x = 0, \tag{4}$$

Problems for the both cases are nonlinear. In order to obtain effective numerical results, new unknown functions are introduced

$$\theta_i(x,t) = \Omega_0 t + \alpha_i(x,t) - \frac{M_0}{G I_{0i}} \left(x - \sum_{k=1}^{i} l_k \right), \quad i = 1, 2, \ldots, N, \tag{5}$$

the variable mass moment of inertia $J_{N+1}(\theta_N)$ is expended in a Taylor series around $\Omega_0 t$, and second and higher order terms are neglected. Introducing the transformations (5), the static deformations corresponding to the nominal torque moment M_0 appearing in the last conditions in formulas (2) and in (4) are separated. This leads to determination of functions α_i by solving N wave equations with linear boundary conditions having variable coefficients.

The solutions for α_i are sought in the form of the sum of the functions representing waves propagating in shafts satisfying identically equations of motion for the shafts. Using this approach we have to solve linear equations with a retarded argument having variable coefficients. These equations are similar to those which one can find in [4, 5] in the case of the first rigid body moving with a constant velocity. The equations in [4, 5], i.e., taking into account condition (3), were solved numerically and analytically. In the case of condition (4) only a numerical way is applied here.

3 Numerical Results

Numerical calculations are focused on the effect of variable inertia on the behavior of the three-mass system shown in Fig. 1 using appropriate equations with a retarded argument with $N = 2$. The system is characterized by the following basic nondimensional parameters, [4, 5],

$$K_r = 0.05, \quad B_1 = B_2 = 1, \quad E_2 = 0.8, \quad l_1 = l_2 = 1, \quad a_0 = 1, \quad a_2 = 0.05 \tag{6}$$

and with the mass moment of inertia described by

$$J_3(\Omega_0 t) = a_0 + a_2 \cos(2\Omega_0 t). \tag{7}$$

For the system with parameters (6) the first four natural frequencies are equal $\omega_1 = 0.222$, $\omega_2 = 0.359$, $\omega_3 = 3.158$, $\omega_4 = 3.183$.

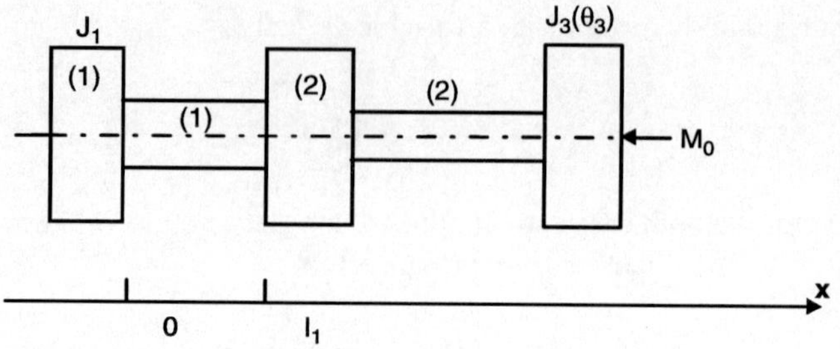

Fig. 1 Three-mass discrete-continuous system

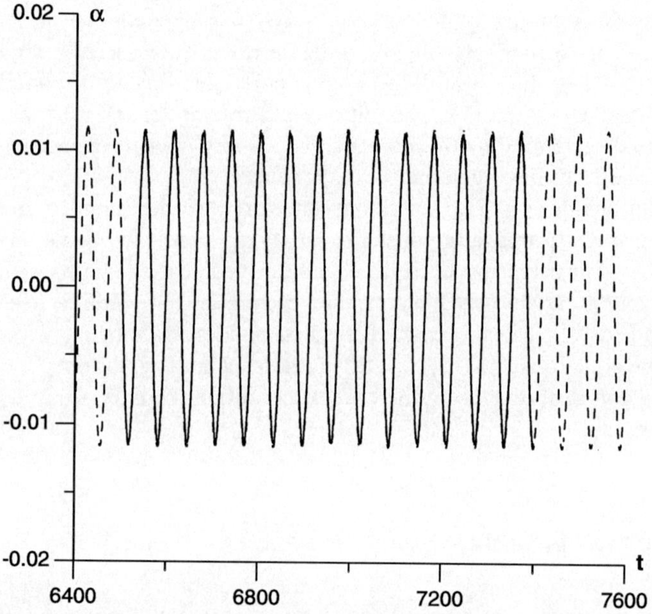

Fig. 2 Comparison of numerical solution (*dashed line*) and analytical solution (*continuous line*)

The efficiency of the applied wave approach is checked in the case of the condition (3) comparing an analytical solution with a numerical one in the steady state of motion. Comparable results for angular displacement α in the cross-section $x=1$ with $\Omega_0 = 0.05$. are given in Fig. 2. The approximate analytical solution for the three-mass system derived by means of the method of a small parameter is presented in [5]. Diagrams in Fig. 2 show a good agreement between results obtained by two different methods.

The analytical solutions in the case when the rigid body representing the motor moving with a constant velocity, enables us to present results in the form of spatial diagrams. Exemplary spatial diagram for the angular displacement α is presented in

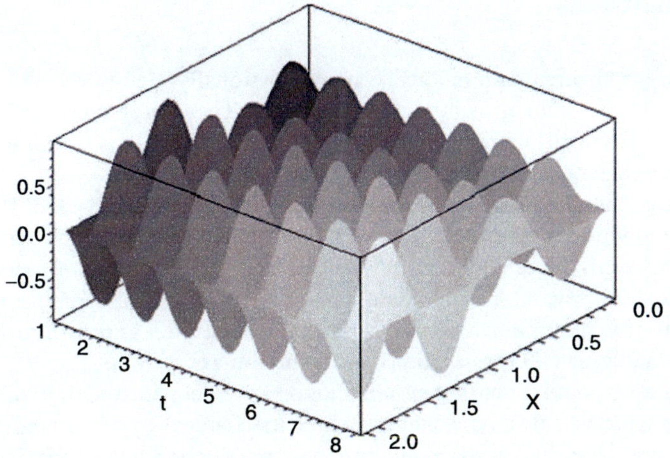

Fig. 3 Spatial diagrams for angular displacement α

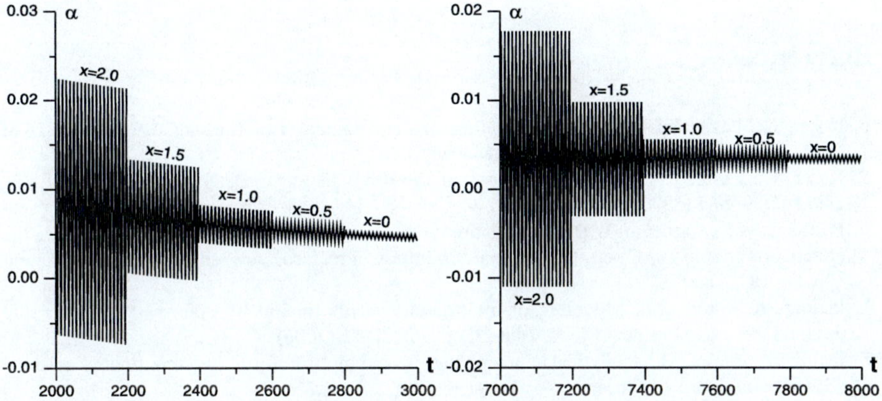

Fig. 4 Angular displacements in the transient and in the steady states

Fig. 3 for $\Omega_0 = 3.14$. The shape of the diagrams for displacements α can be different for different Ω_0. In the case of the assumed value Ω_0 the shape is nonlinear, however for smaller values of Ω_0 appropriate diagrams can be linear along the shaft length, see [5].

Numerical calculations for the system in which the first rigid body moves with a variable speed are performed with the parameters (6), and with the nondimensional parameter K equal to 0.1. Diagrams for angular displacements α in the transient and in the steady states are presented in Fig. 4 for the cross-sections x = 0, 0.5, 1.0, 1.5, 2.0 with $\Omega_0 = 0.3$. The displacement amplitudes decrease with the decrease of x. From Fig. 4 it follows that the steady state is gained for nondimensional time greater than 7,000.

4 Conclusions

Problems for discrete-continuous systems torsionally deformed with position-dependent inertia belong to nonlinear problems. These problems can be reduced to solving partial differential equations with linear boundary conditions having variable coefficients.

It appears that the proposed wave approach can be an effective tool for deriving numerical solutions for the considered system having a rigid body with variable inertia. Diagrams for angular displacements show transient states, steady states for solutions in several cross-sections. Moreover, in the case of the rigid body representing the motor moving with a constant velocity, it was possible to do spatial diagrams including all cross-sections of the three-mass system.

It should be pointed out that systems torsionally deformed with variable inertia play an important role in mechanisms and mechanical systems, and that using the wave approach in the discrete-continues models one can determine required functions in various cross-sections simultaneously.

References

1. Koser, K., Pasin, F.: Continuous modeling of the torsional vibrations of the drive shaft of mechanisms. J. Sound Vib. **188**(1), 17–24 (1995)
2. Koser, K., Pasin, F.: Torsional vibrations of the drive shafts of mechanisms. J. Sound Vib. **199**(4), 559–565 (1997)
3. Pasricha, M.S., Carnegie, W.D.: Formulation of the equations of dynamic motion including the effects of variable inertia on the torsional vibrations in reciprocating engines, Part I. J. Sound Vib. **66**, 181–186 (1979)
4. Pielorz, A., Skóra, M.: Modeling of multimass systems torsionally deformed with variable inertia. Diff. Eq. Nonlinear Mech. **2006**, ID 20758, 1–11 (2006)
5. Pielorz, A., Skóra, M.: Analytical approach in torsional multi-mass discrete-continuous systems with variable inertia. Meccanica **44**, 121–131 (2009)
6. Nadolski, W., Pielorz, A.: Theoretical investigations of nonlinear loads on gear teeth in single gear transmission. Int. J. Mech. Sci. **43**, 299–311 (2001)
7. Pielorz, A.: Nonlinear equations with a retarded argument in discrete-continuous systems. Math. Probl. Eng. **2007**, Article ID 28430, 1–11 (2007). doi:10.1155/2007/28430
8. Pielorz, A.: Vibrations of rod-rigid element systems with a local non-linearity. Int. J. Nonlinear Mech. **45**(9), 888–894 (2010)

On the Dynamics of the Oscillating Moving Point Load Acting on the System Comprising a Pre-stressed Layer and Pre-stressed Half-Space

Surkay D. Akbarov, Nihat İlhan, and Ahmet Temugan

Abstract Within the scope of the piecewise homogeneous body model by the use of the 3D linearized theory of elastic waves in initially stressed bodies the attempt is made for the study of the dynamics of the point located oscillating moving load acting on the system consisting of the pre-stressed covering layer and pre-stressed half-space. The solution method and the algorithm for attaining the numerical results are developed and employed. The numerical results on the stress distribution are presented and discussed. In particular, it is established that the absolute values of the normal stress acting on the interface plane increase with moving velocity of the oscillating point-located load.

Keywords Point located oscillating moving load • Initial stresses • Covering layer • Half-space • Stress distribution

1 Introduction

Within the framework of the piecewise homogeneous body model with the use of the Three-dimensional Linearized Theory of Elastic Waves in Initially Stressed Bodies (TLTEWISB) the critical velocity of the oscillating moving harmonic point load acting on the system comprised a pre-stressed covering layer and a pre-stressed half-plane is studied. This study can be considered as the development of the investigation carried out in the paper [1] for the moving oscillating load, as well as the development of the investigation [2] for the 3D case.

S.D. Akbarov • N. İlhan (✉) • A. Temugan
Yildiz Technical University, Istanbul, Turkey
e-mail: akbarov@yildiz.edu.tr; ilhan@yildiz.edu.tr; temuganim@yahoo.com.tr

J. Náprstek et al. (eds.), *Vibration Problems ICOVP 2011: The 10th International Conference on Vibration Problems*, Springer Proceedings in Physics 139,
DOI 10.1007/978-94-007-2069-5_35,

2 Formulation of the Problem

It is assumed that the covering layer occupies the region $\{-\infty < x_1 < \infty, -h \leq x_2 \leq 0, -\infty < x_3 < +\infty\}$ the related values of which are denoted by upper index (1), and the half-plane occupies the region $\{-\infty < x_1 < +\infty, -\infty \leq x_2 \leq -h$,.. the related values of which are denoted by upper index (2).

The investigation of the considered problem is based on the following field equations of the 3D linearized theory of elastic waves in the pre-stressed elastic bodies [3].

$$\frac{\partial \sigma_{ij}^{(m)}}{\partial x_j} + \sigma_{11}^{(m),0} \frac{\partial^2 u_i^{(m)}}{\partial x_1^2} + \sigma_{33}^{(m),0} \frac{\partial^2 u_i^{(m)}}{\partial x_3^2} = \rho_0^{(m)} \frac{\partial^2 u_i^{(m)}}{\partial t^2}$$

$$\sigma_{ij}^{(m)} = \lambda^{(m)} \theta^{(m)} \delta_{ij} + 2\mu^{(m)} \varepsilon_{ij}{}^{(m)};\ \theta^{(m)} = \varepsilon_{11}{}^{(m)} + \varepsilon_{22}{}^{(m)} + \varepsilon_{33}{}^{(m)},$$

$$\varepsilon_{ij}{}^{(m)} = \frac{1}{2}\left(\frac{\partial u_i^{(m)}}{\partial x_j} + \frac{\partial u_j^{(m)}}{\partial x_i}\right) \quad i, j = 1, 2, 3;\ m = 1, 2 \tag{1}$$

In (1) the conventional notation is used.

The perfect contact conditions and boundary conditions on the upper free surface of the covering layer are written as follows.

$$\sigma_{12}^{(1)}\Big|_{x_2} = 0 = 0,\ \ \sigma_{22}^{(1)}\Big|_{x_2=0} = P_0 \delta(x_1 - Vt) e^{i\omega t} \delta(x_3),$$

$$\sigma_{i2}^{(1)}\Big|_{x_2} = -h = \sigma_{i2}^{(2)}\Big|_{x_2=-h},\ \ u_i^{(1)}\Big|_{x_2=-h} = u_i^{(2)}\Big|_{x_2=-h} \tag{2}$$

where V is a velocity of the point-located moving load, t is a time.

In addition to these, it is supposed that there are boundedness conditions

$$\left|u_i^{(2)}\right|, \left|\sigma_{ij}^{(2)}\right| < M = \textbf{constant} \text{ as } x_2 \to -\infty. \tag{3}$$

This completes the formulation of the problem.

3 Method of Solution

We attempt to use the Lame´ representations for displacements

$$\mathbf{u} = \nabla \cdot \phi + \nabla \times \boldsymbol{\psi},\ \nabla \cdot \boldsymbol{\psi} = 0,\ \mathbf{u} = (u_1, u_2, u_3),\ \boldsymbol{\psi} = (\psi_1, \psi_2, \psi_3) \tag{4}$$

In representation (4) the symbols $\times$ and $\cdot$ show the vector and scalar products of vectors, respectively.

Using the moving coordinate system $x_1' = x_1 - Vt$, $x_2' = x_2$, $x_3' = x_3$ and presenting the sought values as $g(x'_1, x'_2, x'_3, t) = \bar{g}(x'_1, x'_2, x'_3)\, e^{i\omega t}$ we obtain the following equations for functions $\phi^{(m)}$ and $\psi_n^{(m)} (n = 1, 2, 3)$ from equations given in (1).

$$\nabla^2\phi^{(m)} + \frac{\sigma_{11}^{(m),0}}{\lambda^{(m)} + 2\mu^{(m)}}\frac{\partial^2\phi^{(m)}}{\partial x_1^2} + \frac{\sigma_{33}^{(m),0}}{\lambda^{(m)} + 2\mu^{(m)}}\frac{\partial^2\phi^{(m)}}{\partial x_3^2}$$
$$= \frac{1}{\left(c_1^{(m)}\right)^2}\left(V^2\frac{\partial^2\phi^{(m)}}{\partial {x_1}^2} - 2i\omega V\frac{\partial\phi^{(m)}}{\partial x_1} - \omega^2\phi^{(m)}\right)$$
$$\nabla^2\psi_n^{(m)} + \frac{\sigma_{11}^{(m),0}}{\mu^{(m)}}\frac{\partial^2\psi_n^{(m)}}{\partial x_1^2} + \frac{\sigma_{33}^{(m),0}}{\mu^{(m)}}\frac{\partial^2\psi_n^{(m)}}{\partial x_3^2}$$
$$= \frac{1}{\left(c_2^{(m)}\right)^2}\left(V^2\frac{\partial^2\psi_n^{(m)}}{\partial {x_1}^2} - 2i\omega V\frac{\partial\psi_n^{(m)}}{\partial x_1} - \omega^2\psi_n^{(m)}\right)$$
$$\frac{\partial\psi_1^{(m)}}{\partial x_1} + \frac{\partial\psi_2^{(m)}}{\partial x_2} + \frac{\partial\psi_3^{(m)}}{\partial x_3} = 0 \tag{5}$$

where $c_1^{(m)} = \sqrt{\lambda^{(m)} + 2\mu^{(m)}/\rho_0^{(m)}}$ is speed of the dilatation wave, $c_2^{(m)} = \sqrt{\mu^{(m)}/\rho_0^{(m)}}$ is speed of the distortion wave.In Eq. 5, the upper prime in x_1 and x_2 and the over bar in $\phi^{(m)}$ and $\psi^{(m)}$ are omitted.

Now we consider the solutions to Eq. 5. For this purpose, as in the paper [1] we employ the exponential Fourier transformation with respect to the x_1 and x_3 coordinates defined as

$$f_{13F}(s_1, x_2, s_3) = \int_{-\infty}^{+\infty}\int_{-\infty}^{+\infty} f(x_1, x_2, x_3)e^{-i(s_1x_1+s_3x_1)}dx_1dx_3 \tag{6}$$

As a result of this transformation we may obtain from (5) the following equation with respect to $\phi_{13F}^{(m)}(s_1, x_2, s_3)$ and $\psi_{n13F}^{(m)}(s_1, x_2, s_3)$:

$$\phi_{13F}^{(1)} = A_1^{(1)}(s_1, s_3)e^{\gamma_1^{(1)}(s_1,s_3)x_2} + A_2^{(1)}(s_1, s_3)e^{-\gamma_1^{(1)}(s_1,s_3)x_2}$$
$$\phi_{13F}^{(2)} = A_1^{(2)}(s_1, s_3)e^{\gamma_1^{(2)}(s_1,s_3)x_2}$$
$$\psi_{n13F}^{(1)} = B_{1n}^{(1)}(s_1, s_3)e^{\gamma_2^{(1)}(s_1,s_3)x_2} + B_{2n}^{(1)}(s_1, s_3)e^{-\gamma_2^{(1)}(s_1,s_3)x_2}$$
$$\psi_{n13F}^{(2)} = B_{1n}^{(2)}(s_1, s_3)e^{\gamma_2^{(2)}(s_1,s_3)x_2} \tag{7}$$

where

$$\left(\gamma_1^{(m)}\right)^2 = s_1^2\left[1+\frac{\sigma_{11}^{(m),0}}{\lambda^{(m)}+2\mu^{(m)}}\right]+s_3^2\left[1+\frac{\sigma_{33}^{(m),0}}{\lambda^{(m)}+2\mu^{(m)}}\right]$$

$$+\frac{\left(c_2^{(1)}\right)^2}{\left(c_1^{(m)}\right)^2}\left[-V^2s_1^2-2\Omega V s_1+\Omega^2\right]$$

$$\left(\gamma_2^{(m)}\right)^2 = s_1^2\left[1+\frac{\sigma_{11}^{(m),0}}{\mu^{(m)}}\right]+s_3^2\left[1+\frac{\sigma_{33}^{(m),0}}{\mu^{(m)}}\right]+\frac{\left(c_2^{(1)}\right)^2}{\left(c_2^{(m)}\right)^2}\left[-V^2s_1^2-2\Omega V s_1+\Omega^2\right] \tag{8}$$

In (8) the following notation is used $\Omega = \omega h/c_2^{(1)}$

From the algebraic equations we find the aforementioned unknowns and, employing the inverse transform

$$f(x_1,x_2,x_3) = \frac{1}{4\pi^2}\int_{-\infty}^{+\infty}\int_{-\infty}^{+\infty} f_{13F}(s_1,x_2,s_3)e^{i(s_1x_1+s_3x_3)}ds_1ds_3 \tag{9}$$

we determine the displacements and stresses in the components of the considered system.

4 Numerical Result and Discussions

It is known that the stresses in the near vicinity of the point located force have a singularity with order $O(r^{-2})$ where r is a distance from the point at which the force acts. Therefore the foregoing solution will not be available for the cases where $r \to 0$ and this statement is taken into consideration under obtaining numerical results.

The following materials are selected for the considerations: Aluminum (shortly Al) with properties $\rho_0 = 2700\,kg/m^3$, $\nu = 0.35$ $c_1 = 6420m/s, c_2 = 3110m/s$; Steel (shortly St) with properties $\rho_0 = 7860\,kg/m^3$, $\nu = 0.29$, $c_1 = 5890m/s$, $c_2 = 3210m/s$ where the density, Poisson's ratio, the speed of dilatation and distortion waves are denoted by ρ_0, ν, c_1 and c_2 respectively. The numerical investigations have been carried out for the case where the material of the covering layer (half-space) is Al (St). For estimation of the initial stresses in the components of considered system we introduce the parameters

$\eta_{11}^{(m)} = \sigma_{11}^{(m),0}/\mu^{(m)}$ and $\eta_{33}^{(m)} = \sigma_{33}^{(m),0}/\mu^{(m)}$.

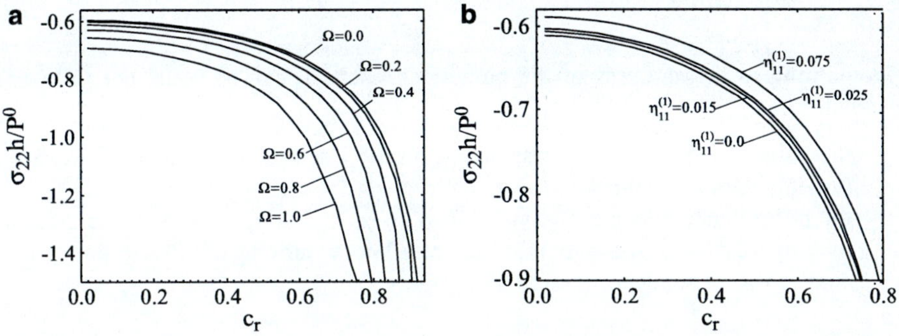

Fig. 1 The dependence between $\sigma_{22}(0)h/P_0$ and c_r for the case where: (**a**) for various Ω under $\eta_{11}^{(1)} = \eta_{33}^{(1)} = \eta_{11}^{(2)} = \eta_{33}^{(2)} = 0.0$; (**b**) for various $\eta_{11}^{(1)}$ under $\Omega = 0.4$, $\eta_{33}^{(1)} = \eta_{11}^{(2)} = \eta_{33}^{(2)} = 0.0$

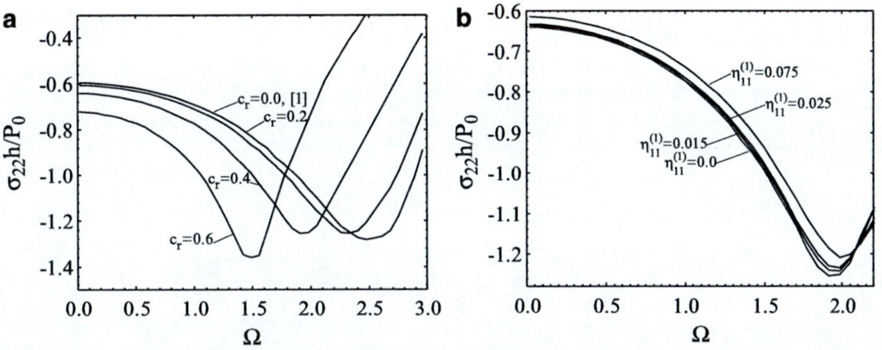

Fig. 2 The dependence between $\sigma_{22}(0)h/P_0$ and Ω for the case where: (**a**) for various c_r under $\eta_{11}^{(1)} = \eta_{33}^{(1)} = \eta_{11}^{(2)} = \eta_{33}^{(2)} = 0.0$; (**b**) for various $\eta_{11}^{(1)}$ under $c_r = 0.4$, $\eta_{33}^{(1)} = \eta_{11}^{(2)} = \eta_{33}^{(2)} = 0.0$

Figure 1 shows the influence of the dimensionless frequency Ω on the graphs of the dependence between $\sigma_{22}h/P_0(= \sigma_{22}^{(1)}(0, -h, 0)h/P_0)$ and dimensionless moving velocity $c_r(= V/c_R^{(2)})$ where $c_R^{(2)}$ is a Rayleigh wave velocity of the half space material. The influence of the dimensionless moving velocity of the external loading on the graphs of the dependencies between $\sigma_{22}h/P_0$ and dimensionless Ω is given in Fig. 2. Note that the graphs given in Figs. 1a and 2a are constructed in the case where $\eta_{11}^{(1)} = \eta_{33}^{(1)} = \eta_{11}^{(2)} = \eta_{33}^{(2)} = 0.0$. But the graphs given in Fig. 1b (Fig. 2b) are constructed in the case where $\Omega = 0.4$ $(c_r = 0.4)$, $\eta_{33}^{(1)} = \eta_{11}^{(2)} = \eta_{33}^{(2)} = 0.0$ and these graphs characterize the influence of the initial stretching of the covering layer in the moving direction of the external loading on the values of the considered stress. The graph obtained in the case $c_r = 0$ (Fig. 2a) coincides with corresponding one obtained in a paper by [1].

5 Conclusions

According to the analyses of the numerical results it can be made the following conclusions:

– the values of the normal stress acting on the interface plane increase with the moving velocity of the oscillating load;
– the dependence between the mentioned stress and frequency of the oscillating moving load has non-monotonic character, but the moving of this oscillating load change this dependence quantitatively, only.

References

1. Emiroglu, I., Tasci, F., Akbarov, S.D.: Dynamical response of two axially pre-strained system comprising of a covering layer and a half space to rectangular time-harmonic forces. Comput. Model. Eng. Sci. **49**(1), 47 (2009)
2. Akbarov, S., Ilhan, N.: Dynamics of a system comprising an orthotropic layer and orthotropic half-plane under the action of an oscillating moving load. Int. J. Solids Struct. **46**, 3873 (2009)
3. Guz, A.N.: Elastic Waves in Bodies with Initial (Residual) Stresses [in Russian], "A.C.K.", Kiev, Ukraine (2004)

On the Stress Field Caused by the Time Harmonic Force in the System Consisting of the Piezoelectric Covering Layer and Piezoelectric Half-Plane

Surkay D. Akbarov and Nihat İlhan

Abstract The Lamb problem is considered for the system consisting of the piezoelectric covering layer and orthotropic half-plane. The investigation is carried out within the framework of the piece-wise homogeneous body model by utilizing the exact equations of motion of the linear theory of electro-elasticity.The plane strain state is considered and it is assumed that between the covering layer and half-plane the perfect contact conditions satisfy. The boundary value problem under consideration is solved by employing Fourier exponential transformation with respect to the coordinate directed along the interface line. The algorithm proposed in a paper (Akbarov S, Ilhan N, Int J Solids Struct 46:3873, 2009) is developed to obtain numerical results on the normal stress distribution on the interface plane. Concrete results are presented and discussed for the piezoelectric materials PZT-5A, PZT-5H, PZT-4 and PZT- 7A and orthotropic half-plane. The mechanical properties of the half-plane material are given through the mechanical constants of the PZT- 5A, PZT-5H, PZT-4 and PZT- 7A. It is established that the piezoelectricity of the covering layer causes to decrease of the absolute values of the normal stresses acting on the interface surface.

Keywords Piezoelectric materials • Time harmonic load • Covering layer • Half-plane • Stress distribution

1 Introduction

The Lamb problem is considered for the system consisting of the piezoelectric covering layer and orthotropic half-plane. The investigation is carried out within the framework of the piece-wise homogeneous body model by utilizing the exact

S.D. Akbarov • N. İlhan (✉)
Yildiz Technical University, Istanbul, Turkey
e-mail: akbarov@yildiz.edu.tr; ilhan@yildiz.edu.tr

J. Náprstek et al. (eds.), *Vibration Problems ICOVP 2011: The 10th International Conference on Vibration Problems*, Springer Proceedings in Physics 139,
DOI 10.1007/978-94-007-2069-5_36,

equations of motion of the linear theory of electro-elasticity. Note that the corresponding investigations in the case where the materials of the constituents are orthotropic ones were made in [1].

2 Formulation of the Problem

The plane strain state is considered in a plane Ox_1x_2. It is assumed that between the piezoelectric covering layer $\{-\infty < x_1 < +\infty, -h < x_2 < 0\}$, the related values of which are denoted by upper index (1), and the orthotropic half-plane $\{-\infty < x_1 < +\infty, -\infty < x_2 < -h\}$, the related values of which are $-\infty < x_3 < \infty\}$ denoted by upper index (2), perfect contact conditions satisfy. On the covering layer the time harmonic point-located force acts.

According to [2], the governing field equations are taken as follows.

$$\frac{\partial \sigma_{ij}^{(k)}}{\partial x_i} = \rho^{(k)} \frac{\partial^2 u_i{}^{(k)}}{\partial t^2}, \quad \frac{\partial D_i^{(k)}}{\partial x_i} = 0$$
$$\sigma_{ij}^{(k)} = C_{ijkl}^{(k)} \gamma_{kl}^{(k)} - e_{kij}^{(k)} E_k^{(k)}, \quad D_i^{(k)} = e_{ikl}^{(k)} \gamma_{kl}^{(k)} - \varepsilon_{il}^{(k)} E_l^{(k)}$$
$$\gamma_{ij}^{(k)} = \frac{1}{2}\left(u_{i,j}^{(k)} + u_{j,i}^{(k)}\right), \quad E_i^{(k)} = -\phi_{,i}^{(k)} \tag{1}$$

where $C_{ijkl}^{(k)}$ is a component of the stiffness matrix of the k-th material, $D_i^{(k)}$ is a component of the electric displacement vector, $E_i^{(k)}$ is a component of the electric field vector, $\phi^{(k)}$ is an electric potential, $e_{kij}^{(k)}$ and $\varepsilon_{il}^{(k)}$ are piezoelectric and dielectric constants, $u_i^{(k)}$ is a component of the displacement vector, $\gamma_{ij}^{(k)}$ is a component of strain tensor, $\sigma_{ij}^{(k)}$ is a component of the stress tensor. We introduce a notation $x = x_1$ and $z = x_2$.

The perfect contact conditions and boundary conditions on the upper free surface of the covering layer are written as follows.

$$\sigma_{xz}^{(1)}\big|_{z=0} = 0; \; \sigma_{zz}^{(1)}\big|_{z=0} = -Pe^{i\omega t}\delta(x); \; \varphi^{(1)}\big|_{z=0} = 0;$$
$$\sigma_{xz}^{(1)}\big|_{z=-h} = \sigma_{xz}^{(2)}\big|_{z=-h}; \; \sigma_{zz}^{(1)}\big|_{z=-h} = \sigma_{zz}^{(2)}\big|_{z=-h};$$
$$u^{(1)}\big|_{z=-h} = u^{(2)}\big|_{z=-h}; \; w^{(1)}\big|_{z=-h} = w^{(2)}\big|_{z=-h}; \; \varphi^{(1)}\big|_{z=-h} = 0 \tag{2}$$

In addition to these, it is supposed that there are boundedness conditions

$$\left|u_i^{(2)}\right|, \left|\sigma_{ij}^{(2)}\right| < M = \textbf{constant} \text{ as } z \to -\infty. \tag{3}$$

This completes the formulation of the problem.

3 Method of Solution

Representing the sought values as $g(x', z', t) = \bar{g}(x', z')\, e^{i\omega t}$ we obtain the following equations of the motion in terms of displacement from Eq. 1:

$$\frac{C_{11}^{(k)}}{C_{44}^{(k)}}\frac{\partial^2 u^{(k)}}{\partial x^2} + \frac{C_{13}^{(k)}}{C_{44}^{(k)}}\frac{\partial^2 w^{(k)}}{\partial x \partial z} + \frac{e_{31}^{(k)}}{C_{44}^{(k)}}\frac{\partial^2 \varphi^{(k)}}{\partial x \partial z} + \left(\frac{\partial^2 u^{(k)}}{\partial z^2} + \frac{\partial^2 w^{(k)}}{\partial x \partial z}\right)$$
$$+ \frac{e_{15}^{(k)}}{C_{44}^{(k)}}\frac{\partial^2 \varphi^{(k)}}{\partial x \partial z} = -\omega^2 \frac{1}{\left(c_2^{(k)}\right)^2} u^{(k)},$$
$$\left(\frac{\partial^2 u^{(k)}}{\partial x \partial z} + \frac{\partial^2 w^{(k)}}{\partial x^2}\right) + \frac{e_{15}^{(k)}}{C_{44}^{(k)}}\frac{\partial^2 \varphi^{(k)}}{\partial x^2} + \frac{C_{13}^{(k)}}{C_{44}^{(k)}}\frac{\partial^2 u^{(k)}}{\partial x \partial z} + \frac{C_{33}^{(k)}}{C_{44}^{(k)}}\frac{\partial^2 w^{(k)}}{\partial z^2}$$
$$+ \frac{e_{33}^{(k)}}{C_{44}^{(k)}}\frac{\partial^2 \varphi^{(k)}}{\partial z^2} = -\omega^2 \frac{1}{\left(c_2^{(k)}\right)^2} w^{(k)}$$
$$\frac{e_{15}^{(k)}}{C_{44}^{(k)}}\left(\frac{\partial^2 u^{(k)}}{\partial x \partial z} + \frac{\partial^2 w^{(k)}}{\partial x^2}\right) - \frac{\varepsilon_{11}^{(k)}}{C_{44}^{(k)}}\frac{\partial^2 \varphi^{(k)}}{\partial x^2} + \frac{e_{31}^{(k)}}{C_{44}^{(k)}}\left(\frac{\partial^2 u^{(k)}}{\partial x \partial z}\right)$$
$$+ \frac{e_{33}^{(k)}}{C_{44}^{(k)}}\left(\frac{\partial^2 w^{(k)}}{\partial z^2}\right) - \frac{\varepsilon_{33}^{(k)}}{C_{44}^{(k)}}\frac{\partial^2 \varphi^{(k)}}{\partial z^2} = 0 \tag{4}$$

where $c_{12}^{(k)} = \sqrt{\mu_{12}^{(k)}/\rho^{(k)}}$

In Eq. 4, the over bar in $u^{(k)}$, $w^{(k)}$ and $\varphi^{(k)}$. are omitted.

Now we consider the solutions to Eq. 4. For this purpose, as in the paper [1] we employ the exponential Fourier transformation with respect to the x coordinate defined as

$$f_F(s, z) = \int_{-\infty}^{+\infty} f(x, z) e^{-isx} dx \tag{5}$$

As a result of this transformation we may obtain from (4) the following equation with respect to $u_F^{(k)}(s, z)$, $w_F^{(k)}(s, z)$ and $\varphi_F^{(k)}(s, z)$:

$$\frac{d^2 u^{(k)}}{dz^2} + \left(\frac{C_{13}^{(k)}}{C_{44}^{(k)}} + 1\right)(is)\frac{dw^{(k)}}{dz} + \left(\frac{e_{31}^{(k)} + e_{15}^{(k)}}{C_{44}^{(k)}}\right)(is)\frac{d\varphi^{(k)}}{dz}$$
$$+ \left(\frac{C_{11}^{(k)}}{C_{44}^{(k)}}(-s^2) + \Omega^2 \frac{\left(c_2^{(1)}\right)^2}{\left(c_2^{(k)}\right)^2}\right) u^{(k)} = 0$$

$$\frac{C_{33}^{(k)}}{C_{44}^{(k)}}\frac{d^2w^{(k)}}{dz^2}+\frac{e_{33}^{(k)}}{C_{44}^{(k)}}\frac{d^2\varphi^{(k)}}{dz^2}+\left(\frac{C_{13}^{(k)}}{C_{44}^{(k)}}+1\right)(is)\frac{du^{(k)}}{dz}$$

$$+\left((-s^2)+\Omega^2\frac{\left(c_2^{(1)}\right)^2}{\left(c_2^{(k)}\right)^2}\right)w^{(k)}+\frac{e_{15}^{(k)}}{C_{44}^{(k)}}(-s^2)\varphi^{(k)}=0$$

$$\frac{e_{33}^{(k)}}{C_{44}^{(k)}}\frac{d^2w^{(k)}}{dz^2}-\frac{\varepsilon_{33}^{(k)}}{C_{44}^{(k)}}\frac{d^2\varphi^{(k)}}{dz^2}+\left(\frac{e_{15}^{(k)}+e_{31}^{(k)}}{C_{44}^{(k)}}\right)(is)\frac{du^{(k)}}{dz}$$

$$+\frac{e_{15}^{(k)}}{C_{44}^{(k)}}(-s^2)w^{(k)}-\frac{\varepsilon_{11}^{(k)}}{C_{44}^{(k)}}(-s^2)\varphi^{(k)}=0 \tag{6}$$

In (6) the following notation is used $\Omega^2=\left(\omega h/c_2^{(1)}\right)^2$. The displacements and electric potential in Eq. 6 are presented as follows.

$$u^{(k)}=A^{(k)}e^{k^{(k)}z},\quad w^{(k)}=B^{(k)}e^{k^{(k)}z},\quad \varphi^{(k)}=C^{(k)}e^{k^{(k)}z} \tag{7}$$

Taking relation (7) into account we obtain the following relations between the unknown constants in (7).

$$A^{(k)}\left(k^2+\left(\frac{C_{11}^{(k)}}{C_{44}^{(k)}}(-s^2)+\Omega^2\frac{\left(c_2^{(1)}\right)^2}{\left(c_2^{(k)}\right)^2}\right)\right)+B\left(\frac{C_{13}^{(k)}}{C_{44}^{(K)}}+1\right)(is)k$$

$$+C\left(\frac{e_{31}^{(k)}+e_{15}^{(k)}}{C_{44}^{(k)}}\right)(is)k=0$$

$$A\left(\frac{C_{13}^{(k)}}{C_{44}^{(k)}}+1\right)(is)k+B\left(\frac{C_{33}^{(k)}}{C_{44}^{(k)}}k^2+\left((-s^2)+\Omega^2\frac{\left(c_2^{(1)}\right)^2}{\left(c_2^{(k)}\right)^2}\right)\right)$$

$$+C\left(\frac{e_{33}^{(k)}}{C_{44}^{(k)}}k^2+\frac{e_{15}^{(k)}}{C_{44}^{(k)}}(-s^2)\right)=0$$

$$A\left(\left(\frac{e_{15}^{(k)}}{C_{44}^{(k)}}+\frac{e_{31}^{(k)}}{C_{44}^{(k)}}\right)(is)k\right)+B\left(\frac{e_{33}^{(k)}}{C_{44}^{(k)}}k^2+\frac{e_{15}^{(k)}}{C_{44}^{(k)}}(-s^2)\right)$$

$$+C\left(-\frac{\varepsilon_{33}^{(k)}}{C_{44}^{(k)}}k^2-\frac{\varepsilon_{11}^{(k)}}{C_{44}^{(k)}}(-s^2)\right)=0 \tag{8}$$

Using Eqs. 7 and 8 and the conditions (2) and (3) we determine the functions $u_F^{(k)}(s,z)$, $w_F^{(k)}(s,z)$ and $\varphi_F^{(k)}(s,z)$. To determine the original of these functions we employ the algorithm developed in the paper [1] is used.

4 Numerical Results

It is known that the stresses in the near vicinity of the point located force have a singularity with order $O\left(r^{-1}\right)$ where r is a distance from the point at which the force acts. Therefore the foregoing solution will not be available for the cases where $r \to 0$ and this statement is taken into consideration under obtaining numerical results.

Concrete numerical results are obtained for the materials the values of the mechanical, piezoelectric and dielectric constants of which, according a book are given in Table 1.

Consider numerical results obtained for the dependence between the stress $\sigma_{22}h/P\left(=\sigma_{22}^{(2)}(0,-h)h\right)/P)$ and dimensionless frequency of the oscillation of the external force. The graphs of these dependencies are illustrated in Fig. 1. In this case the graphs given in Fig. 1a–d are constructed for the cases where the material of the covering layer is PZT- 5A, PZT-5H, PZT-4 and PZT- 7A, respectively. In each noted case the material of the half-plane is indicated in figures field on the graphs. In the mentioned figure the solid lines correspond the case where $e_{kij}^{(1)} \neq 0$ and $\varepsilon_{il}^{(1)} \neq 0$, but $e_{kij}^{(2)} = 0$ and $\varepsilon_{il}^{(2)} = 0$, i.e. the solid lines correspond the case where the piezoelectricity and dielectricity of the covering layer material are taken into account, but the piezoelectricity and dielectricity of the half-plane material are not taken into account. The graphs given through the dashed lines correspond the case where $e_{kij}^{(1)} = 0$, $\varepsilon_{il}^{(1)} = 0$, $e_{kij}^{(2)} = 0$ and $\varepsilon_{il}^{(2)} = 0$.

Table 1 Properties of materials

	C_{44}*10^{10} N/m^2	C_{11}*10^{10} N/m^2	C_{13}*10^{10} N/m^2	C_{33}*10^{1} N/m^2	ρ kg/m^3
PZT-5A	2.11	12.1	7.54	11.1	7,750
PZT-5H	2.3	12.6	8.39	11.7	7,500
PZT-4	2.56	13.9	7.40	11.5	7,500
PZT-7A	2.53	14.8	8.13	13.1	7,600
	e_{31} C/m^2	e_{33} C/m^2	e_{15} C/m^2	ε_{11}*10^{-8} C/Vm	ε_{33}*10^{-8} C/Vm
PZT-5A	−5.4	15.8	12.3	0.811	0.735
PZT-5H	−6.5	23.3	17	1.505	1.302
PZT-4	−5.2	15.1	12.7	0.646	0.562
PZT-7A	−2.1	9.5	9.2	0.407	0.208

Consequently, the difference between the graphs illustrated by solid and dashed lines shows the influence of the piezoelectricity and dielectricity on the character of the considered stress-frequency relation.

5 Conclusions

According the numerical results given in Fig. 1, one can make the following conclusions:

- piezoelectricity and dielectricity of the covering layer material causes to decrease the absolute values of the normal stress acting on the interface plane. This decreasing can be explained with increasing of the stiffness of the covering layer material by taking the piezoelectricity and dielectricity of that into account,
- the relation between the considered stress and oscillation frequency of the external force, as in [1], has non-monotonic character. But in the case under consideration the values of the oscillating frequency of the external force under which the absolute values of the stress has its maximum, decrease.

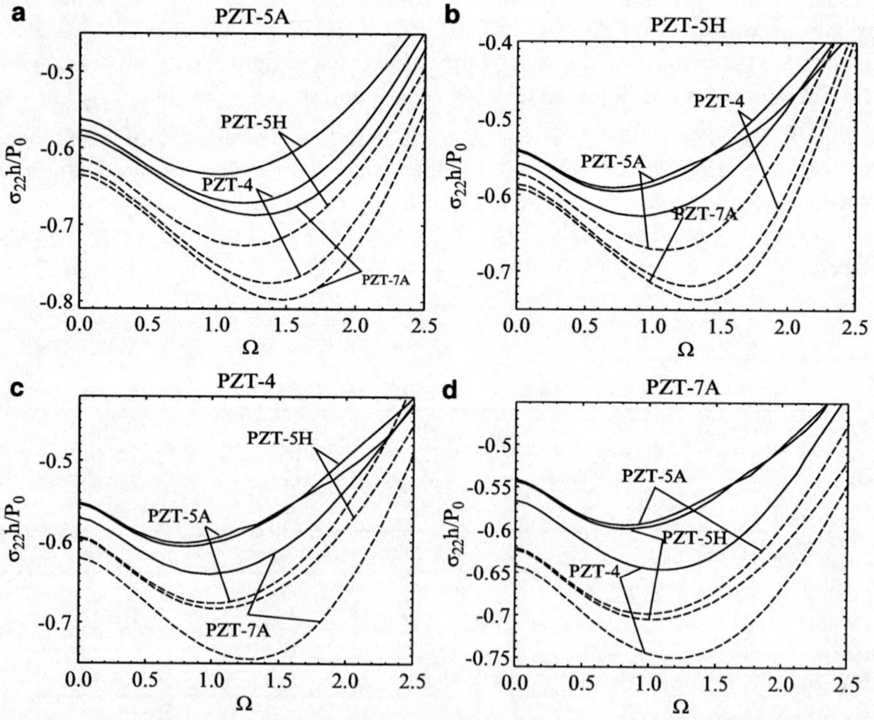

Fig. 1 The dependence between $\sigma_{22}(0)h / P_0$ and Ω

References

1. Akbarov, S., İlhan, N.: Dynamics of a system comprising an orthotropic layer and orthotropic half-plane under the action of an oscillating moving load. Int. J. Solids Struct. **46**, 3873 (2009)
2. Yang, J.: An Introduction to the Theory of Piezoelectricity. Springer, New York, USA (2005)

Contribution to Modal and Spectral Interval Finite Element Analysis

Milan Sága, R. Bednár, and M. Vaško

Abstract Our paper deals with a non-probabilistic computational approach for mechanical systems with structural uncertainties. Uncertainties are considered as bounded possible values – intervals. The main goal is to propose algorithms for modal and spectral interval computations on FE models. An application of the chosen approaches is presented, i.e. the first one a simple combination of only inf-values or only sup-values; the second one presents full combination of all inf-sup values; the third one uses the optimization process as a tool for finding out a inf-sup solution and last one is Monte Carlo technique as a comparison tool.

Keywords Uncertain structural parameters • MATLAB • Monte Carlo • Interval arithmetic

1 Introduction

Generally, it is possible to say that each engineering problem encounters uncertainties in various forms, i.e. in geometrical parameters, material constants, loads, etc. Many of those uncertainties are based on physical imperfections, the general diversity and complexity of natural phenomena and of course our ignorance or inability to precisely describe characteristics of the investigated problem.

Uncertain parameters appear mostly as random variables and thus are described in the terms of stochastic approach. But without the knowledge of the probability density and the nature of distribution we are forced to use another approach, which could describe the parameters with the mentioned restrains and at the same time contain sufficient information about the character of the uncertainty.

M. Sága (✉) • R. Bednár • M. Vaško
University of Žilina, Žilina, Slovak Republic
e-mail: milan.saga@fstroj.uniza.sk; robert.bednar@fstroj.uniza.sk; milan.vasko@fstroj.uniza.sk

J. Náprstek et al. (eds.), *Vibration Problems ICOVP 2011: The 10th International Conference on Vibration Problems*, Springer Proceedings in Physics 139,
DOI 10.1007/978-94-007-2069-5_37, © Springer Science+Business Media B.V. 2011

Alternately to the use of probability methods we can use imprecise probabilities [1, 2] and the possibility theory, which involves the theory of interval numbers [2–5], fuzzy numbers and fuzzy sets [6, 7]. Without the information of the relevance of the data on the interval, we cannot use the fuzzy approach, but are still able to use the interval approach to describe the uncertain parameters, which are considered as unknown but bounded with lower and upper bounds.

Our short study proposes algorithms for modal and spectral interval computations of FE models and their effectivity analysis in view of the input uncertainty degree (2%, 5%, 10%, 20%).

2 Computational Methods for Interval Spectral Analysis

If we want to use interval arithmetic approach, an uncertain number is represented by an interval of real numbers [8, 9]. The interval numbers derived from the experimental data or expert knowledge can then take into account the uncertainties in the model parameters, model inputs etc. Complete information about the uncertainties in the model may be included by this technique and one can demonstrate how these uncertainties are processed by the calculation procedure in MATLAB [10].

Classic interval arithmetic application for solution of the numerical mathematics and mechanical problems in the engineering practice can encounter the problem known as the overestimate effect. Elimination of this problem is possible only in the case of meeting specific assumptions, mainly related to the time efficiency of the computing procedures.

Considering uncertain parameters in interval form, some solution approaches already used or proposed by the authors are analyzed [11]. Our goal is to present algorithm description and comparison study of the following numerical methods:

- Monte Carlo method (MC) - as a comparison tool,
- a simple combination of only inf-values or only sup-values (COM1),
- a full combination of all inf-sup values (COM2),
- a method which uses an optimisation process as a tool for finding out a inf-sup solution (OPT).

The structural uncertainty parameters are usually written into vector $\mathbf{x} = [\underline{\mathbf{x}}, \bar{\mathbf{x}}]$ and the interval modal FE analysis may be formulated as follows [12]

$$[\mathbf{K}(\mathbf{x}) - \lambda_j \cdot \mathbf{M}(\mathbf{x})] \cdot \mathbf{v}_j = \mathbf{0} \text{ or } \left([\underline{\mathbf{K}},\overline{\mathbf{K}}] - [\underline{\lambda_j}, \overline{\lambda_j}] \cdot [\underline{\mathbf{M}},\overline{\mathbf{M}}]\right) \cdot [\underline{\mathbf{v}_j},\overline{\mathbf{v}_j}] = \mathbf{0}, \quad (1)$$

where $\underline{\lambda_j}, \overline{\lambda_j}$ and $\underline{\mathbf{v}_j},\overline{\mathbf{v}_j}$ are the j-th eigenvalue with corresponding eigenvector, $\underline{\mathbf{K}}$, $\overline{\mathbf{K}}$, $\underline{\mathbf{M}}$, $\overline{\mathbf{M}}$ are of course the infimum and supremum of the mass and stiffness matrices. The application of the classic interval arithmetic for FE analysis is very limited. Its "overestimation" grows with the problem size (the dimension of the system matrices) and has not a physical foundation in the reality. Therefore, it is efficient to apply the previous numerical methods.

Application of the Monte Carlo method in IFEA may be realized as follows

1. step: generation of the random matrix (uniform distribution)

$$\mathbf{X}_{\mathrm{MC}} = [\mathbf{x}_1, \ldots..\mathbf{x}_{\mathrm{m}}], \quad (\mathrm{m} \approx 5,000 \div 100,000),$$

2. step: solution of $\boldsymbol{\lambda}_{\mathrm{j\_MC}} \rightarrow [\mathbf{K}(\mathbf{x}_{\mathrm{j}}) - \boldsymbol{\lambda}_{\mathrm{j\_MC}} \cdot \mathbf{M}(\mathbf{x}_j)] \cdot \mathbf{V}_{\mathrm{j}} = \mathbf{0}$ for $\mathrm{j} = 1 \ldots \mathrm{m}$,
3. step:
   - infimum calculation of the *i-th* eigenvalue $\underline{\lambda_i} = \inf(\mathrm{i}^{\mathrm{th}}$ row of $\boldsymbol{\lambda}_{\mathrm{MC}})$,
   - supremum calculation of the *i-th* eigenvalue $\overline{\lambda_i} = \sup(\mathrm{i}^{\mathrm{th}}$ row of $\boldsymbol{\lambda}_{\mathrm{MC}})$.

In the case of COM1, the numerical approach implementation to IFEA is following

- infimum calculation $\underline{\boldsymbol{\lambda}} \rightarrow [\mathbf{K}(\underline{\mathbf{x}}) - \underline{\boldsymbol{\lambda}} \cdot \mathbf{M}(\underline{\mathbf{x}})] \cdot \underline{\mathbf{V}} = \mathbf{0}$,
- supremum calculation $\overline{\boldsymbol{\lambda}} \rightarrow [\mathbf{K}(\overline{\mathbf{x}}) - \overline{\boldsymbol{\lambda}} \cdot \mathbf{M}(\overline{\mathbf{x}})] \cdot \overline{\mathbf{V}} = \mathbf{0}$.

COM1 doesn't give the correct results every time. We can obtain more proper results using COM2. Its computational process for IFEA is

1. step: calculation of realizations matrix $\mathbf{X}_2$, i.e. 2^{n} inf-sup combinations, $\mathbf{X}_{\mathrm{COM2}} = [\mathbf{x}_1, \ldots..\mathbf{x}_{\mathrm{m}}]$, $(\mathrm{m} = 2^{\mathrm{n}})$, n – number of uncertain system parameters,
2. step: solution of

$$\boldsymbol{\lambda}_{\mathrm{j\_COM2}} \rightarrow [\mathbf{K}(\mathbf{x}_{\mathrm{j}}) - \boldsymbol{\lambda}_{\mathrm{j\_COM2}} \cdot \mathbf{M}(\mathbf{x}_{\mathrm{j}})] \cdot \mathbf{V}_{\mathrm{j}} = \mathbf{0} \quad \text{for } \mathrm{j} = 1 \ldots \mathrm{m},$$

3. step:
   - infimum calculation of the *i-th* eigenvalue $\underline{\lambda_i} = \inf(\mathrm{i}^{\mathrm{th}}$ row of $\boldsymbol{\lambda}_{\mathrm{COM2}})$,
   - supremum calculation of the *i-th* eigenvalue $\overline{\lambda_i} = \sup(i^{\mathrm{th}}$ row of $\boldsymbol{\lambda}_{\mathrm{COM2}})$.

Generally, the infimum or supremum are not found only in the boundary points (COM1, COM2) but also in the inner domain of the solution set (OPT). To find the inf-sup solution using the approach OPT means to solve the optimizing problem described as follows

- infimum calculation of the *i*-th eigenvalue

$$\underline{\lambda_{\mathrm{i}}}(\mathbf{x}_{\mathrm{OPT}}) \rightarrow \text{ minimize value of } \lambda_{\mathrm{i}} \text{ for eq.: } [\mathbf{K}(\mathbf{x}) - \boldsymbol{\lambda}_{\mathrm{i}} \cdot \mathbf{M}(\mathbf{x})] \cdot \mathbf{v}_{\mathrm{i}} = \mathbf{0},$$

- supremum calculation of the *i-th* eigenvalue

$$\overline{\lambda_{\mathrm{i}}}(\mathbf{x}_{\mathrm{OPT}}) \rightarrow \text{ maximize value of } \lambda_{\mathrm{i}} \text{ for eq.: } [\mathbf{K}(\mathbf{x}) - \boldsymbol{\lambda}_{\mathrm{i}} \cdot \mathbf{M}(\mathbf{x})] \cdot \mathbf{v}_{\mathrm{i}} = \mathbf{0}.$$

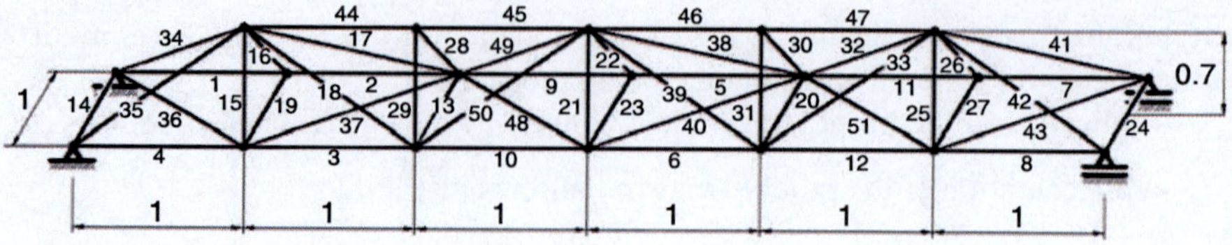

Fig. 1 Analyzed truss structure, dimensions in [m]

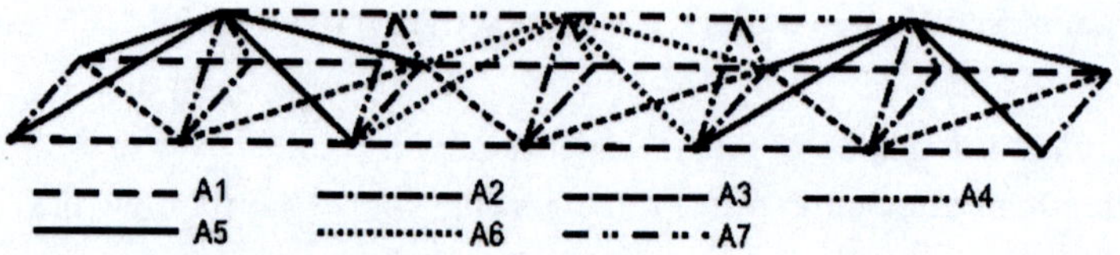

Fig. 2 Truss structure split into 7 cross-sectional groups

It should be noted that it is possible to realize the searching process by a comparison optimizing method (e.g. Nelder-Mead simplex algorithm) or by using genetic algorithm as a robust tool of global optimization.

3 Interval Analysis of Truss Structure

The three-dimensional truss structure has been analyzed (Fig. 1). The interval modal-spectral analyses in the range of the first two modal shapes were performed. As the interval uncertain parameters were the cross-sections of the trusses considered [13]. Because of the computation memory and time demands, 51 bars have been split into 7 cross-sectional groups (Fig. 2). All other parameters were considered as certain. Certain parameters: $E = 2 \cdot 10^{11}\ Pa$, $\mu = 0.3$, $\rho = 7800\ kg \cdot m^{-3}$, $\delta = 10^{-5}$.

Uncertain parameters: $\mathbf{xf} = [0.02,\ 0.05,\ 0.10\ ,\ 0.20]$,

$$A_1 = 3500 \cdot 10^{-6} \cdot (1 + \mathbf{xf}_i)\ m,\ A_2 = 3000 \cdot 10^{-6} \cdot (1 + \mathbf{xf}_i)\ m,$$
$$A_3 = 2500 \cdot 10^{-6} \cdot (1 + \mathbf{xf}_i)\ m,\ A_4 = 2000 \cdot 10^{-6} \cdot (1 + \mathbf{xf}_i)\ m,$$
$$A_5 = 1800 \cdot 10^{-6} \cdot (1 + \mathbf{xf}_i)\ m,\ A_6 = 1500 \cdot 10^{-6} \cdot (1 + \mathbf{xf}_i)\ m,$$
$$A_7 = 1000 \cdot 10^{-6} \cdot (1 + \mathbf{xf}_i)\ m.$$

The chosen results of our short comparison study are presented in Tables 1, 2 and graphically on Fig. 3.

4 Conclusions

The paper presents the interval arithmetic application on modal and spectral FE analysis. The use of the interval arithmetic provides a new possibility of the quality and reliability appraisal of analyzed objects. In the paper authors investigated

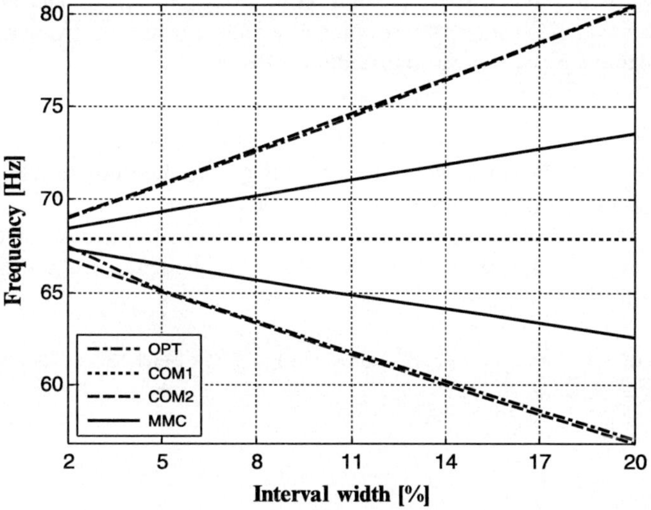

Fig. 3 Inf-sup modal solution for the 2nd natural frequency

Table 1 Inf-sup results of the first two natural frequencies

		Method			
Inputs interval width (%)	Freq.no.	MMC [Hz] <inf sup>	OPT [Hz] <inf sup>	COM1 [Hz] <inf sup>	COM2 [Hz] <inf sup>
2	1	<36.5 37.0>	<36.3 37.3>	<36.8 36.8>	<36.2 37.4>
	2	<67.4 68.4>	<67.5 69.0>	<67.9 67.9>	<66.7 69.1>
5	1	<36.0 37.4>	<35.5 38.1>	<36.8 36.8>	<35.2 38.4>
	2	<66.6 69.3>	<65.1 70.8>	<67.9 67.9>	<65.0 70.9>
10	1	<35.4 38.2>	<34.2 39.4>	<36.8 36.8>	<33.7 40.0>
	2	<65.4 70.7>	<62.3 73.8>	<67.9 67.9>	<62.2 74.0>
20	1	<34.0 39.7>	<31.4 42.4>	<36.8 36.8>	<30.8 43.4>
	2	<62.8 73.4>	<57.0 80.4>	<67.9 67.9>	<56.8 80.4>

Table 2 Results of the interval width of the natural frequencies

		Outputs interval width [%]			
Inputs interval width (%)	Freq. no.	MMC	OPT	COM1	COM2
2	1	0,74	1,41	0,00	1,69
	2	0,80	1,09	0,00	1,72
5	1	1,91	3,56	0,00	4,22
	2	1,95	4,19	0,00	4,31
10	1	3,78	7,14	0,00	8,45
	2	3,84	8,38	0,00	8,62
20	1	7,71	14,69	0,00	16,91
	2	7,80	17,01	0,00	17,24

possibilities of the modal-spectral solution of truss structure model with interval cross-sectional areas. It shows the solution efficiency for solving problems including uncertain parameters with a various width of the interval.

The analyses results can be summarized as follows

- COM1 method is not recommended for this kind of analyses,
- COM2 method provides decent results, but it is limited due to the exponential growth of the analyses number for complicated problems,
- OPT method provides good results and is suitable for complicated problems because it does not need so many analyses as in the cases of the MC or COM2 methods.

Acknowledgments This work has been supported by VEGA grant No. 1/0125/09 and VEGA grant No. 1/0727/10.

References

1. Segľa, Š.: Dynamic and quasistatic analysis of the elevating work platform. J. Mech. Manipulat. **6**(1), 81–86 (2007)
2. Chen, S.H., Yang, X.W.: Interval finite element method for beam structures. Finite Elem. Anal. Des. **34**, 75–88 (2000)
3. Neumaier, A.: Interval Methods for Systems of Equations. Cambridge University Press, Cambridge (1990)
4. Elishakoff, I., Duan, D.: Application of mathematical theory of interval analysis to uncertain vibrations. In: Proceedings of NOISE-CON'94, Ft. Lauderdale, pp. 519–524 (1994)
5. Forssén, P.: Interval methods I. http://www.tdb.uu.se/kurs/optim-mn1/ht01/lectures/lec14_2.pdf
6. Dekýš, V., Sapietová, A., Kocúr, R.: On the reliability estimation of the conveyer mechanism using the Monte Carlo method. In: Proceedings of COSIM2006, Krynica - Zdroj, pp. 67–74 (2006)
7. Zhang, H.: Nondeterministic Linear Static Finite Element Analysis: An Interval Approach. School of Civil and Environmental Engineering, Georgia Institute of Technology, Atlanta (2005)
8. Kulpa, Z., Pownuk, A., Skalna, I.: Analysis of linear mechanical structures with uncertainties by means of interval methods. Comput. Assis. Mech. Eng. Sci. **5**, 443–477 (1998) (Poland)
9. Moore, R.E.: Interval Analysis. Prentice Hall, Englewood Cliffs New Jersey (1966)
10. Rump, S.M.: Intlab. http://www.ti3.tu-harburg.de/~rump/intlab/
11. Vaško, M.: Application of genetic algorithms for solving of mechanical systems with uncertain parameters. PhD. thesis, Žilina (2007)
12. Zhang, H., Muhanna, R.L.: Finite element analysis for structures with interval parameters. In: Proceedings of the 9th ASCE Joint Specialty Conference on Probabilistic Mechanics and Structural Reliability, Albuquerque, NM, USA (2004)
13. Bednár, R.: Analysis of the effectiveness of selected methods for solving mechanical systems with uncertain parameters. Diploma thesis, Žilina (2009)

FEM Modeling and Dynamical Behavior of a Flexible Cracked Linkage Mechanism with Clearance

Mihai Dupac and S. Noorozi

Abstract In this study the vibration characteristics and dynamical behavior of a general planar mechanism with a flexible link that contain cracks and a revolute joint with clearance at the slider is analyzed. The link flexibility is modelled using the finite element method, and the local flexibilities representing cracks are modelled using massless rotational springs. The equations of motion are developed for the non impacting case. The equations of motion in integrated form (called the impact equations) are used to model impact at the joint. The influence of the angular velocity, number of cracks, and clearance of the slider revolute joint on the flexural response of the flexible rod is investigated.

Keywords Vibration • Cracks • Clearance • Impact • FEM

1 Introduction

The effect of clearance, flexibility and cracks on the dynamic behavior of mechanical systems is considerable when high precision and alignment are important. The dynamical effect of clearances for planar mechanical systems has been studied. Rotational massless spring models, with stiffness related to the crack have been used. Different definitions for the spring constants have been presented. Research to simulate the free vibration of beams with multiple cracks has been developed. The design and/or dynamics of impact systems with lumped masses have been studied in [1–4]. Crack initiation and growth occurring under cyclic motion is the reason for

M. Dupac (✉)
School of Design, Engineering and Computing, Bournemouth University, Talbot Campus, Fern Barrow, Poole Dorset, BH12 5BB, UK
e-mail: mdupac@bournemouth.ac.uk

S. Noorozi
Bournemouth University, Dorset, UK
e-mail: snoorozi@bournemouth.ac.uk

J. Náprstek et al. (eds.), *Vibration Problems ICOVP 2011: The 10th International Conference on Vibration Problems*, Springer Proceedings in Physics 139,
DOI 10.1007/978-94-007-2069-5_38,

many connecting rods failure. Fatigue crack propagation and fracture behaviour of rods subjected to various loading and boundary conditions has been discussed.

In this paper the dynamics of a mechanism with a flexible rod with cracks clearance is studied. The rod flexibility is modeled using finite element method. The cracks are modeled by massless rotational springs. Impact equations are used to study the impact of the system. The influence of cracks and clearance of the slider revolute joint on the flexural response of the flexible rod is investigated.

2 Theoretical Model and Governing Equations of Motion

To study the effect of flexibility, cracks and clearance, a slider-crank (Fig. 1) with a flexible link with cracks and joint clearance is considered. The length of the crank is L and the length of the flexible link is l. For the general case there are i_n cracks (modeled with rotational springs denoted by K_i) on the flexible link.

For the clearance model presented in Fig. 2 the journal and sleeve are considered as two impacting bodies, having radius r and R. The difference in radius R-r defines the radial clearance b of the joint. The next distinct states appear between the journal and sleeve: (a) no contact, (b) impact, and (c) permanent contact.

For a beam element (Fig. 3) there are two degrees of freedom at each node $v_{eL_i} = \Omega_1 + \Omega_2 x + \Omega_3 x^2 + \Omega_4 x^3$ and $v_{eR_i} = \Lambda_1 + \Lambda_2 x + \Lambda_3 x^2 + \Lambda_4 x^3$. The relations $v_{eL_i} = Y_{eL_i}$, $v'_{eL_i} = \Phi_{eL_i}$, and $v_{eR_i} = Y_{eR_i}$, $v'_{eR_i} = \Phi_{eR_i}$, are related to the transverse deflections Y_{eL_i}, Y_{eR_i} and nodal slope deflections Φ_{eL_i}, Φ_{eR_i}. The boundary conditions are (i) $v_{eL_i} = v_{eR_i}$ - displacements, (ii) $EI/K_i.v''_{eL_i} + v'_{eL_i} - v'_{eR_i} = 0$ – slopes difference, (iii) $v''_{eL_i} = v''_{eR_i}$ - bending moments, and (iv) $v'''_{eL_i} = v'''_{eR_i}$ - shear forces. Once the interpolating functions are calculated and stiffness matrix elements $[K_{eL_i}]$, $[K_{eR_i}]$ obtained [5, 6] the equations are expressed as $v_{eL_i} = \{\phi_{eL_i}\}[K_{eL_i}]\{x\}$ and $v_{eR_i} = \{\phi_{eR_i}\}[K_{eR_i}]\{x\}$. The spring constant K_i is computed as in [7] $K_i = EJ/(40.8h\beta_j f_1(\beta_j) f_2(\beta_j))$.

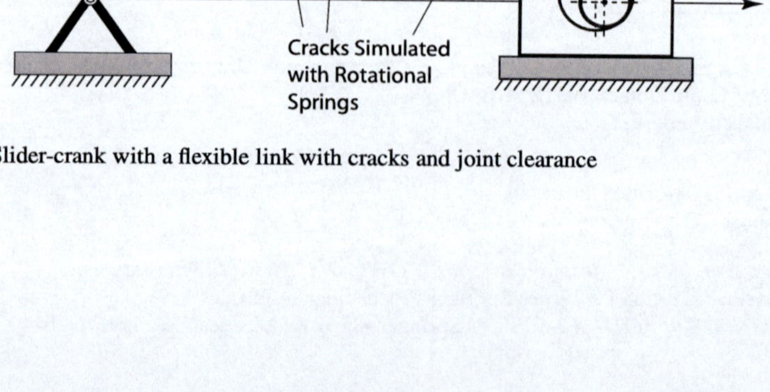

Fig. 1 Slider-crank with a flexible link with cracks and joint clearance

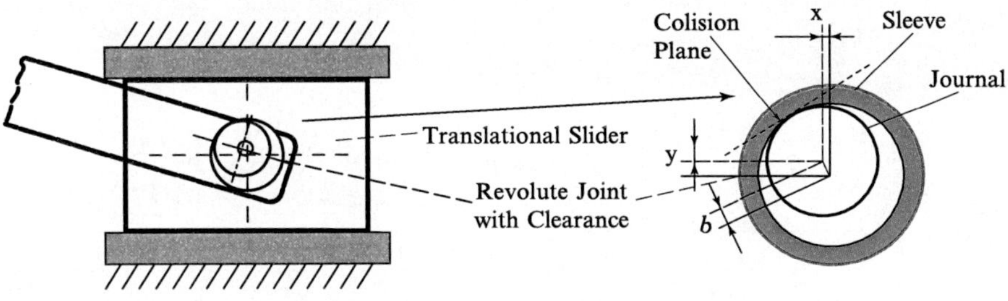

Fig. 2 Revolute joint with clearance (clearance is exaggerated to make it visible)

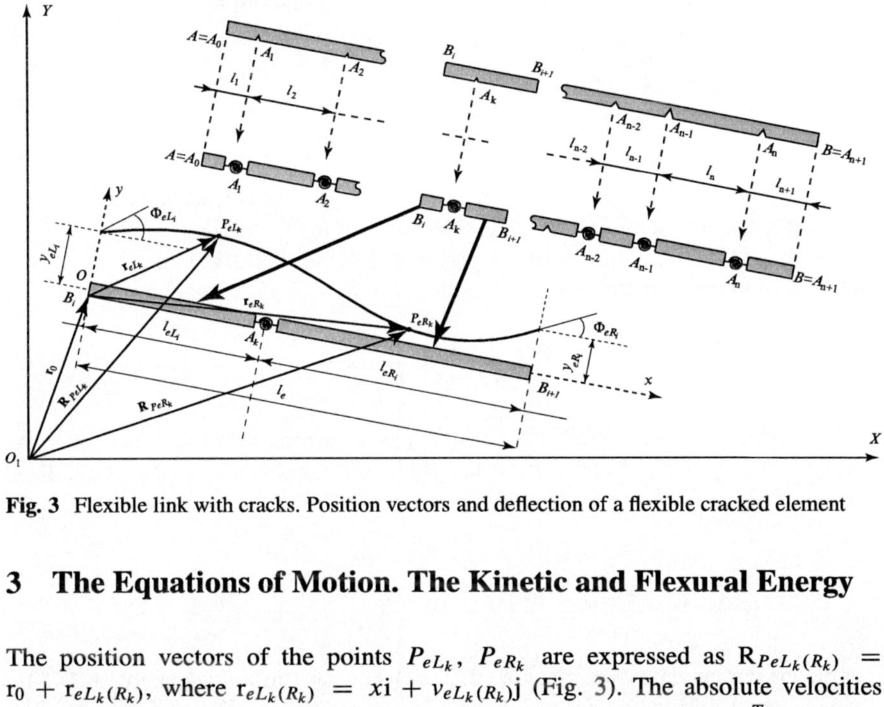

Fig. 3 Flexible link with cracks. Position vectors and deflection of a flexible cracked element

3 The Equations of Motion. The Kinetic and Flexural Energy

The position vectors of the points P_{eL_k}, P_{eR_k} are expressed as $\mathbf{R}_{PeL_k(R_k)} = \mathbf{r}_0 + \mathbf{r}_{eL_k(R_k)}$, where $\mathbf{r}_{eL_k(R_k)} = x\mathbf{i} + v_{eL_k(R_k)}\mathbf{j}$ (Fig. 3). The absolute velocities $\mathbf{v}_{PeL_k(R_k)}$are given by $\{v_{ox} - \dot{\gamma}v_{eL_k(R_k)}\ v_{oy} + \dot{\gamma}\mathrm{x} + v_{eL_k(R_k)}\ 0\}\{\mathrm{i\ j\ k}\}^T$, where the angular velocity is $\dot{\gamma}$ and $v_{P_{eL_k}}$ is the velocity of the point P_{eL_k}. The kinetic energy of the flexible link is

$$T = \frac{1}{2}\rho AL^3 \sum_{i=1}^{n} \left(\int_{x=x_{i-1}}^{\beta_i} \left[(v_{ox} - \dot{\gamma}v_{eL_i})^2 + \left(v_{oy} + \dot{\gamma}\mathrm{x} + \dot{v}_{eL_i}\right)^2 \right] dx \right.$$
$$\left. + \int_{x=\beta_i}^{x_i} \left[(v_{ox} - \dot{\gamma}v_{eL_i})^2 + \left(v_{oy} + \dot{\gamma}\mathrm{x} + \dot{v}_{eL_i}\right)^2 \right] dx \right) \quad (1)$$

where ρ is the mass density, and A is the cross-sectional area.

The flexural strain V energy due to beam deformation can be expressed as $V = V_1 + V_2$, where V_1 is

$$V_1 = \frac{1}{2}\frac{E \cdot I_z}{L}\sum_{i=1}^{n}\left(\int_{x=x_{i-1}}^{\beta_i}\left(\frac{\partial^2 v_{eL_i}}{\partial x^2}\right)^2 + \int_{x=\beta_i}^{x_i}\left(\frac{\partial^2 v_{eR_i}}{\partial x^2}\right)^2\right)dx$$
$$+\frac{1}{2}\sum_{i=1}^{n}K_i \cdot \left(\left.\frac{\partial v_{eL_i}}{\partial x}\right|_{\beta_i} - \left.\frac{\partial v_{eR_i}}{\partial x}\right|_{\beta_i}\right)^2 \tag{2}$$

and the total strain energy V_2 of the beam due to longitudinal loads is

$$V_2 = \sum_{i=1}^{n}\left(\int_{x=x_{i-1}}^{\beta_i} P_x\left(\frac{\partial v_{eL_i}}{\partial x}\right)^2 dx + \int_{x=\beta_i}^{x_i} P_x\left(\frac{\partial v_{eR_i}}{\partial x}\right)^2 dx\right) \tag{3}$$

In Eq. 3 the longitudinal loads P_x in an element having an external load P_R acting at its right end and undergoing transverse deflection was considered $P_x = P_R + \rho L^2 A\,(x-1)\,(\dot{v}_{Ox} - \dot{\gamma} v_{Ox}) - \frac{1}{2}\rho L^2 \dot{\gamma}^2\,(A + x^2)$ as in [8]. The Lagrange differential equation of motion (where L is the Lagrangian) for a single finite element is

$$\frac{\mathbf{d}}{\mathbf{dt}}\left(\frac{\partial L}{\partial q_j}\right) - \frac{\partial L}{\partial q_j} = 0 \tag{4}$$

The equation of motion of the system may be expressed in a matrix form as $[M]\{\ddot{\Phi}\} + [K_s + K_d]\{\Phi\} = F$, where M is the mass is and F the generalized forces.

4 Equations of Motion with Impact

For the present model it was assumed that impact occurs instantaneously [9–14], no change occurs during contact, the coefficient of restitution e quantifies the energy dissipation, and the contact force is as in [11, 14]. The Newton–Euler equations with frictionless bilateral constraints can be written as [12]

$$M\dot{u} - f = 0 \quad \text{and} \quad \dot{q} = u, \tag{5}$$

where $M(q,\,t)$ is the mass, q the generalized coordinates, $\dot{q} = u$ the velocities, $\ddot{q} = \dot{u}$ the accelerations, and f the sum of all generalized forces. The Newton–Euler equations of motion in integrated form (equations of motion with impact) are

$$Mu^{+} - Mu^{-} = P \tag{6}$$

where $u^{\pm}(t) = \lim_{\Delta t \downarrow \uparrow 0} \frac{q(t+\Delta t) \pm q(t)}{\Delta t}$ are the generalized post- and pre-impact velocities, $P = s_N d_N$ the generalized impulsive forces having only a vertical component, s_N the impulsive force in normal direction, $d_N == (\partial \dot{r}_C / \partial \dot{q})^T \cdot n$ the normal force vector, and $\dot{r}_C$ the velocity at the contact point. Using the coefficient of restitution e as in [12, 14] the impact equations can be written in a compact form as

$$\begin{bmatrix} M & \dot{d}_N^T \\ \dot{d}_N & 0 \end{bmatrix} \begin{bmatrix} u^{+} \\ s_N \end{bmatrix} = \begin{bmatrix} Mu^{-} \\ -ed_N u^{-} \end{bmatrix} \tag{7}$$

The contact impulse and velocity after impact can be expressed as

$$s_N = (1+e)\frac{\dot{d}_N u^{-}}{\dot{d}_N M^{-1} \dot{d}_N^T}, \quad u^{+} = u^{-} - M^{-1}\dot{d}_N^T s_N \tag{8}$$

5 Results

In this paper the vibration response of a mechanism with a flexible rod with cracks and clearance was studied. Numerical simulations have been performed for a flexible link with one and two cracks (located at 1/3 and 1/2 of the link length), with the Young's modulus $E = 2.08 \cdot 10^{11}$ [N/m^2], Poisson's ratio $\nu = 0.32$, and restitution coefficient $e = 0.47$. The crank and link lengths are 0.11 and 0.32 m respectively. Angular velocities of 100 and 150 rpm have been considered.

Since the journal and the sleeve exhibit impacts that give rise to discontinuous forces in the joint and jumps in the system velocity, the joint clearance can be treated as small perturbation on the system.

The influence of the angular velocity, cracks and clearance to the dynamics of the flexible link can be observed in Fig. 4a. It was observed that the main variable

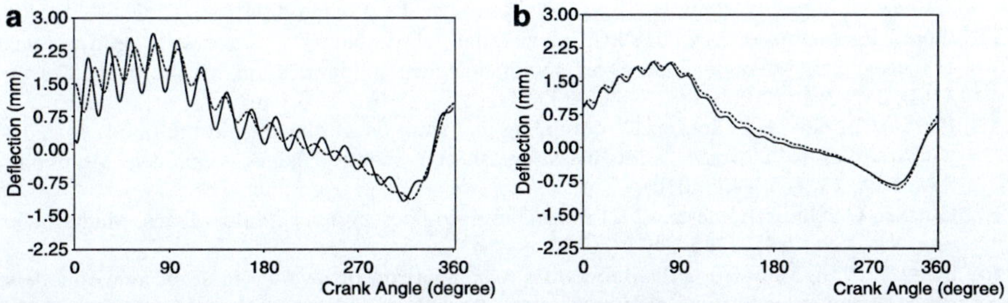

Fig. 4 Flexural deflection at the mid-span of the flexible rod for (**a**) —— one crack, clearance and ω_1, - - - - two cracks, clearance and ω_2, and (**b**) —— no cracks, clearance and ω_1, - - - - no cracks, clearance and ω_2

affecting the link vibration was an increased angular velocity combined with cracks. The influence effect of angular velocity and clearance parameters can be observed in Fig. 4b.

The interactions between the variables (clearance, angular velocity, cracks) showed that the stability tended to be affected for medium to high values of the associated parameters.

6 Conclusions

In this paper the dynamics of a flexible rod with cracks and clearance was studied. The rod flexibility was modeled using FEM and rotational springs. It was observed that medium to high values of the clearance, cracks, and angular velocity has a substantial influence over the dynamics of the mechanism.

References

1. Dupac, M., Beale, D.G.: Dynamic analysis of a flexible linkage mechanism with cracks and clearance. Mech. Mach. Theory **45**(12), 1909–1923 (2010)
2. Dupac, M., Marghitu, D.B.: Nonlinear dynamics of a flexible mechanism with impact. J. Sound Vib. **289**(4–5), 952–966 (2006)
3. Marghitu, D.B., Sinha, S.C., Diaconescu, C.: Control of a parametrically excited flexible beam undergoing rotations and impacts. Multibody Syst. Dyn. **3**, 47–63 (1999)
4. Rubinstein, D.: Dynamics of a flexible rod and a system of rigid rods, fully inverse boundary conditions. Comput. Methods Appl. Mech. Eng. **175**, 87–97 (1999)
5. Skrinar, M., Plibersek, T.: New finite element for transversely cracked slender beams subjected to transverse loads. Comput. Mater. Sci. **39**, 250–260 (2007)
6. Skrinar, M.: Elastic beam finite element with an arbitrary number of cracks. Finite Elem. Anal. Des. **45**, 181–189 (2009)
7. Skrinar, M., Umek, A.: Plane line finite element of beam with crack. Gradbeni Vestnik **1/2**, 2–7 (1996)
8. El-Dannanh, E.H.: Flexural vibrational behaviour of an elastic linkage containing a transverse crack. Mech. Mach. Theory **30**(2), 285–297 (1995)
9. Flores, P.: Modeling and simulation of wear in revolute clearance joints in multibody systems. Mech. Mach. Theory **44**, 1211–1222 (2009)
10. Flores, P., Ambrósio, J., Claro, J.P.: Dynamic analysis for planar multibody mechanical systems with lubricated joints. Multibody Syst. Dyn. **12**, 47–74 (2004)
11. Flores, P., Ambrósio, J., Claro, J.C.P., Lankarani, H.M., Koshy, C.S.: A study on dynamics of mechanical systems including joints with clearance and lubrication. Mech. Mach. Theory **41**(3), 247–261 (2006)
12. Flores, P., Leine, R., Glocker, C.: Modeling and analysis of planar rigid multibody systems with translational clearance joints based on the non-smooth dynamics approach. Multibody Syst. Dyn. **23**, 165–190 (2010)
13. Glocker, C.: On frictionless impact models in rigid-body systems. Philos. Trans. Math. Phys. Eng. Sci. **359**(1789), 2385–2404 (2001)
14. Schwab, A.L., Meijaard, J.P., Meijers, P.: A comparison of revolute joint clearance models in the dynamic analysis of rigid and elastic mechanical systems. Mech. Mach. Theory **37**, 895–913 (2002)

Simulation of a Non-smooth Continuous System

Jan Clauberg, Markus Schneider, and Heinz Ulbrich

Abstract The current paper gives a short overview of the multiphysics simulation tool *MBSim*, developed at the Institute of Applied Mechanics - TU München, and its application on a non-smooth continuous system. Starting at the mathematical equations describing the dynamical behavior of a non-smooth multibody system with uni- and bilateral frictional constraints, the used time-stepping integration method is shortly outlined. The application of *MBSim* is demonstrated by the simulation of a continuous coil spring with non-smooth contact mechanics. To show the quality of the simulation model, a short validation of the coil spring with experimental data is shown.

Keywords Flexible multibody dynamics • Non-smooth contact mechanics • Helicoil springs • Experimental validation • Finite elements

1 Introduction

The simulation tool *MBSim* has been developed at the Institute of Applied Mechanics at the Technische Universität München [1] and is available under the GNU Lesser General Public License [2]. The program is written in C++ and provides a framework for modelling and simulation of dynamic systems in various physical domains (e.g. multibody dynamics, hydraulics, electronics, control theory).

J. Clauberg (✉) • M. Schneider • H. Ulbrich
Institute of Applied Mechanics - TU, München, 85747 Garching, Germany
e-mail: clauberg@amm.mw.tum.de; schneider@amm.mw.tum.de; ulbrich@amm.mw.tum.de

J. Náprstek et al. (eds.), *Vibration Problems ICOVP 2011: The 10th International Conference on Vibration Problems*, Springer Proceedings in Physics 139,
DOI 10.1007/978-94-007-2069-5_39,

2 Postprocessing Framework

The simulation results are written in the hierarchical HDF5 file format, which also enables the efficient treatment of large dynamical systems. These files can be read by H5PlotSerie [3] or Matlab for plotting or by OpenMBV [4] for visualisation.

3 Simulation Framework

In this chapter the mathematical formulation of the dynamic system and the set-valued force-laws are described.

3.1 Mathematical Formulation

Basis of the formulation of the equations of motion of a dynamic system with bi- and unilateral constraints is the finite dimensional measure differential equation [5, 6]

$$\mathbf{M}\,\mathrm{d}\mathbf{u} = \mathbf{h}\,\mathrm{d}t + \mathbf{W}\,\mathrm{d}\underline{\Lambda}. \tag{1}$$

$\mathbf{M} = \mathbf{M}(\mathbf{q})$ denotes the symmetric and positive definite mass matrix. The vector $\mathbf{h} = \mathbf{h}(\mathbf{u}, \mathbf{q}, t)$, depending on the generalized positions $\mathbf{q}$, generalized velocities $\mathbf{u}$ and the time t, contains all external, internal and gyroscopic forces. The directions of the set-valued contact reactions $\mathrm{d}\underline{\Lambda}$ are contained in the matrix $\mathbf{W} = \mathbf{W}(\mathbf{q})$.

The acceleration measure $\mathrm{d}\mathbf{u} = \dot{\mathbf{u}}\,\mathrm{d}t + \left(\mathbf{u}^+ - \mathbf{u}^-\right)\mathrm{d}\eta$ and the measure for impulses $\mathrm{d}\underline{\Lambda} = \underline{\lambda}\,\mathrm{d}t + \underline{\Lambda}\,\mathrm{d}\eta$ is split into a continuous part ($\dot{\mathbf{u}}\,\mathrm{d}t$, $\underline{\lambda}\,\mathrm{d}t$) and an atomic part ($\left(\mathbf{u}^+ - \mathbf{u}^-\right)\mathrm{d}\eta$, $\underline{\Lambda}\,\mathrm{d}\eta$) with the left and right limit $\mathbf{u}^-$, $\mathbf{u}^+$ and the Dirac point measure $\mathrm{d}\eta$.

Integrating (1) yields the equations of motion for a constrained system

$$\mathbf{M}\dot{\mathbf{u}} = \mathbf{h} + \mathbf{W}\,\underline{\lambda} \tag{2}$$

of smooth dynamics as well as the impact equations

$$\mathbf{M}_i\left(\mathbf{u}_i^+ - \mathbf{u}_i^-\right) = \mathbf{W}_i\,\underline{\Lambda}_i \qquad \forall i \in N \tag{3}$$

being valid at times t_i of impact.

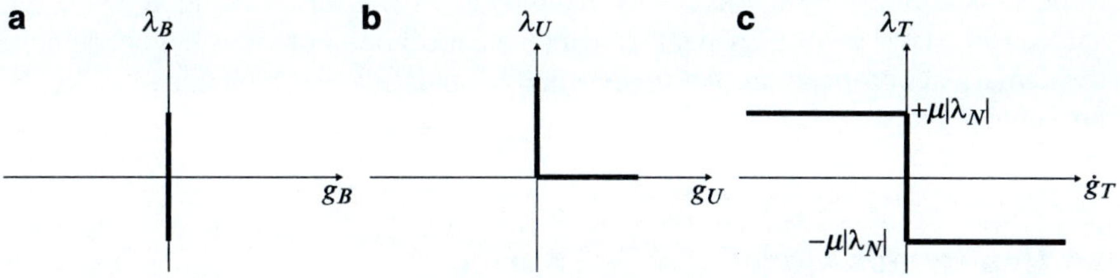

Bilateral constraint. Unilateral constraint. Planar COULOMB friction.

Fig. 1 Force laws for bi- and unilateral contacts and friction

3.1.1 Set-Valued Force Laws

The computation of the accelerations $\dot{\mathbf{u}}$ in (2) and the post-impact velocities $\mathbf{u}_i^+$ in (3) requires the knowledge of the unknown reactions $\underline{\lambda}$ and $\underline{\Lambda}_i$ governed by the set-valued force laws $(\mathbf{q}, \mathbf{u}, \underline{\lambda}, \underline{\Lambda}_i, t) \in \mathcal{N}$. Therefore, additional conditions must be formulated.

In *MBSim* three different basic kinds of set-valued force laws can be considered, representing *bilateral*, *unilateral* and *frictional* constraints. They are depicted in Fig. 1. A bilateral force law, which is always closed, implies a bilateral constraint of the form

$$g_B = 0, \qquad \lambda_B \overset{<}{\underset{>}{=}} 0. \tag{4}$$

Therein g_B denotes the normal distance of the interacting bodies in the contact point and λ_B the corresponding force. The second type of force law represents a contact in mechanical systems. It is given by the *Signorini–Fichera*-condition

$$g_U \geq 0, \quad \lambda_U \geq 0, \quad g_U\,\lambda_U = 0, \tag{5}$$

with the normal distance g_U of the contact contours. As a third force law, friction can be considered. Therefore, the force of a single contact point is decomposed in a component $\lambda_N \in \{\lambda_B, \lambda_U\}$ normal to the contact plane and a tangential component $\underline{\lambda}_T$ in friction direction. With the relative tangential velocity $\dot{\mathbf{g}}_T$ and the friction coefficient μ, *Coulomb*'s friction law is given by

$$\begin{aligned} &\dot{\mathbf{g}}_T = \mathbf{0} \Rightarrow |\underline{\lambda}_T| \leq \mu\,|\lambda_N| \\ &\dot{\mathbf{g}}_T \neq \mathbf{0} \Rightarrow \underline{\lambda}_T = -\frac{\dot{\mathbf{g}}_T}{|\dot{\mathbf{g}}_T|}\,\mu\,|\lambda_N|\,. \end{aligned} \tag{6}$$

An impact influences all relations between the objects concerning the post impact generalized velocity $\mathbf{u}^+$. Internally, the impact laws are formulated on velocity level, substituting g by $\dot{g}^+$ and λ by Λ in (4) and (5). The formulation of the set-valued force laws are self-evident from a mechanical point of view, but not suitable for numerical computation. An appropriate formulation can be attained using the proximal point to a convex set [5].

4 Continuous Model of a Coil Spring

The simulation of continuous coil springs is still a challenging topic within multibody simulations. In most cases it is sufficient to describe the behavior of springs with the spring stiffness c and the spring damping d:

$$F = cx + d\dot{x} \tag{7}$$

However in highly dynamic cases it is not negligible that the dynamic response of a spring significantly differs from the static response. In multibody simulations mainly three methods to describe coil springs are used: multi-mass-models, modal–models and multi-beam-models [7]. In this paper a continuous spring model based on a curved beam is introduced and applied to a coil spring with constant radius R, elliptical cross section and non-constant pitch.

4.1 *Equations of Motion*

As shown in Fig. 2a cartesian coordinate system (x, y, z) is introduced so that the z-axis coincides with the central axis of the spring. The spring wire is parametrized by the two variables s (length of spring wire) and Θ (angle of spring wire), in which u_z denotes the compression of the spring. The spring is approximated as a curved beam (cross section area A, radius R) and the governing equations are derived in accordance with regular beam theory [8]. Furthermore, an isotropic hookean material with Young's modulus E, lateral contraction coefficient ν, density ρ, shear modulus G and torsional constant J is used. The relationship between strains and stresses can be stated by Hook's generalized law. The cross section is assumed to be elliptical (Fig. 2). Finally, the equations of motion can be derived resulting in six coupled differential equations, which can also be obtained from *Wittrick*'s equations [9]. The purpose of the spring model described in this paper is to be applied in multibody simulations. For this reason, the full equations of motion are reduced to the extensional degree of freedom (z-axis), resulting in a hyperbolic partial differential equation, also known as the one-dimensional wave equation, with f_z the external load of the spring.

$$\rho A \frac{d u_z}{dt^2} - \frac{GJ}{R^2}\frac{d^2 u_z}{ds^2} = f_z \,. \tag{8}$$

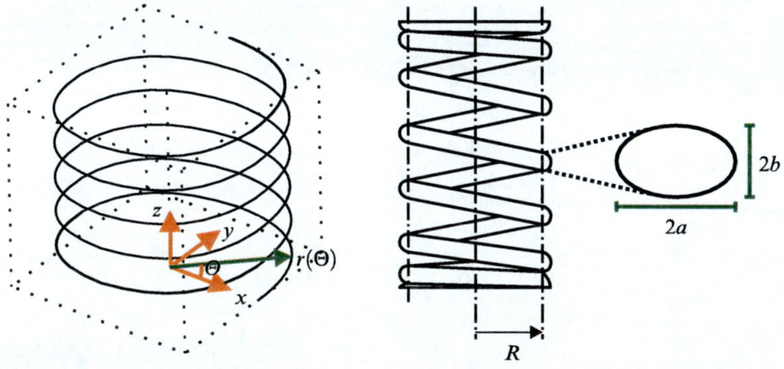

Fig. 2 Coil spring with coordinate systems

To gain an appropriate form for multibody simulations (9), the Finite Element approach is used. In (9) **M** denotes the mass-matrix, **D** the damping-matrix, **K** the stiffness-matrix and **h** the vector of external forces.

$$\mathbf{M}\,\ddot{\mathbf{q}} + \mathbf{D}\,\dot{\mathbf{q}} + \mathbf{K}\,\mathbf{q} = \mathbf{h}(\dot{\mathbf{q}}, \mathbf{q}, t). \tag{9}$$

Applying the *Galerkin–Bubnov-Method* to Eq. 8 yields the discretized equations of motion

$$\underbrace{\int_k^l \rho A \mathbf{N}\mathbf{N}^T \, ds \cdot \frac{d^2\mathbf{q}}{dt^2}}_{\mathbf{M}} + \underbrace{\int_k^l \frac{GJ}{R^2}\frac{d\mathbf{N}}{ds}\frac{d\mathbf{N}^T}{ds}\, ds}_{\mathbf{K}} \cdot \mathbf{q} = \underbrace{[\mathbf{N}\cdot T_z]_k^l + \int_k^l \mathbf{N}\cdot f_z \, ds}_{\mathbf{F}} . \tag{10}$$

Lagrange Polynomials of order 2 are applied as shape-functions **N**. To allow interaction between adjacent coils, contacts between the coils are introduced. These contacts are solved by means of non-smooth contact mechanics. The main advantage of this approach is that the penetration of adjacent coils is minimal, whereby the energy dissipation within the contacts reaches a maximum (rigid contacts are considered as full plastic impacts). To be able to validate the model with experimental data, damping is introduced as Rayleigh damping (mass-proportional parameter α, stiffness-proportional parameter β (11) and within the contacts.

$$\mathbf{D}_{Rayleigh} = \alpha\mathbf{M} + \beta\mathbf{K} \tag{11}$$

4.2 *Example of Experimental Validation*

In this section an example for the experimental validation of the spring model in the time domain is shown. A coil spring with seven coils, a non-constant pitch and

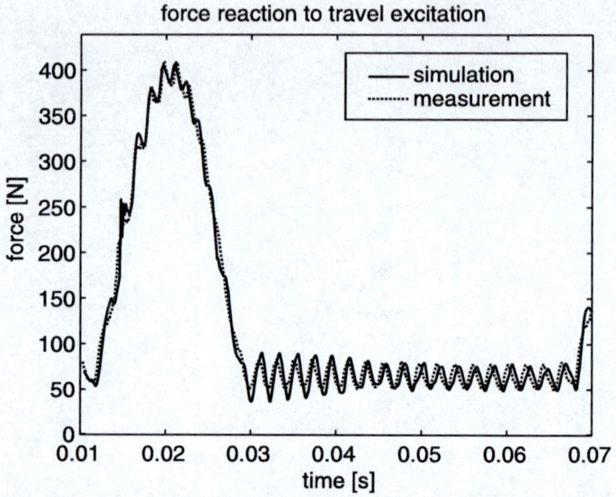

Fig. 3 Force at the clamped end of a excited spring

an elliptical cross section is taken into account. The simulation model has seven elements (Lagrange Polynomials of order 2) and seven contacts between the coils. The experimental data is gained from a test rig, where the spring is excited by a cam profile (analogue to the excitation in a combustion engine). Figure 3 shows the data from the simulation and the experimental data.

5 Conclusions

A simulation model for helicoil springs based on a curved beam is presented. This model is discretized with the Finite Element approach and implemented in the multibody simulation tool *MBSim*. The contacts between the interacting coils of the spring are modeled by means of non-smooth contact mechanics. By validating the simulation model with experimental data gained from a spring test rig, the quality of the spring model is shown. The experimental measured and simulated dynamics (force levels and frequencies) at high and low prestresses of the helicoil spring reveal a great matching. For this purpose only about 15 degrees of freedom and seven contacts between the interacting coils are necessary.

References

1. Förg, M.: Mehrkörpersysteme Mit Mengenwertigen Kraftgesetzen – Theorie und Numerik. Technische Universität München, München (2008)
2. MBSim - Multi-Body Simulation Software. GNU Lesser General Public License. http://mbsim.berlios.de
3. HDF5Serie - A HDF5 Wrapper for Time Series. http://hdf5serie.berlios.de

4. OpenMBV - Open Multi Body Viewer. GNU Lesser General Public License. http://openmbv.berlios.de
5. Glocker, C.: Set-Valued Force Laws. Springer, Berlin (2001)
6. Moreau, J., Panagiotopoulos, P.: Nonsmooth Mechanics and Applications. Springer, Berlin (1988)
7. Huber, R., Clauberg, J.: Ulbrich H an efficient sping model based on a curved beam with non-smooth contact-mechanics for valve train simulations. SAE Int. J. Engines **3**(1), 28–34 (2010)
8. Volmer, K.J.: Finite-Elemente-Methoden, 2nd edn. Springer, Berlin (2002)
9. Wittrick, W.H.:On elastic wave propagation in helical springs. 8th Intern. J. Mech. Sci. **8**(1), 25–47 (1966)

Characterisation of the Dynamic Behaviour of Laminated Sheet Glass in Steel-Glass Façades

C. Colomer Segura and M. Feldmann

Abstract The use of steel-glass façades has become a very standard feature in modern design of buildings, very often including laminated sheet glass panes as a load bearing element. While these elements are essentially subjected to dynamic loading due to wind, its design is still based on static load assumptions. A more exact approach is not possible due to the uncertainties regarding the transfer function of the glass panes to the structure. In this paper, basic modal identification techniques are used to characterise the dynamic behaviour of laminated sheet glass and to give a first approach to account for dynamic effects on steel-glass façades.

Keywords Damping • Laminated glass • Façade • Transfer function • Impulse excitation

1 Introduction

Current investigations in the field of glass construction are being performed in the latest years in order to achieve a better understanding of its mechanical properties, allowing structural engineers to start employing glass elements as load bearing elements. The newest building designs are very often conceived as slender, irregular shaped structures wrapped by glass façades. These construction trend aims for bigger and bigger spans between the glass supports allowing the building to give a more slender light-weight appearance.

The classical first assumption in the design of façades consisting of a statically loaded glass element transmitting its loads to the supporting structure is no longer valid in cases where big glass elements start showing low eigenfrequencies that can be easily excited by normal wind conditions.

C.C. Segura (✉) • M. Feldmann
RWTH Aachen University, Institute for Steel Structures, Mies van der Rohe Str. 1, 52070 Aachen, Germany
e-mail: colomer@stb.rwth-aachen.de

J. Náprstek et al. (eds.), *Vibration Problems ICOVP 2011: The 10th International Conference on Vibration Problems*, Springer Proceedings in Physics 139,
DOI 10.1007/978-94-007-2069-5_40, © Springer Science+Business Media B.V. 2011

In these situations, the dynamic properties of the load transfer between the glass pane and its fixation elements becomes a decisive question in order to predict the dynamic behaviour of wind (dynamically) excited glass façades. So far laminated glass (most widely used in these type of constructions) has been assumed to account for a higher damping than single sheet glass, because of the presence of an intermediate PVB-sheet. The absence of widely extended codes regulating the use of glass as an structural element together with the lack of knowledge of its damping properties are the starting point for this investigation.

Some basic identification techniques [2, 3] of experimental modal analysis (EMA) basing on the Rational Fraction Polynomials (RFP) Method [4] and the Polyreference Frequency Domain Method (PRFD) [6] have been implemented along with a Single Input – Multiple Output (SIMO) experiment with impact hammer to estimate the modal parameters of the glass elements from the measured Frequency Response Functions (FRF).

The numerical implementation of the identification methods as well as the generation of artificial FRFs to validate the implementation have been performed with Matlab©.

2 Experimental Modal Analysis

In this section the used methods for identifying the dynamic properties of the glass elements are briefly introduced. An implementation of the original RFP method according to [2, 4], as well as the original PRFD method according to [2, 6], has been extended with a user interface to allow for a user-guided system identification for the evaluation between the physical and non-physical nodes. There exist more advanced techniques for kind of evaluation [1], but the presence of some strong discontinuities in the coherence plots of the FRFs hinted at non-linearities [2] for which a user interactive implementation was preferred.

The employed methods require a controlled excitation of the structure in order to determine its FRF. In the field of civil engineering, this is often a hard task to carry out due to the dimensions of the structures. In order to avoid this limitation, new methods of extracting modal properties from the structural response under unknown excitation have been developed [5, 7]. For the goal of this investigation and considering the dimensions of our test specimens, classical identification methods were appropriated.

2.1 Rational Fraction Polynomials Method

The RFP is a Single Input – Single Output (SISO) Method [4] for modal identification in the frequency domain. It is based on the viscously damped model and describes the FRF of the system as a ratio between two polynomials:

$$H(\omega) = \left. \frac{\sum_{k=0}^{m} a_k s^k}{\sum_{k=0}^{n} b_k s^k} \right|_{s=j\omega} \quad (1)$$

By minimizing the error between the measures FRF values and the model, the coefficients a_k and b_k are obtained, from which the modal parameters can be retrieved. The original implementation of this method can be found in [4].

2.2 *Polyreference Frequency Domain Method*

The PRFD method is a Multiple Input – Multiple Output (MIMO) Method [6] that has been implemented in order to evaluate the modal shapes of the detected modes. For a detailed description of its implementation, we refer to [2].

3 Experimental Investigations

The investigated specimens consisted of rectangular 1,100 × 360 mm glass plates simply supported on both ends with a span of 1,000 mm.

The response of the plate under impulsive excitation was measured in terms of accelerations, by sensors placed at positions 1–4. The sensors 1 and 2 were placed at the middle of the plate, while sensors 3 and 4 at a distance to the supports of respectively 1 / 3 and 1 / 4 of the span as illustrated in Fig. 1.

The impulse excitation was applied in the center of the plate (Position 2) with a calibrated impulse hammer, allowing for a broad-band excitation of the glass panel (see Fig. 2). The measurements were conducted at a sampling rate of 10,000 Hz, in order to avoid any aliasing effects. The frequencies of interest lied between 0 and 1,000 Hz.

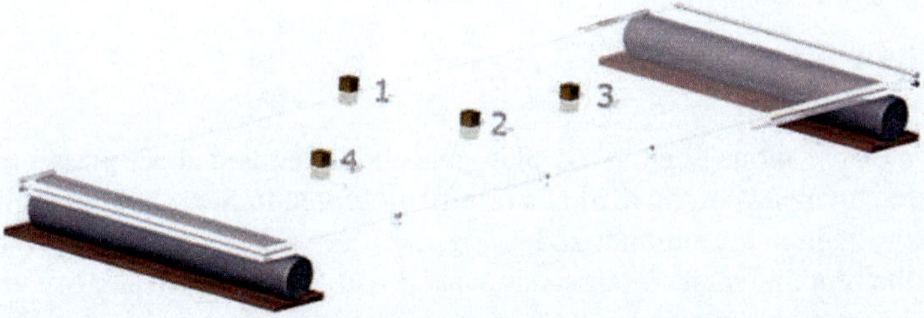

Fig. 1 Test set-up and positions of the sensors, exemplarily laminated sheet glass

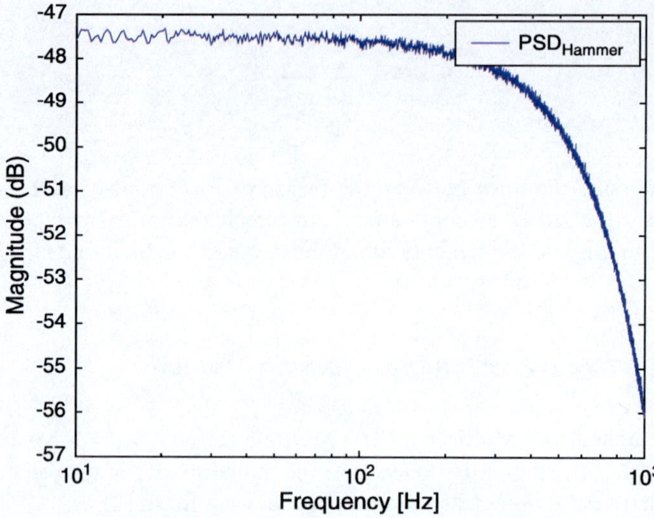

Fig. 2 PSD of the impulse excitation

3.1 Measurement Results

A total of 32 measurements with repeated impulse excitation where performed for a mono-sheet glass plate and for a laminated sheet glass plate. The spectral analysis and determination of FRF was done by means of the PSD using Welchs method with Hamming windowing. A frequency resolution of 0.076 Hz was achieved. The obtained FRFs are given in Fig. 3.

In order to evaluate the linearity of the response and hence the presence of non-linearities in the measured data, the coherence function of the FRFs was calculated (see Fig. 4). During the identification of the modal properties, these results were considered as an additional criteria to eventually distinguish the physical modes from the non-physical.

4 System Identification

The FRFs of the mono-sheet glass plate and the laminated sheet glass plate were investigated by means of the methods briefly presented in Sect. 2. The results of the modal identification are summarized in Fig. 5.

With the implemented methods it was possible to easily identify the main vibration modes of the system. The second natural vibration mode could not be

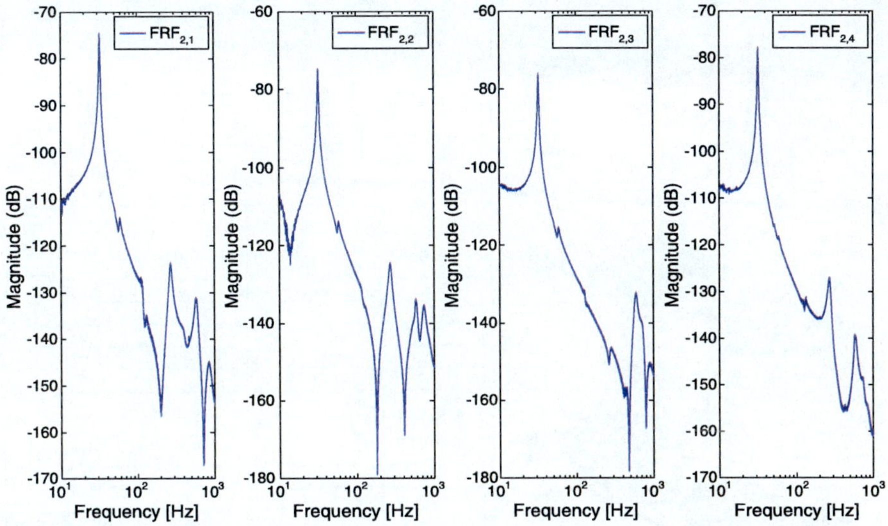

Fig. 3 Frequency response functions for the four measuring positions, laminated sheet glass

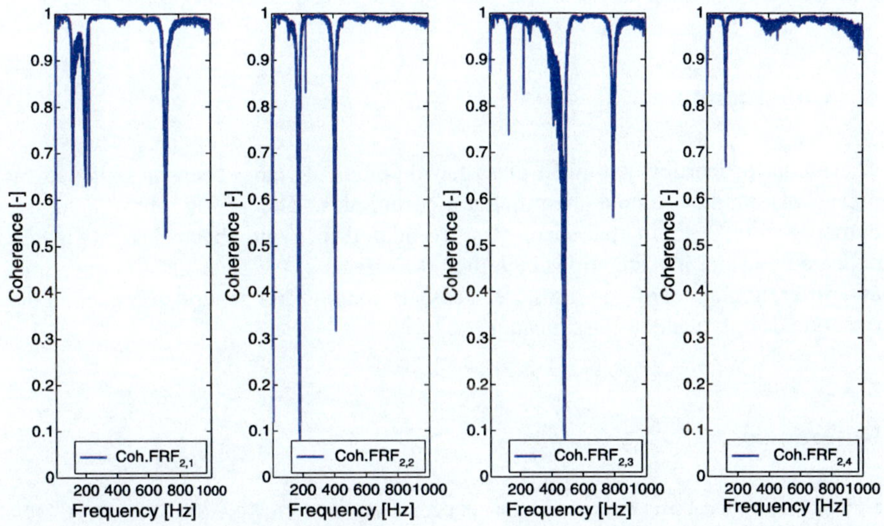

Fig. 4 Coherence of the FRFs from Fig. 3

identified, since the excitation happened in the middle of the plate, where this mode shows no amplitude. Higher frequency modes could be identified as well, although the results (specially for the damping value and modal shape) varied strongly between simulation runs.

Identified Modal Shape	Mono-Sheet Glass		Laminated Sheet Glass	
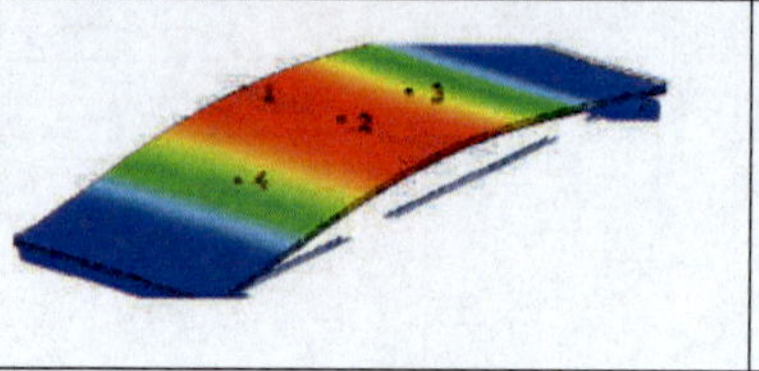	$f_1 = 18{,}46\ Hz$ $\xi_1 = 0{,}0038$		$f_1 = 32{,}02\ Hz$ $\xi_1 = 0{,}0092$	
	Pos.	Disp.	Pos.	Disp.
	1	1	1	1
	2	0.97	2	0.96
	3	0.83	3	0.84
	4	0.66	4	0.68
	$f_1 = 165{,}13\ Hz$ $\xi_1 = 0{,}0188$		$f_1 = 273{,}26\ Hz$ $\xi_1 = 0{,}0466$	
	Pos.	Disp.	Pos.	Disp.
	1	1	1	1
	2	0.91	2	0.92
	3	0.00	3	0.02
	4	-0.72	4	-0.72
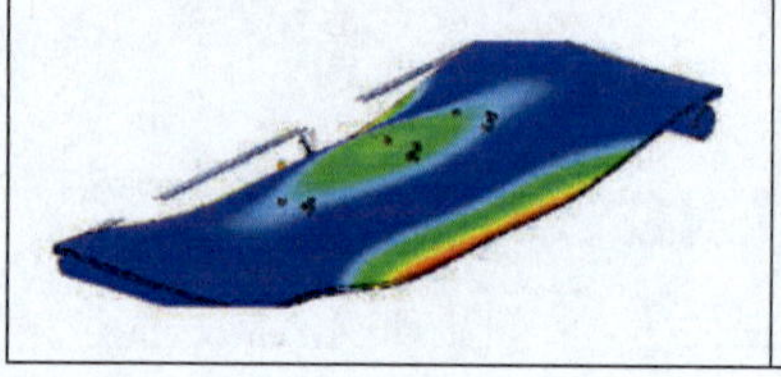	$f_1 = 366{,}06\ Hz$ $\xi_1 = 0{,}148$		$f_1 = 589{,}97\ Hz$ $\xi_1 = 0{,}0346$	
	Pos.	Disp.	Pos.	Disp.
	1	1	1	1
	2	-0.48	2	-0.51
	3	-0.59	3	-0.97
	4	-0.53	4	-0.33

Fig. 5 Results of the first three identified modes for both mono-sheet and laminated sheet glass

5 Conclusions

The modal properties of typical glass façade elements have been investigated by means of standard modal identification techniques. The results show a higher damping ratio in laminated sheet glass compared to mono-sheet glass, as it was expected. Further investigations including the effects of the steel-glass connections and different glass configurations are necessary to aim for a general approach in the dynamic design of steel-glass façades.

References

1. Fasana, A.: Modal parameters estimation in the z-domain. Mech. Syst. Signal Pr. **23**(1), 217–225 (2009)
2. Maia, N.: Theoretical and Experimental Modal Analysis. (1997)
3. Maia, N., Silva, J.: Modal analysis identification techniques. Phil. Trans. R. Soc. Lond. **2001**(359), 29–40 (2001)
4. Richardson, M.H., Formenti, D.L.: Parameter estimation from frequency response measurements using rational fraction polynomials. In: Proceedings of the 1st International Modal Analysis Conference, Orlando, FL (1982)

5. Zhang, Y., Zhang, Z., Xu, X., Hua, H.: Modal parameter identification using response data only. J. Sound Vib. **282**(1–2), 367–380 (2005)
6. Zhang, L., Kanda, H., Brown, D., Allemang, R.: A polyreference frequency domain method for modal parameter identification. ASME **1985**(No.85-DET-106), 1–6 (1985)
7. Zhang, L., Wang, T., Tamura, Y.: A frequency-spatial domain decomposition (fsdd) method for operational modal analysis. Mech. Syst. Signal Pr. **24**(5), 1227–1239 (2010)

Damping Optimization for Linear Vibrating Systems Using Dimension Reduction

Peter Benner, Zoran Tomljanović, and Ninoslav Truhar

Abstract We consider a mathematical model of a linear vibrational system described by the second-order system of differential equations $M\ddot{x}+D\dot{x}+Kx = 0$, where M, K and D are positive definite matrices, called mass, stiffness and damping, respectively. We are interested in finding an optimal damping matrix which will damp a certain part of the undamped eigenfrequencies. For this we use a minimization criterion which minimizes the average total energy of the system. This is equivalent to the minimization of the trace of the solution of a corresponding Lyapunov equation. In this paper we consider an algorithm for the efficient optimization of the damping positions based on dimension reduction techniques. Numerical results illustrate the efficiency of our approach.

Keywords Vibrating system • Lyapunov equation • Energy minimization • Dimension reduction

1 Introduction

The aim of this paper is the determination of optimal damping for the following linear vibrational system:

$$M\ddot{x} + D\dot{x} + Kx = 0, \tag{1}$$

P. Benner (✉)
Max Planck Institute for Dynamics of Complex Technical Systems, Sandtorstr. 1, D-39106 Magdeburg, Germany
e-mail: benner@mpi-magdeburg.mpg.de

Z. Tomljanović • N. Truhar
Department of Mathematics, University J.J. Strossmayer in Osijek, Trg Lj. Gaja 6, 31000 Osijek, Croatia
e-mail: ztomljan@mathos.hr; ntruhar@mathos.hr

J. Náprstek et al. (eds.), *Vibration Problems ICOVP 2011: The 10th International Conference on Vibration Problems*, Springer Proceedings in Physics 139,
DOI 10.1007/978-94-007-2069-5_41,

where M and K (called mass and stiffness, respectively) are real, symmetric positive definite matrices of order n. The damping matrix is defined as $D = C_u + C_{ext}$, where C_{ext} is the external damping. The internal damping C_u is usually taken to be a small multiple of the critical damping, more precisely [1]

$$C_u = \alpha_c C_{crit}, \text{ where the critical damping is } C_{crit} = 2M^{1/2}\sqrt{M^{-1/2}KM^{-1/2}}M^{1/2}. \tag{2}$$

For the sake of simplicity, we will use parameter $\alpha = 2\alpha_c$. Equation 1 can be transformed to phase space which yields a system of first order differential equations. For this purpose, let Φ be a matrix that simultaneously diagonalizes M and K, that is

$$\Phi^T K \Phi = \Omega^2 = \text{diag}(\omega_1^2, \ldots, \omega_n^2) \quad \text{and} \quad \Phi^T M \Phi = I. \tag{3}$$

For the internal damping defined in Eq. 2 it holds that $\Phi^T\ C_u \Phi = \alpha\Omega$. The positive numbers $\omega_1, \omega_2, \ \ldots, \omega_n$ are the eigenvalues of the undamped system, also called undamped eigenfrequencies. Then, we can write the differential equation (1) in phase space as

$$\frac{d}{dt}y = Ay, \quad \text{where} \quad A = \begin{bmatrix} 0 & \Omega \\ -\Omega & -\Phi^T D \Phi \end{bmatrix}, \quad y = \begin{bmatrix} y_1 \\ y_2 \end{bmatrix}, \tag{4}$$

for more details see [1–4]. The first order differential equation given above has the solution $y = e^{At}\,y_0$, where y_0 contains the initial data. It has been shown in [3, 5] that A from Eq. 4 is a stable matrix, that is, the eigenvalues of A are in the open left half of the complex plane.

The main aim is to determine the "best" damping matrix D which will insure optimal evanescence of each component of y. For this purpose, we will use the criterion of minimization of the total energy of the system, that is

$$\int_0^\infty E(t)dt \quad \rightarrow \quad \min, \tag{5}$$

where $E(t)$ is the total energy of the system at a given time t, as a sum of kinetic and potential energies. In [3] it is shown that by taking the average over all initial states of the unit total energy and a given frequency range, the minimization criterion (5) is equivalent to

$$\text{trace} X \quad \rightarrow \quad \min, \tag{6}$$

where X is the solution of the Lyapunov equation

$$AX + XA^T = -GG^T,$$

with A as in Eq. 4. The matrix G depends on the eigenfrequencies which have to be damped. If we are interested in damping of all undamped eigenfrequencies, then $G = I$, while in the case of damping of just the first s eigenfrequencies of the undamped system, the matrix G has the following form [3]:

$$G = \begin{bmatrix} I_s & 0 & 0 & 0 \\ 0 & 0 & I_s & 0 \end{bmatrix}^T. \tag{7}$$

This optimization problem has been intensively considered, see for example [1–4, 6–10]. Optimal damping aims at optimization of the damping positions as well as the corresponding viscosities. In this paper we will be mainly interested in the optimal dampers' positions. The optimal viscosities can then be determined by an additional optimization procedure.

Let the external damping be given by

$$C_{ext} = v_1 e_{i_1} e_{i_1}^T + v_2 e_{i_2} e_{i_2}^T + \cdots + v_k e_{i_k} e_{i_k}^T, \tag{8}$$

where i_j, $j = 1, \ldots, k$, corresponds to the damping positions with viscosities v_j, $j = 1, \ldots, k$. It follows directly from Eq. 8 that it is sufficient to find the optimal positions such that $1 \leq i_1 < i_2 < \ \ldots \ < i_k \leq n$. Since we are interested in determination of the optimal damping positions and viscosities, we will use a new notation for trace$X$ which is now a function of the damping positions $(i_1, \ldots, i_k)$ and the corresponding viscosities $(v_1, \ldots, v_k)$. Thus, let $X(C(v_1, \ldots, v_k; i_1, \ldots, i_k))$ be the solution of the Lyapunov equation

$$A\, X(C(v_1, \ldots, v_k; i_1, \ldots, i_k)) + X(C(v_1, \ldots, v_k; i_1, \ldots, i_k))\, A^T = -GG^T, \tag{9}$$

where $(i_1, \ldots, i_k)$ are the damping positions and $(v_1, \ldots, v_k)$ the corresponding viscosities. The matrix G is given in Eq. 7, while the matrix A equals

$$A = \begin{bmatrix} 0 & \Omega \\ -\Omega & -\alpha \Phi^T C_{crit} \Phi - C(v_1, \ldots, v_k; i_1, \ldots, i_k) \end{bmatrix}, \tag{10}$$

where $C(v_1, \ldots, v_k; i_1, \ldots, i_k) = \Phi^{\,T} C_{ext} \Phi$ and Φ is the matrix given in Eq. 3.

For a given mass matrix M, stiffness matrix K, internal damping C_u and k dampers, we are interested in determining the optimal positions $i_1^{opt}, \ldots, i_k^{opt}$ and corresponding viscosities $v_1^{opt}, \ldots, v_k^{opt}$ such that trace$X(C(v_1, \ldots, v_k; i_1, \ldots, i_k))$ is minimal.

In the next section we will discuss the main difficulties in the process of damping optimization and we suggest a new approach for efficient damping optimization.

2 Damping Optimization by "Discrete to Continuous" Optimization Approach

This section will be mainly devoted to the calculation of the optimal damping positions. The problem of determining the optimal damping is extremely demanding, because numerous Lyapunov equations have to be solved. Furthermore, for systems with large dimensions, even solving a single Lyapunov equations with direct solvers (such as the Bartels–Stewart algorithm [11]) can become very demanding.

One approach for determination of the optimal damping positions is the "direct" approach, which includes viscosity optimization for all possible damping configurations. For the external damping given by Eq. 8, we need to optimize the viscosity at all different configurations of dampering positions such that $1 \le i_1 \le i_2 \le \ldots \le i_k \le n$. Then the optimal positions are those corresponding to the minimal trace$X$.

One heuristic optimization approach for the determination of the optimal damping positions is presented in [9]. In this paper, the authors group the possible damping positions in order to optimize the viscosities with respect to a smaller number of damping positions.

Here we will introduce the "discrete to continuous" heuristical approach which relies on the optimization of functions of real variables. First, we will define an additional function which will be used in the optimization procedure. We want to determine optimal damping for k dampers with different viscosities. Thus, for $\mathcal{D} \subset \mathbb{R}^{2k}$ we define a function $f : \mathcal{D} \to \mathbb{R}$ by

$$f(v_1, \ldots, v_k; i_1, \ldots, i_k) = \mathrm{trace}(X(C(v_1, \ldots, v_k; [i_1], \ldots, [i_k]))), \tag{11}$$

where $[\,\cdot\,]$ stands for the rounding (we use the MATLAB®; function `round`) and the matrix $X(C(v_1, \ldots, v_k; [i_1], \ldots, [i_k]))$ is the solution of the Lyapunov equation (9). Here i_k is considered is as a continuous variable and the damping positions $[i_1], [i_2], \ldots, [i_k]$ with corresponding viscosities $v_1, v_2, \ldots, v_k$ determine the matrix C.

Now, we reduce our optimization problem to the minimization of the function (11) with continuous domain.Thus, for minimization of this function we can use standard methods like Nelder–Mead [12] or, for example, Newton like methods. When we determine the minimum of the function (11) we will denote the point where the minimum is achieved by $(\hat{v}_1, \hat{v}_2, \ldots, \hat{v}_k; \hat{i}_1, \hat{i}_2, \ldots, \hat{i}_k)$. Then the optimal positions are $[\hat{i}_1], [\hat{i}_2], \ldots, [\hat{i}_k]$ with corresponding optimal viscosities equal to $\hat{v}_1, \hat{v}_2, \ldots, \hat{v}_k$.

Apart from the above mentioned minimization procedure like the Nelder–Mead method (implemented in the MATLAB function `fminsearch`) or the Newton-like methods (implemented in the MATLAB function `fmincon` or `fminunc`), one can also use a genetic algorithm (implemented in the MATLAB function `ga`). In the optimization process we will use the Nelder–Mead method which is much more robust than the other mentioned methods for our minimization problem.

Algorithm 1 ("Discrete to continuous" approach for determination of optimal positions)

Require: d_3, d_4 – parameters which determine the first and the second grid;
Ensure: Optimal dampers' positions $[i_1^{opt}], \ldots, [i_k^{opt}]$ with optimal viscosities $v_1^{opt}, \ldots, v_k^{opt}$.
1: **for** $i_1^s = 1 + d_4 : d_3 : n - d_4$ **do**
2:     **for** $i_2^s = i_1^s + d_4 : d_3 : n - d_4$ **do**
3:         ...
4:         **for** $i_k^s = i_{k-1}^s + d_4 : d_3 : n - d_4$ **do**
5:             Using starting points $(v_1^s, \ldots, v_k^s; i_1^s, \ldots, i_k^s)$ calculate (for example with Nelder–Mead algorithm)
$$\min_{\substack{(v_1,\ldots,v_k)\in\mathbb{R}^k \\ 1\le i_1<i_2<\ldots<i_k\le n}} f(v_1, \ldots, v_k; i_1, \ldots, i_k).$$
6:         **end for**
7:         ...
8:     **end for**
9: **end for**
10: The parameters which correspond to the minimal value calculated in Step 5, are returned as optimal parameters $(v_1^{opt}, \ldots, v_k^{opt}; i_1^{opt}, \ldots, i_k^{opt})$.

A further question in minimization with a Nelder–Mead method is the choice of good starting points. For this purpose we will define a grid of starting points for damping positions, which will correspond to the starting points generated in Steps 1–4 of Algorithm 1. Some fixed values $v_1^s, v_2^s, \ldots, v_k^s$ will be taken as starting viscosities.

First, we have to define the parameters d_3 and d_4 which determine the grid of starting points $(i_1^s, \ldots, i_k^s)$. As can be seen from Algorithm 1, the parameter d_3 determines the difference between points inside the region, while the parameter d_4 defines the distance to the edge of the region where the optimal position is to be found.

Note that Nelder–Mead [12] is an unconstrained multidimensional optimization method. In numerical experiments the optimization procedure could require an evaluation at the points that are not in the domain (for example, viscosities may become negative). Thus, at points that are outside the domain where the optimization is performed, in our optimization procedure we set the function value to some constant large enough. With this, our optimization procedure will always return a minimum which is inside the domain of our value function.

3 Damping Optimization Based on Dimension Reduction and Continuous Minimization

We have introduced the "discrete to continuous" approach which can be applied to vibrating systems of moderate dimensions. However, for large systems solving of the corresponding Lyapunov equation is quite demanding itself. Thus, we propose a new approach which will combine approximation algorithms that use dimension reduction techniques with the "discrete to continuous" approach introduced in the previous section.

Algorithm 2 (Approximation of the function $f(\nu_1, \ldots, \nu_k; i_1, \ldots, i_k)$)

```
Require: κ ≥ 1:
    tol_start – tolerance for the first approximation;
    ε – tolerance for bounding the approximation error;
    c1 – a positive constant for scaling the tolerance (c1 < 1).
Ensure: f_approx – approximation of the function f(ν1, ..., νk; i1, ..., ik).
tol = tol_start
while tol > 10^4 u do
    Calculate an approximation of function f(ν1, ..., νk; i1, ..., ik) with [7, Algorithm 2] using
    tolerance tol, and denote the approximation by f_approx.
    Calculate the right-hand side of the bound given in [7, Theorem 3.1] and denote it by b
    if b < ε then
        return f_approx
        break
    else
        tol = c1 · tol
    end if
end while
```

In case we intent to damp all undamped eigenfrequencies, we will use the approximation algorithms introduced in [6]. Contrary to this, in the case when we damp a selected part of the undamped eigenfrequencies, we will use the dimension reduction approach derived in [7].

In the minimization process with the "discrete to continuous" approach we cannot just directly apply the algorithms from [6, 7], since these algorithms optimize viscosities at given damping positions, while in the "discrete to continuous" approach we change the damping positions during the optimization process. Thus, we have to modify our algorithms which use a dimension reduction technique. This modification includes checking of the corresponding error bound at the each step of the optimization procedure. More precisely, for the approximation of the function given in Eq. 11, we have to treat two cases which depend on the eigenfrequencies which have to be damped.

In the case of damping of selected eigenfrequencies, in Step 5 of Algorithm 1, we need to calculate an approximation of the function $f(\nu_1, \ldots, \nu_k; i_1, \ldots, i_k)$. Algorithm 2 gives an approximation with a given tolerance ε. In Algorithm 2, the parameter $\mathbf{u}$ represents the machine precision. For the purpose of simplification, on the input we give just parameters that are essential for the understanding of the algorithm. Similarly, if we damp all eigenfrequencies, in Step 5 of Algorithm 1 we calculate approximation of the function $f(\nu_1, \ldots, \nu_k; i_1, \ldots, i_k)$ using the approximation algorithms introduced in [6].

In the following example we will demonstrate the performance of damping optimization using approximation algorithms and the "discrete to continuous" approach.

Example 1. We consider the n-mass oscillator or oscillator ladder with two dampers, shown in Fig. 1, which describes the mechanical system of n masses and

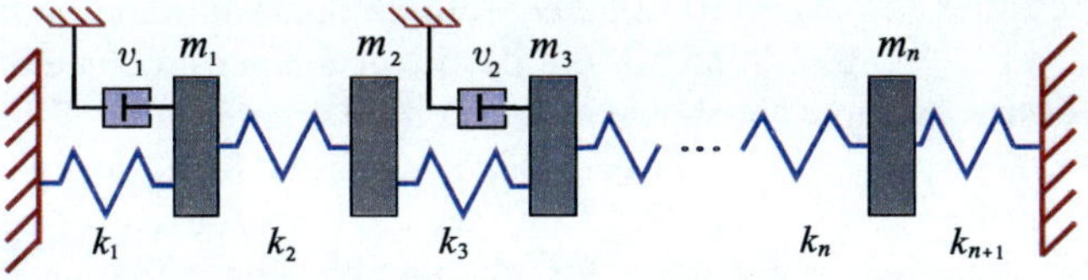

Fig. 1 n-Mass oscillator with two dampers of different viscosities

$n+1$ springs with corresponding stiffnesses. For such a mechanical system the mathematical model is given by $M\ddot{x} + D\dot{x} + Kx = 0$, where for the stiffness and mass matrices we have

$$K = \begin{pmatrix} k_1 + k_2 & -k_2 & & & \\ -k_2 & k_2 + k_3 & -k_3 & & \\ & \ddots & \ddots & \ddots & \\ & & -k_{n-1} & k_{n-1} + k_n & -k_n \\ & & & -k_n & k_n + k_{n+1} \end{pmatrix}, \quad M = \operatorname{diag}(m_1, m_2, \ldots, m_n),$$

where $m_i > 0$ for $i = 1 \ldots, n$ are the masses and $k_i > 0$ for $i = 1 \ldots, n+1$ are the stiffnesses. We will consider the following configuration for $n = 1{,}600$:

$$m_i = 120 - (i-1)/5, \quad i = 1, \ldots, 100;$$
$$m_i = i, \quad i = 101, \ldots, n; \qquad k_i = 4, \quad i = 1, \ldots, n+1.$$

For the numerical test, we demand to damp all the undamped eigenfrequencies that are smaller than 0.005 by magnitude, yielding $s = 34$ (s determines the matrix G given in Eq. 7).

The damping matrix is $D = C_u + C_{ext}$, where the internal damping C_u is defined as in Eq. 2 with $\alpha_c = 0.001$. Since we will consider two dampers of different viscosities, the external damping is defined by $C_{ext} = v_1 e_i e_i^T + v_2 e_j e_j^T$ for $1 \leq i < j \leq n$, where e_i is the ith canonical basis vector, and v_1, v_2 are the viscosities of the dampers applied on the ith and jth mass, respectively.

As we mentioned above, the considered optimization process is extremely demanding, because it requires solving the Lyapunov equation (9) numerous times. In order to compare the performance of our approach, instead of performing optimization in all damping positions (this corresponds to "direct" approach), we will optimize the viscosities on the following equidistant mesh of damping positions:

$$i = 51 : 50 : n, \quad j = i + 51 : 50 : n, \tag{12}$$

which will give 465 different damping positions. We obtain that the optimal position with respect to the mesh (12), is the position $(i,j) = (651,\ 1{,}352)$ with optimal viscosities $(v_1, v_2) = (107.03009,\ 150.49333)$, while the optimal trace is $\text{trace}(X(v_1, v_2)) = 993,\ 067.32851$. More details of the performance of dimension reduction for this example are shown in [7]. For the optimization of the viscosities on the above mesh we needed 104 days, which means that for viscosity optimization at one point of the mesh we need $5.4\,\text{h}$ on average. Recall that a "direct" approach requires the viscosity optimization for all possible configurations of damping positions. As there are $n(n-1)/2$ different positions, it is impossible to apply direct optimization here. But optimization becomes possible with our approach which combines an approximation algorithm with the "discrete to continuous" approach.

Now, we will present results obtained with the "discrete to continuous" approach presented in Algorithm 1, but in Step 5 we will use the function approximation given by Algorithm 2.

In Algorithm 1 we use the following configuration:

$$d_3 = 160; \quad d_4 = 80; \quad v_1^s = v_2^s = 50.$$

In Algorithm 2 we use the following configuration:

$$tol_{start} = 0.02; \quad \varepsilon = 0.05; \quad c_1 = 0.5.$$

The parameters d_3 and d_4 define the grid with 45 different points in Algorithm 1. The function is minimized with the MATLAB function `fminsearch` and for a termination tolerance for the function value we take 0.1, which determines the absolute error. Thus the relative error has magnitude $\mathcal{O}(10^{-7})$. For the termination tolerance of the optimization variables we have used 0.01 (this also determines the absolute error). We have obtained that the optimal damping positions equal to (730, 1,274) with optimal viscosities equal to $(120.47387,\ 120.38917)$. For these parameters the value of our penalty function equals $987,\ 258.34332$. This value was calculated using an algorithm without dimension reduction. Note that combining Algorithm 1 with dimension reduction technique, we obtain a smaller trace. That is we have obtained the optimal positions which are not included in the mesh (12). This is not a surprise since the mesh (12) includes just a small number of positions.

For calculating the approximation of the optimal damping with this approach we only needed 0.532 days. The obtained results with corresponding CPU times were calculated using an Intel(R) Core(TM) i7 CPU 920 with 12 GB of RAM and 8 MB cache.

Another possibility for optimization is the application of the heuristic introduced in [9] with the help of the approximation algorithms which use the mentioned dimension reduction techniques.

4 Conclusions

Damping optimization in a mechanical vibrating system is a very demanding problem due to the numerous Lyapunov equations which have to be solved. In this paper, we have introduced the "discrete to continuous" approach which considerably reduces the number of Lyapunov equations which have to be solved. Furthermore, we have proposed a new approach which is based on dimension reduction and continuous minimization.

We can conclude that combining the approximation algorithm with the "discrete to continuous" approach we have significantly accelerated the time needed for the calculation of the approximation of optimal damping.

References

1. Truhar, N., Veselić, K.: An efficient method for estimating the optimal dampers' viscosity for linear vibrating systems using Lyapunov equation. SIAM J. Matrix Anal. Appl. **31**(1), 18–39 (2009)
2. Brabender, K.: Optimale Dämpfung von linearen Schwingungssystemen. PhD thesis, Fernuniversität, Hagen (1998)
3. Nakić, I.: Optimal damping of vibrational systems. PhD thesis, Fernuniversität, Hagen (2002)
4. Truhar, N., Veselić, K.: On some properties of the Lyapunov equation for damped systems. Math. Commun. **9**(2), 189–197 (2004)
5. Tisseur, F., Meerbergen, K.: The quadratic eigenvalue problem. SIAM Rev. **43**(2), 235–286 (2001)
6. Benner, P., Tomljanović, Z., Truhar, N.: Dimension reduction for damping optimization in linear vibrating systems. Z. Angew. Math. Mech. **91**(3), 179–191 (2011). doi:10.1002/zamm.201000077
7. Benner, P., Tomljanović, Z., Truhar, N.: Optimal damping of selected eigenfrequencies using dimension reduction. Preprint 1/2011. Department of Mathematics, J. J. Strossmayer University of Osijek, Osijek (2011)
8. Truhar, N.: An efficient algorithm for damper optimization for linear vibrating systems using Lyapunov equation. J. Comput. Appl. Math. **172**(1), 169–182 (2004)
9. Truhar, N., Tomljanović, Z.: Estimation of optimal damping for mechanical vibrating systems. Intl. J. Appl. Math. Mech. **5**(5), 14–26 (2009)
10. Veselić, K., Brabender, K., Delinić, K.: Passive control of linear systems. In: Rogina, M., et al. (eds.) Applied Mathematics and Computation, pp. 39–68. Department of Mathematics, University of Zagreb, Zagreb (2001)
11. Bartels, R., Stewart, G.: A solution of the matrix equation $AX + XB = C$. Comm. ACM **15**(9), 820–826 (1972)
12. Nelder, J.A., Mead, R.: A simplex method for function minimization. Comput. J. **7**, 308–313 (1965)

Dynamics Modelling of Tensegrity Structures with Expanding Properties

Musa Abdulkareem, M. Mahfouf, and D. Theilliol

Abstract Given the prestress level of a tensegrity structural system obtained from any form-finding method, an important step in the design process is to develop mathematical models that describe the behaviour of the system. Moreover, tensegrity structures are strongly dependent on their geometric, or kinematic, configurations. As such, except for small scale tensegrity structures with a few structural members, resorting to the use of computational techniques for analysis is a necessity. Because tensegrity structures are kinematically and statically indeterminate structures, a free standing tensegrity structure has at least one rigid body mode apart from the six rigid body modes that can be eliminated, for example, by applying boundary conditions assuming the structure is attached to a base. In this paper, a new general tool (applicable to small and large systems) for systematic and efficient formulation of structural models for tensegrity systems is proposed. Current tools are limited to structures with a few degrees of freedom (DOF), however, this new tool simplifies the analyses of tensegrity structures with several DOFs and provides a new insight into the behaviour of these interesting and yet challenging structures, at least from a control systems' viewpoint.

Keywords Tensegrity • Stiffness matrix • Dynamics • Finite element method • State-space • Singular systems

M. Abdulkareem (✉) • M. Mahfouf
Department of Automatic Control and Systems Engineering, The University of Sheffield, Sheffield, United Kingdom
e-mail: m.abdulkareem@sheffield.ac.uk

D. Theilliol
Faculté des Sciences et Techniques, Nancy-Université, Paris, France

J. Náprstek et al. (eds.), *Vibration Problems ICOVP 2011: The 10th International Conference on Vibration Problems*, Springer Proceedings in Physics 139,
DOI 10.1007/978-94-007-2069-5_42,

1 Introduction

A system that consists of a given set of cables connected to a configuration of rigid bodies which can be put in a state of static equilibrium as a result of internal forces of the cables alone in the absence of external forces is called a tensegrity structure. In tensegrity related research, bars refer to the rigid bodies and a class k tensegrity system has as many as k bars in contact at the nodes [1]. In the technological quest for stronger, stiffer, flexible, and yet, lightweight structural systems, tensegrity systems have the following additional features, among others: mass efficiency, modularity, scalability, deployability, accurate modelling, and, as a consequence, precision control. Form-finding of tensegrity structures involves using the mathematical properties of tensegrity structures to search and define a configuration that satisfies the conditions of static equilibrium of the structural system. The main bottleneck hindering thorough investigation into the dynamic properties of tensegrity structures is the lack of tools that would efficiently solve the equations of motion of tensegrity systems for larger systems with several degrees of freedom. Moreover, most current analysis tools use algebraic formulation and are, as such, limited to tensegrity structures with a few structural members and a simple connectivity. Thus, research in tensegrity dynamics is still an emerging field and this paper attempts to provide tools and insight into the dynamic properties of such structures.

2 Mathematical Formulations

Consider a tensegrity structure with n nodes and b structural members, the forces of tension (for cables) and compression (for bars), the lengths of structural members and the nodal forces in three-dimensional Euclidean space form vectors $f \in \Re^b$, $l \in \Re^b$ and $p \in \Re^{3n}$, respectively. Let the nodal coordinates of all points in 3-dimensional Euclidean space be assembled into column vectors $x \in \Re^n$, $y \in \Re^n$ and $z \in \Re^n$, where (x_i, y_i, z_i) represents the coordinates of node i, then a matrix of nodal coordinates $N \in \Re^{3\times n}$ may be defined as follows:

$$N = [x \ y \ z\ ]^T = [n_i]_{1\times n} \tag{1}$$

where $n_i \in \Re^3$, given by $n_i = \left[x_i \ y_i \ z_i\right]^T$, is the nodal coordinates of node i. Furthermore, a branch-node connectivity matrix [2, 3] denoted C may be defined with the aid of a connectivity graph; for the structural member i connected to two matched nodes numbered $j(i)$ and $h(i)$ (where $j(i) < h(i)$), $c_{i,j}$ is written as follows:

$$c_{i,j} = \begin{cases} +1 \ \ for\ j(i) = 1 \\ -1 \ \ for\ h(i) = 1 \\ 0 \ \ otherwise \end{cases} \tag{2}$$

Equation 2 makes it easy to write the coordinate differences of the connected points as $u = Cx$, $v = Cy$, and $w = Cz$. The i'th structural member with nodes h and j can be uniquely described by a Euclidean row vector $b_i \in \Re^3$, given by $b_i = \left[x_h - x_j \; y_h - y_j \; z_h - z_j \right]$ and has the length $\|b_i\| = \|n_h - n_j\|$ Thus, all members of the structural system can be assembled into matrix $B \in \Re^{b\times 3}$ as follows:

$$B = [u \quad v \quad w] = [b_i]_{b\times 1} = C\ N^T \tag{3}$$

2.1 Derivation of the Stiffness and Mass Matrices

Tensegrity structures are in a state of minimum potential energy and this energy (assumed to be due to straining alone) can be written for the whole structural system as follows:

$$\pi = \sum_{i=1}^{b} \pi_i(\|b_i\|) \tag{4}$$

where $\pi_i(\|b_i\|)$ denotes π_i is a function of $\|b_i\|$. The stiffness matrix of a tensegrity structure is given as follows [1]:

$$K(N) = \sum_{i=1}^{b} K_i(N) = K_q(N) + K_s(N) \tag{5}$$

where $K_i(N) = \left(C_i C_i^T \otimes x_i\right)$, C_i is the i'th row vector of C, $\otimes$ is the Kronecker product, $x_i = \frac{x_i'' b_i b_i^T}{\|b_i\|^2} + q_i \left(I_3 - \frac{b_i b_i^T}{\|b_i\|^2}\right)$, $q_i = \frac{\pi_i'}{\|b_i\|}$ is the tension coefficient of the i'th structural member, $K_s(N) = (C^T \otimes I_3) diag\left[[(\pi]_i'' - q_i \frac{b_i b_i^T}{\|b_i\|^2}\right](C^T \otimes I)$, $K_q(N) = C^T Q C \otimes I_3$, $Q = diag\,(q_i)$, π_i' and π_i'' are first and second derivatives of π_i with respect to nodal displacements, respectively. K_q is called the prestress (or geometric) stiffness matrix and it is mainly a function of tension coefficients, while K_B is a called the elastic stiffness matrix and it is mainly a function of material properties of the structural members. Let the i'th structural member have mass density ρ_i and cross-sectional area A_i; if the member is assumed to deform linearly in the axial direction only, the consistent mass matrix in local coordinates system is given as follows [4]:

$$m_i = \frac{\rho_i A_i l_i,}{6} \begin{bmatrix} 2 & 1 \\ 1 & 2 \end{bmatrix} \tag{6}$$

The transformation to the global coordinate system is obtained as follows:

$$m_i' = \boldsymbol{\lambda}^T m_i \boldsymbol{\lambda},\ \boldsymbol{\lambda} = \frac{1}{l_i} \begin{bmatrix} 1 & 0 \\ 0 & 1 \end{bmatrix} \otimes b_i,\ M = \underline{\sum_{i=1}^{b} m_i'} \tag{7}$$

where m_i' is the global stiffness matrix, $\lambda \in \Re^{6\times6}$ is a transformation matrix and $\underline{m'}_i$ is obtained by identifying the locations of the i'th structural member in the global system and including zeros in the remaining location and M is the stiffness matrix of the entire structure.

3 Numerical Results

For the tensegrity structures shown in Fig. 1, we assume the length l, tension coefficient q, Young's modulus E, cross-sectional area A, and mass densities ρ of each structural member is as given in Table 1.

Figure 2 shows the dynamic simulation of the tensegrity structure of Fig. 1 when three vertically downward loads, each of 300 N, are placed on nodes 10, 11 and 12. Nodes 1, 2, and 3 are made rigid (constrained). The damping constant $\zeta_j = 0.02$ and the integration step-size is 0.02 s. It can be seen in Fig. 2 that there is no displacement in the x-direction of node 4. Although, only the results of the current loading configuration is given, numerous simulations reveal that at least one nodal coordinate remain unaffected by any loading effect irrespective of the loading configuration due to the singular nature of the system. More generally, tensegrity structures are singular systems. Singular systems are systems that satisfy the following state differential equations [5]:

$$E\dot{x} = A\,x + B\,u \tag{8}$$

where E is a singular matrix with dimension equals to the number of degrees of freedom n and rank $n - r$ and r is the number of nodes unaffected by the external forces (zero energy modes). If E is matrix defined as follows:

$$E = \begin{bmatrix} \mathrm{I_e} & 0 \\ 0 & 0_\mathrm{r} \end{bmatrix} \tag{9}$$

It is easy to see how Eq. 8 can be extended to tensegrity systems.

4 Conclusions

Modal analysis shows that tensegrity structures, even after the six rigid-body modes are eliminated, include at least one zero-energy. In general, tensegrity structures are singular systems and formulating the dynamic equations using the state-space theory provides an opportunity of extending many control system concepts to these structural systems. In particular, unlike the traditional techniques where mass and stiffness matrices are reduced by focusing only on few nodal coordinates and which

Table 1 Lengths and tension coefficients of tensegrity structures shown in Fig. 1

Member no.	1–3	4	5	6	7	8	9	10–12	13–15	16–18	19–21	22–24
l (m)	10.00	7.38	7.38	7.38	7.38	7.38	7.38	11.67	11.67	10.00	16.71	16.71
q (N/m)	31.06	30.15	49.09	30.15	49.09	30.15	49.09	43.46	27.30	7.423	−43.46	−20.30
E (GPa)	117	117	117	117	117	117	117	117	117	117	200	200
ρ (kg/m^3)	8,920	8,920	8,920	8,920	8,920	8,920	8,920	8,920	8,920	8,920	7,850	7,850
A ($\times 10^{-6}$ m^2)	1.5	1.5	1.5	1.5	1.5	1.5	1.5	1.5	1.5	1.5	6	6

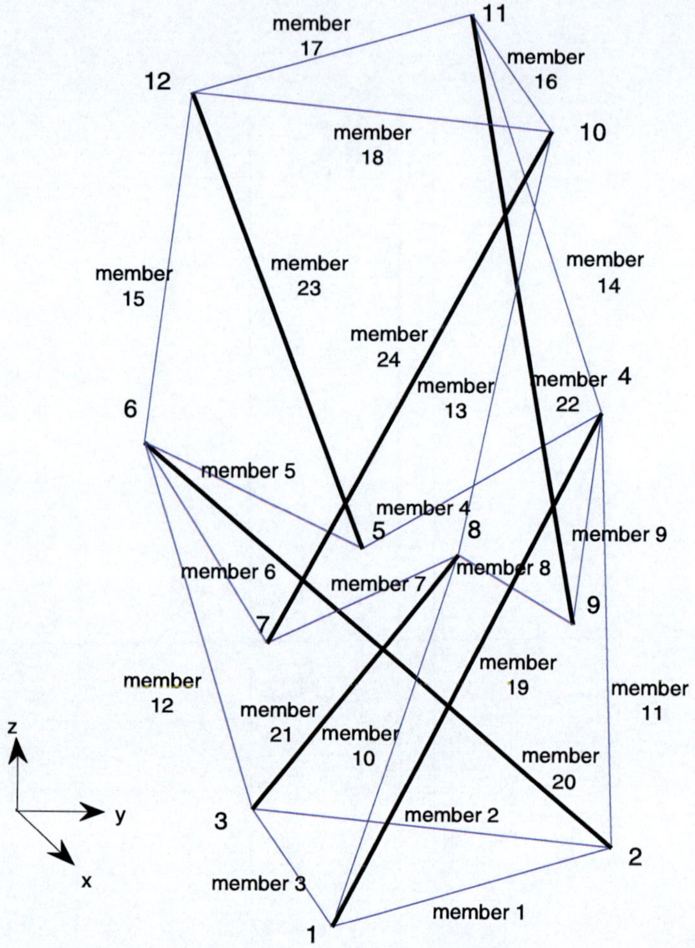

Fig. 1 A 2-stage tensegrity structure (*bars* and *cables* are drawn as *thick* and *thin lines*, respectively)

lead to models that are strongly dependent on the initial choices, the singular state-space representation provides a mechanism for enabling the detection of structural modes with zero energy which is very useful for model reduction and further analysis. Near future work will focus on employing this model for synthesizing controllers for tensegrity structures.

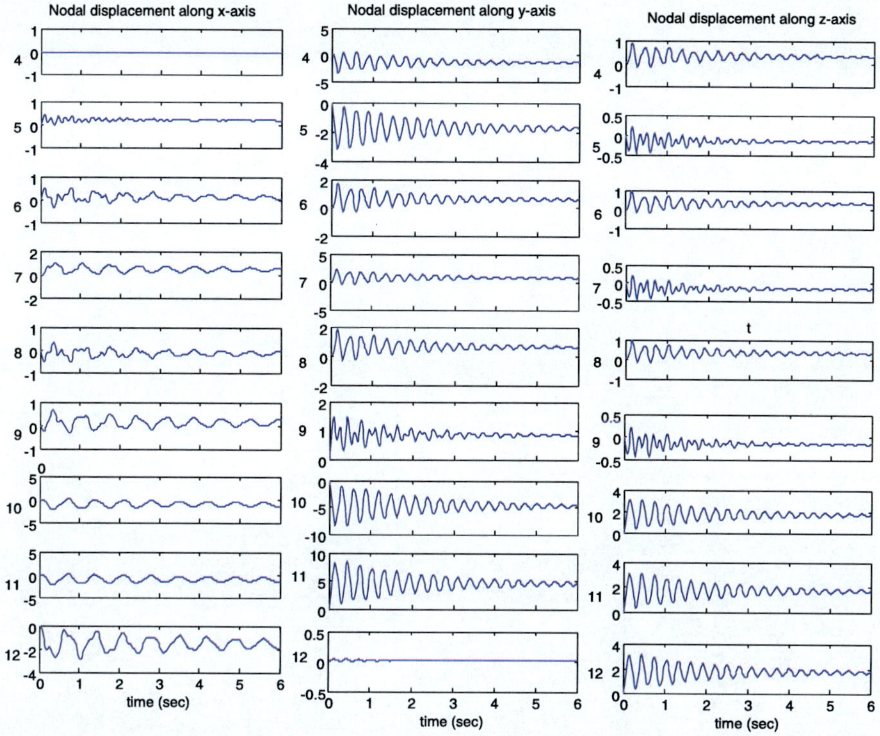

Fig. 2 Dynamic response of the 2-stage tensegrity structure to three *vertically downward* loads of 300N on nodes 10, 11, and 12

References

1. Skelton, R.E., Oliveira, M.C.: Tensegrity Systems. Springer, Dordrecht/Heidelberg/London/NewYork (2009)
2. Schek, H.: The force density method for form finding and computation of general networks. Comput. Methods Appl. Mech. Eng. **3**, 115–134 (1974)
3. Argyris, J.H.: Recent advances in matrix methods of structural analysis. In: Progress in Aeronautical Science, vol. 4. Pergamon Press, New York (1964)
4. Rao, S.S.: The Finite Element Method in Engineering. Butterworth-Heinemann, Boston (1999)
5. Verghese, G., Levy, B., Kailath, T.: A generalized state-space for singular systems. IEEE Trans. Autom. Control **26**, 811–831 (1981)

Transient Response of FGM Pressure Vessels

Hakan Pekel, Ibrahim Keles, Beytullah Temel, and Naki Tutuncu

Abstract The present study aims to investigate the transient behavior of thick-walled cylinders under dynamic internal pressure. Analytical solutions are possible only for simple time-dependent pressure functions. The solution procedure presented is general in the sense that the pressure applied may be an arbitrary continuous function of time, impulsive or given in a discrete form. The material considered is isotropic and heterogeneous with properties varying in the radial direction termed as Functionally Graded Material (FGM). Laplace transform method is used and the inversion into the time domain is performed using the modified Durbin's method. Verification of the numerical procedure is performed by comparing the results with those of an analytical solution available in the literature for a simple exponentially-varying pressure. The inhomogeneity constant in the material property model is shown to have a significant effect on the transient response.

Keywords Functionally graded materials • Vibration • Pressure vessel

1 Introduction

Functionally graded materials (FGM) have seen increasing applications in engineering fields. The continuous spatial variation of physical properties in such materials has eliminated some adverse effects such as stress concentration and

---

H. Pekel (✉) • B. Temel • N. Tutuncu
Cukurova University, Adana, Turkey
e-mail: hpekel@cu.edu.tr; btemel@cu.edu.tr; ntutuncu@cu.edu.tr

I. Keles
Ondokuz Mayis University, Samsun, Turkey
e-mail: kelesibrahim@gmail.com

J. Náprstek et al. (eds.), *Vibration Problems ICOVP 2011: The 10th International Conference on Vibration Problems*, Springer Proceedings in Physics 139,
DOI 10.1007/978-94-007-2069-5_43,

delamination encountered in ordinary composites. Exact vibration analyses of homogeneous cylinders using Laplace transform date back as early as 1940s (e.g. see Tranter [1]). A more recent detailed study on the subject is due to Ghosh [2] where axisymmetric vibration of thick cylinders under various continuous dynamic pressures is investigated.

As for the dynamic response of heterogeneous cylinders; the works by Loy et al. [3] and Pradhan et al. [4] include studies using Rayleigh-Ritz method on the vibration of cylindrical shells made of functionally graded materials composed of stainless steel and nickel or zirconia. The transient waves in a functionally graded cylinder are analyzed by Han et al. [5] using a hybrid numerical method. A finite-element vibration analysis of FGM thick hollow cylinders is done by Shakeri et al. [6] where the cylinder is assumed to be made of many subcylinders.

The present paper presents solutions for the axisymmetric forced vibrations of a functionally graded isotropic thick-walled cylinder. The solutions are obtained in the Laplace domain and inverse transformation to the time domain is done by the modified Durbin's method. For an internal dynamic pressure continuous with time analytical solutions may be obtained through the calculus of residues. This analysis is viable only for simple cases of pressure functions. A benchmark solution given by Keles [7] is used to verify the numerical procedure. In cases of impulsive and piecewise continuous pressures it becomes inevitable to resort to numerical inversion schemes. The method chosen in the present work, Durbin's numerical inverse Laplace transform method, has been efficiently implemented in vibration analysis of structural elements (e.g. see Temel [8]). If the pressure is given in a point-by-point manner its closed-form Laplace transform will not be available and, for such cases, Durbin's method is also a valid, efficient procedure.

Elastic modulus and the density of the cylinder are assumed to vary radially according to a power-law and, for the sake of simplicity, constant Poisson's ratio is assumed throughout. Dynamic response is presented for various values of the inhomogeneity parameter. The analysis presented is readily applicable to FGM disks and hollow spheres.

2 Governing Equations

Consider a thick-walled hollow cylinder (or a disk) of inner radius a and outer radius ka where k is a constant. The cylinder is isotropic and made of functionally graded material. Modulus of elasticity and the density vary through the thickness as $E(r) = E_0 r^{\beta}$ and $\rho(r) = \rho_0 r^{\beta}$, respectively, where the exponent β essentially represents the degree of inhomogeneity. Under axisymmetric conditions using the strain-displacement relations, constitutive equations for plane strain and equation of motion in the radial direction gives the governing differential equation for the radial displacement $\bar{v}$ in the Laplace domain as

$$\frac{d^2\bar{v}}{dx^2} + \frac{m_1}{x}\frac{d\bar{v}}{dx} + \left(\frac{m_2}{x^2} - p^2\right)\bar{v} = 0 \tag{1}$$

where p is the Laplace parameter. The boundary conditions are

$$\bar{\sigma}_r|_{x=1} = -\overline{P}(p) \qquad \bar{\sigma}_r|_{x=k} = 0 \tag{2}$$

The solution of Eq. 1 is given in terms of Bessel functions as

$$\bar{v}(x,p) = x^{\varphi}(C_1 I_n(px) + C_2 K_n(px)) \tag{3}$$

where I_n and K_n are Bessel functions of first and second kind, respectively, of order n with $\varphi = -\frac{\beta}{2}$, $n = \sqrt{1 - \frac{\mu\beta}{1-\mu} + \frac{\beta^2}{4}}$

The constant C_1 and C_2 are evaluated using the boundary conditions. Making use of the recurrence formulae [9] the complete solution in the Laplace domain is written as

$$\bar{v}(x,p) = -\frac{\bar{P}(p)}{C_{11}} x^{\varphi} \frac{F(p)}{G(p)} \tag{4}$$

where

$$\begin{aligned} F(p) = {} & [K_n(pk)S_1 + pkK_{n-1}(pk)]I_n(px) \\ & - [I_n(pk)S_1 - pkI_{n-1}(pk)]K_n(px) \end{aligned} \tag{5}$$

and

$$\begin{aligned} G(p) = {} & [K_n(pk)S_1 + pkK_{n-1}(pk)][I_n(p)S_1 - pI_{n-1}(p)] \\ & - [K_n(p)S_1 + pK_{n-1}(p)][I_n(pk)S_1 - pkI_{n-1}(pk)] \end{aligned} \tag{6}$$

where $S_1 = (n - m - \varphi)$ and $m = C_{12}/C_{11}$

To find the final solution in real time domain, the inverse Laplace transform needs to be performed on Eq. 4. For the internal pressure $P(\tau) = P_0(1-e^{\gamma\tau})$, a benchmark solution given by Keles [7] will be used for comparison purposes.

3 Solutions by the Modified Durbin's Method

For impulsive and piecewise continuous time-dependent pressures the calculus of residues method seizes to be a viable option; numerical methods are inevitably employed.

A numerical inverse Laplace transform technique to obtain the values in the time domain is Durbin's inverse Laplace transform technique based on the Fast Fourier Transform (FFT) algorithm. Durbin's formulation for inverse Laplace transform is summarized as follows [10]:

The function $f(t)$ at time t_j is given by

$$f(t_j) \cong \frac{2Exp[aj\Delta t]}{T}\left[-\frac{1}{2}\text{Re}\left\{\bar{F}(a)\right\} + \text{Re}\left\{\sum_{k=0}^{N-1}(\bar{F}(p_k)L_k)Exp\left[i\left(\frac{2\pi}{N}\right)jk\right]\right\}\right]$$
$$(j = 0, 1, 2, \ldots, N-1) \tag{7}$$

where $\bar{F}(p_k)$ is the Laplace transform of $f(t)$. The kth Laplace parameter is defined as $p_k = a + ik2\pi/T$. The number N is $N = T/\Delta t$ where T is the solution interval and Δt is the time increment which is taken as $\Delta T = 0.1$. The selection of constant a is done by assigning a value to αT. It is suggested that the value of αT be in the range 5–10. For the numerical examples presented in this paper this value is taken as 6. Finally, the results are modified by multiplying each term in the summation by Lanczos factor L_k as suggested in [11]

$$L_k = \sin\left(\frac{k\pi}{N}\right) \Big/ \left(\frac{k\pi}{N}\right), \quad (L_0 = 1) \tag{8}$$

If the Laplace transform of the function $f(t)$ is not given in closed-form such as in the case of point-by-point definition, the discrete values need first to be transformed into the Laplace domain as follows:

$$\bar{F}(p_k) = \Delta t \sum_{n=0}^{N-1} \left[f(t_n)e^{-at_n}\right]e^{-i\frac{2\pi nk}{N}} \tag{9}$$

For various pressures only the term $\overline{P}(p)$ is changed in the solution given by Eq. 4. To verify the numerical procedure, the benchmark solution is resolved by numerical inversion and virtually an exact match is obtained within six-digit accuracy by using the first 20 natural frequencies inherent in the solution.

4 Results and Conclusions

Forced vibration analysis of FGM hollow cylinders has been performed using Laplace transform technique. The material is assumed to be graded with respect to a power-law whose exponent represents the degree of inhomogeneity. A constant Poisson's ratio of 0.3 has been used throughout. For a rectangular wave pressure

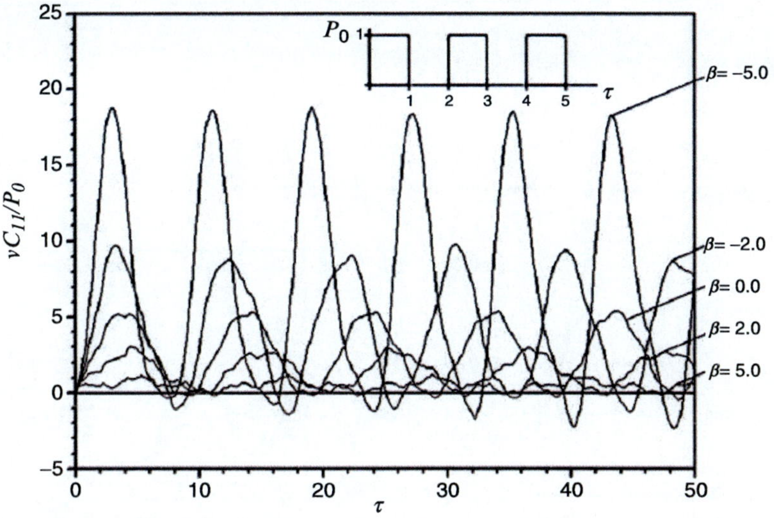

Fig. 1 Radial displacement on inside surface subjected to periodic rectangular pressure for $k = 2.0$, $\beta = -5.0, -2.0, 0.0, 2.0, 5.0$

$P(t) = 2P_0 \sum_{k=0}^{\infty} (-1)^k U(t - 2k)$, where $U(t - 2k)$ is the Heaviside's function, the Laplace transform is given in closed-form as $\overline{P}(p) = 2P_0/[p(1 + Exp(-2p))]$. This pressure is substituted in Eq. 4, and the numerical inversion into the real-time domain is performed by Durbin's method. The radial displacement on the inner surface is presented in Fig. 1 as a function of the non-dimensional time parameter. As mentioned in the preceding sections the pressure may be impulsive and given discretely. As an example of this case, consider the impulsive inner pressure which is present only up to $\tau = 12.0$ as shown in Fig. 2. After this point, there is no pressure and only free oscillations are present. Using Eq. 9 the real-time pressure values at time t_n read from the graph are transformed into the Laplace domain. Substituting these values into Eq. 7 yields the discrete solution in the time domain. The results are displayed in Fig. 2.

The values of the radial displacement decrease when β increases. A positive inhomogeneity constant refers to increasing stiffness in the radial direction which leads to the conclusion that it provides a stress shielding effect. It is obvious that a negative β would create a stress amplification effect.

The inhomogeneity constant, which includes continuously varying volume fraction of the constituents, is empirically determined and is a useful parameter from a design point of view in that it can be tailored for specific applications to control the stress distribution.

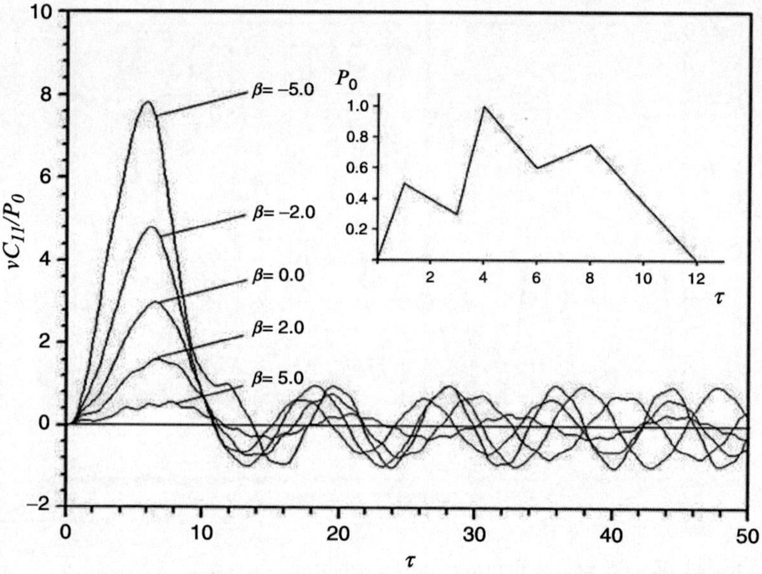

Fig. 2 Radial displacement on inside surface subjected to impulsive pressure given discretely. $k = 2.0$, $\beta = -5.0, -2.0, 0.0, 2.0, 5.0$

References

1. Tranter, C.-J.: The applications of the Laplace transformation to a problem on elastic vibrations. Philos. Mag. **33**, 614–622 (1942)
2. Ghosh, A.-K.: Axisymmetric vibration of a long cylinder. J. Sound Vib. **186**, 711–721 (1995)
3. Loy, C.-T., Lam, K.-Y., Reddy, J.-N.: Vibration of functionally graded cylindrical shells. Int. J. Mech. Sci. **41**, 309–324 (1999)
4. Pradhan, S.-C., Loy, C.-T., Lam, K.-Y., Reddy, J.-N.: Vibration characteristics of functionally graded cylindrical shells under various boundary conditions. Appl. Acoust. **61**, 111–129 (2000)
5. Han, X., Liu, G.-R., Xi, Z.-C., Lam, K.-Y.: Transient waves in a functionally graded cylinder. Int. J. Solids Struct. **38**, 3021–3037 (2001)
6. Shakeri, M., Akhlaghi, M., Hoseini, S.-M.: Vibration and radial wave propagation velocity in functionally graded thick hollow cylinder. Compos. Struct. **76**, 174–181 (2006)
7. Keles I.: Elastic response of FGM and anisotropic thick-walled pressure vessels under dynamic internal pressure. PhD thesis, Cukurova University (2007)
8. Temel, B.: Transient analysis of viscoelastic helical rods subject to time-dependent loads. Int. J. Solids Struct. **41**, 1605–1624 (2004)
9. Watson, G.-N.: A Treatise on the Theory of Bessel Functions. Cambridge University Press, Cambridge (1996)
10. Durbin, F.: Numerical inversion of Laplace transforms: an efficient improvement to Dubner and Abate's method. Comput. J. **17**, 371–376 (1974)
11. Narayan G.-V.: Numerical operational methods in structural dynamics. PhD thesis, University of Minnesota (1979)

Homogenized Perforated Interface in Acoustic Wave Propagation – Modeling and Optimization

Eduard Rohan and Vladimír Lukeš

Abstract We consider acoustic wave propagation described by Helmholtz equation and involving homogenized transmission conditions imposed along a thin *perforated interface* which separates two halfspaces occupied by the acoustic medium. On this interface homogenized transmission conditions are imposed; they were obtained by the two-scale homogenization of a layer with an immersed perforated slab designed as a periodic structure. Design of the perforation can modify significantly acoustic wave propagation, therefore, optimization of the perforation can contribute to development of noise-isolation devices and to other practical acoustic applications. We developed the sensitivity of coefficients involved in the homogenized transmission conditions with respect to the design of the "holes" in the slab. Using the two-scale approach we treat the acoustic waves propagating in a duct which is equipped with the perforated interface. Using numerical examples we illustrate, how the perforation geometry influences the acoustic transmission. The optimization problem is solved with a few parameters describing shape of the perforation.

Keywords Wave transmission • Homogenization • Structural optimization • Sensitivity analysis

1 Introduction

Optimization of noise transmission in the acoustic fluid belongs to important merits of the acoustic engineering. Sieve-like structures are classical elements employed in noise-reducing devices. For example, in the exhaust silencers of the combustion

E. Rohan (✉) • V. Lukeš
Department of Mechanics, Faculty of Applied Sciences, University of West Bohemia, Univerzitní 8, 30614 Pilsen, Czech Republic
e-mail: rohan@kme.zcu.cz

J. Náprstek et al. (eds.), *Vibration Problems ICOVP 2011: The 10th International Conference on Vibration Problems*, Springer Proceedings in Physics 139,
DOI 10.1007/978-94-007-2069-5_44,

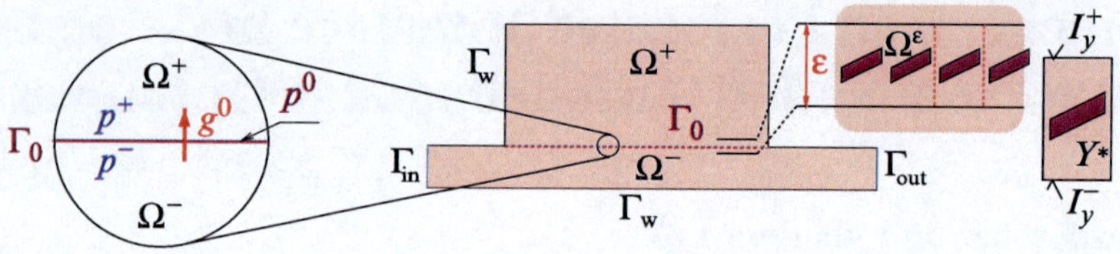

Fig. 1 *Left*: illustration of the transmission coupling – the acoustic pressure jump is proportional to the transverse acoustic velocity g^0. *Center*: the domain and boundary decomposition of the global acoustic problem considered. *Right*: perforated interface and the representative periodic cell $Y = Y* \cup \overline{S}$

engines the gas flows through ducts equipped with various sieves which in part may influence the transmission losses associatedwith acoustic waves propagating in the exhaust gas. In aerospace and automotive industry there are many applications related to acoustic waves and fluid flow where optimal design of the sieves (perforated slabs) is a challenging problem.

In the paper we deal with the acoustic transmission through a perforated slab – the *perforated interface*, cf. [1, 2]. In [5], using the asymptotic analysis we developed the homogenized transmission conditions to be imposed on the interface plane representing the slab with a periodic perforation which, in general, is designed by obstacles having arbitrarily complicated shapes.

The limit model of such an interface plane involves some homogenized impedance coefficients depending on the so-called microscopic problems; these are imposed in the *reference periodic cell* embedding an obstacle which represents the perforation. The two-scale modeling approach allows for an efficient treatment of complicated designs of perforations. Acoustic response to the global acoustic problem involving the transmission conditions is subject to the sensitivity analysis. Namely the total variation of an objective function depending on the acoustic pressure w.r.t. the obstacle shape at the "microlevel" is derived.

An abstract optimization problem is formulated at three levels: at the "global" one the pressure field is controlled by the model variables (acoustic momentum) associated with the homogenized transmission condition; at the "interface" level these interface variables are solutions of the non-local transmission conditions depending on the homogenized impedance parameters; finally, at the "microscopic level" these parameters depend on solutions of auxiliary local problems featured by the shape of perforations.

We consider the global problem of the wave propagation in a duct $\Omega \subset \mathbb{R}^3$ filled by the acoustic fluid. Ω is subdivided by perforated plane Γ_0 in two disjoint subdomains Ω^+ and Ω^-, so that $\Omega = \Omega^+ \cup \Omega^- \cup \Gamma_0$, see Fig. 1 (obviously, much more general setting is possible). The acoustic pressure field p is discontinuous in general along Γ_0. In a case of no convection flow (the linear acoustics), the waves propagating in Ω are described by the following equations where κ is the wave number (i.e. frequency $\omega = \kappa c$)

$$\nabla^2 p + \kappa^2 p = 0 \quad \text{in } \Omega^+ \cup \Omega^-,$$

$$\text{transmission conditions} \quad \mathscr{G}(\kappa, [p]_-^+, [\partial p/\partial n]_-^+) = 0 \quad \text{on } \Gamma_0,$$

$$r i\kappa p + \frac{\partial p}{\partial n} = s 2 i\kappa \bar{p} \quad \text{on } \partial\Omega, \tag{1}$$

where s, r and $\bar{p}$ are given data, $[\cdot]_-^+$ is the jump across Γ_0. $\frac{\partial p}{\partial n} = n.\nabla p$ is the normal derivative on Γ_0. The homogenized transmission conditions $\mathscr{G} = 0$ developed in [5] introduce two internal variables on Γ_0: the "in-layer" acoustic potential p^0 and the "trans-layer" acoustic velocity g^0, which is coupled with the "off-layer" fields through: $\partial p/\partial n^\pm = \pm i\kappa g^0$, so that $[\partial p/\partial n]_-^+ = 0$. Boundary $\partial\Omega = \Gamma_w \cup \Gamma_{\text{in}} \cup \Gamma_{\text{out}}$ of the duct is split into walls and the input/output parts; by the constants r, s in $(1)_3$ different conditions on $\partial\Omega$ are respected: $r = s = 0$ on the duct walls Γ_w, whereas $r = s = 1$ on Γ_{in} and $r = 1, s = 0$ on Γ_{out}.

2 Acoustic Problem with Homogenized Sieve

We now formulate the *state problem* describing acoustic waves in open bounded domain Ω with immersed homogenized sieve represented by non-local transmission conditions. We need the following notation:

$$a_\Omega(p, q) = \int_\Omega \nabla p \cdot \nabla q, \quad (p, q)_\Omega = \int_\Omega pq, \quad \langle p, q \rangle_{\Gamma_0} = \int_{\Gamma_0} pq.$$

The problem is defined at two levels:

At the *global level* the interface conditions involve three geometrical parameters A, B, F which characterize the design of the sieve perforation; we define (summation $\alpha, \beta = 1, 2$)

$$\mathscr{A}(p, q) = \int_{\Gamma_0} A_{\alpha\beta} \partial_\beta p \partial_\alpha q, \quad \mathscr{B}(g, q) = \int_{\Gamma_0} B_\alpha g \partial_\alpha q, \quad \mathscr{F}(g, h) = \int_{\Gamma_0} Fgh. \tag{2}$$

The *global* problem is to find $(p.p^0, g^0) \in H^1(\Omega \setminus \Gamma_0) \times H^1(\Gamma_0) \times L^2(\Gamma_0)$ such that

$$a_\Omega(p, q) - \kappa^2 (p, q)_\Omega + i\kappa \langle p, q \rangle_{\Gamma_{\text{in-out}}} - i\kappa \langle g^0, [q]_-^+ \rangle_{\Gamma_0} = 2i\kappa \langle \bar{p}, q \rangle_{\Gamma_{\text{in}}}$$
$$\mathscr{A}(p^0, \phi) - \kappa^2 \varsigma^* \langle p^0, \phi \rangle_{\Gamma_0} + i\kappa \mathscr{B}(g^0, \phi) = 0,$$
$$-i\kappa \mathscr{B}(\psi, p^0) - \kappa^2 \mathscr{F}(g^0, \psi) + i\kappa \frac{1}{\varepsilon^0} \langle [p]_-^+, \psi \rangle_{\Gamma_0} = 0, \tag{3}$$

for all $(q, \phi, \psi) \in H^1(\Omega \setminus \Gamma_0) \times H^1(\Gamma_0) \times L^2(\Gamma_0)$, where ε_0 is the real thickness of the layer.

At the *local* level the geometrical parameters A, B, F are determined upon solving "microscopic problems". The perforation design is characterized by computational cell $Y = \Xi \times]-1/2, +1/2[$ with $\Xi =]0, b_1[\times]0, b_2[$, where the fluid occupies domain $Y*$ and $S = Y \setminus \overline{Y*}$ represents a rigid obstacle, see Fig. 1. Further $I_y^{\pm} = \Xi \pm (0, 0, 1)$ are the "lower" and "upper" faces of Y. In (3), $\varsigma^* = |Y*|/|\Xi|$ is the porosity. Below the space $H^1_\#(Y^*)$ contains all Ξ-periodic functions in the Sobolev space $H^1(Y*)$. The local problems read: find $\pi^\beta, \xi \in H^1_\#(Y^*)$ such that

$$(\nabla_y \pi^\beta, \nabla_y \psi)_{Y*} = -\int_{Y*} \partial^y_\beta \psi\,, \quad \beta = 1, 2,$$

$$(\nabla_y \xi, \nabla_y \psi)_{Y*} = -\left(\int_{I_y^+} \psi - \int_{I_y^-} \psi\right), \tag{4}$$

for all $\psi \in H^1_\#(Y^*)$, where $\nabla_y = (\partial/\partial y_\beta)$ and $(,)_{Y*}$ is the inner product in $L^2(Y*)$. Using the local responses, the geometrical parameters can now be computed, see (2) and Fig. 1:

$$A_{\alpha\beta} = \frac{1}{|\Xi|}(\nabla_y(\pi^\beta + y_\beta), \nabla_y(\pi^\alpha + y_\alpha))_{Y*},$$

$$B_\alpha = \fint_{Y*} \partial^y_\alpha \xi = \fint_{I_y^+} \pi^\alpha - \fint_{I_y^-} \pi^\alpha, \qquad F = -\left(\fint_{I_y^+} \xi - \fint_{I_y^-} \xi\right), \tag{5}$$

where $\fint = |\Xi|^{-1} \int$. Note $F > 0$ and A is positive definite.

3 Optimal Design Problem

One of the most frequently used criteria of optimality in acoustics is related to transmission loss (TL) evaluated using two pressures $p^a = p(x = a)$, $p^b = p(x = b)$, where p satisfies the *state problem* (3). In our numerical tests we observed some remarkable sensitivity of TL on the perforation design [5].

When the coupling coefficient $B_{\rm ff}$ is nonvanishing, surface acoustic stationary waves propagate along Γ_0 being described by amplitudes p^0, see (3). Thus, it is legitimate to look for an optimal design of the perforation, such that the $|p^0|$ is maximized in $L^2(\Gamma_0)$ norm. We shall considered the following objective functions:

$$\Phi_{\rm TL}(p) = \hat{\Phi}(p^a, p^b) = 20\, log\left(\frac{|p^a|}{|p^b|}\right), \qquad \Phi_0(p^0) = \int_{\Gamma_0} |p^0|^2\,. \tag{6}$$

Let the perforation design be controlled by design variables $\mathbf{d}$ which describe the shape of obstacle S and, thereby, the shape of domain $Y*$, so that $\mathbf{d}$ influences the homogenized coefficients A, B, F involved in (3). Let us recall that these coefficients are integrals of functions π^β, ξ which are solutions of the microscopic problems (4) posed in Y^*. At the global level, $\mathbf{d}$ influences the overall acoustic fields (p, p^0, g^0).

We can now define the *optimal perforation design problem*:

$$\min_{\mathbf{d} \in D_{adm}} \Phi(p, p^0, g^0)$$

$$\text{subject to:} \quad (p, p^0, g^0) \text{ solves (3), where } A, B, F \text{ are given by (4),(5),} \tag{7}$$

where D_{adm} is the set of admissible designs, constraining shape regularity of ∂S and typically some other features, like the size of the obstacle (thickness), or porosity of the interface.

To solve (7) using gradient-based methods, the sensitivity of Φ w.r.t. the design $\mathbf{d} = (d_i)$ must be supplied at any iteration (Φ can be substituted by Φ_{TL} or $-\Phi_0$, for instance). For this, any component d_i is associated with the *design velocity* field $\mathcal{V}^i$ which is constructed e.g. by solving an auxiliary elasticity problem in domain Y^*, whereby $\mathcal{V}^i \not\equiv 0$ is given on ∂S and $\mathcal{V}^i = 0$ on ∂Y. Then the shape sensitivities $\delta A_{\alpha\beta}(\mathcal{V}^i), \delta B_\beta(\mathcal{V}^i), \delta F(\mathcal{V}^i)$ and $\delta \zeta^*(\mathcal{V}^i)$ of coefficients A_α, B_β, F and ζ^* can be obtained, as described in [4], using the general approach based on the material derivative.

The total design sensitivity $\delta\Phi(p, p^0, g^0; \mathcal{V}^i) = \frac{\partial}{\partial d_i}\Phi$ is obtained by formula

$$\delta\Phi(p, p^0, g^0; \mathcal{V}^i) = 2\Re\left\{\int_{\Gamma_0} \delta A_{\alpha\beta}(\mathcal{V}^i)\partial_\beta p^0 \partial \tilde{p}^0 - \kappa^2 \int_{\Gamma_0} \delta F(\mathcal{V}^i) g^0 \tilde{g}^0 \right.$$
$$\left. -\kappa^2 \delta\zeta^*(\mathcal{V}^i)\int_{\Gamma_0} p^0 \tilde{p}^0 + \mathrm{i}\kappa \int_{\Gamma_0} \delta B_\alpha(\mathcal{V}^i)\left(\partial_\alpha \tilde{p}^0 g^0 - \partial_\alpha p^0 \tilde{g}^0\right)\right\}, \tag{8}$$

where $(\tilde{p}, \tilde{p}^0, \tilde{g}^0) \in H^1(\Omega \setminus \Gamma_0) \times H^1(\Gamma_0) \times L^2(\Gamma_0)$ is the *adjoint state*, cf. [3], satisfying the *adjoint equation*, see [4] for details,

$$a_\Omega(\tilde{p}, q) - \kappa^2(\tilde{p}, q)_\Omega + \mathrm{i}\kappa\langle \tilde{p}, q\rangle_{\Gamma_{\text{in-out}}} - \mathrm{i}\kappa\langle \psi, [\tilde{p}]_-^+\rangle_{\Gamma_0} + \mathrm{i}\kappa\frac{1}{\varepsilon^0}\langle [q]_-^+, \tilde{g}^0\rangle_{\Gamma_0}$$
$$+ \mathscr{A}(\tilde{p}^0, \phi) - \kappa^2\varsigma^*\langle \tilde{p}^0, \phi\rangle_{\Gamma_0} + \mathrm{i}\kappa\mathscr{B}(\psi, \tilde{p}) - \mathrm{i}\kappa\mathscr{B}(\tilde{g}^0, \phi) - \kappa^2\mathscr{F}(\tilde{g}^0, \psi)$$
$$= -\frac{1}{2}\left(\partial_{\Re(p,p^0,g^0)}\Phi(p, p^0, g^0; q, \phi, \psi) - \mathrm{i}\partial_{\Im(p,p^0,g^0)}\Phi(p, p^0, g^0; q, \phi, \psi)\right), \tag{9}$$

for all $(q, \phi, \psi) \in H^1(\Omega \setminus \Gamma_0) \times H^1(\Gamma_0) \times L^2(\Gamma_0)$, where $\Re$ and $\Im$ is the real and the imaginary part, respectively.

4 Conclusions

For illustration we consider a 2D problem of acoustic waves in a duct equipped with a perforated sieve, see Fig. 1, designed by repeating a slanted rectangle of thickness h, width b and angle of declination α. In the optimal design problem we allow b and

Table 1 A test of the sensitivity analysis, design gradient $\nabla^{\mathbf{d}} = (\partial_1{}^{\mathbf{d}}, \partial_2{}^{\mathbf{d}})$ and relative error $(\text{rel.error})_{1,2}{}^{\mathbf{d}}$, when compared with central finite difference, where $\mathbf{d} = (b, \alpha)$ is the design vector. Parameterization of the obstacle (perforation) by $\mathbf{d} = (b, \alpha)$, see the illustration below

	Coeff.	$\partial_1{}^{\mathbf{d}}(\cdot)$	$(\text{Rel.error})_1{}^{\mathbf{d}}$	$\partial_2{}^{\mathbf{d}}(\cdot)$	$(\text{Rel.error})_2{}^{\mathbf{d}}$
A	0.729947	−0.312262	5.8e-10	−0.359917	−5.2e-7
B	−0.244708	−0.611555	4.3e-9	−0.326677	1.7e-7
F	2.083840	2.623334	−7.3e-9	−1.632226	2.0e-6
ζ^*	0.840000	−0.160000	−6.2e-9	−1e-18	∼1
Φ_{TL}		1.616879	5.7e-8	0.620776	4.1e-7

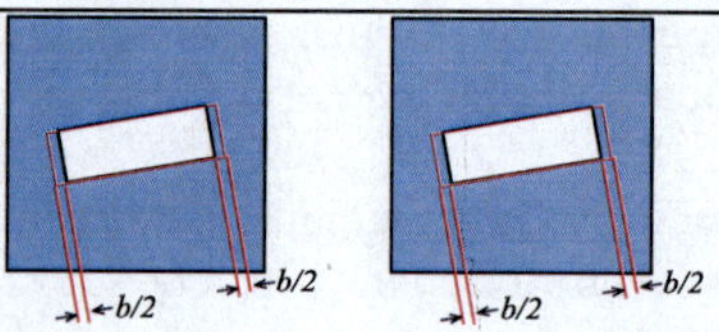

case	2-point TL	Ini.	Opt.
1	TL_{ab}	8.363	8.282
2	TL_{ab}	8.529	8.583
	TL_{ac}	11.459	11.437

Case 1: minimization of $\Phi(p^a, p^b) = TL_{ab}$,
Case 2: minimization of

$$\Phi(p^a, p^b, p^c) = \beta TL_{ab} + \gamma TL_{ac}$$

with $\beta < 0$ and $\gamma > 0$.

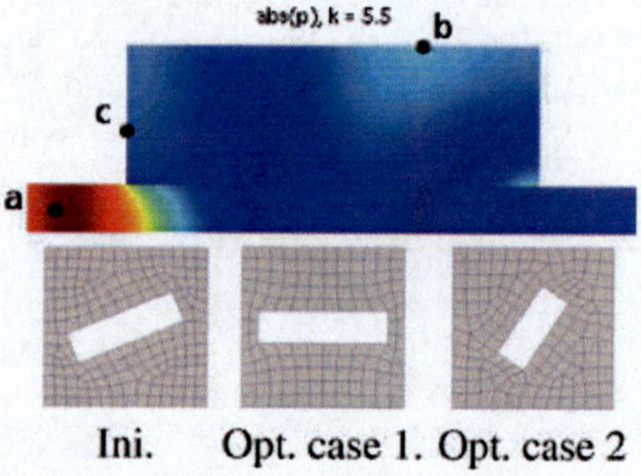

Fig. 2 Optimization of the perforation design. *Left*: two cases, initial and optimized TL values. *Right*: global acoustic problem, three points a, b, c marked, displayed $|p|$ in Ω (*up*) and the cell Y^* for the initial and the two optimized layouts (*bottom*)

α to change, so that the rectangle in the cell Y can rotate and be extended; we state $\mathbf{d} = (b, \alpha)$.

In Table 1 we report on testing the sensitivity analysis; the design sensitivity of transmission coefficients A, B, F and ς^* were compared with central differences (step 1e-3). Standard accuracy was observed also when testing the sensitivity of the two-point objective function $\hat{\Phi}(p^a, p^b)$.

We introduce an *academic example*, to demonstrate the influence of the two parameters, b and α on the global acoustic field. In general, we consider function $\Phi(p^a, p^b, p^c) = \beta\hat{\Phi}(p^a, p^b) + \gamma\hat{\Phi}(p^a, p^c)$, where weights $\beta, \gamma \in \mathbb{R}$ can be selected, and $p^x = p(x)$ is the "nodal value" of the numerical solution. In the first case we take $\gamma = 0$ and $\beta > 0$, so that the transmission loss between a and b is minimized. In the second case $\gamma > 0$ and $\beta < 0$, thus a combination of both transmissions a-b and a-c are considered, see Fig. 2. The results (local minima) were obtained by the SQP algorithm with box constraints which secure the "mesh deformation" during the design iterations. The acoustic transmission model was implemented in our in-house FEM code SfePy, which allows for 3D simulations. We intend to consider non-academic examples with spline parameterization of the obstacle shapes.

Acknowledgments The research and this publication was supported by research projects GAČR 101/07/1471 and MSM 4977751303 of the Czech Republic.

References

1. Bonnet-Bendhia, A.S., Drissi, D., Gmati, N.: Mathematical analysis of the acoustic diffraction by a muffler containing perforated ducts. Math. Models Methods Appl. Sci. **15**(7), 1059–1090 (2005)
2. Chen, K.T.: Study on the acoustic transmission loss of a rigid perforated screen. Appl. Acoust. **47**(4), 303–318 (1996)
3. Feijóo, G.R., Oberai, A.A., Pinsky, P.M.: An application of shape optimization in the solution of inverse acoustic scattering problems. Inverse Probl. **20**, 199–228 (2004)
4. Rohan, E., Lukeš, V.: Sensitivity analysis for the optimal perforation problem in acoustic transmission. Appl. Comput. Mech., UWB Pilsen **3**, 111–120 (2009)
5. Rohan, E., Lukeš, V.: Homogenization of the acoustic transmission through perforated layer. J. Comput. Appl. Math. **234**(6), 1876–1885 (2010)

Reducing Ground-Borne Micro-Vibrations by Plate Foundations

L. Auersch

Abstract The soil-structure interaction of elastic plates with horizontally propagating waves through the soil is investigated by a combined finite-element boundary-element method (FEBEM). The frequency-dependent reduction of the soil amplitudes by the plate and the amplitude distribution along the plate are presented. The following parameters are varied, the length and the width of the plate, the stiffness and the mass of the plate or, as a combination of both, the height of the plate. The results show that a strong reduction can be achieved for higher frequencies, at some distance from the edge of the plate, and for stiff (high) plates.

Keywords Plate-soil interaction • Wave excitation • Vibration reduction • Finite-element boundary-element method

1 Introduction

Production facilities for micro- and nano-technologies must fulfil strong vibration criteria to guarantee a safe production. In order to reduce the ambient ground vibration, thick foundation plates are used (Fig. 1). As a basis for the design of these foundations, the soil-structure interaction of elastic plates with horizontally propagating waves through the soil is investigated by a combined finite-element boundary-element method (FEBEM).

L. Auersch (✉)
Federal Institute of Materials Research and Testing, Berlin, Germany
e-mail: lutz.auersch-saworski@bam.de

J. Náprstek et al. (eds.), *Vibration Problems ICOVP 2011: The 10th International Conference on Vibration Problems*, Springer Proceedings in Physics 139,
DOI 10.1007/978-94-007-2069-5_45,

Fig. 1 Reduction of ground-borne vibration by thick foundation plates

2 Problem of Wave-Field Excited Plate-Soil Interaction

A concrete plate with dimensions length $L = 60$ m*, width $a = 20$ m*, height $h = 0.7$ m*, elastic modulus $E = 3\ 10^{10}$ N/m^2, Poisson's ratio $\nu = 0.15$, and mass density $\rho = 2{,}500$ kg/m^3* is lying on a layered soil of the material constants shear modulus $G = 8\ 10^7$ N/m^2, $\nu = 0.33$, and $\rho = 2{,}000$ kg/m^3 for each layer. The values are given for the standard system of a plate on a homogeneous half-space, an asterisk stands for the parameters that are varied here. The condensed parameters (1) bending stiffness B, mass per area m'', and shear wave velocity V_s

$$B = \frac{Eh^3}{(1-\nu^2)12}, \qquad m'' = \rho h, \qquad v_S = \sqrt{\frac{G}{\rho}} \tag{1}$$

are sometimes advantageous. The surface of the soil and the plate are coupled either in full contact or in a relaxed contact where only the vertical displacements are the same for plate and soil. The plate is excited by a harmonic wave which is travelling along its length (x-axis) and which has a constant amplitude $\mathbf{u_0}$, the frequency f (or $\omega = 2\pi f$), and wave velocity v where the wave velocity and amplitudes of the Rayleigh wave of the soil are used here. In case of the relaxed contact, only the vertical wave amplitude u_0 is effective. The calculated response of the plate u is presented in relation to the vertical excitation amplitude u_0 as $u/u_0(x, f)$. The reduction (or amplification) of the plate amplitudes are studied for the different design parameters of the plate.

3 Finite-Element Boundary-Element and Integral Methods

The finite element method is used to calculate the plate. The Green's functions of the homogeneous or layered soil are used to establish a dynamic stiffness matrix of the soil [1] which is introduced in the FE-code. The soil matrix is fully populated as the forces at each point depend on the displacements of all other points. Its inversion is usually the most time consuming task of the soil-structure solution. Building elements with smaller foundation areas such as high and low walls, beams and rigid strip foundations have been analysed first [2]. With higher computational power available, now plates with a large foundation area of 60 × 20 m are calculated. The use of relaxed boundary conditions facilitates the time consuming solution.

As an alternative, infinitely long plates can be considered using the integration in wavenumber domain k. The dynamic stiffness (2) of the plate in the frequency-wavenumber domain

$$K_P = Bk^4 - m''\omega^2 \tag{2}$$

and the dynamic stiffness (3) of the soil ([3] where the compliance H(k,ω) of the soil is integrated over wavenumber k)

$$K_s(k,\omega) = \frac{1}{H_s(k,\omega)} = 2\pi \Big/ \int_{-\infty}^{+\infty} H(k,\omega)\left(\frac{\sin ka}{ka}\right)^2 dk \tag{3}$$

are combined for the plate-soil stiffness $K_{PS} = K_P + K_S$. The response (4) to the free-field wave excitation

$$\frac{u}{u_0} = \frac{K_S}{K_P + K_S} \tag{4}$$

is a function of frequency f, if the wavenumber k is replaced by $2\pi f/v$, and independent from x due to the infinite extension of the plate. Similar wavenumber integral methods have been used for the force excitation of plates [4].

4 Results of the Coupled Boundary Element Method (FEBEM)

At first, the amplitude distribution along the 60 m long standard plate is shown in Fig. 2 for different frequencies. At the beginning and the end of the plate, higher amplitudes can be observed which reach the value $u/u_0 = 1$ for the low frequencies. After a sharp decay, a moderate decrease of amplitudes follows. At some distance, the minimum amplitude seems to be reached which is smaller with increasing frequency. It could also be concluded that the final reduction is reached at a longer distance for lower frequency. Although the reduction clearly depends on the distance, it was found that it is independent from the length of the plate. That means that the displacements at $x = 30$ m are the same for a 40 m and a 60 m long plate. Moreover, an influence of the width of the plate is only found for narrow plates of 2 or 4 m width.

The influence of the most important parameter, the bending stiffness of the plate, is shown in Fig. 3. The frequency dependent transfer functions between the free-field amplitudes of the soil and the response of the plate start with $u/u_0 = 1$ and hold this value for the low frequencies. At a certain frequency a strong reduction starts and low amplitudes are reached within the next 10 Hz. The cut-off frequency clearly depends on the bending stiffness of the plate.

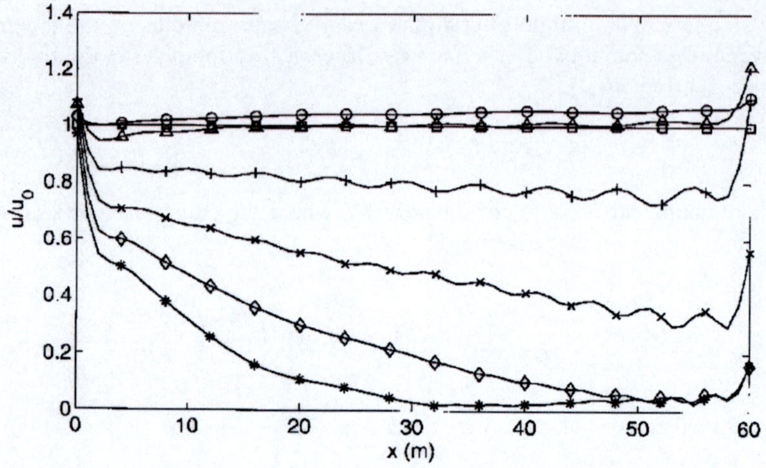

Fig. 2 Amplitude distribution for a standard plate and frequencies $f =$ □ 0, ○ 5, △ 10, + 15, × 20, ◇ 25, * 30 Hz

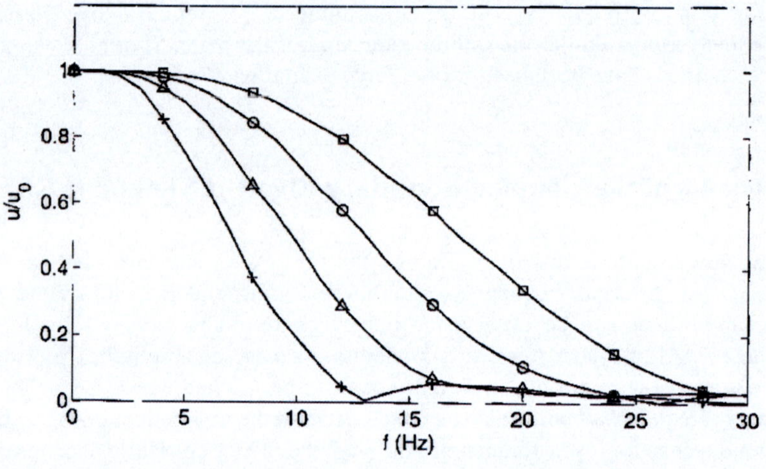

Fig. 3 Plate response for different bending stiffnesses $B =$ □ 0.35, ○ 1, △ 3, + 10 Nm ($L = 60$ m, $m'' = 0$, at $x = 30$ m, relaxed contact)

The results in Fig. 3 are presented for a plate with no mass. The influence of the mass is shown in Fig. 4 for plates without any bending stiffness. The mass is responsible for amplifications of the amplitudes at low frequencies. Without bending stiffness, amplifications of more than $u/u_0 = 2$ are possible if an additional building mass $\rho/\rho_0 = 2$ or 4 is supposed. In these cases, high-frequency reductions are also observed which are caused by the mass and not by the stiffness of the plate.

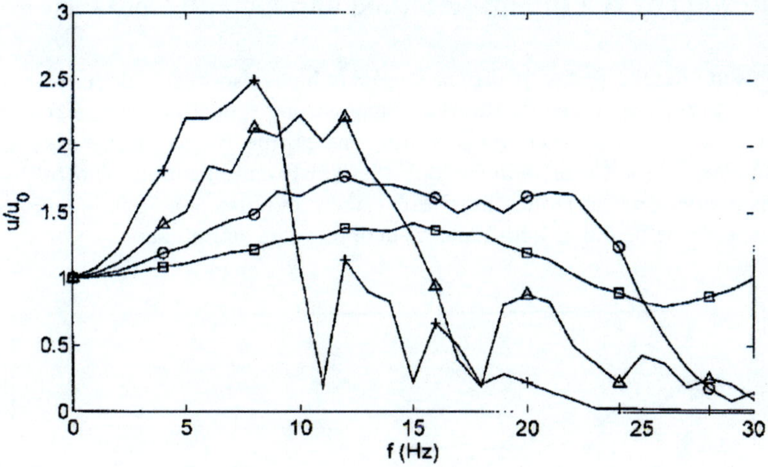

Fig. 4 Plate response for different mass densities $m'' =$ □ 1.8, ○ 3.7, △ 7.5, + 15 10^3 kg/m^2 ($L = 60$ m, $B = 0$, at $x = 30$ m, relaxed contact)

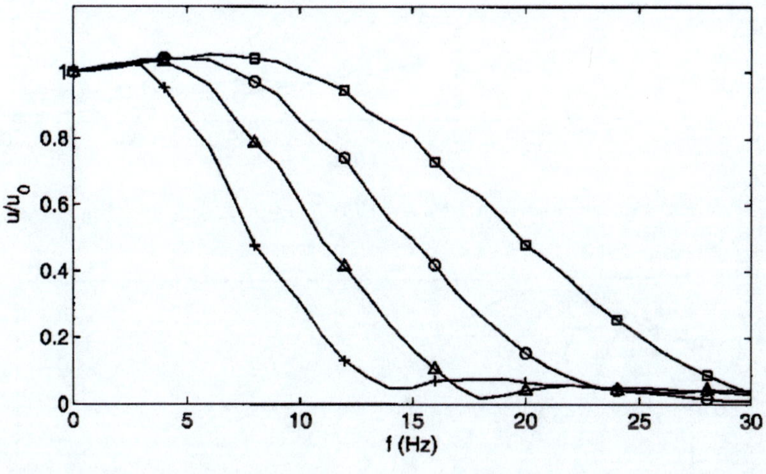

Fig. 5 Plate response for different heights $h =$ □ 0.5, ○ 0.7, △ 1.0, + 1.5 m ($L = 60$ m, at $x = 30$ m, relaxed contact)

If the height of the plate is varied in Fig. 5, stiffness and mass are increased. Nevertheless, the increase of the stiffness is dominating and the results are similar to those in Fig. 3. The strong reduction of the amplitudes is ruled by the bending stiffness. The mass of the different plates is too small to create strong resonance effects. The low-frequency amplitudes are increased only from $u/u_0 = 1.0$ to 1.05. For stiff plates with additional building mass (no figure), the amplifications are also limited to $u/u_0 < 1.2$.

5 Results by the Integral Method and for Full Contact

The results for the plates of different heights have also been calculated with full contact (Fig. 6). The results show the same strong reduction at high frequencies as for the relaxed contact. At low frequencies, the full contact yields smaller amplitudes, a small reduction instead of a small amplification. The full contact means almost zero horizontal amplitudes under the plate, an additional constraint for the wave field, and an additional reduction for the amplitudes.

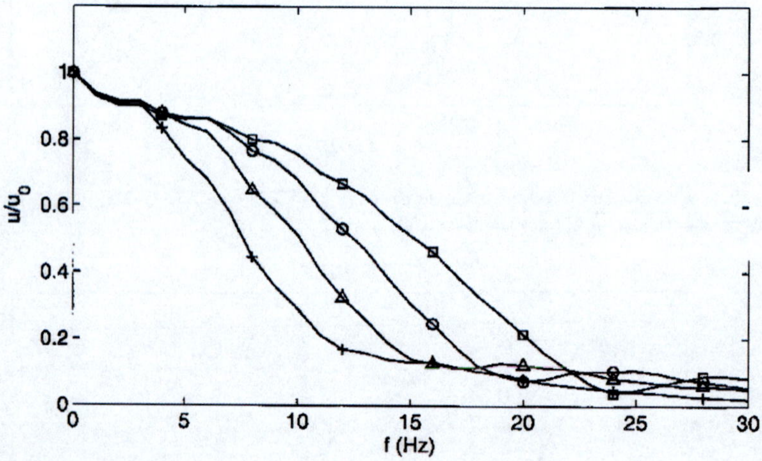

Fig. 6 Plate response for different heights $h = \square$ 0.5, $\circ$ 0.7, $\triangle$ 1.0, $+$ 1.5 m ($L = 60$ m, at $x = 30$ m, full contact)

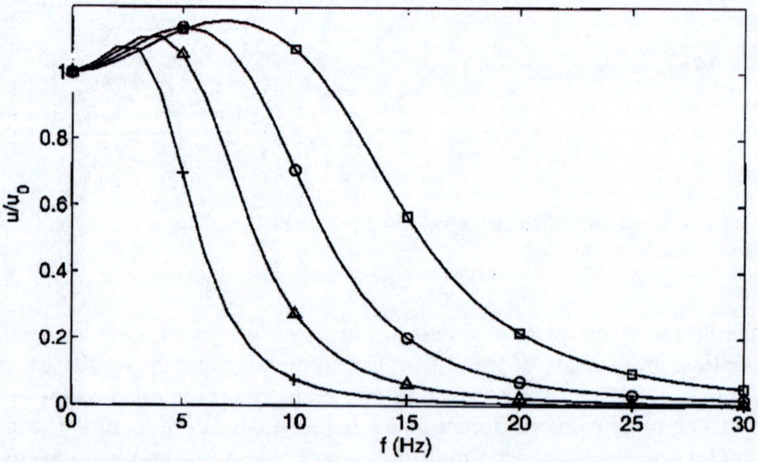

Fig. 7 Response of infinitely long plates for different heights $h = \square$ 0.5, $\circ$ 0.7, $\triangle$ 1.0, and $+$ 1.5 m (relaxed contact)

The response of an infinitely long plate is shown in Fig. 7. The transfer functions for different heights are quite similar to the curves on Fig. 5 for finite plates. Small differences are found for the low-frequency amplitudes, which are amplified up to $u/u_0 = 1.2$, and for the high-frequency amplitudes which are stronger than for the plate of finite length. The results show some similarity to the reduction by finite or infinite beams for which an approximate reduction $u/u_0 \sim f^{-4}$ has been found [1, 5].

6 Conclusion

Both methods (wavenumber integral and combined finite-element boundary-element method) show consistently the strong reduction of $u/u_0 < 0.1$ of the ground vibration by thick plates. The most important parameter for the strong reduction is the stiffness of the plate and the effect increases with frequency and distance from the beginning of the plate.

References

1. Auersch, L.: Wave propagation in the elastic half-space due to an interior load and its application to ground vibration problems and buildings on pile foundations. Soil Dyn. Earthq. Eng. **30**, 925 (2010)
2. Auersch, L., Schmid, G.: A simple boundary element formulation and its application to wavefield excited soil-structure interaction. Earthq. Eng. Struct. Dyn. **19**, 931 (1990)
3. Auersch, L.: Dynamic interaction of various beams with the underlying soil – finite and infinite, half-space and Winkler models. Eur. J. Mech. A Solids **27**, 933 (2008)
4. Auersch, L.: Dynamic plate-soil interaction – finite and infinite, flexible and rigid plates on homogeneous, layered or Winkler soil. Soil Dyn. Earthq. Eng. **15**, 51 (1996)
5. Auersch, L.: Wechselwirkung starrer und flexibler Strukturen mit dem Baugrund insbesondere bei Anregung durch Bodenerschütterungen. BAM Forschungsbericht 151, Dissertation, BAM, Berlin (1988)

Estimation of Parameters of Soil Vibration Due to Impact Loading

Svetlana Polukoshko, V. Gontza, and S. Sokolova

Abstract The results of the finite element analysis of impact action on structure and soil are presented in this work. Finite element analysis is executed by means of Plaxis program, which enables to take into account soil – structure interaction. The next problems are discussed: vertical impact on concrete foundation, modeled with axisymmetric models, and horizontal impact on the sheet-piling and on the concrete bridge pier, simulated with plane strain models.

Keywords Plaxis • Soil vibration • Impulse force • Impact time • Rayleigh damping

1 Introduction

The soil vibrations could cause the damages of nearby building, therefore estimation of its level and character is actual problem. A solitary impact on the construction contiguous with soil results in vibrating of medium. The impact of the moving body at constructive element it is possible to describe using impulse of force: $S = \int_0^{\tau} P\,dt = m(v_0 - v_1)$, where P –impact force, m – mass of striking body, v_0, v_1 – pre- and post-impact velocity of moving body, τ –impact duration. Since duration of impact may be different, the identical impulse may cause different reaction of construction and soil. During the time of the impact force action the construction together with soil displace on some distance, which is determined by the impulse

S. Polukoshko (✉)
Ventspils University College, Ventspils, Latvia
e-mail: pol.svet@inbox.lv

V. Gontza • S. Sokolova
Riga Technical University, Riga, Latvia
e-mail: vladimirs.gonca@rtu.lv; svetavla@inbox.lv

J. Náprstek et al. (eds.), *Vibration Problems ICOVP 2011: The 10th International Conference on Vibration Problems*, Springer Proceedings in Physics 139,
DOI 10.1007/978-94-007-2069-5_46, © Springer Science+Business Media B.V. 2011

intensity, construction properties and resistance of soil. When impulse becomes equal to zero the deformation does not end: at some moment deformation reaches maximal values and due to the elasticity of soil damped vibration of structure with soil is beginning; impact process ends when vibration diminished. In this work the finite element analysis of impact action on structure and soil is executed using dynamic module of Plaxis 2D v8 program, which enables to take into account soil – structure interaction, to use different models of soil and different configuration of impulse. In Plaxis special boundary conditions are used in order to absorb waves, reaching the boundaries. The paper deals with next problems: vertical impact onconcrete foundation, simulated with axisymmetric models, horizontal impact on the sheet-piling wall and on the massive concrete construction of the bridge pier, simulated with plane strain models. For each problem different models of soil are examined: linear-elastic body (LE), Mohr-Coulomb model (MC) and hardening soil model (HS). Impact impulse of sinusoidal form is applied. The purpose of work is to examine the behavior of the system soil-construction under impact loading and to define the parameters of vibrations of soil surface points – displacements, velocities and accelerations, duration of vibrations fading depending on soil properties.

2 Scientific Background

The basic equation for the time-dependent movement of a volume under the influence of a dynamic load is:

$$\boldsymbol{M}\underline{\ddot{u}} + \boldsymbol{C}\underline{\dot{u}} + \boldsymbol{K}\underline{u} = \boldsymbol{F} \tag{1}$$

where: $\boldsymbol{M}$ – the mass matrix (soil, water and construction), $\boldsymbol{C}$ – the damping matrix, $\boldsymbol{K}$ – the stiffness matrix, $\boldsymbol{F}$ – the load vector, u –$\dot{u}$-$\ddot{u}$- displacement, velocity and acceleration vectors. Here the theory is described on the base of linear elasticity. However, all models in Plaxis can be used for dynamic analysis. The matrix $\boldsymbol{C}$ represents the material damping, which is caused by friction or by irreversible deformations (plasticity or viscosity). If elasticity is assumed, damping can still be taken into account using the matrix $\boldsymbol{C}$. In finite element formulations, $\boldsymbol{C}$ is often formulated as a function of mass and stiffness matrices (Rayleigh damping) as:

$$\boldsymbol{C} = \alpha_R \boldsymbol{M} + \beta_R \boldsymbol{K} \tag{2}$$

This limits the determination of damping matrix to the Rayleigh coefficient α_R and β_R. Here, when the contribution of $\boldsymbol{M}$ is dominant, more of low frequency vibrations are damped, and when the contribution of $\boldsymbol{K}$ is dominant more of high-frequency vibrations are damped. Liquefaction of soil is not considered in Plaxis Dynamics. A water-saturated porous soil can bring in the viscous damping on high-frequencies.

In the case of perfectly elastic continuous medium for determination of amplitudes of surface waves on comparatively large distance from a source it is possible to use a formula:

$$A_r = A_0 \sqrt{r_0/r^{e^{-\alpha(r-r_0)}}} \tag{3}$$

where A_r and A_0 – amplitude of the soil vibration on the distance r and r_0; α– damping coefficient, for sand $\alpha = 0.04 \div 0.06\ \text{m}^{-2}$, for clay $\alpha = 0.06 \div 0.10\ \text{m}^{-2}$.

3 Numerical Examples of Computation

3.1 Vertical Impact on Concrete Foundation

Concrete foundation for sledge hammer is examined by means of axisymmetrical model. Height of foundation is 1.7 m, radius –1.5 m; distributed impact load is imposed in the circle with radius 0.5 m (Fig. 1). Mass of falling part of hammer is m = 3.5 t, impact velocity –7 m/s, impact impulse S = 25 t·m/s, distributed impulse s = 32 t·m/s·m^2. Impulse action has a duration from 0.1 to 0.001 s, impulse is assumed sinusoidal Material model of concrete is linear-elastic, non-porous, unit weigh $\gamma = 25$ kN/m^3, Young's modulus $E_{ref} = 3{\cdot}10^7$ kN/m^3, Poisson's ratio $\nu = 0.2$. Material model of soil is LE drained sand, $\gamma = 21$ kN/m^3, $E_{ref} = 50{,}000$ kN/m^3, $\nu = 0.2$. In Fig. 1 geometry and finite element mesh is presented, control points A,B,C,D,E, F,G,H on the surface are located in the coordinates r = 1.53; 2.0; 3.0; 4.5; 6.0; 8.0; 10; 12.5;15.0;18.0 m, point O-under center of foundation.

Results of influence of solitary impact impulse with t = 0.01 s, applied to the concrete foundation, on its surrounding soil are presented bellow. In Fig. 2 plots of changing with depth of the modulus of maximal amplitude of vertical vibrations of soil in control points are given.

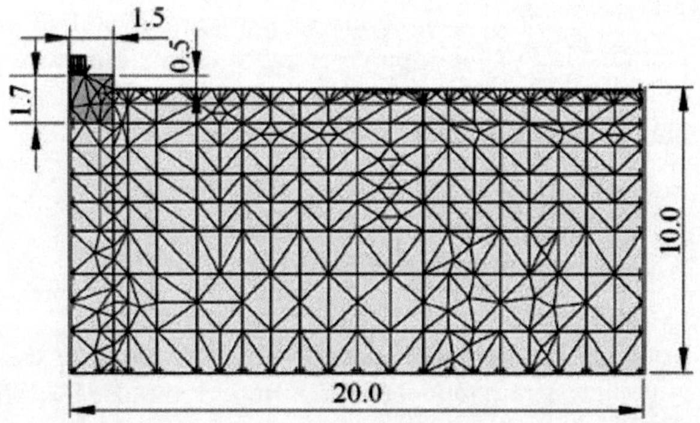

Fig. 1 Finite element mesh and geometry

Fig. 2 Plots of displacement $u_{ymax} = f(y)$

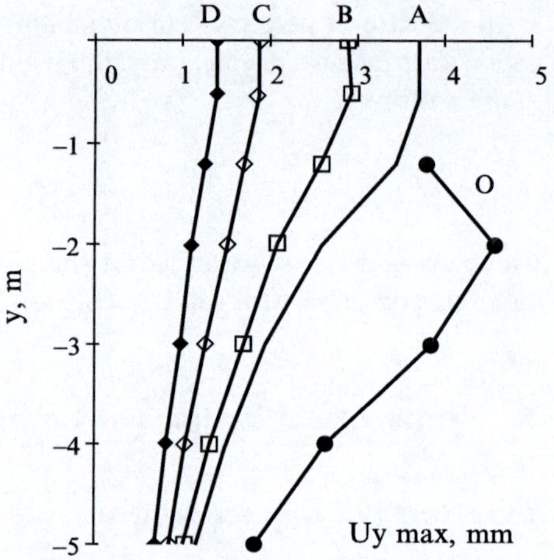

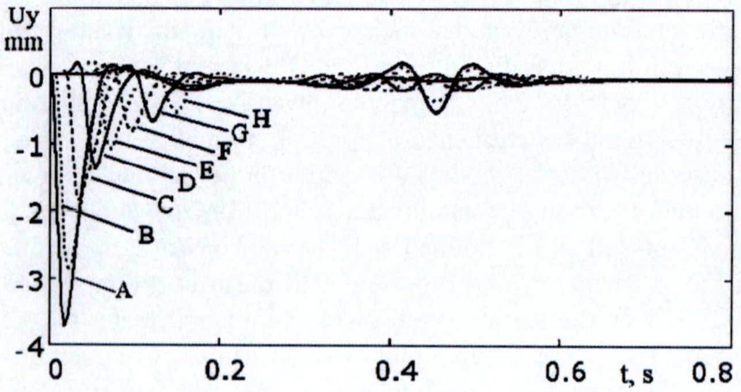

Fig. 3 Vertical displacement $u_y = f(t)$ of control points

In Fig. 3 plots of soil particle vertical displacement in control points depending on time are shown; in Figs. 4–6 modulus of the maximal amplitudes of vibrations, velocities and accelerations of particles of soil in control points are presented. Diminishing of these parameters in a direction away from foundation takes place in accordance with exponential law. MC soil model is also used for comparison of the results of different impulse duration action (Table 1).

3.2 Horizontal Impact to Bridge Pier

Concrete bridge pier of length 12 m, imposed to the vessel impact, is examined with plane strain model (Fig. 7). Impact impulse S = 550 tm/s. impact time duration t = 0.01 s, impact force amplitude of sinusoidal pulse q = 3,600 kN/m^2. Material

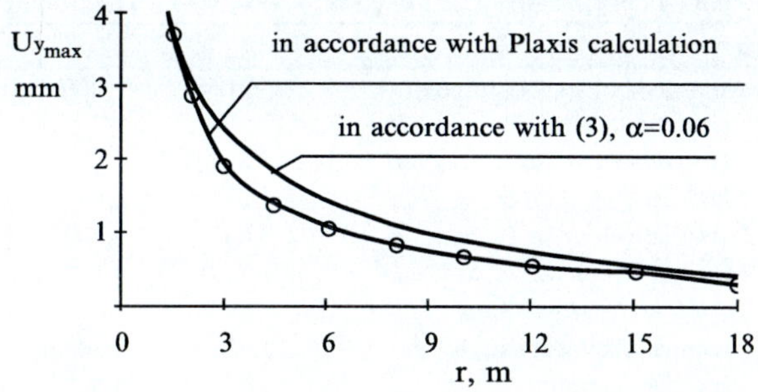

Fig. 4 Amplitude u_{ymax} of control points

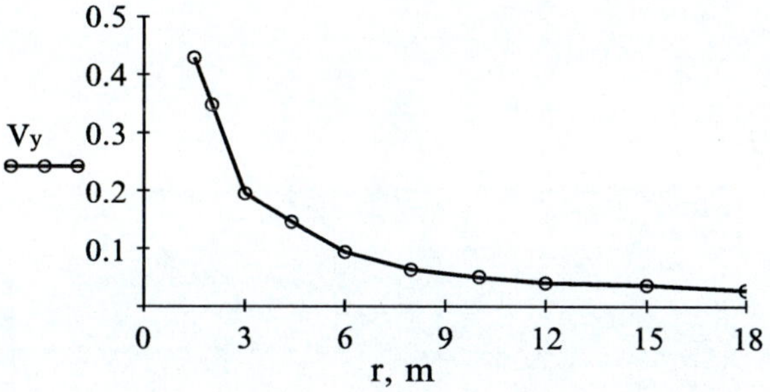

Fig. 5 Amplitudes of velocity V_{ymax}, m/s of control points on the surface

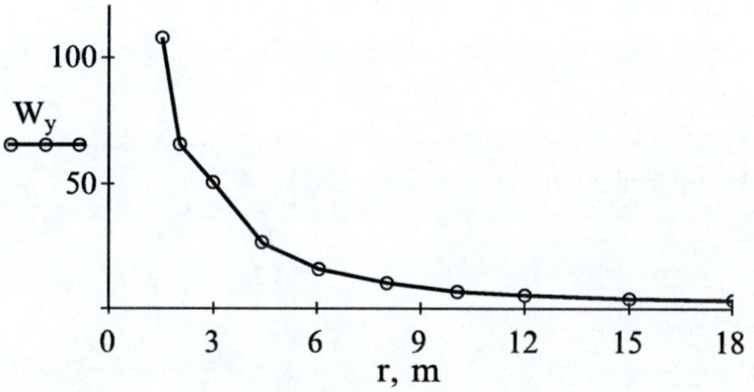

Fig. 6 Amplitudes of acceleration W_{ymax}, m/s^2 of control points

model of concrete is linear-elastic, $E_{ref} = 3 \cdot 10^7$ kN/m^3, $\nu = 0.2$. Material model of soil is LE drained sand, $\gamma = 21$ kN/m^3, $E_{ref} = 50{,}000$ kN/m^3, $\nu = 0.3$. Rayleigh damping coefficients are: $\alpha_R = 0.010$, $\beta_R = 0.1$; it is assumed that $\beta_R > \alpha_R$ in order to take into consideration the water presence. Calculation is executed also for MC soil model, the results are close similar. In Figs. 8 and 9 plots of dependence of vertical u_y and horizontal u_x displacement of surface points on time are presented for LE model.

The vibrations relax in ~1 s, the residual horizontal deformation is 4.5 mm, max vertical velocity 0.06 m/s, acceleration-14 m/s^{-2}.

Table 1 Parameters of soil in points A (r = 1.53 m) depending on duration of impact

Time interval t, s	0.1	0.05	0.01	0.005	0.001
Amplitude of impact force of sinusoidal impulse, q, kN/m^2	500	1,000	5,000	10,000	50,000
For linear-elastic model of soil					
Maximal displacement, u_r, mm	1.57	2.61	3.63	3.67	3.67
Maximal velocity, Vy, m/s	0.042	0.102	0.428	0.604	1.144
Maximal acceleration, W_y, m/s^2	1.772	9.117	115.0	281.2	242.9
For Mohr-Coulomb model of soil					
Maximal displacement, u_r, mm	2.58	3.36	3.84	3.86	3.81
Maximal velocity, Vy, m/s	0.063	0.129	0.326	0.390	0.468
Maximal acceleration, W_y, m/s^2	4.12	19.06	152	198	950

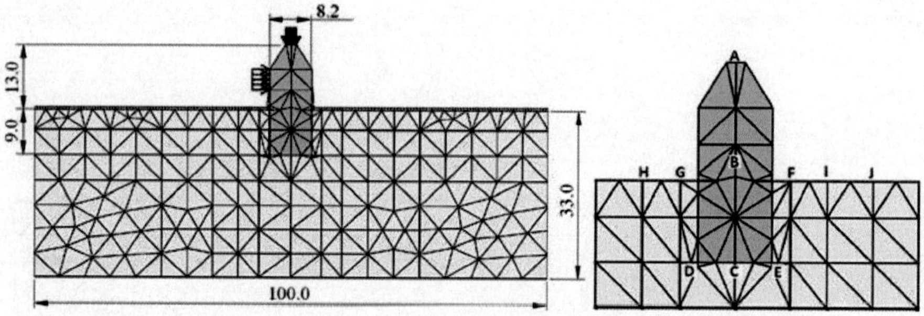

Fig. 7 Finite element mesh for plain strain model and scheme of control point location

Fig. 8 Plots of $u_y = f(t)$ for surface points

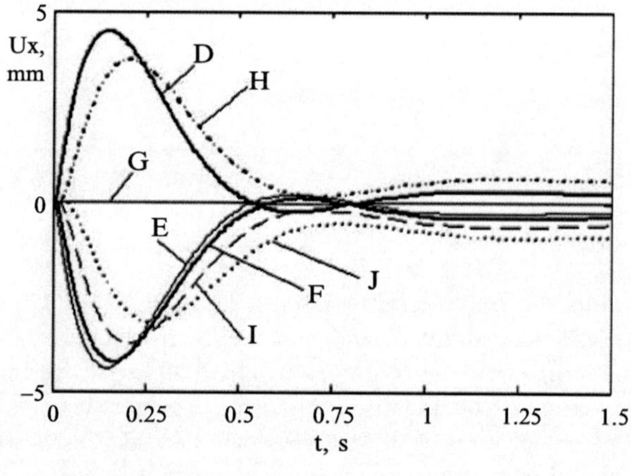

3.3 *Horizontal Impact on Sheet Pile Wall*

Sheet pile wall is often used for the berthing; it consists of wall elements, connected to the anchor wall by means of high-strength steel roads. Sheet piles wall is

Fig. 9 Plots $u_x = f(t)$ for surface points

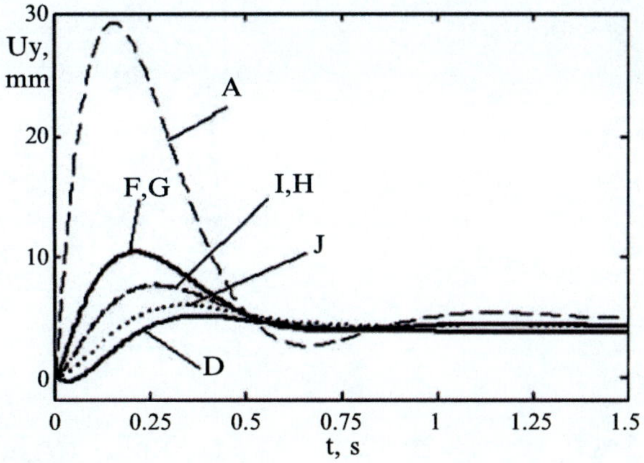

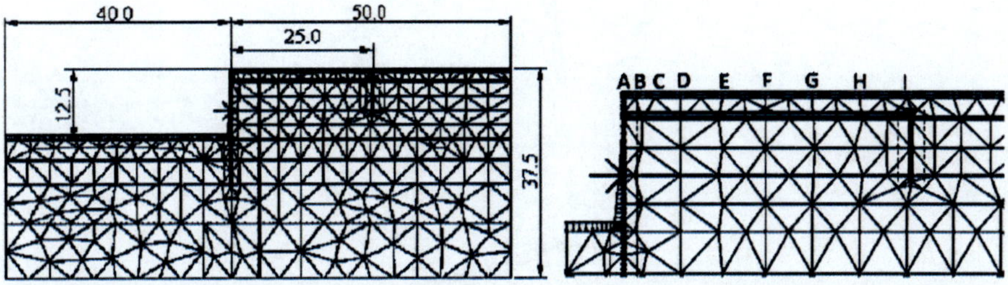

Fig. 10 Finite element mesh for plain strain model and scheme of control point location

imposed to impact of vessel, with duration $t = 0.01$ s, impact force amplitude of sinusoidal impulse $q = 6{,}000$ kN/m (Fig. 10). Material properties of wall are: normal stiffness $EA = 5.96 \cdot 10^6$ kN/m, flexural rigidity $EI = 9.13 \cdot 10^5$ kNm^2/m, weight $w = 2.21$ kN/m, material model is linear-elastic. Material model of soil is LE sand, drained, $\gamma = 21$ kN/m^3, $E_{ref} = 35{,}000$ kN/m^3, $\nu = 0.3$. Rayleigh damping coefficient $\beta_R = 0.1$, $\alpha_R = 0.01$. MC model of soil with friction angle $\varphi = 34°$, cohesion $C = 1$ kN/m^2, $E = 35{,}000$ kN/m^2 and hardening soil model are also examined and received results of the first and the second model are similar, the third is differs.

In Figs. 11 and 12 plots of dependence of vertical displacement u_y on time are presented for LE model and horizontal displacement u_x for all models. Residual horizontal deformation of surface points is close for MC and LE modes and more less for HS model. The vibrations relax in 0.6 s, the residual horizontal deformation is 45 mm, max vertical velocity 0.31 m/s, max horizontal velocity 1.68 m/s.

Max acceleration $-W_x$=973 m/s^{-2} (under force imposing point), W_y=171 m/s^{-2}.

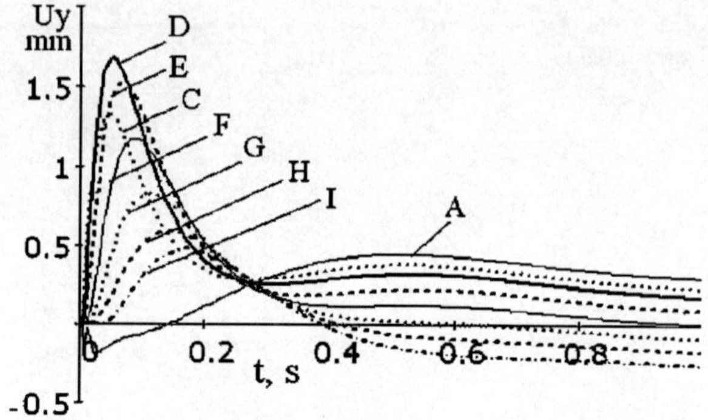

Fig. 11 Plots of $u_y = u_y$ (t) for surface points

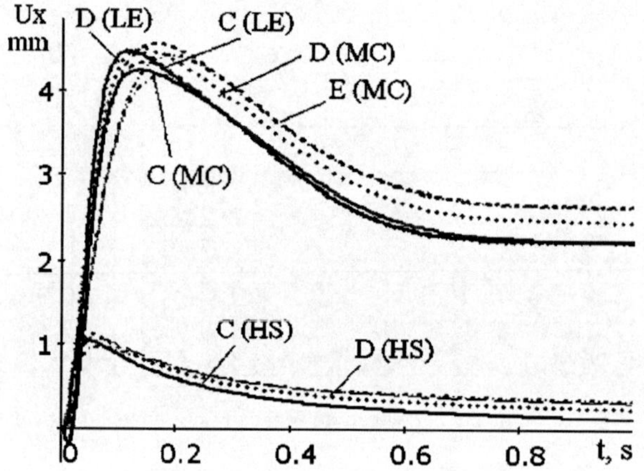

Fig. 12 Plots of $u_x = u_x$(t) for surface points

4 Conclusions

The results of the finite element analysis of impact action on soil and structure are presented in this work. Impact on massive foundation and on a flexible construction is examined. Results of calculation, executed using LE and MC models are differ not much. In all cases of application of the nonlinear hardening soil model the displacements and velocities of soil particles are less, but time of damping is longer, accelerations are higher. Duration of impulse influences on a result, if time of impulse $\tau$is comparable with the natural period of system. The form of impulse does not influence on results. This approach may be used for the analysis of accidents results.

References

1. Verrujit, A.: Soil Dynamics. Delft University of Technology, Papendrecht (2008)
2. Holleman, A.: Soil behavior under vibratory driving. In: Proceedings of the International Conference on Vibratory Pile Driving and Deep Soil Compaction, Transvib, pp. 3–19 (2002)
3. Plaxis 2D-v8, Plaxis b.v. Delft, Netherland (2007)
4. Tzitovich, N.A.: Soil Mechanics. Moscow (1983)

Finite Element Model Updating and Validation of the Uniovi Structure

José Luis Zapico-Valle, M. García-Diéguez, M.P. González-Martínez, and J. Abad-Blasco

Abstract The aim of this paper is the finite element modelling, updating and validation of the Uniovi Structure. This is a middle-size steel frame. The updating is based on natural frequencies, which are obtained from the free vibration of the structure. Three uncertain structural parameters are updated on the basis of the natural frequencies of two different configurations through an adaptive stochastic algorithm. Additional configurations corresponding to mass modifications were used to validate the updated model. It is found that the updated model can reproduce accurately the first two natural frequencies in all the configurations.

Keywords Model updating • Semi-portal finite element model • Free vibration test • Model validation

1 Introduction

The Uniovi Structure is a benchmark intended for many purposes: novelty detection [1], nonlinear identification [2], damage identification and other future applications. The present work is framed within a methodology for overall seismic damage identification in buildings. The methodology has been previously developed by the authors [3] and tested by numerical simulations, but it need to be validated with experimental data. The methodology relies on a reference finite element model.

J.L. Zapico-Valle (✉) • M. García-Diéguez • M.P. González-Martínez
Universidad de Oviedo, Campus de Gijón 7.1.16, 33203 Gijón, Spain
e-mail: jzapico@uniovi.es

J. Abad-Blasco
Universidad de Zaragoza, Campus Río Ebro, Zaragoza, Spain

J. Náprstek et al. (eds.), *Vibration Problems ICOVP 2011: The 10th International Conference on Vibration Problems*, Springer Proceedings in Physics 139,
DOI 10.1007/978-94-007-2069-5_47, © Springer Science+Business Media B.V. 2011

The objective of this work is to develop an appropriate finite element model capable to reproduce the main dynamic features of the structure from the available experimental data.

2 Experimental Part

2.1 *Description of the Structure, Exciter and Instrumentation*

The Uniovi Structure is a four-storey steel frame with two bays in the longitudinal direction and one bay in the transversal one. The overall dimensions of the structure are 4 m long, 1.5 m wide and 7.3 m height. All columns and beams are HEA-120 and IPN-100, respectively, of steel grade S-275. The floors of the frame are steel sheets 4 mm thick connected to the beams through discontinuous weld. The foundations are two continuous concrete beams lying on the floor of the laboratory.

Each column consisted of two pieces. They are spliced through end plates connected by four bolts 12 mm in diameter. The columns are welded to 20 mm thick plates. The beams corresponding to the transversal direction are directly connected to the web of the columns by a welded-all-around fillet. In the longitudinal direction, however, the beams are connected to the flange of the columns. The connection was designed with a 8 mm thick end plate and four bolts 10 mm in diameter.

A pendulum-like exciter was used to excite the frame. It consists of an eccentric mass connected to the shaft of an electrical motor. The excitation was carried out by varying harmonically the angular position of the mass. The frequency and amplitude of the excitation were achieved by controlling the angular position of the shaft through an electronic regulator.

The regulator is feedbacked by an encoder. The frequency range of the exciter is from 0 to 100 Hz and the angle is from 1° to 20°. The total mass of the pendulum is 1 kg.

The structure was provided with four seismic Brüel & Kjær accelerometers with a sensibility of 10 V/g. Each was screwed at the middle of the end beam on each floor pointed to the longitudinal direction. The hardware used to record the signals was a dSPACE RTI1104 data acquisition card. The signal processing was carried out by MATLAB.

2.2 *Test Planning*

Six different mass configurations of the structure were tested. The additional masses corresponding to each configuration are shown in Table 1. All the experiments consisted of a 20 s harmonic excitation at frequencies close to the resonances

Table 1 Mass configurations. m_i: additional mass on floor i

Configuration	m_1 [kg]	m_2 [kg]	m_3 [kg]	m_4 [kg]
Initial 0	0	0	0	0
Mass 1	43.4	43.2	43.5	43.3
Mass 2	43.4	43.2	43.5	86.7
Mass 3	43.4	43.2	86.4	86.7
Mass 4	43.4	86.7	86.4	86.7
Mass 5	87.3	86.7	86.4	86.7

Table 2 Identification results. f_i^e: experimental results. f_i^a: analytical results

Configuration	f_1^e [Hz]	f_1^a [Hz]	Error [%]	f_2^e [Hz]	f_2^a [Hz]	Error [%]
Initial 0	4.5836	4.5836	−0.0006	15.9551	15.9126	0.3206
Mass1	4.4367	4.4367	0.0009	15.3734	15.4149	−0.2698
Mass 2	4.3655	4.3629	0.0586	15.2495	15.3220	−0.4757
Mass 3	4.3227	4.3226	0.0025	15.2094	15.2637	−0.3568
Mass 4	4.3070	4.3070	−0.0011	14.9769	15.0359	−0.3938
Mass 5	4.3031	4.3049	−0.0410	14.9471	14.9678	−0.1385

and then 30 s of free vibration. From the measured free responses the first two natural frequencies were identified. For this purpose, a time-domain identification procedure was used [4].

2.3 *Data Preprocessing and Nonlinear Modal identification*

The free vibration analogue signals were converted into digital ones at a sampling frequency of 1,000 Hz. As a signal offset and a sort of periodic high frequency noise were found in the signals, a band-pass filter was used to remove the offsets and the noise. It was a four order Butterworth filter with cut-off frequencies of 3 and 20 Hz.

A softening model was adopted for the modal stiffness. This nonlinear model was defined with respect to the modal displacements. The acceleration corresponding to the free vibration of the model was obtained by numerical integration of the nonlinear equation of motion as a function of the model parameters and initial conditions. The identification was posed as an optimization issue. An error function accounting for the discrepancies between analytical and experimental accelerations was defined and minimized with respect to the model parameters and initial conditions. More details about the algorithm can be found in Zapico et al. [4]. The results of the identification are a softening parameter and the dynamic modal properties of the underlying linear model. More details can be found in [2]. The results of this modal identification on each mode and configuration are shown in Table 2 labeled as experimental results.

3 Analytical Part

3.1 Modelling

The shear model is common in seismic analysis of regular buildings due to its simplicity. In this case, the structure was modelled with a semi-portal model (see Fig. 1) in order to take into account the stiffness of the beams, which will be used for damage identification purposes in future applications [3].

The columns of the model and the structure have the same length, while the beams of the model have half length than the actual ones. The sum of all the stiffnesses of columns of each floor was selected as the stiffness of the corresponding column of the model. They were modelled as Timoshenko beam elements to take into account the shear deformations. The column splice is modelled with two additional nodes (9 and 10) and a different beam element. The beams were modelled with beam elements. All the columns and beams of the model were modelled as massless elements. The total mass of each floor was lumped at the corresponding nodes (1, 2, 3 and 4) (see Fig. 1). Only the degrees of freedom corresponding to the horizontal and vertical translation and the rotation around normal direction are considered for all the nodes except nodes 5 to 8 where vertical translations were fixed. The model was coded in MATLAB [5] using the Structural Dynamics Toolbox [6].

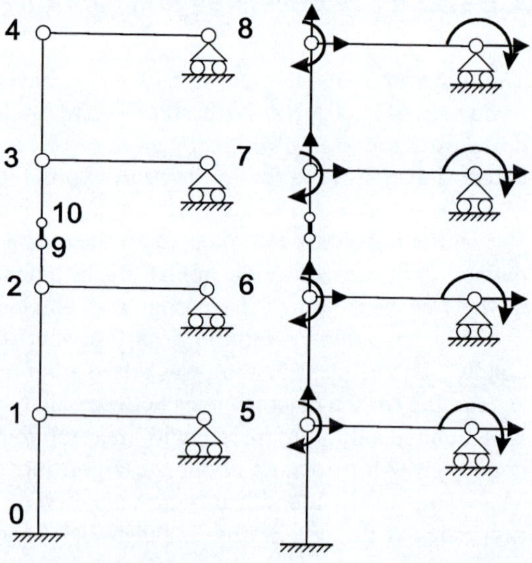

Fig. 1 *Left*: Semi-portal model. *Right*: Degrees of freedom

3.2 Updating

The selection of parameters is a critical task on the updating process; only uncertain parameters that the response is sensitive to should be selected [7]. The Young's modulus and columns second moment of area were set to their standard values ($E = 2.07 \times 10^{11}$ Pa and $I_c = 6.06 \times 10^{-6}$ m^4) because only small discrepancies are expected in these parameters. Other parameters, however, are much more uncertain. This is the case of the stiffness of the column splice and the stiffness of the beams, due to the contributions of the floors and handrails. It is assumed that all the beams have the same stiffness. Thus, the splice second moment of area I_s and the beam second moment of area I_b were selected as parameters to be updated. In addition, a mass factor d was selected as a global parameter aimed to correct not only the effects of the mass errors, but indirectly the global stiffness and geometric errors. The stiffness of the column-foundation joint was modelled as fixed due to the insensitivity of the response to this parameter.

The updating is posed as the minimization of the following error function

$$\varepsilon = \frac{100}{4} \sum_{i=0}^{1} \sum_{j=1}^{2} \frac{|f_{ij}^{e} - f_{ij}^{a}|}{f_{ij}^{e}}, \tag{1}$$

where subscript i represents the number of configuration, in this case two configurations were selected (Initial 0 and Mass 1), and subscript j represents the mode, in this case the two first ones. The Eq. 1 represents the normalized mean squared error. The optimization was carried out through a novel adaptive stochastic method explained in [4].

The algorithm was run 10 times in order to check the converged of the results. The maximum number of iterations was set to $N_s = 2{,}000$. The updated parameters were $I_s = 1.055 \times 10^{-5}$ m^4, $I_b = 1.182 \times 10^{-5}$ m^4 and $d = 1.0461$, which correspond to a fitting error of 0.2960%. As expected, the second moment of area of the splice, Is, is around three times lower than that of the column, due to the flexibility of the joint. The obtained second moment of area of the beams, I_b, is higher than that of the standard section, as a result of the contribution of the floors and other elements. The global parameter, d, is close to one with a discrepancy lower than 5%, which indicates that the rest of the physical and geometrical properties of the model are close to their nominal values. The physical meaning of the results and the low fitting error guarantee, to some extent, the updating process; it is not a mere mathematical optimization, but it has physical meaning.

4 Validation

In order to validate the updated model, other mass configurations (Mass 2 to Mass 5) were used. The results for these configurations along with the discrepancies with respect the experimental frequencies for each configuration and mode are shown in Table 2. As can be seen, the model reproduces quite accurately the first two frequencies. In the first one, the model prediction errors are up to 0.06%, and in the second one, they are up to 0.48%.

5 Conclusions

A semi-portal finite element model has been proposed to model the linear component of the dynamic behaviour of the Uniovi Structure. The most uncertain parameters of this model have been selected and updated on the basis of the first natural frequencies of the structure. These were previously identified in the modal domain through a softening nonlinear model from the measured free vibration of the structure. An adaptive stochastic optimization procedure has been used to update the uncertain parameters. Additional configurations of the structure consisting in mass modifications have been used to validate the updated model. It is found that the optimization procedure, which was originally intended for time-domain data, is effective and robust when using natural frequencies. Moreover, the updated model can reproduce accurately the first two natural frequencies in all the considered configurations with discrepancies lower than 0.5%.

Acknowledgments The financial support given by the Spanish Ministry of Education and Innovation trough the project BIA2006-15266-C02-01 is gratefully appreciated.

References

1. Zapico-Valle, J.L., García-Diéguez, M., González-Martínez, M.P., Worden, K.: Experimental validation of a new statistical process control feature for damage detection. Mech. Syst. Signal Process.. doi:10.1016/j.ymssp.2011.02.007
2. García-Diéguez, M., Zapico-Valle, J.L., González-Martínez, M.P., Abad-Blasco, J.: Nonlinear modal identification of the Uniovi Structure. In: Proceedings of the International Conference on Structural Engineering Dynamics, Tavira-Portugal, June 2011
3. González, M.P., Zapico, J.L.: Seismic damage identification in buildings using neural networks and modal data. Comput. Struct. **86**, 416–426 (2008)
4. Zapico-Valle, J.L., Alonso-Camblor, R., González-Martínez, M.P., García-Diéguez, M.: A new method for finite element model updating in structural dynamics. Mech. Syst. Signal Process. **24**, 2137–2159 (2010)
5. Users Manual Version 6.5 MATLAB. The Math Works, Inc., Natick (2002)
6. Balmès E.: Structural Dynamic Toolbox Manual. Scientific Software Group, Sèvres, France (1997)
7. Friswell, M.I., Mottershead, J.E.: Finite Element Model Updating in Structural Dynamics. Kluwer Academic Publishers, Dordrecht, The Netherlands (1999)

Design of Symmetric Positive Definite Vibrating Systems via Inverse Eigenvalue Methods

Ladislav Starek and Daniel Starek

Abstract This contribution considers the inverse eigenvalue problem for linear vibrating systems described by a vector differential equation with constant coefficient matrices. The inverse problem of interest here is that of determining real symmetric positive definite mass normalized velocity and position coefficient matrices, which create the vibrating system with the prescribed (in advance) complex eigenvalues and complex eigenvectors.

Keywords Eigenvalue • Eigenvector • Spectral • Modal matrix • Inverse formulas

1 Introduction

Here we consider linear lumped parameter system (1) which can be modeled by a vector differential equation in the second order form given by

$$M\ddot{q}(t) + D\dot{q}(t) + Kq(t) = f_1(t) \tag{1}$$

where $q(t)$ is an n vector of time-varying elements representing the displacement of the masses in a lumped mass model of some structure or device. The vectors $\dot{q}(t)$ and $\ddot{q}(t)$ represent the velocities and accelerations, respectively. The n vector $f_1(t)$ represents applied external forces and is time varying. The coefficients M, D and

L. Starek (✉)
Faculty of Mechanical Engineering, Slovak University of Technology, Bratislava, Slovakia
e-mail: ladislav.starek@stuba.sk

D. Starek
Faculty of Civil Engineering, Slovak University of Technology, Bratislava, Slovakia
e-mail: daniel.starek@stuba.sk

J. Náprstek et al. (eds.), *Vibration Problems ICOVP 2011: The 10th International Conference on Vibration Problems*, Springer Proceedings in Physics 139,
DOI 10.1007/978-94-007-2069-5_48,

K are $n \times n$ matrices of constant real elements representing the various physical parameters of mass, damping, and stiffness. The matrices M, D and K could be either asymmetric or symmetric and if symmetric M is positive definite.

If M is positive definite and symmetric it has a positive definite matrix square root, with a symmetric, positive definite inverse denoted by $M^{-1/2}$. Let us then consider the transformation $q(t) = M^{-1/2}v(t)$. Substitution of this change of coordinates into Eq. 1 yields

$$\ddot{v}(t) + \tilde{D}\dot{v}(t) + \tilde{K}v(t) = p(t) \tag{2}$$

where $\tilde{K} = M^{-1/2}KM^{-1/2}$ and $\tilde{D} = M^{-1/2}DM^{-1/2}$ are necessarily symmetric if D and K are and $p(t) = M^{-1/2}f_1(t)$. The matrices $\tilde{D}$ and $\tilde{K}$ are refered to here as the mass normalized damping and stiffness matrices. The eigenvalue problem for the systems described by (2) is defined by

$$(\lambda^2 I + \lambda\tilde{D} + \tilde{K})v = 0 \tag{3}$$

where v is a nonzero vector of constants, called the eigenvector, and λ is a scalar, called the eigenvalue. From the spectral theory of matrix polynomials it is well known that the solutions of the system (2) are intimately connected with the algebraic properties of the matrix polynomials [1] of the form

$$L(\lambda) = \lambda^2 I + \lambda\tilde{D} + \tilde{K}. \tag{4}$$

Here a scalar λ and a nonzero vector x are again called an eigenvalue and associated (right) eigenvector of $L(\lambda)$ if $\det L(\lambda) = 0$ and $L(\lambda)\,x = 0$, respectively. This forms an obvious connection.

2 Inverse Formulas

The eigenvalues (and their multiplicities) of a lumped linear system described by real coefficient matrices are symmetric with respect to the real axis of the complex plane. This implies that there is a Jordan matrix for such a system with the block diagonal form

$$\Lambda = (J_C, J_R, \bar{J}_C) \tag{5}$$

where J_C is a matrix with all its eigenvalues in the open upper half of the complex plane, J_R is real matrix and the entries of $\bar{J}_C$ are the complex conjugate of those in J_C. Here the dimensions of J_C and J_R are determined by the number of real roots. If there are $2m$ complex roots, then the dimension of J_C and $\bar{J}_C$ is $2m \times 2m$ and the dimension of J_R is $2(n - m)$. The modal matrices V and W are partitioned in a compatible way as Eq. 5, i.e.

$$V = (V_C, V_R, \bar{V}_C), \quad W = (W_C, W_R, \bar{W}_C). \tag{6}$$

Here the overbar denotes the complex conjugate. Note that the matrix of eigenvectors must be arranged in columns such that the complex eigenvectors appear first, followed by the real eigenvectors and the complex conjugate eigenvector.

Inverse formulas and conditions for given spectral and modal data to determine symmetric coefficient matrices are given in [2]. Inverse formulas specify the coefficient matrices $\tilde{K}$ and $\tilde{D}$ in terms of the spectral matrices V, W and Λ. We obtain

$$\tilde{D} = -V\Lambda^2 W^T \ \tilde{K} = \tilde{D}^2 - V\Lambda^3 W^T \tag{7}$$

with the conditions

$$VW^T = 0 \text{ and } V\Lambda W^T = I. \tag{8}$$

The formulas (7) determine the two desired coefficient matrices of the system (2), if the third is chosen with spectral and modal properties, which must satisfy the conditions (8). Unfortunately, there is no guarantee at this point, that the formulas given by Eq. 7 will result in $\tilde{D}$ and $\tilde{K}$ being symmetric.

If the inverse formulas are to generate real symmetric coefficient matrices, then the specified spectral and modal matrices must fulfill some further requirements. From the theory of matrix polynomials [1] it is known that if a monic matrix polynomial is self-adjoin, then there exist a Jordan triple $(U, \Lambda, P_\Lambda U^*)$, where P_Λ is a permutation matrix and U^* denotes the complex conjugate transpose of the matrix U. Hence in the case of interest here the spectral matrix Λ and modal matrices V and W will generate hermitian coefficient matrices if the left modal matrix is of the form

$$W^T = P_\Lambda V^* \tag{9}$$

where the real valued matrix P_Λ is given by the formula [1]

$$P_\Lambda = \begin{pmatrix} 0 & 0 & P_C \\ 0 & P_R & 0 \\ P_C & 0 & 0 \end{pmatrix}. \tag{10}$$

Then, if $X = \begin{pmatrix} V & \Lambda V \end{pmatrix}^T$, conditions (8) takes the form

$$\begin{pmatrix} V & \Lambda V \end{pmatrix}^T P_\Lambda V^* = \begin{pmatrix} 0 & I \end{pmatrix}^T. \tag{11}$$

The conditions given by Eq. 11 specify vectors that will generate real symmetric coefficient matrices for the given spectral matrix Λ. Note that this also depends on the structure of the state matrix A. State matrix$A$ is assumed to be simple or of general Jordan structure and whether or not the dynamic system is overdamped, underdamped, critically damped, or exhibits mixed damping.

3 A Symmetric Positive Definite Inverse Vibration Problem for Non-proportional Underdamped Systems

Theorem [3] A Jordan pair (X, J) corresponds to a self adjoint monic matrix polynomial, if and only if, there is a matrix $T \in A_I(J)$ and a canonical matrix P_C such that

$$\begin{pmatrix} X \\ XJ \\ \vdots \\ XJ^{n-1} \end{pmatrix} \left(T \; P_C \; T^*\right) X^* = \begin{pmatrix} 0 \\ 0 \\ \vdots \\ I \end{pmatrix} \tag{12}$$

$T \in A_I(J)$ denotes the subalgebra of invertible matrices that commute with J. When the canonical matrix takes the form of (10) and with $T = I$, $n = 2$ and for the case of simple underdamped system (no real eigenvalues), Eq. 12 takes the form

$$V_C P_C V_C^T + \overline{V_C} P_C V_C^* = 0 \tag{13}$$

$$V_C J_C P_C V_C^T + \overline{V_C}\,\overline{J_C} P_C V_C^* = I. \tag{14}$$

For any $Z \in C^{n \times n}$ the matrix polynomial $L(\lambda) = (I\lambda - Z^*)(I\lambda - Z)$ is nonnegative, if $x^* L(\lambda) x \geqslant 0$. Furthermore, $L(\lambda)$ is positive, if and only if Z has no real eigenvalues. For such polynomials the Jordan structure of $L(\lambda)$ is simply determined by that of Z.

If $Z = X_Z J_Z X_Z^{-1}$, where $J_Z = J_C$ is in Jordan normal form and has no real eigenvalues and the matrix $X_Z = X_C$ is a matrix of corresponding eigenvectors for J_Z, will ensure that $L(\lambda)$ has real symmetric coefficients and the pair $J = diag\left(J_C, \bar{J}_C\right)$ and $X = \left(X_C, \bar{X}_C\right)$ determine a self adjoint triple for $L(\lambda)$.

If the pair $(X,\ J)$ determines a self adjoint polynomial, then by (5) and (6) (or by Theorem) it is real valued.

When the all conditions for generating symmetric positive definite coefficient matrices via theory of matrix polynomials are fulfill, then all others inverse formulas given by (7), where the entries to the inverse formulas are given by Eq. 8 can be used for determining coefficient matrices.

Let the imaginary part of the modal matrix be represented as the product $V_i = V_r C$, of real part of the modal matrix V_r and the matrix C which is real valued and nonsingular. Assume underdamped system and substitute the value for V_i and after same manipulations, we yield the simplified form of conditions (8) [2].

$$2V_r E V_r^T = 0 \text{ and } 2V_r F V_r^T = I \tag{15}$$

where $E = P_C - C P_C C^T$ and $F = J_r P_C - C J_i P_C - J_i P_C C^T - C J_r P_C C^T$. If we will assume vibrating systems of the simple structure, then $P_C = I$. In the case $E = 0$ the matrix C must be orthonormal and upon further examination F must be symmetric. If J_i is chosen to be in the lower part of the complex plane, then the matrix F is seen to be positive definite with the appropriate choise of entries of the matix C (according to the choise of eigenvalues).

Then from (15) can we consider the matrix L defined by

$$L = 2V_r^c F^c \left(V_r^c\right)^T \tag{16}$$

It is possible to approve that matrix L is symmetric and positive definite (for the above condition). Since the matrix L is positive definite it has a Cholesky decomposition, i.e., there exists a nonsingular matrix T, such that $L = T^T T$ and then we have $T^T T = 2V_r^c F^c \left(V_r^c\right)^T$. Ending this equation yields

$$2\left(T^T\right)^{-1} V_r^c F^c \left(V_r^c\right)^T T^{-1} = I. \tag{17}$$

A comparison of Eqs. 17 and 15 is

$$V_r = \left(T^T\right)^{-1} V_r^c \tag{18}$$

then V_r is a matrix of real part of eigenvectors such that the choise of $\tilde{D}$ and $\tilde{K}$, given by equations $\tilde{D} = -Z - Z^*$ and $\tilde{K} = Z * Z$, will be symmetric and positive definite [4].

The procedure for determining real symmetric positive definite matrices $\tilde{D}$ and $\tilde{K}$ is summarized as follows:

1. Select the required eigenvalues (with the negative real parts and imaginary part chosen to be in the lower part of the complex plane) of the system described by Eq. 2 (they are given by the matrix J_Z).
2. Choose the appropriate orthonormal matrix C according to the desired designed system (diagonal for systems with proportional damping and non diagonal for systems with non proportional damping).
3. Choose a real part of the matrix of eigenvectors V_r^c and check the condition (16). If the matrix $L \neq I$, then correct the real part of chosen eigenvectors by (18) to fullfil the condition (14).
4. Compute the desired coefficient matrices.

4 Conclusions

This paper presents a solution to the inverse eigenvalue problem for the case that the desired coefficient matrices be symmetric and positive definite and the system can have an arbitrary damping (proportional or nonproportional damping). In [2] the conditions (15) for given spectral and modal data have been defined under which the inverse formulas determine real symmetric coefficient matrices $\tilde{D}$ and $\tilde{K}$. However, that solution does not preserve given eigenvectors.

The method has been outlined which allows the synthesis of a symmetric, underdamped linear system, having desired eigenvalues and eigenvectors. The new result gives the alternative solution to the inverse eigenvalue problems in vibration described in [2] and [4].

As was stated, by sequence modification Eq. 15, base on conditions, which define underdamped system, we yield relation for matrix L Eq. 16, that is real valued, symmetric and positive definite. Hence is possible to apply a Cholesky decomposition, from which it is possible to derive corrected eigenvectors of the system and these have been used on determination coefficient matrices of system $\tilde{D}$ and $\tilde{K}$.

Acknowledgments The authors gratefully acknowledges the support of the Slovak Ministry of Education Grant Agency VEGA # 1/0744/10.

References

1. Gohberg, I., et al.: Matrix Polynomials. Academic, New York (1982)
2. Starek, L., Inman, D.J., Kress, A.: A symmetric inverse vibration problem. ASME J. Vib. Acoust. **114**, 564–568 (1992)
3. Lancaster, P., Maroulas, J.: Inverse eigenvalue problems for damped vibrating systems. J. Math. Anal. Appl. **123**(1), 238–261 (1987)
4. Starek, L., Inman, D.J.: Design of nonproportional systems via symmetric positive inverse problems. ASME J. Vib. Acoust. **126**, 212–219 (2004)

Characteristic Response Functions (CRFs)

Nuno M.M. Maia

Abstract Characteristic Response Functions (*CRFs*) are functions that are typical of a given structure. Frequency Response Functions (*FRFs*) are commonly used with many different objectives, relating output and input and can be expressed either in terms of poles and residues or in terms of natural frequencies, damping ratios and modal constants. Frequencies and damping ratios are global properties, whereas modal constants are local properties, i.e., the former are independent on the locations of the excitation and measurement co-ordinates and the latter vary with those locations. Therefore, *FRFs* vary along the structure. However, one can use the *FRFs* to build other functions that do not depend on the modal constants. Such functions, that depend only on the global properties are called Characteristic Response Functions (*CRFs*), as they remain the same no matter where the measurements are taken. Such feature may be quite useful in some applications. These kind of functions are developed here. Comparisons are made through the use of some numerical simulations to illustrate their performance and usefulness.

Keywords Frequency response functions • Modal analysis • System identification

1 Introduction

The study of the dynamic behavior of a structure often implies the establishment of a theoretical model from measurements either in the laboratory or in the field, namely when a numerical model is not easy to develop. Those measurements are usually expressed in terms of frequency response functions (*FRFs*), which depend

N.M.M. Maia (✉)
IDMEC-IST, Technical University of Lisbon, Lisbon, Portugal
e-mail: nmaia@dem.ist.utl.pt

J. Náprstek et al. (eds.), *Vibration Problems ICOVP 2011: The 10th International Conference on Vibration Problems*, Springer Proceedings in Physics 139,
DOI 10.1007/978-94-007-2069-5_49,

on the natural frequencies, damping ratios and also mode shapes. The measured *FRFs* vary along the structure, according to the co-ordinates that have been taken as inputs and outputs. However, some properties of the structure remain unchanged, like the natural frequencies and damping ratios, while others, related to the mode shapes vary along the structure. If one manages to "filter out" the effect of the modal constants, one obtains something independent of the measured co-ordinates and thus expresses a characteristic of the dynamic behavior of that structure. The advantages of such type of formulation can be various, like to have a kind of "signature" of the structure, to have a clearer picture of the number of modes, to avoid the influence of modes outside the frequency range of interest, to easily identify the global properties, etc. The first proposal was given by the author and some colleagues in 1994 [1] and further developments were presented later [2, 3]. The goal of this paper is to derive and study such type of functions.

2 Theoretical Formulation

Assuming that the model of our structure is based on the hysteretic type of damping, the *FRF* relating co-ordinates p and q is given by the expression [4, 5]:

$$H_{pq}(i\omega) = \sum_{r=1}^{N} \frac{\varphi_p^{(r)} \varphi_q^{(r)}}{\omega_r^2 - \omega^2 + i\eta_r\omega_r^2} = \sum_{r=1}^{N} \frac{C_{pq}^{(r)}}{\omega_r^2 - \omega^2 + i\eta_r\omega_r^2} \tag{1}$$

where N is the total number of degrees of freedom, $\varphi_p^{(r)}$ is the pth element of the rth mode shape, $\varphi_q^{(r)}$ is the qth modal participation factor, ω_r is the rth natural frequency and η_r the rth damping ratio. Often, $\varphi_p^{(r)}\varphi_q^{(r)}$ is called the modal constant ($C_{pq}^{(r)}$). As the response is dominated by each mode in a narrow band frequency around it, it is usual to approximate expression (1) in the following way:

$$H_{pq}(i\omega) = \frac{C_{pq}^{(r)}}{\omega_r^2 - \omega^2 + i\eta_r\omega_r^2} + R_{pq} \tag{2}$$

where R_{pq} represents the contribution of the neighboring modes. This is acceptable if the modes are not too close together. Simplifying, Eq. 2 becomes:

$$H_k = \frac{C_k}{D - \Omega} + R_k \tag{3}$$

where $D = \omega_r^2 + i\eta_r\omega_r^2$, $\Omega = \omega^2$ and k is the index for each pair pq. Differentiating once and twice with respect to Ω, the effect of the residual term disappears:

$$\frac{\partial H_k}{\partial \Omega} = \frac{C_k}{(D - \Omega)^2}; \quad \frac{1}{2}\frac{\partial^2 H_k}{\partial \Omega^2} = \frac{C_k}{(D - \Omega)^3} \tag{4}$$

Dividing Eq. 4 between them leads to a new function, the *CRF* – say β – independent of the effect of neighboring modes and of the modal constant C_k:

$$\beta = \frac{\partial^2 H_k}{\partial \Omega^2} \Big/ \left(2\frac{\partial H_k}{\partial \Omega}\right) = \frac{1}{D - \Omega} = \frac{1}{\omega_r^2 - \omega^2 + i\eta_r\omega_r^2} \tag{5}$$

3 Numerical Implementation and Discussion

Equation 5 shows that β has the same form as the *FRF* of a single degree-of-freedom with unit modal constant, and so, it must look like such an *FRF*, around each natural frequency. Therefore, β should be the same, no matter where one measures the *FRF*. It is the Characteristic Response Function (*CRF*). From Eq. 2, and apart the residual term, the response *FRF* at each natural frequency is dominated not only by the damping ratioη_r, but also by the modal constant $C_{pq}^{(r)}$. A small modal constant may hide a resonance peak (Fig. 1). However, the *CRF* will show that peak, as the *CRF* is independent of the modal constants (Fig. 2). The *CRF* can thus act as a valuable mode indicator. The identification of ω_r and η_r is easy to perform from Eq. 5. To have a clear picture of the behavior of the structure, one should superimpose various *CRFs*. From three *FRFs* (Fig. 3), the three *CRFs* are shown in Fig. 4. The three curves coincide around each ω_r. Adding random noise may lead to something like in Fig. 5, where the right peaks emerge.

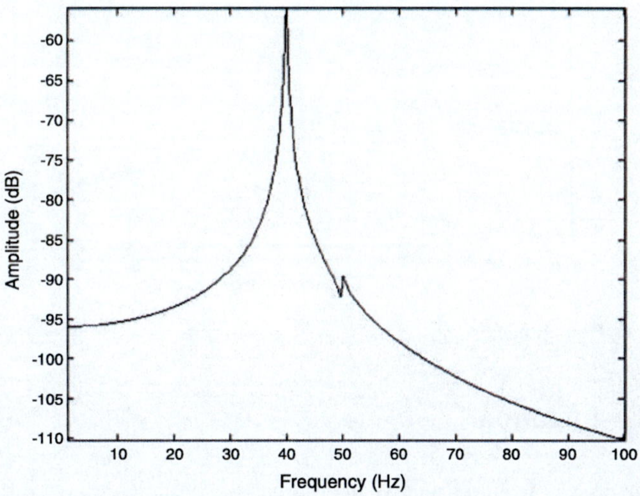

Fig. 1 *FRF* with 2 resonance peaks, with $\eta_1 = 0.01$, $\eta_2 = 0.01$, $C_1 = 1$, $C_2 = 0.01$

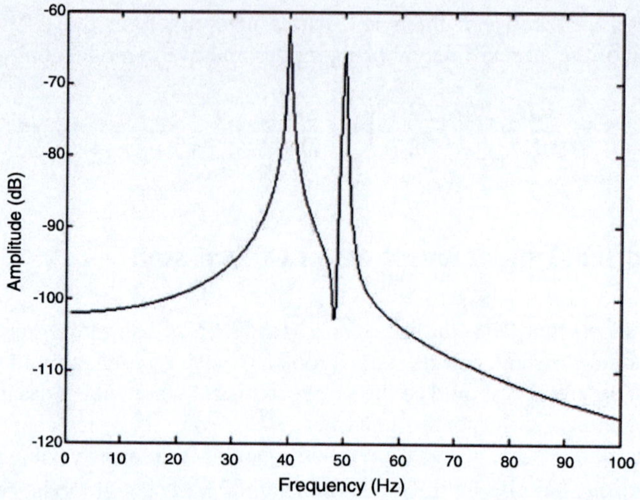

Fig. 2 *CRF* corresponding to the *FRF* of Fig. 1

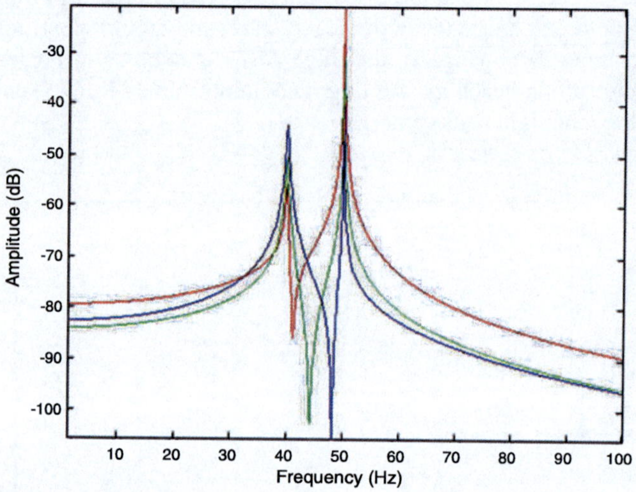

Fig. 3 Three *FRF* amplitudes

4 New Formulation

The *CRF* can also be used to identify the modal properties, using the inverse of β:

$$\operatorname{Re}\left(\frac{1}{\beta}\right) = \omega_r^2 - \omega^2 \quad \text{and} \quad \operatorname{Im}\left(\frac{1}{\beta}\right) = \eta_r \omega_r^2 \tag{6}$$

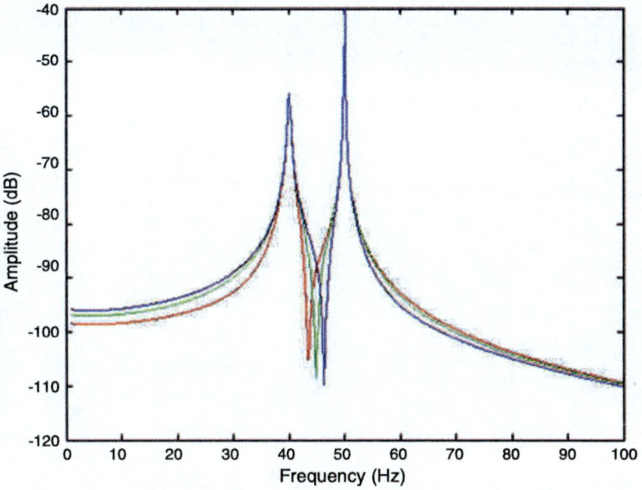

Fig. 4 *CRF* amplitudes corresponding to the *FRFs* of Fig. 3

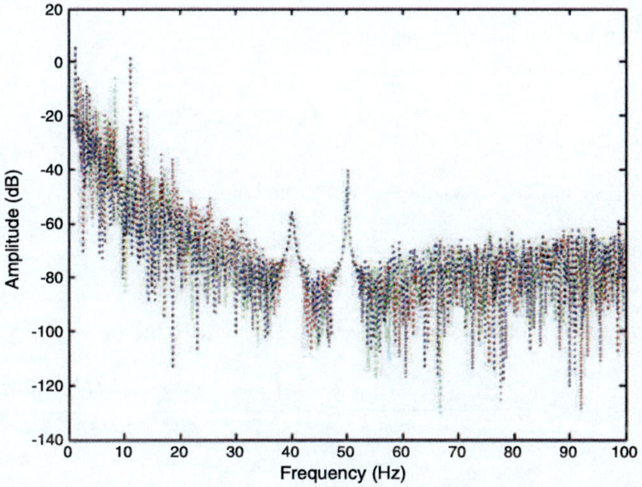

Fig. 5 Same results as in Fig. 4 with some added random noise

An important step to take is to consider that the influence of the residual modes is not a simple constant. The low frequency residual modes are represented by a term inverse to the square of the frequency, and the upper frequency residual modes by a constant [4]. A more accurate model for the *FRF* is therefore:

$$H_k = \frac{C_k}{D - \Omega} + \frac{R_k}{\Omega} + S_k \tag{7}$$

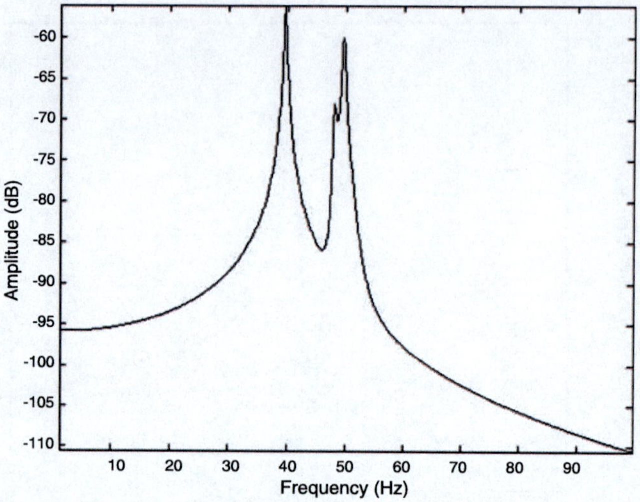

Fig. 6 New *CRF* corresponding to the *FRF* of Fig. 1

The first and second derivatives are:

$$H_k' = \frac{\partial H_k}{\partial \Omega} = \frac{C_k}{(D-\Omega)^2} - \frac{R_k}{\Omega^2}; \quad H_k'' = \frac{\partial^2 H_k}{\partial \Omega^2} = \frac{2C_k}{(D-\Omega)^3} + \frac{2R_k}{\Omega^3} \tag{8}$$

Extracting C_k and R_k from Eqs. 8 in 7, one has:

$$H_k = H_k' D - 2H_k'\Omega + \frac{1}{2}H_k'' D\Omega - \frac{1}{2}H_k''\Omega^2 + S_k \tag{9}$$

Differentiating once more with respect to Ω, leads to the new *CRF*:

$$\beta = \frac{1}{\omega_r^2 - \omega^2 + i\eta_r\omega_r^2} = \frac{1}{D-\Omega} = \frac{3H_k'' + H_k'''\Omega}{6H_k' + 3H_k''\Omega} \tag{10}$$

Using the same data as in Fig. 1 the result is in Fig. 6. An extra peak appears, but taking another *FRF* (e.g., $C_2 = 1$), the result is in Fig. 7. The middle peak moves and so it is not a genuine resonance. More FRFs can be used for this goal.

5 Identification

Example 1. Comparing the performance of the new formula with the previous one, with $\omega_1 = 2\pi \times 40$ rad/s, $\omega_2 = 2\pi \times 50$ rad/s, $\eta_1 = 0.01$, $\eta_2 = 0.01$, $C_1 = 1$, and $C_2 = 0.01$ (the same data used in Fig. 1), the results are presented in Table 1.

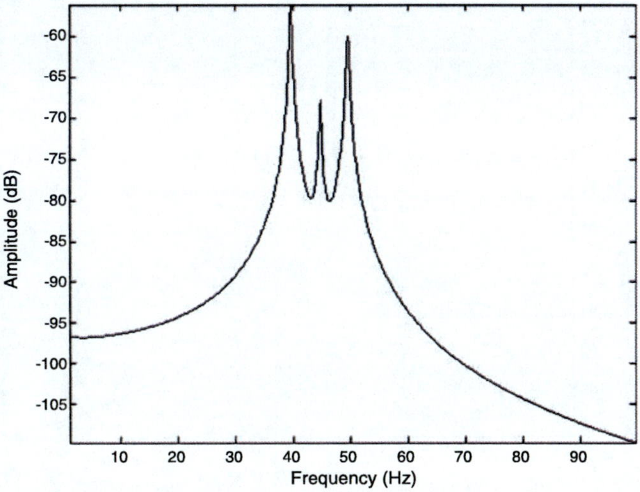

Fig. 7 Equivalent to Fig. 6, with $C_2 = 1$

Table 1 Example 1. Results of the identification of the natural frequencies and damping ratios

	$\omega_1/2\pi$ (Hz)	ε (%)	$\omega_2/2\pi$ (Hz)	ε (%)	η_1	ε (%)	η_2	ε (%)
CRF	40.257	0.64	50.240	0.48	0.0107	7	0.0099	1
New *CRF*	39.769	0.58	49.628	0.74	0.0101	1	0.0099	1

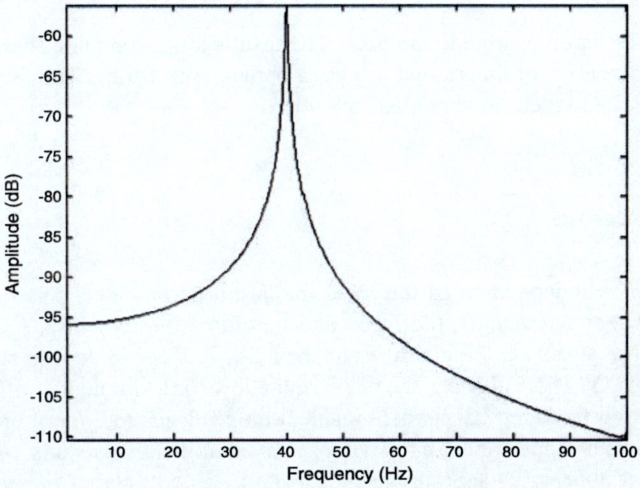

Fig. 8 *FRF* of example 2

Example 2. For two closer modes, say 40 and 43 Hz (the other data remain the same), the *FRF* and *CRF* are shown in Figs. 8 and 9. The *CRF* detects clearly the

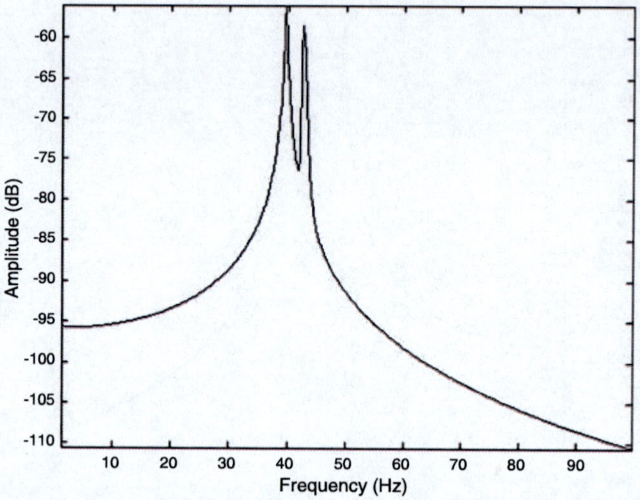

Fig. 9 New *CRF* of example 2, corres-ponding to the *FRF* of Fig. 8

Table 2 Example 2. Results of the identification of the natural frequencies and damping ratios

	$\omega_1/2\pi$ (Hz)	ε (%)	$\omega_2/2\pi$ (Hz)	ε (%)	η_1	ε (%)	η_2	ε (%)
CRF	40.215	0.54	43.468	1.09	0.0099	1	0.0155	55
New *CRF*	39.789	0.53	42.282	1.67	0.0101	1	0.0077	23

second mode, barely visible in the *FRF*. The results are in Table 2. There is a slight increase in the error of the second ω_r, but a better result for η_r. This is a very hard example, as the modes are very close together.

6 Conclusions

- The *CRF* is independent of the local modal properties, i.e., it is the same no matter where one excites the force and measures the response; however, this is only true around each natural frequency. So, it is convenient to superimpose several *CRFs*. This enhances the peaks and allow for a simple identification;
- when taking frequency dependent residual modes leads to a more precise *CRF*, with a similar appearance, but giving better identification results. A few extra peaks may appear, but superimposing several *CRFs* will clarify the problem.
- both functions are quite sensitive to noise and further refinements are needed.

Acknowledgements The current investigation had the support of FCT, under the project POCI2010.

References

1. Maia, N.M.M., Ribeiro, A.M.R., Silva, J.M.M.: A new concept in modal analysis: the characteristic response function (CRF). Int. J. Anal. Exp. Modal Anal. **9**(3), 191–202 (1994)
2. Maia, N.M.M., Silva, J.M.M., Ribeiro, A.M.R.: A new modal identification method using an advanced characteristic response function. In: Proceedings of IMAC XIII, Nashville, vol. II, pp. 1069–1075 (1995)
3. Silva, J.M.M., Maia, N.M.M., Ribeiro, A.M.R.: Structural dynamic identification with modal constant consistency using the Characteristic Response Function (CRF) concept. Mach. Vib. **5**(2), 83–88 (1996)
4. Ewins, D.J.: Modal Testing: Theory, Practice and Application, 2nd edn. Research Studies Press Ltd, Baldock (2000)
5. Maia, N.M.M., Silva, J.M.M., HE, J., Lieven, N.A.J., Lin, R.M., Skingle, G.W., To, W.M., Urgueira, A.P.V.: Theoretical and Experimental Modal Analysis. Research Studies Press Ltd, Taunton (1997)

Part VI
Active Vibration Control

Active Vibration Suppression by the Receptance Method: Partial Pole Placement, Robustness and Experiments

R. Samin, M. Ghandchi Tehrani, and John E. Mottershead

Abstract The receptance method in active vibration suppression is presented. Robust eigenvalue assignment is achieved by selecting controller parameters that not only place the eigenvalues at desired locations but also minimize the sensitivity of the eigenvalues to measurement uncertainty. Partial pole placement is demonstrated experimentally, in which case certain closed-loop eigenvalues are assigned while others are deliberately selected to be unchanged by control action.

Keywords Active vibration suppression • Receptance method • Robust and partial pole placement

1 Introduction

Numerous techniques and algorithms have been developed for active vibration control, many of which are described in well-known texts including Inman [1], Preumont [2], Fuller et al. [3] and the review paper of Mottershead and Ram [4]. Ram and Mottershead [5] developed a method based entirely on measured vibration data, having significant modeling advantages over conventional matrix methods, including no requirement to know or to evaluate the **M**, **C**, **K** matrices, no need for the estimation of the unmeasured state and no need for model reduction. This approach was developed in a series of papers [6–9] including full implementation in experiments. Datta et al. [10] proposed partial pole assignment as a solution

R. Samin • J.E. Mottershead (✉)
Centre for Engineering Dynamics, University of Liverpool, Liverpool, UK
e-mail: j.e.mottershead@liv.ac.uk

M.G. Tehrani
Institute of Sound and Vibration Research, University of Southampton, Southampton, UK
e-mail: M.Ghandchi-Tehrani@soton.ac.uk

J. Náprstek et al. (eds.), *Vibration Problems ICOVP 2011: The 10th International Conference on Vibration Problems*, Springer Proceedings in Physics 139,
DOI 10.1007/978-94-007-2069-5_50, © Springer Science+Business Media B.V. 2011

to the problem of spillover in structural control problems and this was developed for the receptance-based approach by Tehrani et al. [8]. In [10] those eigenvalues that remained unchanged from the open-loop condition were made unobservable whereas in [8] they became uncontrollable. Full state feedback was necessary to the method of Datta [10] whereas in [8] one set of eigenvalues were assigned, a second set were rendered uncontrollable and a third set were neither assigned nor unchanged, but they took values that could be computed. In practice the third set of eigenvalues would have high frequencies so, for structures with significant mass, they would be difficult to destabilize.

2 Brief Review of the Theory

We consider the case of single-input state feedback, and write the second-order matrix equation as,

$$\left(\mathbf{M}s^2 + \mathbf{C}s + \mathbf{K}\right)\mathbf{x}(s) = \mathbf{b}(s)u(s) + \mathbf{p}(s) \tag{1}$$

where

$$u(s) = -\begin{pmatrix}\mathbf{f}^T & \mathbf{g}^T\end{pmatrix}\begin{pmatrix}s\mathbf{x}\\ \mathbf{x}\end{pmatrix} = -(s\mathbf{f} + \mathbf{g})^T\mathbf{x} \tag{2}$$

is the control input and $\mathbf{p}(s)$ is an external disturbance. The force location vector $\mathbf{b}(s)$ is written as a general function of s. One particular form results in the control force as a proportional plus integral term,

$$\mathbf{b}(s) = \mathbf{b}_1 + \frac{\mathbf{b}_2}{s} \tag{3}$$

By combining Eqs. 1 and 2 it is seen that,

$$\left(\mathbf{M}s^2 + \left(\mathbf{C} + \mathbf{b}(s)\,\mathbf{f}^T\right)s + \left(\mathbf{K} + \mathbf{b}(s)\,\mathbf{g}^T\right)\right)\mathbf{x}(s) = \mathbf{p}(s) \tag{4}$$

which amounts to a rank-1 modification to the dynamic stiffness matrix.

Modal controllability and observability conditions may be derived from,

$$\boldsymbol{\varphi}_k^T\left(\mathbf{M}\lambda_k^2 + \mathbf{C}\lambda_k + \mathbf{K}\right)\boldsymbol{\varphi}_k = -\left(\boldsymbol{\varphi}_k^T\mathbf{b}\left(\lambda_k\right)\right)\left(\left(\mathbf{g}^T + \lambda_k\mathbf{f}^T\right)\boldsymbol{\varphi}_k\right) \tag{5}$$

The right-hand-side vanishes whenever $\left(\boldsymbol{\varphi}_k^T\mathbf{b}\left(\lambda_k\right)\right) = 0$, the *uncontrollability* condition, or $(\mathbf{g}^T + \lambda_k\mathbf{f}^T)\boldsymbol{\varphi}_k = 0$, the *unobservability* condition [11]. We notice that under either of these conditions the eigenvalue λ_k remains unchanged by control action. In this article, partial pole placement is achieved by invoking the uncontrollability condition.

The Sherman-Morrison formula [12] gives the inverse of a matrix with a rank-1 modification in terms of the inverse of the original matrix. Thus the closed-loop receptance matrix is found to be,

$$\hat{\mathbf{H}}(s) = \mathbf{H}(s) - \frac{\mathbf{H}(s)\mathbf{b}(s)\left(\mathbf{g} + s\mathbf{f}\right)^T \mathbf{H}(s)}{1 + \left(\mathbf{g} + s\mathbf{f}\right)^T \mathbf{H}(s)\mathbf{b}(s)} \tag{6}$$

where,

$$\mathbf{H}(s) = \left(\mathbf{M}s^2 + \mathbf{C}s + \mathbf{K}\right)^{-1} \tag{7}$$

The characteristic polynomial of the closed-loop system is $1+\left(\mathbf{g} + s\mathbf{f}\right)^T \mathbf{H}(s)\,\mathbf{b}(s)$ and the problem of assigning the poles of the system to predetermined values $\left\{\mu_1\ \mu_2 \cdots \mu_{2n}\right\}$ may then be achieved by the procedure now to be explained.

We begin by denoting,

$$\mathbf{r}_k(\mu_k) = \mathbf{H}(\mu_k)\mathbf{b}(\mu_k) \tag{8}$$

so that for each characteristic equation,

$$\mathbf{r}_k^T(\mu_k) \times (\mathbf{g} + \mu_k \mathbf{f}) = -1,\ k = 1, \ldots, 2n \tag{9}$$

or

$$\mathbf{r}_k^T \mathbf{g} + \mu_k \mathbf{r}_k^T \mathbf{f} = -1,\ k = 1, \ldots, 2n, \tag{10}$$

The set of $2n$ equations with $2n$ unknowns may be written in matrix form,

$$\mathbf{G}\begin{pmatrix}\mathbf{g}\\ \mathbf{f}\end{pmatrix} = \begin{pmatrix}-1\\ -1\\ \vdots\\ -1\end{pmatrix}, \quad \mathbf{G} = \begin{bmatrix}\mathbf{r}_1^T & \mu_1 \mathbf{r}_1^T\\ \mathbf{r}_2^T & \mu_2 \mathbf{r}_2^T\\ \vdots & \vdots\\ \mathbf{r}_{2n}^T & \mu_{2n} \mathbf{r}_{2n}^T\end{bmatrix} \tag{11}$$

which allows the determination of $\mathbf{g}$ and $\mathbf{f}$ by inversion of the matrix $\mathbf{G}$.

Robustness of the closed-loop poles to measurement uncertainty may be achieved by choosing $\mathbf{g}$ and $\mathbf{f}$ that minimize the sensitivities, typically [9]

$$\frac{\partial \mu_k}{\partial h_{pq}} = \frac{-\left(\mathbf{g} + \mu_k \mathbf{f}\right)^T \mathbf{e}_p \mathbf{e}_q^T \mathbf{b}(\mu_k)}{\mathbf{f}^T \mathbf{H}(\mu_k)\,\mathbf{b}(\mu_k) + \left(\mathbf{g} + \mu_k \mathbf{f}\right)^T \left(\left.\frac{\partial \mathbf{H}}{\partial s}\right|_{s=\mu_k} \mathbf{b}(\mu_k) + \mathbf{H}(\mu_k) \left.\frac{\partial \mathbf{b}}{\partial s}\right|_{s=\mu_k} \right) - \left(\mathbf{g} + \mu_k \mathbf{f}\right)^T \mathbf{e}_p \left.\frac{\partial h_{pq}}{\partial s}\right|_{s=\mu_k} \mathbf{e}_q^T \mathbf{b}(\mu_k)} \tag{12}$$

3 Numerical Example – Sequential Multi-input Robust Pole Placement

We consider the three degree of freedom system described by,

$$\mathbf{M} = \begin{bmatrix} 2 & & \\ & 1 & \\ & & 3 \end{bmatrix}, \mathbf{C} = 0.5 \begin{bmatrix} 1 & 0 & 0 \\ 0 & 1 & -1 \\ 0 & -1 & 1 \end{bmatrix}, \mathbf{K} = \begin{bmatrix} 6 & -2 & -1 \\ -2 & 4 & -2 \\ -1 & -2 & 3 \end{bmatrix}$$

The open-loop poles are given by,

$$\lambda_{1,2} = -0.0166 \pm 0.5516\mathrm{i}$$
$$\lambda_{3,4} = -0.1890 \pm 1.6044\mathrm{i}$$
$$\lambda_{5,6} = -0.2528 \pm 2.2289\mathrm{i}$$

The closed-loop poles are assigned to ellipses having centres,

$$(x_{\mathrm{c}1,2}, y_{\mathrm{c}1,2}) = (-0.2, \pm 0.8)$$
$$(x_{\mathrm{c}3,4}, y_{\mathrm{c}3,4}) = (-0.5, \pm 2)$$
$$(x_{\mathrm{c}5,6}, y_{\mathrm{c}5,6}) = (-1, \pm 2.5)$$

with semi-axes $a = 0.1$ and $b = 0.2$.

The assignment is applied sequentially. The first pole is assigned followed by the second and then the third. At each step the previously assigned poles are deliberately made uncontrollable by the choice of $\mathbf{b}(s)$.

Random perturbation of the measured $\mathbf{H}(s)$ then results in eigenvalue distributions for both sequential multi-input without minimisation (denoted by grey points) and with minimisation of the sensitivities (black points) as shown in Fig. 1. The improvement in robustness to measurement noise is clear. The deterministic solution is given by the white point for each pole.

4 Experimental Example – Partial Pole Placement

The Liverpool modular structure in the 'H' configuration is shown in Fig. 2. Details of this structure, the test configuration, and measurements for estimating the receptance matrix may be found in [8]. The receptance transfer function $\mathbf{H}(s)$ was determined from the measured $\mathbf{H}(\mathrm{i}\omega)$ by curve-fitting using the PolyMAX routine [13].

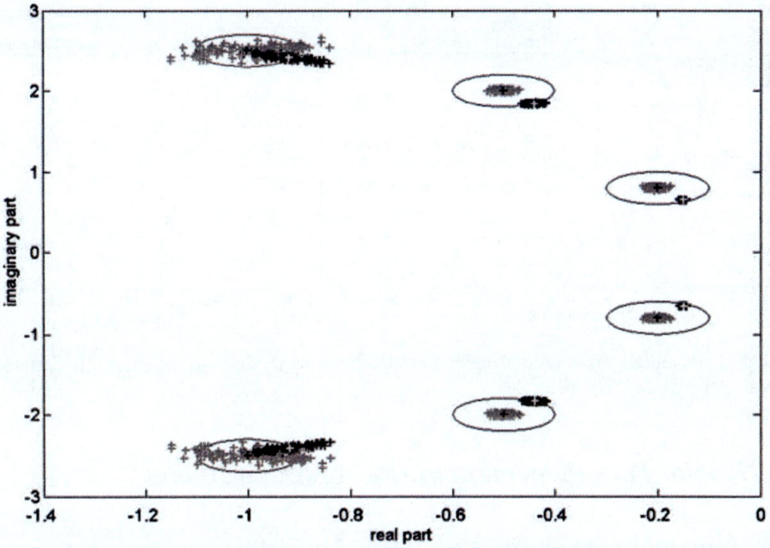

Fig. 1 Sequential multi-input robust assignment of poles (Reprinted from [9]. With permission from Elsevier)

Fig. 2 Modular test structure – 'H' configuration (Reprinted from [8]. With permission from Elsevier)

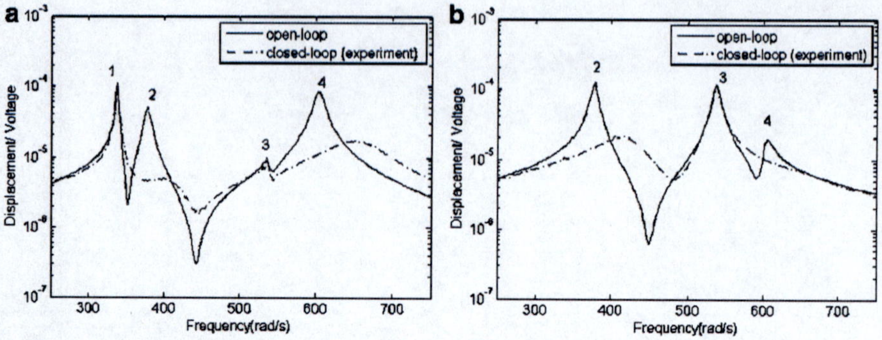

Fig. 3 Experimental partial pole placement of the torsional modes: **(a)** h_{11} **(b)** h_{33} (Reprinted from [8]. With permission from Elsevier)

4.1 Partial Pole Placement of the Torsional Modes

Poles were to be assigned at $\mu_{3,4} = -12 \pm 410i$ and $\mu_{7,8} = -65 \pm 640i$. The control force distribution $\mathbf{b} = \begin{bmatrix} 1 & -1 & 1 & -1 \end{bmatrix}^T$ was chosen so that the two torsional modes were easily excited. The control gains were found to be $\mathbf{g} = \begin{bmatrix} 20000 & -20000 & 840 & -840 \end{bmatrix}^T$ and $\mathbf{f} = \begin{bmatrix} -55 & 55 & 16 & -16 \end{bmatrix}^T$. Figure 3 shows the measured displacement/input voltage for the open-loop system as the full lines and the closed-loop system as dashed-dotted lines, for h_{11} and h_{33}. The results show that the torsional mode poles were assigned to the prescribed values, while the bending modes were rendered uncontrollable. The mode at 54 Hz disappears in h_{33} (and also in h_{44},) since the upper arms are at a node of the bending mode labeled '1'. Similarly, the mode at 85 Hz almost disappears in h_{11} (and also in h_{22}), since the lower arms are very close to a node of the bending mode labeled '3'.

5 Conclusions

The receptance method in active vibration suppression is described. Robust pole placement and partial pole placement are demonstrated in numerical and experimental examples. It is seen that the eigenvalues of physical structures may be adjusted significantly in both frequency and damping.

Acknowledgments R Samin wishes to acknowledge the support of the Ministry of Higher Education, Malaysia. JE Mottershead and M Ghandchi Tehrani acknowledge the support provided by EPSRC grant EP/F008724/1.

References

1. Inman, D.J.: Vibration with Control, Measurement and Stability. Prentice Hall, New York (1989)
2. Preumont, A.: Vibration Control of Active Structures, 2nd edn. Kluwer Academic Publishers, Dordrecht (2002)
3. Fuller, C.R., Elliot, S.J., Nelson, P.A.: Active Control of Vibration. Academic, San Diego (1996)
4. Mottershead, J.E., Ram, Y.M.: Inverse eigenvalue problems in vibration absorption: passive modification and active control. Mech. Syst. Signal Process. **20**(1), 5–44 (2006)
5. Ram, Y.M., Mottershead, J.E.: Receptance method in active vibration control. Am. Inst. Aeronaut. Astronaut. J. **45**(3), 562–567 (2007)
6. Mottershead, J.E., Ghandchi Tehrani, M., James, S., Ram, Y.M.: Active vibration suppression by pole-zero placement using measured receptances. J. Sound Vib. **311**(3–5), 1391–1408 (2008)
7. Mottershead, J.E., Ghandchi Tehrani, M., Ram, Y.M.: Assignment of eigenvalue sensitivities from receptance measurements. Mech. Syst. Signal Process. **23**(6), 1931–1939 (2009)
8. Ghandchi Tehrani, M., Elliott, R.N.R., Mottershead, J.E.: Partial pole placement in structures by the method of receptances: theory and experiments. J. Sound Vib. **329**, 5017–5035 (2010)
9. Ghandchi Tehrani, M., Mottershead, J.E., Shenton, A.T., Ram, Y.M.: Robust pole placement in structures by the method of receptances. Mech. Syst. Signal Process. **25**(1), 112–122 (2011)
10. Datta, B.N., Elhay, S., Ram, Y.: Orthogonality and partial pole assignment for the symmetric definite quadratic pencil. Linear Algebra Appl. **257**, 29–48 (1997)
11. Hamdan, A.M.A., Nayfeh, A.H.: Measures of modal controllability and observability for first- and second-order linear systems. AIAA J. Guidance Control Dyn. **12**(3), 421–428 (1989)
12. Golub, G.H., Van Loan, C.F.: Matrix Computations. John Hopkins University Press, Baltimore (1983)
13. Peeters, B., Lowet, G., Van der Auweraer, H., Leuridan, J.: A new procedure for modal parameter estimation. J. Sound Vib. **38**(1), 24–29 (2004)

Attitude Control of Micro- and Nanosatellites Using Multi-degree-of-Freedom Piezoelectric Actuators

Ramutis Bansevicius and V. Kargaudas

Abstract The paper deals with new approach to control attitude of small satellites by using high resolution multi-degree-of-freedom piezoelectric motors and piezoelectric transducers, generating travelling wave oscillations in the plane, perpendicular to satellite's rotation axis. It leads to the increase of the resolution of system to enable orienting the payload in a precise direction, extends the range of temperatures to the direction of both positive and negative values and significantly reduces the cost and simplifies the design. Theoretical analysis of described attitude control mechanisms is presented.

Keywords Microsatellite • Nanosatellite • Attitude control • Piezoelectric actuator • Traveling waves

1 Introduction

The major concern of the control of small satellites is to reduce the spinning rate of the satellite after the launch and, if it is technically possible, to orient the payload in a precise direction. The attitude determination system of the satellite has to determine the position of the captured image with a precision of $\pm 5°$ in latitude, $\pm 7.5°$ in longitude (which can be correlated to the solar local time at zenith) and ± 700 km

R. Bansevicius (✉)
Mechatronics Centre for Research, Studies and Information, Kaunas University of Technology, Kaunas, Lithuania
e-mail: ramutis.bansevicius@ktu.lt

V. Kargaudas
Faculty of Civil Engineering and Architecture, Kaunas University of Technology, Kaunas, Lithuania
e-mail: vytautas.kargaudas@ktu.lt

J. Náprstek et al. (eds.), *Vibration Problems ICOVP 2011: The 10th International Conference on Vibration Problems*, Springer Proceedings in Physics 139,
DOI 10.1007/978-94-007-2069-5_51,

in altitude to guarantee that the atmosphere limb is within the telescope's field view [1–4]. Two directions of R&D are covered to realize these aims: (a) The application of multi-degree-of-freedom piezoelectric motors in orienting systems, giving as advantages dramatically increased resolution, temperature range and reliability, simplifying the design and reducing weight and cost; (b) The application of monolithic piezoelectric transducers in specific travelling waves oscillation modes, ensuring extremely high resolution, reliability and low energy consumption.

2 Attitude Control by Using Multi-degree-of-Freedom Piezoelectric Motors

Radial traveling wave is generated in piezoelectric ring (Fig. 1a), resulting in rotation of the passive mass. The specifics of this 1*D* attitude control mechanism lies in high resolution (less than 0.001°) and small response time, not to mention small weight and dimensions. Even more effective system with 3 DOF is shown in Fig. 1b [5, 6], where passive sphere is contacting with piezoelectric cylinder in 3 contact areas; axial traveling wave of the cylinder (Fig. 2) results in controllable rotation of the sphere around vertical axis; rotation around two perpendicular axes are generated by exciting asymmetrical oscillations of the cylinder by connecting signal generators to electrodes 5 [7, 8]. The modification of this system is shown in Fig. 3, here passive sphere 1 is placed on two piezoelectric disks through contact points 3, and control of rotation direction is realized by connecting signal generator to electrodes 4.

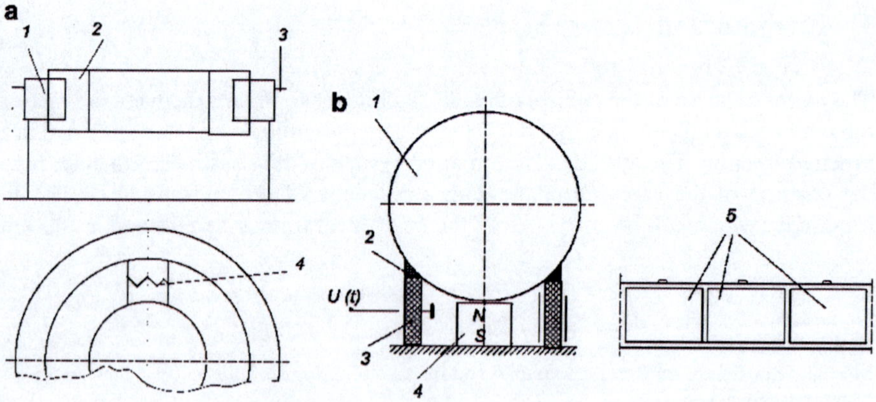

Fig. 1 1*D* and 3*D* attitude control by one DOF (**a**) and three DOF (**b**) piezoelectric motors. (**a**): *1* – piezoelectric ring with sectioned electrodes, *2* – passive mass, *3* – elastic fixing element, *4* – spring to ensure contact between piezoelectric ring and passive mass. (**b**): *1* – passive ferromagnetic sphere, *2* – contact points, *3* – piezoelectric cylinder, *4* – permanent magnet, *5* – electrodes

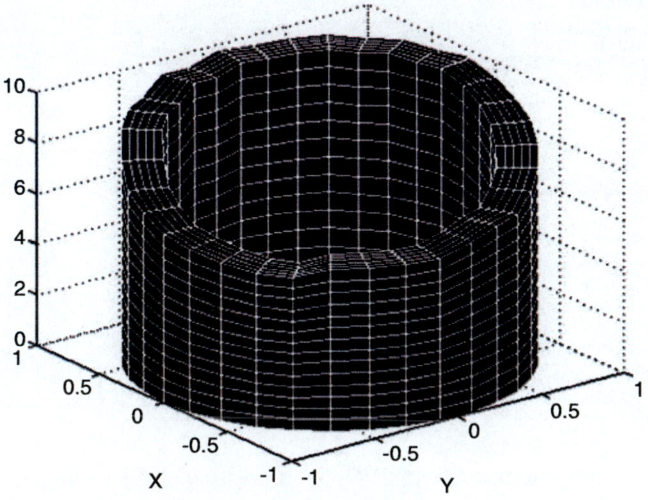

Fig. 2 Axial traveling wave in piezoelectric cylinder

Fig. 3 Symmetric 3 DOF system

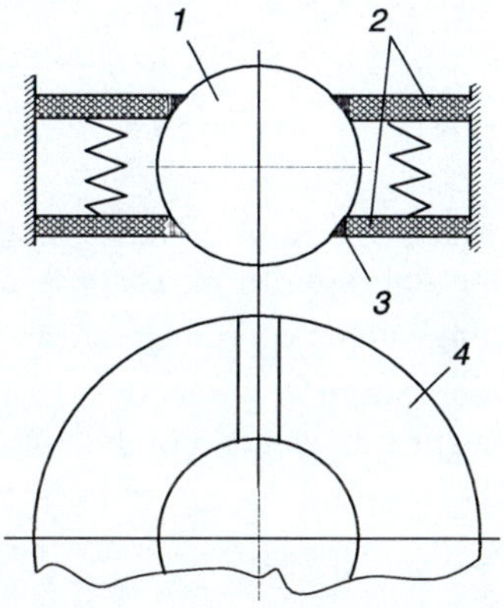

3 Attitude Control by Using Bending Traveling Waves in Piezoelectric Transducers

Mass m_1 of the satellite is assumed to be much more than the mass m_2 of the piezoelectric rod. If C_1, C_2 are the mass centers of the satellite and the rod correspondingly, the distance $C_1C = r_1 \ll r_2 = CC_2$, where C is the mass center of composite body of the satellite and the rod (Fig. 4). The rod of length l performs traveling wave oscillations around an axis $O_1\,O_2$, parallel to the axis z, and the mass

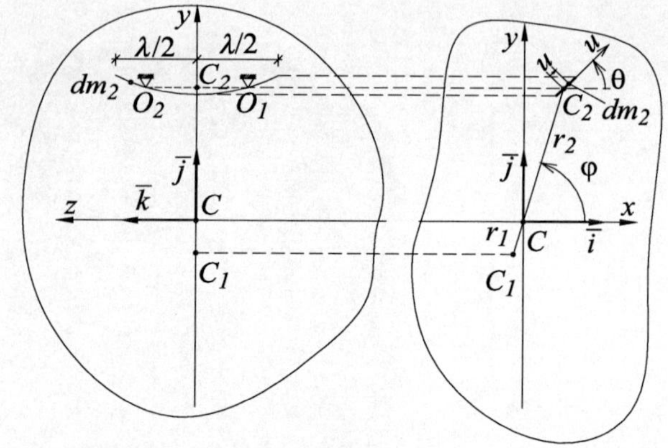

Fig. 4 Traveling bending waves in piezoelectric rod and the satellite

center C_2 is assumed to be on the axis O_1O_2. If position of the satellite is determined by angle ϕ and traveling wave oscillations of the rod by angle θ, coordinates of a rod particle dm_2 are

$$x_2 = r_2 \cos\phi + u \cos\theta,$$

$$y_2 = r_2 \sin\phi + u \sin\theta,$$

$$z_2 = const,$$

where u is displacement of the particle from the axis O_1O_2. If $|u| \ll l$ and the rod is uniform, $dm_2 = \rho ds \approx \rho dz$, where ρ is mass per length unity, s – longitudinal coordinate of the rod. As $\int_{(V)} u dm_2 = \rho \int_{-l/2}^{+l/2} u dz = 0$ and angular momentum of the particle is $d\bar{L}_{C2} = \left[-z_2\dot{y}_2\bar{i} + z_2\dot{x}_2\bar{j} + (x_2\dot{y}_2 - y_2\dot{x}_2)\bar{k}\right] dm_2$, angular momentum of the rod

$$\bar{L}_{C2} = \left(m_2 r_2^2 \dot{\phi} + J_2\dot{\theta}\right)\bar{k}, \tag{1}$$

where J_2 is moment of inertia of the rod about the axis O_1O_2. Derivatives $\dot{\phi} = d\phi/dt$, $\dot{\theta} = d\theta/dt$ are angular velocities of the satellite and the rod. Symmetry of the rod deformation $u(-z) = u(+z)$ is taken into account. Motion of an every cross-section of the rod about the axis $O_1 O_2$ is translation, when any particle of the same cross-section moves in a circle of the same radius.

Angular momentum of the satellite $\bar{L}_{C1} = \bar{L}_{01} + \bar{L}^*_{C1}$, where

$$L_{0z} = m_1 r_1^2 \dot{\phi} \tag{2}$$

is angular momentum z component of the mass center C_1. The next term is angular momentum of the satellite in its motion about the mass center

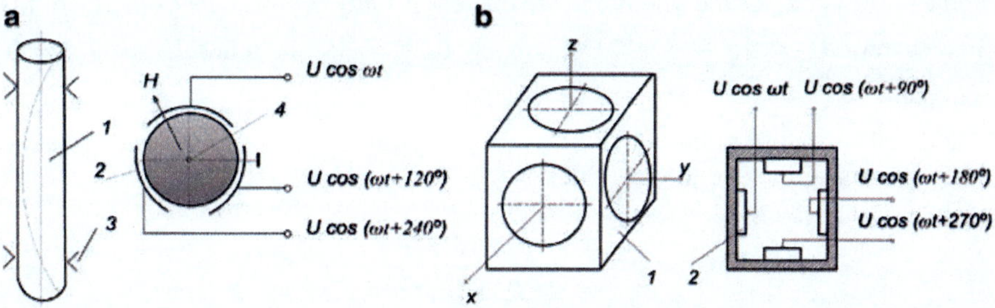

Fig. 5 1*D* (**a**) and 3*D* (**b**) attitude control systems with bending waves. (**a**): *1* – piezoelectric rod with central electrode, *2* – external electrodes, *3* – fixing elements, positioned at the nodes, *4* – central electrode. (**b**): *1* – case, *2* – piezoelectric transducers

$$
\begin{aligned}
L^*_{C1x} &= \omega_x J_x - \omega_y J_{xy} - \omega_z J_{xz},\\
L^*_{C1y} &= \omega_y J_y - \omega_z J_{yz} - \omega_x J_{yz},\\
L^*_{C1z} &= \omega_z J_z - \omega_x J_{zx} - \omega_y J_{zy},
\end{aligned}
\tag{3}
$$

where $\bar{\omega}$ is angular velocity, $\{J_{ij}\}$ – inertia tensor of the satellite. The angular momentum of the whole mechanical system of satellite and rod is constant, and this constant is zero. Consequently, the vector sum $\bar{L}_{C2} + \bar{L}^*_{C1} = 0$ and the angular momentum of the satellite are parallel to the axis of traveling wave oscillation. The vector $\bar{L}_0$, its component (2) and the other two, are neglected as $m_1 r_1^2 \ll J_z = m_1 i_z^2$. The radius of gyration i_z of the satellite is approximately as the distance $r_2 \gg r_1$, and $\dot{\phi} = \omega_z$.

Angular velocity of the satellite can be deduced from (1) and (3)

$$
\omega_z = -\frac{J_2}{J_z - \mu_x J_{zx} - \mu_y J_{yz}}\dot{\theta},\ \omega_x = \mu_x \omega_z,\ \omega_y = \mu_y \omega_z
$$

where $\mu_x = \frac{J_{xy}J_{yz}+J_{xz}J_y}{J_xJ_y-J_{xy}^2}$, $\mu_y = \frac{J_{xy}J_{xz}+J_{yz}J_x}{J_xJ_y-J_{xy}^2}$.

If the axes *x, y, z* are the principal satellite axes of inertia and consequently $J_{xy} = J_{yz} = J_{zx} = 0$, then $\omega_x = \omega_y = 0$ and $\omega_z = -\frac{J_2}{J_z}\dot{\theta}$.

Figure 5 show two practical cases of generating traveling wave oscillations. Piezoelectric rod *1* is polarized radially, using central electrode (Fig. 5a); connecting voltage supply with shifted phases to sectioned electrodes generates bending traveling wave (with two nodes) and velocity ω. This is the example of one DOF attitude control; three DOF mechanism is shown in Fig. 5b. Here traveling wave is realized in planes *xz, xy* or *yz* by connecting multiphase signal generator to six corresponding piezoelectric transducers.

If satellite is a solid cube of mass m_o with edges of length a and *x, y, z* – the principal axes, which intersect in the geometric center of the cube, then $J_x = J_y = J_z = m_o a^2/6$, $J_{xy} = J_{yz} = J_{zx} = 0$. Let a small in size mass m_p is fixed to some point y_p, z_p of the satellite in the plane y, z. Only one product of inertia of the new system $J'_{yz} = \left[\frac{\vartheta^2}{(1+\vartheta)^2} + \vartheta\right] y_p z_p m_o \neq 0$, $\vartheta = \frac{m_p}{m_o}$, the other $J'_{xy} = J'_{xz} = 0$. Moments of inertia J'_x, J'_y, J'_z will be slightly changed and $\mu_x = 0$, $\mu_y = \frac{J'_{yz}}{J'_y}$. The axis of satellite rotation will be in the y, z plane. If α – angle between rotation axes of the satellite and the rod, then

$$\tan\alpha = \mu_y = \frac{\vartheta^2 + \vartheta(1+\vartheta)^2}{\frac{(1+\vartheta)^2}{6} + (\vartheta^2 + \vartheta)\frac{z_p^2}{a^2}} \frac{y_p z_p}{a^2}.$$

When $y_p = z_p = 0.5a$ and $\vartheta = 0.1$, then $\tan\alpha = 0.143$. But when the mass m_p moves along x or y axis, then $\tan\alpha = \mu_y = 0$, that is the satellite is rotating around the axis of the oscillating rod.

4 Conclusions

The applications of multiDOF piezoelectric motors for attitude control of micro- and nanosatellite's can significantly increase the resolution. Applying piezoelectric transducers in traveling wave mode effectively reduce the cost and simplify the design, increasing the reliability of orienting systems.

References

1. Samir A.R.: Passive attitude stabilization for small satellites. Thesis, The Graduate School University of Kentucky, Lexington (2009)
2. Ortega, P., et al.: Miniaturized two axis sun sensor for attitude control of nano-satellites. Sens. J. IEEE **10**(10), 1623–1632 (2010) (ISSN: 1530-37X)
3. Kundu, P., Bhattacharyya, T.K., Das, S.: A monopropellant hydrazine MEMS thruster for attitude control of nanosatellites. In: Students' Technology Symposium (TechSym), 2010 IEEE, IEEE Explore Digital Library (IEEE Xplore), 3–4 April 2010, pp. 137–141. ISBN: 978-1-4244-5975-9
4. Iai, M., Durali, M., Hatsuzawa, T.: Attitude control of nanosatellites by paddle motion using elastic hinges actuated by shape memory alloy. J. Space Eng. **3**(1), 38–51 (2010)
5. Bansevicius R., Ahmed S.: Bearings and support. US Patent 6,262,514 (2001)
6. Bansevicius R., Ahmed S.: Drive mechanism. Great Britain Patent 2 332 090 (2001)
7. Tumasonienė, I., Kulvietis, G., Mazeika, D., Bansevicius, R.: The eigenvalue problem and its relevance to the optimal configuration of electrodes for ultrasound actuators. J. Sound Vib. **308**(3-5), 683–691 (2007) (Elsevier Science, London. ISSN 0022-460X)
8. Bansevicius, R., Drukteiniene, A., Kulvietis, G.: Trajectory planning method of rotating mobile piezorobot. J. Vibroeng. **11**(4), 690–696 (2009) (Lithuanian Academy of Sciences, Vilnius: Vibromechanika. ISSN 1392-8716)

Vibration Control via Positive Delayed Feedback

Zaihua Wang and Qi Xu

Abstract Based on the method of stability switches and the calculation of the real part of the rightmost characteristic root, this brief review paper demonstrates, through two control problems via delayed position feedback and delayed acceleration feedback, that from the viewpoint of stability, a positive delayed feedback can be superior to the corresponding negative delayed feedback and the delayed acceleration feedback can be superior to the delayed position feedback.

Keywords Time-delayed feedback • Positive feedback • Vibration control • Stability switch • Rightmost characteristic root

1 Introduction

One common issue encountered in active control is the time delay in controllers, filters and actuators. It is frequently the time delay that deteriorates the system stability and performance. Thus, time delay is often considered as a negative factor to system modeling, control and analysis. On the other hand, delayed feedback control has been shown superior to conventional techniques in some active control problems, such as in reducing sway of cranes [1, 2]. In addition, a delayed feedback control can be used to achieve total absorption in a harmonically excited damped system, such as in delayed resonator [3, 4]. In applications, one usually use delay-line to analog a delayed feedback, or use recording-reproduction to realize a delayed feedback digitally [5]. A large delay requires more delay elements in controller, which makes the controller complicated. Thus, a small delay is usually preferable.

Z. Wang (✉) • Q. Xu
Institute of Vibration Engineering Research, Nanjing University of Aeronautics and Astronautics, 210016 Nanjing, China
e-mail: zhwang@nuaa.edu.cn

J. Náprstek et al. (eds.), *Vibration Problems ICOVP 2011: The 10th International Conference on Vibration Problems*, Springer Proceedings in Physics 139,
DOI 10.1007/978-94-007-2069-5_52, © Springer Science+Business Media B.V. 2011

Though acceleration sensor is widely used in vibration control due to its small volume, lower cost and easy installation, delayed acceleration feedback is used very few in literature, compared to the large number of works that use delayed position feedback. In addition, a feedback control with positive gain is seldom used in vibration control. This paper shows that from the viewpoint of stability, a positive delayed feedback can be superior to the corresponding negative delayed feedback and the delayed acceleration feedback can be superior to the delayed position feedback, through two control problems for a single degree of freedom of vibration system.

2 Stabilization to a Free SDOF Vibration

To make the exposition as simple as possible, let us consider the stabilization problem to a free vibration in *dimensionless form* described by

$$\ddot{x}(t) + 2\xi\dot{x}(t) + x(t) = u \tag{1}$$

where ξ is negative, u is a control to ensure that the stability of $x = 0$ is stabilized [6]. The case with small $\xi > 0$ can be dealt with similarly. Two cases are considered.

Firstly, let $u = kx(t-\tau)$, then the closed-loop reads

$$\ddot{x}(t) + 2\xi\dot{x}(t) + x(t) = kx(t-\tau) \tag{2}$$

The solution $x = 0$ is asymptotically stable if and only if the characteristic equation

$$\lambda^2 + 2\xi\lambda + 1 - ke^{-\lambda\tau} = 0 \tag{3}$$

have roots with negative real parts only. When $\tau = 0$, Eq. 3 has two unstable roots due to $\xi < 0$. As τ increases from 0, the stability of $x = 0$ keeps unchanged for small $\tau > 0$ and it can be changed only if τ passes through some points where Eq. 3 has a pair of conjugate roots on the imaginary axis. Letting $\lambda = \pm i\omega$ ($i^2 = -1$) and separating the real and imaginary parts of Eq. 3 give

$$\sin(\omega\tau) = -(2\xi\omega)/k, \qquad \cos(\omega\tau) = (1-\omega^2)/k \tag{4}$$

and $F(\omega) = 0$, where $F(\omega) = \omega^4 + (4\xi^2 - 2)\omega^2 + 1 - k^2$. In order that the unstable $x = 0$ is stabilized, it is necessary that $F(\omega)$ has two different positive roots $0 < \omega_- < \omega_+$ [6], which is guaranteed if and only if $4\xi^2 - 2 < 0,\ 1 - k^2 > 0,\ (4\xi^2 - 2)^2 - 4(1-k^2) > 0$. With $\omega = \omega_-$ or $\omega = \omega_+$, one obtains the unique $\theta_- \in (0, 2\pi),\ \theta_+ \in (0, 2\pi)$ from $\sin(\theta) = -2\xi\omega/k,\ \cos(\theta) = (1-\omega^2)/k$ respectively. The critical delay values can be expressed by

$$\tau_{\pm,j} = (\theta_\pm + 2j\pi)/\omega_\pm, \quad (j = 0, 1, 2, \cdots) \tag{5}$$

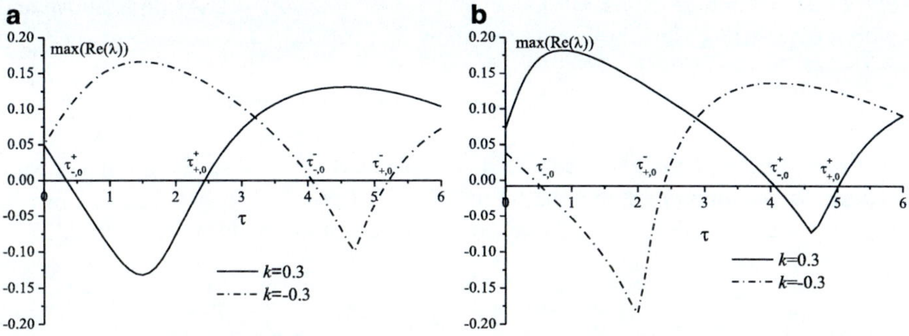

For Eq. (2) with ξ = –0.05 For Eq. (8) with ξ = –0.05

Fig. 1 The real part of the rightmost characteristic root of the closed loop vs delay

The unstable $x = 0$ can be stabilized, it is necessary that [6]

$$\tau_{-,0} < \tau_{+,0} \tag{6}$$

And if $\tau_{-,0} < \tau_{+,0}$, then unstable $x = 0$ is stabilized for $\tau \in (\tau_{-,0}, \tau_{+,0})$.

Moreover, let $\tau^{+}_{\pm,j}$ and $\tau^{-}_{\pm,j}$ denote the above defined critical delay values for the positive feedback ($k > 0$) and the corresponding negative feedback ($k < 0$) respectively, then straightforward computation shows that [6]

$$\tau^{+}_{-,j} < \tau^{-}_{-,j}, \quad \tau^{+}_{+,j} - \tau^{+}_{-,j} > \tau^{-}_{+,j} - \tau^{-}_{-,j}, \quad (j = 0, 1, 2, \cdots) \tag{7}$$

This implies that from the viewpoint of stability, a *positive* delayed position feedback is *superior* to the corresponding *negative* delayed position feedback. For small delay $\tau > 0$, as shown in Fig. 1a, a negative feedback deteriorates the stability of the closed-loop, but a positive feedback improves its stability, and $0 < \tau^{+}_{-,0} \ll \tau^{-}_{-,0}$, $\tau^{+}_{+,0} - \tau^{+}_{-,0} > \tau^{-}_{+,0} - \tau^{-}_{-,0}$. At $\tau \approx 1.5$, the positive delayed feedback leads to optimal stability. Now, if the feedback is taken as $u = k\ddot{x}(t - \tau)$, then the closed-loop reads

$$\ddot{x}(t) + 2\xi\dot{x}(t) + x(t) = k\ddot{x}(t - \tau) \tag{8}$$

When $|k| < 1$, the solution $x = 0$ is asymptotically stable if and only if the roots

$$\lambda^2 + 2\xi\lambda + 1 - k\lambda^2 e^{-\lambda\tau} = 0 \tag{9}$$

have negative real parts only. With $\lambda = \pm i\omega$, Eq. 9 gives

$$\sin(\omega\tau) = (2\xi\omega)/(k\omega^2), \qquad \cos(\omega\tau) = -(1 - \omega^2)/(k\omega^2) \tag{10}$$

and $F(\omega) = 0$, where $F(\omega) = (1-k^2)\omega^4 + (4\xi^2 - 2)\omega^2 + 1$. As done above and with the notations used above, one can show that [6]

$$\tau^+_{-,j} > \tau^-_{-,j}, \quad \tau^+_{+,j} - \tau^+_{-,j} < \tau^-_{+,j} - \tau^-_{-,j}, \quad (j = 0, 1, 2, \cdots) \tag{11}$$

Unlike the delayed position feedback, here a *negative* delayed acceleration feedback is *superior* to the corresponding *positive* delayed position feedback, as shown in Fig. 1b for Eq. 8 with $\xi = -0.05$, where at $\tau \approx 2$, optimal stability is achieved.

3 Total Absorption to a Harmonic Excited Vibration

For a SDOF vibration system with mass m, damping coefficient c and elastic coefficient k and excitation $f(t)$, the combined resonator system is described by

$$\begin{cases} m_a\ddot{x}_a(t) + c_a\dot{x}_a(t) + k_a x_a(t) - c_a\dot{x}(t) - k_a x(t) = u \\ m\ddot{x}(t) + (c_a + c)\dot{x}(t) + (k_a + k)x(t) - c_a\dot{x}_a(t) - k_a x_a(t) = -u + f(t) \end{cases} \tag{12}$$

where m_a, c_a, k_a are the mass, damping coefficient, elastic coefficient of the resonator, and u is a feedback control to the resonator, see Fig. 2a. When total absorption occurs, namely $x(t) \to 0$ as $t \to +\infty$, one has

$$\begin{cases} m_a\ddot{x}_a(t) + c_a\dot{x}_a(t) + k_a x_a(t) - u = 0 \\ -c_a\dot{x}_a(t) - k_a x_a(t) + u = f(t) \end{cases} \tag{13}$$

It follows that $m_a\ddot{x}_a(t) = f(t)$. If $f(t)$ takes simply as $f(t) = f_0 \sin(\omega_0 t)$, then

$$x_a(t) = f_0 \sin(\omega_0 t)/(m_a\omega_0^2) \tag{14}$$

In what follows, two cases will be considered.

Firstly, let $u = -gx_a(t-\tau)$ [3], then due to the first equation in (13), the characteristic function of the resonator via delayed position feedback reads

$$m_a\lambda^2 + c_a\lambda + k_a + ge^{-\lambda\tau} = 0 \tag{15}$$

In order that the resonator vibrates with frequency ω_0, it requires that Eq. 15 has a pair of conjugate roots $\lambda = \pm \mathrm{i}\omega_0$. Let $\lambda = \pm \mathrm{i}\omega_0$, it follows from Eq. 15 that

$$-m_a\omega_0^2 + k_a + g\cos(\omega_0\tau) = 0, \quad c_a\omega_0 - g\sin(\omega_0\tau) = 0 \tag{16}$$

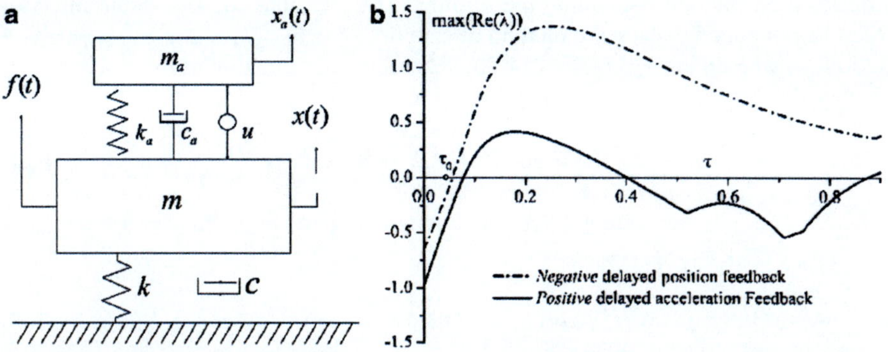

The combined resonator system

Comparison between two feedback controls

Fig. 2 The delayed resonator and the delay effect on the stability of the combined resonator system

It follows that the feedback gain g must equal to one of the following two values

$$g_{po} = +\sqrt{(k_a - m_a\omega_0^2)^2 + (c_a\omega_0)^2} \quad \text{or} \quad g_{po} = -\sqrt{(k_a - m_a\omega_0^2)^2 + (c_a\omega_0)^2} \tag{17}$$

Let θ_{po} be the unique solution of $\sin(\theta_{po}) = c_a\omega_0/g_{po}$, $\cos(\theta_{po}) = (m_a\omega_0^2 - k_a)/g_{po}$ in $(0, 2\pi)$, then the critical values of delay can be expressed by

$$\tau_j = (\theta_{po} + 2j\pi)/\omega_0, \quad j = 0, 1, 2, \cdots \tag{18}$$

For both cases of $m_a\omega_0{}^2 - k_a > 0$ and $m_a\omega_0{}^2 - k_a < 0$, a positive g_{po} (corresponding to a negative feedback) yields a smaller τ_0 (the minimal critical delay) than the negative one (corresponding to a positive feedback). That is to say, for the delayed resonator system, the delayed position feedback should be

$$u = -g_{po}x_a(t - \tau_0) := -\sqrt{(k_a - m_a\omega_0^2)^2 + (c_a\omega_0)^2}\, x_a(t - \tau_0) \tag{19}$$

where τ_0 is the minimal critical delay in (18) for the delayed negative feedback.

With $g = g_{po}$ and $\tau = \tau_0$, the corresponding characteristic function defined in (15) has infinite number of roots including the conjugate pair $\pm\,\mathrm{i}\omega_0$ and the other characteristic roots have negative real parts only.

Moreover, let $p(\lambda)$ be the characteristic function of Eq. 12 with $f(t) \equiv 0$, then

$$p(\lambda) = \det\begin{bmatrix} m_a\lambda^2 + c_a\lambda + k_a + g_{po}\mathrm{e}^{-\lambda\tau} & -c_a\lambda - k_a \\ -c_a\lambda - k_a - g_{po}\mathrm{e}^{-\lambda\tau} & m\lambda^2 + (c_a + c)\lambda + (k_a + k) \end{bmatrix} \tag{20}$$

In order to exhibit total absorption, it is necessary that the combined delayed resonator system described by Eq. 12 with $f(t) \equiv 0$ is asymptotically stable for delay in an interval containing the minimal critical value τ_0. The stable intervals can be obtained by using the method of stability switches [7].

Now, when $u = -g\ddot{x}(t-\tau)$ [4], the characteristic function of the resonator via delayed acceleration feedback is given by

$$p_{ac}(\lambda) = m_a\lambda^2 + c_a\lambda + k_a + g\lambda^2 e^{-\lambda\tau} \tag{21}$$

With $\lambda = \pm i\omega_0$, separating the real and imaginary parts of $p_{ac}(i\omega_0) = 0$ with $g_{ac} = g_{po}/\omega_0^2$ or $g_{ac} = -g_{po}/\omega_0^2$ gives

$$\sin(\omega_0\tau) = -(c_a\omega_0)/(g_{ac}\omega_0^2), \qquad \cos(\omega_0\tau) = -(m_a\omega_0^2 - k_a)/(g_{ac}\omega_0^2) \tag{22}$$

Unlike the resonator via delayed position feedback, here a negative g_{ac} (corresponding to a positive delayed feedback) yields a smaller minimal critical delay than the positive one. Namely, in order to get a smaller minimal critical delay value, the delayed acceleration feedback should be

$$u = -g_{ac}\ddot{x}_a(t-\tau_0) := g_{po}\,\ddot{x}_a(t-\tau_0)/\omega_0^2, \quad (g_{po} > 0) \tag{23}$$

where the minimal critical delay τ_0 of the resonator, determined from (22), is *the same* as the one obtained for delayed position feedback, and $|\, g_{ac}\,| < g_{po}$ if $\omega_0 > 1$.

One interesting observation is, as shown in Fig. 2b, with $m = 1kg,\ c = 2kg/s,\ k = 100N/m,\ m_a = 0.2kg,\ c_a = 0.5kg/s,\ k_a = 10N/m,\ \omega_0 = 10rad/s$, that both the stable intervals $\tau \in [0, 0.0565)$ and $\tau \in [0, 0.0779)$ of the combined system cover the minimal critical delay $\tau_0 = 0.0464$ of the resonator, but $g_{po} = 11.180N/m$ and $g_{ac} = -0.1118N/m$ satisfying $|\, g_{ac}\,| \ll g_{po}$, and the delayed acceleration resonator (with smaller negative real part of the rightmost characteristic root of the combined system) makes the total absorption occurring in a shorter time than the delayed position resonator. In this sense, a delayed resonator via *positive* delayed acceleration feedback is *superior* to the one via *negative* delayed position feedback.

4 Conclusions

Time-delay feedback control is a kind of control with memory, and it has some special features different from delay-free control. Though a positive delay-free feedback is seldom used in control applications, this paper shows that a positive delayed feedback not only works in vibration control, but also can be superior to the corresponding negative delayed feedback, from the viewpoint of stability. In addition, it shows that a delayed acceleration feedback can be superior to the delayed position feedback. Realization of this observation is left for future investigation.

Acknowledgements This work was supported by NSF of China under Grants 11032009, 10825207.

References

1. Masoud, Z.N., Nayfeh, A.H., Al-Mousa, A.: Delayed position feedback controller for the reduction of payload pendulations of rotary cranes. J. Vib. Control **9**, 257–277 (2003)
2. Masoud, Z.N., Nayfeh, A.H., Nayfeh, N.A.: Sway reduction on quayside container cranes using delayed feedback controller: Simulations and experiments. J. Vib. Control **11**, 1103–1122 (2005)
3. Olgac, N., Holm-Hansen, B.T.: A novel active vibration absorption technique: Delayed resonator. J. Sound Vib. **176**(1), 93–104 (1994)
4. Olgac, N., Elmali, H., Hosek, M., et al.: Active vibration control of distributed systems using delayed resonator with acceleration feedback. ASME J. Dyn. Syst., Meas., Control **119**, 380–389 (1997)
5. Liu, B., Hu, H.Y.: Stabilization of linear un-damped systems via position and delayed position feedbacks. J. Sound Vib. **312**(3), 509–525 (2008)
6. Wang, Z.H., Li, J.Y.: New features of time-delayed positive feedbacks in vibration control. Chin. J. Theor. Appl. Mech. **42**(5), 933–942 (2010)
7. Wang, Z.H., Hu, H.Y.: Stabilization of vibration systems via delayed state difference feedback. J. Sound Vib. **296**, 117–129 (2006)

Optimization of Semi-active Seat Suspension with Magnetorheological Damper

Stefan Segla, J. Kajaste, and P. Keski-Honkola

Abstract The paper deals with modeling, control and optimization of semiactive seat suspension with pneumatic spring and magnetorheological damper. The main focus is on isolating vertical excitation from the cabin of a bucket-wheel excavator in order to protect the excavator driver against harmful vibration. Three different control algorithms are used to determine the desired semi-active damping force: skyhook control, balance control and combination of balance and skyhook controls. The dynamic behavior of the semi-active system is optimized using genetic algorithms. As the objective function the effective value of the seat (sprung mass) acceleration is used.

Keywords Magnetorheological damper • Vibration isolation • Optimization • Random vibration

1 Introduction

Most suspension units are still passive ones which do not require any external source of power. The vibration isolation characteristics of most passive suspensions are rather limited as a consequence of the trade off in the choice of the spring rate and damping characteristics in order to achieve acceptable behavior over the whole range of working frequencies. It is well known from linear systems theory that

S. Segla (✉)
Technical University of Liberec, Liberec, Czech Republic
e-mail: stefan.segla@tul.cz

J. Kajaste • P. Keski-Honkola
Aalto University School of Engineering, Aalto, Finland
e-mail: jyrki.kajaste@tkk.fi; petri.keski-honkola@tkk.fi

J. Náprstek et al. (eds.), *Vibration Problems ICOVP 2011: The 10th International Conference on Vibration Problems*, Springer Proceedings in Physics 139,
DOI 10.1007/978-94-007-2069-5_53,

low damping gives good vibration isolation at high frequencies but poor resonance characteristics. Higher damping results in good resonance characteristics but the high frequency performance is poor.

The necessity of compromising between the conflicting requirements has motivated the investigation of controlled suspension systems [1–3], where the elastic and damping characteristics are controlled in closed-loop. The conflicting requirements led to using optimization techniques [4, 5].

The control strategies can be divided into two main categories: active and semi-active. In the case of active systems optimum transmissibility has no resonance amplification and the suspension system performance is superior to any passive suspension system. Regardless of their indisputable advantages they are still more complex, expensive and less reliable than the passive suspensions. Further problems are caused by their high energy consumption (an external source of power is needed) and possible instability. Semi-active control systems provide reliability comparable to that of passive devices, yet maintaining the versatility and adaptability of fully active systems, without requiring large power sources and without possible instability. The amount of damping can be tuned in real time. Examples of such devices are variable orifice dampers, controllable friction devices and dampers with controllable fluids (e.g., electrorheological and magnetorheological fluids). A variety of control algorithms have been developed in the past 30 years both for active and semi-active suspensions.

Magnetorheological (MR) dampers have attracted interest in using controllable actuators for their quick time response and low energy consumption. They belong to the most promising semi-active devices used nowadays in automotive engineering and are employed in a number of passenger cars and other types of vehicles. The key feature of an MR damper is the magnetorheological oil whose rheological properties can be substantially altered by applying a magnetic field. Variable damping can be produced by controlling the magnetic field by the electrical current in a solenoid.

2 Dynamic and Mathematical Model of the Seat Suspension

Figure 1 shows a dynamic model of the semi-active seat suspension containing a passive air spring and a controllable MR damper. The rheological properties (mainly yield stress) of the MR fluid inside the damper can be substantially altered with the application of a magnetic field which is produced by the solenoids placed around the orifices between chambers of the damper. The solenoids supply a controllable current which changes the rheological properties of the MR fluid and in consequence the damping force is controlled.

The equation of motion of the semi-active seat suspension with parallelogram mechanism was derived by using Lagrange's equations of the second kind. After linearization for $\varphi_0 = 9°$, $\beta_0 = 45°$, see Fig. 1, the equation of motion has the following form

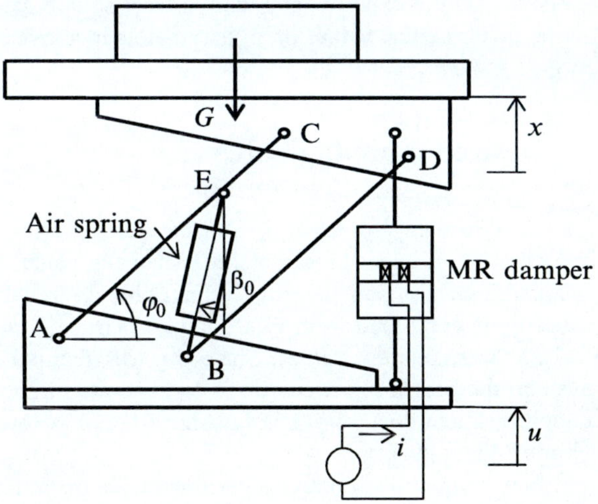

Fig. 1 Dynamic model of the seat suspension with air spring and MR damper

$$\frac{m}{\cos^2\varphi_0}\ddot{x} + \frac{k\,c^2\cos^2(\varphi_0+\beta_0)}{h^2\cos^2\varphi_0}(x-u) + F_{sa} = \frac{m\,[1+\cos\varphi_0\cos(\pi-\varphi_0)]}{\cos^2\varphi_0}\ddot{u}, \tag{1}$$

where m is the mass of the upper part of the seat including the driver, k is the air spring stiffness coefficient, F_{sa} is the damping force, $c = \overline{\mathrm{AC}}$, $h = \overline{\mathrm{AE}}$, see Fig. 1. The cabin random excitation u was measured from the platform of the cabin (in vertical direction).

The simplified model of the MR damper LORD RD −1005-3 described in [6] is used in this paper. It is more accurate than e.g. the Bingham model. In this simplified model the mass inertia effects of the fluid in the channels of the piston are neglected. Consequently the hysteresis loop of the MR damper is neglected and the semi-active damping force is given by

$$F_{sa} = F_{sa}(\dot{x}-\dot{u}, I) = c_f k_3(\dot{x}-\dot{u}) + F_y(I)\tanh[\beta\,k_3(\dot{x}-\dot{u})]\;, \tag{2}$$

where I is the electric current (input signal), $\dot{x}-\dot{u}$ is the relative velocity between the upper part of the seat and its base, c_f expresses viscous damping of the MR fluid, β influences the rate of change of the friction force in the vicinity of zero relative velocity $\dot{x}-\dot{u}$ and $k_3 = -k_2/k_1$ where k_1 and k_2 describe compressibility of individual hydraulic and pneumatic parts of the damper.

The parameters of the MR damper described above were experimentally identified [6] and the following approximation functions hold

$$\begin{aligned} c_f(I) &= c_1 I + d_1, \quad F_y(I) = a_2 I^3 + b_2 I^2 + c_2 I + d_2,\\ k_1(I) &= b_3 I^2 + c_3 I + d_3, \quad k_2(I) = b_4 I^2 + c_4 I + d_4. \end{aligned} \tag{3}$$

Based on the experimental data, a least-square method was employed to determine the numerical values of all parameters in above equations [6].

3 Desired Damping Force

The desired damping force F_d can be realized only if this force and the relative velocity $\dot{x} - \dot{u}$ have the same sign. Then the calculated input current of the MR damper, depending on the actual value of the desired damping force and the relative velocity, is computed from Eqs. 2 and 3. The input current varies in the range of 0–1.25 A (maximum value). Since the MR damper cannot supply power to the system, the best thing it can do is to generate minimum possible force when the condition mentioned above is not met (damping force in MR damper cannot be eliminated totally).

Many semi-active control algorithms have been developed so far. According to the balance logic, the damping force F_d is as follows

$$F_d = F_{balance} = \begin{cases} K\,|x_1|\,\mathrm{sgn}\,\dot{x}_1 \;\; (\text{or} - K x_1) & x_1\dot{x}_1 \le 0, \\ 0 & x_1\dot{x}_1 > 0. \end{cases} \tag{4}$$

where $x_1 = x - u$ means the relative displacement between the upper part of the seat and its base and $\dot{x}_1 = \dot{x} - \dot{u}$ means the relative velocity.

The ideal skyhook is given by the following control logic

$$F_d = F_{skyhook} = \begin{cases} b_{sky}\,\dot{x} & \dot{x}_1\dot{x} > 0, \\ 0 & \dot{x}_1\dot{x} \le 0. \end{cases} \tag{5}$$

In this paper we also apply the control logic which is a combination of the control algorithms based on the balance control and the skyhook control. The desired damper force is defined as follows

$$F_d = K_1\dot{x} + K_2\,(x - u)\,, \tag{6}$$

where K_1 is the proportionality factor of absolute velocity feedback loop and K_2 is the proportionality factor of relative displacement feedback loop.

4 Optimization and Results

The dynamic behavior of the semi-active system is optimized using genetic algorithms [7]. As the objective function the effective value of the seat acceleration is used [3, 8]

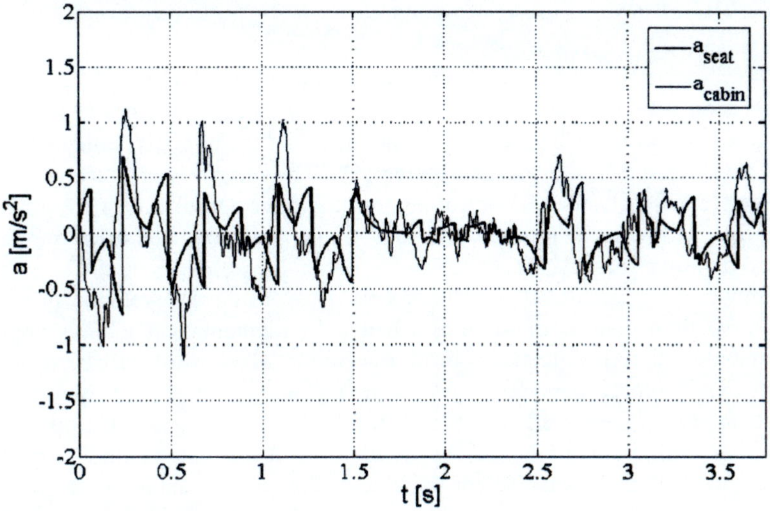

Fig. 2 Comparison of seat acceleration a_{seat} and cabin acceleration a_{cabin} (combined control)

$$\ddot{x}_{ef} = \sqrt{\frac{1}{T}\int_0^T \ddot{x}^2(t)\mathrm{d}t}. \tag{7}$$

Optimization parameters correspond to individual control algorithms:

(a) *Balance control*: two optimization parameters are used: K, see Eq. 4, and k (air spring stiffness). Their optimum values are: $K_{\text{opt}} = 20.1925{\cdot}10^3$ N/m, $k_{\text{opt}} = 15.01{\cdot}10^3$ N/m. Objective function: $\ddot{x}_{\text{ef,min}} = 1.2056{\cdot}10^{-2}$ m/s^2.
(b) *Skyhook control*: two optimization parameters are used: b_{sky}, see Eq. 5, and k. Their optimum values are: $b_{\text{sky,opt}} = 2.927{\cdot}10^3$ N.s/m, $k_{\text{opt}} = 15.07{\cdot}10^3$ N/m. Objective function: $\ddot{x}_{\text{ef,min}} = 1.0954{\cdot}10^{-2}$ m/s^2.
(c) *Combined control*: three optimization parameters are used: K_1, K_2, and k. Their optimum values are: $K_{1,\text{opt}} = 1.5015{\cdot}10^3$ N.s/m, $K_{2,\text{opt}} = 1.6097{\cdot}10^4$ N/m, $k_{\text{opt}} = 15.12{\cdot}10^3$ N/m. Objective function: $\ddot{x}_{\text{ef,min}} = 1.0843{\cdot}10^{-2}$ m/s^2.

Figure 2 shows the sprung mass acceleration in comparison with the bucket wheel excavator cabin acceleration in the vertical direction for the combined control. Despite significant improvements (about 60%), which can be seen from comparing the effective value of the cabin vertical acceleration ($\ddot{x}_{\text{ef,cabin}} = 2.7124{\cdot}10^{-2}$ m/s^2) and the effective values of the seat acceleration for cases (a), (b) and (c), the conventional implementation of all control algorithms used in the paper introduces a sharp increase (jump) in damping force, which, in turn, causes a jump in sprung mass acceleration. This acceleration jump, or jerk, causes a significant reduction in isolation benefits that can be offered by semi-active suspensions. Figure 2 clearly illustrates the problem of acceleration jumps.

5 Conclusions

The semi-active seat suspension with MR damper effectively reduces vibration acceleration acting on the excavator driver as it can be seen from comparing the effective value of the cabin vertical acceleration and the effective values of the seat acceleration given in the previous paragraph. The improvements are about 60%. The differences between all control algorithms (after optimization) are practically negligible. However, it is important to optimize the parameters determining the desired damping force.

Simplified mathematical models of MR dampers were developed in order to allow real time control of systems when their mathematical models are to be implemented directly into the control algorithms. Next work will be devoted to complete mathematical models of MR dampers taking into account compressibility and also the mass inertia effects of the MR fluid in the channels of the piston.

Further improvements of the results are associated with modifications to the formulations of the control algorithms such that the damping force jumps are eliminated.

Acknowledgments The research reported herein was supported by a grant from the Ministry of Industry and Trade of the Czech Republic, No. FT-TA5/102.

References

1. Guglielmino, E., et al.: Semi-active Suspension Control, Improved Vehicle Ride and Road Friendliness. Springer, London (2008)
2. Ballo, I.: Comparison of the properties of active and semi-active suspension. Vehicle Syst. Dyn. **45**, 1065–1073 (2007)
3. Segla, S., Reich, S.: Optimization and comparison of passive, active, and semi-active vehicle suspension systems. In: Proceedings of the 12th IFToMM World Congress on Mechanism Science, Besancon (2007)
4. He, Y., McPhee, J.: Multidisciplinary design optimization of mechatronic vehicles with active suspensions. J. Sound Vib. **283**, 217–241 (2005)
5. Sága, M. et al.: Selected Methods of Analysis and Synthesis of Mechanical Systems (in Slovak). VTS ŽU, Žilina (2009)
6. Úradníček, J.: Multidisciplinary optimization of a vehicle suspension model with semi-active magnetorheological damper (in Slovak). PhD Thesis, Slovak Technical University in Bratislava, Bratislava (2008)
7. The Mathworks Inc.: Genetic Algorithm and Direct Search Toolbox – User's Guide. The Mathworks Inc., Natick (2004)
8. Ostertagová, E.: Applied Statistics (in Slovak). ELFA, Košice (2011)

Dynamics of a Rotor Suspended on Gimbals

H.I. Weber and D.H.Z. Carrera

Abstract The Magnus demonstration gyroscope is an instrument that allows a series of amazing experiences when the rotor loses its axial symmetry. It consists of a rotor on two gimbals whose inertia and energy loss will be considered in the modeling. A normalization procedure to reduce the number of parameters is established. Unlimited rotation around the center is possible for all bodies. A pointer on the rotor shaft will allow to define two distinct hemispheres which may capture for some instants the motion, when the energy decreases due to damping. In changing hemispheres the usual singularities occurring with Euler angles are crossed challenging the numerical methods. For non symmetric rotors a minimum kinetic energy will be needed to eject the motion from Lyapunov stable orbits to a sequence of jumps between both defined hemispheres. In this condition the movement does not remain inside any basin of attraction, it continues tumbling without regard to the loss of kinetic energy.

Keywords Free body rotation • Tumbling • Gimbals • Stability

1 Introduction

Thanks to the straightforward way to describe analytically the motion of a free body in space this is "the" classical problem in rotational dynamics and is found in all the timeless literature of Mechanics. Centuries passed and the need to gather analytical

H.I. Weber (✉)
PUC-Rio, Rio de Janeiro, Brazil
e-mail: hans@puc-rio.br

D.H.Z. Carrera
ITA, São José dos Campos, Brazil
e-mail: hernan@ita.br

J. Náprstek et al. (eds.), *Vibration Problems ICOVP 2011: The 10th International Conference on Vibration Problems*, Springer Proceedings in Physics 139,
DOI 10.1007/978-94-007-2069-5_54, © Springer Science+Business Media B.V. 2011

solutions was substituted by numerical integrations. The visualization by graphical representation of the body motion in space substituted the abstraction imagining a body cone rolling on a fixed cone, Poinsot 1834. Nevertheless rotation in space is an advanced subject, rich in unexpected movements, that still brings trouble to our young engineers. But, since inertial sensors to obtain attitude (angular position in space) are getting much cheaper, lots of new applications are possible and an effort should be made to enhance the feeling (intuition) of engineers to problems related to rotation in space. The first author of this paper remembers well classes in the end 1960s given by K. Magnus using the demonstration gyroscope he developed and is produced by Phywe [5] where a rotating body in space could be approached by the gyroscope on gimbals, since the inertia of the gimbals were small compared to the rotor of the gyroscope. One of the most fascinating experiments is the behavior of a naturally unstable rotor, which occurs when it turns around of the axis of intermediate moment of inertia. Repeating these experiments through the last four decades the first author got the help of the second author to look for answers to questions that intrigued us for a good time.

These questions refer to regions of operation that usually are outside of the practical range in common engineering problems like wide amplitude motion. An unlimited rotation for a body on gimbals seems to be good for training astronauts; you find it also in science fiction movies and even as fitness training apparatus. Following topic will be discussed: considering a non-axisymmetric rotor, which rotates around the axis of minimum or maximum moment of inertia, when it suffers a disturbance that is big enough, it will start a tumbling motion. Then, what is the behavior and how is the influence of the inertia of the gimbals and the damping in the joints. Since this implies in the description of the motion crossing the linear stability borders, as well as the singularity condition of the Euler angles, what is the real physical meaning of these angles in terms of the rotor motion?

2 Equations of Motion

Three bodies compose the gyroscope: the outer gimbal will be positioned using the angle α, while β is for the internal gimbal and γ represents the rotor spin. There may be additional masses at the rotor, without changing the position of the center of mass but eliminating its axi-symmetry. There may also exist other masses fixed on the extremities of the rotor shaft, flattening or elongating its inertia ellipsoid (Fig. 1).

The compact representation (1) for sequential rotations follows: the angles below are associated to 3 elementary sequential rotations making use of 4 reference frames (RF), (*F*) fixed in space, (*Q*) to the outer gimbal, (*R*) to the inner and (*S*) to the rotor. The *x* axis of (*F*) is vertical.

$$\underset{(x,y,z)}{F} \xrightarrow{\alpha\ (x)} \underset{(x,y',z')}{Q} \xrightarrow{\beta\ (y')} \underset{(x'',y',z'')}{R} \xrightarrow{\gamma\ (z'')} \underset{(x''',y''',z'')}{S} \tag{1}$$

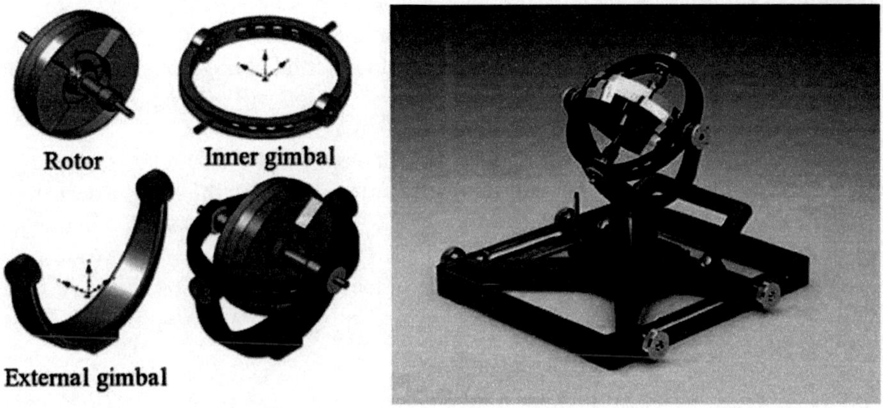

Fig. 1 Gyroscope (Phywe) and its main components

Total kinetic energy is obtained by adding the energies of each body, which can be calculated in the respective RF. For the analysis, a specific problem was considered: with resting gimbals, the rotor is initially turning at constant angular speed when, suddenly, an external impact force acts on one of the axes, x or y of RF (R). This impact generates an instantaneous change in the angular momentum and the spin axis of the rotor starts a conical motion around the angular momentum vector. The non axi-symmetric condition of the rotor's inertia leads to large amplitude displacements if it was originally turning around the axis of intermediate inertia; in both other alternatives there is a threshold for the impact force before this unbound rotation is attained. Small damping will cause a gradual change in the resulting motion.

Newton-Euler law leads to equations of motion that include the damping moments between the three bodies due to the low friction ball bearings. Reaction moments are eliminated. Non-dimensional formulations are more adequate in nonlinear dynamics: we introduce an equivalent angular speed ν obtained by division of the resulting angular momentum h_G after the impact by the principal moment of inertia of the rotor around axle z'' that supports it (I_3). The time in the analysis will be changed to the non-dimensional value $\tau = \nu t$ where $\nu = h_G / I_3$.

Also: $I^A_{(xx)}$ is the moment of inertia of the external gimbal corresponding to axis x of (Q), I_p and I_d are the principal (polar and diametric) moments of inertia of the internal gimbal ring in (R), and I_1, I_2 and I_3 are the principal moments of inertia of the rotor in axes x, y, z of (S) respectively. Parameters (2) are introduced:

$$\mu_x = I^A_{(xx)} / I_3 \quad \mu_1 = I_1 / I_3 \quad \mu_2 = I_2 / I_3 \quad \mu_p = I_p / I_3 \quad \mu_d = I_d / I_3 \quad (2)$$

Viscous friction is assumed in the bearings of the gyroscope; $M_{[fric]AF} = C_{AF}\,\dot{\alpha}$ (normalized as $k_{AF} = C_{AF}/h_G$) is the resistive moment of the bearing in the fixed base of the outer gimbal and equivalently to the other joints. This is a good model for the present case but an investigation on the change of behavior due to other contact conditions could make good use of the results of [1]. Four additional parameters *a, b, c, d* are dependent on the inertia of the rotor and together with the rotor angle γ are used to describe the non axi-symmetry of the inertial configuration of the rotor in (*R*). Due to lack of space the final equations of motion will not be presented here. Please refer to [2, 3] or contact the authors by e-mail.

Parameter values for the simulation are obtained from the Magnus gyroscope. Initial conditions for the rotor will be $\gamma'(0) = 1,\ \gamma(0) = 0$, the impact occurring along *y* axis leads to $\alpha'(0) = 0,\ \alpha(0) \neq 0,\ \beta'(0) \neq 0,\ \beta(0) = 0$ while the impact occurring along *x* axis leads to $\alpha'(0) \neq 0,\ \alpha(0) = 0,\ \beta'(0) = 0,\ \beta(0) \neq 0$. The impact strength will define the angular velocity along the impact axis and the deviation (angle) of the other axis.

3 Results and Comments

A sample from the results of Carrera [2] is presented next.

While axi-symmetric rotors present no surprises in the dynamic behavior, even if the gimbals are considered as part of the system, the lack of symmetry and motion through the nonlinear range present a great challenge to the understanding. An example for a flattened rotor ($\mu_1, \mu_2 < 1$) is presented in Fig. 2 (left) considering a conservative system: representing the angles α and β correspond to observe the conical motion of the rotor shaft and will be called orbit. More impact intensity will increase the radius of this orbit and increase the kinetic energy present in the system, as it can be observed in the figure. After a certain minimum level of energy the rotor starts jumping from one side to the other (changing hemispheres), looking for other equilibrium positions, but it does not stop in these large excursion motion. These jumps will cause high acceleration in the angle coordinates, which represent the motion of the external and internal gimbal respectively. Careful analysis of this problem leads to following results: the minimum impact impulse to remove the

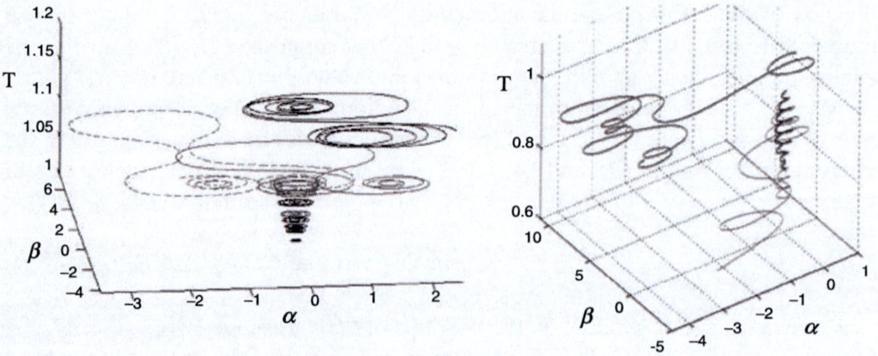

Fig. 2 Orbits of the conservative (*left*) and non-conservative (*right*) three body system for different values of kinetic energy T

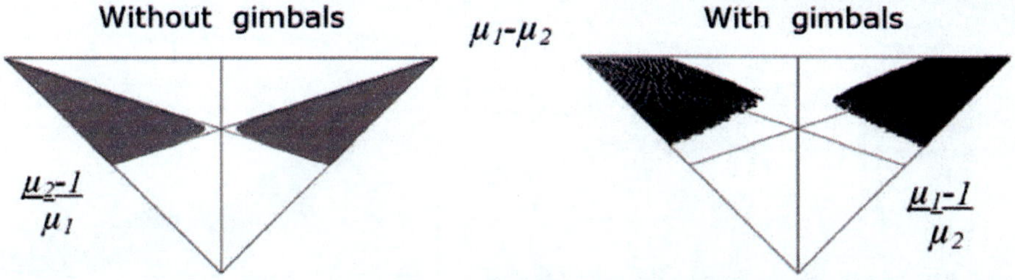

Fig. 3 Instability areas for a flat rotor without gimbals and with gimbals

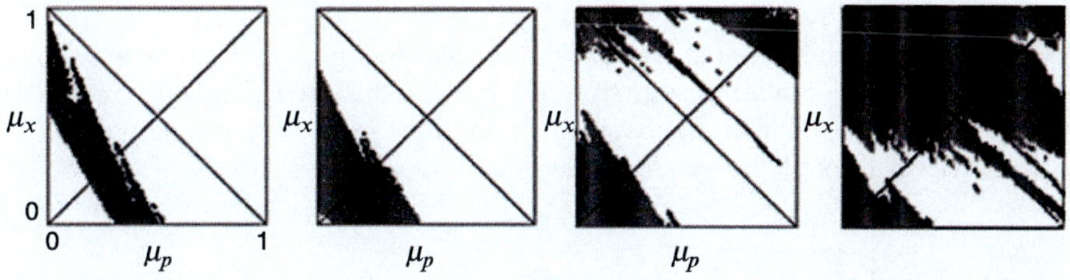

Fig. 4 Instability areas for a flat rotor on gimbals, for increasing impact disturbance from *left*

system from stable oscillations condition can be obtained analytically; infinitesimal changes in initial impact condition result in other motion geometry, which means chaos; if the gimbals have no inertia (free body problem), the only possible jump to the opposite hemisphere is in the α angle (this can be shown analytically); near to the jump, in the free body problem the orbit tends to a square while the orbit of the rotor on gimbals stays more like a circle. The non-conservative system in Fig. 2 (right) shows that if the initial kinetic energy exceeds the minimum for the tumbling of the rotor, it keeps tumbling till the complete dissipation of the energy.

Figure 3 presents the natural instability regions in the Magnus triangle [4], which is constructed using only with the moments of inertia of the rotor. Observe that the gimbals change the system stability, even if they have small inertia and damping.

In Fig. 4 stability results for flattened rotors on gimbals are presented, increasing the impact impulse from left to right. The bottom left corner of each square represents the free body problem (0,0), showing that between the second and third square the limit condition was reached. The change in greyscale responds to the energy present in the system after impact. One can observe that a stable free body may become unstable due to the gimbals (left), and also that a naturally unstable flattened rotor may be stabilized through the gimbals (white regions).

In Fig. 5 (right) the friction in the body connections is considered; a bifurcation diagram is represented considering the viscous damping (equal for all connections) as a control parameter in the range k (0:0.02) and representing the kinetic energy variation in a time interval $t = 5$. The system presents a change in behavior after a certain value for the damping, showing that the energy loss may be slower even if the damping is higher. Figure 5 (left) shows that the rotor and the inner gimbal are

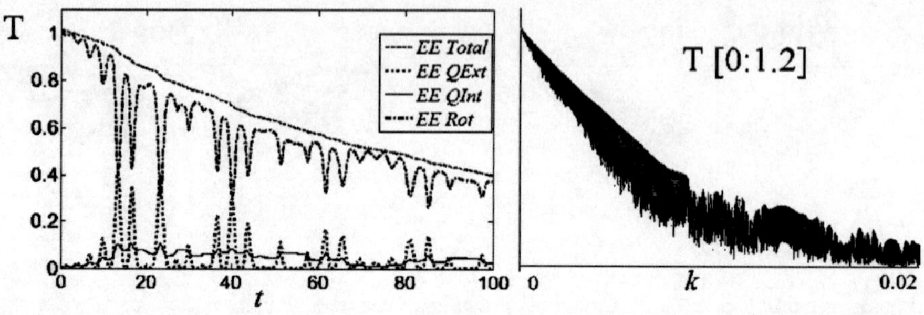

Fig. 5 Energy and energy loss of a flat rotor on gimbals considering friction

the main participants in the energy exchange. The outer gimbal keeps small values: remember that the motion passes from the rotor to the outer gimbal by means of the inner gimbal. The inner gimbal is the head of the great orchestra in the walk of kinetic energy through the gyroscopic system.

4 Conclusions

The minimum energy required to start tumbling motion can be radically altered by the presence of gimbals. Nonlinear tools are essential for the understanding of complex rotation conditions. To complete the analysis presented here, [2] investigates Poincaré maps, quaternions for the free body problem, Poincaré maps for quaternion coordinates, phase plane analysis and the use of Magnus triangle for stability representation. Low-cost satellites demand a great care not to be lost by some instability; besides numerical simulation the use of instruments like the Magnus gyroscope are of great value in the understanding of phenomena. We could modify the behavior of the gyroscope, using the torque between the components. This is a step towards active gimbals.

Acknowledgements This research work was supported by CNPq, Brazilian Council for Research and Technological Development.

References

1. Awrejcewicz, J., Supel, B., Lamarque, C.-H., Kudra, G., Wasilewski, G., Olejnik, P.: Numerical and experimental study of regular and chaotic motion of triple physical pendulum. Int. J. Bifurcation Chaos **18**(10), 2883–2915 (2008)
2. Carrera, D.H.Z.: Libration and tumbling of a rigid body. Doctorate Thesis, Department of Mechanical Engineering, Pontifícia Universidade Católica do Rio de Janeiro, Brazil (2010)
3. Carrera, D.H.Z., Weber, H.I., Morrot, R.: Alternatives of long term behavior of a gyroscope considering damping. In: PACAM XI, Foz do Iguaçu, PR, Brazil, January 04–08, 2010
4. Magnus, K.: Der Kreisel. Springer, Berlin (1971)
5. Phywe Catalogue Physics, "Cardanic Gyroscope, Magnus Type", prod. nr. 02550-00, pp. 51 (2011)

Experiment Study on Fuzzy Vibration Control of Solar Panel

Dongxu X. Li, Rui Xu, and Jiangjian P. Jiang

Abstract Some flexible appendages of spacecraft are cantilever plate structures, such as solar panels. These structures usually have very low damping ratios, high dimensional order, low modal frequencies and parameter uncertainties in dynamics. Their unwanted vibrations will be caused unavoidably, and harmful to the spacecraft. To solve this problem, the dynamic equations of the solar panel with piezoelectric patches are derived, and an accelerometer based fuzzy controller is designed. In order to verify the effectiveness of the vibration control algorithms, experiment research was conducted on a piezoelectric adaptive composite honeycomb cantilever panel. The experiment results demonstrate that the accelerometer-based fuzzy vibration control method can suppress the vibration of the solar panel effectively, the first bending mode damping ratio of the controlled system increase to 1.64%, and that is 3.56 times of the uncontrolled system.

Keywords Vibration control • Fuzzy logic control • Piezoelectric smart structure • Solar Panel • Experiment study

1 Introduction

In order to meet limited launch weight and costs, and more power requirement, space craft solar panels are becoming lager and more flexible, and they usually have very low damping ratios, high dimensional order, low modal frequencies and parameter uncertainties in dynamics [1]. Their unwanted vibrations will be caused unavoidably [2], and harmful to the spacecraft. Vibration suppression must

D.X. Li (✉) • R. Xu • J.P. Jiang
National University of Defense Technology, Changsha, China
e-mail: dongxuli@263.net; xurui@nudt.edu.cn; jianping202@163.com

J. Náprstek et al. (eds.), *Vibration Problems ICOVP 2011: The 10th International Conference on Vibration Problems*, Springer Proceedings in Physics 139,
DOI 10.1007/978-94-007-2069-5_55,

be implemented for high precision pointing spacecrafts. In the past decades, many theoretical and simulated works have been presented. Hence, we should pay more attention on experiment study of vibration control.

In recent year, some experimental works on vibration control of beam and panel structure are presented. Paolo Gaudenzi et al. proposed position and velocity control algorithms to perform vibration control of an active cantilever beam, with the experimental tests and a good correspondence has been obtained [3]. Guillaume Barrault et al. implemented a H_∞ controller design for high-frequency control over a specified bandwidth using a reduced order plant with corrective terms, experiments on a cantilever beam demonstrated the effectiveness of the proposed approach in controlling vibration due to vibration modes within the high-frequency bandwidth of interest [4]. Z. Mohamed performed experimental investigations into the development of feed forward and feedback control schemes for vibration control of a very flexible and high-friction manipulator system [5]. M.A. Rastgaar et al. introduced orthogonal eigenstructure control for active vibration cancellation in a plate that is clamped along its four edges, experiment result show that this method can achieve a better isolation across the entire range of frequency of the disturbance [6].

In the present work, experiment research was conducted on a piezoelectric adaptive composite honeycomb cantilever panel. The control algorithm is designed based on fuzzy logic control theory, and receives the acceleration and acceleration rate of the panel as inputs. Then the controller was run on dSPACE Real-Time System. At last, the experimental comparison research was conducted.

2 Mathematic Modeling

A typical single-panel solar panel structure made of composite material with rectangular cross section having length L, width b, and thickness h is considered. The schematic diagram of the experiment system is shown in Fig. 1. The accelerometer is fixed on the center of the panel tip, and it is used as sensor. And there are eight PZT patches (length l_A, width b_A, thickness h_A) bonded perfectly on the top surface of the panel, and they are used as one-channel actuator. The patches are close to the clamped side of the panel to make better control effect.

The mathematical model is based on mechanics of thin walled structures [7, 8] and liner theory of piezoelectricity. Furthermore, quasi-static motion is assumed, which means that the mechanical and electrical forces are balanced at any given instant [9]. When the panel is vibrating, the stress of the panel and PZT patches are

$$\begin{cases} \sigma_p(z) = E_p \varepsilon_p = E_p z \rho \\ \sigma_A(z) = E_A \varepsilon_A = E_A (z\rho - \Lambda) \end{cases} \tag{1}$$

where $E_p,\ E_A$ are the elastic modulo of the panel and the PZT patches, $\varepsilon_p,\ \varepsilon_A$ are the x-direction strain of the panel and the PZT patches, $\sigma_p,\ \sigma_A$ are the x-direction

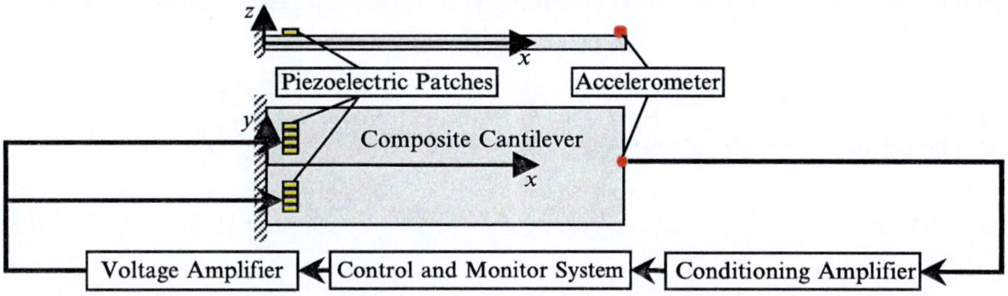

Fig. 1 Schematic diagram of the experimental system

stress of the panel and the PZT patches, ρ is the curvature of the panel in the certain section, $\rho = m/E_pI_p$, where m is the bending moment in the certain section, I_p is the inertia moment of the certain section, Λ is the unconstraint strain of the PZT patches, it can be calculated by the following equation:

$$\Lambda = d_{31} \cdot U_A/h_A \tag{2}$$

where U_A is the control voltage loaded on the PZT patches, d_{31} is the piezoelectric constant of the PZT patches.

Then, expressing and solving the moment equilibrium about the centroid of the beam section

$$\int_{-\frac{1}{2}h}^{\frac{1}{2}h} \sigma_p bhzdz + 8\int_{\frac{1}{2}h}^{\frac{1}{2}h+h_A} \sigma_A b_A h_A zdz = 0 \tag{3}$$

we can obtain

$$\rho = \frac{48d_{31}U_AE_Ab_Ah_A\,(h+h_A)}{E_pbh^4 + 24E_Ab_Ah^2h_A^2 + 48E_Ab_Ahh_A^3 + 32E_Ab_Ah_A^4} \tag{4}$$

So the corresponding nominal bending moment induced in the panel by the actuators can be calculated as

$$M = ml_A = \rho E_pI_pl_A = \frac{48d_{31}E_AE_pI_pb_Ah_A\,(h+h_A)\,l_A}{E_pbh^4 + 24E_Ab_Ah^2h_A^2 + 48E_Ab_Ahh_A^3 + 32E_Ab_Ah_A^4} \cdot U_A \tag{5}$$

By applying Hamilton's principle, the finite element formulation of the solar panel structure can be derived. The equations of equilibrium governing the linear dynamic response of a system of finite elements are

$$\mathbf{M}\ddot{\mathbf{q}} + \mathbf{C}\dot{\mathbf{q}} + \mathbf{K}\mathbf{q} = \mathbf{F}_\mathbf{f} - \mathbf{F}_{\mathbf{control}} \tag{6}$$

where **M**, **C**, **K** are the mass, damping and stiffness matrix respectively, $\mathbf{q}$, $\dot{\mathbf{q}}$, $\ddot{\mathbf{q}}$ are the displacement, velocity and acceleration vectors of the finite element assemblage, $\mathbf{F_f}$, $\mathbf{F_{control}}$ are the disturb force and control force matrixes, and $\mathbf{F_{control}} = \left\{M\left(U_A\right)0\cdots 0\right\}^T$.

Damping matrixes is calculated by Rayleigh damping model as follow:

$$\mathbf{C} = a\mathbf{M}+b\mathbf{K} \tag{7}$$

3 Fuzzy Controller Design

Since the sufficiently precise process mathematic model is hard to get and it presence of non-linearity in this issue, fuzzy controller will be very suitable. In this paper, a Mamdani-type fuzzy inference system, which consists of two inputs and one output, was constructed by using the Fuzzy Toolbox of Matlab. The controller receives the acceleration (a) and acceleration rate $(\dot{a})$ of the solar panel tip as inputs, while gives the control voltage (V) as output. Triangular and trapezoidal shape membership functions were chosen both for inputs and output, the membership function plots is shown in Fig. 2. Forty-nine rules were used to describe the present system-controller. All rules have weights equal to 1 and use the AND-type logical operator, show in Table 1.

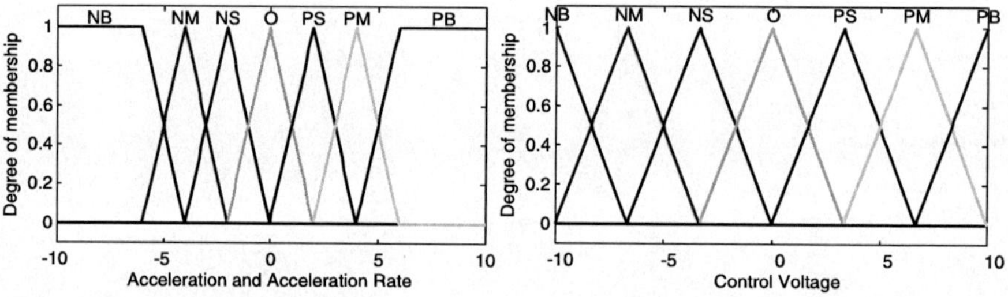

Fig. 2 Membership function plots

Table 1 Fuzzy IF-THEN rule base

a/V/$\dot{a}$	NB	NM	NS	O	PS	PM	PB
NB	NB	NB	NB	NB	NM	O	O
NM	NB	NB	NB	NM	NM	O	O
NS	NM	NM	NM	NS	O	PS	PS
O	NM	NS	O	O	O	PS	PM
PS	NS	NS	O	PS	PM	PM	PM
PM	O	O	PM	PM	PB	PB	PB
PB	O	O	PM	PB	PB	PB	PB

4 Experiments

In order to verify the effectiveness of the proposed fuzzy controller, experimental research was conducted on a piezoelectric adaptive carbon fiber/aluminum honeycomb cantilever panel. The accelerometer is fixed at the tip of the panel to measure the vibration. The PZT patches bonded perfectly on the top surface of the panel.

The control system is implemented by using dSPACE DS1005. The accelerometer signal is amplified by conditioning amplifier to the voltage range of -10 to $+10$ V, and the control voltage of the PZT actuator is amplified by voltage amplifier to the voltage range of 0–200 V. The sampling period of the control system is selected as 0.001 s.

Since the first order frequency of the panel is 3 Hz. To excite the first bending mode of the panel, an excitation voltage $\varphi = 100\sin(3 \times 2\pi t) + 100$ was loaded on the PZT patches. The vibration of the panel reaches the maximum amplitude after 20 s excitation, and then the excitation voltage was deactivated. The resonance and damping response of the panel is shown in Fig. 3a, and the excitation voltage is shown in Fig. 3b.

In order to verify the effectiveness of the fuzzy controller, the experimental comparison research was conducted. To make the same comparison condition, the same sine wave excitation voltage was loaded on the PZT patches for 20 s, and then fuzzy controller turned on. The attenuation of the vibration effects with controlled and uncontrolled are compared in Fig. 4a. The control voltage is shown in Fig. 4b.

From Fig. 4a we can get that: the decay time of the first bending mode vibration with fuzzy logic control effect is about 12 s, when compared with the open-loop decay time (more than 50 s), it reveals a great improvement on the response attenuation. And the logarithmic attenuation rates of every adjacent period are calculated, shown in Fig. 5a. The damping ratio of the controlled system is 1.64%, it's 3.56 times of the uncontrolled system (0.46%). The frequency response of the controlled and uncontrolled system is shown in Fig. 3. The first order resonance experiment Fig. 5b.

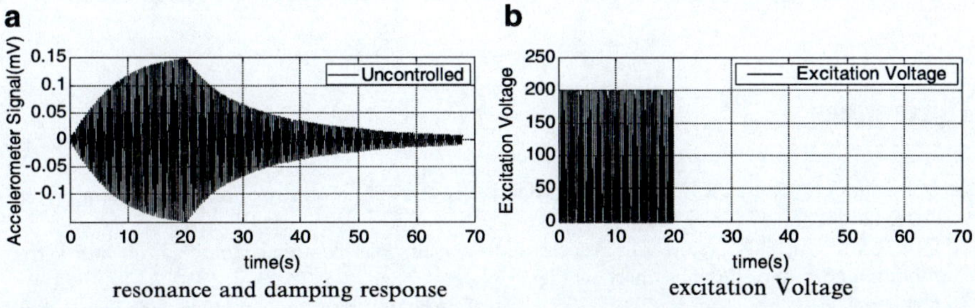

Fig. 3 The first order resonance experiment. (**a**) Resonance and damping response. (**b**) Excitation voltage

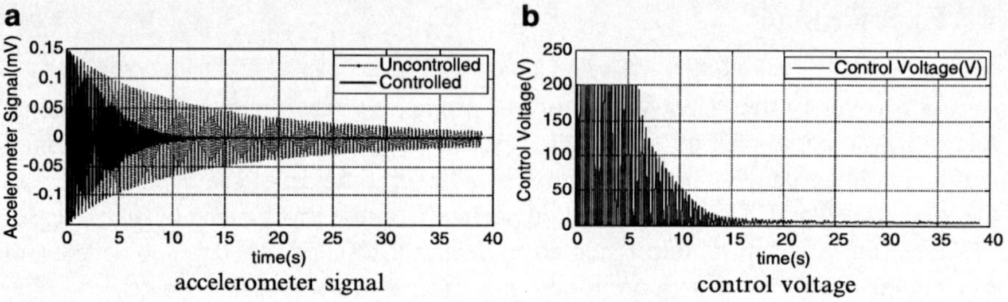

Fig. 4 Fuzzy vibration control experiment. (**a**) Accelerometer signal. (**b**) Control voltage

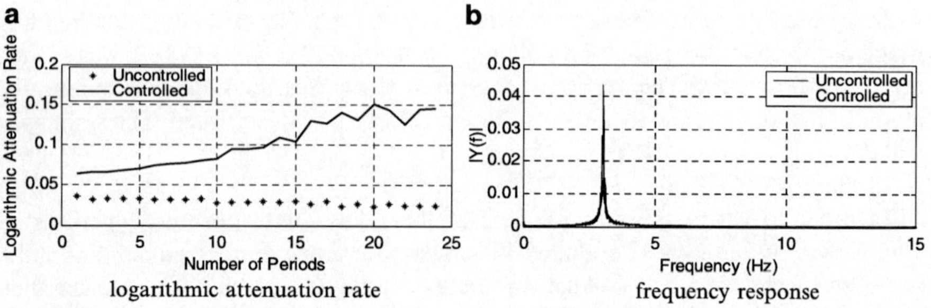

Fig. 5 Vibration control effectiveness analyze. (**a**) Logarithmic attenuation rate. (**b**) Frequency response

5 Conclusions

The problem of vibration control for an active cantilever panel has been considered in the present research. An accelerometer based fuzzy control system has been used for its solution. By implemented dSPACE Real-Time System, the control system was established. Experiment results demonstrate that the fuzzy controller proposed in this paper can suppress the bending vibration of the composite panel effectively.

References

1. Maganti, G.B., Singh, S.N.: Simplified adaptive control of an orbiting flexible spacecraft. Acta Astronautica **61**, 575–589 (2007)
2. Qiu, Z.C., Wu, H.X., Ye, C.D.: Acceleration sensors based modal identification and active vibration control of flexible smart cantilever plate. Aerosp. Sci. Technol. **13**, 277–290 (2009)
3. Gaudenzi, P., Carbonaro, R., Benzi, E.: Control of beam vibrations by means of piezoelectric devices: theory and experiments. Compos. Struct. **50**, 373–379 (2000)
4. Barrault, G., Halim, D., Hansen, C.: High frequency spatial vibration control using method. Mech. Syst. Signal Process. **21**, 1541–1560 (2007)

5. Mohamed, Z.: Vibration control of a very flexible manipulator system. Control. Eng. Pract. **13**, 267–277 (2005)
6. Rastgaar, M.A., Ahmadian, M., Southward, S.C.: Experimental application of orthogonal eigenstructure control for structural vibration cancellation. J. Sound Vibration **329**, 3873–3887 (2010)
7. Andrianov, I.V., Awrejcewicz, J., Manevitch, L.I.: Manevitch Asymptotical Mechanics of Thin Walled Structures. Springer, Berlin/Heidelberg/New York (2004)
8. Awrejcewicz, J., Krysko, V.A., Vakakis, A.F.: Nonlinear Dynamics of Continuous Elastic Systems. Springer, Berlin/New York (2004)
9. Marinaki, M., Marinakis, Y., Stavroulakis, G.E.: Fuzzy control optimized by PSO for vibration suppression of beams. Control. Eng. Pract. **18**, 618–629 (2010)

Seismic Vibration Attenuation of a Structural System Having Actuator Saturation with a Delay-Dependent H_∞ Controller

H. Yazici, Rahmi Guclu, and I.B. Kucukdemiral

Abstract This paper deals with the design of state-feedback delay-dependent H_∞ controller for active vibration control problem of seismic-excited structures having actuator delay, L_2 disturbances and actuator saturation. First sufficient delay-dependent criteria are developed by choosing a Lyapunov-Krasovskii functional candidate for a stabilizing H_∞ synthesis involving a matrix inequality conditions. Then actuator saturation phenomenon is added to the controller design using LMI constraints. The sufficient conditions for designing such controller are obtained in terms of delay-dependent bilinear matrix inequalities (BMIs). To overcome nonlinear optimization problem involved in the delay-dependent conditions, a cone complementary linearization method is used to find a feasible solution set. Using proposed method, a suboptimal controller with maximum allowable delay bound, minimum allowable disturbance attenuation level under actuator saturation constraints can be obtained simultaneously by a convex optimization technique. A four-degree-of-freedom structural system subject to Kobe Earthquake excitations is used to illustrate the effectiveness of the approach through simulations. Simulation results show that the proposed controller is very effective in reducing vibration amplitudes of storeys and guarantees stability at maximum actuator delay bound under actuator saturation constraints.

Keywords Active vibration control • H_∞ control • Actuator delay • Actuator saturation • Linear matrix inequalities

---

H. Yazici • R. Guclu (✉) • I.B. Kucukdemiral
Yildiz Technical University, Istanbul, Turkey
e-mail: hyazici@yildiz.edu.tr; guclu@yildiz.edu.tr; beklan@yildiz.edu.tr

J. Náprstek et al. (eds.), *Vibration Problems ICOVP 2011: The 10th International Conference on Vibration Problems*, Springer Proceedings in Physics 139,
DOI 10.1007/978-94-007-2069-5_56, © Springer Science+Business Media B.V. 2011

1 Introduction

Theoretical and experimental results show that active control methods can reduce the maximum response of building structure against dynamic excitations successfully [1]. However, one important issue of active structural control is the existence of time delay phenomenon. In active control process, unavoidable time-delay may appear especially in control channel which mainly results from on-line data acquisition from long distance sensors at different location of the structure, computing the control forces, transmitting data and signals to actuator and applying control forces to the structure. Due to time-delay in control input, unsynchronized control forces are applied to the structure and this may cause some amount of degradation in control efficiency or even instability of system. Apart from actuator delay, one of the main sources of instability is the actuator saturation phenomenon for the active vibration control problem of structural systems [2]. Due to the stochastic nature of seismic and wind loadings, it is conceivable that the required control force may exceed the capacity of the actuator in active vibration control of structural systems application, resulting in actuator saturation. Actuator saturation may cause serious deterioration in the performance of the closed-loop system, also and may lead to instability. In this study, an approach based on LMI and cone complementary algorithm is presented for designing delay-dependent H_∞ controller which provides best performance depending on the actuator saturation limit.

2 Formulation of the Problem

Consider a class of time delayed systems with input saturation and external disturbance as follows:

$$\begin{aligned} \dot{x}(t) &= Ax(t) + B_h \mathrm{sat}(u(t-h)) + B_w w(t) \\ z(t) &= Cx(t), \quad x(t) = 0, \quad t \in [-\bar{h}, 0] \end{aligned} \tag{1}$$

where $x(t) \in \Re^n$ is the state vector. $u(t) \in \Re^{m_u}$ is the control input, $w(t) \in \Re^{m_w}$ is the disturbance input acting on the system, $z(t) \in \Re^p$ is the controlled output. Then A, B_h, B_w, and C are known real constant state space matrices with appropriate dimensions. Also, sat(·) is the standard saturation function with unity saturation level, i.e, $\mathrm{sat}(u) = [\mathrm{sat}(u_1)\mathrm{sat}(u_2)\ldots\mathrm{sat}(u_{m_u})]^T$, where $\mathrm{sat}(u) = \mathrm{sign}(u_i)\min\{1, |u_i|\}$. Here we mildly harm the notation by using sat(·) to denote both the scalar valued and vector valued saturation functions. On the other hand, the delay h is assumed to be a fixed time delay which satisfies $0 < h \le \bar{h}$. Here, $\bar{h}$ are known positive constants. Throughout this work, we assume that the disturbance signals acting on system have bounded energy, i.e., $W_\delta := \{w : \Re_+ \to \Re^{m_w}; \int_0^\infty w^T(t)w(t)dt \le \delta < \infty\}$ for some known constant $\delta > 0$. Then our goal is to find a suitable state-feedback control law in the form of $u(t) = Kx(t)$, such that the closed-loop system exhibits a globally asymptotic stable behaviour and minimum H_∞ gain from w(*t*) to z(*t*) where disturbances distribute from the set W_δ. The following lemma is useful in providing the main results of this paper [3].

Lemma 1. Let $u, v \in \Re^m$ with $u = [u_1 u_2, \ldots, u_m]^T$ and $v = [v_1 v_2, \ldots, v_m]^T$. Suppose that $|v_i| \leq 1$ for all $i \in [1, m]$. Then, $\text{sat}(u) \in \text{co}\{D_i u + D_i^- v : i \in [1, 2^m]\}$ where co denotes the convex hull. From Lemma 1, we can write $\text{sat}(Kx(t-h)) = D_i Kx(t-h) + D_i^- Hx(t-h), i = [1, 2^{m_u}]$. And applying the variable change, $B_s := B_h(D_i K + D_i^- H)$ the closed-loop system can be obtained as below.

$$\dot{x}(t) = Ax(t) + B_s(x(t-h)) + B_w w(t)$$

$$z(t) = Cx(t), \quad x(t) = 0, \quad t \in [-\bar{h}, 0] \tag{2}$$

3 Simulation Study

In this section, simulations are carried out in order to illustrate the effectiveness of the proposed controller in reducing the effect of earthquakes on structures. For simulation study, a four-storey structure is modelled (Fig. 1) [1]. The structural system has been simulated against the ground motion of Kobe Earthquake (Fig. 1) [8]. During the simulation the mass, damping and stiffness coefficients for storey are assumed to be as follows: $m_1 = 450$, $m_2 = m_3 = m_4 = 345$ ton, $c_1 = 26.170$, $c_2 = 490$, $c_3 = 467$, $c_4 = 410$ KNs/m, $k_1 = 18.050$, $k_2 = 340{,}000$, $k_3 = 326{,}000$, $k_4 = 280{,}000$ KN/m.

Figure 2 shows the time responses, frequency responses for the considered structure and time history of the applied control force saturated $\bar{h} = 0.0407$. Solving the control problem by using Yalmip parser with Sedumi solver [9, 10] under Matlab,

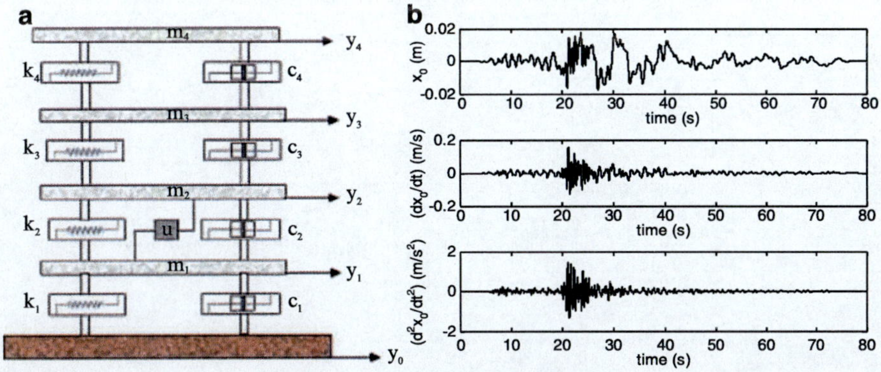

Fig. 1 (**a**) Physical model of the structural system and (**b**) time responses of Kobe Earthquake excitation

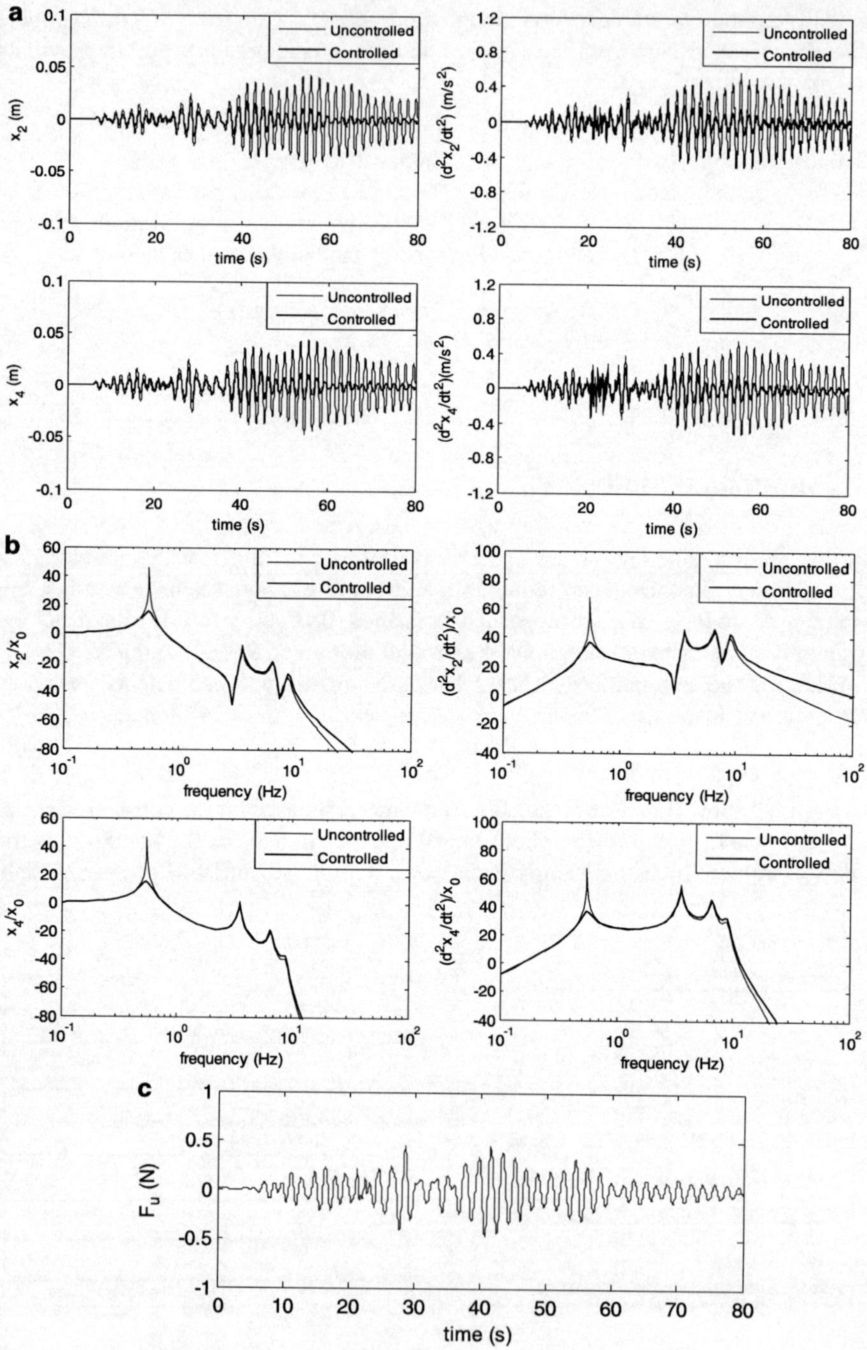

Fig. 2 (**a**) Time responses, (**b**) frequency responses of the second and fourth storeys and (**c**) time history of the applied control force

we achieve the following optimal values: the allowable upper actuator delay bound $\bar{h} = 0.0407$ s and minimum allowable disturbance attenuation level $\gamma = 444{,}39$ using cone complementary algorithm and the controller gain is obtained as $\mathrm{K} = \bar{\mathrm{L}}\mathrm{X}^{-1} = \begin{bmatrix} -0.1937 & 0.3433 & 0.0109 & 0.0891 & -2.1265 & -1.7608 & -1.8466 & -1.8975 \end{bmatrix}$.

As it can be observed from Fig. 2, satisfactory vibration suppression is achieved by the proposed delay-dependent H_∞ controller under actuator saturation. Since the system has four degrees of freedom, there are four resonance frequency points at 0.54, 3.55, 6.56, and 8.86 Hz. Note that the first mode is the most dangerous for structures during an earthquake. However, it is obvious that this mode is successfully suppressed by the use of proposed controller.

4 Conclusions

This paper presents an approach for designing a delay dependent state feedback H_∞ controller to attenuate the vibration of seismic excited structural system with actuator delay and actuator saturation. The performance of the proposed control approach and the system stability are demonstrated by numerical simulation results. This results show that, in spite of the actuator saturation, the designed controller is all effective in reducing vibration amplitudes of storeys and guarantees stability at maximum allowable actuator delay bound.

Acknowledgments This work was supported by The Scientific and Technological Research Council of Turkey (TUBITAK) under Grant 109M523.

References

1. Guclu, R., Yazici, H.: Vibration control of a structure with ATMD against earthquake using fuzzy logic controllers. J. Sound Vib. **318**(1–2), 36–49 (2008)
2. Du, H., Lam, J.: Energy-to-peak performance controller design for building via static output feedback under consideration of actuator saturation. Comput. Struct. **84**, 2277–2290 (2006)
3. Hu, T., Lin, Z.: Control Systems with Actuator Saturation: Analysis and Design. Birkhkauser, Boston (2001)
4. Moon, Y.S., Park, P., Kwon, W.H., Lee, Y.S.: Delay-dependent robust stabilization of uncertain state-delayed systems. Int. J. Control **74**(14), 1447–1455 (2001)
5. Boyd, S., Ghaoui, L.E.Feron, E., Balakrishnan, V.: Linear Matrix Inequalities in System and Control Theory, Society for Industrial and Applied Mathematics. SIAM, Philadelphia (1994)
6. Ghaoui, L., El Qustry, F., Ait Rami, M.: A cone complementarity linearization algorithm for static output-feedback and related problems. IEEE Trans. Autom. Control **42**(8), 1171–1176 (1997)
7. Parlakci, M.N.A., Kucukdemiral, I.B.: Robust delay-dependent H_∞ control of time-delay systems with state and input delays. Int. J. Robust Nonlinear Control (2010). doi:10.1002/rnc.1637
8. Kasimzade, A.A.: Structural Dynamics. Birsen Publication, Istanbul (2004)
9. Löfberg, J.: Yalmip: a toolbox for modeling and optimzation in MATLAB. In: Proceedings of the CACSD Conference, Taipei (2004)
10. Strum, J.F.: Using SeDuMi 1.02 a Matlab for optimization over symmetric cones. Optim. Methods Softw. **11**(2), 625–653 (1999)

Active Vibration Control of Seismic-Excited Structural System with LMI Based Dynamic Output-Feedback H_∞ Controller

H. Yazici, Rahmi Guclu, and G. Keskin

Abstract This paper is concerned with the active vibration control of multi-degree-of-freedom-structure, which is effected by earthquake. To obtain desired time history and frequency responses for solution of active vibration control problem, linear matrix inequality (LMI) based output-feedback H_∞ controller is designed in this study. The time history of ground motion of the Northridge earthquake, which is a disturbance input, is applied to modeled structure. At the end of the study, the time history of the storey displacements, velocities and frequency responses of both controlled and uncontrolled cases are presented and results are discussed. Performance of the designed controller has been shown for the different loads and disturbance using ground motion of the Northridge Earthquake.

Keywords Active control • LMI • Dynamic H_∞ control • Output-feedback • Vibration • Structure

1 Introduction

In recent years, remarkable progress has been made in the field of active vibration control of engineering structures subjected to earthquakes and strong winds. Proposed techniques to minimize the structural vibrations, in general, consist of two categories, namely passive control systems and active control systems [1]. Passive systems add damping to the structure or isolate it from the source of environmental excitation, thus reducing vibration. Active systems have the advantage of strong capacity. Active devices can be designed to influence a number of vibration modes. Hence, active control is most suited for a multi-degree-of-freedom structure, which

H. Yazici • R. Guclu (✉) • G. Keskin
Yildiz Technical University, Istanbul, Turkey
e-mail: hyazici@yildiz.edu.tr; guclu@yildiz.edu.tr; gkhnkskn@gmail.com

J. Náprstek et al. (eds.), *Vibration Problems ICOVP 2011: The 10th International Conference on Vibration Problems*, Springer Proceedings in Physics 139,
DOI 10.1007/978-94-007-2069-5_57,

the response can be influenced by a number natural mode. The effects of the active control are obviously superior to the passive control in decreasing the response of structure vibration. Theoretical and experimental results show that active control methods can reduce the maximum response of structural systems against dynamic excitations. Kose et al. designed static output feedback controller to attenuate the structural vibrations against earthquake excitation [2]. Guclu and Yazici applied a Fuzzy PID controller design in order to mitigate Seismic-Vibration of a Non-linear Structural System with an ATMD [3].Yazici and Guclu designed LMI based mixed H_2 /H_∞ state and output feedback controller for active vibration control of seismic excited structural system [4].

In this study LMI based dynamic output-feedback H_∞ controller is designed for active vibration attenuation problem. In order to show effectiveness of the proposed approach, a five-degree-of freedom structural system is modelled using a spring-mass-damper subsystem. The system is then simulated against the real ground motions of Northridge earthquake. Simulation results exhibit that efficient vibration suppressions are achieved by use of the proposed LMI based dynamic output-feedback H_∞ controller.

2 Dynamic Model of Structural System

In this study, a five-storey structure is modeled in Fig. 2a. Since the destructive effect of earthquakes is mainly the result of horizontal vibrations, the degrees of freedom of the structure have been assumed only in this direction. During an earthquake, the maximum inter-storey shear force occurs on the first storey. Assuming equivalent storey stiffness and ultimate capacities, the destructive effect of an earthquake is expected to be the largest on the first storey. Besides, it is well known that the maximum displacements and velocities are expected at the top storey of structures during an earthquake. Therefore, active control devices (actuators), which supplies energy to suppress seismic vibrations, are installed on the first and top storey of the structure. The modeled structural system is shown in Fig. 1. Here, m_1 is movable mass of the ground storey, the mass of each storey is m_2, m_3, m_4, respectively. x_1, x_2, x_3, x_4 are the horizontal displacements and x0 is the earthquake-induced ground motion disturbance to the structure. The masses, damping and stiffnesses for each storey are assumed to be identical, and the realistic structural parameters are given as $m_1 = 450$, $m_2 = m_3 = m_4 = m_5 = 345$ ton, $c_1 = 26.170$, $c_2 = 490$, $c_3 = 467$, $c_4 = 410$, $c_5 = 390$ KNs/m, $k_1 = 18{,}050$, $k_2 = 340{,}000$, $k_3 = 326{,}000$, $k_4 = 280{,}000$, $k_5 = 240{,}000$ KN/m. All springs and dampers are acting in horizontal direction. The equations of motion of the system can be obtained easily using Lagrange equations as below:

$$M\ddot{X} + C\dot{X} + KX = F_u + F_w \tag{1}$$

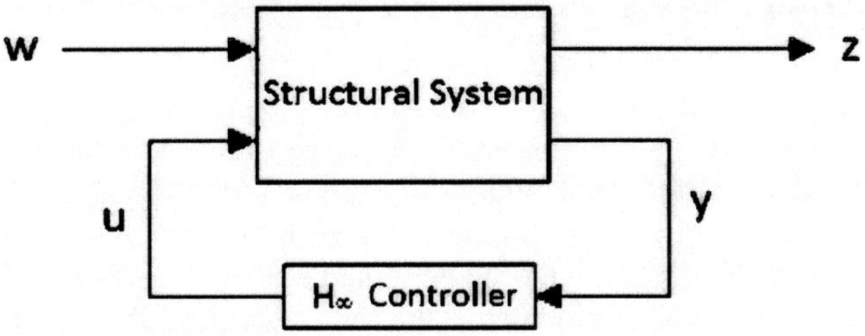

Fig. 1 Generalized output-feedback control system

Here, $[M]$, $[C]$, $[K]$ are the mass, damping and stiffness matrices of the structure, respectively. Also, F_u is control force and F_w represents disturbance input.

3 Dynamic Output-Feedback H_∞ Controller

In the last two decades, robust control problems have been studied effectively in many fields of control engineering. Active vibration is one of the main topic in these research works and still remains attractive for new control design schemes. H_∞ control depends on minimizing the infinitive norm of transfer function matrix which is written from controlled output to disturbance input in order to avoiding the disturbance input to affect the system. Therefore, H_∞ control is very suitable control algorithm for the structural systems which are under affect of disturbance inputs with unknown magnitude as earthquakes.

Equation of motion of the structural system (1) arranges in state-space form for H_∞ control design. Figure 1 shows a general plant.

In generalized system chosen matrices and vectors for controller design is given in equality (2).

$$\begin{aligned} \dot{x} &= Ax + B_1 w + B_2 u \\ z &= C_1 x + D_{11} w + D_{12} u \\ y &= C_2 x + D_{21} w + D_{22} u \end{aligned} \tag{2}$$

Here, A, B_1, B_2, C_1, C_2, D_{11}, D_{12}, D_{21}, D_{22} is respectively system matrix, disturbance input matrix, control input matrix, performance output matrix, measurement matrix, disturbance input matrix for performance output, control input matrix for performance output, disturbance input matrix for measurement, control input matrix for measurement. Also, $x \in \Re^n$, $w \in \Re^{m_1}$, $u \in \Re^{m_2}$, $z \in \Re^{p_1}$, $y \in \Re^{p_2}$ show state vector, disturbance input vector, control input vector, performance output vector, and measurement vector, respectively.

4 Simulation Results

The structural system has been simulated against the ground motion of Northridge earthquake. Earthquake ground motions are used as input to a building structure. The corresponding earthquake motions are shown in Fig. 2b. The horizontal displacements of the related storeys are obtained by on-line integration of storey accelerations which are caused by the earthquake effect, see Fig. 3a. This figure indicates displacements and velocities of the time responses of the second and fifth storeys of the considered structure, respectively, for both controlled and uncontrolled cases. As can be seen from the figure, satisfactory vibration suppression is achieved by the designed H_∞ controller. During an earthquake, the maximum displacements are expected at the top storey of a structure. The displacements of the fifth storey are minimized successfully using H_∞ controller. Figure 3b shows the frequency responses of the second and fifth storey displacements and velocities, respectively, for both controlled and uncontrolled cases. Since the system has five degrees of freedom, there are five resonance frequency points at 0.4852, 2.8116, 5.2554, 7.3001 and 8.9607 Hz. As expected, the upper curves belong to the uncontrolled system. When the response plots of the structural systems with uncontrolled and controlled cases are compared, a superior improvement in the mitigation of the resonance values is observed with the proposed controller. Thc first mode is the most dangerous during an earthquake and it is suppressed successfully by the use of H_∞ controller. As a result, Figs. 3 and 4 show the robustness of the proposed H_∞ controller and control force. It is desired that the controller remains stable and effective when the structure is subjected to seismic excitation.

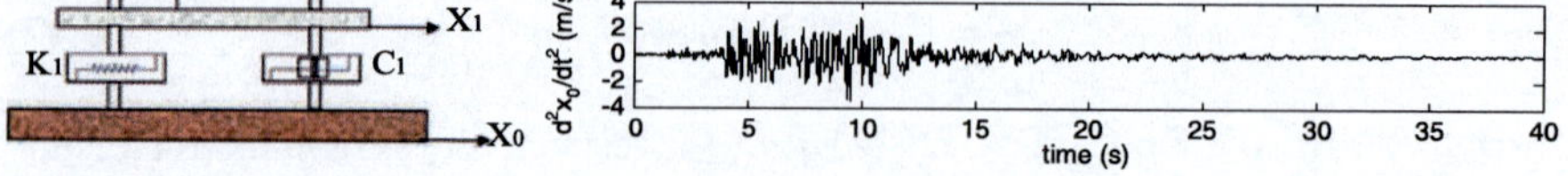

Fig. 2 (**a**) Physical model of n degree of freedom structural system and (**b**) time responses of Northridge Earthquake excitation

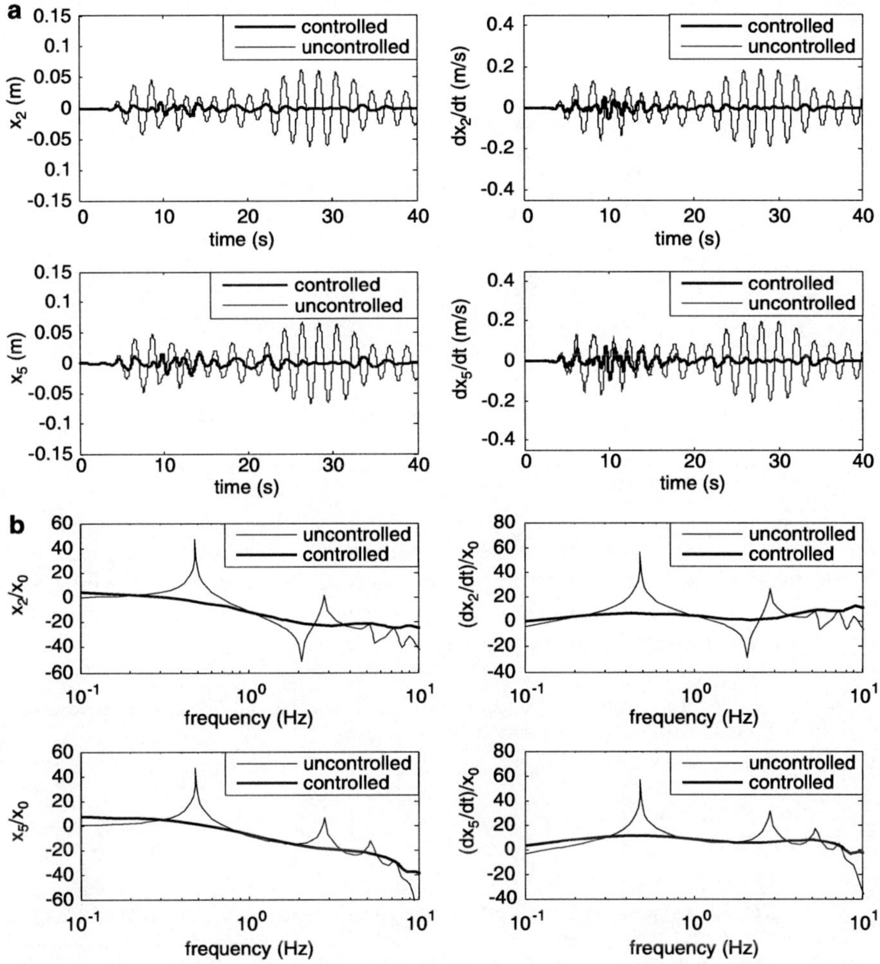

Fig. 3 Controlled and uncontrolled displacement and velocities (**a**) time responses and (**b**) frequency responses of the second and fifth storeys

5 Conclusions

In this study, H_∞ controller have been designed for a multi-degree-of-freedom structural system having the parameters of a real building and simulation results have been presented. The main idea behind proposing H_∞ controller has great potential in active structural control. The system is modeled including the two actuators which are installed on first and fifth storey. The structural system is then subjected to Northridge Earthquake vibrations effects. The simulation results exhibit that the implementation of H_∞ controller shows a good response as far as absorbing

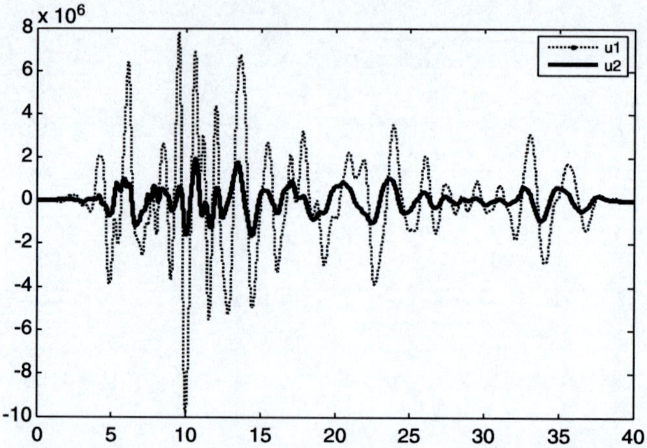

Fig. 4 Time history of the applied control force scaled from -10×10^6N to 8×10^6N

the vibration due to earthquake effects. The improvement in resonance values and the decrease in vibration amplitudes support this result. Robustness of the proposed H_∞ controller is desired that the controller remains stable and effective when the structure is subjected to seismic excitation.

Acknowledgments This work was supported by The Scientific and Technological Research Council of Turkey (TUBITAK) under Project 109 M523.

References

1. Soong, T.T., Constantinou, M.C. (eds.): Passive and Active Structure Vibration Control in Civil Engineering, CISM Courses and Lectures 345, International Centre for Mechanical Sciences. Springer, New York (1994)
2. Kose, I.E., Schmitendorf, W., Jabbari, F., Yang, J.: H_∞ active seismic response control using static output feedback. J. Eng. Mech. **122**(7), 651–659 (1996)
3. Guclu, R., Yazici, H.: Seismic-vibration mitigation of a non-linear structural system with an ATMD through a fuzzy PID controller. Nonlinear Dyn. **58**(3), 553–564 (2009)
4. Yazıcı, H., Guclu, R.: Active vibration control of seismic excited structural system using LMI-based mixed $H2/H_\infty$ state feedback controller. Turk. J. Electr. Eng. Comput. Sci. (2010). doi:10.3906/elk-1007-592
5. Masubuchi, I., Ohara, A., Suda, N.: LMI-based controller synthesis: a unified formulation and solution. Int. J. Robust Nonlinear Control **8**(8), 669–686 (1998)
6. Jabbari, F.: Output-feedback controllers for systems with structured uncertainty. IEEE Trans. Autom. Control. **42**(5), 715–719 (1997)

Active Vibration Control of Hydrodynamic Journal Bearings

J. Tůma, J. Šimek, J. Škuta, J. Los, and J. Zavadil

Abstract Rotor instability is one of the most serious problems of high-speed rotors supported by sliding bearings. With constantly increasing parameters, new machines problems with rotor instability are encountered more and more often. Even though there are many solutions based on passive improvement of the bearing geometry to enlarge the operational speed range of the journal bearing, the paper deals with a working prototype of a system for the active vibration control of journal bearings with the use of piezoactuators. The actively controlled journal bearing consists of a movable bushing, which is actuated by two piezoactuators. It is assumed that the journal vibration is measured by a pair of proximity probes. Force produced by piezoactuators and acting at the bushing is controlled according to error signals derived from the proximity probe output signals. The active vibration control was tested with the use of a test rig, which consists of a rotor supported by two controllable journal bearings and driven by an inductive motor up to 23,000 rpm. As it was proved by experiments the active vibration control extends considerably the range of the rotor operational speed.

Keywords Journal bearings • Hydrodynamic bearings • Active vibration control • Oil film instability

---

J. Tůma (✉) • J. Škuta • J. Los • J. Zavadil
Faculty of Mechanical Engineering, VSB – Technical University of Ostrava, Ostrava, Czech Republic
e-mail: jiri.tuma@vsb.cz; jaromir.skuta@vsb.cz; jaroslav.los.st@vsb.cz; jaromir.zavadil.st1@vsb.cz

J. Šimek
TECHLAB Ltd., Prague, Czech Republic
e-mail: j.simek@techlab.cz

J. Náprstek et al. (eds.), *Vibration Problems ICOVP 2011: The 10th International Conference on Vibration Problems*, Springer Proceedings in Physics 139,
DOI 10.1007/978-94-007-2069-5_58,

1 Introduction

Both the VSB – Technical University, Faculty of Mechanical Engineering and TECHLAB Ltd., Prague, are focused on long-term research in the field of rotor dynamics. One of the most serious problems is instability of high-speed rotors due to the journal bearing oil film. To study possibilities of affecting rotor behavior by controlled movement of bearing bushings, a test stand was designed, manufactured and assembled. Even though there are many solutions based on passive improvements of the bearing geometry to enlarge the operational speed range of the journal bearing, such as a lemon bore, pressure dam, tilting pad, etc., the approach to preventing the journal bearing instability, presented in the paper, is based on the use of the active vibration control.

Many authors pay attention to the active control of sliding bearings with the use of magnetic actuators as for example [1, 2]. Piezoactuators as a tool to control of rotating machines have been intensively investigated in the literature since the end of 1980s. One of the first original contributions dated from the beginning of the 1990s [3]. These papers did not study the effect of the oil film on the onset of instability and its suppression using the active vibration control. Worth mentioning are papers [4] and [5] dealing with the problem of the rotor instability. Because of the lack of information, it was decided to start research of methods suppressing sliding journal bearing instability by the active vibration control. The research work was granted by the Czech Science Foundation as a part of the research project No. 101/07/1345 Active control of journal bearings aimed at suppressing the rotor instability". The control system adds an electronic feedback to the rotor-bearing system actuating the position of a movable bushing. The current passive damper changes into an active component of the system with controllable properties. The laboratory test facilities, including the journal bearing equipped with the movable bushing was designed by TECHLAB Ltd., Prague. The research group of Technical University of Ostrava developed the control system, which is based on piezoactuators, and put the system into operation [6, 7].

2 Test Stand

The photos of a controllable journal bearing arrangement, which is implemented for the active vibration control, are shown in Fig. 1. The test rig consists of a rigid shaft of 30 mm diameter supported in two cylindrical hydrodynamic journal bearings. Radial clearance is 45 μm. Bearing span is 200 mm. It is possible to put one or two discs on the shaft, thus increasing bearing load and rotor mass. However, lowest stability limit should be achieved with the minimum bearing load, i.e. with hollow shaft without discs. The test shaft is driven by high-frequency motor through an elastic membrane coupling, constituting two joints, so that the shaft is decoupled from motor and free to move. The test rig was designed for speeds up to 23,000 rpm.

Fig. 1 Arrangement of the test rig and the controllable journal bearing

Bearing bushings are supported in rubber "O" rings, which ensure sealing of oil inlet and at the same time enable movement of bushings within the clearance in bearing casing. Bearing bushings can be excited by means of piezoactuators oriented in vertical and horizontal directions and fastened to the frame. The preloaded open-loop LVPZT piezoactuators are of the P-842.40 at the coupling and P-844.60 type at the shaft free end. Both the piezoactuator types require a low voltage amplifier with the 100 V peak value at the output. The pushing force produced by the P-842.40 type is of 800 N and the pulling force only 300 N. The piezoactuator travel range is up to 90 μm. The same travel range reaches the LVPZT piezoactuator of the P-844.60 type while the pushing force is up to 3,000 N and the pulling force is up to 700 N in contrast to the P-842.40 type. Concerning the lubricant, it was initially used the hydraulic oil of the VG 32 grade and then bearing special oil of the OL-P03 grade for high-speed grinder spindle bearing.

Shaft movement is measured by two pairs of proximity probes. Firstly the eddy current sensors IN-085, supplied by Brüel & Kjær Company, were tested, but after some problems with the measurement errors the capacitive sensors of the capaNCDT CS05 type supplied by the Micro Epsilon Company, were installed.

3 Active Vibration Control

There is a critical angular velocity Ω_{CRIT} at which the rotating shaft becomes unstable due to the properties of the oil film [2]. If a feedback between the shaft position and bushing position with respect to the bearing housing is introduced then the critical speed is enlarged to the value

$$\Omega_{MAX} = \Omega_{CRIT}\sqrt{K_P + 1}\,. \tag{1}$$

where K_P is the gain of the open control loop. The control system does not stabilize the behavior of the journal bearing directly by changing the position of the bearing bushing, but indirectly by changing force that acts on this bushing. Except of

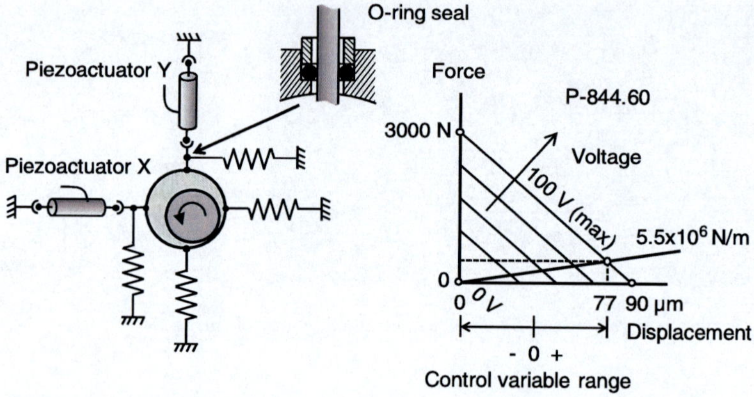

Fig. 2 Arrangement of piezoactuators and dependence of their force on displacement

the controller gain, the displacement of the bushing depends on stiffness of its connection with the bearing body through rubber seal rings as it is shown in Fig. 2. The dependence of the piezoactuator travel and force on electrical voltage and the dependence of clamping force on bushing displacement are depicted on the right side of Fig. 2. The working point of the electromechanical system results from the voltage which is supplied to the piezoactuator. The gain K_P of the open control loop results not only from the setting up of the controller, but from the property of the bushing clamping as well. Properties of the piezoactuator of the P-844.60 type (catalogue values) and measured stiffness of clamping (5.5×106 N/m) gives the bushing travel range which is equivalent to the control variable range. The range of the rotor stable rotational speed is limited by the travel range of piezoactuators and measurement errors of the proximity probes.

As was mentioned earlier, the signal from the proximity probes is connected to the dSpace signal processor. The output of the signal processor is connected to the input of the amplifier that powers the piezoactuator. The electronic feedback (see Fig. 3) in the below presented experiments was of the proportional controller type. Although improved dynamic properties of the control loop require adding a derivative component, the noisy signal produced by the proximity probes is the reason, for which the derivative feedback was not used [8]. Even if the sensors based on the electrical capacity principle have a smaller error than the eddy current ones, the active vibration control has been tested with sensors based on eddy currents.

The shaft rotational speed during the tests under active control (ON) and without active control (OFF) was increasing at the ramp rate of 7,000 rpm per a minute for all measurements. For the oil of the VG 10 grade the onset of instability starts at 4,300 rpm. Because the piezoactuator travel range cannot cover the change of the shaft position from the very beginning up to the level of the bushing centre position, the active vibration control is switched ON when the shaft lifts up into the stabilized position, which corresponds approximately to 3,000 rpm. Due to the measurement

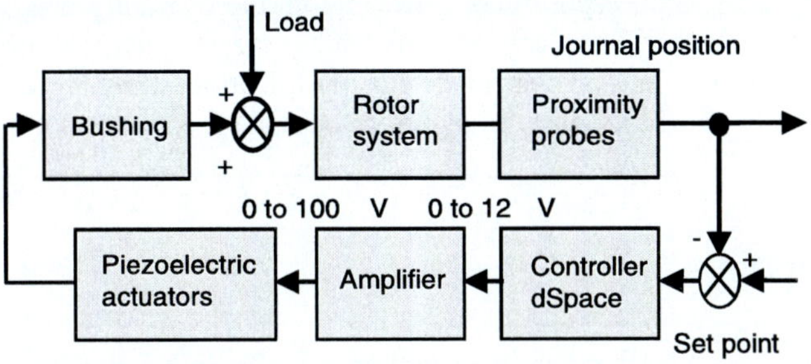

Fig. 3 Active vibration control system

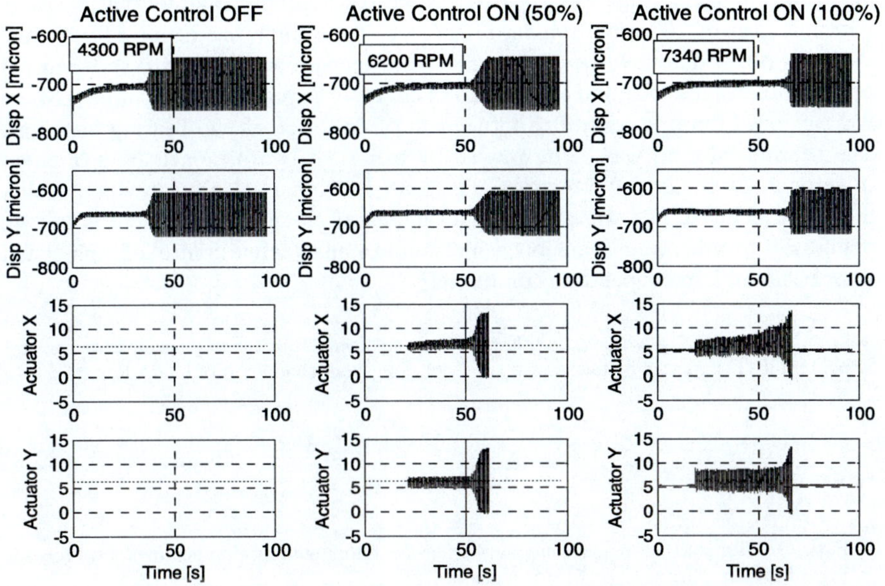

Fig. 4 Time history of the rotor rotation speed when the active vibration control is ON and OFF

error the controller output voltage starts to oscillate with a limited magnitude. As is clear from Fig. 4, if the active control is switched ON during the rotor run-up, the onset of instability is changed to 7,300 rpm. This increasing of the limit speed corresponds to the controller gain. The result of measurements at half the open-loop gain (50%) is shown in the middle part of Fig. 4. The onset of instability occurs at about 6,200 rpm. The active vibration control is immediately switched OFF after starting the unstable vibration with frequency, which is 0.475 multiple of the shaft rotation frequency. This phenomenon is called "whirl" due to the oil film and is different from the first natural frequency of the shaft, which is equal to a constant value.

As is demonstrated in Fig. 4, the active vibration control significantly extends the range of operating rotational speed. With active control ON, onset of instability is increased by about 3,000 rpm in comparison with the operating range without the active vibration control. The electronic feedback is clearly seen as a complementary way to the traditional journal bearing design modification, which prevent instability or shift the rotor instability onset to higher rotational speed.

4 Conclusions

The test rig for experimental investigation of affecting behavior of the rotor supported in sliding bearings by external excitation was put into operation. The rotational speed of the rotor is up to 23,000 rpm. The test rig consists of a rigid shaft of 30 mm diameter supported in two cylindrical hydrodynamic journal bearings with the span of 200 mm. Standard behavior of the rotor was achieved with low viscosity oil, with which the oil film had insufficient load capacity to shift journal centre into unstable position at the bearing centre. The proposed goal of the project was achieved through controlled movement of only one of two bearing bushings. The bearing bushing was being moved by means of two piezoactuators oriented in vertical and horizontal directions. The shaft rotational velocity, at which occurs the onset of instability, was substantially increased. It seems, that there is a large potential for further improvements, which could lead to active control of high-speed rotor behavior in real operating conditions.

Acknowledgments The presented results have been obtained during the solving o research project SP201118/2011 supported by the Ministry of Education, Youth and Sport of Czech Republic.

References

1. Fürst, S., Ulbrich, H.: An active support system for rotors with oil-film bearings. In: Proceedings of IMechE, Series C, pp 61–68, paper 261/88 (1988)
2. Tůma, J., Šimek. J., Škuta, J., Los, J. Zavadil, J.: Active vibration control of journal bearings with the use of piezoactuators. In: Proceedings of Ninth International Conference on Vibrations in Rotating Machines 2011, Darmstadt, Germany, 21–23 Feb 2011, Paper-ID: 54. pp. 11
3. Palazzolo, A.B., Lin, R.R., Alexande, R.M., Kascak, A.F., Montague, G.: Test and theory of Piezoactuators – active vibration control of rotating machinery, ASME Trans. J. Vib. Accoustics **113**(2), 167–175 (1991)
4. Carmignani, C., Forte, P., Rustighi, E.: Active control of rotor vibrations by means of Piezoelectric Actuators. In: Proceedings of DETC2001 18th Biennial Conference on Mech Vibration and Noise, Pitts-burgh, PA (2001)
5. Rho, B-H., Kim, K-W.: The effect of active control on stability characteristics of hydrodynamic journal bearings with an axial groove. In: Proceedings of the Institution of Mechanical Engineers, Part C: Journal of Mechanical Engineering Science, vol. 216, Number 9/2002, pp. 939–946 (2002)
6. Tůma, J., Škuta, J., Klečka, R., Los, J. Šimek. J.: A laboratory test stand for active control of journal bearings. In: Proceedings of Colloquium Dynamics of Machines 2010, Institute of Thermomechanics, Prague, 2–3 Feb 2010, pp. 95–100

7. Šimek, J., Tůma, J., Škuta, J., Klečka, R.: Unorthodox behavior of a rigid rotor supported in sliding bearings. In: Proceedings of Colloquium Dynamics of Machines 2010, Institute of Thermomechanics, Prague, 2–3 Feb 2010, pp. 85–90
8. Tůma, J., Šimek, J. Víteček, A.: Simulation study of a rotor system response to kinematic perturbation. Acta Mechanica Slovaca, 3/2008
9. Awrejcewicz, J., Dzyubak, L.P.: 2-dof non-linear dynamics of a rotor suspended in the magneto-hydrodynamic field in the case of soft and rigid magnetic materials. Int. J. Non-Linear Mech. (2010). doi:10.1016/j.ijnonlinmec. 2010.01.006

Vibrations Reduction of Industrial Sewing Machines

P. Šidlof and V. Votrubec

Abstract This paper deals with balancing of industrial sewing machines. We have proposed an eccentric balancing mechanism that balances crank mechanism of the needle bar. It significantly reduces vibrations. Big advantage is that the variation of drive shaft angular velocity only slightly raises. To create a new computational apparatus we began to use and supplement new simulating software MathModelica. Optimization, some calculations and evaluations are then executed in software Mathematica.

Keywords Sewing machine • Balancing • Ununiform angular velocity • MathModelica

1 Introduction

Industrial sewing machines have high-performance parameters and consequently high dynamic forces, vibrations and noise. The main dynamic forces source of most of sewing machines is the needle bar mechanism due to the fact that the needle bar has relatively high stroke and mass. Less influence has the thread feeder mechanism and possible rotary imbalances. Many methods for reduction of dynamic forces were patented, but none of them is practically used, because they are expensive, require large space or do not reduce vibrations sufficiently. In VÚTS an industrially implementable method using eccentric balancing mechanism that balances crank mechanism of the needle bar was patented. Kinematically it is similar to a crank mechanism with very small stroke. Inertial forces are balanced almost entirely and imbalanced torque is relatively small. According to small stroke it is a big advantage

P. Šidlof (✉) • V. Votrubec
VÚTS Liberec, Plc., Liberec, Czech Republic
e-mail: pavel.sidlof@vuts.cz; vlastimil.votrubec@vuts.cz

J. Náprstek et al. (eds.), *Vibration Problems ICOVP 2011: The 10th International Conference on Vibration Problems*, Springer Proceedings in Physics 139,
DOI 10.1007/978-94-007-2069-5_59,

that the balancing mechanism only slightly raises the variation of drive shaft angular velocity, which influences other mechanisms adversely.

For optimal balancing of the whole machine which contains a number of other mechanisms, it is necessary to use appropriate software. One of the best and most universal software is software Mathematica by Wolfram Research Company. Recently we began to use new simulating software MathModelica for creation of a new computational apparatus of mechanisms. Optimization and some calculations are then executed in Mathematica.

2 Balancing of Crank Mechanism of the Needle Bar

Dynamic forces of the crank mechanism are balanced by many methods that are used e.g. in combustion engines. The simplest method is partial balancing. Usually only the first or second harmonic component is balanced by means of rotary balancers placed on shafts that rotate in opposite direction in the same or twice angular velocity considering the crank shaft. The reduction of the first harmonic force component in the direction of needle bar motion is possible to achieve by a balancer on crank shaft which is always used but the force component in the orthogonal direction rises. Another method is balancing through passive vibratory dumpers. Total balancing of dynamic forces is possible to achieve simply by inverse crank mechanism and for total dynamic torque balancing it is necessary to use other balancing crank mechanisms.

In our proposed method we used at first two balancing mechanisms on the prototype (main and auxiliary), but only main mechanism was implemented in the production due to economic reasons [1]. The principle of the main balancing mechanism is obvious from the Fig. 1. Parts of the machine frame are depicted in the darkest color. On the drive shaft the inner cam is rotated by 180° against the crank of needle bar mechanism. The inner cam is embedded in outer cam that substitutes connecting rod of the mechanism. The balancer is guided by two flat springs so that it conducts movable motion in the opposite direction than the needle bar. Its trajectory is close to a parabola. Mounting on springs is without clearance and cheap to manufacture. It is possible to use it because the stroke of the balancer is multiple times smaller than the stroke of the needle bar. The mass of the balancer is bigger in the similar ratio. The outer cam that substitutes the connecting rod is the main construction problem, because its angular acceleration regarding to geometric similarity is as large as the connecting rod angular acceleration of the needle bar mechanism.

The auxiliary balancing mechanism of the machine prototype had similar construction and it was placed on the opposite end of the drive shaft. Its balancer moved in the same sense as the needle bar, but it had significantly smaller dimensions and stroke than the main balancing mechanism.

The balancing was evaluated by measurement on the machine using accelerometers. The measurements were sampled as a function of main shaft angle of rotation

Fig. 1 Schema of the needle bar mechanism, thread feeder mechanism and main eccentric balancing mechanism

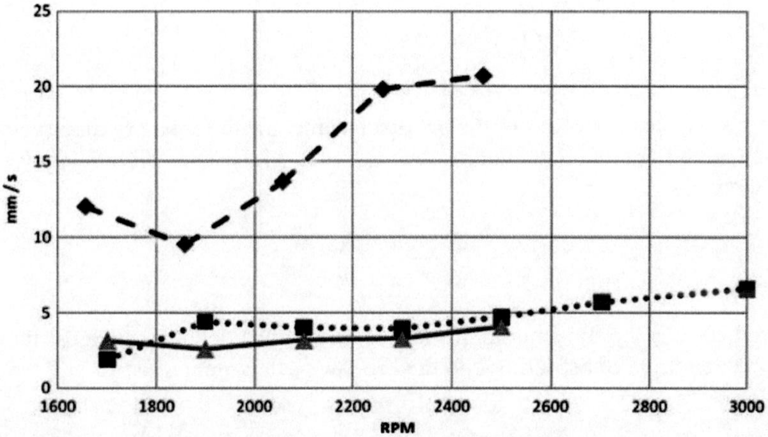

Fig. 2 Effective velocities of vertical vibrations of sewing machine arm as a function of drive shaft revolutions (*solid line* – machine with main and auxiliary balancing mechanism, *dotted line* – machine only with main balancing mechanism, *dashed line* – a competitive machine of a renowned producer)

so that quality statistic processing would be possible. Vibrations were evaluated by transferring to time domain. A simple and often used method of vibration check is calculation of effective velocities of vibrations. On the Fig. 2 there is a comparison of results of measurement on a machine with balancing mechanisms (dotted line) and on a machine from a renowned producer (dashed line), which had similar parameters of the needle bar mechanism but more massive frame and bigger mass. It's allowed revolutions were only 2,500 RPM. Machines were embedded in silent blocks in the working desk. Effective velocities were evaluated in a frequency range 5–2,200 Hz. It is clear that application of the balancing mechanisms leads to a significant reduction of vibrations.

Bad influence of additional balancing mechanisms may be the increase of irregularity of drive shaft rotation that is caused by addition of masses with reversible motion.

3 Irregularity Increase of Drive Shaft Rotation due to the Balancing Mechanism

Masses with reversible motion cause angular velocity change of the drive shafts. Inside the sewing machine other important mechanisms work near dead centers of the needle bar and due to the stopping of the needle bar, the shaft velocity is maximal near dead centers. This increases the dynamic stress of mechanisms and it also changes the frequency content of motions. For qualitative appreciation the simplest model based on energy conservation principle [2] with only the mass of the needle bar m and the constant total crank shaft moment of inertia I is used:

$$\frac{1}{2} m\, \dot{s}^2 + \frac{1}{2} I\, \omega^2 = \frac{1}{2} I\, \omega_{\max}^2 \tag{1}$$

Here $\dot{s}$ denotes derivative of the bar displacement with respect to time (velocity), ω is the main shaft angular velocity and $\omega_{\max}$ is its highest velocity. Velocity can be expressed as

$$\dot{s} = \frac{ds}{dt} = \frac{ds}{d\varphi}\frac{d\varphi}{dt} = s'\, \omega \tag{2}$$

Symbol s' is the transmission ratio; the apostrophe denotes the derivative with respect to angle φ. Substituting into the Eq. 1 we can write

$$\omega(\varphi) = \frac{\dot{s}}{s'} = \frac{\omega_{\max}}{\sqrt{1 + \frac{m}{I} s'^2}} = \omega_{\max} \left(1 - \frac{m\, s'^2}{2I} + \frac{3m\, s'^4}{8I} - \ldots \right) \tag{3}$$

Sliding mass of the balancing mechanism does the same movement as the needle bar, but with $p \times$ smaller stroke and $p \times$ bigger mass. The adjunction of the balancing mechanism consequently causes that the element $\frac{m}{I} s'^2$ extends $\left(1 + \frac{1}{p}\right)$ times. In our case it was $p = 9.5$ and the increase of the rotation irregularity is then practically insignificant.

From the Eq. 3 it is possible to estimate relation between the frequency content of the geometric velocity (transmission ratio) of the needle bar $s'$and the angular velocity ω. Because the first most significant member of the series is proportional to the second power of s', frequencies of the main harmonic components of the angular velocity ω are twofold to frequencies of s'. That can be dangerous in terms of vibrations excitation. The mentioned effect has a certain analogy in rectification of alternative current in electrical engineering.

4 The Method of Design and Calculation of Machine Balancing

The balancing of the sewing machine has to work in large range of operating speed and a level of wear. If mechanisms are balanced through rotary balancers and balancing mechanisms, it is convenient to use a kinetostatic method for design and optimization of main parameters. Software capable of calculating with flexibilities and clearances (e.g. Adams), that are expensive and inconvenient for optimization are then used for solving design details. One of the best universal computational software is software Mathematica from Wolfram Research Company. There the calculation of all mechanisms and the optimization of parameters were realized. Ten parameters were optimized there, e.g. dimensions and angular rotation of four rotary balancers on the upper and the lower shaft and dimensions of the balancer of the main and auxiliary balancing mechanism. The criterion was weighted maximum of total dynamic force components and torque to the center of mass of the machine. The optimization of the total dynamic force can be favorably figured by polar diagrams. An example of the calculation of the balancing that comprises five main mechanisms is on the Fig. 3.

Computational schemas of mechanisms in the software Mathematica lead to relatively complicated equations with many constants and variables and it is easy to make lots of mistakes. Therefore we began to use graphic oriented software MathModelica which is also able to cooperate with the software Mathematica.

5 Application of Software MathModelica

Software MathModelica uses Modelica language which is based on object oriented modeling. Its big advantage is describing the model directly by a system of equations, not as algorithm of solving these equations. Large libraries allow modeling of complex problems from different fields [3].

Modeling of the mechanism is done by compiling elements. In this software it is possible to edit and create elements according to user needs. Because the definition of elements is quite general, it is necessary to make some modifications. The main

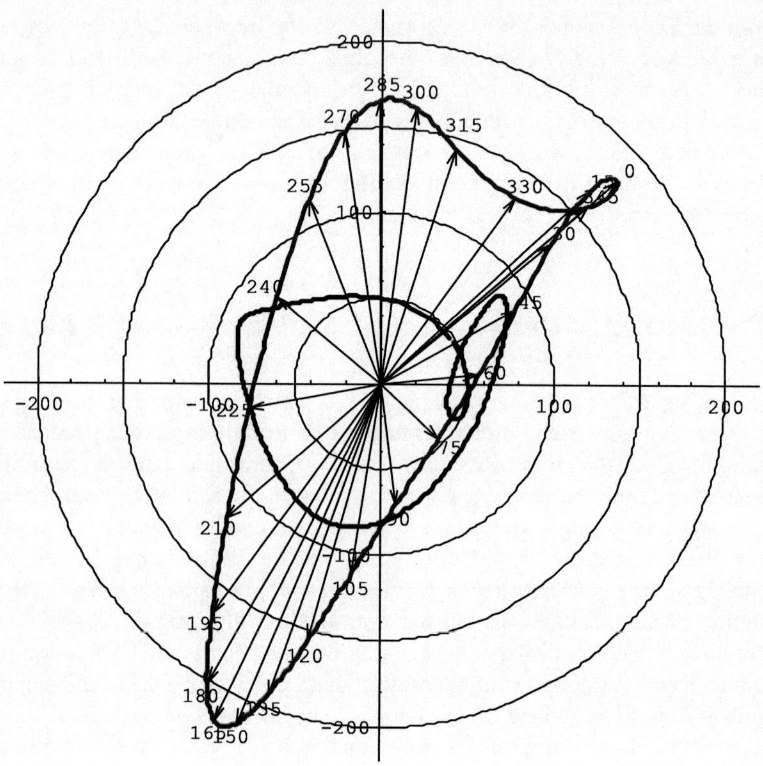

Fig. 3 Polar diagram of total dynamic force vectors [N] before the balancing (*outer curve*) and after the balancing (*inner curve*) as a function of angular rotation [deg] of the crank. Angles of main shaft are subjoined to the vectors of force (Total force maximum before the balancing is 220 [N], the maximum after the balancing is 69.7 [N])

requirement during the calculation is the course of inertial force and torque of individual bodies, then courses of kinematic quantities. For these and other reasons a new element of rigid body was created.

The advantage of software MathModelica is its compatibility with Mathematica. Although MathModelica allows the simulation of the model, some calculations are more convenient to execute in Mathematica. An example can be optimization of balancers in the sewing machine. In Mathematica the function is created that includes definition of parameters for optimizing (e.g. masses of balancers), simulation of the model in MathModelica, finding the maximum value of total inertial force etc. This function is then part of the operator for finding the minimum. It is possible to proceed similarly in other cases. This method eliminates most drawbacks, which occurred in Mathematica, and makes the calculation more effective.

6 Conclusions

To reduce vibrations of industrial sewing machines balancing eccentric mechanism was used. This mechanism almost completely balances dynamic forces of the needle bar mechanism. In the total balancing we must consider other mechanisms and rotary balancers. Overall calculation and optimization were executed in software Mathematica and a significant reduction of vibrations was achieved. In order to facilitate the calculation, mechanisms are solved in software MathModelica and optimization and some other calculations are executed in Mathematica.

References

1. Šidlof, P., Škop, P.: Equipment for vibrations reduction of industrial sewing machines. CZ Patent 301,075 B6
2. Šidlof, P., Čejka, V., Škop, P.: Method for vibration and noise reduction of loom shed mechanisms. In: Proceedings of ICOVP – 2003, Liberac, pp. 74 (2003)
3. Fritzson, P.: Principles of Object-Oriented Modeling and Simulation with Modelica 2.1. Wiley-IEEE Press, Chichester (2003). ISBN 0-471-47163-1

Monitoring of Roof Truss Girder Vibration Using Piezoelectric Sensors – FEM Simulation

Jan Freundlich and Marek Pietrzakowski

Abstract In this paper modelling and numerical simulations of the roof truss girder with the surface mounted piezoelectric sensors applied to monitoring purposes are presented. In the first stage, a simply supported beam with piezoelectric sensors, excited by time varying vertical displacements of the supports, is examined using the finite element method (FEM) and an analytical approach. The aim of calculations is to validate the applied finite element (FE) model of the system. The compared results are in a good agreement and confirm the applied FE model correctness. In the second stage, a simplified FE model of the roof truss girder is investigated. Dynamic responses of the roof truss girder with two piezoelectric sensors are achieved using FEM simulations. As previously, the roof truss girder is excited by the time varying vertical movement of the supports. The influence of a local failure of the structure on its dynamic behaviour is tested as well. A defect is modelled by decreasing stiffness of a selected joint of the bars in the truss girder. The amplitude-frequency characteristics are calculated and compared with those obtained for the healthy structure.

Keywords Piezoelectric sensors • Vibration • FEM simulation • Monitoring

1 Introduction

Roof structures used in building construction are exposed to detrimental influence of environment and various types of vibration generated by heavy vehicle traffic, which cause alternate loads. Hence, the alternate values of stress, deflection and acceleration appearing in the important elements ought to be monitored to detection

---

J. Freundlich (✉) • M. Pietrzakowski
Warsaw University of Technology, Warszawa, Poland
e-mail: jfr@simr.pw.edu.pl; mpi@simr.pw.edu.pl

J. Náprstek et al. (eds.), *Vibration Problems ICOVP 2011: The 10th International Conference on Vibration Problems*, Springer Proceedings in Physics 139,
DOI 10.1007/978-94-007-2069-5_60,

and identification of failures in the early stage of evolution. Monitoring of dynamic behaviour of structures can be realised, among other techniques, using piezoelectric distributed transducers. A lack of additional power supply is a quite important advantage in the case of piezoelectric sensors implementation. In order to improve the operational performance of piezoelectric transducers numerical simulations are often required. A great deal of theoretical and experimental research efforts concentrate on actuation (cf [1]) and active damping of slender structures like beams where piezoelectric sensors/actuators are mainly of the extension-type with transverse poling direction (cf [2, 3]). In this paper numerical simulations of the roof truss girder with the surface mounted piezoelectric sensors applied to monitoring purposes are presented.

2 Comparative Calculations of the Piezoelectric Sensor Performance

A simply supported beam with the piezoelectric sensor, excited by time varying vertical movements of the supports (Fig. 1), is examined using the finite element method (FEM) and an analytical approach. The aim of the calculations is to validate the applied finite element (FE) model of the system. The piezoelectric sensor is assumed to be perfectly bonded to the beam surface. The simplified coupling model is applied, in which the mass and longitudinal inertia forces of piezoelectric elements are ignored. The beam is modelled according to the Bernoulli-Euler theory, namely neglecting rotary inertia and shear deformation. The viscoelastic material properties are approximated by the Kelvin-Voigt model.

The equation of the beam transverse vibration $w(x, t)$ excited by displacements $y_1(t)$ and $y_2(t)$ and referred to the undeformed beam axis can be written as

$$E_b J_b \left(\frac{\partial^4 w}{\partial x^4} + \mu \frac{\partial^5 w}{\partial x^4 \partial t} \right) + \rho_b A_b \frac{\partial^2 w}{\partial t^2} = -\rho_b A_b \left(\left(1 - \frac{x}{l}\right) \ddot{y}_1(t) + \frac{x}{l} \ddot{y}_2(t) \right) \quad (1)$$

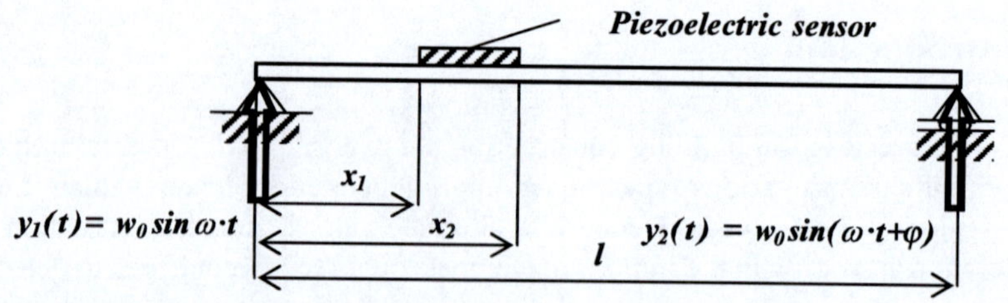

Fig. 1 Scheme of the considered simply supported beam with the piezoelectric sensor

where: E_b and ρ_b are the Young's modulus and the mass density of the beam material, respectively, J_b and A_b are the moment of inertia and area of the beam cross-section, respectively, μ is the damping coefficient.

The solution to the governing Eq. 1 has to satisfy the simply supported boundary conditions.

The voltage generated by the sensor can be calculated from the formula (cf [4])

$$V_s = \frac{d_{31}^s E_s (h_b + h_s)}{2C} \int_0^l \frac{\partial^2 w}{\partial x^2} b_s(x) dx \tag{2}$$

where: $b_s(x)$ and h_s indicate the width distribution and thickness of the sensor, respectively, E_s and d_{31}^s are the Young's modulus and piezoelectric constant of the sensor material, C is the sensor capacitance.

The system steady-state response is expressed in frequency domain in terms of a transfer function. The transfer function (in the complex form) relating the beam deflection to left side support displacement is given by

$$H_{wy_1} = \frac{w(x,\omega)}{y_1(\omega)} = \frac{2}{\pi} \sum_{n=1}^{\infty} \frac{\omega^2}{n\left(\omega_n^2 - \omega^2 + i\mu\omega_n^2\omega\right)} \sin\frac{n\pi}{l}x \tag{3}$$

The transfer function relating the output piezoelectric voltage to the left side support displacement has the form

$$H_{Vy_1} = \frac{V_s(\omega)}{y_1(\omega)} = -\frac{d_{31}^s E_s (h_b + h_s) b_s}{Cl} \sum_{n=1}^{\infty} \frac{\omega^2\left(\cos\frac{n\pi}{l}x_2 - \cos\frac{n\pi}{l}x_1\right)}{\omega_n^2 - \omega^2 + i\mu\omega_n^2\omega} \tag{4}$$

The analogous transfer functions can be obtained relating to the right support.

The achieved transfer functions enable amplitude-frequency characteristics to be calculated. The calculations are performed using symbolic calculation software "Mathematica". Vibration and generated voltage amplitudes versus excitation frequency for phase angle $\varphi = 0$ and $\varphi = \pi$ are calculated. The comparable calculations are performed using FE model of the considered system. The appropriate FE model of the beam with the piezoelectric sensor is prepared. The examined beam is modelled using 2-nodes, Bernoulli beam elements (B33 [5]). The FE model of the beam consists of 40 elements. The piezoelectric sensor is modelled using 3D-solid, second order, piezoelectric 20-nodes elements i.e. C3D20E elements [5]. These elements have additionally electric potential as the nodal quantity. The FE model of the sensor consists of 24 elements. Piezoelectric coupling is provided by introducing the piezoelectric and dielectric material coefficients. The compared voltage frequency characteristics (Fig. 2) are similar for analytical and FEM approaches and confirm the applied FE model correctness.

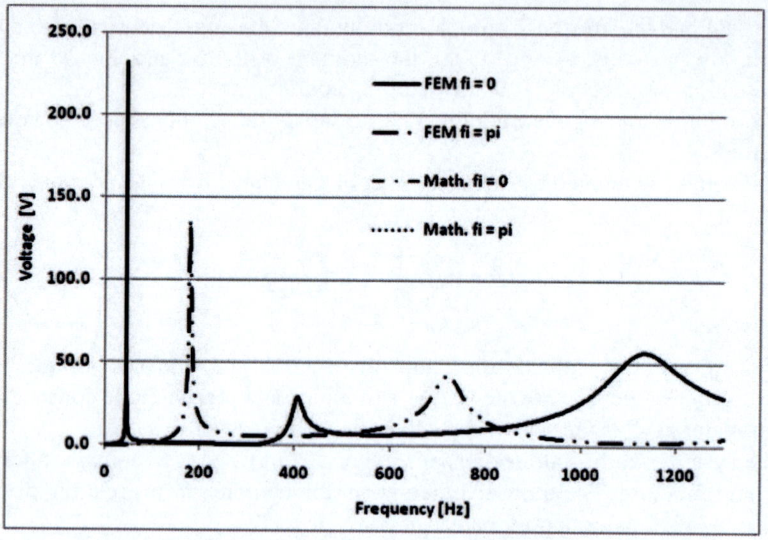

Fig. 2 Voltage amplitude-frequency characteristics for the piezoelectric sensor

3 The FEM Calculations of a Roof Truss Girder

A simplified FE model of the roof truss girder established in MONIT project [1] is investigated. Dynamic response of the roof truss girder with two piezoelectric sensors is examined using FEM simulations. The first sensor is located on a bottom chord whereas the second is located on a diagonal post (web) (Fig. 3). As previously, the roof truss girder is excited by the time varying vertical movement of the supports. The scheme of the considered structure is shown in Fig. 3. The amplitude-frequency characteristics related to the node displacements and the piezoelectric sensor voltage for the in-phase ($\varphi = 0$) and out-of-phase ($\varphi = \pi$) excitation are calculated.

In the next step, the influence of a local failure of the truss girder structure on its dynamic behaviour is tested. The defect is modelled by decreasing stiffness of a selected joint of the bars. The sensors are in the same position as in the healthy structure. The amplitude-frequency characteristics for the two cases of excitation ($\varphi = 0$ and $\varphi = \pi$) are calculated and compared with those obtained for the undamaged structure. Examples of the voltage amplitude-frequency characteristics of the sensor, which is situated on the bottom chord, are calculated for the roof truss without and with failure are shown in Fig. 4.

Decreasing the stiffness of the joint causes additional resonant amplitudes of the roof girder, which produce the voltage resonant peaks in the piezoelectric sensor. Decreasing of the joint stiffness also results in a shift of the additional resonance frequencies to lower values (Fig. 4). The changes in the sensor voltage response

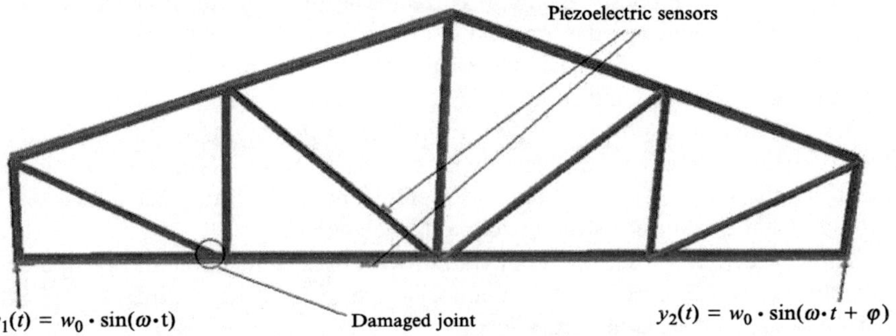

Fig. 3 Scheme of the considered roof truss with the sensors location

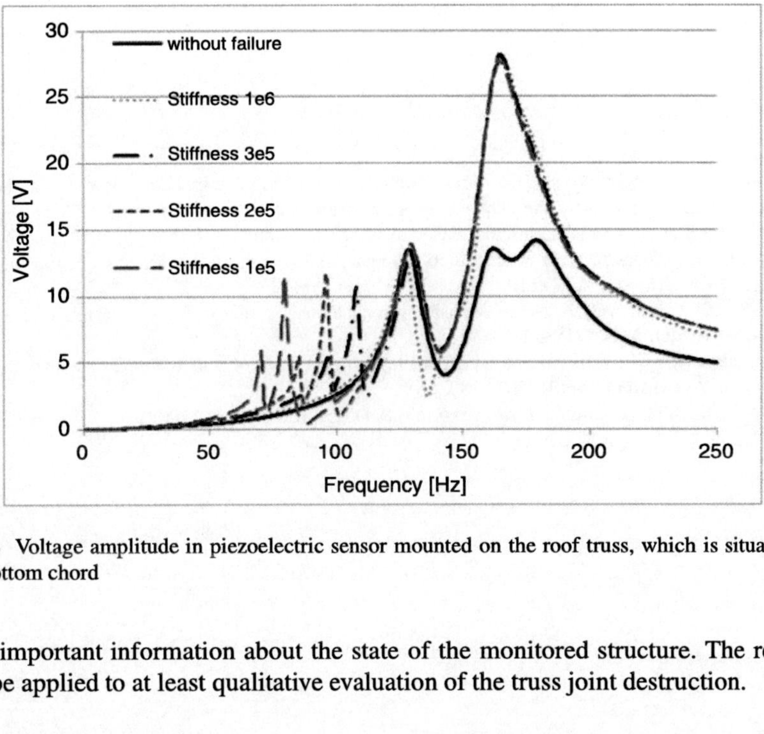

Fig. 4 Voltage amplitude in piezoelectric sensor mounted on the roof truss, which is situated on the bottom chord

give important information about the state of the monitored structure. The results can be applied to at least qualitative evaluation of the truss joint destruction.

4 Conclusions

The FE model of the truss structure with piezoelectric sensors is established. As a validation of the proposed FE model of the truss girder, the amplitude-frequency characteristics of the simply supported beam with the piezoelectric sensor is numerically investigated using the FEM and the analytical approach. The obtained results prove the applied FE model correctness. In the case of the truss girder structure the influence of the local failure on its dynamic behaviour is tested. The voltage amplitude-frequency characteristics for the healthy and locally damaged structure are calculated and compared.

The dynamic responses achieved from FEM calculations of the healthy and locally damaged roof girder demonstrate sensitivity and usefulness of the applied piezoelectric sensor for the failure detection and monitoring the effects of its progress on the examined structure dynamic behaviour.

Thus, piezoelectric sensors cooperating with data acquisition and computation systems give an opportunity for monitoring the state of truss girders to avoid a danger of the roof structure breakdown.

Acknowledgments This publication was supported by the European Union project "Monitoring of technical state of construction and evaluation of its lifespan" (MONIT).

References

1. Tylikowski A., Pietrzakowski M., Freundlich J., Monitoring of vibration processes in the truss structures using piezoelectric sensors, Proc. of the II Seminar of the Project "Monitoring of Technical State of Construction and Evaluation of its Lifespan" (in Polish), 2010, pp. 65
2. Crawley, E.F., Anderson, E.H.: Detailed models of piezoceramic actuation of beams. J. Intell. Mater. Syst. Struct. **1**, 4 (1990)
3. Dosch, J.J., Inman, D.J.: A self-sensing piezoelectric actuator for collocated control. Int. J. Solids Struct. **3**, 166 (1992)
4. Pietrzakowski, M.: Experiment on a cantilever beam control and theoretical approximation. J. Theor. Appl. Mech. **40**(3), 667 (2002)
5. ABAQUS HTML documentation: version 6.8, Dassault Systèmes (2008)

An Application of Electromagnetic Induction in Vibration Control

Radoslav Darula, George Juraj Stein, and Sergey Sorokin

Abstract Excessive vibration of machines and/or structures can be controlled passively (e.g. introducing resilient elements) and/or actively (e.g. using elements capable to adjust their properties for actual state of the vibration). An electromagnet, as a vibration control element, can be implemented in active (or semi-active) control strategy. This approach is analyzed in the present paper. The electromagnetic induction occurs in a magnetic circuit exposed to variable magnetic flux, which can be obtained e.g. by changing reluctance (magnetic resistance) of the system due to a variable air gap, as the result of armature vibration. The lumped parameter mathematical model of the coupled electro-magneto-mechanical system is formulated. The performance of the model is analyzed, assuming harmonic forced vibration. By the induction, mechanical energy of vibration is converted into electrical one and dissipated in the shunt resistance. Two concepts are investigated further – electromagnet behaves as (a) a spring element (reduction of equivalent system stiffness); (b) a damping element.

Keywords Electro-magneto-mechanical system • Stiffness modulation • Shunt damping • Semi-active vibration control

---

R. Darula (✉) • S. Sorokin
Aalborg University, Aalborg, Denmark
e-mail: dra@m-tech.aau.dk; svs@m-tech.aau.dk

G.J. Stein
Slovak Academy of Sciences in Bratislava, Bratislava, Slovak Republic
e-mail: stein@savba.sk

J. Náprstek et al. (eds.), *Vibration Problems ICOVP 2011: The 10th International Conference on Vibration Problems*, Springer Proceedings in Physics 139,
DOI 10.1007/978-94-007-2069-5_61,

1 Introduction

When a yoke of an electromagnet (Fig. 1a) is exposed to vibratory motion, the variable magnetic flux induces voltage in the closed electric circuit, as a result of Faraday's law of induction [1]. The induced voltage pushes current through the closed electric circuit and thus mechanical energy from vibrating yoke is transformed via magnetic field into electric energy, which is then dissipated in the shunt resistance. The concept of shunted electromagnetic transducer is analyzed in [2]. Implementation of electromagnet, as an active inertial control actuator, is presented in [3]. Some ideas how to combine the passive springs and active vibration isolation using 'negative stiffness' properties of electromagnetic elements can be found in [4]. The eddy currents induced by electromagnets and their utilization in vibration control are analyzed e.g. in [5].

The concept of implementation of induced currents dissipation in mechanical structures using electromagnets was analyzed and proved experimentally in [6].

2 Electro-Magneto-Mechanical Model

The electromagnet is a complex device, which is mathematically described using a coupled electro-magneto-mechanical model.

The magnetic circuit is composed of three reluctance terms: (a) an air gap of width $2{\cdot}d(t)$; (b) an iron core with flux line length l_C and (c) a permanent magnet (PM) of flux line length l_{PM}. A coil with N_w turns is located around the core. Then, from Ampere's law [7] can be written

$$N_w i_i(t) = \oint_C \vec{H} \cdot d\vec{l} = H_{PM} l_{PM} + H_C l_C + H_g \left[2d(t)\right] \tag{1}$$

where $i_i(t)$ is the current induced in the electric circuit; H is the magnetic field intensity. In general, the induced current $i_i(t)$ is negative and behaves as an energy sink.

Assuming that the PM is operating in linear region of demagnetization curve (Fig. 1b), then the magnetic field density at PM can be expressed in the form [8]

$$B_{PM} = B_r + \mu_{PM} H_{PM} \tag{2}$$

where the permeability of PM is $\mu_{PM} = B_r/H_c'$, with B_r remanent magnetization and H_c' apparent coercivity. Both parameters are material characteristics of the PM.

Neglecting the fringing effects of the magnetic flux and assuming constant cross-sectional area throughout the magnetic circuit S_C, i.e. the same magnetic flux throughout the circuit ($B_{PM} = B_g = B_C = B(t)$), the magnetic flux density becomes

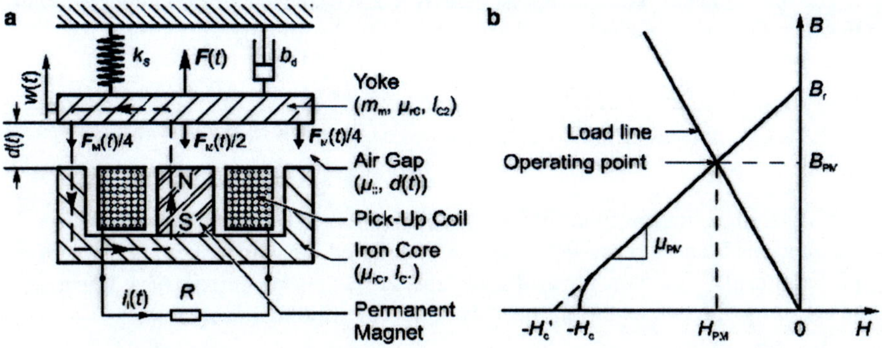

Fig. 1 (**a**) Electromagnetic semi-active controller; (**b**) Demagnetization curve of PM after [8]

$$B(t) = \frac{\mu_0}{2d_0} \frac{[M_{PM} + N_w i_i(t)]}{[(1+\delta) + \varepsilon(t)]} \tag{3}$$

where the scaled parameters are: $d_C = l_C/(2\mu_{rC})$, $d_{PM} = l_{PM}/(2\mu_{rPM})$, $\delta = (d_C + d_{PM})/d_0$, dimensionless displacement $\varepsilon(t) = w(t)/d_0$ and magnetization $M_{PM} = H_c' l_{PM}$.

From Faraday's law [1], the variation in the magnetic flux $\Phi(t)$, due to change of air gap reluctance in the system shown in Fig. 1a, induces voltage in the coil

$$u_i(t) = -\frac{d\Phi(t)}{dt} = -N_w S_C \frac{dB(t)}{dt} = -N S_C \left(\frac{\partial B(t)}{\partial \varepsilon} \frac{d\varepsilon(t)}{dt} + \frac{\partial B(t)}{\partial i_i} \frac{di_i(t)}{dt} \right) \tag{4}$$

Substituting the Eq. 3 into Eq. 4, connecting the shunt circuit to the coil and using Ohm's law for the electric circuit, $u_i(t) = Ri_i(t)$

$$\frac{L_0}{(1+\delta)} \frac{(M_{PM} + N_w i_i(t))}{[1 + \varepsilon(t)/(1+\delta)]^2} \frac{d\varepsilon(t)}{dt} = Ri_i(t) + L(\varepsilon(t)) \frac{di_i(t)}{dt} \tag{5}$$

where $L_0 = \mu_0 S_C N_w{}^2/(2d_0(1+\delta))$ is the self-inductance of the coil for d_0 and $L(\varepsilon(t)) = L_0/(1 + \varepsilon/(1+\delta))$ is the inductance of the coil for $d(t)$.

Equation 5 is the governing equation of the electrical system.

The electromagnetic coupling force between the mechanical and electrical system is derived using concept of Maxwell's pulling force [9]

$$F_M(F_M(t)) = 2\frac{1}{2\mu_0} B^2(t) S_C = \frac{L_0}{2d_0 N_w^2 (1+\delta)} \frac{[M_{PM} + N_w i_i(t)]^2}{[1 + \varepsilon(t)/(1+\delta)]^2} \tag{6}$$

The mechanical system is represented by a SDOF mass-spring-damper oscillator (Fig. 1a). Governing equation is derived using the D'Alembert principle [10]

$$F^*(t) - m_{\mathrm{m}}\frac{d^2\varepsilon(t)}{dt^2} - b_{\mathrm{d}}\frac{d\varepsilon(t)}{dt} - k_{\mathrm{s}}\varepsilon(t) - \frac{L_0}{2d_0^2 N_{\mathrm{w}}^2(1+\delta)}\frac{[M_{\mathrm{PM}} + N_{\mathrm{w}} i_{\mathrm{i}}(t)]^2}{[1+\varepsilon(t)/(1+\delta)]^2} = 0 \tag{7}$$

with a scaled excitation force $F^*(t) = F(t)/d_0$.

Equations 5 and 7 represent a system of coupled, non-linear equations, with the coupling terms – the non-linear displacement $\varepsilon(t)$ and the current $i_{\mathrm{i}}(t)$. Because of the coupling, the two equations need to be solved simultaneously.

3 Operation of the Vibration Controller

Let us assume that the displacement $\varepsilon(t)$ is sufficiently small, so we can linearize the non-linear terms in Eqs. 5 and 7 using Taylor's expansion. Then we get a system of coupled linear equations in the form:

$$m_{\mathrm{m}}\frac{d^2\varepsilon(t)}{dt^2} + b_{\mathrm{d}}\frac{d\varepsilon(t)}{dt} + \left(k_{\mathrm{s}} - \frac{L_0 {M_{\mathrm{PM}}^*}^2}{d_0^2(1+\delta)^2}\right)\varepsilon(t) + \frac{L_0 M_{\mathrm{PM}}^*}{d_0^2(1+\delta)} i_{\mathrm{i}}(t) = F^*(t)$$

$$\frac{L_0 M_{\mathrm{PM}}^*}{(1+\delta)}\frac{d\varepsilon(t)}{dt} = L_0\frac{di_{\mathrm{i}}(t)}{dt} + Ri_{\mathrm{i}}(t), \tag{8}$$

with the scaled magnetization $M_{\mathrm{PM}}{}^* = M_{\mathrm{PM}}/N_{\mathrm{w}}$. The static term of magnetic force (which is not considered in Eq. 8) is responsible for pre-stress of the mechanical spring k_{s}. Depending on the air gap width d_0 and the PM properties, permanent fixation of armature to the core can occur, which is unwanted. The limitations on air gap width d_0 are analyzed in [11].

Solving the system of Eq. 8 for harmonic force excitation $F^*(t) = F_0{}^* e^{j\omega t}$, introducing $i_i(t) = I_0 e^{j\omega t}$, $\varepsilon(t) = E_0 e^{j\omega t}$ (where amplitudes $F_0{}^*$, I_0, E_0 are complex valued) and neglecting higher order and cross terms, we obtain

$$\left(k_{\mathrm{s}} - \frac{L_0 {M_{\mathrm{PM}}^*}^2}{d_0^2(1+\delta)^2}\right) - \omega_{\mathrm{a}}^2 m_{\mathrm{m}} + \mathrm{j}\omega_{\mathrm{a}} b_{\mathrm{d}} + \frac{L_0 {M_{\mathrm{PM}}^*}^2}{d_0^2(1+\delta)^2}\frac{\omega_{\mathrm{a}}^2}{(\omega_{\mathrm{a}}^2 + R^2/L_0^2)}$$

$$+\mathrm{j}\frac{{M_{\mathrm{PM}}^*}^2}{d_0^2(1+\delta)^2}\frac{R\omega_{\mathrm{a}}}{(\omega_{\mathrm{a}}^2 + R^2/L_0^2)} = \frac{F_0^*}{E_0}. \tag{9}$$

Analyzing the expression on left hand side of Eq. 9, two coupling terms are present, one real and the other one imaginary (the last two terms). Taking the

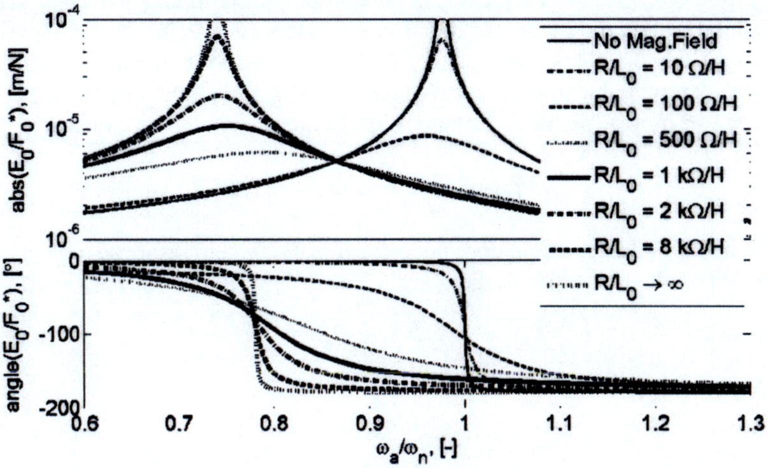

Fig. 2 Response of the system for various values of shunt resistance

excitation frequency (ω_a) and shunt resistance (R) as parameters, based on the denominator ($\omega_a{}^2 + R^2/L_0{}^2$) of the two coupling terms, we can distinguish three operation regimes of the controller:

- If $R/L_0 \gg \omega_a$ (high resistance value and/or excitation frequency within sub-resonance region), then

$$\left(k_s - \frac{L_0 M_{PM}^{*\,2}}{d_0^2(1+\delta)^2}\right) - \omega_a^2\left(m_m - \frac{L_0^3 M_{PM}^{*\,2}}{d_0^2(1+\delta)^2}\frac{1}{R^2}\right) + j\omega_a\left(b_d + \frac{L_0^2 M_{PM}^{*\,2}}{d_0^2(1+\delta)^2}\frac{1}{R}\right) = \frac{F_0^*}{E_0} \tag{10}$$

The electromagnet behaves as an inertia reduction element (the second term in Eq. 10). Magnetic damping, inversely proportional to the resistance R, is added to the system, too (the third term).

If the resistance is very high ($R \gg L_0\, M_{PM}{}^*$), or if the shunt is disconnected, then the Eq. 10 can be simplified further

$$\left(k_s - \frac{L_0 M_{PM}^{*\,2}}{d_0^2(1+\delta)^2}\right) - \omega_a^2 m_m + j\omega_a b_d = \frac{F_0^*}{E_0} \tag{11}$$

It means that without the shunt resistor just modulation of natural frequency, due to downwards acting magnetic force, is obtained (Fig. 2, for $R/L_0 \to \infty$).

- If $R/L_0 \approx \omega_a$ (neither resistance, nor frequency are dominant), then the system is strongly coupled and the Eq. 9 remains unsimplified. The magnetic terms are function of excitation frequency, i.e. the amount of damping as well as stiffness modulation is dependent on excitation frequency (Fig. 2 for $R/L_0 = 100\ \Omega/\mathrm{H}$ to $R/L_0 = 2\ \mathrm{k}\Omega/\mathrm{H}$). This operation region is of particular interest, since amount of added damping due to the electro-magnetic subsystem influence is the largest.

- If $R/L_0 \ll \omega_a$ (low resistance and/or high frequency excitation), then

$$k_s - \omega_a^2 m_m + j\omega_a b_d = F_0^* / E_0 \tag{12}$$

The overall magnetic field does not influence the system response, which is fully controlled by the properties of the mechanical system (Fig. 2, for $R/L_0 = 10\ \Omega/\mathrm{H}$).

4 Conclusions

The linearized model of vibration controller, which implements electromagnetic induction phenomena to convert mechanical energy into electrical and dissipate it within a shunt circuit, was investigated. Analyzing the Eq. 9, it was shown, that the controller operates in three regimes:

- Mechanical system is coupled only to static magnetic field (for dominant resistance/inductance term, i.e. $R/L_0 \gg \omega_a$). The controller is capable to modulate natural frequency (behaves as a spring element, Eq. 11), no damping is added.
- Mechanical system is fully coupled to the electrical field (when $R/L_0 \approx \omega_a$), where the coupling is strong. Added damping, as well as stiffness, are dependent on excitation frequency (Eq. 9). This regime is of particular interest.
- Mechanical system is fully de-coupled (for very low resistance, high inductance or large excitation frequency, i.e. $R/L_0 \ll \omega_a$). There is no influence of either magnetic, or the electric circuit, i.e. the controller behaves passively (Eq. 12).

For the region with strong coupling ($R/L_0 \approx \omega_a$), there is a shunt resistance value with maximal displacement attenuation ($R/L_0 \approx 500\ \Omega/\mathrm{H}$, Fig. 2). In general, this value depends on the ratio of shunt resistance R and coil inductance L_0.

Acknowledgments The contribution is a result of projects No.2/0075/10 of the Slovak VEGA Grant Agency and the InterReg IV A – the 'Silent Spaces' project at Aalborg University.

References

1. Cheng, D.K.: Field and Wave Electromagnetic. Addison-Wesley, Boston (1989)
2. Fleming, A.J., Moheimani, S.O.R.: Inertial vibration control using a shunted electromagnetic transducer. IEEE/ASME Trans. Mechatronics **11**(1), 84 (2006)
3. Paulitsch, C., Gardonio, P., Elliott, S.J.: Active vibration control using an inertial actuator with internal damping. J. Acoust. Soc. Am. **119**(4), 2131 (2006)

4. Mizuno, T., et al.: Vibration isolation system combining zero-power magnetic suspension with springs. Control. Eng. Pract. **15**, 187 (2007)
5. Sodano, H.A., Inman, D.J.: Non-contact vibration control system employing an active eddy current damper. J. Sound Vib. **30**(5), 596 (2007)
6. Darula, R.: Multidisciplinary analysis and experimental verification of electromagnetic SAVC. Master Thesis, Institute of Mechanical Engineering, Aalborg University (2008)
7. Krause, P.C., Wasynczuk, O.: Electromechanical Motion Devices. McGraw-Hill, New York (1989)
8. Furlani, E.P.: Permanent Magnet and Electromechanical Devices. Materials, Analysis, and Applications. Academic, San Diego (2001)
9. Mayer, D., Ulrych, B.: Electromagneticke aktuatory (Electromagnetic actuators – *in Czech*). BEN, Prague (2009)
10. Rao, S.S.: Mechanical Vibrations. Pearson Prentice Hall, Upper Saddle River (2004)
11. Stein, G.J., Darula, R., Chmúrny, R.: A clamped-clamped beam static sag limits under perpendicular magnetic force., Proc. of the 17th International Conference "Engineering Mechanics 2011", 2011, pp. 567–570.

Suppression of the Work-Piece Vibrations in Milling Using Active Clamp System

A. Parus, M. Hoffmann, and T. Okulik

Abstract The machining is always accompanied by vibration. In certain cases the level of vibration is very high and may cause shortening of the tool life, poor quality of machined surface. Operational speed and machined surface depend on dynamic stability of three components of the machine tool-cutting system: the cutting tool, the machine tool structure, the work-piece and the clamping system. To assure stable machining, parameters of the cutting process have to be tuned and frequently the machining productivity is decreased. For this reasons different types of systems are developed for suppressing the work-piece vibration. In some cases an additional modification of the work-piece is allowed and mounting the vibration absorber is possible. The paper describes a modification of the work-piece dynamic properties using active clamp system. In comparison to the vibration absorbers this solution has a great advantageous – adaptation of the work-piece is not necessary. In the paper the simulation results of different variants of milling process with work-piece mounted using the active clamp are presented. Piezo actuators are used in order to assure active influence on the work-piece. The aim of the state space feedback control system is to minimize the amplitude of the vibration during machining process.

Keywords Active clamp system • Vibration suppression • Stability

1 Introduction

The most undesirable phenomena in machining is chatter. This results in an impaired surface texture of the work-piece. To alleviate the problem different type of solutions can be found in literature. During designing stage of a machine tool its dynamics

A. Parus (✉) • M. Hoffmann • T. Okulik
West Pomeranian University of Technology, Szczecin, Poland
e-mail: Arkadiusz.Parus@zut.edu.pl; Marcin.Hoffmann@zut.edu.pl; Tomasz.Okulik@zut.edu.pl

J. Náprstek et al. (eds.), *Vibration Problems ICOVP 2011: The 10th International Conference on Vibration Problems*, Springer Proceedings in Physics 139,
DOI 10.1007/978-94-007-2069-5_62,

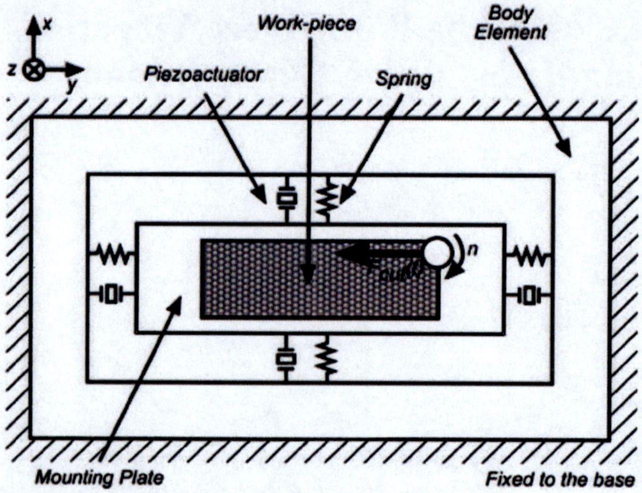

Fig. 1 Model of the work-piece mounted in the active clamp

can be shaped to elevate the stability limit. However, in the case of existing machine tools other methods are to be used. These may be classified into three groups. The first of them includes methods based on reducing the effect of the phase shift between the outer and inner modulation through relevant varying the spindle speed [1, 2]. The next group includes methods based on adaptive control of the feed rate [3, 4]. A separate group is composed of methods to control the energy flow in the MT-CP system through e.g. changing the tool geometry [5], using smart materials [6] and employing different passive and active vibration absorbers [7]. Dynamics of the mass-damping-spring (MDS) system may be shaped by introducing an additional active element. On the basis of this assumption a concept of the work-piece active clamp system is proposed. When the work-piece exhibits one dominant vibration direction the vibration absorber could be used. In some cases the vibration scenario is more complex and therefore the efficiency of suppression vibration is low. In such cases the active clamp system acting in X and Y direction could be very effective. The model of proposed system is presented in Fig. 1.

The work-piece is attached to the mounting plate which can be moved in XY plane by the piezoactuators. The cutting force $F_{\mathrm{cut}}(t)$ acting on the work-piece cause vibration, which can be suppressed by a proper control of piezoactuators.

2 Mathematic Model of the Active Clamp with Control System

To control system presented in Fig. 1 the LQG controller is used. Designing of this type control is based on the model of the MDS system with actuators. Model of the system is defined by sets of Eqs. 1–3.

$$\dot{\mathbf{x}} = \mathbf{Ax} + \mathbf{Bu}$$
$$\mathbf{y} = \mathbf{Cx} + \mathbf{Du}. \tag{1}$$

$$\mathbf{A} = \begin{bmatrix} 0 & 0 & 1 & 0 & 0 & 0 & 0 & 0 \\ 0 & 0 & 0 & 1 & 0 & 0 & 0 & 0 \\ \frac{-K_1}{m_p} & 0 & \frac{-C_1}{m_p} & 0 & \frac{K_{ox}}{m_p} & 0 & \frac{C_{ox}}{m_p} & 0 \\ 0 & \frac{-K_2}{m_p} & 0 & \frac{-C_2}{m_p} & 0 & \frac{K_{oy}}{m_p} & 0 & \frac{C_{oy}}{m_p} \\ 0 & 0 & 0 & 0 & 0 & 0 & 1 & 0 \\ 0 & 0 & 0 & 0 & 0 & 0 & 0 & 1 \\ \frac{K_{ox}}{m_o} & 0 & \frac{C_{ox}}{m_o} & 0 & \frac{-K_{ox}}{m_o} & 0 & \frac{-C_{ox}}{m_o} & 0 \\ 0 & \frac{K_{oy}}{m_o} & 0 & \frac{C_{oy}}{m_o} & 0 & \frac{-K_{oy}}{m_o} & 0 & \frac{-C_{oy}}{m_o} \end{bmatrix} \tag{2}$$

$$\mathbf{B} = \begin{bmatrix} 0 & 0 & 0 & 0 \\ 0 & 0 & 0 & 0 \\ 0 & 0 & \frac{K_a l_p d_{33}}{m_i} & 0 \\ 0 & 0 & 0 & \frac{K_a l_p d_{33}}{m_i} \\ 0 & 0 & 0 & 0 \\ 0 & 0 & 0 & 0 \\ \frac{1}{m_o} & 0 & 0 & 0 \\ 0 & \frac{1}{m_o} & 0 & 0 \end{bmatrix} \quad \mathbf{C} = \begin{bmatrix} 0 & 0 \\ 0 & 0 \\ 0 & 0 \\ 0 & 0 \\ 1 & 0 \\ 0 & 1 \\ 0 & 0 \\ 0 & 0 \end{bmatrix}^T \quad \mathbf{D} = [\mathbf{0}] \tag{3}$$

$$\mathbf{x} = \left[x_{px}\ x_{py}\ \dot{x}_{px}\ \dot{x}_{py}\ x_{ox}\ x_{oy}\ \dot{x}_{ox}\ \dot{x}_{oy} \right]^T,$$

$$K_1 = K_a + K_p + K_{ox},\ K_2 = K_a + K_p + K_{oy},\ C_1 = C_p + C_{ox},\ C_2 = C_p + C_{oy}$$

Where subscript p – denotes moving plate, o – work-piece object, x, y – moving direction, a – actuator, m, K, C – mass, stiffness and damping coefficients, l_p – number of elements in piezo stack, d_{33} – piezo coefficient. The inputs are piezo voltage in both direction $\mathbf{u}_1 = [U_x\ U_y]$ and cutting force $\mathbf{u}_2 = [F_{cx}\ F_{cy}]$ considered as a disturbance.

In this model there is no coupling effect between axis x and y. Therefore LQG controller for all axis was designed separately in similar way to [8]. However in manufactured system this assumption could be not fulfilled. Moreover, the properties of the active clamp system (e.g. stiffness, damping etc.) in x direction could be different than in y direction. In Fig. 2 the frequency response function (FRF) between piezo voltage and work-piece displacement is presented with and without LQG controller in x and y direction.

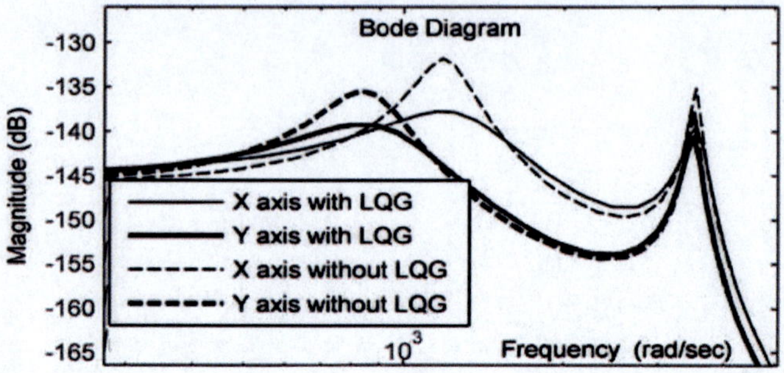

Fig. 2 FRF of the work-piece mounted in active clamp with and without LQG control. The work-piece stiffness $k_{ox} = 2k_{oy}$

Damping increase in the vicinity of natural frequencies with active LQG controller can be observed in Fig. 2. In results the vibration amplitude is lower than in the case, when the work-piece is mounted without active clamp.

3 Cutting Force Model

In order to assess efficiency of vibration suppression a set of milling simulation for different machining condition was performed. In the simulation the mechanistic model of cutting force was employed. Equations describing the cutting forces acting on single insert are presented in a linear form (4) [9, 10]. The location of component forces and geometrical parameters which are included in this model are shown on Fig. 3a.

$$\begin{aligned} F_t(\varphi) &= K_{tc} A(\varphi) + K_{te} b \\ F_r(\varphi) &= K_{rc} A(\varphi) + K_{re} b \\ F_a(\varphi) &= K_{ac} A(\varphi) + K_{ae} b \end{aligned} \tag{4}$$

Where K_{tc}, K_{rc}, K_{ac} – factors of specific resistance for a linear model, respectively in the tangential, radial and axial directions, K_{te}, K_{re}, K_{ae} – coefficients are constants of the cutting tool edge [9], $A(\varphi)$ – the area of cut, b – width of the cut.

The experimental identification of the coefficients was conducted on the basis of recorded cutting forces during milling with DIN 845-B K-N HSS milling shank cutter (Fig. 3b). The experiment was performed on a CNC milling machine, DMU-60 Monoblock (Fig. 3c). The cutting forces were measured using Kistler dynamometer. The work-piece was a rectangular block of '45' steel. The dynamometer and the work-piece were mounted on the machine table. Cutting forces were recorded using a d-SPACE system. Estimation of cutting force model coefficients was performed using method described in [11, 12].

The comparison of simulation using the parameters of a mechanistic linear model with experimental results are presented in Fig. 4.

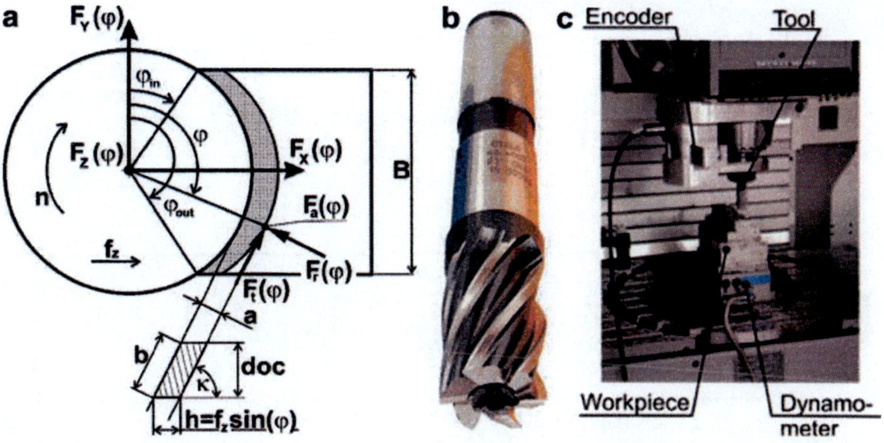

Fig. 3 (**a**) Cutting forces in end milling, (**b**) Milling shank cutter DIN 845-B K-N HSS, (**c**) Photography of a workstation for the measurement of cutting forces on the milling machine DMU 60 monoblock

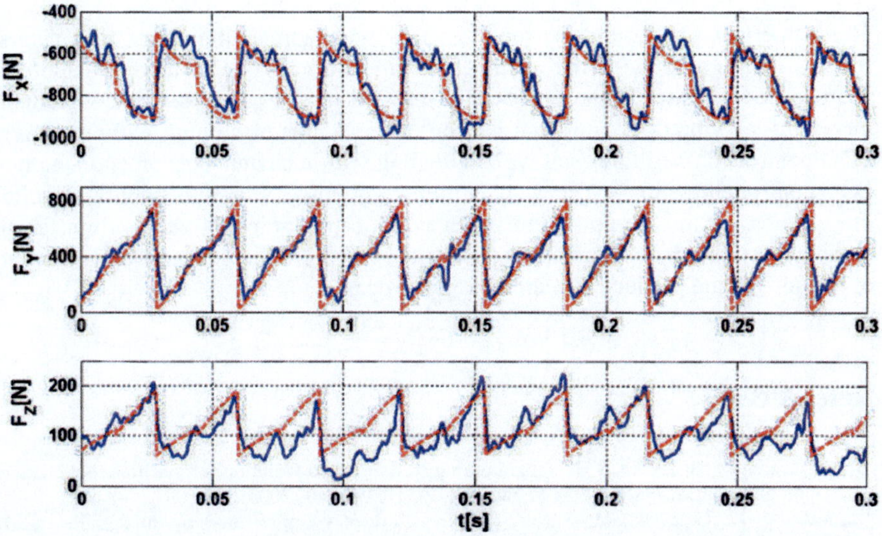

Fig. 4 Actual cutting force components (F_X, F_Y, F_Z) (*solid line*) and forces synthesized using estimated coefficients (*dashed line*) for material constants $[K_{tc}, K_{te}, K_{rc}, K_{re}, K_{ac}, K_{ae}] = [1800, 40.5, 816, 25, 396, 3.7]$, (machining parameters: $doc = 3$ mm, $n = 256$ rev./min., $f_z = 0.12$ mm/tooth, $v = 20$ m/min., $B = 12.5$ mm, $z = 6$)

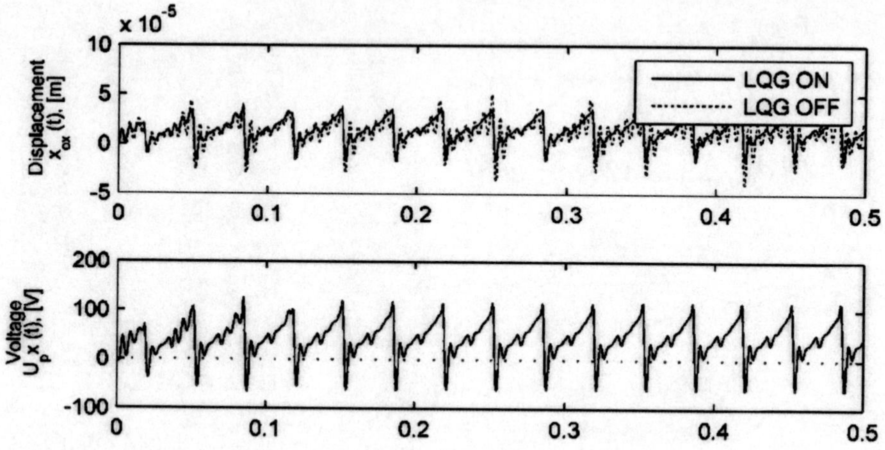

Fig. 5 Work-piece displacement and piezo control signal for x axis n = 300 rpm ap = 2.3 mm

4 Conclusions

The numerical simulations performed for a wide range of milling parameters show that usage of the active clamp system is an effective method to suppress work-piece vibration. In the Fig. 5 only selected results are presented. A gradual increasing of vibration amplitude can be observed for machining without active LQG control (dashed line) and we are dealing with a chatter phenomenon. active LQG control allows to perform stable milling with lower vibration amplitude (solid line). Moreover the absolute limit width of cut (ap) significantly grows up. From obtained results can be assumed that using active workholder leads to improvement of the quality and productivity and increase process.

References

1. Al-Regib, E., Ni, J., Lee, S.H.: Programming spindle speed variation for machine tool chatter suppression. Int. J. Mach. Tools Manuf. **43**(12), 1229–1240 (2003)
2. Kaliński, K., Galewski, M., Sawiak, S.: Reduction of vibration level by the spindle speed optimization during ball end milling of flexible details. In: Fifth International Conference on High Speed Machining, vol. L, Metz, France, pp. 479–488, 14–16 Mar 2006
3. Alter, D.M., Tsao, T.C.: Stability in turning process with actively controlled linear motor feed drives. ASME J. Eng. Ind. **116**, 298–307 (1994)
4. Shiraishi, M., Yamanaka, K., Fujita, H.: Optimal control of chatter in turning. Int. J. Mach. Tools Manuf. **31**(1), 31–43 (1991)
5. Lee, W.Y., Kim, K.W., Sin, H.C.: Design and analysis of a milling cutter with the improved dynamic characteristics. Int. J. Mach. Tools Manuf. **42**, 961–967 (2002)
6. Wang, M., Fei, R.: Chatter suppression based on nonlinear vibration characteristic of electrorheological fluids. Int. J. Mach. Tools Manuf. **39**, 1925–1934 (1999)

7. Miguélez, M.H., Rubio, L., Loya, J.A., Fernández-Sáez, J.: Improvement of chatter stability in boring operations with passive vibration absorbers. Int. J. Mech. Sci. **52**(10), 1376–1384 (2010)
8. Parus, A., Marchelek, K., Domek, S., Hoffmann, M.: Vibrostability improvement in milling process using piezoelectric eliminator. (In Polish: Podniesienie wibrostabilności w procesie skrawania z zastosowaniem eliminatora piezoelektrycznego). Modelowanie inżynierskie **39**(8), 159–170 (2010)
9. Altintas, Y.: Manufacturing Automation. Cambridge University Press, Cambridge (2000)
10. Ehmann, K., Kapoor, S., Devor, R., Lazoglu, I.: Machining process modeling: a review. ASME J. Manuf. Sci. Eng. **119**, 655–663 (1997)
11. Hoffmann, M., Powałka, B., Berczyński, S., Pajor, M.: Identification of cutting forces in frequency domain for milling. Adv. Manuf. Sci. Technol. **34**(1) pp. 5–20 (2010)
12. Jayaram, S., Kapoor, S.G., Devor, R.E.: Estimation of the specific cutting pressures for mechanistic cutting force models. Int. J. Mach. Tool Manuf. **41**, 265–281 (2001)

Modelling of Piezoactuator Edge Delamination in Active Beam Systems

Marek Pietrzakowski

Abstract The objective of the study is to develop the modelling of piezoactuator edge delamination and the analysis of effects of the progressive damage process on the active beam dynamic behaviour. Delamination is described as a significant reduction of the bonding interlayer shear stiffness. It is assumed that the damaged region extends uniformly across the actuator from its ends to the centre. The beam is divided into sections for which the governing equations are formulated separately. The steady-state solution is obtained taking into account boundary and continuity conditions at the borders of the beam and actuator sections. The influence of the length and equivalent coupling stiffness of the damaged region on the dynamic characteristics and the control effectiveness is numerically investigated.

Keywords Vibration control • Piezoactuator • Delamination • Bonding interlayer

1 Introduction

Techniques based on piezoelectric control show great advantages for thin walled structures to improve their operational behaviour and ability to reduce unwanted vibration. Relatively large deformations of piezoelectric actuator patches, generated during the vibration control, create severe interfacial shear stresses. Alternating in time loading and also environmental conditions initiate geometrical and material degradation, which may progress and finally lead to a failure of the control system. The geometrical degradation is mainly introduced as delamination, which refers especially to adhesive interlayer damage. The adhesive degradation is characterized

M. Pietrzakowski (✉)
Warsaw University of Technology, Warsaw, Poland
e-mail: mpi@simr.pw.edu.pl

J. Náprstek et al. (eds.), *Vibration Problems ICOVP 2011: The 10th International Conference on Vibration Problems*, Springer Proceedings in Physics 139,
DOI 10.1007/978-94-007-2069-5_63,

by a local concentration of micro-cracks, which leads to reduction of the glue shear stiffness and in consequence decreases significantly the coupling performance of piezoactuators and their operational effectiveness. In the field of damage detection well established are techniques based on vibration responses whose a literature review is given by [1].

This paper develops the model of piezoactuator edge delamination, considered previously as a gap between interfaces by [2], including now a bonding interlayer whose shear stiffness is modified depending on the delamination process phase.

2 Formulation of the Problem and Analysis

The system considered herein is a simply supported beam with piezoelectric patches bonded to its both the upper and lower sides and operating as a collocated sensor/actuator pair. The beam is modelled according to the Bernoulli-Euler theory supposing viscoelastic material properties approximated by the Kelvin-Voigt relation. The closed loop control with velocity feedback is applied to reduce transverse vibration excited by a time-dependant point force $F(t)$.

For analysis the beam is divided into five sections due to the acting force cross-section (x_1), the location of piezoelectric transducers (x_2, x_4) and the actuator delamination length (x_3) (Fig. 1).

The delamination process is considered as a local shear stiffness reduction of a massless viscoelastic bonding layer between the piezoceramic actuator and the host structure. It is assumed that the damaged section with the constant shear stiffness of the adhesive layer extends uniformly across the actuator to its centre.

In the case of a relatively thin piezopolymer sensor the perfect bonding assumption is reasonable.

Taking into account the actuator extension, to which the inertial forces also contribute, and the shear stresses transmitted by the bonding layer, the motion of

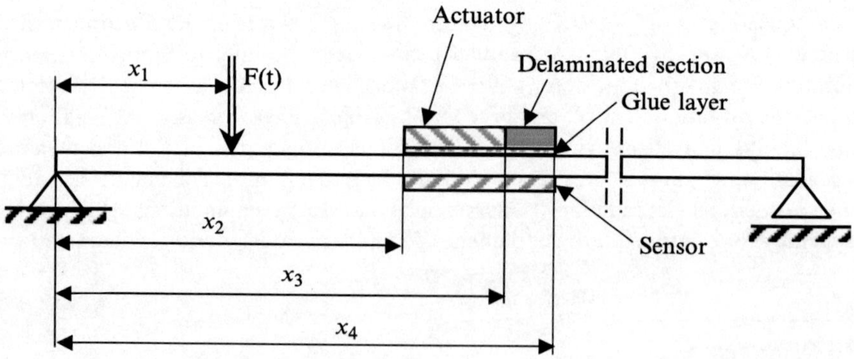

Fig. 1 Model of the beam with partially delaminated piezoactuator

both the undamaged and damaged activated sections is described by two coupled equations. They can be expressed in terms of actuator strains ε_a and beam surface strains ε_b in the following form

$$E_a t_a \frac{\partial^2 \varepsilon_a}{\partial x^2} - \frac{G_k}{t_g}(\varepsilon_a - \varepsilon_b) - \rho_a t_a \frac{\partial^2 \varepsilon_a}{\partial t^2} = 0 \quad x \in (x_2, x_3)$$

and

$$\frac{\tilde{E} t_b^2}{12} \frac{\partial^4 \varepsilon_b}{\partial x^4} + \frac{t_b}{4} \frac{G_k}{t_g} \left(\frac{\partial^2 \varepsilon_a}{\partial x^2} - \frac{\partial^2 \varepsilon_b}{\partial x^2} \right) + \tilde{\rho} \frac{\partial^2 \varepsilon_b}{\partial t^2} = 0 \quad x \in (x_3, x_4) \tag{1}$$

where: t_a, t_b and t_g – actuator, beam and bonding layer thickness, respectively, $\tilde{\rho}$ – equivalent mass density of the activated beam section, ρ_a – actuator mass density, E_a – Young's modulus of the actuator material, $\tilde{E}$ – equivalent Young's modulus of the activated beam section, G_k – Kirchhoff's modulus referred to the undamaged ($k = u$) or damaged ($k = d$) actuator section, respectively.

The motion of other beam sections is described by the classical equation

$$\frac{E_b t_b^2}{12} \frac{\partial^4 \varepsilon_b}{\partial x^4} + \rho_b \frac{\partial^2 \varepsilon_b}{\partial t^2} = 0 \tag{2}$$

where: E_b, ρ_b – Young's modulus and mass density of the beam, respectively.

Supposing a viscoelastic material of the beam and the bonding layer, Young's moduli E_b, $\tilde{E}$ and Kirchhoff's moduli G_u, G_d are complex.

The details of the governing equations formulation one can find in [2, 3].

The governing Eqs. 1 and 2 have to satisfy boundary conditions at the beam ends at $x = 0$ and $x = l$ for a simply-supported beam, continuity of beam deflection, slope, curvature and transverse force at the borders of the sections at $x = x_1$, x_2, x_3, x_4, free edge conditions at the actuator ends at $x = x_2$, x_4 and continuity of the actuator longitudinal displacements at the border between its undamaged and damaged sections at $x = x_3$. The requirements related to the actuator, according to the stress-strain relation and the displacement-strain relation, yield

$$\varepsilon_a\left(x_2^+\right) = \varepsilon_a\left(x_4^-\right) = \varepsilon_p \quad \text{and} \quad \left.\frac{\partial \varepsilon_a}{\partial x}\right|_{x=x_3^-} = \left.\frac{\partial \varepsilon_a}{\partial x}\right|_{x=x_3^+} \tag{3}$$

The strain $\varepsilon_p(t)$ is induced by the external voltage $V(t)$ applied to the unconstrained transversally polarised actuator and is given by the equation $\varepsilon_p = d_{31} V / t_a$ where d_{31} indicates the piezoelectric constant.

The steady-state responses of the active system discussed are harmonic single frequency functions with an angular velocity of excitation ω and can be written in the general form as

$$\begin{bmatrix} \varepsilon_a(x,t) \\ \varepsilon_b(x,t) \end{bmatrix} = \begin{bmatrix} \varepsilon_a(x) \\ \varepsilon_b(x) \end{bmatrix} \exp(i\omega t) \tag{4}$$

The spatial functions $\varepsilon_a(x)$ and $\varepsilon_b(x)$, formulating using the modal superposition, have the form dependent on the section of the beam or the actuator, which is determined by the boundary and continuity conditions. In order to solve the boundary-value problem for a viscoelastic system the complex moduli are used for material properties description.

3 Results

Numerical calculations are performed for the beam of dimensions 380 × 40 × 2 mm loaded by the harmonic force $F(t)$ of amplitude equal to unity and acting at $x_1 = 75$ mm. The sensor/actuator pair is located between $x_2 = 76$ mm and $x_4 = 114$ mm with its centre placed on the fourth mode line. The thickness of the PZT (lead-zirconate-titanate) actuator is $t_a = 0.2$ mm. The PVDF (polyvinylidene fluoride) sensor is of thickness $t_s = 0.04$ mm. The material properties of the beam and piezoelectric transducers are listed in Table 1.

The actuator bonding layer within the undamaged section is of the shear stiffness parameter $G_u/t_g = 5 \times 10^{11}$ N/m^3. The beam and glue layer material dumping of retardation time $\mu_b = 10^{-7}$ s and $\mu_g = 5 \times 10^{-5}$ s, respectively, is applied to limit the resonant amplitudes.

Figures 2 and 3 show the beam deflection and shear stress distribution along the actuator, respectively, which are induced by almost a static voltage ($\omega = 0.1$ s^{-1}) of the amplitude $V_0 = 100$ V applied to the actuator. The diagrams demonstrate effects of the bonding layer stiffness degradation comparing with the performance of the undamaged actuator. A constant length of the damaged section given by the relative length parameter $\delta = 30\%$ is assumed.

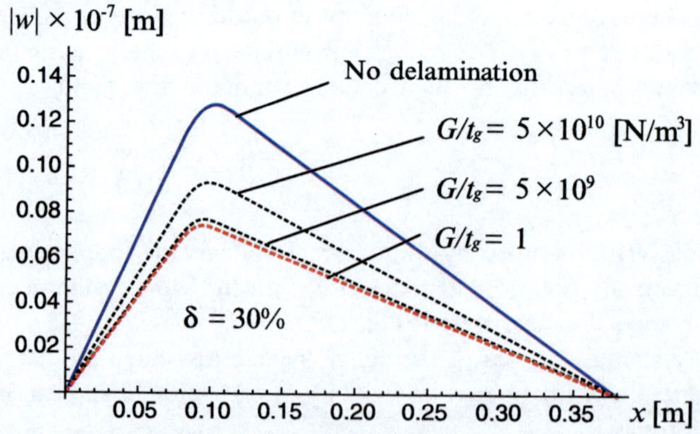

Fig. 2 Influence of the bonding layer stiffness parameter G/t_g within the damaged actuator section on the beam deflection for a quasistatic voltage loading ($\omega = 0.1$ s^{-1})

Table 1 Material properties

Material parameter	Beam	Actuator (PZT)	Sensor (PVDF)
Mass density ρ [kg/m^3]	7800	7280	1780
Young's modulus E [N/m^2]	2.16×10^{11}	6.3×10^{11}	2.0×10^9
Piezoelectric constant d_{31} [m/V]	–	1.9×10^{-10}	3.3×10^{-11}
Dielectric constant $\in_{33}$ [F/m]	–	–	1.06×10^{-10}

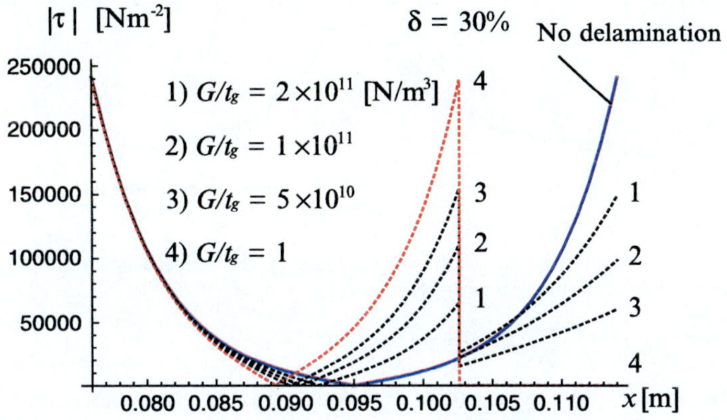

Fig. 3 Effects of variation in the bonding layer stiffness parameter G/t_g within the damaged actuator section on the shear stress distribution for a quasistatic voltage loading ($\omega = 0.1\ s^{-1}$)

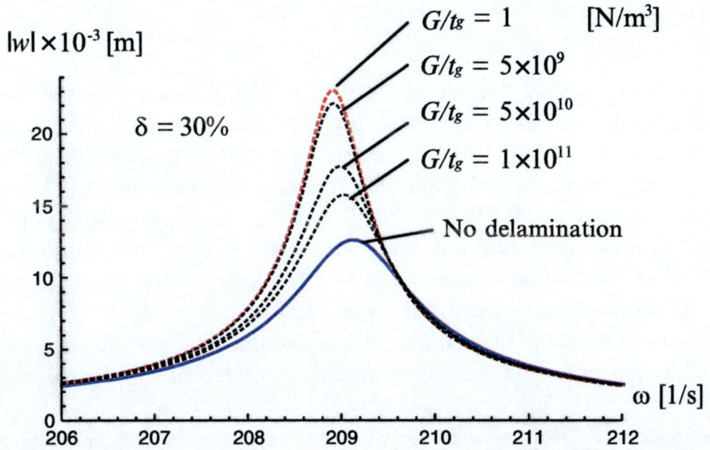

Fig. 4 Effects of variation in the bonding layer stiffness parameter G/t_g within the damaged actuator section on the active beam response in the first resonance region ($\delta = 30\%$)

As an example of vibration characteristics the frequency response functions calculated at the activated field point $x = 90$ mm are shown in Figs. 4 and 5 in the first beam resonance. The influence of the shear stiffness degradation within the 30% length of the edge delamination is shown in Fig. 4. The effect of variation in

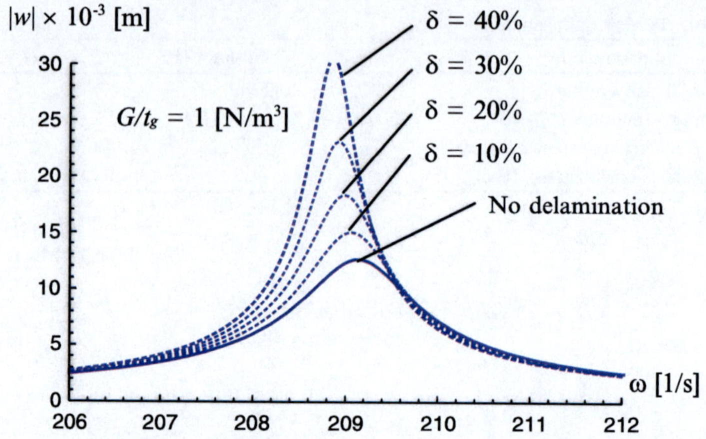

Fig. 5 Effects of variation in the delamination length on the active beam response in the first resonance region for a minimal bonding layer stiffness ($G/t_g = 1$ [N/m^3])

the relative length of delamination (δ%) assuming an extreme reduction of the glue layer stiffness ($G/t_g = 1$ N/m^3) is presented in Fig. 5.

4 Conclusions

The developed model of the actuator delamination basing on the bonding layer stiffness degradation is formulated and analysed. The numerical simulations show the influence of the delamination parameters on the transmitted shear force distribution, beam deflection and amplitude-frequency characteristics. The increased delamination length as well as the local bonding interlayer softening result in a disadvantageous modification of transmitted shear forces which diminishes the control system effectiveness, and the natural frequency shift related to the global stiffness reduction of the system is also noticed.

The proposed model of the actuator failure and the results obtained can be useful for the failure detection and evaluation of damage effects in active structures.

Acknowledgments The work reported here was supported by the State Committee for Scientific Research (KBN), grant no 0673/B/T02/2008/35

References

1. Zou, Y., Tong, L., Steven, G.P.: Vibration-based model-dependent damage (delamination) identification and health monitoring for composite structures – review. J. Sound Vibration **230**(2), 357–378 (2000)
2. Tylikowski, A.: Effects of piezoactuator delamination on the transfer functions of vibration control systems. Int. J. Solids Struct. **38**, 2189–2202 (2001)
3. Pietrzakowski, M.: Active damping of beams by Piezoelectric System: effects of bonding layer properties. Int. J. Solids Struct. **38**, 7885–7897 (2001)

Application of Proportional Velocity Feedback Control to Attenuate the Vibrations of a Flexible Plate Using Piezoceramic Patch Actuators

S. Kulah, U. Boz, and I. Basdogan

Abstract This paper presents a theoretical and experimental study on the control performance of proportional velocity feedback control with rectangular piezoceramic patch actuators to attenuate the vibrations of a thin flexible plate. For this purpose, first, frequency response funciton of the plate is obtained based on the experimental frequency sweep data. Then, a state space model was fitted to the measured frequency response to be used in the simulations to represent the plant dynamics. The controlled response of the plate is investigated via simulations using MATLAB/SIMULINK. Control performance of the controller is investigated and discussed for various feedback gains.

Keywords System identification • Active vibration control • Piezoelectric actuator • Velocity feedback • Finite element modeling

1 Introduction

In recent years, many researches showed that Active Vibration Control (AVC) can successfully reduce vibration of structures at low frequency. Due to this fact, AVC has found many application areas. One of the first applications is reduction of sound radiation and transmission by using piezo-actuators on ships' and submarines' hull [1]. Then active control of interior sound radiation at aircraft fuselages and vibration control at wing surfaces are added to the application field [2]. Piezo-actuators are imbedded directly into the structure and can defeat many point force drawbacks [1]. They are not heavy and they are easy to implement and use. Dimitriadis et al. [3] studied the excitation of 2-D structures by using piezo-actuators.

S. Kulah • U. Boz • I. Basdogan (✉)
Mechanical Engineering Department, Koç University, Istanbul, Turkey34450
e-mail: ibasdogan@ku.edu.tr

J. Náprstek et al. (eds.), *Vibration Problems ICOVP 2011: The 10th International Conference on Vibration Problems*, Springer Proceedings in Physics 139,
DOI 10.1007/978-94-007-2069-5_64,

Fig. 1 Experimental setup

However it is really important to choose control points of the plate. Control points and measurement points have vital role at controlling vibration [4]. There are many studies about selecting the control points for the piezo-actuators [5]. The main objective of this paper is to establish general guidelines concerning the use of piezoceramic patch actuators for the implementation of proportional velocity feedback control loop on a thin plate. The aim is to investigate the effect of the piezoceramic patch location and also the control performance of the proportional velocity feedback loop when the gain is altered. There are studies in the literature [6, 7], the works of Elliott et al. and Bianchi et al., which have shown that the response and sound radiation of a thin plate could be effectively reduced.

1.1 System Introduction

In this study, we demonstrate the vibration control of a flexible plate using a velocity feedback control loop. For this purpose, we first measure the velocity of the flexible plate using a Laser Doppler Vibrometer (LDV) and then multiply it with a gain and feed the signal back to the piezo-patches to change its effective damping.

The thin flexible plate ($1 \times 1 \times 0.001$ m) separates two empty cavities. This setup is mainly constructed to demonstrate some control methodologies that can be applied for automotive applications. The flexible plate is clamped in all four sides.

The flexible plate acts like a firewall that separates the engine and passenger compartments in a vehicle body. Figure 1 shows the flexible plate with the piezo actuators mounted on it. The shaker is placed at the engine side to excite the flexible plate. A force sensor (PCB 208 C02 ICP) is attached to the shaker end to obtain the excitation data needed for system identification procedure and disturbance information for future control purposes.

The piezo actuators and the velocity measurement points are selected on the passenger cavity side of the flexible plate. The vibration measurements are done using Polytech PDV – 100 Portable Digital vibrometer. Physik instrumente (PI) P-876 DuraAct piezo patches is used to as the actuators. As illustrated in Fig. 1, this piezoceramic patch transducer works as an actuator to implement a proportional feedback control loop with a velocity sensor, which detects the out-of-plane velocity of the panel. Since the aim of this paper is to investigate how the effect of the control performance of the feedback loop, the velocity sensor signal has been remotely measured with a laser vibrometer pointed to the center of the piezoceramic patch actuator, which does not introduce passive dynamic effects to the response of the panel, as would do an accelerometer for example.

1.2 Determination of Sensor Location

Two PI E-413 piezo amplifiers are used with the NetdB DAQ 12 data acquisition system. To determine optimal sensor locations on the 0–160 Hz range (the frequency range of interest for structural borne vibrations for automotive applications), a finite element model of the flexible plate is built and modal analysis is performed. Then using the modal analysis results, plate deformations that occur in the first four modes are drawn on top of each other and intersection of the modes which have the highest displacement amplitude is selected (also have the highest values for velocity) such that the sensor locations do not lie along the nodal lines of the low order natural modes of the panel.

1.3 Determination of Piezopatch Locations

To determine the optimal piezo patch actuator position, frequency response simulations are conducted in ANSYS software. The frequency response functions are obtained between all the grid points and one of the sensor locations obtained in the previous section. It was observed that if the sensor and the actuator is used at the same point then, the maximum amplitudes can be obtained for the frequency range of interest. For that reason, we decided to use collocated sensor and actuator locations for this study.

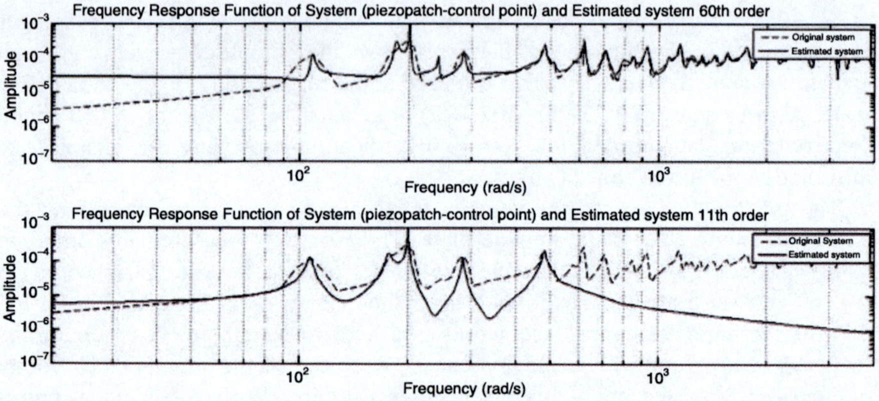

Fig. 2 Frequency Response Function and estimated models (60th and 11th order) between piezopatch (volt) and sensor location (velocity)

2 System Identification

The LDV is used to measure the velocity and the upper piezo patch is used to actuate the panel during the experiments. The obtained data files are imported into MATLAB workspace as in separate complex and real parts and then this data is used in the System identification toolbox GUI of MATLAB [8, 9] to generate the transfer functions between the piezopatch and the velocity sensor.

In the System identification toolbox, initially, a filter between 5 and 150 Hz range is applied and a linear parametric state space model is estimated from the frequency response data. For estimation N4SID [10–12] method is applied. Initially a 60th order model was fit to cover the whole frequency range. However, in order to obtain a reduced model, order estimation is set between 5 and 50 Hz range such that the first three modes of the flexible panel are targeted for the vibration control. The identification procedure is repeated and then a low order 10th order model is estimated. FRF results for both measured and estimated models (10th and 60th order) are shown in Fig. 2.

3 Controller Design and Simulation Results

In this study, a velocity feedback control is applied in order to change the effective damping of the first three plate modes. A SIMULINK® model of the flexible plate and the feedback loop are developed to demonstrate the effect of the feedback loop on the vibrations of the flexible plate (See Fig. 3). After obtaining the state space model of the plate as described in the earlier sections, the velocity gain, G in the SIMULINK® model was altered to test the significance of the velocity gain

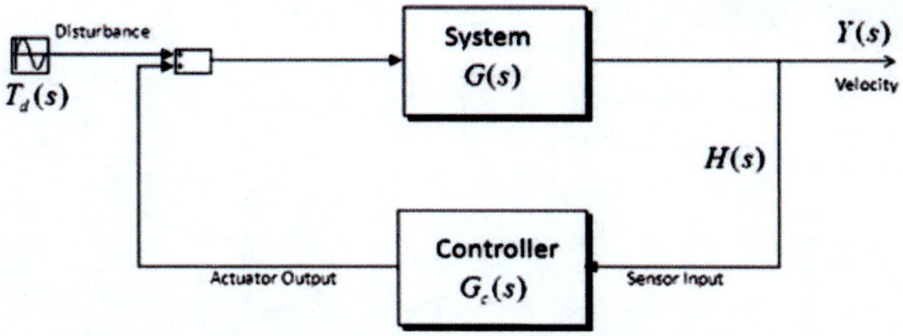

Fig. 3 Block diagram of control system

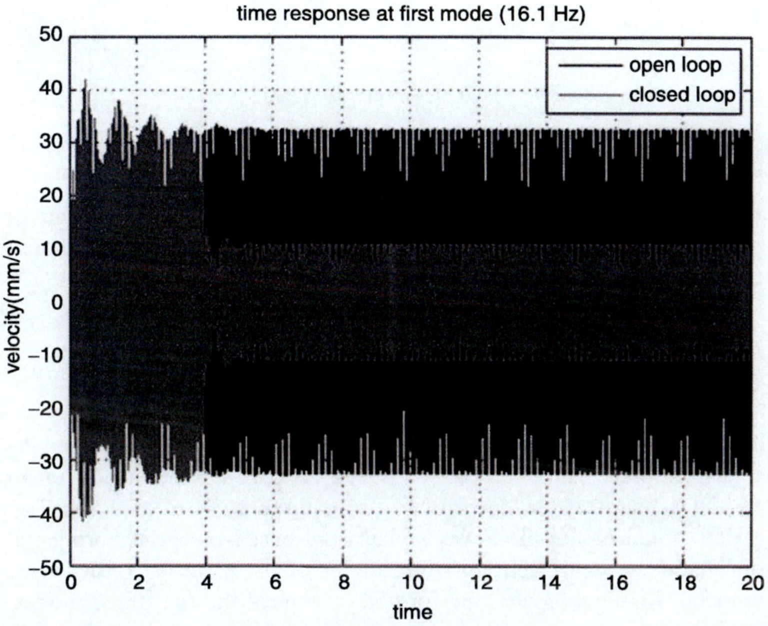

Fig. 4 Time response comparison of the flexible plate

on the vibration characteristics of the flexible plate. The measurement noise was assumed to be negligible. The system is excited by a sinusoidal signal then a sine sweep was performed to see the effect of the velocity gain in the frequency range of interest. The time response of the system is given at 16.1 Hz (first resonance mode of the plate) in Fig. 4. The open loop and closed loop time response of the system is compared and the effect of the control feedback is clearly observed after 4 s. The amplitude of the measured velocity reduces significantly when the piezo patches are actuated. The frequency response of the system is given in Fig. 5 for

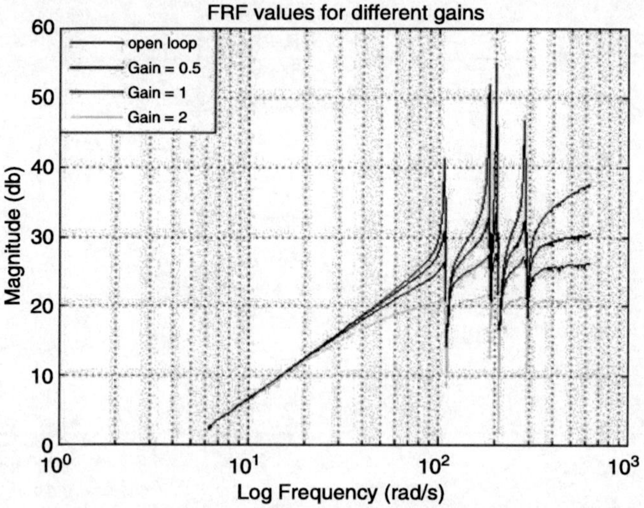

Fig. 5 Frequency response function

different velocity gains. Both figures show that the velocity feedback control can be effectively applied to such systems to control the vibrations. However, these results must be validated with experimental studies in the future.

4 Conclusions

We altered the effective damping of the first three modes to attenuate the vibrations of the flexible plate using velocity feedback control. Optimum positions of the sensors and actuators were determined using finite element analysis. Then, the state space model of the plate was constructed based on the experimental data. We considered the first three vibration modes of the plate to construct the state space model. Simulations were performed to control the first three modes using velocity feedback. Our study shows that all of the modes can be attenuated using the same velocity feedback gain. However, these results should be also validated using experimental studies.

Acknowledgement This work is supported by Scientific and Technological Research Council of Turkey (TUBITAK).

References

1. Lee, J.C., Chen, J.C.: Active control of sound radiation from rectangular plates using multiple piezoelectric actuators. Appl. Acoustics **57**(4), 327–343 (1999)
2. Fuller, C.R., Hansen, C.H.: Active control of interior nose in model aircraft fuselages using piezoelectric actuators. In: Aeroacoustics Conference, AIAA, Tallahassee, pp. 22–24 (1990)

3. Fuller, C.R., Rogers, C.A., Dimitriadis, E.K.: Piezoelectric actuators for distributed vibration excitation. J. Sound Vib., 100–108 (1991)
4. Fahy, F., Gardonio, P.: Sound and Structural Vibration: Radiation, Transmission and Response, 2nd edn. Academic, Oxford (2007)
5. Wang, B.T., Fuller, C.R., Dimitriadis, E.K.: Active control of noise transmission through rectangular plates using multiple piezoelectric or point force actuators. J. Acoust. Soc. Am **90**(5), 2820–2830 (1991)
6. Elliott, S.J., Gardonio, P., Sors, T.C., Brennan, M.J.: Active vibroacoustic control with multiple local feedback loops. J. Acoust. Soc. Am. **111**, 908–915 (2002)
7. Bianchi, E., Gardonio, P., Elliott, S.J.: Smart panel with multiple decentralized units for the control of sound transmission. Part III: control system implementation. J. Sound Vib. **274**, 215–232 (2004)
8. Ljung, L., Zhang, Q., Lindskog, P., Juditsky A.: System Identification Toolbox 7 For Use with MATLAB (2010)
9. Ljung, L.: System Identification – Theory for the User. Prentice Hall, Englewood Cliffs (1987)
10. Favoreel, W., De Moor, B., Van Overschee, P.: Subspace state space system identification for industrial processes. J. Process Control **10**, 149–155 (2000)
11. De Moor, B., Van Overschee, P.: Subspace algorithms for the stochastic identification problem. In: Proceedings of the 30th IEEE Conference on Decision and Control, Brighton, pp. 1321–1326 (1991)
12. De Moor, B., Van Overschee, P.: Subspace Identification for Linear Systems, Theory, Implementation, Applications. Kluwer Academic Publishers, Boston/Dordrecht (1996)

Residual Vibrations in the Drives of Working Links of Electronic Cam Mechanisms

Miroslav Václavík and Petr Jirásko

Abstract In the drives of working links of the mechanisms of processing machines, electronic cams are currently being promoted due to their effective change of motion working functions which represent the kinematic excitation of a system. Electronic cams have their dynamic limitations and provide a compliant electromagnetic constraint between the stator and the rotor into the kinematic chain of a mechanism. This constraint in influenced by the electronic drive control system. The paper generally deals with the dynamic analysis of the drives of electronic cam working motions which describes a system composed of a mechanical part and an electronic part. Thus, the analysis describes the generation of disturbing vibrations and suggests some methods to minimize them.

Keywords Electronic cam • Displacement law • Motion function • Residual vibrations • Residual spectrum

1 Introduction

In the introduction, there are to be defined such terms as *displacement law*, *motion function* and *electronic cam*, which are frequently used in this paper. A function assigning a position quantity of the definite link of a composed cam mechanism to time, we designate it as *motion function* of that link [1, 2]. The motion function of electronic cam (theoretical, actual) is servomotor shaft motion

M. Václavík (✉)
TUL Liberec, VÚTS, a.s., Liberec, Czech Republic
e-mail: miroslav.vaclavik@vuts.cz

P. Jirásko
VÚTS, a.s., Liberec, Czech Republic
e-mail: petr.jirasko@vuts.cz

J. Náprstek et al. (eds.), *Vibration Problems ICOVP 2011: The 10th International Conference on Vibration Problems*, Springer Proceedings in Physics 139,
DOI 10.1007/978-94-007-2069-5_65, © Springer Science+Business Media B.V. 2011

in dependence on time and that motion function kinematically excites the compliant dynamic system of composed mechanism on whose end there is working link [3]. We designate a function assigning a position quantity of another link of composed cam mechanism to the position of the definite link as *displacement law.* The displacement law of electronic cam is a theoretical function of the servomotor shaft position on the virtual shaft position (Master) or virtual rotation. The displacement law transformed by a constant (0. derivative) and the angular velocity of the virtual shaft (1st and 2nd derivatives) is the theoretical motion function.

Electronic cam is a *drive* (*synchronous servomotor* fed by a frequency inverter-*servo inverter* and controlled by a *controller*) which realizes exciting motion function on the output shaft (servomotor rotor). The controller can be programmed in the *PLC* via development environment in the area of contiguous motions. The electronic cam is programmed in the *PLC.* In each passage (*scan*) of the *PLC* programme, *position, velocity and servomotor moment size* are defined by means of output registers. Most of electronic cam producers use a cascade regulation structure of servo inverters, consisting of a position regulator (proportional usually), a speed controller (proportional plus integral) and a moment controller (or current controller, proportional plus integral usually). With some producers (such as Siemens, producers of single-purpose electronics) it is possible to modify the structure of controllers and to use it for special applications (adaptive regulation of dynamic processes) in such a way. From the viewpoint of the engineering applications of mechanism working links, it is meant by the term of *electronic cam* such a use of servomotor (as a servo force link) which is *alternative* to the drives which are possible combinations of articulated (joint) cam mechanisms driven by conventional asynchronous motors.

Residual vibrations are disturbing (spurious) and there are the deviations of the actual motion function of a mechanism working link from its theoretical course in cam resting zones. Due to compliances in the working link drive, undesirable deviations from the theoretical motion function occur not only in rest areas. Thus, all those positional deviations have a periodical character of vibration motions in both displacement areas and the rest areas of displacement laws or motion functions. For the important group of technical applications, the resting dwells of motion functions are essential in which processing technologies are carried out. Thus, it concerns various handling movements performed by turntables and other mechanisms as the case may be. Residual vibrations are generally caused by the compliance of mechanism links and in the case of electronic cams by the compliance of the electromagnetic constraint between the stator and rotor of the electronic cam servomotor as well.

In this paper, sources of those vibrations are divided in the mechanical and the electronic part of the kinematic chain of a working mechanism. Briefly, their source and possible ways of their minimization are depicted. We will describe those parasite vibrations in technically significant non-periodical displacement laws with resting dwells which are also designated as *step* ones.

Not only electronic cams but also serial and parallel combinations with conventional cam articulated (joint) mechanisms are used in the drives of the working links of processing mechanisms. Furthermore, one group will be mentioned only.

2 Residual Vibrations of the Links of Electronic Cam Mechanisms

In principal, electronic cams have two compliances. It is a flexible electromagnetic constraint stator-rotor and a flexible output driven mechanical part, with one natural frequency in the simplest case as to Fig. 1. The features of the electromagnetic constraint are influenced by the cascade regulation structure of the inverter, in the case of Yaskawa's hardware that we use. In Fig. 1, there are two drive variants with/without output mechanical compliance and with/without a speed-reducing gear box behind the servomotor (Spinea reducer).

The depiction of the solution is, as in the case of a conventional cam mechanism, carried out in References [4, 5], including the measurement results on the stand. In Fig. 2, there is mentioned only the result of the numerical solution of the discrete model as to Fig. 1 with parabolic displacement law and a natural frequency of approx. 15 [Hz] in steps of 50 [min^{-1}].

It is worth noting the nature of vibrations on the servomotor output shaft and the nature of the vibrations of working inertia mass. In Fig. 2, there is reducer i = 33 between the servomotor shaft and the rotational working mass. The shape of the deformed (acceleration and positional error) motion function (due to the compliant electromagnetic constraint stator-rotor) of the working inertia mass, which is positioned directly on the servomotor shaft, is in Fig. 3 for P modes (proportional) and PI (proportional plus integral) of speed and moment regulators (controllers). The movement of the working inertia mass as to Fig. 2 and the size of residual vibration amplitudes are not significantly influenced by P/PI modes of speed and moment controllers. The figures are the results of the numerical solution of the discrete model and agree with the response which is acquired from the servomotor encoder or the external sensor as to Fig. 1. Due to the restricted size of the paper, the measurement results are not mentioned, they are contained in References [2, 4].

The amplitudes of disturbing residual vibrations at the end of displacement law, which are caused by the compliance of the stator-rotor constraint, are a

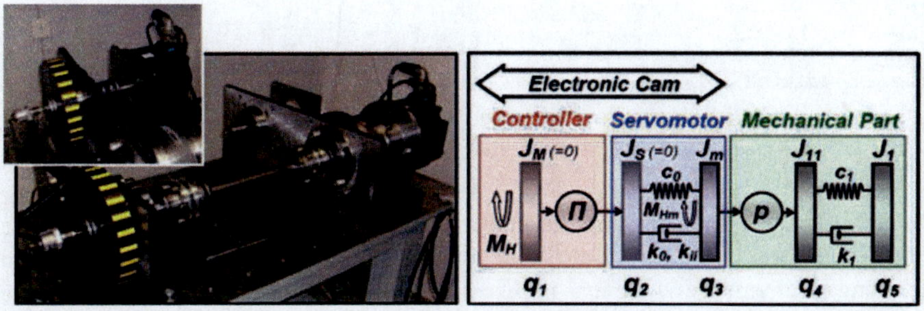

Fig. 1 An electronic cam stand and a discrete computation model

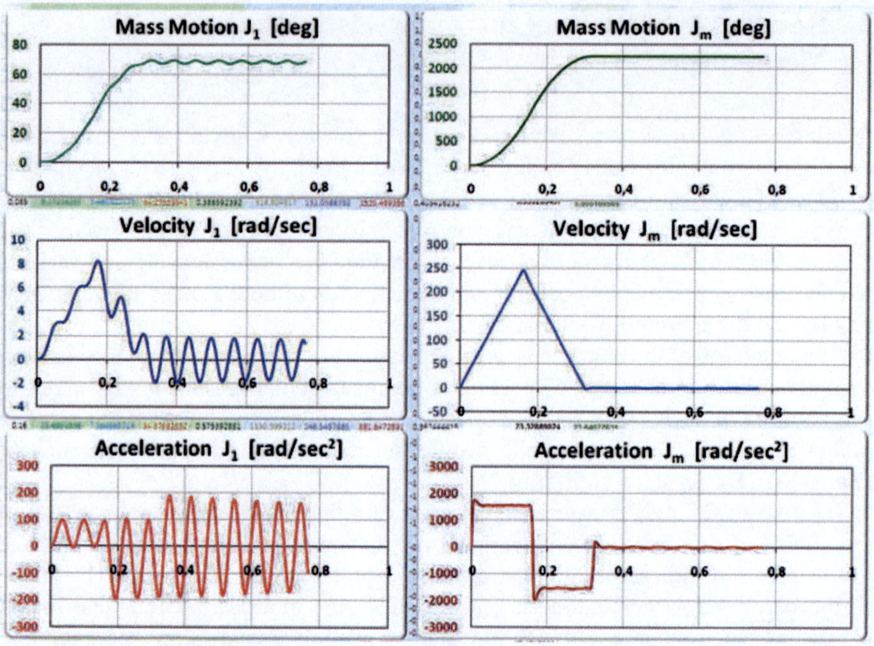

Fig. 2 A solution of the discrete model as to Fig. 1

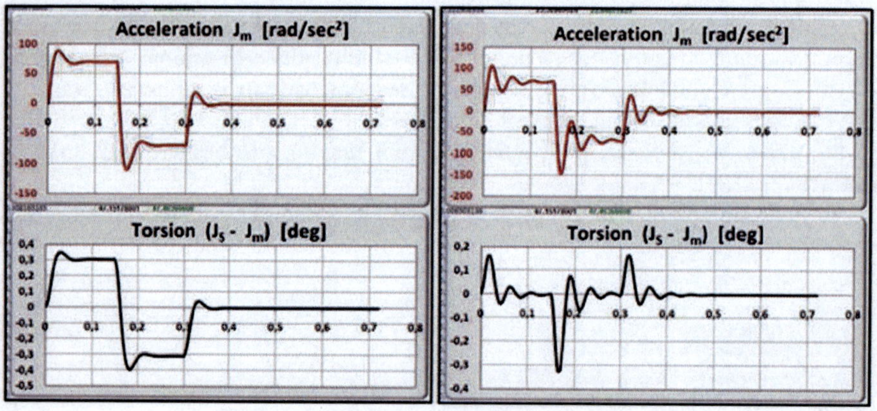

Fig. 3 P and PI regulations (parabolic motion function with resting dwell)

function of the parameters of the controllers in cascade arrangement (Yaskawa HW case). Those vibrations are significant in the applications of electronic cams in handling mechanisms where it is necessary to position by a rotational output at the shortest time possible. The residual vibrations extend the time interval of handling remarkably. The end position is signaled by the appropriate bit variable (red) which

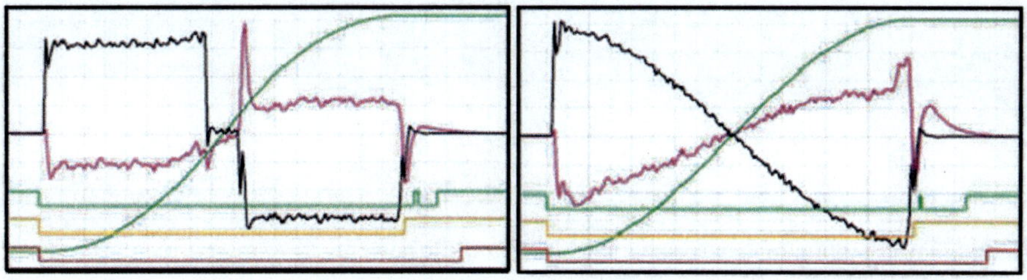

Fig. 4 Parabolic (modification as to VDI 2143) and harmonic motion functions (axes auto)

starts the following technological operations, including the usual blocking of the mechanism. In Fig. 4, as follows, positioning for two different displacement laws (*green* is displacement, *black* is moment, *violet* is positional error) is demonstrated under the same conditions.

Position mechanisms are designed with a high stiffness of the driven mechanical part; therefore, there are the properties of the electromagnetic constraint stator-rotor dominant here. One of the important applications is a spindle drum turning of automatic lathes.

3 Residual Vibrations of the Links of Combined Mechanisms

It is the principle of mechatronic differential drive [2, 6] we have on mind with combined mechanisms. The resulting working motion consists of two inputs whereas one input is the displacement law of electronic cam. In the case of step motions with resting dwells, the resting part of the motion function is a superposition of both inputs whereas the input motion function of electronic cam has its positional deviation. The minimization of this positional error or disturbing vibrations is essential in the differential drive.

4 Methods of Minimizing the Parasite Residual Vibrations

The methods of minimizing the residual vibrations caused by the compliance in the driven mechanical part of the electronic cam are principally divided in the *superposition* of displacement law with a partial harmonic function by the *choice* of suitable *kinetostatic parameters*, methods of *feedback control* and *signal modification* and methods based on *feed forward control*.

The method of *superposition* is a reliable method when a correction harmonic impulse [3, 4] is superimposed to the basic exciting displacement law. This method is illustrated in Fig. 5 schematically. Superposition is programmed in the PLC and this impulse is shifted and scaled in phases by the input parameters of the PLC user function.

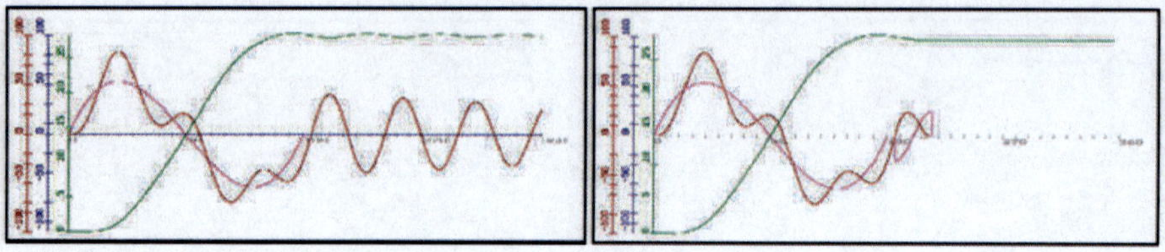

Fig. 5 Superposition of displacement law with harmonic pulse

Methods based on the choice of suitable *kinetostatic quantities* go out from the shapes of the residual spectra of displacement laws [1, 4]. As a choice of kinetostatic quantities it is meant the choice of displacement law as well. The method when the movement of a dynamical system is determined unambiguously by the values of the impulse frequency spectrum in the natural frequencies of the system after completion of an exciting impulse is mentioned in References [7, 8].

Methods based on *feedback control* are completed by a module for identifying the residual vibrations and a correction signal generator which is fed to the input of a speed or current regulator (controller) in a conventional cascade regulation structure [9].

In the methods based on *signal modification* in the direct branch of the regulation structure, the control signal is modified in such a way that residual vibration can be suppressed (input shaping technique [10]).

Some methods of *forward regulation* (input shaping as well) are based on the principle of inverse dynamics. This method is currently being investigated.

5 Conclusions

The residual disturbing vibrations, their description and minimization are a topic issue of the dynamics of electronic cam applications. Currently, we are working on the methodical processing of those issues concerning dividing the investigating methods and evaluating their usability in the application fields as to the definition from Chap. 1.

References

1. Koloc, Z., Václavík, M.: Cam Mechanisms. Elsevier, Amsterdam (1993)
2. Jirásko, P., Bušek, M.: Position accuracy of motion functions of electronic cams. In: X. International Conference IFToMM, Liberec, 2–4 Sept 2008
3. Václavík, M., Jirásko, P.: Research and application of displacement laws of electronic cams. In: 12th IFToMM World Congress, Besançon, France, 18–21 June 2007
4. Jirásko, P.: Methodology of electronic cam applications in drives of working links of mechanisms of processing machines. Dissertation TU Liberec CZ, 2010

5. Jirásko, P., Václavík M.: Residual spectra of displacement of conventional and electronic cams. In: National Conference Engineering Mechanics 2009, Svratka, 11–14 May 2009
6. Václavík, M., Jirásko, P.: Mechatronic drive variants of processing machine working links. In: X. International Conference IFToMM, Liberec, 2–4 Sept 2008
7. Souček, P., Bubák, A.: Vybrané statě z kmitání v pohonech výrobních strojů, ČVUT 2008, ISBN 978-80-01-04048-5
8. Výzkumná zpráva *V10-067*: Pokročilé metody zpětnovazebního řízení pohonů, kapitola Souček P.: Redukce reziduálního kmitání při rozběhu dynamických systemů, 2010, VCSVTT při ČVUT Praha
9. Lindr, D., Rydlo, P.: Two-mass model based vibration suppression feedback control method applied to standard servo control system. ElectroScope [online]. 2010, 2010, I, [cit. 2010-11-15]. Dostupný z WWW: http://electroscope.zcu.cz. ISSN 1802–4564
10. Beneš, P., Valášek, M.: Application of input shaping control to drives of machine tools. In: Proceedings Interaction and Feedbacks 2006, vol. 1, Ústav termomechaniky AV ČR, Praha, pp. 5–12. ISBN 80-87012-02-X (2006)

Synthesis of Multiple Multiloop Digital Controller for Vibroprotection Device with Parallel Structure

L. Rybak, Aleksey Chichvarin, and J. Šklíba

Abstract In this paper presented vibroprotection device with 6 degree of freedom based on Stewart platform. An actuation mechanism with parallel structure consists of pulse drivers and screw gears. Mathematical model of vibroprotection device is represented as equation of state in vector-matrix form. Digital controller designed as multiloop taking into account coordinated work of six actuation mechanisms with feedback control. Feedback is optimized with the use of solving the discrete algebraic Riccati equation. Mathematical modeling results are presented.

Keywords Mechatronics • Riccati equation • Vibroprotection • Optimal control • Stewart platform

1 Introduction

We consider the use of mechatronics system (see Fig. 1a) as an active vibration isolation system designed for the protection of technological objects, as well as a human operator, from low-frequency vibration of the foundation. Disks I (platform) and II (foundation) are connected to one another through six absolutely rigid legs with spherical joints at their ends.

L. Rybak (✉)
Belgorod State Technological University named after V.G. Shoukhov, Belgorod, Russia
e-mail: rl_bgtu@intbel.ru

A. Chichvarin
Stary Oskol Technological Institute, National University of Science and Technology "MISIS", Stary Oskol, Russia
e-mail: ch_alexx79@rambler.ru

J. Šklíba
Technical University of Liberec, Liberec, Czech Republic
e-mail: jan.skliba@tul.cz

J. Náprstek et al. (eds.), *Vibration Problems ICOVP 2011: The 10th International Conference on Vibration Problems*, Springer Proceedings in Physics 139,
DOI 10.1007/978-94-007-2069-5_66,

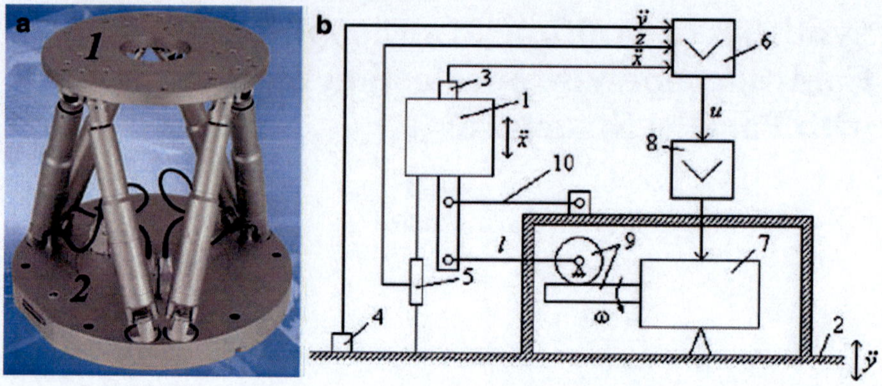

Fig. 1 Parallel-structure mechatronics system (**a**) and scheme of driving mechanism (**b**)

Each leg is an electromechanical driving mechanism shown in Fig. 1b. This mechanism has the following structure. The foundation 2 transmits disturbing vibrations to the object 1. The actuator used to the reducing vibration influence on the object is controlled by regulator 6 and located between the object and foundation.

The actuator consists of the parallelogram mechanism 10, warm gear 9, and electric motor 7. The warm wheel is rigidly connected to the pair of bottom levers of the parallelogram mechanism. Signals from the sensors (accelerometers 3 and 4 on the object and foundation and the relative displacement sensor 5) are processed in the regulator. The output signal of the regulator controls the power amplifier 8, which, in turn, controls the winding of the electric motor. The other five driving mechanisms of the spatial mechatronic system are implemented in the same way.

2 One-Dimensional Optimal Regulator Synthesis

Describe one-dimensional system in state space. Introducing three state variables $x_1 = \dot{x}$, $x_2 = z$, and $x_3 = I_a$, where $z = x - y$, we write state equations of the system in matrix form as

$$\dot{\mathbf{X}} = \mathbf{AX} + \mathbf{B}u + \mathbf{GY} \tag{1}$$

Equation 1 completely describe behavior of the one-dimensional vibroprotection system. Particular models of the electric motor differ by the matrices **A**, **B**, and **G**. The regulator synthesis problem will be solved in a discrete domain. To this end, Eq. 1 are transformed to the discrete form.

The structure of the optimal digital regulator is expressed as a matrix relation $u[i] = -\mathbf{FX}[i]$, where $\mathbf{F} = \begin{bmatrix} f_1 & f_2 & f_3 \end{bmatrix}$ is the matrix of feedback coefficients with respect to the state variables. In fact, the optimal regulator is based of feedback with respect to the state of the system. Synthesis of the optimal regulator is based on feedback with respect to the state of the system. Synthesis of the optimal regulator begins with formulation of the control-variable $\mathbf{Z} = \begin{bmatrix} z_1 \\ z_2 \end{bmatrix} = \begin{bmatrix} x_1 \\ x_2 \end{bmatrix} = \mathbf{DX}$, where $\mathbf{D} = \begin{bmatrix} 1\,0\,0 \\ 0\,1\,0 \end{bmatrix}$ is the matrix of coupling state and control variables.

The general expression adopted for the quality criterion is an integral of quadratic form of the control variables and the control signal, of the form

$$J = \int_0^\infty \left(q_1 z_1^2(t) + q_2 z_2^2(t) + r u^2(t)\right) dt \to \min$$

where $\mathbf{Q}$ is the weighting-factor matrix for the control variables, r is the weighting factors for the square of the control signal and are selected experimentally.

Thus, the control problem may be stated as follows: to minimize the amplitude of the object's velocity (and hence the amplitude of its acceleration) and limit the displacement of the object relative to the base, thereby restricting the scope for control. The matrix of feedback coefficients if $\mathbf{F} = \left(r + \mathbf{B}_\Delta^{\mathrm{T}}\mathbf{P}\mathbf{B}_\Delta\right)^{-1}\mathbf{B}_\Delta^{\mathrm{T}}\mathbf{P}\mathbf{A}_\Delta$, where $\mathbf{P}$ is 3×3 quadratic positive-defined matrix and satisfies the discrete Riccati equation [1]

$$\mathbf{P} = \mathbf{D}^{\mathrm{T}}\mathbf{Q}\mathbf{D} + \mathbf{A}_\Delta^{\mathrm{T}}\mathbf{P}\mathbf{A}_\Delta - \mathbf{A}_\Delta^{\mathrm{T}}\mathbf{P}\mathbf{B}_\Delta\left(r + \mathbf{B}_\Delta^{\mathrm{T}}\mathbf{P}\mathbf{B}_\Delta\right)^{-1}\mathbf{B}_\Delta^{\mathrm{T}}\mathbf{P}\mathbf{A}_\Delta \quad (2)$$

Equation 2 has a solution if the pair $(\mathbf{A}, \mathbf{B})$ is completely controllable, which is the case here.

Solving Eq. 2 in accordance with the recommendations of [2], we obtain the control law of the optimal regulator in the form of feedback with respect to state of the system $u\,[i] = -f_1\dot{x}\,[i] - f_2 z\,[i] - f_3 I\,[i]$

The structure of a vibroprotection system with an optimal regulator is shown in Fig. 2.

For given matrix $\mathbf{F}$, system operation with a perturbing signal ($f = 1\,\mathrm{Hz}$) of various type is modeled, as shown in Fig. 3: (1) acceleration at the base; (2) acceleration at the object; (3) displacement of the object relative to the base. The amplitude of the perturbing signal $A_{\ddot{y}} = g$. In that case, the coefficient of amplitude suppression of the perturbing signal is $A_{\ddot{x}}\,/A_{\ddot{y}} = 0{,}08$.

3 Multiple-Dimensional Optimal Regulator Synthesis

Consider the design of the spatial system depicted in Fig. 4. When one of the discs is loaded by an external force and another is fixed, there appear reaction forces in the rods, which form, generally, a power-transmission screw.

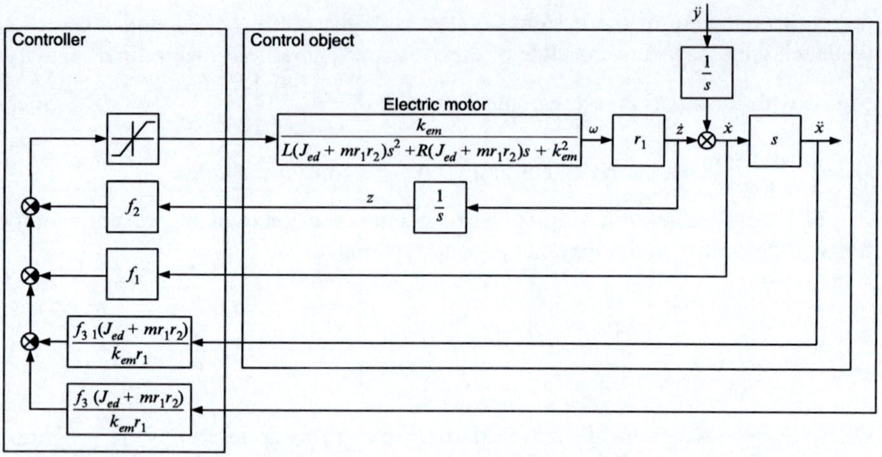

Fig. 2 Structure of one-dimensional vibroprotection system

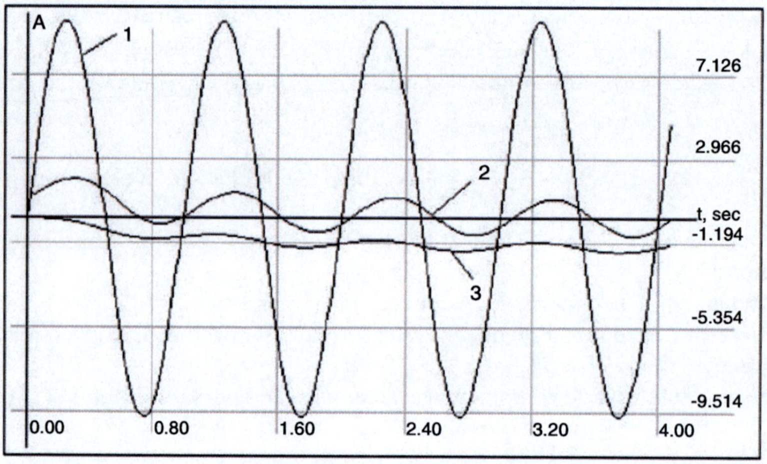

Fig. 3 System operation with a perturbing signal ($f = 1$ Hz) of various type

The interaction of the rigid body and foundation along the leg axes is determined by the Plucker coordinates of these axes [3]. Let α_i, β_i, and γ_i be angles of the leg axis unit vector with axes of some rectangular body frame XYZ (Fig. 4).

Moments of the unit vector with respect to these axes are given by $l_i = \eta_i \cos\gamma_i - \zeta_i \cos\beta_i$, $m_i = \zeta_i \cos\alpha_i - \xi_i \cos\gamma_i$, and $n_i = \xi_i \cos\beta_i - \eta_i \cos\alpha_i$, where ξ_i, η_i, and ζ_i are coordinates of the points where the rods are attached to the rigid body (centers of the spherical joints). Displacement of the rigid body (object) in the space is related to displacements along the six rods by the matrix of the Plucker coordinates $\mathbf{A}_\Gamma$.

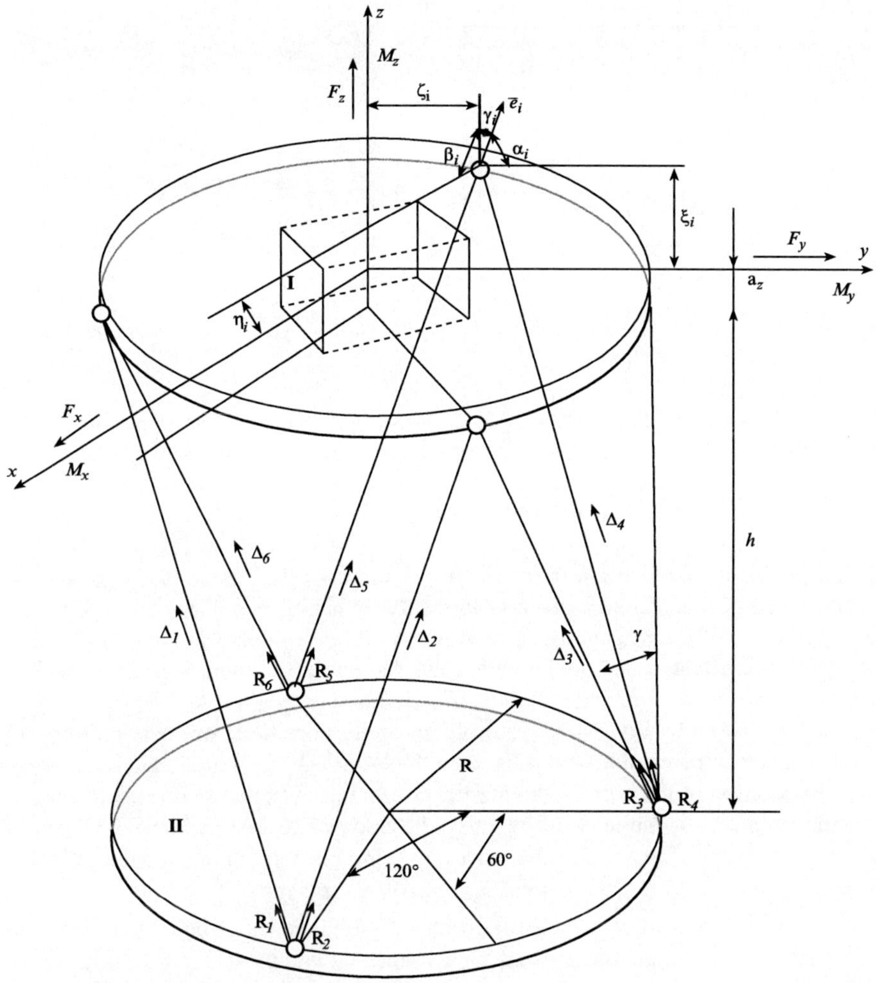

Fig. 4 Design diagram of spatial mechatronic system

Small displacement of the rigid body I relative to foundation II is given by the column vector $\mathbf{X} = \left[x \ y \ z \ \varphi \ \psi \ \chi \right]^{\mathrm{T}}$ where x, y, and z are displacements along axes X, Y, and Z, respectively, and φ, ψ, and χ are rotation angles about these axes. The column vector of speeds is defined similarly as $\dot{\mathbf{X}} = \left[\dot{x} \ \dot{y} \ \dot{z} \ \dot{\varphi} \ \dot{\psi} \ \dot{\chi} \right]^{\mathrm{T}}$.

In the framework of the theory of small probable displacements, we can write $\Delta = \mathbf{A}_\Gamma \mathbf{X}$, $\dot{\Delta} = \mathbf{A}_\Gamma \dot{\mathbf{X}}$ where Δ and $\dot{\Delta}$ are column vectors of relative displacements and speeds along the leg axes. If the rigid body is subjected to forces and moments given by the vector $\mathbf{F} = \left[F_x \ F_y \ F_z \ M_x \ M_y \ M_z \right]^{\mathrm{T}}$, where F_x, F_y, and F_z are forces along axes X, Y, and Z and M_x, M_y, and M_z are moments with respect to these axes,

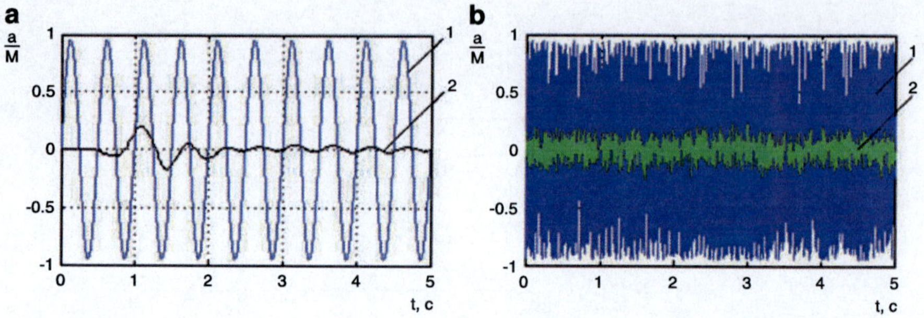

Fig. 5 Reaction of the system to harmonic signal of frequency 1 Hz (**a**) and to white noise (**b**)

we have $\mathbf{F} = \mathbf{A}_\Gamma \mathbf{R}$, where $\mathbf{R} = [R_1, R_2, R_3, R_4, R_5, R_6]^T$ are reactions of the rods subject to the forces F. Solving this equations in Δ, $\dot{\Delta}$, and $\mathbf{R}$, we obtain

$$\mathbf{X} = (\mathbf{A}_\Gamma)^{-1}\Delta;\ \dot{\mathbf{X}} = (\mathbf{A}_\Gamma)^{-1}\dot{\Delta};\ \mathbf{R} = (\mathbf{A}_\Gamma)^{-1}\mathbf{F} \tag{3}$$

From (3), it follows that, if matrix $\mathbf{A}_\Gamma$ is singular, then there is no one-to-one relationship between the displacements of the attachment points of the rods and displacements of the rigid body in the frame *XYZ*, since the inverse matrix does not exist. Displacements $\mathbf{X}$ are not bounded. The same refers to the forces and moments applied to the rigid body and reactions of the rods.

To solve the vibration isolation problem under consideration, it is required to control displacements of body I relative to the immovable foundation in such a way that the position of the body in the inertial coordinate system remains unchanged.

The reactions of the legs in the case of a body of mass m and radii of inertia ρ_x, ρ_y, and ρ_z are found from the equation (which is an analogue of the equation $m\ddot{x} = R_1$) $\left[\ddot{x}\ \ddot{y}\ \ddot{z}\ \ddot{\varphi}\ \ddot{\psi}\ \ddot{\chi}\right]^{\mathrm{T}} = \mathbf{M}^{-1} A_\Gamma \left[R_1\ R_2\ R_3\ R_4\ R_5\ R_6\right]^{\mathrm{T}}$.

For the six driving mechanisms, each of which is described by three state coordinates, we obtain the system of 18 state equations in matrix form $\dot{\mathbf{X}} = \mathbf{AX} + \mathbf{B}u$.

Solving this equation we obtain the 18×6 matrix of the regulator feedback coefficients $\mathbf{K}$. Further, in MATLAB we build model of mechatronic system.

Figure 5a shows reaction of the system (curve 2) to the harmonic excitation (curve 1) in the vertical direction (along z-axis) of frequency 2 Hz. As can be seen, the system ensures efficient damping of the vibration with the coefficient equal to 0.07. Figure 5b shows reaction of the same system (curve 2) to white noise. In this case, the damping coefficient is 0.18.

4 Conclusions

Results of our studies demonstrate that mechatronic systems based on parallel structure mechanisms can efficiently be used for the isolation of objects from low frequency vibration applied to the foundation. The multidimensional system is shown to possess not only a greater damping coefficient compared to the one dimensional system but also, owing to the effect of cross-coupled links, a wider frequency range up to 20 Hz. For the one-dimensional system, the frequency range is limited to 4–8 Hz.

Acknowledgments The research work reported here was made possible by RFBR, grant No. 10-08-01144-a and Ministry of Education and Science of the Russian Federation, state contract No.471 from 04.09.2009.

References

1. Fedosov, E.A., Krasovskii, A.A., Popov, E.P.: Encyclopedia, vol. I–4, Automatic Control: Theory [in Russian]. Machinostroenie, Moscow (2000)
2. Mita, T., Hara, S., Kondo, R.: Introduction to Digital Control [Russian translation]. Mir, Moscow (1994)
3. Dimentberg, F.M.: Screw Method and its Applications in Applied Mechanics [in Russian]. Machinostroenie, Moscow (1971)

Investigation of Contact Point Trajectories of Piezoelectric Actuator with Bidirectional Polarization

Dalius Mazeika, R. Lucinskis, and R. Bansevicius

Abstract Contact point trajectories of the beam type, quadratic cross-sectional piezoelectric actuator with non-uniform polarization are analyzed in the paper. Polarization vectors of the actuator have perpendicular directions on the first and second half of the actuator. Bidirectional polarization is used to achieve flexural oscillations of the actuator in perpendicular planes and to increase the number of degree of freedom of the contact point movement. A particular electrode pattern of the actuator was found and contact point trajectories are analyzed under different excitation regimes. Prototype piezoelectric actuator was build and experimental measurements of contact point trajectories were performed. The results of numerical modeling and experimental study are compared and discussed.

Keywords Piezoelectric actuator • Bidirectional polarization • Trajectory of motion

1 Introduction

Piezoelectric actuators are widely used for high precision mechanical systems such as positioning devices, manipulating systems, control equipment and etc. [1, 2]. Piezoelectric actuators have advanced features such as high resolution, short response time, compact size, and good controllability [2]. A lot of design and operating principles of the actuators are used in different devices [1, 2].

D. Mazeika (✉) • R. Lucinskis
Vilnius Gediminas Technical University, Vilnius, Lithuania
e-mail: Dalius.Mazeika@vgtu.lt; Raimundas.Lucinskis@vgtu.lt

R. Bansevicius
Kaunas University of Technology, Kaunas, Lithuania
e-mail: Ramutis.Bansevicius@ktu.lt

J. Náprstek et al. (eds.), *Vibration Problems ICOVP 2011: The 10th International Conference on Vibration Problems*, Springer Proceedings in Physics 139,
DOI 10.1007/978-94-007-2069-5_67,

A beam type piezoelectric actuator of multi-mode oscillations is analyzed in the paper. This type of actuators usually is a single degree-of-freedom (DOF) system and has limited applicability. Mechanical systems composing several single DOF actuators are used to design multi DOF system that is each degree of freedom uses separate actuator [2]. The accuracy of the system decreases and the control scheme becomes more complex when applying such design principle. Obviously a better way is to use only one actuator that can move or rotate the body in several directions.

A new principle how to design a beam type actuator with three independent DOF is proposed. Numerical modeling of a piezoelectric actuator was carried out to evaluate operating principle by analyzing trajectories of the contact point under different excitation schemes. A prototype actuator has been made and experimental study was performed. Finally, the results from numerical simulation and experimental study were analyzed and discussed.

2 Design and Operating Principle of the Actuator

The contact point of a 3 DOF actuator must produce elliptical motions in three perpendicular planes independently to move or rotate the slider in three different directions. Flexural resonant oscillations in two perpendicular directions and longitudinal resonant oscillation of the beam are employed. Both flexural and longitudinal resonant frequencies of the beam must coincide to achieve appropriate elliptical motion of the contact point. Flexural oscillations of the beam type actuator can be obtained in two perpendicular directions at the same resonant frequency when the structure is symmetric and bending stiffness of the structure in both directions is the same.

Based on these requirements, a bulk piezoceramic beam with quadratic cross-section was chosen to build the actuator. The actuator was polarized in two perpendicular directions on the first and second half of the actuator (Fig. 1). Both vectors of polarization are perpendicular to the longitudinal axis of the beam and d31 effect is used for generating mechanical motion. Quadratic cross-section and bidirectional polarization of the beam allows to make possible to excite actuator in perpendicular planes. Contact point of the actuator was defined at the middle of top surface of the actuator (Fig. 1).

Multi-mode oscillations consisting of the 1st longitudinal and 3rd flexural oscillation modes of the beam were chosen as operating mode of the actuator. These modes define particular dimensions of the beam and can be calculated by equalizing

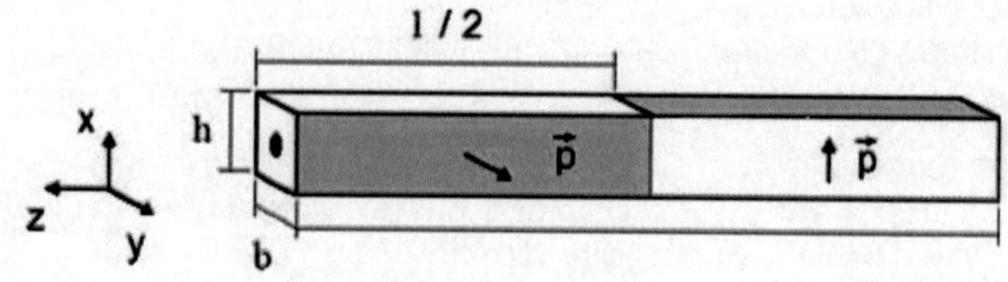

Fig. 1 Principle scheme of the actuator

equations of resonant frequencies of longitudinal and flexural modes [3]. In our case the ratio of height and length of the beam actuator is equal to 0.09.

Electrode pattern is very important for the beam type actuator because the required trajectory of the contact point is obtained by exiting particular electrode sections [4]. The following algorithm how to find the electrode pattern of the actuator was suggested i.e. to find electrode pattern for longitudinal and flexural modes separately and then to combine them by defining intersecting areas.

Electrodes pattern for separate oscillation modes can be found referring to effective work of equivalent mechanical forces of the excited actuator that must be maximized. The effective work of a piezoelectric actuator can be written as follows:

$$A_k^{ef} = \frac{\omega_k}{2\pi} \int_0^{\frac{2\pi}{\omega_k}} \sum_{i=0}^{n} \mathbf{y}_k^T \mathbf{F}_k \sin(\omega_k t)\, z_k dt; \tag{1}$$

$$\mathbf{F} = \mathbf{L}\,\mathbf{T}\,\mathbf{W} \mathrm{sgn}(\phi)\,; \tag{2}$$

where $\mathbf{F}$ is a vector of amplitudes of equivalent mechanical force in the global coordinate system, $\mathbf{y}_k$ – is the eigenvector of k mode, $\mathbf{W}$ is a vector of excitation voltage amplitude, $\mathbf{L}$ is a transformation matrix, $\mathbf{T}$ is an electroelasticity matrix, ω_k is a resonant frequency. Equation 2 shows the relation between electric and mechanical fields. By evaluating the product of the vectors, Eq. 1 could be written as follows:

$$A_k^{ef} = \frac{\omega_k}{2\pi} \int_0^{\frac{2\pi}{\omega_k}} \sum_{i=0}^{n} |\mathbf{y}_k|\; |\mathbf{F}_k| \cos\alpha_k \sin(\omega_k t)\; z_k\, dt; \tag{3}$$

It can be seen from Eq. 3, that effective work of the actuator strongly depends on the angle between the vector of the amplitude of the equivalent mechanical force and the eigenvector of particular mode. So the electrode pattern of the beam actuator can be defined by evaluating values of $\cos\alpha$.

3 Numerical Modelling and Results

Numerical simulation was carried out by employing software ANSYS 11.0. A finite element model of a piezoelectric actuator with bidirectional polarization was build. The following dimensions of a beam actuator were used $b = h = 4$ mm and $l = 44$ mm. SP6 piezoelectric ceramics was used for the modeling. No mechanical constrains were applied in the model. The required natural frequencies were found at the following frequencies: 34.56 kHz (3rd flexural mode) and 35.34 kHz (1st longitudinal mode). Based on Eq. (4), the values of $\cos\alpha$ were calculated

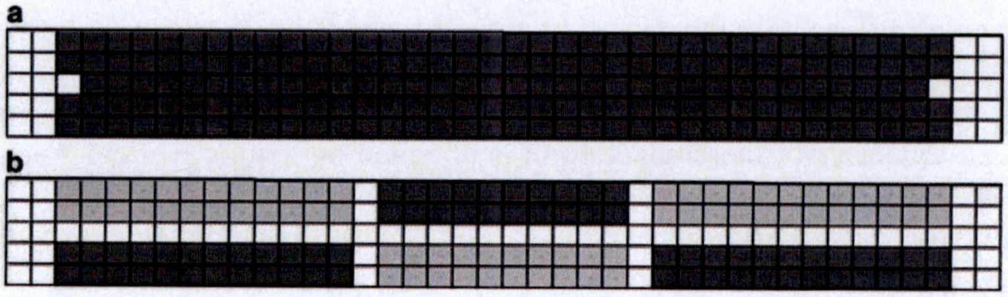

Fig. 2 The electrode pattern of the actuator for 1st longitudinal mode (**a**), and 3rd flexural mode (**b**)

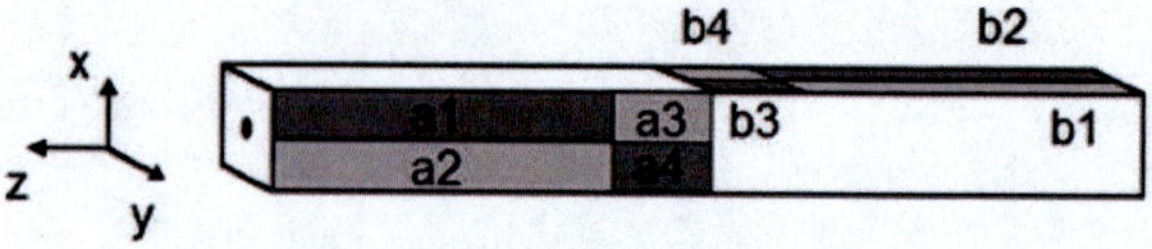

Fig. 3 The final electrode pattern of the piezoelectric actuator

for these modes. The results of calculations are given in Fig. 2, where different grey colors represent positive and negative signs and represent electrode sections for direct and reverse motion. The pattern is given just in one plane for better understanding.

The intersection of these two patterns gives the same electrode configuration as in Fig. 2b. So the final electrode pattern of the beam actuator is given in Fig. 3.

Harmonic response analysis of the actuator was performed at the frequency range from 0–40 kHz with 100 Hz frequency step. The electrodes with numbers a1–a4 were excited (Fig. 3). 100 V harmonic signal was used for excitation. Oscillations of the contact point were analyzed. Elliptical type trajectory is generating during the period of the oscillation that could be written in *xz* plane as follows:

$$\left.\begin{aligned} u_x &= a\cos\tau\ \cos\beta - b\sin\tau\ \sin\beta \\ u_z &= a\cos\tau\ \sin\beta + b\sin\tau\ \cos\beta \end{aligned}\right\} \tag{4}$$

where a and b are the length of major and minor semi-axis respectively, τ is the parameter and β is the rotation angle of the ellipse axis.

Figure 4 shows the dependence of the length of major and minor semi-axes of elliptical motion of the contact point versus frequency. The peaks can be noticed in the graphs where the first and second peaks represent the 1st and 2nd flexural modes respectively and the 3rd peak at 34.75 kHz corresponds to 1st longitudinal and 3rd flexural mode and will be used for further investigation.

Three different excitation schemes were defined: (1) oscillations in *xz* plane are achieved when electrode a1, a3 are excited; (2) oscillations in yz plane are achieved when electrode b1, b3 are excited; (3) oscillations in xyz are achieved when electrode a1, a3, b2, b4 are excited (Fig. 3). The trajectories of the contact point movement are presented in Fig. 5 when all defined schemes are applied. The

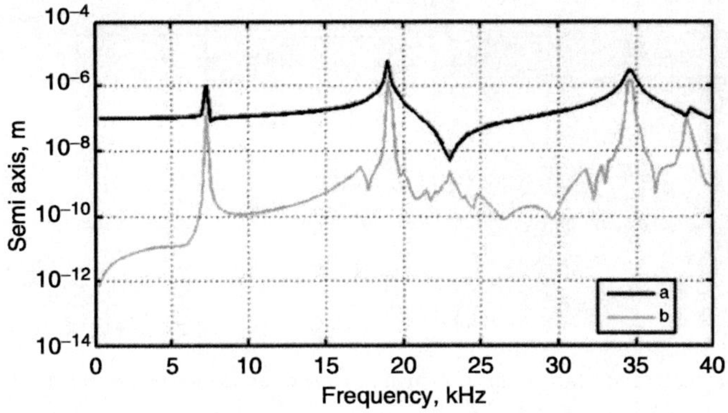

Fig. 4 Length of major and minor semi axes of elliptical trajectory versus excitation frequency

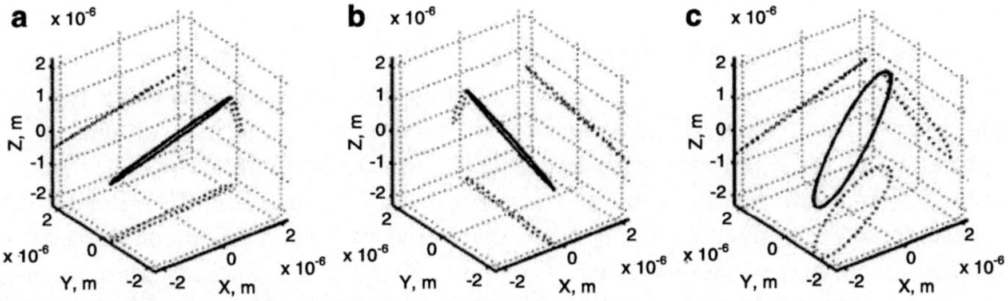

Fig. 5 Simulated trajectories of the actuator's contact point under different exciting schemes: (**a**) 1st excitation scheme, (**b**) 2nd excitation scheme, (**c**) 3rd excitation scheme

Table 1 Numerically and experimentally obtained parameter of the ellipses

	Numerical results			Experimental results		
Parameters	No. 1	No. 2	No. 3	No. 1	No. 2	No. 3
Length of major semi-axis, μm	2.3	2.2	2.7	1.4	1.2	0.5
Length of minor semi-axis, μm	0.4	0.4	1.4	0.27	0.33	0.4
Axis angle in xy plane	4°	85°	61°	6°	88°	67°
Axis angle in xz plane	34°	72°	34°	30°	84°	61°
Axis angle in yz plane	98°	146°	139°	83°	127°	107°

major axis of the elliptical trajectory of the contact point is almost parallel to x and y axes when excitation scheme No. 1 and No. 2 is used respectively. By observing these two ellipses it can be noticed that the parameter of these ellipses are very similar (Table 1). This means that quite stable and controllable motion of the slider must be achieved in xy plane. The rotational motion of the slider will be achieved based on the elliptical motion of the contact point, when the 3rd excitation scheme is used.

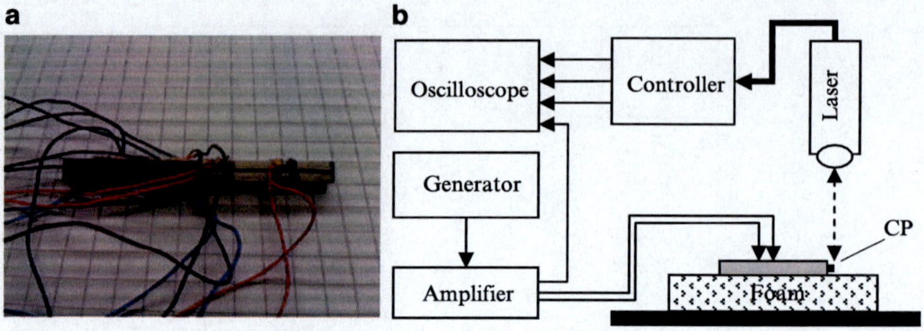

Fig. 6 Prototype actuator (**a**) and the principal scheme of experimental setup (**b**)

4 Experimental Study

A prototype piezoelectric actuator with bidirectional polarization was made and experimental investigation was performed (Fig. 6). The same electrode pattern and excitation schemes of the actuator were used as the numerical simulation. A special stand for experiments was designed (Fig. 6). Oscillations of the actuator contact point were measured by a laser vibrometer (Polytec CLV-3D). Results of measurements are given in Table 1 and compared with the corresponding results obtained from numerical simulation. It can be noticed, that elliptical motion of the contact point was achieved in three different plane. Measured angles of the ellipses axis in different planes are similar as obtained in numerical simulation but lengths of the semi-axis differ. This can be explained as damping influence of the foam that was used to put the actuator during the experimental measurements.

5 Conclusions

Numerical and experimental investigation of the piezoelectric actuator with bidirectional polarization confirm that the elliptical trajectory of contact point motion can be achieved in three directions. An algorithm for electrode pattern calculation was developed and a particular pattern for corresponding multimode oscillations was found. Three separate directions of contact point elliptical motion were obtained during numerical simulation and experimentally. It means that 3DOF motion of the slider can be achieved. Results of numerical simulation and experimental measurements are in good agreement.

Acknowledgments This work has been supported by Research Council of Lithuania, Project No. MIP-122/2010.

References

1. Kenji, U.: Piezoelectric Actuators and Ultrasonic Motors. Kluwer Academic Publishers, Boston (1997). 364 p
2. Uchino, K., Giniewicz, J.: Micromechatronics, p. 504. Marcel Dekker, New York (2003)
3. Bansevičius, R., Kulvietis, G., Mažeika, D.: Numerical identification of modal shapes sequence of multicomponent piezoelectric actuators. J. Vibroeng. vol. 4, pp. 14. Vibromechanika, Vilnius (2000)
4. Calas, H., Moreno, E., Eiras, J., Aulet, A., Figueredo, J., Leija, L.: Non-uniformly polarized piezoelectric modal transducer: fabrication method and experimental results. Smart Mater. Struct. **15**(4), 904–908 (2006)

Part VII
Dynamics of Rotating Systems

Dynamics of High-Speed Rotors Supported in Sliding Bearings

J. Šimek and R. Svoboda

Abstract The higher the operating speed, the more serious are problems with rotor stability. Three basic groups of rotors are analyzed and some methods of suppressing instability are shown. In the first group are classical elastic rotors supported in hydrodynamic bearings. Practically all high-speed rotors now run in tilting pad bearings, which are inherently stable, but in specific conditions even tiling pad bearings may not ensure rotor stability. The second group is composed of combustion engines turbocharger rotors, which are characteristic by heavy impellers at both overhung ends of elastic shaft. These rotors are in most cases supported in floating ring bearings, which bring special features to rotor behaviour. The third group of rotors with gas bearings exhibits special features.

Keywords High-speed rotors • Sliding bearings • Floating ring bearings • Gas bearings • Oil whip instability • Oil whirl • Pneumatic instability

1 Introduction

Many technical papers deal with rotor stability [e.g. 1], but most of them present results of calculation and only few of them bring case histories of stability problems. Rotors supported in fluid film bearings exhibit basically two types of instability, both of which are characterized by subharmonic vibration with big amplitudes. Instability of "oil whirl" type, with frequency dependent on rotational speed, occurs more likely with rigid rotors, paradoxically namely with rotors in gas bearings. "Oil whip" instability, encountered mainly with elastic rotors, is characterized by constant vibration frequency, usually the lowest eigenfrequency of the system. The

J. Šimek (✉) • R. Svoboda
Techlab Ltd., Prague, Czech Republic
e-mail: j.simek@techlab.cz; r.svoboda@techlab.cz

J. Náprstek et al. (eds.), *Vibration Problems ICOVP 2011: The 10th International Conference on Vibration Problems*, Springer Proceedings in Physics 139,
DOI 10.1007/978-94-007-2069-5_68,

source of instability is in most cases the bearing support itself. Cross-coupling terms of stiffness matrix, which promote journal orbit around bearing centre, prevail at some operational conditions (low load, high speed) over direct terms. Special shapes of bearing surface, consisting of several areas with preload relative to the bearing centre, increase stability. Tilting-pad journal bearings are inherently stable, because their cross-coupling terms are at least one order lower than direct terms. However, in cases of strong external excitation, e.g. from labyrinth seals, even tilting-pad bearings may not ensure stability of the rotor.

Special type of bearing – floating ring – is used in turbochargers of combustion engines. The bearing consists of two circular bearings, one of which is rotating with the speed equal to about 0.1 to 0.3 of the rotor speed. Two oil films in series ensure strong damping of rotor vibration. The bearings are relatively near one to another and heavy impellers located at both overhang ends of the rotor produce substantial gyroscopic moments influencing bending critical speeds of the rotor by splitting to branches with co-rotating and counter-rotating precession. Some rotors supported in floating ring bearings exhibit "oil whirl" instability. Special types of instability are encountered in aerostatic bearings, namely pneumatic instability called "air hammer".

2 Oil Whip Instability

As was mentioned earlier, oil whip instability is characterized by violent vibrations with constant frequency equal in most cases to the lowest eigenfrequency of the system, usually the first bending critical speed of the rotor. As an example of this type of instability we can use that of two-stage high-speed steam turbine with nominal output of 1.7 MW [2].

Impeller of this type of turbine is mounted directly on pinion of high-speed gearbox. In this particular case the pinion had operating speed about 19,000 rpm. During function tests of the turbine it appeared that it is impossible to achieve full operating speed because of excessive vibrations. This fact is demonstrated by frequency spectra shown in Fig. 1, where amplitude of subharmonic frequency around 151 Hz grew from about 1.5 μm at 16,100 rpm to more than 40 μm at 18,500 rpm, which led to shut-down by vibro-diagnostic system. Dynamic analysis of the rotor supported in tilting pad journal bearings showed, that measured subharmonic frequency is very near to the calculated first bending critical speed of the pinion. The results of calculation also proved, that stability reserve of pinion is very low and that it decreases with growing turbine output. Stability reserve (SR) at operating speed at idle run was about 24%, which corresponds to logarithmic decrement (LD) of 0.75. With turbine output increased to 1.5 MW the stability reserve fell to 0.2% (LD = 0.006), and with 1.7 MW output the calculation indicated already rotor instability. As the pinion was already supported in tilting pad bearings, it was not possible to achieve stability by use of bearings with better dynamic properties. It was necessary to find the source of destabilizing forces, which was

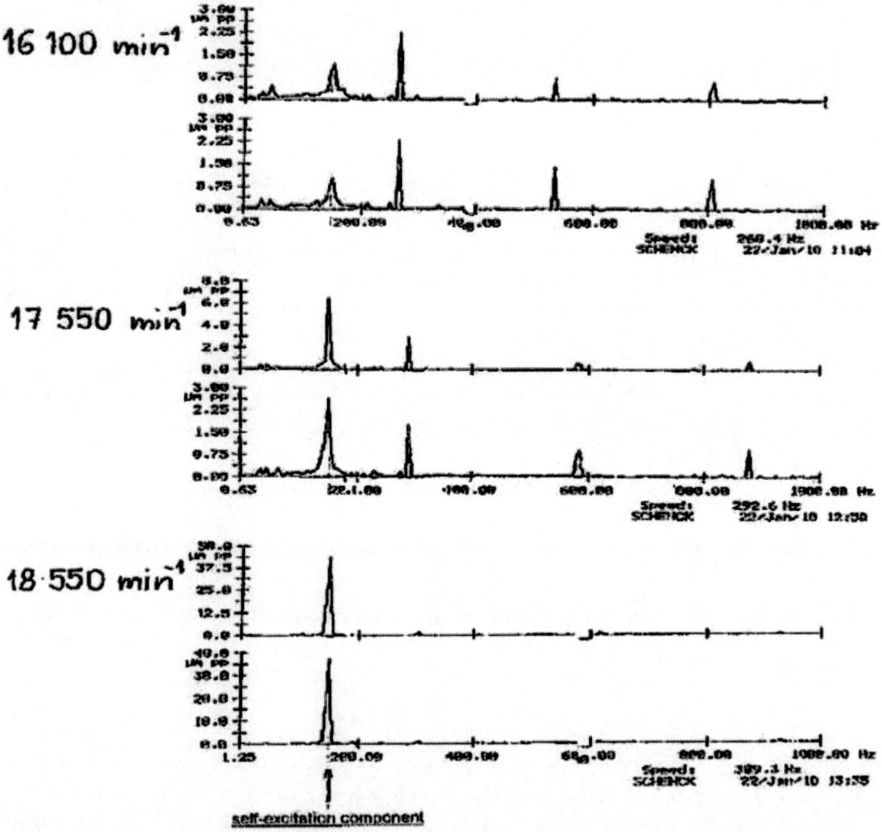

Fig. 1 Evolution of subharmonic vibration component with speed

as usual the labyrinth seal at the back of impeller. Due to great pressure drop and relatively big diameter the labyrinth seal generated great destabilizing forces, which exceeded damping capacity of bearing support. Reduction of destabilizing forces was achieved by installation of barriers between impeller backside and turbine casing, which obstructed circumferential flow in this area, so that the steam entered labyrinth seal with low circumferential velocity component. After this modification it was possible to operate the turbine up to maximum speed and nominal output.

3 Oil Whirl Instability of Rotor in Floating Ring Bearings

Typical example of turbocharger (TCH) rotor is shown in Fig. 2.

Characteristic feature of TCH rotors are heavy impellers at both overhung ends, relatively small shaft diameter and short distance between journal bearings. Though most TCH rotors run with steady small vibration amplitudes, some rotors supported

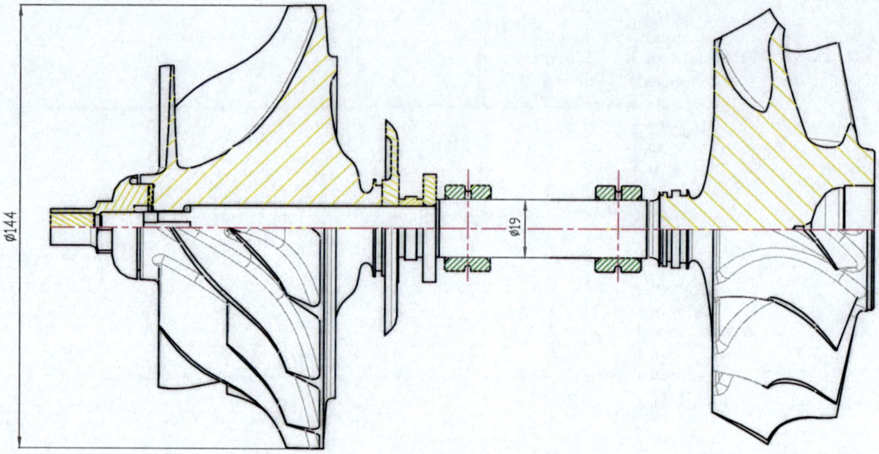

Fig. 2 Typical turbocharger rotor supported in floating ring bearings

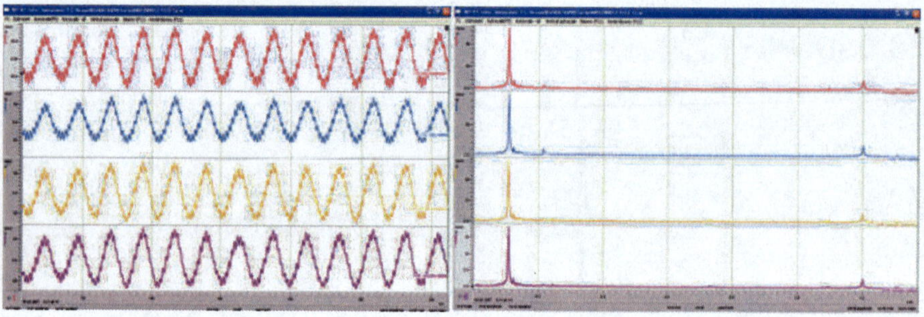

Fig. 3 Instability of outer oil film in floating ring bearing

in rotating floating bushings exhibit instability, which is manifested by subharmonic vibration with amplitudes reaching the whole bearing clearance. Instability is of "oil whirl" type, because its frequency is dependent on rotating speed.

In most documented cases the instability occurred in outer oil film, because dominating subharmonic frequency was equal to one half of bushing rotational speed. An example of instability of outer oil film is shown in Fig. 3 [3].

Top down are signals from:

rotor – compressor side
rotor – turbine side
bushing – compressor side
bushing – turbine side

Recorded signals in time domain (left) show, that both ends of the rotor vibrate in phase and rotor and bushing movement are is also in phase. Frequency spectra

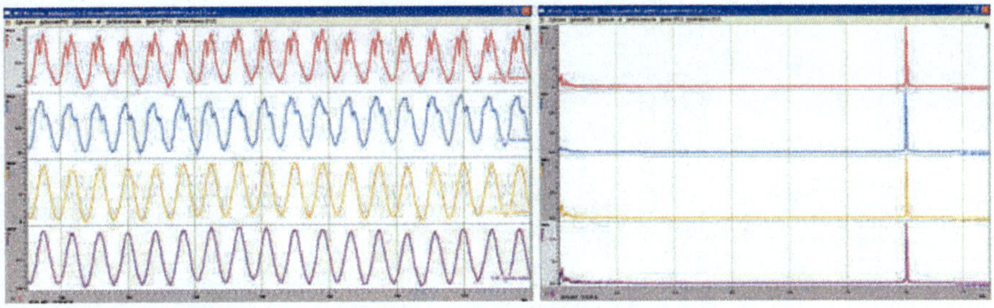

Fig. 4 Stable operation in non-rotating floating bushings

(right) indicate dominating subharmonic frequency of the bushing about 110 Hz, which is one half of bushing speed 220 Hz. Rotational frequency of the rotor 1,200 Hz (72,000 rpm) is hardly distinguishable at the spectra. Double amplitudes of subharmonic vibration of the rotor exceed 100 μm, which is practically the whole bearing clearance (sum of inner and outer diametral clearance).

TCH operation with vibration amplitudes such as in Fig. 3 is very risky and immediate failure does not occur only due to highly nonlinear properties of oil film. As permanent operation in these conditions was not permissible, rotating floating bushings were replaced by non-rotating floating bushings with three- lobbed inner geometry. As is documented by Fig. 4 (with the same signals, as in Fig. 3), this modification ensured stable operation with vibration amplitudes at least one order lower.

4 Pneumatic Instability of the Rotor in Gas Bearing

Rotor instability in gas bearings has in most cases quite fatal consequences, because gasses have no lubricating properties. In case of instability occurring in high speed, contact of sliding surfaces results in their heating to high temperature. Even with very good sliding properties of sliding surface material the contact causes some damage. The process is so rapid, that there is no chance to record its course.

As an example of rotor supported in gas bearings we use that of power gyro intended for stabilization of vibro-isolating system [4] (Fig. 5).

Flywheel two with vertical rotation axis is supported by aerostatic thrust bearing seven and two tilting pad journal bearings. Journal bearing pads three are supported on flexible elements four, which provide additional damping and enable adjustment of basic clearance by means of screws five and nuts six. Flywheel drive, provided by pressurized air, enabled to reach speeds in excess of 14,000 rpm.

Aerostatic thrust bearing exhibited at higher air inlet pressure pneumatic instability called "air hammer", because it is characterized by noise resembling pneumatic hammer. Pneumatic instability is caused by periodic compression and expansion

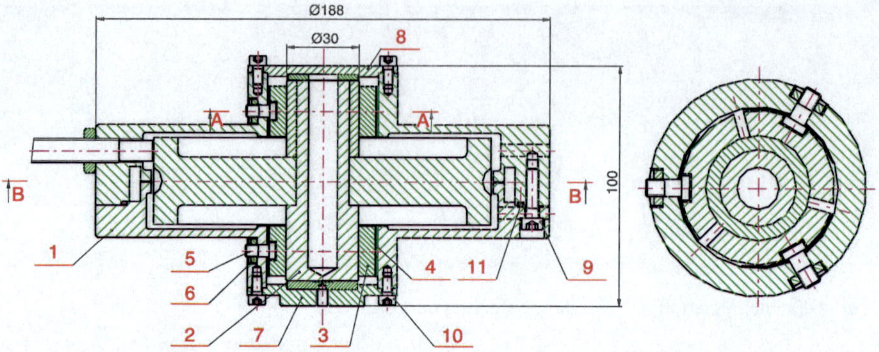

Fig. 5 Air bearing support of power gyro for stabilization of vibro-isolating system

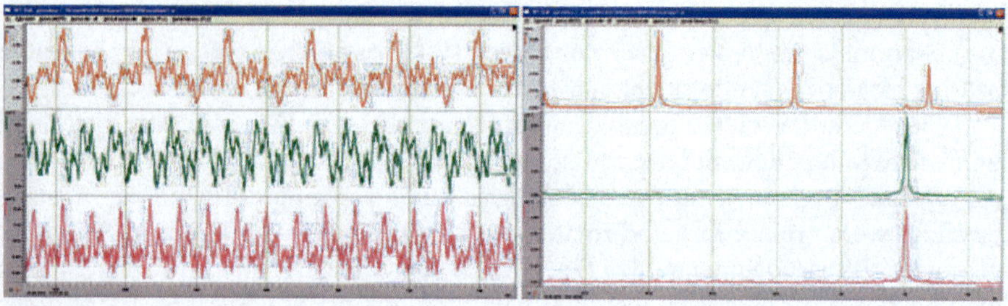

Fig. 6 Pneumatic instability of aerostatic thrust bearing

of air in a chamber, distributing pressure to greater area. Even if the chamber volume is minimized (in given example the chamber depth was only 0.1 mm) in some conditions air hammer sets in. It is illustrated by Fig. 6, which shows top down signals from relative vibration sensor tracing gyro surface, and from two accelerometers fastened to gyro frame in vertical and horizontal directions.

Frequency of pneumatic instability around 120 Hz can be seen at both vibration signals from accelerometers. With gyro mass of 3.5 kg this frequency corresponds to thrust bearing stiffness of $2{\cdot}10^6\ \mathrm{Nm}^{-1}$, which is quite close to the calculated value of $2.48{\cdot}10^6\ \mathrm{Nm}^{-1}$.

5 Conclusions

Three types of rotor bearing support were considered and different kinds of instability were described with help of particular examples. Elastic rotors suffer mostly from "oil whip" instability, while some turbocharger rotors in floating ring bearings and also rigid rotors in aerodynamic bearings exhibit "oil whirl" type of instability with frequency dependent on speed. Aerostatic bearings can suffer in certain conditions from pneumatic instabilities.

Acknowledgments This work was supported by the Czech Science Foundation under project No. 101/091481 "Gyroscopic stabilization of vibro-isolation system".

References

1. Kocur, J.A., Nicholas, J.C., Chester, C.L.: Surveying titling pad journal bearing and gas labyrinth seal coefficients and their effect on stability. In: Proceedings of the 36th Turbomachinery Symposium, Houston (2007)
2. Šimek, J., Svoboda, R.: Calculation of bearing support and dynamics of high-speed rotor of a steam turbine. Technical Report TECHLAB No.10-404 (2010) (in Czech)
3. Šimek, J.: Examination of journal bearing concept influence on a small turbocharger rotor behavior. Technical Report TECHLAB No.07-403 (2007)
4. Šimek, J., Šklíba, J., Prášil, L., Svoboda, R.: Design of power gyro for stretcher stabilization. In: Colloquium Dynamics of Machines, p. 79. Institute of Thermomechanics, Prague (2010)

Smart Machines with Flexible Rotors

Arthur W. Lees

Abstract The concept of smart machinery is of significant current interest. Several technologies are relevant in this quest including magnetic bearings, shape memory alloys (SMA) and piezo-electric activation. Recently a smart bearing pedestal was proposed based on SMAs and elastomeric O-rings. However, such a device is clearly relevant only for the control of rigid rotors, for flexible rotors there is a need for some modification on the rotor itself. In this paper, rotor actuation by piezo-electric patches on the rotor is studied. A methodology is presented for the calculation of rotor behaviour and appropriate control strategies are discussed. It is shown how this form of control is a viable option and an estimate of system requirements is given. It is shown that the most promising option for the control of rotor imbalance is the imposition of a counteracting bend. However, because the bend can be controlled as a function of speed, the usual restrictions in comparing a rotor bend and unbalance do not apply. After outlining the methodology of the calculation, a series of transient simulations are presented. Finally, difficulties and options in the construction of a real machine are discussed.

Keywords Rotor • Imbalance • Piezo-electric • Smart

1 Introduction

The concept of a Smart Structure is now a familiar one referring to a structure which can react to its environment, apply appropriate forces and correct incipient faults. To date far less attention has been devoted to smart rotor or smart machines and yet this seems a natural extension of recent work on Condition Monitoring and fault

---

A.W. Lees (✉)
School of Engineering, Swansea University, Singleton Park, Swansea SA2 8PP, UK
e-mail: a.w.lees@swansea.ac.uk

J. Náprstek et al. (eds.), *Vibration Problems ICOVP 2011: The 10th International Conference on Vibration Problems*, Springer Proceedings in Physics 139,
DOI 10.1007/978-94-007-2069-5_69,

identification. The simplest case to consider is a machine with a rigid rotor, such as occurs in a range of small machines. This is a particularly simple case to consider since the natural frequencies (or more particularly, critical speeds) are controlled entirely by the stiffness of the bearing pedestals. Lees et al. [1] have shown how a controllable bearing pedestal may be designed using Shape Memory Alloys and elastomers, although it is appreciated that there are a number of possible routes to achieve this goal. Other approaches to the introduction of controllable support stiffness have been considered by Zak et al. [2] and Cartmell et al. [3]. Recently Maslen [4] has given a brief review of progress on Smart Machinery, but this too focuses on bearings, with particular emphasis on magnetic bearings: this is not surprising as this technology offers great promise.

However, looking further ahead one is naturally led to consider the possibility of controlling machines with flexible rotors. This is important for two reasons: in machinery generally there is a trend towards the use of flexible rotors and secondly, adequate control would enable the operation of much light machines leading to higher efficiencies and better material utilization. A preliminary study of such a system has been reported recently by Lees [5] and in the present paper the methodology is refined and extended.

2 Theoretical Framework

The effective distinction between an imbalance on a rotor and a bend is well documented: whilst both give rise to synchronous vibration at shaft speed, their respective variations with frequency is quite different [6]. At low frequency and bend will give rise to a vibration signal which is finite, but at high frequency the vibration level will tend to zero. A rotor with imbalance will show the opposite variation: the response will be negligible at low speed by will tend to a finite limit as the speed of the rotor is increased. A consequence of this contrast is that under normal circumstances a the effects of a bend can only be compensated with imbalance at a single shaft speed, a fact well known to plant operators. Figure 1 shows the layout for a simple system with a central piezo-electric patch;

In this work, the objective is precisely the opposite – the compensation of an imbalance by the imposition of a rotor bend induced by the activation of piezo-electric patches. This is rather different to a conventional bend insofar as it can be varied at will by modification of the applied voltages. Lees [5] described a machine with a smart flexible rotor with equations

Fig. 1 Simple rotor with central patch

$$\begin{bmatrix} 0 & 0 & -\mathbf{K} & 0 \\ 0 & 0 & 0 & -\mathbf{K} \\ \mathbf{K} & -\mathbf{KR} & \mathbf{C} & 0 \\ 0 & \mathbf{K}-\Omega^2\mathbf{M}+2\Omega\mathbf{G} & 0 & \mathbf{C_r} \end{bmatrix} \begin{Bmatrix} y \\ b_0 \\ \dot{y} \\ \dot{b}_0 \end{Bmatrix} + \begin{bmatrix} \mathbf{K} & 0 & 0 & 0 \\ 0 & \mathbf{K} & 0 & 0 \\ 0 & 0 & \mathbf{M} & 0 \\ 0 & 0 & 0 & \mathbf{M} \end{bmatrix} \frac{d}{dt} \begin{Bmatrix} y \\ b_0 \\ \dot{y} \\ \dot{b}_0 \end{Bmatrix} = \begin{Bmatrix} 0 \\ 0 \\ ma\Omega^2 e^{j\Omega t} - jma\dot{\Omega}e^{j\Omega t} \\ P(t) \end{Bmatrix} \tag{1}$$

where

$$\mathbf{R} = \begin{bmatrix} r & 0 & \vdots & 0 \\ 0 & r & \vdots & 0 \\ \cdots & \cdots & \ddots & \vdots \\ 0 & 0 & 0 & r \end{bmatrix} \quad r = \begin{bmatrix} \cos\Omega t & \sin\Omega t & 0 & 0 \\ -\sin\Omega t & \cos\Omega & 0 & 0 \\ 0 & 0 & \cos\Omega t & \sin\Omega t \\ 0 & 0 & -\sin\Omega t & \cos\Omega t \end{bmatrix} \tag{2}$$

and the gyroscopic term is given by

$$\mathbf{G} = \frac{d\mathbf{R}}{dt}\mathbf{MR}$$

The equations of motion were developed in a Lagrangian formulation in which the state variables were taken as the deflection and the bend together with their respective time derivatives. The particular example presented in reference 1 studies a rotor comprising 12 elements with 52 degrees of freedom. Using a Runge–Kutta time integration it was necessary to condense this problem using Guyan reduction and 8 degrees of freedom were retained: although this may be considered as the minimum meaningful example it is worth noting that even this requires integration of 32 degrees of freedom. The response was reduced by the application of a voltage to the piezo patch which in turn induced a bend. In reference [5] the optimum level of voltage is established by applying a trial voltage in a procedure similar to single plane balancing. Calculations were presented at speeds below the first critical. Increasing rotor speed presents difficulties because the response of the rotor to the current application is of a similar form to that due to imbalance.

In the present paper, two changes of approach are reported. Firstly, the rotor imbalance can be calculated quickly from the state-space equations of motion without the need for any substantial settling time. Secondly, the difficulties arising from transients as the bend develops may be circumvented by imposing a bend as a function of time, then in a separate calculation, calculate the force required to produce this bend. Not only does this procedure alleviate transient problems, it also

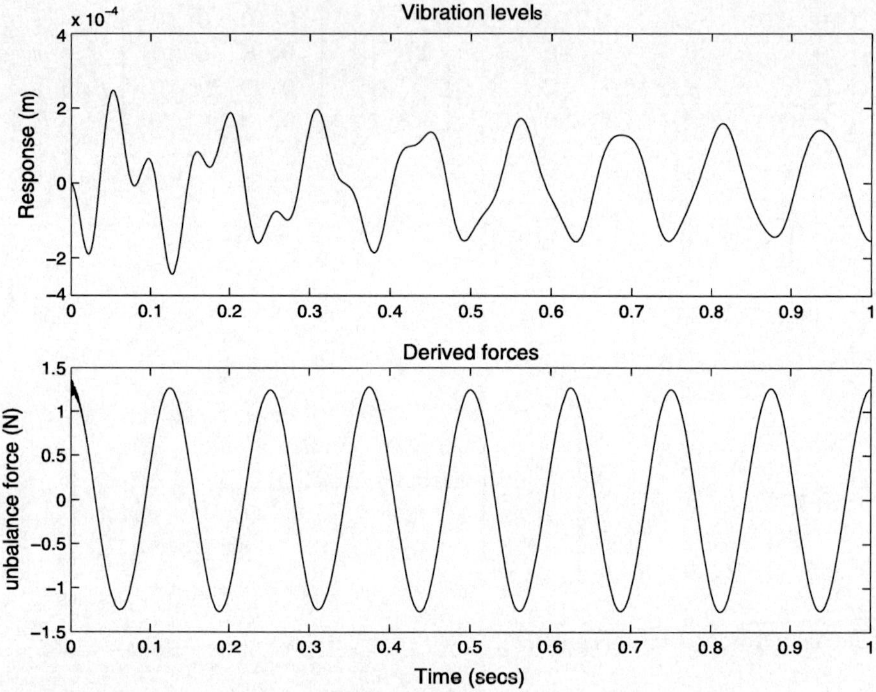

Fig. 2 Variation in time of vibration and forces acting

reduces the dimensionality of the problem by half. In this approach we demand that the bend at time *t* is *b(t)* and then the equations of motion of the rotor are given by

$$\begin{bmatrix} 0 & -\mathbf{K} \\ \mathbf{K} & \mathbf{C} \end{bmatrix} \begin{Bmatrix} x \\ \dot{x} \end{Bmatrix} + \begin{bmatrix} \mathbf{K} & 0 \\ 0 & \mathbf{M} \end{bmatrix} \frac{d}{dt} \begin{Bmatrix} x \\ \dot{x} \end{Bmatrix} = \begin{Bmatrix} 0 \\ ma\Omega^2 e^{j\theta} + b(t)e^{j\phi} \end{Bmatrix} e^{j\Omega t} \tag{3}$$

The time development of the bend *b,* is now a matter of choice depending on the forces applied. Consider the case

$$b(t) = b_0 \left(1 - e^{-\alpha t}\right) \tag{4}$$

A difficulty is in the decision of what *b* should be: in reference [5] this was achieved by a process akin to single plane balancing. It is clear that an appreciable interval is required. There is an alternative approach to achieving this result by using the state- space equation. Figure 2, also taken from reference [5], shows a section of the timetrace as the calculation begins. It is clearly non-sinusoidal as the starting transients decay, but the second trace of the figure shows the unbalance force which is taken from the appropriate component of Eq. 1. Note that this is sinusoidal from the start of the calculation. This implies that a very much quicker estimate of the unbalance may be obtained by using the available knowledge of the system. Note that the match between imbalance and the bend is obtained by considering modal contributions.

From Eq. 1, note that it may be conveniently partitioned into blocks and an estimate for the imbalance is readily obtained for the constant speed case by writing the relevant part of the equations of motion.

$$\mathbf{K}y - \mathbf{KR}b_0 + \mathbf{C}\dot{y} - \mathbf{M}\Omega^2 y + 2\mathbf{G}\Omega\dot{b}_0 = ma\Omega^2 e^{j\Omega t} \quad (5)$$

Note that since all parameters are available on the left hand side of this equation, the imbalance are the rotor is readily calculated. This does assume that the model of the system is sufficiently accurate, but any shortcomings may be overcome by model upating.

3 Sample Performance

In the examples shown, the effective imbalance term is calculated after two complete cycles of the shaft. The examples shown use a rotor system with a first critical speed of 20 Hz. Figure 3 shows the response at a running speed of 10 Hz when the corrective patch force is applied suddenly. Note the very substantial transients. The small remaining amplitude is due to phase error and this may be ammended in more sophisticated calculations. Clearly the rate at which compensation is applied ddictates the force required in the piezo-patch (or other device).

This could overcome by the introduction of a feedback loop to form a closed loop system. At speeds above the first critical speeds as one might anticipate, the response can be much higher but it is again controlled by the bend application.

Obviously there are options to continously update the estimate of the effective imbalance. The other point to note here is that even with a constant estimate of the imbalance, the moments and hence the required voltages to induce a bend will

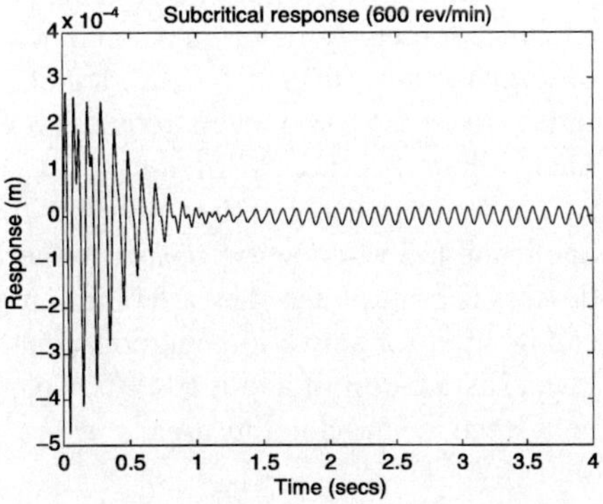

Fig. 3 Response to a sudden patch voltage (at 10 Hz)

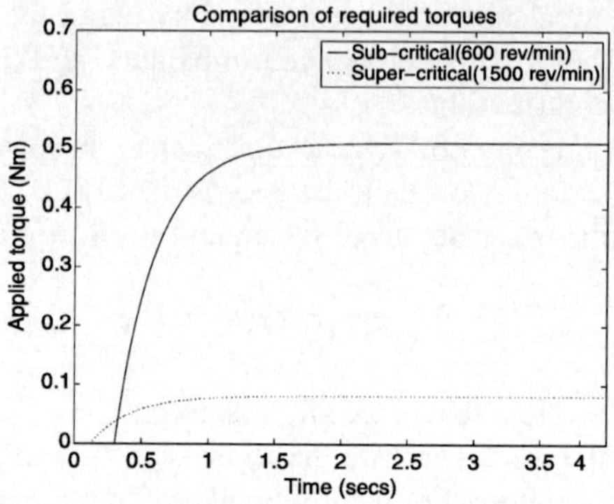

Fig. 4 Comparison of requires torques at different speeds

vary with the rotational speed of the machine. This variation is, at first sight, a little surprising and is shown in Fig. 4: substantially less moment is required in the supercritical case.

4 Practical Issues and Difficulties

The preceeding sections have discussed in detailed the ways in which a rotor can be controled by the application of moments and forces to the rotor itself. There are, of course, some formidable obstacles to be overcome becore such a 'Smart Machine' can be realised. Three areas of difficulty emerge: (1) bonding of the actuator to the rotor, (2) The transmission of the high voltages to the actuator, and (3) The selection of an actuator with suitable dimensions

Piezo-electric actuators have a number of attractions including ease of control, but the do suffer from a major disadvantage in requiring high voltages, often 1KV or more. Since they are predominently capacitive devices, the current requirement is negligible. In view of the ease of rapid control with these devices they must be viewed with some preference over rival technologies. One challenge, however, will be the identification of a suitable patch (or set of patches) which will generate adequate forces. A major question is whether the full voltage should be transmitted through the slip-rings or transmitted at some lower voltage for amplification of the rotor. Given the force levels requred, a completely different approach is the use of Shape Memory Alloys (SMAs).

These are avaiable in the form of wire which would be convenient for the current application. SMAs, being basically thermally, are intrinsically slower than piezo-devices in control and they are highly non-linear. On the other hand, ohmic heating can be used for activation and consequently, the voltages required are dramatically lower (by a factor of about 100). It is clear that some practical trials are needed as a next stage in studying this area.

5 Discussion and Conclusions

Magnetic bearings is probably the most developed aspect of this field as discussed in reference [4], but their primary application to date has been with respect to rigid rotors. For flexible rotors, considerable work is required to establish the optimum control strategy and it is clear that the controllability decreases as the rotor becomes more flexible. For such cases a control on the rotor become necessary since a flexible rotor cannot be adequately controlled from the bearings alone at all speeds.

References

1. Lees, A.W., Jana, S., Inman, D.J., Cartmell, M.P.: The Control of Bearing Stiffness Using Shape Memory. ISCORMA, Calgary (2007)
2. Zak, A.J., Cartmell, M.P., Ostachowicz, W.M.: Dynamics and control of a rotor using and integrated SMA/composite active bearing actuator. In: 5th International Conference on Damage Assessment of Structures, University of Southampton, Southampton, pp. 233–240 (2003)
3. Cartmell, M.P., Leize, T.H., Atepor, L., Inman, D.J., Lees, A.W.: Controlling flexible rotor vibrations by means of an antagonistic SMA/composite smart bearing. In: IoP Conference, Modern Practice in Stress and Vibration Analysis, Bath, July 2006
4. Maslan, E.H.: Smart machine advances in rotating machinery. In: 9th International Conference on Vibrations in Rotating Machinery, I. Mech. E., Exeter, 8–10 Sept 2008
5. Lees, A.W.: Smart machines with flexible rotors. Mech. Syst. Signal Process. **25**, 373–382 (2011)
6. Friswell, M.I., Penny, J.E.T., Garvey, S.D., Lees, A.W., Lees, A.W.: Dynamics of Rotating Machines. Cambridge University Press, Cambridge (2010). ISBN 978-0-521-85016-2

Stability Investigation of the Steady State Response of Flexibly Supported Rigid Rotors

Jaroslav Zapomĕl and Petr Ferfecki

Abstract A new semiactive coupling element working on the principle of squeezing two concentric films formed by classical and magnetorheological liquids has been proposed to achieve the optimum attenuation of the lateral vibration of rotors. The damping effect is controlled by changing the magnetic field induction. The steady state solution of the equations of motion is obtained by application of a collocation method and its stability is evaluated using the Floquet theory. The newly proposed controllable damping device enables to minimize amplitude of the rotor vibration by adapting the damping effect to the current operating conditions.

Keywords Rotors • Semiactive damping elements • Magnetorheological fluid

1 Introduction

To achieve optimum performance of damping devices placed between the rotor and its stationary part, the damping effect must be controllable. For this purpose a new semiactive coupling element working on the principle of squeezing two concentric films of classical and magnetorheological liquids has been proposed.

Modelling of classical and magnetorheological squeeze film dampers is mostly based on assumptions of the classical theory of lubrication [1–5].

J. Zapomĕl (✉)
Institute of Thermomechanics, Prague, Czech Republic
e-mail: jaroslav.zapomel@vsb.cz

P. Ferfecki
VSB-Technical University of Ostrava, Ostrava, Czech Republic
e-mail: petr.ferfecki@vsb.cz

J. Náprstek et al. (eds.), *Vibration Problems ICOVP 2011: The 10th International Conference on Vibration Problems*, Springer Proceedings in Physics 139,
DOI 10.1007/978-94-007-2069-5_70,

In this paper there is presented a method for analysing the influence of a new damping element on the steady state vibration of a rigidrotor flexibly supported and excited by the unbalance forces. The magnetorheological liquid is represented by Bingham material. The steady state response is calculated by means of a collocation method and to evaluate its stability the Floquet theory is applied.

2 Determination of the Damping Force

The proposed damping element consists of two rigid rings mounted with the stationary part and of two ones coupled with the shaft through a rolling element bearing and with the stationary part by a flexible element (squirel spring). This enables their vibration in the radial direction but prevents their rotation together with the shaft (Fig. 1). The gaps between the fixed and flexibly supported rings are filled with lubricants. The inner layer is formed by classical oil, the outer one by magnetorheological fluid. The damping element is equipped with an electric coil generating magnetic field. The change of the magnetic flux passing through the layer of magnetorheological liquid can be used to control the damping effect.

The mathematical model of the proposed damping element is based on assumptions of the classical theory of lubrication. The normal and magnetorheological oils are represented by Newtonian and Bingham materials and the yield shear stress of magnetorheological liquid depends on magnetic induction. The further attention is focused only on the dampers that can be treated as short [1].

The thickness of the oil films depends on the position of the journal centre relative to the stationary part of the rotor system [1]

$$h_{CO} = c_{CO} - e_H \cos(\varphi - \gamma), \quad h_{MR} = c_{MR} - e_H \cos(\varphi - \gamma) \tag{1}$$

h_{CO}, h_{MR} denote the thickness of the films of classical and magnetorheological oils, c_{CO}, c_{MR} are the widths of the gaps between the rings filled with classical oil

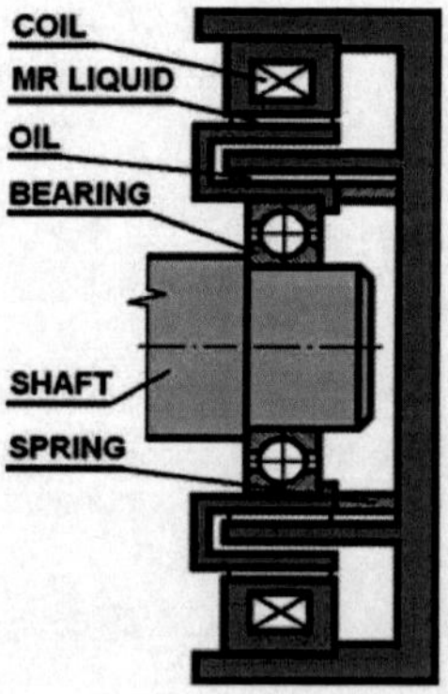

Fig. 1 The proposed damping element

and magnetorheological liquid, e_H denotes the rotor journal eccentricity, φ is the circumferential coordinate and γ denotes the position angle of the line of centres.

The pressure distribution in the layer of classical oil is governed [1]

$$\frac{\partial^2 p_{CO}}{\partial Z^2} = \frac{12\eta}{h_{CO}^3}\dot{h}_{CO}. \tag{2}$$

p_{CO} denotes the pressure in the layer of the classical oil, η is its dynamic viscosity, Z is the axial coordinate and $(\dot{})$ denotes the first derivative with respect to time.

To describe the pressure field in the layer of magnetorheological fluid, the Reynolds equation was modified for the case of Bingham liquid [5]. Assuming the damper symmetry relative to its middle plane it takes the form for $Z > 0$

$$h_{MR}^3 p_{MR}'^3 + 3\left(h_{MR}^2 \tau_y - 4\eta_B \dot{h}_{MR} Z\right) p_{MR}'^2 - 4\tau_y^3 = 0. \tag{3}$$

p_{MR}, p'_{MR} denote the pressure and the pressure gradient in the axial direction in the layer of magnetorheological liquid, η_B is the Bingham dynamical viscosity and τ_y represents the yield shear stress.

Equations 2 and 3 are solved with the boundary conditions expressing that the pressure at the damper's faces is equal to the pressure in the ambient space. In addition the solution of Eq. 3 must fulfill the conditions that the pressure gradient p' is real (not complex), negative and satisfies the relation

$$p'_{MR} < -\frac{2\tau_y}{h_{MR}}. \tag{4}$$

In the simplest design case of the damping element, the rings, between which there is a layer of magnetorheological liquid, can be considered as a divided core of an electromagnet. Then the dependence of the yield shear stress on magnetic induction can be approximately expressed

$$\tau_y = k_y \left(\frac{N_C I}{2 h_{MR}}\right)^{n_y}. \tag{5}$$

k_y and n_y are material constants of the magnetorheological liquid, N_C is the number the coil turns and I is the electric current.

In the areas where the thickness of the lubricating films rises with time ($\dot{h}_{CO} > 0$, $\dot{h}_{MR} > 0$) a cavitation is assumed. The pressure in these areas remains constant and is equal to the pressure in the ambient space. Components of the damping force are then calculated by integration of the pressure distributions around the circumference and along the length of the damping element taking into account the cavitation in the oil films.

3 Vibration and Stability Analysis of the Investigated Rotor

The rotor consists of a shaft carrying one disc (Fig. 2). The rotor is mounted with the rolling element bearings whose outer races are coupled with the stationary part through squirrel springs. Between the springs and the stationary part there are inserted the new coupling elements. The system is symmetric relative to the disc middle plane. The rotor is loaded by its weight and is excited by the disc unbalance. The squirrel springs are pre-bent to eliminate their deflection caused by the rotor weight.

In the mathematical model, the rotor is considered as absolutely rigid. Its lateral vibration is described by two nonlinear equations of motion

$$m_R \ddot{y} + b_P \dot{y} + 2 k_D y = 2 F_{dy} (y, z, \dot{y}, \dot{z}) + m_R e_T \omega^2 \cos (\omega t + \psi_o), \qquad (6)$$

$$m_R \ddot{z} + b_P \dot{z} + 2 k_D z = 2 F_{dz} (y, z, \dot{y}, \dot{z}) + m_R e_T \omega^2 \sin (\omega t + \psi_o). \qquad (7)$$

m_R is the rotor mass, b_P is the coefficient of external damping, k_D is stiffness of the squirrel spring, ω is the angular speed of the rotor rotation, e_T is the eccentricity of the rotor unbalance, t is the time, ψ_o is the phase shift, y, z are displacements of the rotor centre, F_{dy}, F_{dz} are components of the damping force and $(\dot{\ })$, $(\ddot{\ })$ denote the first and second derivatives with respect to time.

As the spring elements are pre-bent, the steady state trajectory of the rotor centre is circular and therefore the time history of the rotor displacements can be described by harmonic functions of time

$$y = r_C \cos \omega t - r_S \sin \omega t, \quad z = r_C \sin \omega t + r_S \cos \omega t \qquad (8)$$

To determine the unknown parameters r_C and r_S a collocation method is applied. Only one collocation point is needed. Then substitution of (8) and their derivatives with respect to time into (6) and (7) yields a set of two nonlinear algebraic equations,

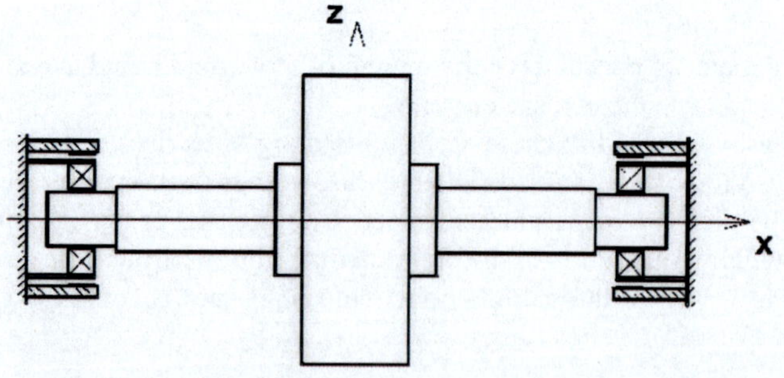

Fig. 2 Scheme of the investigated rotor system

which for the moment of time t equal to zero take the form

$$\left(2k_D - m_R\omega^2\right) r_C - \omega b_P\, r_S - m_R e_T \omega^2 - 2\,F_{dy}\,(r_C, r_S) = 0, \tag{9}$$

$$\omega b_P r_C + \left(2k_D - m_R\omega^2\right) r_S - 2F_{dz}\,(r_C, r_S) = 0. \tag{10}$$

Using the method presented in [6], stability of the steady state vibration is evaluated by means of the Floquet theory. The equations for the time history of deviations Δy, Δz of displacements y, z expressed in the state space have the form

$$\Delta\dot{\mathbf{r}} = \mathbf{W}(t)\,\Delta\mathbf{r}, \tag{11}$$

$$\Delta\mathbf{r} = \left[\Delta\dot{y}\ \ \Delta\dot{z}\ \ \Delta y\ \ \Delta z\right]^T, \tag{12}$$

$$\mathbf{W}(t) = \begin{bmatrix} -\dfrac{b_P - 2\dfrac{\partial F_{dy}}{\partial \dot{y}}}{m_R} & \dfrac{2\dfrac{\partial F_{dy}}{\partial \dot{z}}}{m_R} & -\dfrac{2k_D - 2\dfrac{\partial F_{dy}}{\partial y}}{m_R} & \dfrac{2\dfrac{\partial F_{dy}}{\partial z}}{m_R} \\ \dfrac{2\dfrac{\partial F_{dz}}{\partial \dot{y}}}{m_R} & -\dfrac{b_p - 2\dfrac{\partial F_{dz}}{\partial \dot{z}}}{m_R} & \dfrac{2\dfrac{\partial F_{dz}}{\partial y}}{m_R} & -\dfrac{2k_D - 2\dfrac{\partial F_{dz}}{\partial z}}{m_R} \\ 1 & 0 & 0 & 0 \\ 0 & 1 & 0 & 0 \end{bmatrix}. \tag{13}$$

The transition matrix $\mathbf{H}$ related to Eq. 13 is set up according to the approach described in details in [6]

$$\mathbf{H}(T,0) = e^{(T-t_{N-1})\,\mathbf{W}_N}\, e^{(t_{N-1}-t_{N-2})\,\mathbf{W}_{N-1}} \ldots\ldots e^{(t_1-t_0)\,\mathbf{W}_1},\ \mathbf{W}_k = \frac{1}{t_k - t_{k-1}} \int_{t_{k-1}}^{t_k} \mathbf{W}(t)\,\mathrm{d}t \tag{14}$$

for $t_0 = 0$, $t_N = T$ and $k = 1,\ 2,\ \ldots\ N$. T is the vibration period.

4 Example

The orbits of the rotor journal centre and the time histories of the damping force acting on the rotor in the vertical direction are drawn in Figs. 3 and 4. At rotor revolutions 100 rad/s, the rising current reduces the vibration amplitude and increases magnitude of the damping force. When the rotor revolutions are changed

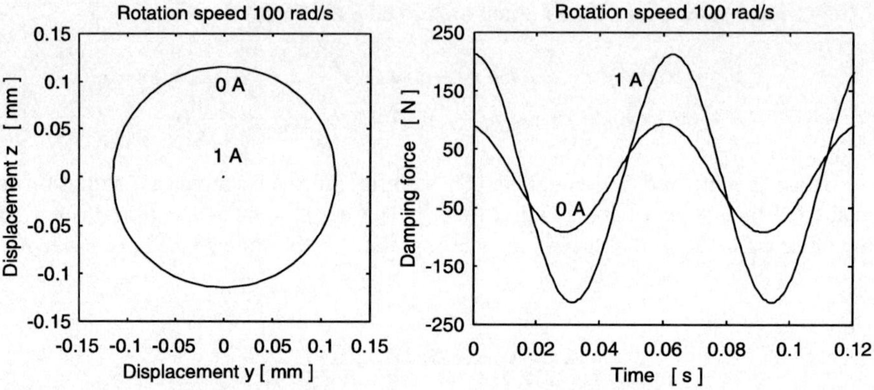

Fig. 3 The rotor centre orbits and the time-force history (100 rad/s)

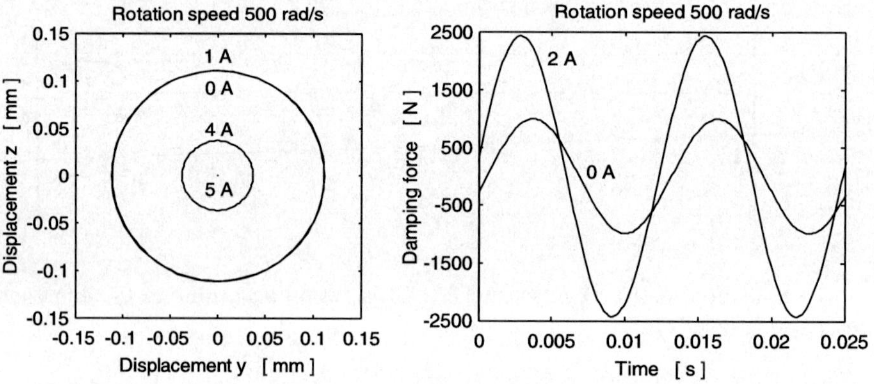

Fig. 4 The rotor centre orbits and the time-force history (500 rad/s)

for 500 rad/s, a small current arrives at increasing amplitude of the vibration instead of its attenuation. To achieve the damping effect, magnitude of the current must be considerably increased and the damping force rises. There were also observed multiple solutions of the equations of motion in some speed intervals.

Such behaviour of the studied damping element is caused by its mutually coupled stiffness and damping properties and by their nonlinear dependence on the electric current and speed of the rotor rotation. This shifts the resonance peaks and can lean the resonance curves to higher excitation frequencies.

Moduli of all eigenvalues of the transition matrix referred to the speed of rotation 100 rad/s are less than one and this implies that the rotor vibration is stable for both magnitudes of the applied current (0 A, 1 A).

5 Conclusions

The developed procedure represents a tool for analysing the steady state vibration and its stability of rigid rotors attenuated by newly proposed semiactive damping elements working on the principle of squeezing two thin films of classical and magnetorheological oils. Advantage of these damping devices is that they always produce some amount of damping, which can be increased if needed, they do not require a complicated and expensive control system and make possible to achieve the optimum performance of the rotating machine by means of adapting the damping effect to the current operating conditions.

Acknowledgments The work reported here has been supported by the research projects P101/10/0209 and AVO Z20760514. The support is gratefully acknowledged.

References

1. Krämer, E.: Dynamics of Rotors and Foundations. Springer, Berlin (1993)
2. Forte, P., Paterno, M., Rustighi, E.: A magnetorheological fluid damper for rotor applications. Int. J. Rotating Mach. **10**, 175 (2004)
3. Wang, G.-J., Feng, N., Meng, G., Hahn, E.-J.: Vibration control of a rotor by squeeze film damper with magnetorheological fluid. J. Intell. Mater. Syst. Struct. **17**, 353 (2006)
4. Zapoměl, J., Ferfecki, P.: Mathematical modelling of a long squeeze film magnetorheological damper for rotor systems. Model. Optim. Phys. Syst. **9**, 97 (2010)
5. Zapoměl, J., Ferfecki, P.: Mathematical modelling of a short magnetorheological damper. Transactions of the VŠB Technical University of Ostrava, Mechanical Series, LV, pp. 289 (2009)
6. Zhao, J.-Y., Linnett, I.-W., McLean, L.-J.: Stability and bifurcation of unbalanced response of a squeeze film damped flexible rotor. ASME J. Tribol. **11**(6), 361 (1994)

Rotor-Model-Based Identification of Foundations in Rotating Machinery Using Modal Parameters

N. Feng and E. Hahn

Abstract The vibration of a rotor bearing foundation system is affected by its foundation which needs to be identified. Outlined is an identification procedure for foundations of turbomachinery installations which avoids rotor removal. Input data requires knowledge of the forces transmitted to the foundation at the bearings and of the motion of the foundation at specified locations. Such data can be obtained from existing monitoring instrumentation provided the dynamic properties of the rotor and the rotor unbalance are known. Numerical experiments show that even with input data truncated to two significant digits, satisfactory identification is possible for a flexibly supported undamped rigid block foundation. It is concluded that this identification technique promises to be applicable in the field.

Keywords Foundation identification • Rotating machinery

1 Introduction

The vibration of a rotor bearing foundation system (RBFS) is affected by its foundation [1]. To include this effect in the overall analysis, one can either use measurements to identify equivalent foundation parameters; or one can model the foundation by finite elements, this latter approach being generally too difficult [1]. Here, the former approach is adopted and a procedure is developed which is applicable to existing turbomachinery installations and does not require rotor removal. The procedure presumes knowledge of the dynamic properties of the rotor; and uses for input data the forces transmitted to the foundation at the bearings and the motion of the foundation at specified locations. Suchdata is obtainable

N. Feng • E. Hahn (✉)
The University of New South Wales, Kensington, Australia
e-mail: n.feng@unsw.edu.au; e.hahn@unsw.edu.au

J. Náprstek et al. (eds.), *Vibration Problems ICOVP 2011: The 10th International Conference on Vibration Problems*, Springer Proceedings in Physics 139,
DOI 10.1007/978-94-007-2069-5_71,

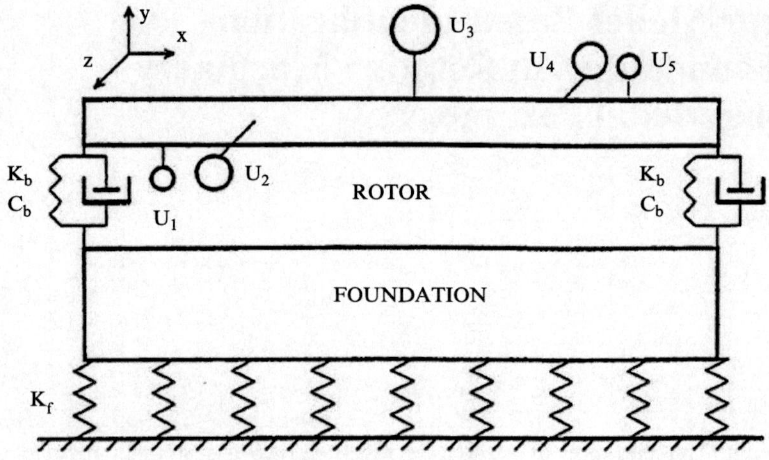

Fig. 1 Unbalanced rotor mounted via bearings on a flexibly supported rigid foundation block

from existing performance monitoring instrumentation [2]. The procedure is here evaluated for flexibly supported rigid foundation blocks as shown in Fig. 1.

Earlier approaches to identify such foundations proved unsatisfactory, either because the identification was poor [3] or the resulting nonlinear equations proved problematic [4]. In ref. [5] the procedure proved successful assuming in-phase harmonic force excitation at the bearing mounts. In this paper, the force excitation is caused by rotor unbalance, taking full account of the attenuating/amplifying effects of the rotor and the nonlinearity inherent in hydrodynamic bearings.

2 Theory

The equations of motion for an n degree of freedom (dof) foundation are:

$$\mathbf{M}\ddot{\mathbf{x}} + \mathbf{C}\dot{\mathbf{x}} + \mathbf{K}\mathbf{x} = \mathbf{f} \tag{1}$$

Here **M**, **C** and **K** are the $n \times n$ mass, damping and stiffness matrices of the foundation; and the elements of **x** are the n dofs chosen to coincide with convenient measurement locations and include the excitation force application points; i.e.

$$\mathbf{x} = \{x_1, x_2, \ldots\ldots x_n\}^T \tag{2}$$

The elements of **f** are the excitation forces transmitted to the foundation at the bearing supports. In the absence of fluid inertia, they are equal and opposite to the forces acting on the rotor at the bearing locations. Assuming that the mass matrix

$\mathbf{M}$ is diagonal, and that the system response due to unbalance excitation is periodic with fundamental frequency Ω equal to the rotor speed, one can write:

$$\left[-\Omega^2\mathbf{I} + i\Omega\mathbf{M}^{-1}\mathbf{C} + \mathbf{M}^{-1}\mathbf{K}\right]\mathbf{X} = \mathbf{M}^{-1}\mathbf{F} \tag{3}$$

Here $\mathbf{X}$ and $\mathbf{F}$ are the corresponding (complex) displacement and force amplitudes and $\mathbf{M}^{-1}$ is diagonal with elements $1/M_1, \ldots, 1/M_n$. The elements of $\mathbf{X}$, viz. $X_1, \ldots, X_n$, are obtained from foundation motion measurements, whereas the elements of $\mathbf{F}$, viz. $F_1, \ldots, F_n$, are calculated from knowledge of the rotor model, the rotor unbalance and rotor motion measurements at the bearing stations [2]. For identification, one needs to determine the elements of the $n \times n$ matrices $\mathbf{M}^{-1}\mathbf{K}$ and $\mathbf{M}^{-1}\mathbf{C}$ and those elements of $\mathbf{M}^{-1}$ which appear in the vector $\mathbf{M}^{-1}$ $\mathbf{F}$.

Selecting a transformation matrix $\mathbf{A}$, so that $\mathbf{X} = \mathbf{A}\mathbf{Q}$, one can write (3) as:

$$\left[-\Omega^2\mathbf{I}\mathbf{A} + i\Omega\mathbf{M}^{-1}\mathbf{C}\mathbf{A} + \mathbf{M}^{-1}\mathbf{K}\mathbf{A}\right]\mathbf{Q} = \mathbf{M}^{-1}\mathbf{F} \tag{4}$$

If the n foundation eigenvalues are $\lambda_1, \ldots, \lambda_n$ and if $\mathbf{A}$ is appropriately related to the system modal matrix, then premultiplying (4) by $\mathbf{A}^\mathbf{T}$ and assuming proportional damping, one obtains [5]:

$$\left(-\Omega^2\mathbf{I} + i\Omega\boldsymbol{\xi} + \boldsymbol{\lambda}\right)\mathbf{A}^T\mathbf{X} - \mathbf{A}^T\mathbf{M}^{-1}\mathbf{F} = \mathbf{0} \tag{5}$$

Here the $n \times n$ matrices ξ , and λ are diagonal. Thus, for $k = 1, \ldots, n$, one has:

$$\left(-\Omega^2 + i\Omega\xi_k + \lambda_k\right)\sum_{j=1}^{n} a_{jk}X_j - \sum_{j=1}^{n} a_{jk}F_j/M_j = 0 \tag{6}$$

The parameters to be identified in the k'th equation are: ξ_k, λ_k, a_{jk} $(j = 1, \ldots, n)$ and $1/M_j$ $(j = 1, \ldots, n)$. Because the elements of $\mathbf{A}$ are relative values, one can assign $a_{kk} = 1$. Sufficient independent simultaneous equations can be obtained by substituting for Ω and the corresponding X_j and F_j into (6). These equations are nonlinear, so an effective solution strategy is required. Having found the unknown parameters in each of the n identification equations, one has, in effect, obtained the equivalent foundation. Various means are then available for using this 'foundation' to obtain the unbalance response of the RBFS. One could, for example, use these identified parameters to obtain the $\mathbf{M}^{-1}\mathbf{C}$ and $\mathbf{M}^{-1}\mathbf{K}$ matrices in (3). Thus, if $\mathbf{k} = \mathbf{A}^\mathbf{T}\mathbf{M}^{-1}\mathbf{K}\mathbf{A}$, and noting that $\mathbf{k} = \lambda\mathbf{m}$ where $\mathbf{m} = \mathbf{A}^\mathbf{T}\mathbf{I}\mathbf{A}$:

$$\mathbf{M}^{-1}\mathbf{K} = \left[\mathbf{A}^T\right]^{-1}\mathbf{k}\mathbf{A}^{-1} = \left[\mathbf{A}^T\right]^{-1}\boldsymbol{\lambda}\mathbf{A}^T \tag{7}$$

A similar expression pertains to $\mathbf{M}^{-1}\mathbf{C}$. Note that should any of the F_j be zero, one cannot determine $1/M_j$. To recover (1), one needs to assign any nonzero number to $1/M_j$ and premultiply (3) by the inverse M of this partly fictitious $\mathbf{M}^{-1}$, to yield the identified mass matrix M and the identified stiffness matrix K $(= M\mathbf{M}^{-1}\mathbf{K})$.

3 Numerical Experiments

The flexibly supported rigid undamped foundation block, previously identified in ref. [4], was selected to evaluate the proposed identification procedure. Here $n = 6$. Figure 2 shows a set of suitable measurement locations on the upper surface of the block. Using these six independent displacements as dofs is referred to as the translational coordinate formulation. It allows for the application of the external force **f** in the x_2 and x_5 directions at the connection point C_1 and in the x_3 and x_6 directions at the connection point C_2. Unfortunately such a choice of dofs does not result in a diagonal mass matrix. To ensure this, one can locate rectangular coordinate axes at the mass centre of the block, aligned in the direction of its principal axes of inertia. A diagonal mass matrix results if one then selects for the dofs the displacements of the mass centre in the directions of these coordinate axes and the rotations of the block about these axes. Using these six independent displacements as dofs is referred to as the mass centre coordinate formulation. In this formulation the equations of motion of the foundation are similar to (1) except the mass, damping and stiffness matrices $\mathbf{M}_c$, $\mathbf{C}_c$ and $\mathbf{K}_c$ have different elements. The vector of displacements is then given by:

$$\mathbf{u} = \{x, y, z, \theta_x, \theta_y, \theta_z\}^T \tag{8}$$

Also, the force vector needs to be modified to **Df** where **D** is a 6×6 matrix. The mass centre and the translational dofs are related according to $\mathbf{u} = \mathbf{G}\mathbf{x}$ and it can be shown that the transpose of the 6×6 **G** matrix equals the inverse of the **D** matrix [4]. Also, the mass matrix **M** in (1) may be obtained from $\mathbf{M}_c$ according to $\mathbf{M} = \mathbf{G}^T\mathbf{M}_c\,\mathbf{G}$, with similar expressions for obtaining the **C** and **K** matrices.

Figure 3 shows the stiffnesses supporting the foundation block and their locations. For W = 317.5 mm, H = 158.75 mm, L = 1,270 mm, and for a mass of 502.49 kg, one obtains the principal moments of inertia $I_x = 5.2765$ kg m^2,

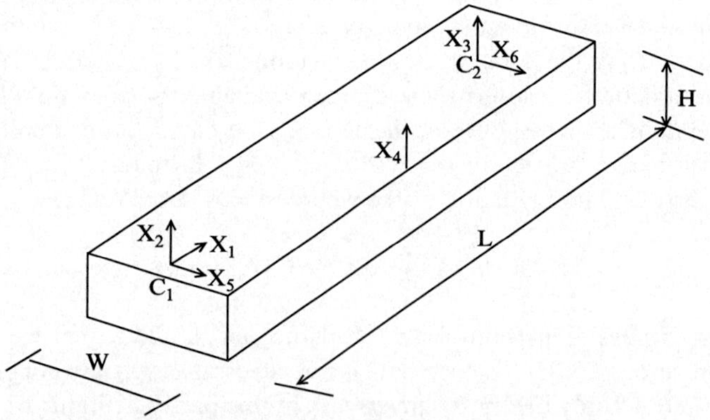

Fig. 2 Measurement locations and directions

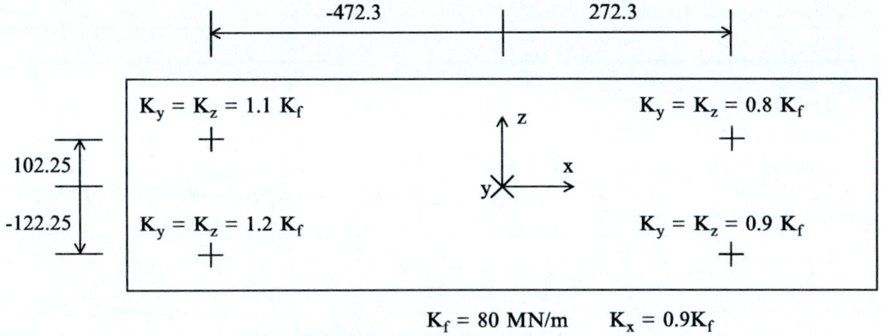

Fig. 3 Support stiffnesses and their locations

$I_y = 71.761$ kg m^2 and $I_z = 68.595$ kg m^2. With respect to the mass centre, the connection points C_1 and C_2 are at (−L/2, H/2, 0) and at (3 L/8, H/2, 0) respectively.

For the above undamped foundation, one can evaluate $\mathbf{M}_c$, $\mathbf{K}_c$ and $\mathbf{D}$ directly, and hence obtain **M** and **K** in (1). The numerical experiments then involved using the above foundation in Fig. 1 with unbalances of $U_1 = 10^{-4}$ kg m, $U_2 = 10^{-5}$ kg m and $U_3 = 10^{-6}$ kg m. Details of the rotor and bearings are given in ref. [3].

Using in-house software [3], the steady state system response was obtained over the frequency range of 300 to 1,450 rad/s in steps of 50 rad/s. The input data 'measurements' needed for identification were then the complex response amplitudes **X**, and the complex force amplitudes **F** at these speeds, resulting in 24 input data sets. Parameter identifications were carried out using both the mass centre and the translational coordinate formulations, assuming that the 'measurements' were available either to 5 significant digits or they were truncated to 2 significant digits. The 5 digit input data results served to evaluate the validity of the identification procedure in principle and to define the achievable accuracy of the adopted computational procedure by minimising the effect of measurement errors. The 2 digit input data better reflected attainable field measurement accuracy.

The solution for the unknowns in (6) involved a nested iterative approach [5]. Each of the six equations yielded 48 simultaneous linear equations with up to seven unknowns which were solved using least squares regression [3]. The converged solutions yielded the identified mass and stiffness matrices ***M*** and ***K*** which were then used to predict the unbalance response of the RBFS in Fig. 1.

4 Results and Discussion

Using the mass centre coordinate formulation, for both 5 and 2 digit data input, there was excellent agreement between the actual and identified foundation natural frequencies and inverse inertia values as shown in Table 1.

Table 1 Natural frequencies ω (rad/s) and inverse inertia values (kg^{-1} or $kg^{-1}\ m^{-2}$)

Mode	1	2	3	4	5	6
Actual ω	765.00	626.74	543.45	1259.0	823.61	1002.2
Actual $1/M_c$	0.00199	0.00199	0.00199	0.190	0.0139	0.0146
5 digit data ω	765.00	626.74	543.45	1259.0	823.62	1002.2
5 digit data $1/M_c$	–	0.00199	0.00199	0.189	0.0139	0.0146
2 digit data ω	765.13	626.24	544.12	1258.9	823.79	1002.3
2 digit data $1/M_c$	–	0.00202	0.00197	0.192	0.0140	0.0147

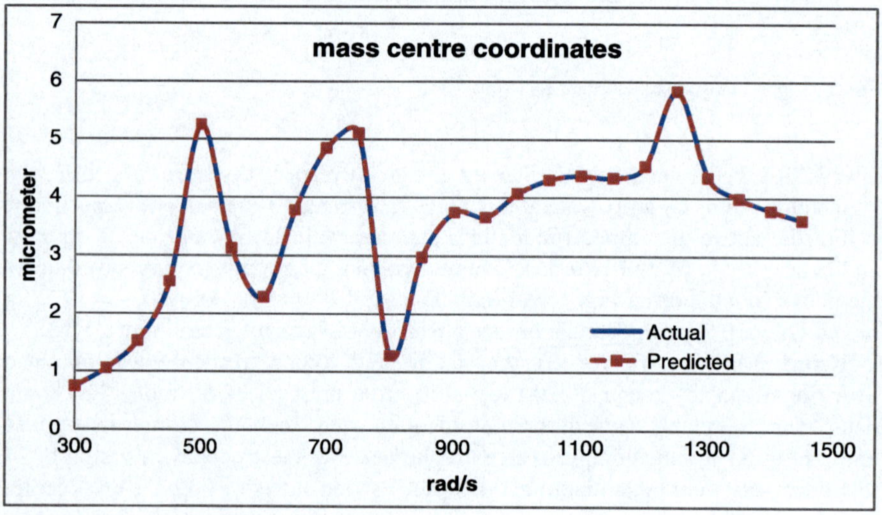

Fig. 4 Comparison of unbalance responses at different running speeds using mass centre coordinates

Figure 4 compares the actual unbalance response amplitudes midway along the rotor of the RBFS in Fig. 1 with that predicted with the identified foundation, using 2 digit input data accuracy. Not surprisingly, with such excellent identification of the natural frequencies and inverse inertia terms, the predicted and actual unbalance responses agree perfectly. Similarly excellent agreement was obtained for the responses at other locations on the rotor as well.

The translational coordinate formulation is far more straightforward, as there is no need to locate the mass centre nor the directions of the principal axes. Unfortunately the assumption of a diagonal mass matrix introduced error. As may be seen in Fig. 5, even with 5 digit input data accuracy, it was not possible to get such good agreement between the actual and predicted unbalance responses.

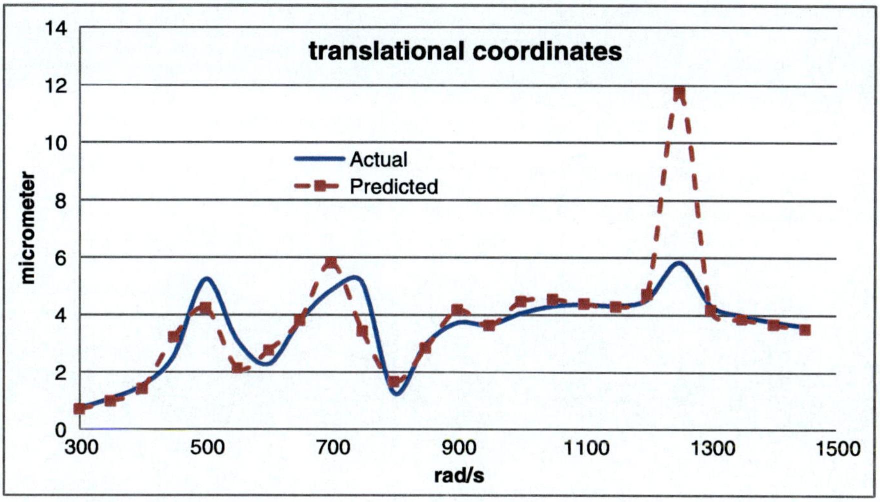

Fig. 5 Comparison of unbalance responses at different running speeds using translational coordinates

5 Conclusions

The proposed identification procedure is robust and promises to be applicable to practical installations where the foundation comprises a flexibly supported rigid foundation block and where measurements to two digit accuracy are feasible. However, the dofs should ideally be chosen to ensure a diagonal mass matrix.

References

1. Lees, A.W., Sinha, J.K., Friswell, M.I.: Model-based identification of rotating machines. Mech. Syst. Signal Process. **2**(3), 1884 (2009)
2. Feng, N.S., Hahn, E.J.: Identification of pedestal parameters in rotor bearing systems using known unbalance. In: Proceedings of the 2001 ASME Design Engineering Technical Conferences, DETC2001/VIB-21553, Pittsburgh (2001)
3. Feng, N.S., Hahn, E.J.: Including foundation effects on the vibration behavior of rotating machinery. Mech. Syst. Signal Process. **9**, 243 (1995)
4. Feng, N.S., Hahn, E.J.: Identification of the modal parameters of a flexibly supported rigid casing. Trans. ASME J. Eng. Gas Turbines Power **125**, 184 (2003)
5. Feng, N.S., Hahn, E.J., Medina L.U.: On the identification of foundations in rotating machinery using modal parameters. In: Proceedings ISMA31 Noise and Vibration Engineering Conference, Leuven (2010)

Magnetic Actuator Modelling for Rotating Machinery Analysis

Ricardo Ugliara Mendes, Hélio Fiori de Castro, Kátia Lucchesi Cavalca, and Luiz Otávio Saraiva Ferreira

Abstract Rotating machines have a wide range of application such as airplanes, factories, laboratories and power plants. Lately, with computer aid design, shafts finite element models including bearings, discs, seals and couplings have been developed, allowing the prediction of the machine behavior. In order to keep confidence during operation, it is necessary to monitor these systems, trying to predict future failures. One of the most applied technique for this purpose is the modal analysis. It consists of applying a perturbation force into the system and then to measure its response. However, there is a difficulty that brings limitations to the excitation of systems with rotating shafts when using impact hammers or shakers, once due to friction, undesired tangential forces and noise can be present in the measurements. Therefore, the study of a non-contact technique of external excitation becomes of high interest. In this sense, the present work deals with the study and development of a finite element model for rotating machines using a magnetic actuator as an external excitation source. This work also brings numerical simulations where the magnetic actuator was used to obtain the frequency response function of the rotating system.

Keywords Magnetic actuator • Rotating machinery • Excitation without contact • Rotor dynamics

R.U. Mendes (✉) • H.F. de Castro • K.L. Cavalca • L.O.S. Ferreira
Laboratory of Rotating Machinery, Faculty of Mechanical Engineering, University of Campinas, Campinas, SP, Brazil
e-mail: rumqld@gmail.com; heliofc@fem.unicamp.br; katia@fem.unicamp.br; lotavio@fem.unicamp.br

J. Náprstek et al. (eds.), *Vibration Problems ICOVP 2011: The 10th International Conference on Vibration Problems*, Springer Proceedings in Physics 139,
DOI 10.1007/978-94-007-2069-5_72, © Springer Science+Business Media B.V. 2011

1 Introduction

Lately, the computer aid design opened doors for a new technique, called finite element model (FEM), to be applied to rotating machinery modeling. In this technique, the continuous shaft is divided into finite elements. The first models used the Euler-Bernoulli beam, whose equations are obtained from the application of the Lagrange equation in a simple beam. A breakthrough in the modeling of rotating machines is shown in [1], who studied a model called the Rayleigh beam, which takes into account the rotational inertia of the shaft and the gyroscopic effect, dependent on the shaft rotational speed. After a representative model of the shaft is obtained, other elements of rotating machines (such as bearings, seals, foundations, rotors) should be incorporated to the model in order to study the phenomena inherent to these systems. Some expressive authors in this area, [2–4], modeled many of the components mentioned above and analyzed their influence on rotor dynamics. A non-linear approach in journal bearings modeling was presented in [5], and an equivalent coefficients approach by numerically solving the Reynolds equation in [6]. Non-symmetric rotors or with considerable anisotropy have precession modes which can be forward or backward. In this sense, a complex modal analysis is proposed to distinguish these two directional modes [7]. Concerning to non-contact excitation source, there are important references in this area [8, 9], which deal with the development of magnetic actuators including the differential linearization, where two actuators are placed radially opposed. Once magnetic actuators can generate only attractive forces, this positioning allows applying forces in both the actuator directions and also linearizes the magnetic force. This work refers to a magnetic actuator model numerically coupled to a rotor-bearing system model to be used in a control system design application.

2 Mechanical System

The model of the mechanical system consists in the finite element model (Fig. 1) of the steel SAE 1030 shaft of 12 mm diameter and 800 mm length, which is split into 21 nodes – 20 elements (represented by the dots). There is also a 94.82 mm diameter and 47.5 mm width disc which is placed in the center of the shaft. The model also considers the actuator journal placed at 62 mm of the first bearing, of 80 mm length and 40 mm diameter. Both the disc and the journal are made of steel SAE 1020. The system is supported by two hydrodynamic bearings (triangles in Fig. 1) symmetrically placed at 600 mm from each other regarding to the middle of the shaft, both with 90 μm clearance, 20 mm width and 30 mm inner diameter.

The equation of motion [1] is given in (1), where Ω is the rotational speed, M is the mass matrix, G the gyroscopic effect matrix, K the stiffness matrix and C is the structural proportional damping matrix [10].

$$[M]\{\ddot{q}(t)\} + ([C] + \Omega[G])\{\dot{q}(t)\} + [K]\{q(t)\} = [F(t)] \tag{1}$$

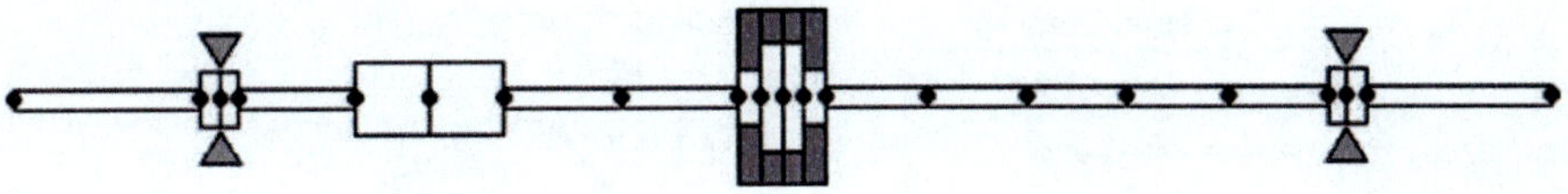

Fig. 1 Finite element model of the mechanical system

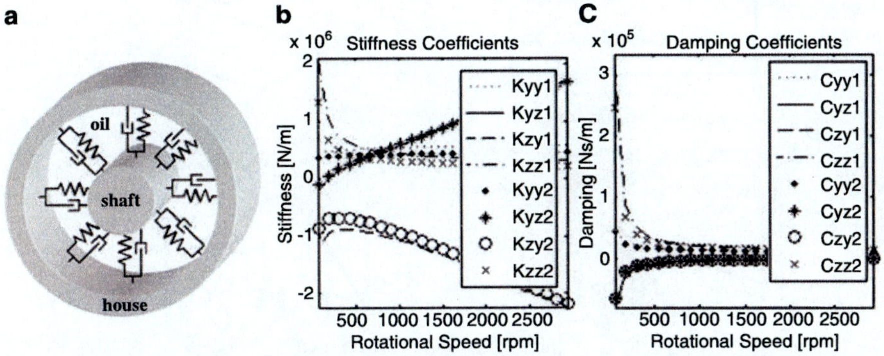

Fig. 2 (**a**) Model of the journal bearing; (**b**) Stiffness coefficients; (**c**) Damping coefficients

$F(t)$ is the external forces vector, containing an unbalanced mass and the actuator force; q is the generalized coordinate vector, with two translations and two rotations for each node. The journal bearings are modeled by equivalent coefficients of damping and stiffness (Fig. 2a) that are added in the corresponding system matrix. The coefficients are evaluated from the pressure distribution, which is obtained by Reynolds equation solution, using the finite-difference method. The coefficients behavior, regarding to the rotational speed, are shown in Fig. 2b, c, [6].

The solution of the homogeneous part of (1) provides the natural frequencies and natural modes of the system. However, it can be seen that (1) depends on the rotational speed, i. e. the properties of the system changes as the rotational speed changes. Therefore, this effect is shown in Fig. 3a, the Campbell Diagram. The points where the natural frequencies cross the 1X line are called critical velocities. In rotor dynamics, the precession movements (also called whirl), can be forward, if they occur in the same direction of the rotor spin, or backward, if in the opposite direction [11]. The natural modes, for a rotational speed of 1,200 rpm, are presented in Fig. 3b, c. Figure 3b presents the first forward bending mode of the shaft, and the first backward bending mode is presented in Fig. 3c. The particular solution of (1), regarding to the unbalance, gives the operational deflection shape (ODS), as shown in Fig. 3d, for the rotational speed of 660 rpm. The ODS presents the relative motion between two or more degrees of freedom from a structure due to an external excitation force [12], and gives an idea of how the system responds in a specific frequency, once the response of the system is a linear combination of its natural modes.

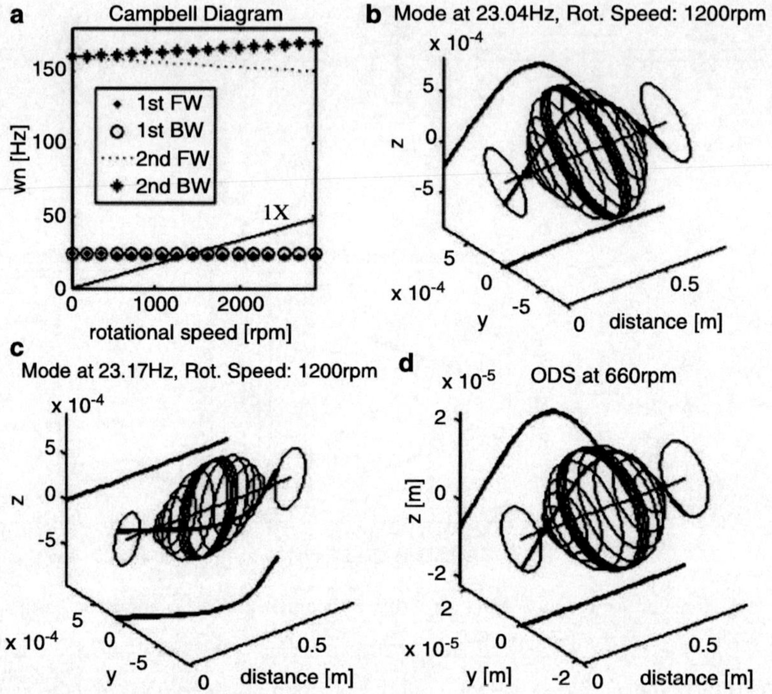

Fig. 3 (**a**) Campbell diagram; (**b**) First forward bending mode; (**c**) First backward bending mode; (**d**) Operational deflection shape

3 Magnetic Actuator

The force obtained by a magnetic actuator (2) is proportional to the area of the actuator pole (A), the magnetic flux density (B), which is given by (3) and it is proportional to the number of turns in the coils (N), the current (i), the length of the air gap (l_g) and the magnetic permeability of the air (μ_g) [8].

$$F_m = \frac{B^2 \cdot A}{\mu_g} \tag{2}$$

$$B = \frac{\mu_g \cdot N \cdot i}{2 \cdot l_g} \tag{3}$$

$$F_m = \frac{A}{\mu_g} 4 B_b B_{ref} \tag{4}$$

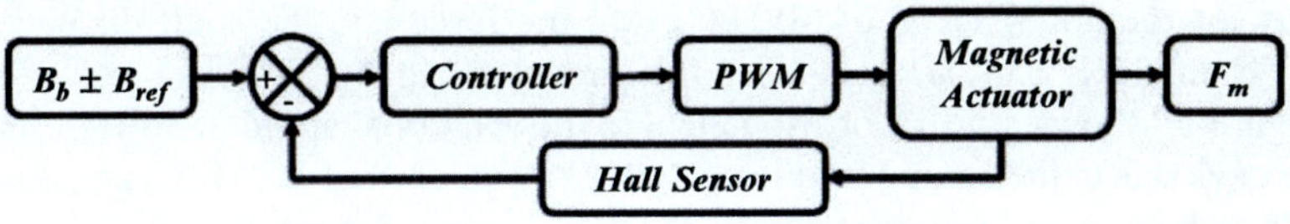

Fig. 4 Magnetic system diagram

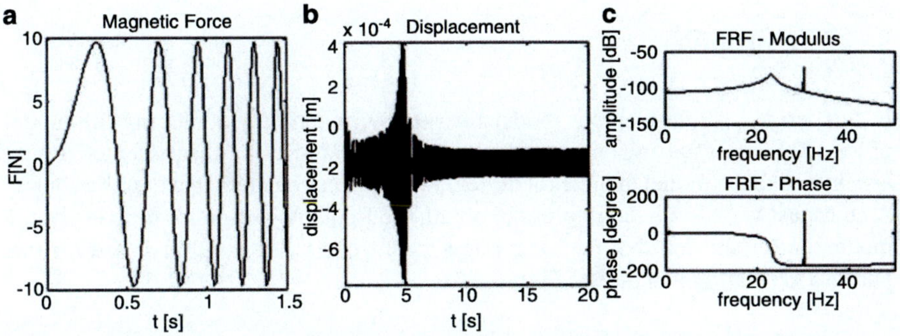

Fig. 5 (**a**) Magnetic force; (**b**) Displacement of the rotor; (**c**) Frequency response function

Therefore, the control of the actuator force can be made through the magnetic field control [13] as shown in Fig. 4, where B_{ref} is the magnetic field correspondent to the reference force which is compared to the magnetic field measurement made by the hall sensor. This comparison is the proportional controller input. The output voltage of the controller is transformed into the correspondent current by the PWM amplifier (which works in the current control mode) and sent to the coils of the actuator, generating the magnetic force F_m. Thus, using the differential assembly with a bias magnetic field (B_b), [8, 9], the linearized resultant force can be written as (4), according to [13]. The PWM presents a linear behavior in the operation range of the system, therefore, its behavior is described by a first order polynomial equation based on experimental tests. The hall sensor behavior is governed by a first order differential equation, and its gain and time constant were provided by its manufacturer.

4 Results

Once the model of the magnetic actuator was integrated to the mechanical system model, a chirp test was accomplished in order to obtain the frequency response function (FRF) of the rotating system. The frequency of the external excitation applied by the magnetic actuator was increased from 0 to 50 Hz, in 20 s. Figure 5a shows a detail of the excitation force frequency increasing. The shaft was kept in a constant rotational speed of 30 Hz. From the force and the shaft displacement (Fig. 5b) the FRF was calculated, and it is shown in Fig. 5c.

Figure 5c shows a peek correspondent to the rotational speed at 30 Hz. There is also a peek due to the natural frequency (with a phase shift of 180°) at approximately 23 Hz, which is in agreement with the first natural frequency of the rotor, as experimentally observed in previous works [5, 13].

5 Conclusions

In this work, a finite element model for rotating machinery, including the model of journal bearings, using an equivalent coefficients approach, was presented. It was also included the model of a magnetic actuator, used as an external excitation source. A chirp test was simulated in order to obtain the frequency response function of the model, and it showed that the magnetic actuator can be successfully used for this purpose in rotating machinery analysis.

Acknowledgments The authors would like to thank Fapesp, CAPES and CNPq for the financial support to this research.

References

1. Nelson, H.D., McVaugh, J.M.: The dynamics of rotor-bearing systems using finite elements. J. Eng. Ind. **98**, 593–600 (1976)
2. Vance, J.M.: Rotordynamics of Turbomachinery. Wiley-Interscience, New York (1988)
3. Kramer, E.: Dynamics of Rotors and Foundations. Springer, Berlin (1993)
4. Childs, D.: Turbomachinery Rotordynamics: Phenomena, Modeling, and Analysis. Wiley-Interscience, New York (1993)
5. Castro, H.F.: Hydrodynamic journal bearing analysis under fluid-induced instabilities. Dissertation (Ph.D.), Faculty of Mechanical Engineering, University of Campinas, São Paulo (2007)
6. Machado, T.H., Cavalca, K.L.: Evaluation of dynamic coefficients of fluid journal bearings with different geometries. In: Proceedings of the Brazilian Congress on Mechanical Engineering (COBEM2009), Belo Hortizonte (2009)
7. Kessler, C.L.: Complex modal analysis of rotating machinery. Dissertation (Ph.D.), University of Cincinnati, Cincinnati (1999)
8. Chiba, A., Fukao, T., Ishicawa, O., Oshima, M., Takemoto, M., Dorrell, D.G.: Magnetic Bearings and Bearingless Drives. Newnes, Amsterdam (2005)
9. Maslen, E.: Magnetic Bearings. University of Virginia, Charlottesville (2000)
10. Weiming, L., Novak, M.: Dynamic behavior of turbine-generator-foundation systems. Earthq. Eng. Struct. Dyn. **24**(3), 339–360 (1996)
11. Lee, C.W.: Vibration Analysis of Rotors. Springer, Dordrecht (1993)
12. Richardson, M.H.: Is it a mode shape, or an operating deflection shape? Sound Vib. Mag. **31**(1), 54 (1997)
13. Furtado, R.M.: A Magnetic actuator development for contactless excitation in rotor systems. Dissertation (Ph.D.), Faculty of Mechanical Engineering, University of Campinas, São Paulo (2008)

Quenching Damping-Induced Instability

Ferdinand Verhulst

Abstract The first complete analytic and geometrical explanation of destabilization of a conservative system by small dissipation was given by Bottema in 1956. We will discuss a simple-looking but generic model for a particle in an rotating vessel studied by L.E.J. Brouwer using averaging-normalization techniques with the purpose to quench such instabilities. Remarkably enough it turns out that for the rotating vessel both in the case of a rotating well and a rotating saddle, quenching is possible.

Keywords Dissipation-induced instabilities • Whitney's umbrella • Brouwer's rotating vessel • Gyroscopic quenching

1 Introduction

A number of rotor systems, when modeled as systems without damping, possess stable equilibria or stable steady motions, but when small damping is introduced, some of these equilibria or steady motions become unstable.

The crucial ideas for the explanation of these destabilization phenomena were formulated by Bottema. In the context of bifurcation theory a geometrical explanation can be given in terms of the so-called 'Whitney umbrella' and singularities of maps, see for references [2].

We consider a physically very simple model studied by Brouwer [1] that can serve as an illustration of destabilization by dissipation. Note however that, although simple looking, the equations that arise are familiar from other applications. For an

F. Verhulst (✉)
Mathematisch Instituut, University of Utrecht, 80.010, 3508 TA, Utrecht, the Netherlands
e-mail: f.verhulst@uu.nl

J. Náprstek et al. (eds.), *Vibration Problems ICOVP 2011: The 10th International Conference on Vibration Problems*, Springer Proceedings in Physics 139,
DOI 10.1007/978-94-007-2069-5_73, © Springer Science+Business Media B.V. 2011

important and typical case of Brouwer's rotating vessel we show that quenching of the instability can be achieved by small modulations of the rotation speed. Several resonances play a part here.

2 Brouwer's Rotating Vessel

In 1918 Brouwer [1] studied the equilibrium position of a point mass in a vessel, described by a smooth surface S, rotating around a vertical axis with constant angular velocity ω. With the equilibrium chosen on the vertical axis at $(x, y) = (0, 0)$ on S, the linearized equations of motion with internal dissipation added are:

$$\ddot{x} - 2\omega\dot{y} + \varepsilon c\dot{x} + (gk_1 - \omega^2)x = 0, \quad (1)$$

$$\ddot{y} + 2\omega\dot{x} + \varepsilon c\dot{y} + (gk_2 - \omega^2)y = 0. \quad (2)$$

The constants k_1 and k_2 are the x, y-curvatures of S at $(0, 0)$, g is the gravitational constant, ε is a positive small parameter, c a positive constant. If $k_1 < 0, k_2 < 0$ we have a 'bump' on S and always instability; we leave this case out. Consider two cases:

1. $k_1 = k_2 = k > 0$ (single-well at equilibrium).Without damping ($\varepsilon = 0$) stability if $0 < \omega^2 < gk$ (slow rotation) or $\omega^2 > gk$ (fast rotation). With damping ($\varepsilon > 0$) stability if $0 < \omega^2 < gk$.
2. $k_1 = k > 0$ and $k_2 = -k$, (saddle at equilibrium). With damping ($\varepsilon > 0$) always instability.

Note that asymmetric damping, putting $c\dot{x} \to c_1\dot{x}, c\dot{y} \to c_2\dot{y}$ does not change the results qualitatively.

Based on earlier studies (see [4]) we choose *modulation of the vessel rotation* as this might produce the combination resonances that will change the stability characteristics. For $\varepsilon = 0$ we have solutions $(x, \dot{x}, y, \dot{y})$ that are used to produce variational equations of the form $\dot{z} = \varepsilon F(z, y)$ with z a 4-vector. The righthand side F is quasi-periodic in t; averaging over the two periods produces averaged-normal forms that yield an $O(\varepsilon)$-approximation valid on the timescale $1/\varepsilon$ (see [3]). The variational equations contain hundreds of terms, the algebraic manipulation program MATHEMATICA is used to compute the normal forms.

3 Quenching Unstable Single-Well Motion

Assume that we have the unstable case $\omega^2 > gk$. The modulation is given by $\omega^2 := \omega^2 + 2a\varepsilon \cos \nu t$ with again ω a constant. We have

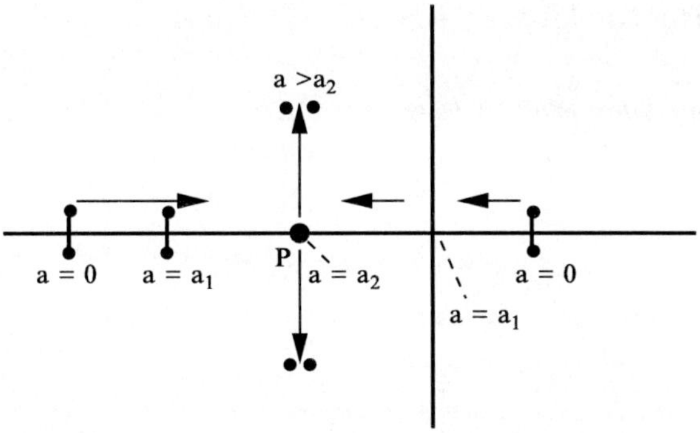

Fig. 1 Movement of the eigenvalues near equilibrium at the single-well with dissipation and small modulation of the rotation ω. The eigenvalues have multiplicity two; for $a=0$ (no modulation) we have two positive and two negative eigenvalues. At $a = a_1 = c^2\omega^2\frac{\omega^2-gk}{gk}$ the two positive eigenvalues arrive at the origin. For $a_1 < a < a_2 (= \frac{c^2\omega^4}{gk})$ four eigenvalues are real and negative (stability). At P (for $a=a_2$) we have four equal eigenvalues, for $a > a_2$ the eigenvalues move into the complex plane with negative real part

$$\ddot{x} - 2\omega\dot{y} - \left(\omega^2 - gk\right)x = \varepsilon\left(-c\dot{x} + \tfrac{2a}{\omega}\dot{y}\cos\nu t + 2ax\cos\nu t\right),$$
$$\ddot{y} + 2\omega\dot{x} - \left(\omega^2 - gk\right)y = \varepsilon\left(-c\dot{y} - \tfrac{2a}{\omega}\dot{x}\cos\nu t + 2ay\cos\nu t\right)$$

Putting $\varepsilon=0$ we obtain for the frequencies near the origin: $\lambda_1 = \omega + \sqrt{gk}$, $\lambda_2 = \omega - \sqrt{gk}$. Using the independent solutions of the unperturbed ($\varepsilon=0$) solutions we introduce the usual variation of constants transformation into the equations of motion. Using matrix inversion and applying averaging-normalization we have to make assumptions about the frequencies $\lambda_1, \lambda_2, \nu$. There is only one (combination) resonance leading to synchronization: $\lambda_1 - \lambda_2 = \nu$ or $2\sqrt{gk} = \nu$. The averaged-normal form equations have in the case of this resonance solutions an $O(\varepsilon)$ error-estimate on the timescale $1/\varepsilon$; in the case of asymptotic stability the estimate is valid for all time (see [3]). For the equilibrium at the origin we have from the normal form equations the eigenvalues, see Fig. 1.

3.1 *Outside the Combination Resonance*

If we have no resonance at first order averaging, the off-diagonal elements vanish to $O(\varepsilon)$, so that the equilibrium remains unstable. The tuning into the combination resonance $\lambda_1 - \lambda_2 = \nu$ is essential to stabilize the system.

4 Quenching the Unstable Saddle Motion

Consider now *modulation of the vessel rotation* in the case of the rotation of a saddle (unstable by dissipation for any rotational velocity ω). Put $k_1 = k > 0$ and $k_2 = -k$ and assume $\omega^2 > gk$. The modulation is given as before. The equations of motion become:

$$\ddot{x} - 2\omega\dot{y} - \varepsilon\frac{2a}{\omega}\cos\nu t\dot{y} + \varepsilon c\dot{x} - \beta^2 x - 2\varepsilon ax\cos\nu t = 0,$$

$$\ddot{y} + 2\omega\dot{x} + \varepsilon\frac{2a}{\omega}\cos\nu t\dot{x} + \varepsilon c\dot{y} - \alpha^2 y - 2\varepsilon ay\cos\nu t = 0.$$

$$\alpha^2 = \omega^2 + gk, \ \beta^2 = \omega^2 - gk.$$

We perform a similar analysis as before producing the following five resonances: $2\alpha = \nu$ (Mathieu resonance); $2\beta = \nu$ (Mathieu resonance); sum-resonance $\alpha = \beta + \nu$; sum-resonance $\alpha + \beta = \nu$; special resonance $\alpha = 3\beta = \beta + \nu$.

Each of these cases correspond with different synchronization scenarios. After computing the normal forms to obtain the stability characteristics it turns out that the two Mathieu-resonances and the fifth, special resonance do not lead to quenching. Consider the two sum-resonances:

The combination resonance $\boldsymbol{\alpha = \beta + \nu}$. Putting $K = gk/\omega^2$ and computing the normal form matrix we have that the modulation coefficient a produces stability with four real negative real eigenvalues if

$$\omega c\frac{\sqrt{4-K^2}}{K} < a < \frac{2c\omega}{K}.$$

The eigenvalues are complex with negative real part, implying stability, if

$$a > \frac{2c\omega}{K}.$$

The combination resonance $\boldsymbol{\alpha + \beta = \nu}$. After normalization we find the same double eigenvalues as in the case $\alpha = \beta + \nu$ and so the same stability conclusions. The two combination resonances enable us to stabilize the gyroscopic system.

In the case of the combination resonances $\alpha = \beta + \nu$ and $\alpha + \beta = \nu$ the value

$$a = \omega c\frac{\sqrt{4-K^2}}{K}$$

corresponds with two real negative eigenvalues and two eigenvalues zero. For the linear system this boundary case gives instability. On adding nonlinear terms we will have a two-dimensional center manifold. For both combination resonances we conclude that for stability the amplitude of the frequency modulation has to be larger than a magnitude proportional to $c\omega^3/gk$.

5 Conclusion

We have revisited the rotating vessel problem formulated by Brouwer that is destabilized by small damping, whatever its magnitude; this can be interpreted by what is now called the Whitney umbrella singularity. Interestingly, this gyroscopic problem can be stabilized again by introducing a small modulation of the rotation speed and tuning into an appropriate combination resonance. It is remarkable that this can even be achieved for the rotating saddle problem. As Brouwer's model is typical for rotating and gyroscopic systems, the results are promising for quenching such systems.

Acknowledgements O.N. Kirillov drew my attention to the paper by Brouwer [1]. In a pleasant cooperation with Theo Ruijgrok, MATHEMATICA was used to handle the normalization procedure in this paper.

References

1. Brouwer, L.E.J.: The motion of a particle on the bottom of a rotating vessel under the influence of the gravitational force. Nieuw Arch. v. Wisk., 2e reeks **12**, 407–419 (1918) (English transl. in collected works, North-Holland Publ. 1976)
2. Kirillov, O.N., Verhulst, F.: Paradoxes of dissipation-induced destabilization or who opened Whitney's umbrella? ZAMM **90**(6), 462–488 (2010)
3. Sanders, J.A., Verhulst, F., Murdock, J.: Averaging methods in nonlinear dynamical systems. Appl. Math. Sci. **59**, Springer (2007, rev. edn.)
4. Tondl, A., Ruijgrok, M., Verhulst, F., Nabergoj, R.: Autoparametric Resonance in Mechanical Systems. Cambridge University Press, Cambridge (2000)

An Equivalent Blisk Model Considering the Influence of the Air Flow on Blade Vibrations of a Mistuned Compressor Blisk

Bernd Beirow, A. Kühhorn, and J. Nipkau

Abstract A discrete blisk model composed by mass-, spring- and dashpot-elements, the so called Equivalent Blisk Model (EBM) is adapted to the mechanical properties of a real high pressure compressor blisk considering measured vibration data. Aiming to integrate the aerodynamic influences in the EBM, aerodynamic influence coefficients derived from aeroelastic computations are employed. Intensive numerical analyses of the forced response behavior of the blisk are performed. In addition, probabilistic mistuning analyses are carried out to assess the influence of increasing mistuning on the response amplification in dependence on the excitation order. Since the coupling between disk and blades is one of the governing parameters affecting the forced response, the influence of the disk's stiffness is studied varying the stiffness of those springs representing the disk.

Keywords Blisk • Blade vibration • Mistuning • Fluid structure interaction

1 Introduction

Analyses of rotating turbomachinery components such as compressor rotors (Fig. 1) are widely based on simulations of perfectly tuned systems. However, due to deviations from the ideal design intention following from geometrical imperfections, material inhomogeneties or wear respectively, a real turbomachinery rotor will always be mistuned no matter how small the tolerances may be. In that regard,

B. Beirow (✉) • A. Kühhorn
Brandenburg University of Technology Cottbus, Cottbus, Germany
e-mail: beirow@tu-cottbus.de

J. Nipkau
Rolls-Royce Deutschland, Blankenfelde-Mahlow, Germany
e-mail: Jens.Nipkau@Rolls-Royce.com

J. Náprstek et al. (eds.), *Vibration Problems ICOVP 2011: The 10th International Conference on Vibration Problems*, Springer Proceedings in Physics 139,
DOI 10.1007/978-94-007-2069-5_74,

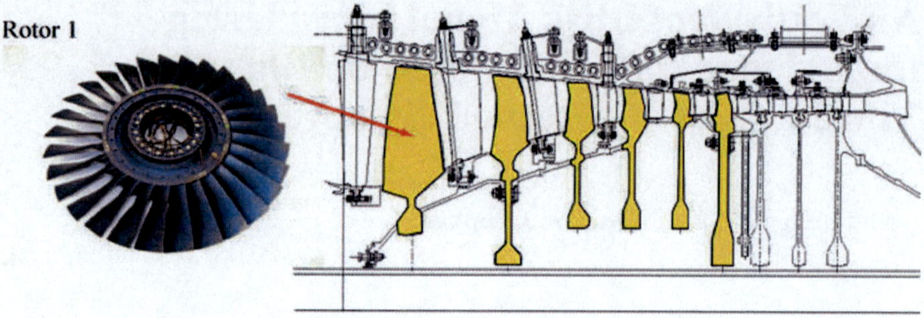

Fig. 1 The 9 stage E3E high pressure compressor [1], six rotors manufactured as blisk

mistuning is referred to as variations of mechanical properties of the rotor's blades. These variations lead to differing blade natural frequencies and a splitting of the orthogonal modes of the slightly damped cyclic symmetric blisk system.

Due to lack of friction at the contact faces of the blades' roots and the disk, compressor rotors manufactured as blisk ('blade integrated disk'), meaning the blades being an integral part of the disk, are characterized by only little mechanical damping. During operation a blisk may face engine order excitation which, in the case of a mistuned system, may lead to so called mode localization and an increase of blade amplitudes ranging between 20% and 402% [2–5] compared to the tuned response of an academically ideal blisk. Whitehead [2] found an upper limit of blade displacement magnification of the forced response due to mistuning given by $1/2(1+\sqrt{N})$. In [5] a mitigation of the blade displacement magnification due to the flow could be shown. In [6, 7] even a reduction of the forced response could be proved in case of large frequency mistuning.

The following work addresses the mistuning problem focusing on the effect of increasing mistuning in dependence on structural coupling under consideration of the flow. A real mistuned blisk of a high pressure compressor is analyzed for that purpose (Fig. 1).

2 Mistuning Identification and Structural Modeling

Aiming at an identification of structural mistuning as introduced in [8, 9] blade by blade measurements are carried out. Impact excitation is employed to each blade step by step applying a miniature hammer and the vibration velocity is measured at the same blade based on laser-Doppler-vibrometry. In order to isolate a blade dominated frequency in the particular FRF all blades except the excited one are detuned with additional mass elements to destroy the slightly disturbed cyclic symmetry of the mistuned blisk completely. As a result distributions of blade dominated frequencies are identified (Fig. 2).

The experimental modal information is used to update low degree of freedom structural models (one for each blade mode shape), the Equivalent Blisk Models [5]

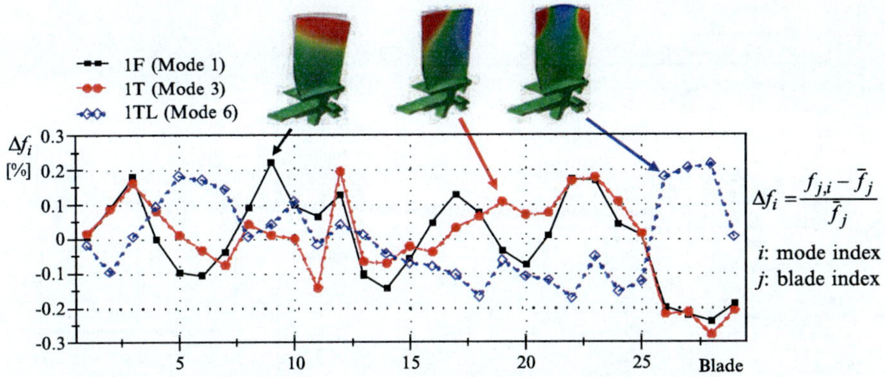

Fig. 2 Normalized blade dominated frequency distributions and blade modes

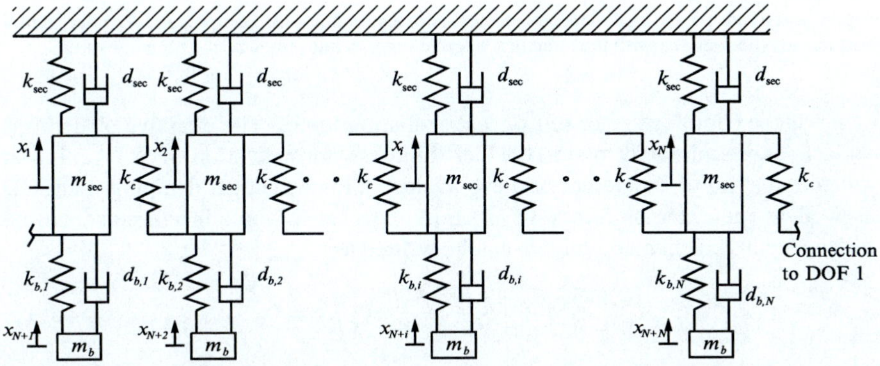

Fig. 3 Equivalent Blisk Model (EBM)

(Fig. 3). The update of each model is based on an adjustment of both the blade stiffness ($k_{b,i}$) and the damping ratio D_i of each blade in an iterative approach requiring that measured and calculated FRF match.

Aiming at a consideration of flow effects, the method of aerodynamic influence coefficients (AIC) as introduced among others by Hanamura et al. [10] and Nipkau [6] is implemented in the EBM with the objective to compute forced responses by means of sensitivity studies. The AIC are determined forcing one reference blade to perform a sine-shaped vibration in a particular mass normalized blade mode shape $\boldsymbol{\Psi}$ in a blade assembly. As a result, flow perturbations and hence an unsteady static pressure distribution arise acting on each blade from which the AIC can be calculated [6] and written in an influence coefficients matrix $\underline{\hat{\mathbf{L}}}^{\Psi}$. Hence, the EBM equation of motion is given with

$$\underline{\mathbf{M}}\ddot{\underline{\mathbf{x}}}(t) + \underline{\mathbf{D}}\dot{\underline{\mathbf{x}}}(t) + \underline{\mathbf{K}}\underline{\mathbf{x}}(t) = \underline{\hat{\mathbf{L}}}^{\Psi} \cdot m_b \underline{\mathbf{x}}(t) + \underline{\mathbf{F}}^F(t). \tag{1}$$

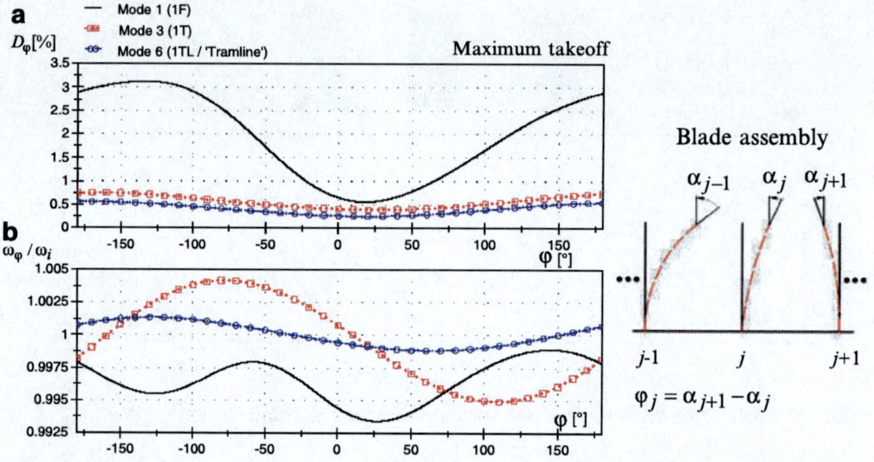

Fig. 4 Interblade phase angle-dependent aeroelastic eigenvalues: **(a)** Damping **(b)** Angular natural frequencies (normalized with the structural angular eigenfrequency ω_i of the *i*th mode)

$\underline{\mathbf{F}}^F$ is the complex vector of external excitation forces. The effective blade mass m_b acts as a scaling factor to consider the mass normalization of $\mathbf{\Psi}$. Aiming at a determination of the aeroelastic eigenvalues the mechanical damping matrix $\underline{\mathbf{D}}$ as well as the external forces $\underline{\mathbf{F}}^F$ in Eq. 1 are set to zero. In consequence the homogeneous eigenvalue problem can be written as

$$\left\{\lambda_\varphi^2\underline{\mathbf{M}} + \left[\underline{\mathbf{K}} - \underline{\hat{\mathbf{L}}}^\Psi \cdot m_b\right]\right\}\hat{\mathbf{x}} = 0 \tag{2}$$

with

$$\lambda_\varphi = -D_\varphi\omega_i + j\,\omega_\varphi \tag{3}$$

naming the aeroelastic eigenvalue λ_φ containing the aerodynamic damping described with the damping ratio D_φ and the angular aerodynamic natural frequency ω_φ. The results in Fig. 4a show that the aerodynamic damping contribution of mode 1 is between 3 and 4.5 times higher compared to modes 3 and 6. In general, clearly higher damping values arise for high absolute interblade phase angles φ. The dependence of ω_φ on φ (Fig. 4b) is inconsistent and its deviation from the angular structural frequency ω_i remains clearly below 1%.

3 Sensitivity Analyses

Applying a forward rotating engine order (EO) excitation, forced response analyses are carried out with respect to increasing mistuning, changes in disk stiffness and hence, the coupling of disk and blade motion. Exemplarily, the blade modes 1 (1F),

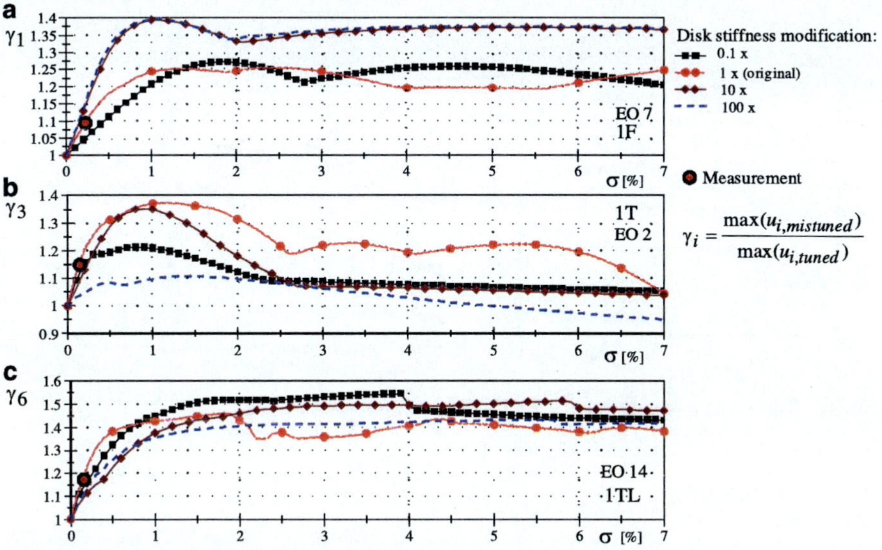

Fig. 5 Amplification of the maximum tuned displacements due to mistuning

3 (1T) and 6 (1 TL) excited by the operationally relevant EO 7 (1F), EO 2 (1T) and EO 14 (1TL) are chosen. The modification of mistuning is accomplished scaling up the measured frequency distributions given in Fig. 2 meaning that the shape of mistuning is maintained and only a modification of the standard deviation σ is assumed.

The results given in Fig. 5 show that apart from one exception a general amplification of the maximum blade displacement compared to a tuned blisk arises. Only for mode 3 and an approximately rigid disk (disk stiffness 100-fold increased, Fig. 5b) a decrease of the mistuned response below the tuned one for standard deviations greater than 5% is obvious. However, considering EO2-excitation also for modes 1 and 6 (Fig. 6) and a rigid disk, a general decrease below the tuned response arises already at $\sigma = 0.4\%$ (mode 1) and $\sigma = 0.5\%$ (mode 6). In case of a flexible disk, only mode 1 leads to $\gamma < 1$ ($\sigma > 1.3\%$). Since mistuning destroys the cyclic symmetry, a particular EO 2 does not only excite a blisk mode with 2 nodal diameters as is it the case for a tuned disk, but a linear combination of multiple EOs depending on the mistuned blisk mode shape (Fig. 6, right). Because the aerodynamic damping is comparatively low for low nodal diameter modes (low φ, see Fig. 4) but high for higher modes in case of mode 1, the aerodynamic damping increases with increasing mistuning which works against the otherwise response-amplifying effect of mistuning. In contrast, an EO14-excition generally increases the tuned response.

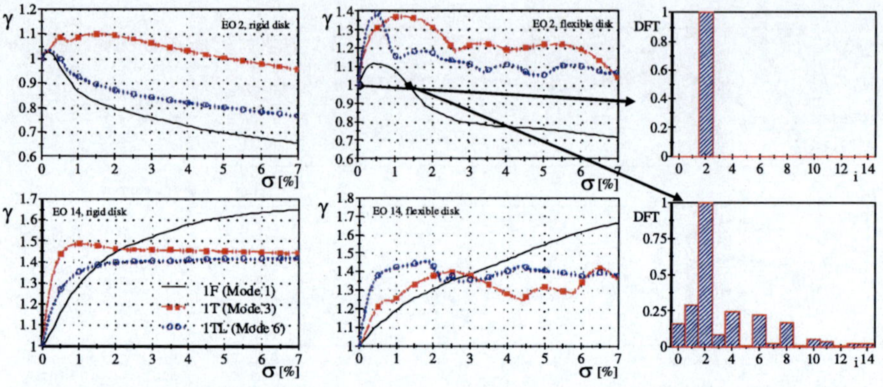

Fig. 6 Displacement amplification γ for EO 2 and 14 for a flexible and rigid disk

4 Conclusions

The paper addresses a methodology to calculate forced responses of a real blisk (Rotor 1, E3E-engine) employing discrete low degree of freedom models introduced as Equivalent Blisk Models (one for each blade mode) with a focus on the consideration of the fluid–structure interaction. Starting from a blade by blade frequency check by which mistuning is taken into account, the structural parts of the actual models are updated. In addition, the effect of increasing mistuning is analysed scaling up the measured distributions. Based on CFD-computations aerodynamic influence coefficients are calculated and implemented in the EBMs, which are henceforward ready for forced response simulations.

It has been shown that the response amplifying effect of mistuning can be attenuated due to the fluid structure interaction. An excitation of low engine orders, high mistuning and low coupling abets the attenuation and can even lead to a reduction of the forced response. Due to the strong dependence of the aerodynamic damping on the interblade phase angle, the fundamental flap mode 1 is most sensitive in this context. Nevertheless, since real frequency mistuning of modern, new blisks commonly does not exceed 1%, the amplifying effect of mistuning still plays a major role with respect to save engine operation.

References

1. Klinger, H., Lazik, W., Wunderlich, T.: The engine 3E core engine. In: Proceedings of the ASME Turbo Expo 2008, Paper GT2008-50679, Berlin, 9–13 June 2008
2. Whitehead, D.S.: Effect of mistuning on the vibration of turbomachine blades induced by wakes. J. Mech. Eng. Sci. **8**, 15–21 (1966)
3. Ewins, D.J.: The effects of detuning upon the forced vibrations of bladed disks. J. Sound Vib. **9**, 65–79 (1969)

4. Petrov, E.P., Ewins, D.J.: Analysis of the worst mistuning patterns in bladed disk assemblies. ASME J. Turbomach. **125**, 623–631 (2003)
5. Beirow, B., Kühhorn, A., Schrape, S.: A discrete model to consider the influence of the air flow on blade vibrations of an integral blisk compressor rotor. In: Proceedings of the ASME Turbo Expo 2008, Paper GT2008-50613, Berlin, 9–13 June 2008
6. Nipkau, J.: Analysis of mistuned blisk vibrations using a surrogate lumped mass model with aerodynamic influences. Ph.D. Thesis, Brandenburg University of Technology (2011)
7. Petrov, E.P.: A method for forced response analysis of mistuned bladed discs with aerodynamic effects included. In: Proceedings of the ASME Turbo Expo, Paper GT2009-59634, Orlando (2009)
8. Beirow, B., Kühhorn, A., Nipkau, J.: On the influence of strain gauge instrumentation on blade vibrations of integral blisk compressor rotors applying a discrete model. In: Proceedings of the ASME Turbo Expo 2009, Paper GT2009-59207, Orlando, 8–12 June 2009
9. Strehlau, U., Kühhorn, A.: Experimental and numerical investigations of HPC blisks with a focus on travelling waves. In: Proceedings of the ASME Turbo Expo, Paper GT2010-22463, Glasgow (2010)
10. Hanamura, Y., Tanaka, H., Yamaguchi, K.: A simplified method to measure unsteady forces acting on the vibrating blades in cascade. Bull. JSME **23**, 880–887 (1980)

Influence of Dry Friction Damping on Bladed Disk Vibration

Luděk Pešek and L. Půst

Abstract The paper deals with influence of dry friction damping on bladed disk vibration. Firstly the 3D FE solution and analysis of the imperfect blade disk dynamics is presented. Then the dry friction effect is studied analytically on linearized spring-dry-friction model consisting of equivalent linear stiffness and damping coefficients. An influence of nonlinearities and unproportional damping distribution on response curves is discussed as well.

Keywords Blade • Disk • Dynamics • Dry friction • Damping

1 Introduction

The flexural vibrations of imperfect bladed circular disk have been solved by analytical and numerical models [1]. The aim of this research is to evaluate the influence of elastic, mass and damping imperfections on the dynamic behavior of turbine disks. Nowadays experiments on the bladed disk model are carried out in the dynamic laboratory of the IT AS CR [2]. The disk imperfection results from implementation of two bunches of five blades with blade-to-blade head contacts into the original free-end beam blading. The bunches are fixed on opposite ends of the disk and introduces imperfections in mass, stiffness and nonlinear damping distribution on the periphery and non-proportional distribution of damping properties, too. In the first part the numerical FE solution presents method for analysis of dynamic properties of non-rotating or rotating disk. Spectral analysis of imperfect disk harmonically excited by turning force field is presented. The circular disk has many double eigenfrequencies. Positions of nodal diameters are arbitrary at

L. Pešek (✉) • L. Půst
Institute of Thermomechanics AS CR, v.v.i., Dolejškova 5, 18200 Praha 8, Czech Republic
e-mail: pesek@it.cas.cz; pust@it.cas.cz

J. Náprstek et al. (eds.), *Vibration Problems ICOVP 2011: The 10th International Conference on Vibration Problems*, Springer Proceedings in Physics 139,
DOI 10.1007/978-94-007-2069-5_75, © Springer Science+Business Media B.V. 2011

perfect disks, but they are fixed with respect to position of imperfection at imperfect disks. The spectrum of resonance peaks is twice denser than that of a perfect disk. Since the computation of the wheel dynamics with dynamic friction contacts is very time consuming the viscous linear damping is considered as approximation for the blade-to-blade friction coupling [1]. The spring-dry-friction model was also proposed and its corresponding equivalent linearized form for different friction and spring parameters was derived. The measured shift of eigenfrequencies at increasing friction is explained by the effect of non-proportional damping.

2 Spectral Analysis of the FE Model Under Rotating Forces

Experimental bladed wheel model consists of a steel disk with 50 beam blades completed with two bunches of blades ended by heads linked by wedge-shaped prisms. The friction couplings between the blades and prisms create damping effect. The disk is fixed in its center overhung on the rotating shaft. For theoretical solution of the wheel dynamics with friction contact a three-dimensional FE-model has been developed in ANSYS environment (Fig. 1). The 57 thousands of eight-node hexagonal elements SOLID45 were used for the finite element mesh. For modal analysis and dynamic response on the harmonic excitation the friction contacts were supplied by fixed connections. Lanczos's method was applied for calculation of eigenvalues and modes and harmonic balance method for evaluation of amplitude-frequency spectra. This paper investigates the eigenfrequencies of the two lowest modes of nodal diameters ($n = 1$ and $n = 2$) and no nodal circle. The lowest pairs of eigenfrequencies of a perfect non-rotating disk with 60 prismatic blades are: $f_1 = 59.21$ Hz and $f_2 = 78.60$ Hz [1, 2]. Due to the bunches of five blades, the bladed disk lost its perfect circular properties and became imperfect with a countable number of axes in this case two or four symmetry axes. The perfect disk has double eigenfrequencies f_1, f_2, which split into two pairs of close eigenfrequencies at the imperfect disks: $f_{1a} = 59.02$ Hz and $f_{1b} = 52.46$ Hz for $n = 1$ and $f_{2a} = 77.98$ Hz and $f_{2b} = 72.70$ Hz for $n = 2$.

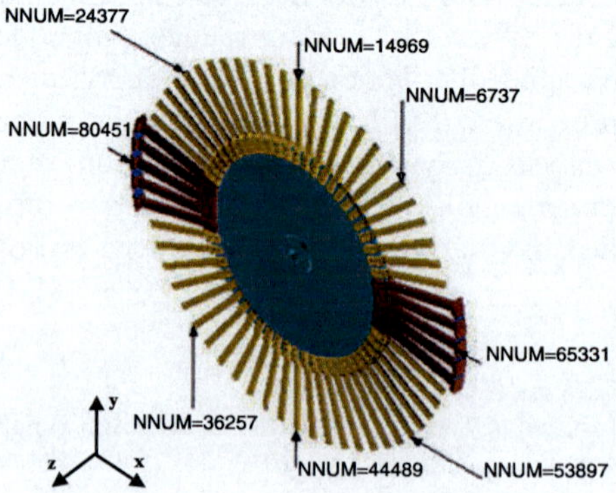

Fig. 1 FE wheel model

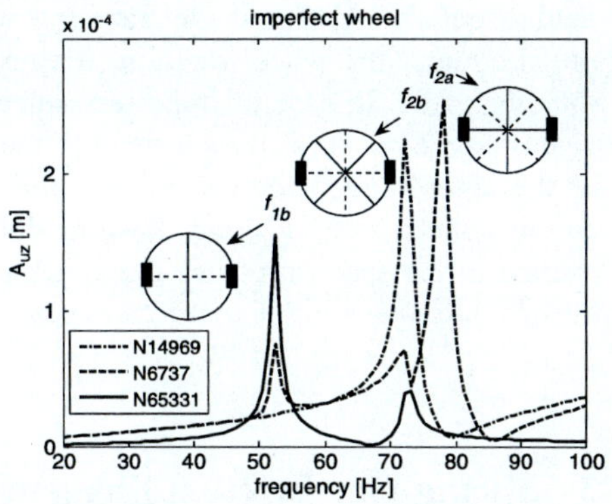

Fig. 2 Amplitude-frequency response spectrum of nodes N14969, N6737 and 65331 of the wheel with association to the modes (n = 1, 2) with nodal (*solid*) and anti-nodal (*dashed line*) diameters

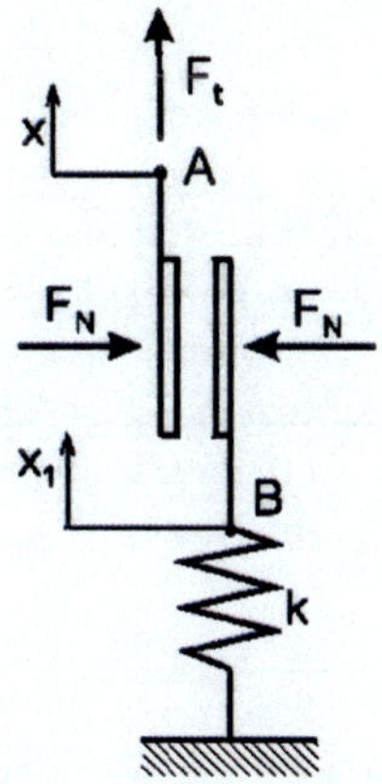

Fig. 3 Spring-damped system

For numerical simulations under rotation the turning force field excited the wheel in steady non-rotating space. Forces of the turning field for optimal excitation of eigen-mode with *n* nodal diameters [1] can be described by

$$F(\theta_{ai}) = k\cos(n\theta_{ai} + \omega t), \quad i = 1, 2, \ldots, nf \tag{1}$$

where k is an amplification coefficient, θ_{ai} is a force circumferential position, ω is excitation frequency, nf number of forces. For excitation of the eigen-mode with $n = 2$, the eight forces ($nf = 8$) placed at the ends of blades with position $\theta_{ai} = (i-1)\pi/4$ ($i = 1, \ldots, 8$) and $k = 1$ were chosen. Then the force field (1) moves in Cartesian co-ordinate system clockwise with a revolution frequency ω/n.

The wheel is embedded in the hub. Damping ration 1% was chosen for the harmonic analysis. The amplitude frequency spectrum of axial displacements uz in three selected nodes for imperfect (with bunches) wheel under designed turning force field is drawn in Fig. 3. For symmetric perfect wheel the only resonance f_2 ($n = 2$) is achieved. This resonance occurs when the rotation speed of the turning field is equal to $f_2/2$ and the traveling wave of the wheel moves with the same speed counter the wheel rotation. It causes the standing wave in the steady non-rotating space. In case of imperfect wheel, however, there are two resonances for eigenfrequencies f_{2a} and f_{2b} instead of the one. Nevertheless due to higher splitting of the eigenfrequencies the corresponding modes are excited separately without building the travelling wave. Besides the extra resonance for the frequency f_{1b} is excited in the spectrum since the associated eigenmode is not fully orthogonal to the excitation vector due to different stiffness of the free-ended blades and bunches.

3 Spring-Dry-Friction Element and Its Effect on Resonance Peak

Friction connection with linear elastic micro-deformation can be modeled by a spring-damper system shown in Fig. 3. Let us suppose a harmonic motion $x = a\cos\omega t$ of point A at initial conditions $\tau = \omega\, t = 0$, $x = a$, $x_1 = a - F_{t0}/k$, where $F_{t0} = fF_N$, is dry friction Coulomb force.

Motions $x(\tau)$, $x_1(\tau)$ during one period of non-dimensional time $\tau \in (0, 2\pi)$ are drawn in Fig. 4. The motion $x(\tau)$ of point A is supposed to be continuous harmonic, but the motion $x_1(\tau)$ of point B is twice interrupted during a period [3].

In interval $0 < \tau < \tau_1$ is x_1 constant

$$x_1 = a - F_{t0}/k, \tag{2a}$$

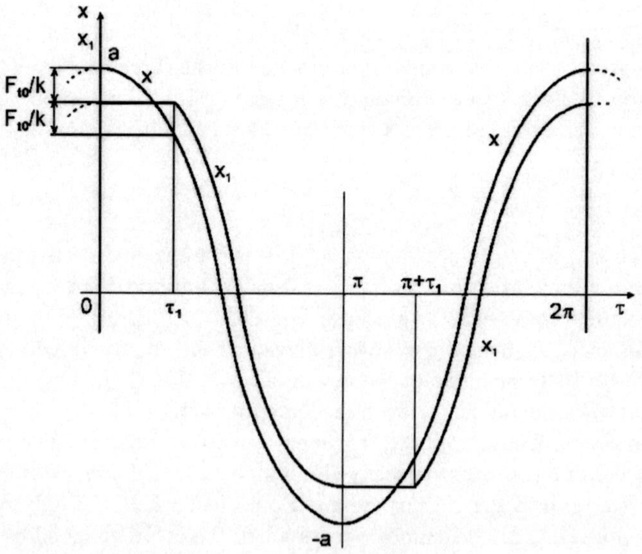

Fig. 4 Time history of motion x, x_1

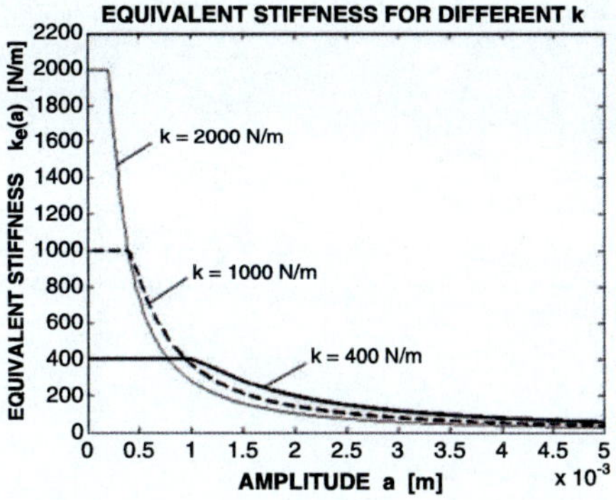

Fig. 5 Equivalent stiffness $k_e(a)$

where the time τ_1 of change is

$$\tau_1 = \arccos(1 - 2F_{t0}/(ka)). \tag{2b}$$

In this time the strength of spring reaches the opposite friction force – F_{t0} and the point B starts motion $x_1(\tau)$ with a constant shift $\Delta = F_{t0}/k$ against the motion x during time interval $\tau_1 < \tau < \pi$: $x_1 = a\cos\tau - F_{t0}/k$.

The same motions $x(\tau)$, $x_1(\tau)$ but with opposite sign occur in the second half period $\pi < \tau < 2\pi$: The friction force F_t (or spring force in point B) is given by the difference $x - x_1$ multiplied by stiffness k:

$$\begin{aligned} &F_t = k(a\cos\tau - a + \Delta) \text{ for } 0 < \tau < \tau_1, \quad F_t = -k\Delta = -F_{t0} \text{ for } \tau_1 < \tau < \tau \\ &F_t = k(-a\cos\tau + a - \Delta) \text{ for } \pi < \tau < \pi + \tau_1,\ F_t = k\Delta = F_{t0} \text{ for } \pi < \tau_1 < \tau < 2\pi. \end{aligned} \tag{3}$$

Force F_t depends both on displacement $x = a\cos\omega t$ and on velocity $\dot{x} = -a\omega\sin\omega t$: $F_t(x, \dot{x})$ and can be replaced by an equivalent linear expression

$$F_t(x, \dot{x}) = k_e(a)x + b_e(a\omega)\dot{x}. \tag{3a}$$

Multiplying function (3) by $\cos\tau$ and integrating we get the equivalent linear stiffness $k_e(a)$ [3]. Similarly we get the equivalent linear damping coefficient $b_e(a\omega)$ by multiplying F_t by function $\sin\tau$ and integrating.

Dependence of equivalent stiffness $k_e(a)$ and damping $b_e(a\omega)$ on amplitude a of vibrations is for $F_{t0} = 0$, 4 N/m shown in Figs. 5 and 6 for three spring stiffness $k = 400$, 1,000, 2,000 N/m. For very small amplitudes $a < F_{t0}/k$, only elastic deformation occurs at constant stiffness value $k_e(a) = k$.

Up to the limit $a = F_{t0}/k$, no energy dissipation exists: $b_e(a\omega) = 0$. After crossing this limit, stiffness quickly decreases and damping coefficient $b_e(a\omega)$ increases, reaches optimal value and then again slowly decreases. Resonance peaks of bladed disk computing model in Fig. 2 as well as in [1, 4] are narrow and a predominant

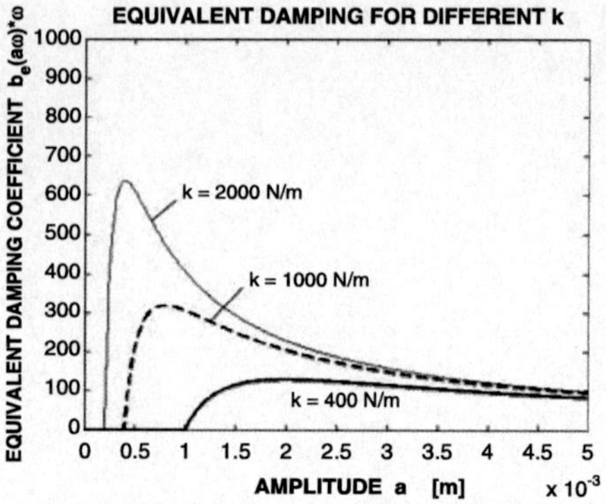

Fig. 6 Equivalent damping $b_e(a)$

sinus form characterizes the resonance vibrations. Therefore the application of equivalent linearization is justifiable. Let us see, how the addition of spring-dry-friction element to the linear vibrating system changes the form, shift and height of resonance peak. Motion in the narrow resonance zones of bladed disk can be described by

$$\ddot{q} + b\dot{q} + k_1 q = F_0 \cos \omega t, \tag{4}$$

where in the first mode: $m = 6, 8$ kg, $k_1 = 740{,}000$ N/m, $b = 5$ Ns/m, $F_0 = 4$ N.

Including equivalent linear damping and stiffness into (4) gives

$$m\ddot{q} + (b + b_e(a\omega))\dot{q} + (k_1 + k_e(a))q = F_0 \cos \omega t. \tag{5}$$

with response $q = a\cos(\omega t - \psi)$, Solution of Eq. 5 gives nonlinear algebraic equation of response curve $a(\omega)$. These curves are drawn e.g. in Fig. 7a. Spring-friction damping element contains two parameters: dry friction force F_{t0} and stiffness k. Influence of stiffness k on response curve of a weakly damped system ($b = 5$, $F_{t0} = 2, 5$ N) is insignificant.

Influence of dry friction force F_{t0} on position and height of resonance peak is demonstrated in Fig. 7a for $F_{t0} = 0$; 1; 2; 3 N at constant spring stiffness $k = 10{,}000$ N/m. Due to increasing friction force F_{t0}, the resonance peak decreases and shifts to higher frequency.

These shifts of resonance are small and are negligible for model of bladed disk IT ASCR. The amplitude fall is important, but it is different from the gained measurement data. Equation 5 is derived at the assumption that the addition damping is uniformly distributed on the whole periphery and therefore the height of resonance peaks continuously decreases at increasing friction force F_{t0} as shown in Fig. 7a. However, the experimental disk IT ASCR in Fig. 1 has damping heads concentrated only in two separated places, which changes this structure into non-proportionally damped system [5]. Computational model of such system is in

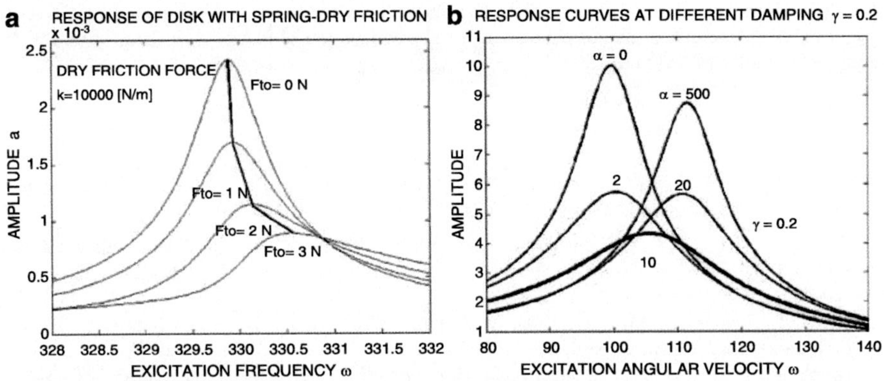

Fig. 7 Comparison of response curves. (**a**) Proportionally damped system, (**b**) non-proportional damping

Fig. 7b. The effect of such damping distribution on response curves is shown in Fig. 7c, where resonance amplitudes increase at very low ($\alpha \cong 0$) and at very high additional damping ($\alpha \cong 500$), but where an optimum damping exists ($\alpha \cong 10$).

4 Conclusion

In the numerical approach of the wheel dynamic the amplitude-frequency spectra of axial displacements for perfect and imperfect wheel under turning force field were presented. This loading simulates the loading of the rotation wheel by forces coming from e.g. static pressure field or permanent magnets. For symmetric case the only resonance occurs when the travelling wave of the wheel moves with the same speed counter the wheel rotation. It causes the standing wave in the steady non-rotating space. In case of imperfect wheel in the spectrum there are two resonances instead of one in the previous case due to the eigenfrequency splitting. Nevertheless due to higher splitting of the eigenfrequencies the corresponding modes are excited separately without building the travelling wave.

The main attention in the theoretical part was focused on the analysis of nonlinear properties of spring-dry-friction element and its linearization by means of equivalent linear stiffness and equivalent linear damping coefficient. It is shown that the influence of local stiffness in contact point on the damping level and in consequence of this also on height of resonance peak is practically small.

More important is the non-uniform distribution of damping elements on the periphery of disk, which change the disk into a non-proportionally damped system. The possibility of damping optimization in such a case is shown.

Acknowledgement This work has been elaborated in a frame of the grant project GA CR 101/09/1166 "Research of dynamic behaviour and optimisation of complex rotating system with non-linear couplings and high damping materials".

References

1. Půst, L., Pešek, L.: Travelling waves in circular disc with imperfections. In: Proceedings of the International Conference on Dynamical System Theory and Applications, Lodž, pp. 345–352 (2009)
2. Pešek, L., Vaněk, F., Půst, L., Bula, V.: Testing of dynamics of blade wheel with double periodicity. Eng. Mech. **17**(3/4), 161–173 (2010)
3. Brepta, R., Půst, L., Turek, F.: Mechanical Vibrations (in Czech – Mechanické kmitání), TP 71. Sobotáles, Praha (1994)
4. Rao, J.S.: Turbomachine Blade Vibration. Wiley Eastern Limited, New Delhi (1991)
5. Půst, L., Pešek, L.: Non-proportional nonlinear damping in experimental bladed disk. Eng. Mech. **17**(3/4), 237–251 (2010)

Damping Capacity of Superelastic Nickel-Titanium Plain Textiles

L. Heller, B. Marvalová, J. Vlach, K. Janouchová, M. Syrovátková, and J. Hanuš

Abstract Presented experimental work analyzes the effect of hysteretic superelastic material behaviour on the vibration damping capacity of plain weft knitted fabrics made of superelastic Nickel-Titanium wires. The damping is evaluated while the fabric is clamped at a circular boundary and subjected to out-of-plane mechanical loadings. Through this experiment, we mainly analyze whether the hysteretic superelastic behavior of Nickel-Titanium wires can effectively increase the damping capacity of weft knitted fabrics. For this purpose, we present and discuss comparison of experimentally identified damping capacity of samples containing, on one hand, wires exhibiting superelasticity and, on the other hand, common elastic stainless steel wires. Moreover, effects of knit size and loading parameters such as vibration amplitude, frequency and mean force are analyzed. The presented study shows clearly that the superelasticity contribute to the overall damping capacity of knitted fabrics. However, this contribution is unexpectedly low as it doesn't exceed 25% of the overall damping capacity. Therefore, it is believed that, within the range of applied loadings and for given structural parameters, elasticity is a dominant deformation mechanism in the Nickel-Titanium wires; whereas, superelastic deformation takes place in a minor volume fraction of wires.

Keywords Nitinol • Superelasticity • Superelastic damping • Weft knitting

L. Heller (✉)
Institute of Physics AS CR, Prague, Czech Republic
e-mail: heller@fzu.cz

B. Marvalová • J. Vlach
Faculty of Mechanical Engineering, Technical university of Liberec, Liberec, Czech Republic
e-mail: bohda.marvalova@tul.cz; vlach.jarmil@seznam.cz

K. Janouchová • M. Syrovátková • J. Hanuš
Faculty of Textile Engineering, Technical university of Liberec, Liberec, Czech Republic
e-mail: katerina.janouchova@tul.cz; martina.syrovatkova@tul.cz; jaroslav.hanus@tul.cz

J. Náprstek et al. (eds.), *Vibration Problems ICOVP 2011: The 10th International Conference on Vibration Problems*, Springer Proceedings in Physics 139,
DOI 10.1007/978-94-007-2069-5_76,

1 Introduction

Commercially available metallic knitted wire meshes made of common metals such as stainless steel, plain-carbon steel, copper, tungsten etc. are used in many applications such as vibration and shock absorbers, engine catalysts, flame arrestors etc. [1].

The vibration damping capacity of metallic knitted wire meshes arises from two effects. First, thanks to structural morphology of wire meshes, friction at wires interlockings and contacts contribute largely to the overall damping capacity. Second, material damping is an additional source of damping.

Shape memory alloys (SMAs) have been used in damping applications as they exhibit so called superelasticity. Mechanism responsible for superelastic deformation of SMAs is a diffusionless displacive martensitic transformation [2], which basically consists in a reversible change of crystal structure upon loading. in SMAs has been recognized as an effective deformation mechanism for two reasons. First, superelasticity allows for large recoverable strains that can reach up to 10%, which is not common to metals. Second, cyclic superelastic deformation shows large hysteresis; therefore, superelasticity can be used for vibration damping. However, this so called superelastic damping of SMAs is highly dependent on three loading conditions – amplitude, mean force, and frequency [3].

Presented work analyzes the influence of the material damping capacity and structural parameters on the damping capacity of metallic plain weft knitted fabrics. First, we describe manufacture technology, structural parameters and wire materials used for all measured samples. Second, experimental set up and the damping capacity evaluation are described. Finally, experimental results in terms of the damping capacity dependence on wire material, structural and loading parameters are presented and discussed.

2 Metallic Weft Knitted Fabrics

Different types of plain weft knitted fabrics were made in order to test the structural and material effects on the damping capacity of weft knitted fabrics. Structural effects were evaluated by testing fabrics with different knit sizes; whereas, material effects were evaluated on fabric samples made of materials having different material properties, namely, Young modulus and material damping. In order to test effects of these two material properties, stainless steel and NiTi weft knitted fabric samples were made.

2.1 Stainless Steel Weft Knitted Fabrics

Stainless steel samples were manufactured from an austenitic steel wire 304 V purchased from Fort Wayne Metals Ltd. The wire was produced via cold drawing manufacture technology. No heat treatment was applied on the wire after the last stage of the cold drawing. Tensile behavior of this wire is characterized by ultimate tensile strength of ~2,100 MPa, ductility of ~1.6% and Young modulus of ~163 GPa. Only one structural type of stainless steel samples was manufactured, which is characterized by a knit size denoted H6. Dimensions of the knit are shown in Fig. 2.

2.2 NiTi Weft Knitted Fabrics

Several types of NiTi weft knitted fabrics were made, differing in the initial material state of NiTi wires, final heat treatments applied on fabrics and knit sizes. All used NiTi wires were of grade NiTi#1 as denoted by wire supplier Fort Wayne Metals Ltd. Wires have chemical composition 55.82 wt% Ni giving the wires superelastic behaviour at temperatures above 10°C.

NiTi wires in two different initial material states were used to manufacture weft knitted fabrics. Samples denoted NiTi HT and NiTi Hard were made of NiTi wires that were not heat treated after the final stage of cold drawing wire manufacture technology. Such so called NiTi hard wires exhibit an elastic tensile behavior characterized by ultimate tensile strength of ~1,800 MPa, ductility of ~4.5% and Young modulus of ~55 GPa. Samples denoted NiTi SA were made of straight annealed NiTi wires, which received a final thermal treatment in a straight shape after the final stage of cold drawing wire manufacture technology. Therefore, such wires exhibit superelastic behavior as depicted in Fig. 2 prior to fabric manufacture. Such a behavior is characterized by an appearance of loading and unloading stress plateaus that delimit the stress-strain hysteresis (see Fig. 2). Hence, samples denoted NiTi HT or NiTi Hard were knitted from an elastically behaving NiTi wires; whereas, samples denoted NiTi SA were knitted from NiTi wires able to undergo superelastic deformation during knitting.

Weft knitted fabrics made of NiTi hard wires were either heat treated in a prestretched shape (samples denoted NiTi HT) in an air furnace at 450°C for 30 min or not heat treated (samples denoted NiTi Hard). Applied heat treatment has two effects. First, it induces the functional behavior of NiTi wires such as superelasticity i.e. tensile behavior of such a wire switches from linear elastic one to the one depicted in Fig. 2. Second, it fixes the wire stress free shape into the one in which the wire is kept during the heat treatment. Therefore, the wire accommodates the knit shape and, consequently, heat treated fabrics are relieved from internal stresses introduced by deformation during knitting. In contrast, fabric samples that were not heat treated after knitting (i.e. samples denoted NiTi Hard and NiTi SA) contained internal stresses. To sum up, the samples denoted NiTi HT were free of internal stresses and the initial microstructure was purely austenitic capable of undergoing the superelastic deformation upon loading. Samples denoted NiTi Hard contained an elasticallyprestressed NiTi wires that don't have the ability to be deformed superelastically (see Fig. 1).

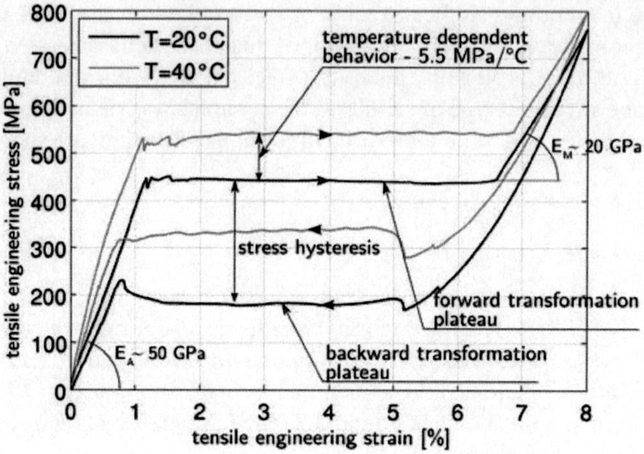

Fig. 1 Cyclic tensile stress-strain curves of a heat treated superelastic NiTi wire at two temperatures

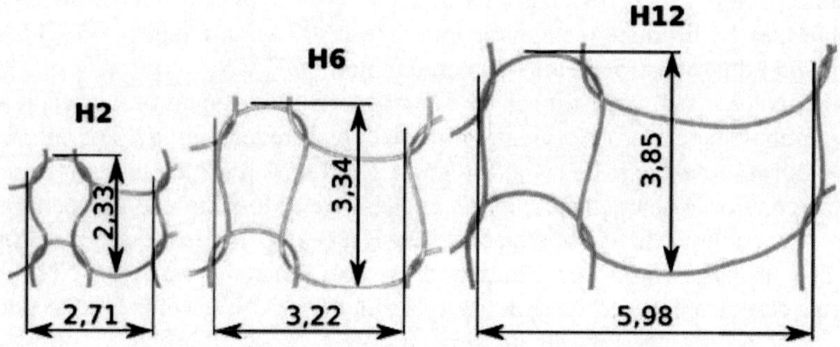

Fig. 2 Comparison of three different knit dimensions of tested weft knitted fabrics

Fabric samples denoted NiTi SA possess a specific material microstructure due to the initial material state of NiTi wires. As samples were knitted from a superelastic NiTi wires, the wires might undergo superelastic deformation in some volume fraction due to bending into knits during the knitting. Therefore, we suppose that the NiTi wire within this samples was in a partially transformed state i.e. microstructure consisted of a mixture of austenite and martensite phase. Consequently, superelastic deformation in these samples could proceed under lower loadings.

Finally, all described samples of NiTi fabrics were made with a knit size denoted H6 (see Fig. 2). In addition, samples of NiTi HT fabrics were made with two other knit sizes denoted H2 and H12 (see Fig. 2) in order to investigate the effect of structural morphology on the superelastic damping. As mentioned above, high strains are needed to trigger superelastic damping. Within a weft knitted fabric, the prevailing deformation mechanism is bending of individual knits. As the knit size

Table 1 Notation and wire material and structural properties of tested weft knitted fabrics

Fabric samples notation	Wire material and its initial state	Fabric heat treatment	Knit size		Young modulus [GPa]
Steel hard	Stainless steel, hard	No treatment	H6	No	163
NiTi hard	NiTi, hard	No treatment	H6	No	55
NiTi HT	NiTi, hard	450°C/30 min	H2/H6/H12	Yes	20–50
NiTi SA	NiTi, straight annealed	No treatment	H6	Yes	20–50

determines the bending rigidity of knits, the knit size also drives the ability of knits to undergo superelastic deformation during loading of the whole fabric. Therefore, three types of NiTi HT weft knitted fabrics were made, differing in the knit size as to identify the influence of knit size on the occurrence of superelastic deformation.

All tested samples, their notation and related initial wire material state, applied heat treatment, knit size, presence of superelastic behaviour and tensile Young modulus are listed in Table 1.

3 Experimental Setup

The damping capacity of weft knitted fabrics was evaluated experimentally under specific loading conditions. A tested fabric was clamped at a circular boundary of diameter 150 mm via a home-made rig and subjected to an out-of-plane pressure through an aluminum spherical cap with base radius of 30 mm and height of 30 mm. Both the clamping rig and the spherical cap were mounted on an Instron tensile machine ensuring mechanical loading. The loading of fabrics applied through the spherical cap consisted of a sequence of displacement controlled harmonic motions with different mean displacement, amplitude and frequency. The mean displacement defined a prestretch of a tested fabric. In order to ensure steady state vibration, each harmonic motion spanned over several tens of cycles. Then, an averaged damping capacity was evaluated using last ten cycles in terms of damping coefficient defined as a ration of the area of the force-displacement loop to the area under the loading part of the force-displacement loop.

4 Experimental Results

The damping capacity of all samples was evaluated within an amplitude range 1–5 mm, mean force range 1–10 N and frequency range 1–10 Hz. However, no considerable sensitivity of the damping capacity on frequency, within the studied range, was identified. Therefore, reported damping capacities represent averaged values of measurements performed at 1 and 2 Hz.

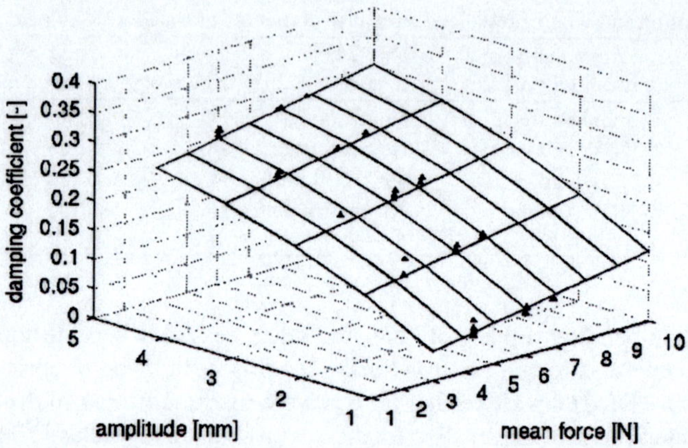

Fig. 3 Amplitude and mean force dependence of the damping coefficient of non-heat treated NiTi fabric NiTi Hard H6

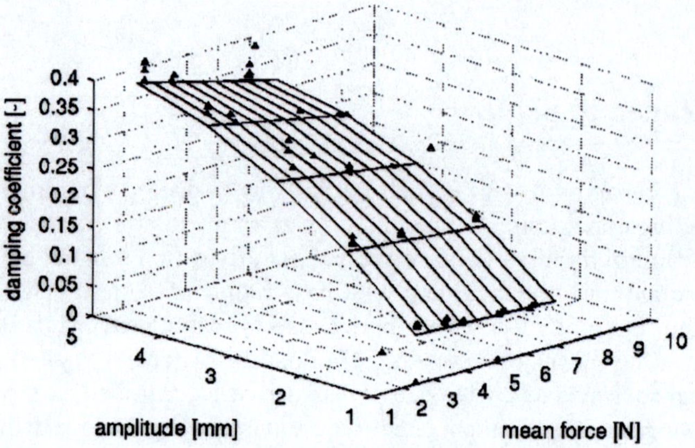

Fig. 4 Amplitude and mean force dependence of the damping coefficient of heat treated NiTi fabric NiTi HT H12

No unique dependence of the damping capacity on mean force was identified. The largest mean force dependence of the damping capacity showed non-heat treated NiTi fabric (NiTi Hard H6) and heat treated NiTi fabrics with the largest knit size (NiTi HT H12). NiTi Hard H6 fabric made of an elastic NiTi wire showed increasing mean force dependence (see Fig. 3); whereas, NiTi HT H12 fabric made of a superelastic NiTi wire showed decreasing mean force dependence (see Fig. 4). Other samples showed only slight dependence of the damping coefficient on mean force.

The damping capacity of all samples displayed increasing dependence on amplitude (see Fig. 5). Such a dependence was more pronounced in the case of samples made of wires exhibiting the superelastic damping i.e. NiTi HT H2/H6/H12, NiTi

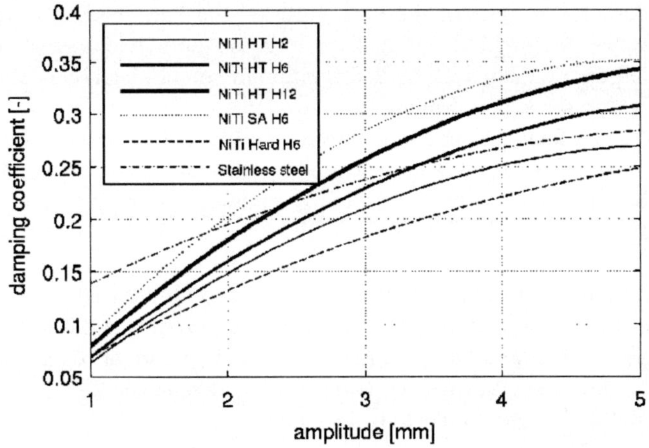

Fig. 5 Comparison of amplitude dependencies of the damping coefficient of all weft NiTi samples identified at mean force of 5N

SA H6. These samples, except for sample NiTi HT H2, showed higher absolute values of the damping capacity, at the highest amplitude, compared to the stainless steel sample. Moreover, the damping capacity at the highest measured amplitude of NiTi SA H6 was higher by 25% compared to the stainless steel sample.

5 Discussion and Conclusions

As postulated in introduction, the damping capacity of NiTi weft knitted fabrics is due to two dissipative mechanisms – (1) friction at wire contacts (2) superelastic damping. Both mechanisms are inherently amplitude dependent. On the other hand, friction is the only dissipative mechanism in the case of metallic weft knitted fabrics made of elastic material such as stainless steel and NiTi hard wires. Therefore, all tested fabrics showed amplitude dependence of the damping coefficient. However, samples incorporating wires exhibiting superelasticity showed stronger amplitude dependence as there are two amplitude dependent dissipative mechanisms acting at the same time.

NiTi SA H6 fabric made of superelastic NiTi wires showed damping capacity superior to other superelastic fabrics. Supposedly, applied mean forces were not sufficient to induce considerably martensitic transformation in NiTi HT samples that were initially in austenitic state. Therefore, superelasticity in NiTi SA H6 could proceed more easily upon applied amplitudes as wires in these samples were in partially transformed state due to deformation applied on them during knitting.

Knit size showed a considerable effect on the damping capacity of superelastic fabrics. The larger knit width the higher damping capacity was identified. Supposedly, it is due to the effect of the knit width on the knit bending rigidity, which decreases with increasing knit width. Therefore, the fabric with largest knit width is able to undergo larger superelastic deformation in bending and, hence, to dissipate more energy through superelasticity.

Only a slight effect of Young modulus on amplitude dependence of the damping capacity was identified from data measured on the stainless steel sample and NiTi hard wire having three times lower Young modulus. In contrast, absolute values of the damping differ considerably, which might be due to a different friction coefficient.

All in all, tested superelastic NiTi weft knitted fabrics showed a higher damping capacity (up to 25%) under certain circumstances in terms of the fabric structure, NiTi wire initial state and heat treatment. The best performance showed the NiTi fabric made from a straight annealed wire, which showed up to 25% higher damping capacity compared to the stainless steel sample.

Acknowledgments The research work reported was supported by Czech Science Foundation through project No. P108/10/1296.

References

1. http://www.metexcorp.com/kwm_intro.htm
2. Otsuka, K., Ren, X.: Physical metallurgy of Ti-Ni-based shape memory alloys. Prog. Mater. Sci. **50**(5), 511–678 (2005)
3. Heller, L., et al.: Factors controlling superelastic damping capacity of SMAs. J. Mater. Eng. Perform. **18**, 603 (2008)

Stress and Dynamic Analysis of Rotating Composite Disc

Rahul Dev, K. Gupta, and S.P. Singh

Abstract The present paper focuses on the dynamic behaviour and stresses in a fiber reinforced composite disc for possible applications to turbomachines. The discs of rotors in turbomachines, in addition to centrifugal loading, are also subjected to flexural vibrations. Variation of bending natural frequencies of composite disc in different modes, with several numbers of nodal diameters and nodal circles, is studied with respect to rotational speed and the various fibre reinforcement configurations. Natural frequencies associated with forward and backward travelling waves are obtained, which give the critical speed(s) of the disc. The semipolar configuration is studied in detail, in which the effect of the inclination of the reinforcement with the radial direction, on the dynamic characteristics of the disc is examined.

Keywords Composite disc • Polar orthotropic disc • Semipolar disc • Critical speed

1 Introduction

With the advent of high performance composites, more and more applications are being explored for effective and light weight designs. Present paper deals with application of composite material to discs, which are an important subcomponent of a rotor system. Researchers have examined the possibility of an all-composite-rotor [1] for turbomachines. Extensive work has been reported on application of fibre reinforced composite rotor shafts [2, 3]. There have been a few studies reported on composite discs,for example, on estimation of stresses and buckling loads in

R. Dev • K. Gupta (✉) • S.P. Singh
Mechanical Engineering Department, Indian Institute of Technology Delhi,
New Delhi 110016, India
e-mail: kgupta@mech.iitd.ernet.in

J. Náprstek et al. (eds.), *Vibration Problems ICOVP 2011: The 10th International Conference on Vibration Problems*, Springer Proceedings in Physics 139,
DOI 10.1007/978-94-007-2069-5_77,

an orthotropic disc [4], and stiffening of an optical disc [5] by circumferential reinforced composite annular ring. Koo [6] has explored the possibility of using composite data disc for larger data transfer rate in computers due to higher critical speeds. However, most of the work on dynamic analysis of composite disc has been limited to polar circumferential (CR) and radial reinforcement (RR) configurations. In the present work the effect of inclination of reinforcement with the radial direction, on the dynamic characteristics of the disc is studied in detail. The energy approach is used to set up the necessary governing equations, which are solved by Rayleigh Ritz method.

2 Theory of Orthotropic Disc

The composite rotating disc has several possibilities of fiber sequences and layer alignment. The polar orthotropic disc has the fibers in circumferential and/or radial directions, as shown in Fig. 1a, b. In Fig. 1c, a different orientation is used in which the fibers are neither radial nor circumferential, but at an angle to the radial direction. This is referred to as semipolar disc.

The stress strain relation for a single lamina, defined along the material axes (One for along fibre and two for perpendicular to the fibre), can be expressed as

$$\{\sigma_1 \ \sigma_2 \ \mathrm{T}_{12}\}^{\mathrm{T}} = [\mathrm{Q}_{\mathrm{ij}}]\{\varepsilon_1 \ \varepsilon_2 \ \gamma_{12}\}^{\mathrm{T}} \tag{1}$$

where $[Q_{ij}]$ is 3×3 stiffness matrix and its elements can be given as,

$$\mathrm{Q}_{11} = \frac{\mathrm{E}_{11}}{1 - \nu_{12}\nu_{21}}, \quad \mathrm{Q}_{22} = \frac{\mathrm{E}_{22}}{1 - \nu_{12}\nu_{21}}, \quad \mathrm{Q}_{12} = \frac{\nu_{12}\mathrm{E}_{22}}{1 - \nu_{12}\nu_{21}} = \frac{\nu_{21}\mathrm{E}_{11}}{1 - \nu_{12}\nu_{21}} \tag{2}$$

The other terms of $[Q_{ij}]$ are zero. For the semipolar configuration of Fig. 1c, in which the fibers are inclined to the radial direction, the necessary transformation along the radial and circumferential direction (r and θ coordinates) is given by

$$\{\sigma\} = \left[\bar{\mathrm{Q}}\right]\{\varepsilon\}, \text{ where } \left[\bar{\mathrm{Q}}\right] = [\mathrm{T}]^{-1}\,[\mathrm{Q}]\,[\mathrm{T}]^{-\mathrm{T}} \tag{3}$$

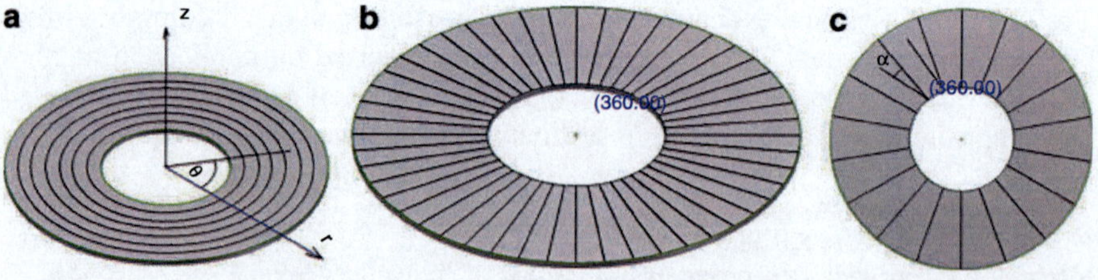

Fig. 1 **(a)** Circumferential **(b)** Radial reinforcement and **(c)** Semipolar configuration

where $[\bar{Q}]$ is transformed stiffness matrix. The matrix [*T*] is given by

$$[T] = \begin{bmatrix} \cos^2\alpha & \sin^2\alpha & 2\sin\alpha\cos\alpha \\ \sin^2\alpha & \cos^2\alpha & -2\sin\alpha\cos\alpha \\ -\sin\alpha\cos\alpha & \sin\alpha\cos\alpha & \cos^2\alpha - \sin^2\alpha \end{bmatrix} \tag{4}$$

In above expression, [*T*] is transformation matrix for a fiber angular position α as in reference [7]. For rotating disc, the strains, including bending as well as inplane membrane effects are written in cylindrical coordinate system (r, θ) as,

$$\{\varepsilon\} = \{\varepsilon_r \ \ \varepsilon_\theta \ \ \gamma_{r\theta}\}^T = \{\varepsilon_r^0 \ \varepsilon_\theta^0 \ \gamma_{r\theta}^0\}^T + z\{\kappa_r \ \ \kappa_\theta \ \ \kappa_{r\theta}\}^T = \{\varepsilon^0\} + z\{\kappa\} \tag{5}$$

where ε^0 and κ are midplane strains and curvatures respectively. From the classical laminate theory, forces and bending moments per unit length for an *N* ply laminate accounting for coupling effects, are

$$\{S_r \ \ S_\theta \ \ S_{r\theta}\}^T = [A_{ij}]\{\varepsilon^0\} + [B_{ij}]\{\kappa\} \text{ and } \{M_r \ \ M_\theta \ \ M_{r\theta}\}^T = [B_{ij}]\{\varepsilon^0\} + [D_{ij}]\{\kappa\} \tag{6}$$

where A_{ij}, B_{ij} and D_{ij} for an *N* ply laminate are given by,

$$A_{ij} = \sum_{k=1}^{N} (\bar{Q}_{ij})_k (z_k - z_{k-1}), \quad B_{ij} = \frac{1}{2}\sum_{k=1}^{N} (\bar{Q}_{ij})_k (z_k^2 - z_{k-1}^2) \tag{7}$$

$$D_{ij} = \frac{1}{3}\sum_{k=1}^{N} (\bar{Q}_{ij})_k (z_k^3 - z_{k-1}^3) \tag{8}$$

Here z_i are position of laminae plane from the midplane. The strain energy associated with bending effect is given as follows [6],

$$U_b = \frac{1}{2}\int_{R_I}^{R_O}\int_{0}^{2\pi} \{\sigma_r \ \ \sigma_\theta \ \ \sigma_{r\theta}\}\{\varepsilon_r \ \ \varepsilon_\theta \ \ \gamma_{r\theta}\}^T \, r dr d\theta \tag{9}$$

The disc is rotating at a speed of ω_r, and the strain energy due to centrifugal loading is given by Koo [6],

$$U_c = \frac{1}{2}\int_{R_I}^{R_O}\int_{0}^{2\pi} \left\{ S_r\left(\frac{\partial w}{\partial r}\right)^2 + S_\theta\left(\frac{\partial w}{r\partial\theta}\right)^2 + 2S_{r\theta}\frac{\partial w}{\partial r}\frac{\partial w}{r\partial\theta} \right\} r dr d\theta \tag{10}$$

Here, S_r, S_θ and $S_{r\theta}$ are inplane centrifugal loads per unit length induced because of rotation ω_r, which are found by solving the equilibrium equation of motion. For nonrotating disc, the inplane membrane strains in (5) are neglected. The kinetic energy of rotating disc for only transverse deflection w, is given by

$$V = \frac{1}{2}\int_{R_1}^{R_0}\int_0^{2\pi} \rho t_k \left(\frac{\partial w}{\partial t}\right)^2 r dr d\theta \tag{11}$$

For symmetric fiber orientation, coupling effect is also not present which eliminates [B] matrix. The solution for w is assumed in product form as follows,

$$w = W(r)\cos(n\theta)\, e^{j\omega t} \tag{12}$$

For the Rayleigh Ritz method, the function $W(r)$ is assumed as a series solution which consists of polynomial terms multiplied with unknown constants. Each polynomial term satisfies at least the geometric boundary conditions of the disc. The disc is fixed at inner radius and free at the outer radius. For rotating disc, the total strain energy is updated by the inclusion of strain energy due to inplane centrifugal loading as given in (10). The values of S_r and S_θ are obtained from the solution of equilibrium equations as in (13), while $S_{r\theta}$ vanishes for axisymmetric problems.

$$\frac{\partial}{\partial r}(rS_r) + \frac{\partial}{\partial \theta}(S_{r\theta}) - S_\theta + \rho r\,\omega_r^2 = 0, \quad \text{and} \quad \frac{\partial}{\partial r}(rS_{r\theta}) + \frac{\partial}{\partial \theta}(S_\theta) + S_{r\theta} = 0 \tag{13}$$

Substituting solution of S_r and S_θ in energy expression (10) and applying the Rayleigh Ritz method, stiffness matrix [K_c], because of energy due to membrane stresses is obtained while [K_b] is bending stiffness matrix. The eigenvalue problem for the case of rotating disc becomes of the form,

$$\left[-\omega^2[M] + [K_b] + [K_c]\right]\{X\} = \{0\} \tag{14}$$

3 Results

A disc with the same dimensions as in reference [6] is considered. The outer and inner radii are 60 mm and 15 mm respectively with uniform thickness of 1.2 mm. A composite disc with four layers and any of the two materials CFRP (carbon fibre reinforced plastic) or GFRP (glass fibre reinforced plastic) is taken up for analysis (Material properties are given in Table 1).

For rotating disc, the critical speed is that rotational speed at which the backward whirl frequency becomes zero. These backward whirl frequencies are plotted for CR (circumferential) and RR (radial) discs in Fig. 2. CR configuration gives higher critical speed than RR configuration. As shown in Fig. 2, GFRP polar CR orthotropic disc backward whirl frequency in [0/2] mode approaches zero at

Table 1 Material properties

Material	E_1 (GPa)	E_2 (GPa)	G_{12} (GPa)	ν_{12}	ρ (kg/m^3)
GFRP	38.6	8.27	4.14	0.26	1,800
CFRP	181.0	10.3	7.17	0.28	1,600

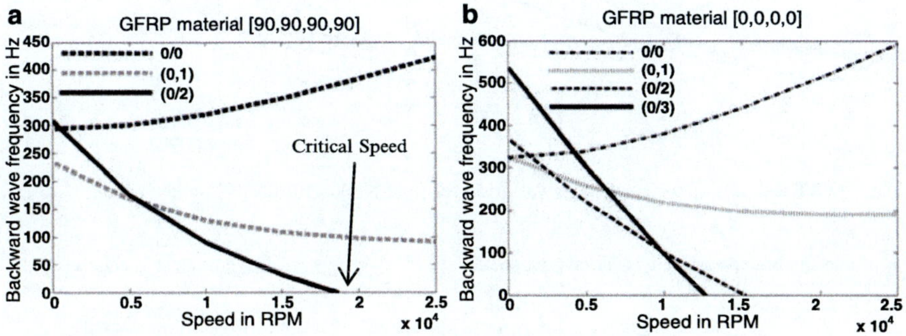

Fig. 2 GFRP polar orthotropic disc backward wave frequencies (CR (**a**) and RR (**b**))

Table 2 Summary of results of GFRP and CFRP rotating disc

Fiber orientation	Critical speed (RPM) GFRP	Critical speed (RPM) CFRP	Critical mode (NC/ND)
[0] (radial only)	12,450	19,300	[0/3]
[90] (circumferential only)	18,650	39,900	[0/2]
$[0/90]_s$	15,475	27,750	[0/3]
$[45]_4$	18,100	34,500	[0/3]

18,650 RPM. The results were in close match with results obtained by Koo [6]. In a disc mode, [NC/ ND] are number of nodal circles and nodal diameters respectively. In addition to CR and RR types, the disc is also analyzed with $[0/90]_s$ and semipolar $[45]_4$ configurations and results are shown in Table 2. Higher critical speeds for CFRP material with similar fibre configurations are listed in Table 2.

The effect of centrifugal loading on natural frequencies of disc for various fibre configurations was studied. As shown in Fig. 3 for umbrella mode [0/0], as well as for [0/1] mode, the stiffening effect of centrifugal loading in RR disc is maximum, while for the CR disc, it is minimum. For the semipolar $[45]_4$, and combined CR and RR $[0/90]_s$ configurations, the centrifugal stiffening effect lies in between these two. To check the effect of fibre orientation in semipolar configurations, the natural frequencies are obtained for various semipolar fibre orientations from pure radial to pure circumferential and results are shown in Fig. 4a. For ND = 1, natural frequency decreases continuously with fibre angle. For ND = 0, the natural frequency decreases upto 45° followed by marginal increase with higher fiber angle, for ND = 2 the trend is opposite. The variation of inplane radial and hoop (circumferential) stresses with fibre orientation obtained from the solution of

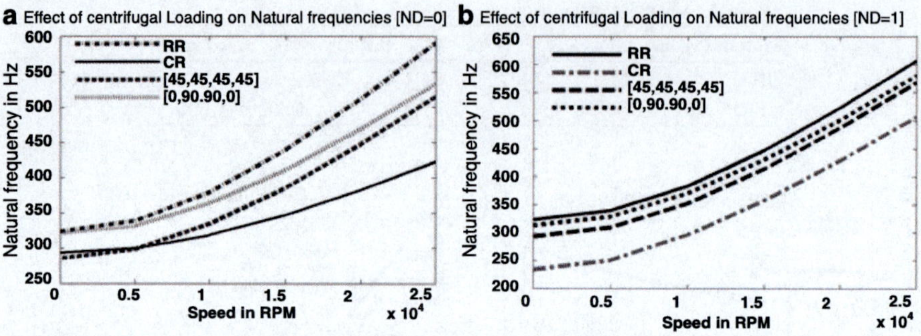

Fig. 3 GFRP disc natural frequencies variation for [NC/ND] = [0/0]&[0/1] with speed

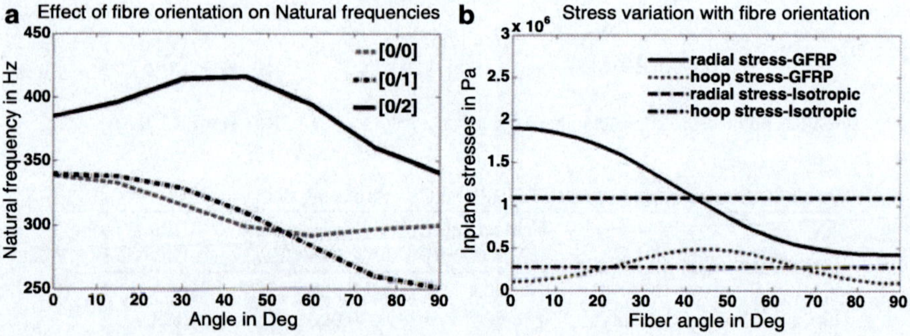

Fig. 4 Variation of natural frequencies and stresses with fibre orientation, (**a**) Disc natural frequencies at 5,000 RPM, (**b**) Inplane radial and hoop stresses at inner radius at 5,000 RPM

equilibrium equation (13) are shown in Fig. 4b for a typical case of 5,000 RPM at inner radius of disc. Due to difference in moduli of composite disc in the radial and circumferential direction the predicted stress varies with the fibre orientation. This is in contrast with isotropic material disc where centrifugal stresses are independent of longitudinal modulus.

4 Conclusions

Results show that centrifugal stiffening effect on bending natural frequencies of composite disc is most dominant for the radial fibre (RR) reinforcement, while its effect is less for the circumferential direction (CR) fibre reinforcement. From the analysis of the semipolar fibre reinforcement configuration, it is observed that the effect of fibre angle (with the radial direction) on natural frequencies of the rotating disc has different trends in modes with different nodal diameters. Three modes with nodal diameters equal to 0, 1 and 2 are analyzed.

References

1. Gupta, K.: Some issues relating to design and development of an all composite aero gas turbine engine rotor. Def. Sci. J. **4**(8), 93 (1998)
2. Singh, S.P., Gubran, H.B.H., Gupta, K.: Developments in dynamics of composite shafts. Int. J. Rotat. Mach. **3**, 189 (1997)
3. Singh, S.P., Gupta, K.: Composite shaft rotor dynamic analysis using layerwise theory. J. Sound Vib. **191**, 739 (1996)
4. Fu, L., Waas, A.M.: Buckling of polar and rectilinearly orthotropic annuli under uniform internal or external pressure loading. Comput. Struct. **22**, 47 (1992)
5. Koo, K.-N.: Vibration and critical speeds of composite-ring disks for data storage. J. Sound Vib. **329**, 833 (2010)
6. Koo, K.-N.: Vibration analysis and critical speeds of polar orthotropic annular disks in rotation. Compos. Struct. **76**, 67 (2006)
7. Jones, R.M.: Mechanics of Composite Materials. Taylor & Francis, Michigan (1999)

Identification of Missing Dynamics in Rotor Systems Using Robust Control Theory Approach

Jerzy T. Sawicki and Ryan Madden

Abstract Analytical models only approximate the true dynamics of analyzed rotating machines, due to the presence of components that are inherently difficult to model. Such models of rotating machines are driven by the best engineering knowledge and experience, and very often are updated based on experimental results. The problem of un-modeled or missing dynamics can be exacerbated in the presence of rotor structural damage such as a transverse crack on a shaft. This paper will present an effective approach for model updating using advanced tools developed in the robust control theory, specifically μ-synthesis. The methodology will be applied to the identification of minute changes in the dynamics of the rotor due to the presence of a transverse crack on a shaft. Experimental data collected from the cracked rotor rig will be utilized to validate the developed approach.

Keywords Identification • Dynamics • Rotor • Structural damage • Robust control

1 Introduction

Despite continuous improvements in the area of modeling of rotordynamic systems, some components still remain difficult to model. Examples of such components are bearings, seals, shrink fits, rotor-bearing interaction or foundations. An intricate system, containing multiple instances of parts and physical effects that are difficult to model, leads to an engineering model that can behave much differently than the actual system. The field of model updating has arisen to correct the engineering model by driving its output to match that of the experimental data.

J.T. Sawicki (✉) • R. Madden
RoMaDyC, Cleveland State University, Cleveland, OH, USA
e-mail: j.sawicki@csuohio.edu; r.madden90@csuohio.edu

J. Náprstek et al. (eds.), *Vibration Problems ICOVP 2011: The 10th International Conference on Vibration Problems*, Springer Proceedings in Physics 139,
DOI 10.1007/978-94-007-2069-5_78,

Maslen et al. [1] first developed the method of model reconciliation which adjusts analytic rotordynamic models to make them consistent with experimental data. The undertaken strategy by Maslen et al. [1] uses the basic approach of model-based identification, i.e., controlling the nominal model in order to minimize the error between its response and that of the experimentally identified system. The assumption in model-based identification is that the basic model structure is correct, but there is a missing dynamics that is left out due to modeling errors or unknown physical phenomena related to specific parts of the system. This method is innovative in that it distinguishes between the parts of the system which are not certain in modeling and the easily modeled portions. A controller is applied only at the uncertain locations and the systems are driven to minimize the modeling error. When the two responses match, the controller, or its dynamics, mimics the unmodeled dynamics that is missing in the nominal model. Vazquez et al. [2] applied model-based identification to identify the dynamics of magnetic journal bearings. Wang and Maslen [3] advanced the technique by the direct use of frequency response functions for the nominal and engineering systems, as well as by implementing a μ-controller to drive the model correction.

The objective of this paper is to apply the method of model-based identification to extract structural damage-induced dynamics in a rotor system. The structural damage is caused by a minute transverse crack on a rotor. This paper will present an outline of the robust control tools which are utilized in finding the unmodeled dynamics. Next, the experimental apparatus will be introduced. Subsequently, the approach for finding the locally induced crack dynamics will be presented. Finally, the results of the experiment will be shown along with an analysis of the resulting unmodeled fault dynamics.

2 Robust Control Tools for Model Updating

The model-based identification technique presented in this paper utilizes a controller synthesized using μ-synthesis [4]. The generated controller is a linear time invariant state-space controller which minimizes the structured singular value, μ, of the closed loop system. The necessary and sufficient condition of a successful μ-controller is that the maximum singular value of the closed-loop system, *clp*, is less than one [5]:

$$\sigma_{\max}(clp(j\omega)) \leq 1.0 \quad \forall \omega \in \mathbb{R} \tag{1}$$

This condition requires that the closed-loop input, w, and output, z, are weighted such that:

$$|\hat{z}_i| < 1.0 \quad : \quad |\hat{w}_i| < 1.0 \tag{2}$$

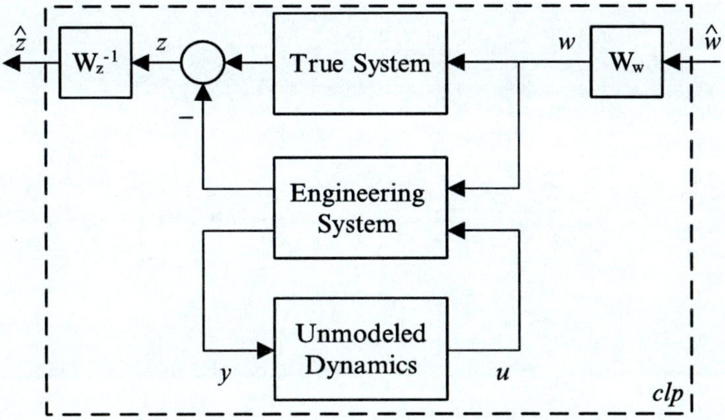

Fig. 1 μ-controlled model-based identification schematic for finding unmodeled dynamics

Such weighting introduces two performance criteria, W_z and W_w, which place limits on the norms of the input and output signals. When the criteria in Eqs. 1 and 2 are met, it means that a controller has been generated which meets the performance criteria.

Figure 1 illustrates the combination of μ-synthesis and model-based identification into a single control schematic. The unmodeled dynamics is found to minimize the weighted difference between the true and engineering system outputs due to a weighted external excitation. The output weighting, W_z, indicates the maximum allowable error between the true system and engineering system responses to the external input. This value will be decreased as much as possible while keeping $\mu \leq 1$ in order to ensure extracting of the most accurate unmodeled dynamics.

3 Application to Structural Damage Detection in Rotating Machines

The model identification method has been proven to be able to identify a known difference in dynamics in a simple example [3]. The developed approach now will be applied to identify a difference in dynamics induced by structural damage. Test data taken from a damage detection experimental rotor setup will be utilized to create a new model for the change in dynamics caused by the presence of a minute transverse crack in a rotor. The experimental rotor test rig employed in this study is shown in Fig. 2. The facility was described in detail in [6, 7]. The radial magnetic bearing serves as the exciter for a sine sweep system ID test. A crack is located at the bearing midspan and has a depth of 40% of the shaft.

The experiment of extracting the dynamics induced by a transverse crack on a shaft utilizes the same unmodeled dynamics control schematic as shown in Fig. 1.

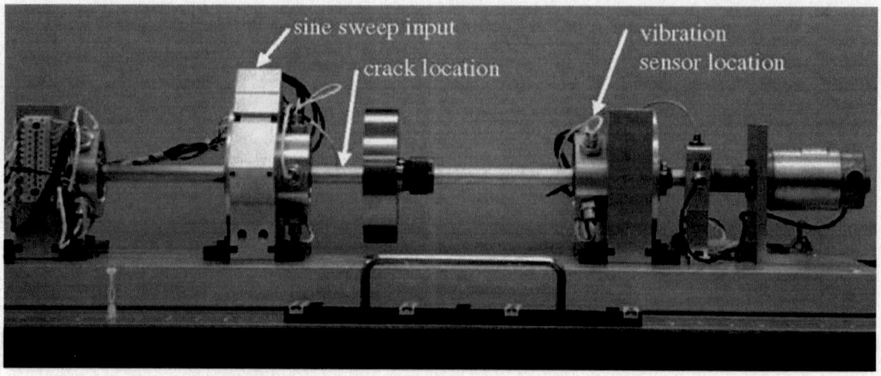

Fig. 2 Rotor crack detection test rig

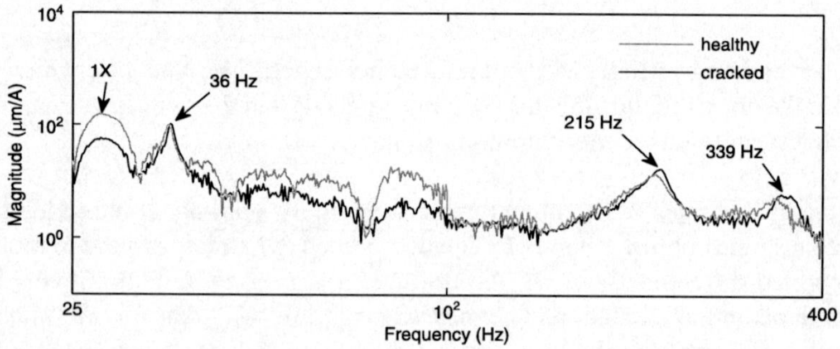

Fig. 3 Comparison of sine sweeps of healthy and cracked rotors running at 1,680 RPM (28 Hz)

In this case, the cracked rotor is taken as the true system, *truesys*, the healthy rotor is taken as the engineering system, *engsys*, and the dynamics induced by the presence of a transverse crack is the unmodeled dynamics. The output weighting, W_z, controls the maximum allowable error between the cracked and controlled healthy rotor vibrations at the sensor location.

Both the healthy and cracked models are constructed from identical sine sweep tests, with the input on the exciter bearing and the output measured at the conical MB rotor. The sine sweeps are taken while the rotor is rotating at 28 Hz (1,680 RPM), which is done to allow the crack to breathe, i.e., open and close during each shaft revolution. The results of these sine sweeps are shown in Fig. 3, with indicated running speed and critical speeds. The developed finite element (FE) based models are tuned to match the sine sweep trials. The inputs and outputs of the sine sweep trial are appropriately located in the FE model and an input and output at the crack location are included as well. This allows for the identification of the local change in dynamics due to the presence of a minute transverse crack, providing additional insight that may not be available by simply analyzing the results shown in Fig. 3.

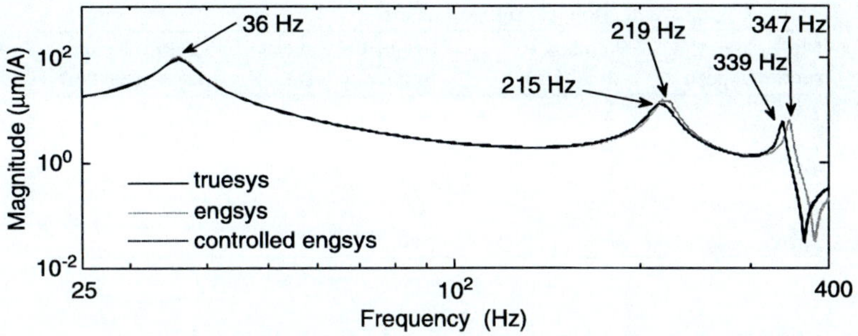

Fig. 4 Rotor model identification system responses

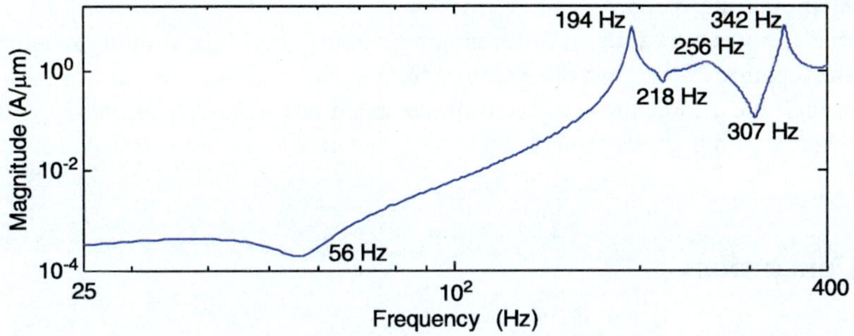

Fig. 5 Identified transverse crack dynamics

In the final step, the μ-controlled model-based identification code is run with a 5.25 μm maximum allowable error between the cracked and controlled healthy rotor vibrations at the sensor location. A corresponding μ-bound of 0.9926 is obtained, confirming that the performance specifications are met. An uncertainty of 1% is placed on the sensor for both the true and engineering systems. The true, engineering, and controlled engineering responses are shown in Fig. 4. As shown, the difference between the healthy (*engsys*) and cracked (*truesys*) engineering responses is eliminated by the addition of the controller on the healthy model (*controlled engsys*). The controller is so effective that the two plots overlap over the entire frequency range. The associated controller plot is shown in Fig. 5. This controller plot can be interpreted as a model for the local change in dynamics brought on by the presence of a minute transverse crack on a shaft at the bearing midspan.

Analysis of Fig. 5 reveals interesting conclusions about the localized effects of the crack. The controller response shows a more intricate behavior of the crack dynamics than just a flat reduction of stiffness. Natural frequencies are shown at 194, 256 and 342 Hz, along with anti-resonances at 56 and 307 Hz. Additionally, a valley occurs at 218 Hz though it is not anti-resonance. Explanations for the controller response points of interest can be found in the cracked and healthy system responses

Table 1 Crack detection test rig parameters

Multiple of the running speed	Frequency value (Hz)	Controller natural frequency (Hz)	Controller anti-resonance (Hz)
2×	56	–	56
7×	196	194	–
9×	252	256	–
11×	308	–	307
12×	336	342	–

in Fig. 3 and the associated models in Fig. 4. The valley at 218 Hz appears to be caused by the healthy system peak at 219 Hz and an attempt of the controller to dampen it. The natural frequencies and anti-resonances all appear to be a result of the 28 Hz running speed.

Table 1 shows the values of different integer multiples of the running speed along with the location of the controller natural frequencies and anti-resonances. The data shown in Table 1 indicates that the running speed has a close relationship with the local change in dynamics brought on by the presence of a minute transverse crack.

4 Conclusions

This paper shows the potential of the μ-controlled model-based identification strategy to determine the change in dynamics induced by the uncharacterized physical phenomena. The presented results address the identification of the structural damage induced dynamics for a possible enhancement of fault detection technologies. The extracted transverse crack dynamics demonstrates strong nonlinear characteristics and influence on the running rotor speed. The approach can be useful in the development of more accurate mathematical description for crack behavior and its incorporation into FE based rotordynamic codes.

Acknowledgments The research work reported here was partially funded by the NASA Aviation Safety and Security Program, "Research Opportunities in Aeronautics", Grant No. NNX08AC31A.

References

1. Maslen, E.H., Vazquez, J.A., Sortore, C.K.: Reconciliation of rotordynamic models with experimental data. ASME J. Eng. Gas Turbines Power **124**, 351 (2002)
2. Vazquez, J.A., Maslen, E.H., Ahn, H.-J., Han, D.-C.: Model identification of a rotor with magnetic bearings. ASME J. Eng. Gas Turbines Power **125**, 149 (2003)
3. Wang, Q., Maslen, E.H.: Identification of frequency-dependent parameters in a flexible rotor system. ASME J. Eng. Gas Turbines Power **128**, 670 (2006)
4. Zhou, K., Doyle, J.C.: Essentials of Robust Control. Prentice Hall, Upper Saddle River (1998)

5. Maslen, E.H., Sawicki, J.T.: Mu-synthesis for magnetic bearings: Why use such a complicated tool? In: Proceedings of the ASME 2007 International Mechanical Engineering Conference and Exposition (IMECE2007), Seattle, pp. 1 (2007)
6. Sawicki, J.T., Storozhev, D.L., Lekki, J.: Exploration of NDE properties of AMB supported rotors for structural damage detection. ASME J. Eng. Gas Turbines Power **133**, 1 (2011)
7. Sawicki, J.T., Friswell, M.I., Kulesza, Z., Wroblewski, A., Lekki, J.D.: Detecting cracked rotors using auxiliary harmonic excitation. J. Sound Vib. **330**, 1365–1381. Elsevier Ltd. (2011)

An Analysis on the Supporting Structure Representative Model in Rotating Systems

Felipe Wenzel da Silva Tuckmantel, Kátia Lucchesi Cavalca, Hélio Fiori de Castro, Patrick Felscher, and Richard Markert

Abstract Power generation systems are composed by several rotating system, which are supported by bearings, and are installed on foundation structure. For this reason, a representative model needs to take into account the components effects. Rotating systems numerical models are well known by the scientific community specialized in rotor-dynamics to predict the system dynamic behaviour. Instead, the foundation numerical model is more sensitive to uncertainties. Experimental models can be the solution to the foundation representation. In this case, some difficulty in identifying the modal parameters, mainly the damping factors, can influence the results. This work proposes a comparison between complete system response using both experimental models of mechanical impedance and mixed coordinates to represent a test-rig foundation. A classical approach is the assembly of the impedance matrix of the supporting structure directly from the flexibility matrix inversion. However, this technique can be limited by the number of degrees of freedom associated to the rotor-structure connecting elements. Therefore, a solution based on modal parameters of mass, stiffness and damping to represent the foundation is also applied and both models are associated to a rotor-bearing system model for comparison.

Keywords Rotor dynamics • Foundation or Supporting structure • Mechanical Impedance • Mixed coordinates

---

F.W. da Silva Tuckmantel
Schaeffler Brasil Ltda, Sorocaba, Brazil
e-mail: tuckmfli@schaeffler.com

K.L. Cavalca (✉) • H.F. de Castro
University of Campinas, Campinas, Brazil
e-mail: katia@fem.unicamp.br; heliofc@fem.unicamp.br

P. Felscher • R. Markert
TUD Technische Universität Darmstadt, Darmstadt, Germany
e-mail: felscher@sdy.tu-darmstadt.de; markert@sdy.tu-darmstadt.de

J. Náprstek et al. (eds.), *Vibration Problems ICOVP 2011: The 10th International Conference on Vibration Problems*, Springer Proceedings in Physics 139,
DOI 10.1007/978-94-007-2069-5_79,

1 Introduction

The goal of the rotor dynamics is to analyze how the vibration levels can be kept as small as possible. Therefore, its design and optimization become essential to study the behavior of the system. Therefore, to study the dynamic behavior of this system, it is necessary to determine the interaction of all components for understanding the phenomena involved.

Previous works were developed to simulate the behavior of a foundation [1–6]. In 1992, Stephenson [7] included the effects of the foundation into a rotating machine model through its modal parameters. However, this method always requires a number of identified modes equal to the number of measured points to keep the eigenvector matrix square, facilitating its inversion. To overcome this limitation, Cavalca [8, 9] used the method of mixed coordinates to include the effect of the foundation on the rotors, using physical coordinates for the rotor and generalized (or modal) coordinates for the foundation. The method has the advantage of allowing the usage of a variable number of modes, since it is not necessary to invert the eigenvector array. Therefore, the aim of this work is to compare the complete system response using mechanical impedance and mixed coordinates methods to represent a test-rig foundation. Finally, the critical points to be considered in each method can be highlighted.

2 Mathematical Modeling

The complete rotor-bearings-structure system is analysed as two subsystems separately: rotor-bearings and supporting structure [10]. Thus, each subsystem is analyzed and the response of the complete system is obtained by unifying the dynamic response of the subsystems. The equation of motion of the complete system considers the rotor inertia, mass, damping and stiffness of the shaft and the structure, as well as the dynamic coefficients of the bearings, which are obtained by numerical solution of Reynolds equation.

$$\begin{bmatrix} [M_{rr}] & [0] \\ [0] & [M_f] \end{bmatrix} \begin{Bmatrix} \ddot{q}_r(t) \\ \ddot{q}_f(t) \end{Bmatrix} + \begin{bmatrix} [C_{rr}]+[C_{ff}] & [C_{rf}] \\ [C_{fr}] & [C_f]+[C_{ff}] \end{bmatrix} \begin{Bmatrix} \dot{q}_r(t) \\ \dot{q}_f(t) \end{Bmatrix}$$
$$+ \begin{bmatrix} [K_{rr}]+[K_{ff}] & [K_{rf}] \\ [K_{fr}] & [K_f]+[K_{ff}] \end{bmatrix} \begin{Bmatrix} q_r(t) \\ q_f(t) \end{Bmatrix} = \begin{Bmatrix} \{F_r\} \\ \{0\} \end{Bmatrix} \tag{1}$$

Where, $\ddot{q}_r, \dot{q}_r, q_r$ are the vectors of acceleration, velocity and displacement of the rotor, $\ddot{q}_f, \dot{q}_f, q_f$ are the vectors of acceleration, velocity and displacement of the foundation at the connecting points with the rotor, M_{rr}, C_{rr}, K_{rr} are the matrices of mass, damping and stiffness of the rotor, M_f, C_f, K_f are the matrices of mass, damping and stiffness of the foundation, C_{rf}, C_{fr}, C_{ff}, K_{rf}, K_{fr}, K_{ff} are the equivalent coefficients of damping and stiffness of the bearings, F_r is the external force due to the rotor unbalance. The index r corresponds to the rotor, and the index f corresponds to the connecting points with the foundation. The matrix of mechanical impedance $[I(\Omega)]$ is dependent on the rotation frequency Ω and it contains the mass, damping and stiffness of the foundation, being associated to the physical coordinate system, as given:

$$[I\ (\Omega)] = -\Omega^2 \left[M_f\right] + j\Omega \left[C_f\right] + \left[K_f\right] \text{where } j = \sqrt{-1} \tag{2}$$

Based on the method of mechanical impedance used to simulate the complete system, the equation of motion in frequency domain is given by Eq. 3:

$$\left(-\Omega^2 \begin{bmatrix} [M_{rr}] & [0] \\ [0] & [0] \end{bmatrix} + j\Omega \begin{bmatrix} [C_{rr}] + [C_{ff}] & [C_{rf}] \\ [C_{fr}] & [C_{ff}] \end{bmatrix} + \begin{bmatrix} [K_{rr}] + [K_{ff}] & [K_{rf}] \\ [K_{fr}] & [K_{ff}] + [I(\Omega)] \end{bmatrix} \right) \begin{Bmatrix} q_{ro} \\ q_{fo} \end{Bmatrix} = \begin{Bmatrix} \{F_{ro}\} \\ \{0\} \end{Bmatrix} \tag{3}$$

The mixed coordinates method consists in describing the displacement vector of the structure at the connecting nodes as independent variables through a modal approach [8]. Physical coordinates are used to the rotor and generalized coordinates to the foundation, to describe the behavior of the complete system. Thus, only the identified modes that effectively contribute in the system response are considered, avoiding the need of a number of modes equal to the number of DOF of the supporting structure at the connecting points with the rotor. The vector of generalized coordinates $\{p\}$, using the modal approach, is:

$$\{q_f\} = [\Phi]\{p\} \tag{4}$$

q_f is the displacements vector of the nodes of the foundation connecting points in physical coordinates, p is the vector of generalized coordinates and $[\Phi]$ is the modal matrix (or the eigenvector matrix of the foundation).

The equation of motion for the complete system, in mixed coordinates [8] is obtained:

$$\begin{bmatrix} [M_{rr}] & [0] \\ [0] & [m_f] \end{bmatrix} \begin{Bmatrix} \ddot{q}_r(t) \\ \ddot{p}(t) \end{Bmatrix} + \begin{bmatrix} [C_{rr}] + [C_{ff}] & [C_{rf}][\Phi] \\ [\Phi]^t [C_{fr}] & [c_f] + [\Phi]^t [C_{ff}][\Phi] \end{bmatrix} \begin{Bmatrix} \dot{q}_r(t) \\ \dot{p}(t) \end{Bmatrix} + \begin{bmatrix} [K_{rr}] + [K_{ff}] & [K_{rf}][\Phi] \\ [\Phi]^t [K_{fr}] & [k_f] + [\Phi]^t [K_{ff}][\Phi] \end{bmatrix} \begin{Bmatrix} q_r(t) \\ p(t) \end{Bmatrix} = \begin{Bmatrix} \{F_r\} \\ \{0\} \end{Bmatrix} \tag{5}$$

Once the modal parameters of the foundation are founded, it can be represented only by the most meaningful modes within the range of frequencies been analyzed, regardless of the number of degrees of freedom associated with connection points (bearings). The generalized coordinates allow the diagonalization of mass, stiffness and damping matrices of the foundation when its structural damping can be considered proportional.

3 Results

The foundation structure was tested with the plate supported by four columns at its edges (Fig. 1). The structure was excited with an electromechanical actuator (shaker). Accelerometers measured the response of the foundation due to the excitation at the bearing supports, besides other points located on the metallic plate and between the bearings. The excitation was applied in horizontal, vertical and axial directions at the bearings housing, and in horizontal and vertical directions in the support plate. The transfer functions obtained in the tests were analyzed in the modal analysis *Modan* 3.0 software [11], given the vibration modes, natural frequencies and modal damping. Table 1 shows the critical frequencies in the range of 0 up to 100 Hz (the operating range of the rotor is up to 60 Hz), which were ranked in order of importance by the Global Mode Rank method [12]. The modes were selected by the mean square of the imaginary components. The bar graph shown in Fig. 2 represents the modal matrix with 8 modes in the frequency range. The four degrees of freedom considered are the horizontal and vertical displacements at the connection points of the foundation (bearings). This chart shows the influence of the DPFs on each vibration mode. As the modes are coupled, the first step is to verify if there is a strongly or weakly modal coupling. This analysis can indicate if it is

Fig. 1 Testing the foundation structure of experimental bench

Table 1 Frequency of the 8 modes in the range of 0 up to 100 Hz in order of importance

Importance	Frequency[Hz]
1	7.81
2	79.52
3	72.15
4	83.38
5	42.72
6	67.26
7	47.54
8	46.06

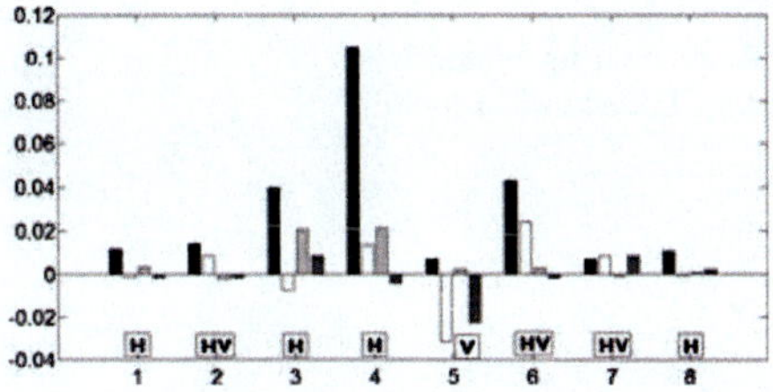

Fig. 2 Influence of the degrees of freedom on the vibration modes: bearing 1 horizontal (*black*), bearing 1 vertical (*white*), bearing 2 horizontal (*light gray*) and bearing 2 vertical (*dark gray*)

possible to consider them uncoupled and, consequently, to assume a proportional damping matrix to the foundation. As shown in Fig. 2, modes 1, 3, 4 and 8 are predominantly horizontal due to the highest horizontal components regarding to the vertical ones. Instead, mode 5 is strongly vertical.

Modes 2, 6 and 7 present horizontal and vertical components of same magnitude in both bearings 1 and 2. If the coupling is weak, the modes can be uncoupled in vertical and horizontal modes at the same frequencies, which are 47.54, 67.26 and 79.52 Hz. Initially, the mechanical impedance was obtained in mixed coordinates, considering the eight coupled modes, so every mode can influence the vibration of the rotating system in both directions. Figure 3 shows bearing 1 response to unbalance, which has been chosen due to the fact that the effects of 8 modes of the foundation were more significant in this bearing. The response is shown in logarithmic scale, in order to highlight the effects of vibrating modes of the foundation in the system response. The peak of amplitude at 19 Hz is due to the natural frequency of the shaft, while the other peaks are related to the foundation. Afterwards, the 8 modes were uncoupled in 16 modes (8 horizontal modes and 8 vertical modes). This approach presented no expressive changes in the system response (Fig. 3), which indicates the foundation modes are weakly coupled. As modes 1, 3, 4 and 8 are prominent in the horizontal direction, mode 5 in the vertical direction, and modes 2, 6 and 7 in both directions, 11 uncoupled modes can be considered, being 7 horizontal and 4 vertical (repeating modes 2, 6 and 7 for both directions). Figure 4 shows the effect of this reduction is negligible in the system response. At this point, the less significant modes (mode 8–46.06 Hz – horizontal and mode 7–47.57 Hz – horizontal and vertical) were disregarded successively until a significant change in the system response.

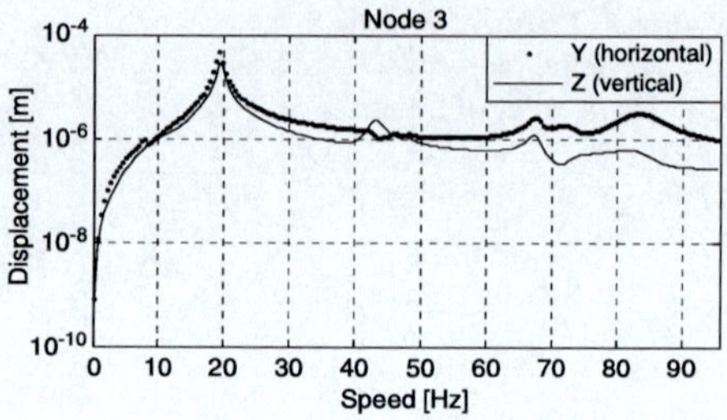

Fig. 3 Unbalance response (bearing 1): Rotor-Bearing-Foundation with 8 coupled modes

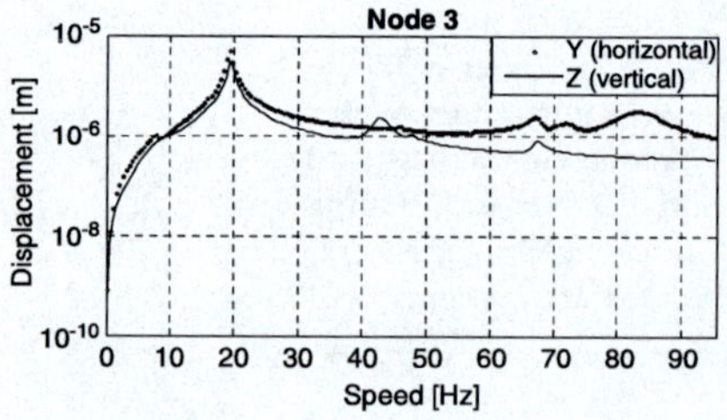

Fig. 4 Unbalance response (bearing 1): Rotor-Bearing-Foundation with 11 uncoupled modes

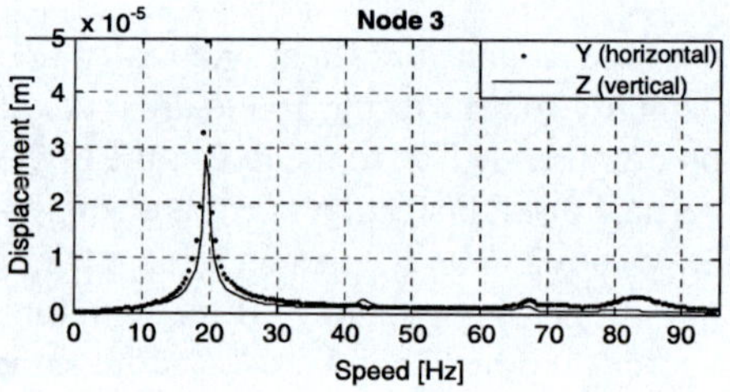

Fig. 5 Unbalance response (bearing 1): Rotor-Bearing-Foundation with 8 uncoupled modes (linear scale)

The first meaningful change occurs in the elimination of the mode at the frequency of 67.26 Hz. Therefore, for the uncoupled system, at least 8 uncoupled modes should be considered to avoid a lack of information in the response (Fig. 5). The most significant modes are: 5 horizontal modes (7.81 Hz, 79.52 Hz, 72.15 Hz, 83.38 Hz and 67.26 Hz) and 3 vertical modes (79.52 Hz, 42.72 Hz and 67.26 Hz).

The response for the foundation by the method of mechanical impedance (Eq. 3) is then obtained (Fig. 6). There are differences in the responses shown in Figs. 5 and 6. However, the results are in good agreement in general. The main differences are due to the approach used in the method of mixed coordinates when considering uncoupled and, consequently, in the evaluation of the damping factor. The adjustment of the damping factor at 42.72 Hz (vertical mode) can be seen in Fig. 7. So, an accurate identification of the modal damping is required when applying the mixed coordinates method.

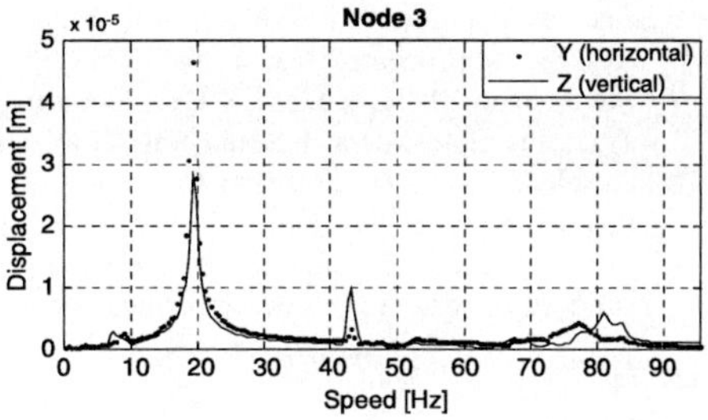

Fig. 6 Unbalance response (bearing 1): Rotor-Bearing-Foundation by Mechanical Impedance (linear scale)

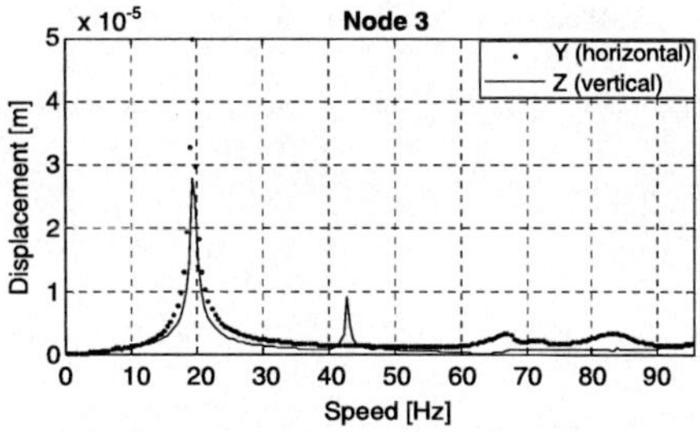

Fig. 7 Unbalance response (bearing 1): Rotor-Bearing-Foundation with 8 coupled modes, modified damping (linear scale)

4 Conclusions

Mechanical impedance and mixed coordinates method have been tested. The supporting structure was represented by coupled and uncoupled modes and in both cases only the most meaningful modes can be considered in the system response. In practice, there are limitations in identifying the vibrating modes and their corresponding damping factors. Consequently, the modal approach must be carefully carried out when applied to the supporting structure representation in a rotating system.

Acknowledgments The research work reported here was supported by CAPES-DAAD Probral program, CNPq and Fapesp.

References

1. Weber, H.: Über das gemeinsame Schwingungsverhalten von Welle und Fundament bei Turbinenanlagen. VDI-Berichte **48**, 55–62 (1961)
2. Wilson, R.R., Brebbia, C.A.: Dynamic behavior of steel foundations for turbo-alternators. J. Sound Vib. **18**(3), 405–416 (1971)
3. Morton, P.G.: Analysis of rotors supported upon many bearings. J. Mech. Eng. Sci. **14**(1), 25–33 (1972)
4. Gasch, R.: Vibration of large turbo-rotors in fluid film bearings on an elastic foundation. J. Sound Vib. **47**, 53–73 (1976)
5. Craggs, A.: A component mode method for modeling dynamics of turbo-generator sets. J. Sound Vib. **117**(2), 277–288 (1987)

6. Lees, A.W., Sinha, J., Friswell, M.I.: Model based identification of rotating machines. Mech. Sys. Signal Process. **23**(6), 1884–1893 (2009)
7. Stephenson, R.W., Rouch, K.E.: Generating matrices of the foundation structure of a rotor system from test data. J. Sound Vib. **154**(3), 467–484 (1992)
8. Cavalca, K.L.: L'Interazione tra rotori e struttura portante: metodologie per la sua modellazione. Ph.D. Thesis, Politecnico di Milano, Italy (1993)
9. Cavalca, K.L., Cavalcante, P.F., Okabe, E.P.: An investigation on the influence of the supporting structure on the dynamics of the rotor system. Mech. Syst. Signal Process. **19**(1), 157–174 (2006) (Cambridge, UK)
10. Weiming, L., Novak, M.: Dynamic behaviour of turbine-generator-foundation systems. Earthq. Eng. Struct. Dyn. **24**(3), 339–360 (1996)
11. Modan – Logiciel d´analyze modale: Notice d'utilization Versión 3.0, Laboratoire de Mécanique Appliquée R. Chaléat Université de Franche-Comté, UMR CNRS 6604, France (2009)
12. Global mode Rank: Proto-Dynamique V4.0, Intespace et LMARC, France (1997)

Part VIII
Experimental Techniques, Signal Processing and Analysis

Adaptable Vibration Monitoring

Vitalijus Volkovas

Abstract In order to optimize monitoring and diagnostics procedures, the data collection periodicity should be altered considering equipment technical condition, after identifying specific defect and having information about its development. The paper presents the methodology and algorithm of adaptive vibration monitoring and diagnostics of rotor system bearings based on vibration monitoring data archive analysis and the modified vibration's power spectrum diagnostics method.

Keywords Technical state • Vibration • Monitoring • Adaptability

1 Introduction

Vibration measurements are widely applied to control condition of various technical systems. Regular measuring of this parameter and estimation of changes in time allows taking relevant actions based on the results to extend the lifetime of the tested object, avoid accidents or emergencies. Such functions are performed by vibration monitoring and diagnostics systems.

Technical condition of the objects and their systems is frequently related to parameters of dynamics (different constructions), of bearing (rotor systems) vibro-acoustic processes and their alterations [1, 2]. Vibration monitoring of operating objects is widely used in practice. It is divided into two groups. In the first group stationary measurement (with short time between measurements) and analysis systems are used while in the second case mobile and most frequently modern vibration measurement (with suitably long time between measurements) and analysis devices

V. Volkovas (✉)
Kaunas University of Technology, Kaunas, Lithuania
e-mail: vitalijus.volkovas@ktu.lt

J. Náprstek et al. (eds.), *Vibration Problems ICOVP 2011: The 10th International Conference on Vibration Problems*, Springer Proceedings in Physics 139,
DOI 10.1007/978-94-007-2069-5_80,

are applied periodically. In first case information collection is performed so frequent that only real time measurements are relevant, and in the second case the periodicity is under consideration in each case.

Both above mentioned monitoring methods (real time or stationary monitoring and mobile monitoring with suitably long intervals between objects' testing) are not optimal. The first one is relatively costly and a considerable amount of inexpedient information is processed while the second method is subject to a high-level risk of missing a defect. Therefore monitoring research, the periodicity of which depends on the existing technical condition of the controlled object, is highly significant in respect of science, economy and practice.

Adaptive bearing vibration monitoring using data trend characteristics is analyzed in the monograph [2]. In terms of a concept and idea it was a new attitude towards monitoring as to an optimized procedure but non-considered issues are related to specific defects of rotor systems though the technical condition depends exactly on them. This means that varying measurement intervals did not depend on specific defects impacting different rate of rotor system deterioration.

A diagnostics method when trend characteristics are not applied to technical state identification is familiar [3]. It may be modified and then used in the algorithm of adaptive monitoring measurement interval alterations subject to assessment of specific defects in the systems.

This study presents the methodology and algorithm of adaptive vibration monitoring and diagnostics of rotor system bearings which are generated on the basis of vibration monitoring data archive analysis and the modified diagnostics method.

2 Concept and Principles of Adaptive Monitoring

The theory and practice of vibration monitoring and diagnostics show that these procedures are hardly distinguished; assessment of the object condition is a complex process during which both procedures influence one another. This was clarified when termination of the object functioning has been recently conducted based on technical condition and not according to the schedule of routine service or non-sanctioned event. Regardless of the fact that diagnostics may be carried out at the "off-line" mode monitoring data is still used and diagnostics results often require a new part of data of measured quantities. If the amount of the data is considerably higher, the probability of inaccurate diagnostics may be expected to be lower [4]. In this case trends of measured quantities may be determined, the technical condition of the object may be forecasted and periodicity of data collection and analysis may be objectively selected. However, the trend is integral characteristics of changed technical condition which does not identify a particular defect. The defect may develop abruptly and the technical condition may change radically (cannot be remedied) until the next data collection and analysis moment. The same applies to mobile monitoring systems but upon assessment of their significantly greater periodicity of data collection and analysis a probability of emergency that may occur between the object control moments highly increases. This means that in the event of optimization of monitoring and diagnostics procedures, the data collection periodicity should be altered in respect of technical condition (diagnostics result), specific defect identified and information on its development.

Periodicity of vibration measurements in monitoring and diagnostics systems may be basically changed if the tendencies of the controlled object technical condition change and development are known or if by means of the diagnostics procedure a specific defect with the known symptoms is detected. Considering the technical state of object, optimal monitoring expenses and change tendencies of vibration, it is possible to optimize periodicity of system vibration measurement.

2.1 Principle of Technical Condition Change Tendency

Evaluation of the tendencies of vibration magnitude variations provided new possibilities related to objective results and existing condition of the object (for instance, rotor system) as well as reasonable variable interval of vibration measurements, which reduces monitoring costs. When the condition of monitored system is satisfactory the interval may be sufficiently long and constant. If the tendency of vibration variations occurs, but the standard requirements (or allowable vibration levels) do not require shutdown of object of monitoring, the interval between data collection (measurements) should be decreased in order not to miss the defect and avoid emergency situations between measurements [2]. Thus, we have two distinctive tendency stages of the parameter change where measurement periodicity can be different and objectively grounded. Of course, this applies only to rather slowly developing faults of the monitored object, which often prevail in practice. Practically, the state in the first stage ($t < t_1$) is stable and a root-mean-square values of vibration (r.m.s.) do not reach permissible values. In this case, periodical measurements are characterized by (m_i, ${\sigma_i}^2$) and $m_i \approx$ const, ${\sigma_i}^2 \approx$ const as well as a very slow change of r.m.s. (a change is described by the equation linear regression), and they are performed with a constant period T. In the second stage ($t > t_1$), vibration r.m.s. measurements are distinguished for the fact that $m_i =$ var, ${\sigma_i}^2 =$ var, the fault develops considerably quicker and changes of r.m.s. average approximate in a non-linear function. Then collection of the data to be monitored is better to be organized with a variable periodicity, using data changing tendencies and r.m.s. levels, evaluating the risk of omission of a sudden change of the technical state and help to avoid the economic consequences. The consequent algorithm [2] is rather simply implemented in a specialized data analysis program of mobile monitoring device.

2.2 Principle of Technical Condition Vector Variation

The technical condition of the object is defined by vector $\mathbf{G} = \{X_j\}^T$, $j = 1, \ldots, m$, and its elements are generalized defect characteristics (symptoms) q_j. Vector **G** depends on object performance mode $\mathbf{G}_o$ (as well as on defects which occurred after assembly or during the process and were eliminated and on the quality of repair) and operation $\mathbf{G}(t)$ components. Mathematical model of the technical condition change may be composed on the basis of the sensitivity theory [5]. It may be written down as follows:

$$\mathbf{G} = \mathbf{G}_0 + \mathbf{G}(t) = \mathbf{G}(n_0, p_0) + \sum_{j=1}^{m} \frac{\partial \mathbf{G}}{\partial q_j} \Delta q_j(t) \tag{1}$$

when $\mathbf{G}_0 = \mathbf{G}(n_0, p_0)$ – a component attributed to the object with deterioration of almost zero (new) and operated at nominal rate (n_0) and load mode (p_0);

$\Delta q_j = X_j$ – object failure j symptom which is detected when analyzing lifetime of the object when defect is present during manufacturing ($t = 0$);

$\partial \mathbf{G}/\partial q_j$ – technical condition vector **G** sensitivity in relation to the defect according to corresponding generalized defect characteristic (symptom) $q_j(t)$ as dependent on time within a certain observed interval.

The concept of adaptive monitoring which is based on evaluation of technical condition alteration according to the formula (1) requires determining symptoms of known defects of the monitoring object and deterioration of the technical condition by means of applying the methods of diagnostics. For this reason the statistic analysis of JSC *Lietuvos Elektrinė* monitoring system measurement data and numerical modeling of defect detection possibilities were conducted [6].

3 Numerical Modeling of Defect Detection

Based on JSC *Lietuvos Elektrinė* periodical monitoring and diagnostics system measurement data collected during 8 years a bearing deterioration model was analyzed.

Data is presented in Fig. 1 where black measurement points stand for high-frequency peak noise of rolling bearings and blue curve means general bearing noise (the SPM methodology to assess bearing condition is used). During the analyzed period non-allowable (exceeding 35 dB) peak bearing noises were twice detected for pump S-008. In September 2003 and October 2007 the bearings were replaced with the new ones. After replacement the noise decreased to the allowable limit and did not exceed 35 dB. Meanwhile r.m.s. measurements of the same bearings remained stable and did not exceed the allowable limit. In order to conduct numerical modeling of these defects and determine adaptive monitoring periodicity the tendencies and models of bearing deterioration were defined in a form of a regression equation.

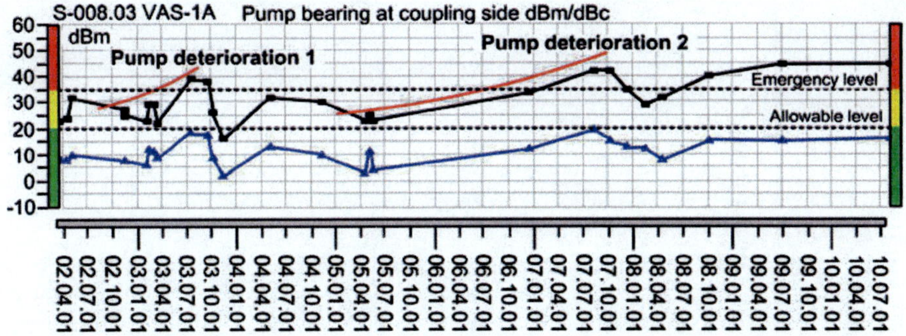

Fig. 1 Pump bearing noise measurements

Deterioration model of the first defect (September 2003) of the pump bearing:

$$y = 22.354e^{0.0534x}, r^2 = 0.3799. \quad (2)$$

Deterioration model of the second defect (October 2007) of the pump bearing:

$$y = 19.758e^{0.0646x}, r^2 = 0.4319. \quad (3)$$

Here r^2 is a determination coefficient of regression equation.

Due to modeling of the monitoring object technical condition alteration the most common defects of pump electric motors were selected [6].

Every structural failure (defect), not only in rotor systems bearings, has its own characteristics in the form of spectrum components f_{sp} [1, 2, 4]. According to the description of a defect provided in the study [4], each defect is evaluated only in consideration with energy of its characteristic frequencies f_{sp}, existing during failure, and vibration magnitude $V_{r.m.s.}$, which completely defines vibration energy, is received in [6]. This way the object defect will be estimated by relative rate R:

$$R = \frac{f_{sp}}{V_{r.m.s.}}. \quad (4)$$

$V_{r.m.s.}$ value is received after vibrations are measured at the time. During the initial stage of the defect the components that are characteristic to specific defects occur in vibration spectrum, therefore magnitude f_{sp} is calculated using these vibration components as follows:

$$f_{sp} = \sqrt{\sum s\left(f_k\right)^2}. \quad (5)$$

Here $s(f_k)^2$ are components of vibrations spectrum of defected system.

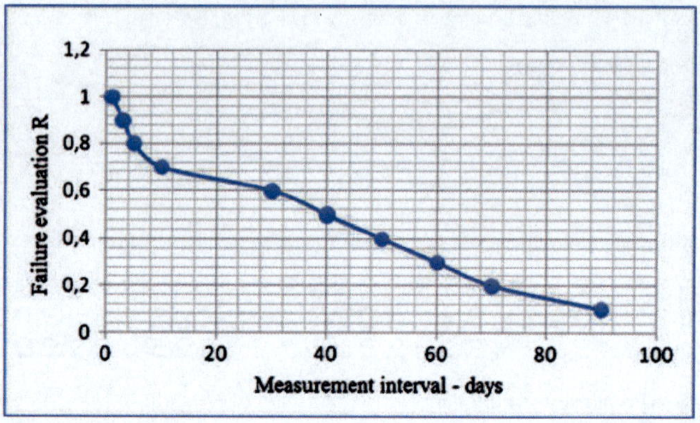

Fig. 2 Measurement (time) interval correction curve

The standard criteria did not evaluate vibration magnitude proximity to non-allowable limits. When the defect is under development diagnostics additionally requires rate R which assesses failure so as the defect is not missed in the early stage. According to vibration measurement spectra rate R which assesses failure is deducted from relatively different frequencies f_k. The results[6] shown, when the likelihood of $R \leq 0.5$ defect occurrence is low, when $0.5 \leq R \leq 0.8$ is the defect development feature, and if $R \approx 1$ defect is progressive and it may endanger further operation of equipment. According to R the periodicity of measurement is changed in real time based on dependence, which is illustrated in Fig. 2.

The achieved results allow suggesting the methodology of adaptive vibration monitoring when the periodicity of data collection depends on technical condition of the object.

4 Conclusion

On the basis of vibration measurement data bank statistical analysis the research of failure and vibration monitoring data correlation has been conducted, vibration spectrum frequencies characteristic to a specific defect have been determined, and it has been suggested to evaluate the technical condition applying the modified diagnostics method which allows to assess developing defects by means of criteria offered by standard procedures. In terms of identification of developing defects the methodology of adaptive (varied vibration measurement periodicity) monitoring has been suggested on the basis of spectrum characteristic constituents and their energy ratio with general vibration energy.

References

1. Adams, M.L.: Rotating Machinery Vibration. Marcel Dekker, New York (2001)
2. Jonusas, R., Jurkauskas, A., Volkovas, V.: Rotor Systems Dynamics and Diagnostics. Technologija, Kaunas (2001)
3. Kawabe, Y., Toyota, T.: New technology of machine diagnosis without using the trend data. Int. J. COMADEM **7**(2), 1–23 (2004)
4. Eidukevičiūtė, M., Volkovas, V.: Measurement uncertainty in vibromonitoring systems and diagnostics reliability evaluation. J. Sound Vib. **308**(3–5), 1–625 (2007)
5. Cacuci, D.: Sensitivity and Uncertainty Analysis. CRC/Chapman & Hall, Boca Raton (2003)
6. Volkovas, V., Perednis, A.: Adaptable vibration monitoring in rotor systems. J. Vibroeng. **12**(4), 1–396 (2010)

Design of an Optical-Fiber Accelerometer Based on Polarization Variation Due to Crushing of the Fiber

Pierre Tihon, Nicolas Linze, Olivier Verlinden, and Marc Wuilpart

Abstract This paper presents the design of a novel kind of accelerometer. The sensor is based on the modification of the polarization state of the light inside the fiber by the means of a mechanical transducer. The transducer presented uses crushing to deform the fiber. Measurements were done with an acceleration varying from 1 to 40 m/s² and a frequency range varying from 100 to 1,000 Hz. The measured output is the ratio between the variable part and the constant part of the optical power at the end of the fiber. A sensitivity of the order of 6E-4/ms² was found. Numerical simulations were also performed with the Abaqus software. The obtained sensitivity is quite greater than the measured one which can be explained by the hypothesizes we made for the simulations.

Keywords Accelerometer • Optical fiber • Polarization • Crushing

1 Introduction

The measurement of accelerations is of great importance in civil and mechanical engineering. The vibrations of a structure (bridge, building, airplane wing) learn about its ageing and the need of maintenance.

Different studies have been carried out to develop optical fiber accelerometer [1–3]. The use of an optical fiber enables distributed or quasi distributed sensing along the fiber.

Up to now, those studies have especially focused on the distributed aspect of the sensor. The vibrations are produced by winding a fiber around a piezoelectric stretcher [2, 3]. The design of a mechanical transducer able to measure actual

P. Tihon (✉) • N. Linze • O. Verlinden • M. Wuilpart
University of Mons – UMONS, Mons, Belgium
e-mail: pierre.tihon@umons.ac.be; Nicolas.Linze@umons.ac.be; Olivier.Verlinden@umons.ac.be; Marc.Wuilpart@umons.ac.be

J. Náprstek et al. (eds.), *Vibration Problems ICOVP 2011: The 10th International Conference on Vibration Problems*, Springer Proceedings in Physics 139,
DOI 10.1007/978-94-007-2069-5_81, © Springer Science+Business Media B.V. 2011

structure vibrations is essential to build a sensor for industrial applications. This paper proposes a home-made accelerometer based on the crushing of the fiber.

2 Experimental Set-Up

The experimental set-up is given in Fig. 1. Polarized light produced by a laser with a wavelength of 1,550 nm passes through a polarization controller so that the initial state of polarization (SOP) gives a maximal sensitivity.

The light then goes along an optical fiber where a mechanical transducer deforms the fiber. The mechanical transducer is placed on a shaker. The fiber is attached between two arms of an aluminium U-shape piece (on Fig. 2). One of the arms is fixed to the shaker, the second one has a mass attached to it and is free to move. A rubber end is fixed to the crushing part of the arm in order to avoid damaging the fiber. Its vibrations deform the fiber, causing some birefringence to appear. Two different arms were made, one with a length of 30 mm, the second with a length of 60 mm.

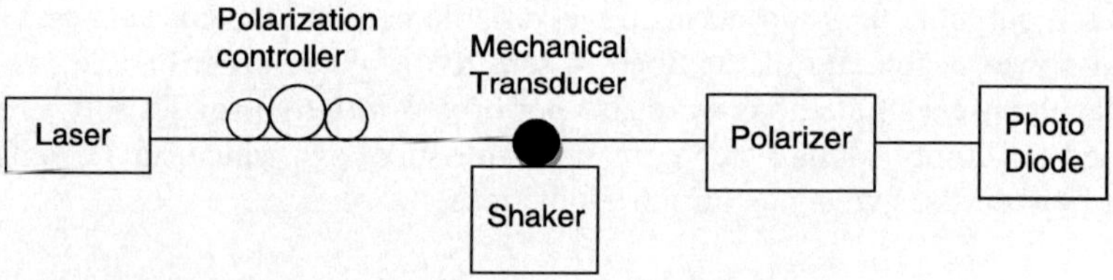

Fig. 1 Experimental set-up

Fig. 2 Mechanical transducer

The light passes through a polarizer and arrives to a photodiode producing a tension proportional to the incident optical power. Thanks to the polarizer, the power at the photodiode varies if the SOP of the fiber varies.

3 Theoretical Background

Light in a single mode optical-fiber has two orthogonal modes of propagation characterized by two orthogonal states of polarization. Those modes are degenerated in a perfect circular symmetric fiber. If the circular symmetry is broken (e.g. because of internal stresses) the two modes are no more degenerated and propagate with different velocities. The refraction indices are then different for the two modes. Their difference is called the birefringence β and induces a phase shift δ. They are given by Eq. (1) [4]

$$\beta = k_x - k_y \approx k(\delta n_x - \delta n_y); \quad \delta = \beta L \tag{1}$$

where β is the birefringence, δn_i the variation of the refractive index in the direction i and k is 2π divided by the wave length of the light. L is the length of the birefringent part of the fiber.

The theory of photoelasticity gives the link between the deformation and the birefringence [4]

$$\delta n_i = -\frac{n^3}{2} \sum p_{ij} \varepsilon_j \tag{2}$$

where n is the refractive index of the core, ε_j the deformation in the direction j and the p_{ij} are the elasto-optic coefficients. For silica, $p_{ij} = 0.121$ if $i = j$ and $p_{ij} = 0.270$ otherwise.

When the phase shift is known it is possible to calculate at each point of the path of the light its state of polarization thanks to the Stokes formalism [5].

In this formalism the state of an elliptically polarized light is represented by a vector

$$s = \begin{bmatrix} 1 \\ \cos 2\chi \cos 2\varphi \\ \cos 2\chi \sin 2\varphi \\ \sin 2\chi \end{bmatrix} \tag{3}$$

where χ characterizes the ellipticity of the SOP and φ is the orientation of the linear part of the SOP with regard to a reference axis.

When a fiber is crushed it becomes a linear retarder with a Mueller matrix given by

$$M_1 = \begin{bmatrix} 1 & 0 & 0 & 0 \\ 0 & \cos^2\frac{\delta}{2} + \sin^2\frac{\delta}{2}\cos 4q & \sin^2\frac{\delta}{2}\sin 4q & -\sin\delta\sin 2q \\ 0 & \sin^2\frac{\delta}{2}\sin 4q & \cos^2\delta - \sin^2\delta\cos 4q & \sin\delta\cos 2q \\ 0 & \sin\delta\sin 2q & -\sin\delta\cos 2q & \cos\delta \end{bmatrix} \quad (4)$$

δ is the phase shift and q the angle between the fastest axis of the light and the reference axis.

If the linear polarizer at the end of the fiber has an azimuth θ its Mueller matrix is given by

$$M_2 = \frac{1}{2}\begin{bmatrix} 1 & \cos 2\theta & \sin 2\theta & 0 \\ \cos 2\theta & \cos^2 2\theta & \sin 2\theta\cos 2\theta & 0 \\ \sin 2\theta & \sin 2\theta\cos 2\theta & \sin^2 2\theta & 0 \\ 0 & 0 & 0 & 0 \end{bmatrix} \quad (5)$$

The state of polarization at the end of the crushed fiber is given by

$$s_{out} = M_2 M_1 s_{in} \quad (6)$$

The first component of this vector gives the power at the end of the fiber. We get

$$s_{out}(1) = \frac{1}{2}(1 + \cos 2\theta\cos 2\chi\cos 2\varphi + \sin 2\theta\cos 2\chi\sin 2\varphi + \delta\sin 2\theta\sin 2\chi)) \quad (7)$$

In this expression the term in δ is maximal if χ and θ are equal to $\pm 45°$ (this corresponds to an initial circular SOP and to a polarizer with an azimuth of $\pm 45°$). In this case the sensitivity of the sensor is maximal and we get

$$s_{out}(1) = \frac{1}{2}(1 + \delta) \quad (8)$$

The phase shift δ varies at the same frequency as the mechanical vibration. The optical output at the photodiode will then have a part varying at the same frequency as the vibration added to a constant part.

We see that the ratio of the variable power at the end of the fiber divided by the constant part is δ.

4 Numerical Simulations

The deformations in the fiber were simulated with the finite element Abaqus software.

Three parts are considered in the fiber: the core through which the light passes, the cladding and the buffer. The core and the cladding are in fused silica, the buffer in polymer.

The method to calculate the deformation in the crushed fiber is the following: a model of the fiber alone is first made and crushed with static displacement in order to measure the stiffness of the fiber. A model of the U-shaped mechanical transducer is made. The fiber is then replaced by a spring of the calculated stiffness and a modal dynamic simulation is performed. The calculated displacements at the extremities of the spring are statically applied to the fiber to find the core deformations.

When the core deformations along the crushed portion of the fiber are known it is necessary to find the induced birefringence. Matlab is used to calculate at each point of the core the induced birefringence thanks to the equations of photoelasticity and then the state of polarization of light at the end of the fiber with the Stokes formalism. The assumption is made that we are in the most favorable case with regards to the initial polarization and to the polarization at the final polarizer.

5 Results and Discussion

A first experiment is performed, exciting the shaker at frequencies of 110 and 1,010 Hz with increasing accelerations. Both arms are used. The optical output measured is the ratio between the amplitude of the variable part of the signal to the constant part. The results are presented in Figs. 3 and 4.

We can see that in all cases the signal evolves linearly with the acceleration in the range 3–40 m/s^2 which means that the sensitivity is constant. The measured sensitivities are nearly the same at 110 and 1,010 Hz. They are presented in Table 1.

The simulated sensibilities are 2–6 times greater than the measured ones. An explanation is the fact that the simulated results are computed with the hypothesis that the light polarization is the best at both the mechanical transducer and the final polarizer. Nevertheless we only have one polarization controller in our experimental set-up which is placed at the beginning of the fiber. We then cannot set the light at the best polarized state at the mechanical transducer and the final polarizer independently but we can only set a best initial polarized state.

Another difference between simulated and experimental results is that the simulated optical signal at 1,010 Hz is greater than at 110 Hz while the experimental results show that this signal is the same for the 6 cm beam and is slightly smaller for the 3 cm beam. In order to study the influence of frequency on the sensitivity, measurements are performed at different frequencies between 110 and 1,800 Hz at a constant acceleration of 15 m/s^2. The results are presented in Fig. 5. We can

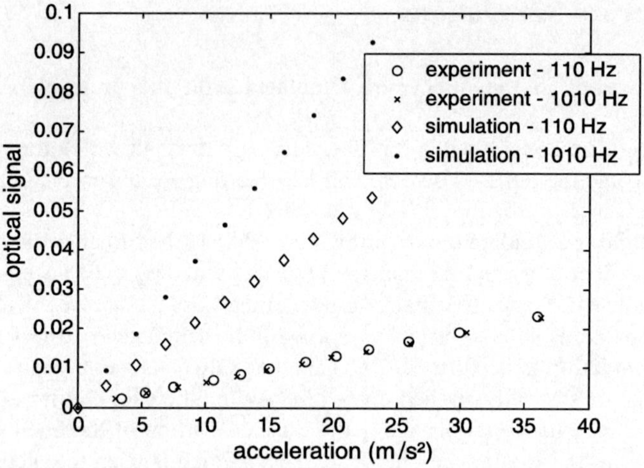

Fig. 3 Optical output for the 60 mm beam

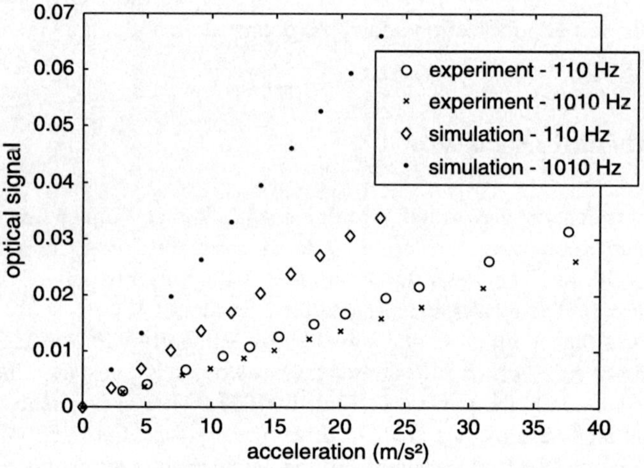

Fig. 4 Optical output for the 30 mm beam

Table 1 Sensibilities of the 60 mm beam (*left column*) and the 30 mm beam (*right column*) at frequencies of 110 and 1,010 Hz

	Sensitivity (1/ms²)	Sensitivity (1/ms²)
Experiment – 110 Hz	6.4E-4	8.3E-4
Experiment – 1,010 Hz	6.4E-4	6.7E-4
Simulation – 110 Hz	2.3E-3	1.5E-3
Simulation – 1,010 Hz	4E-3	2.9E-3

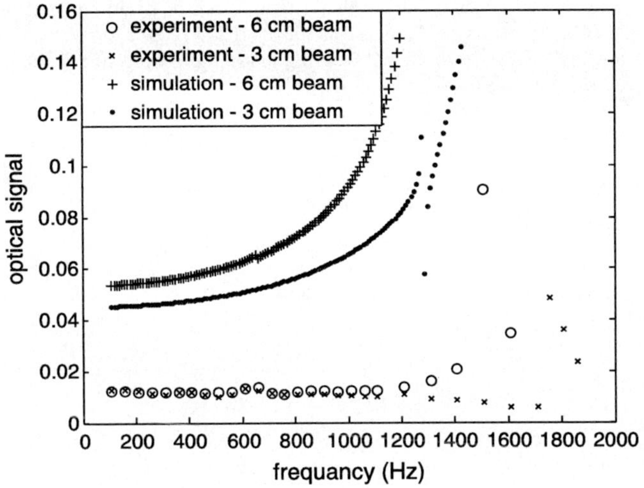

Fig. 5 Simulated and measured optical data for both beams

see that the measured sensitivity is constant in the range 110–1,000 Hz with an exception around 600 Hz where it becomes slightly greater. Then the influence of the first resonance mode occurs and the sensitivity grows quickly. The simulated results present this growth but it comes at lower frequencies.

6 Conclusion

We have proposed a mechanical transducer to develop an accelerometer based on the variation of the polarization state of the light. We have shown the possibility to use a device crushing the fiber to modify its state of polarization. Our device presents a linear behavior at least in the range 3–40 m/s^2 and in a frequency range going from 110 to 1,000 Hz. The simulations performed give results with the same order of magnitude as the measured ones. Work must still be done to get those simulations closer to the measured results.

References

1. Liu, J.-G., Schmidt-Hattenberger, C., Born, G.: Dynamic strain measurement with a fiber Bragg grating sensor system. Measurement **32**, 151–161 (2002)
2. Zhang, Z., Bao, X.: Continuous and damped vibration detection based on fiber diversity detection sensor by Rayleigh Backscattering. J. Lightwave technol. **26**, 832–836 (2008)

3. Zhang, Z., Bao, X.: Distributed optical fiber vibration sensor based on spectrum analysis of Polarization-OTDR system. Opt. Express **16**, 10240–10247 (2008)
4. Ulrich, R., Rasleigh, S.C., Eickhoff, W.: Bending-induced birefringence in single-mode fibers. Opt. Lett. **5**, 273–275 (1980)
5. Chah, K., Caucheteur, C., Verlinden, O., Mégret, P., Wuilpart, M.: Polarimetric optical fiber sensor for vibration measurements. OPTIMESS2009, Antwerp (2009)

Experimental Research on Brake Squeal

Daniel Wallner and Stefan Bernsteiner

Abstract Since many years experimental researches have been performed in order to get an insight into the issue of brake squeal. This work presents an innovative test setup for analyzing friction-excited vibrations and squeal triggering mechanisms. The investigated brake system showed brake squeal issues at certain operating points. The purpose was to analyse the brake system in detail on a test rig. Because the well-known methods of investigation such as laser vibrometer were not possible, the aim was to develop and generate new measuring points, which are close to the contact area of disc and pad. Therefore the brake calliper has been modified; hence the guide are replaced by modified ones. At these pins the friction force can be measured. Because of an optimized shape it is possible to measure the vibration of the friction force. The measured signal correlates with a parallel measured microphone signal. Next the brake disc will be assembled with triaxial accelerometers. Thus it is possible to determine the operating deflection shape of the disc. With the results of this work a new contact algorithm should be developed which can be used e.g. in Finite Element calculations.

Keywords Brake squeal • Vibration measurement • Friction force measurement • Friction-excited vibration

D. Wallner (✉)
Institute of Automotive Engineering, Graz University of Technology, A-8010 Graz, Austria
e-mail: daniel.wallner@tugraz.at

S. Bernsteiner
Graz University of Technology, Graz, Austria
e-mail: stefan.bernsteiner@tugraz.at

J. Náprstek et al. (eds.), *Vibration Problems ICOVP 2011: The 10th International Conference on Vibration Problems*, Springer Proceedings in Physics 139,
DOI 10.1007/978-94-007-2069-5_82,

1 Introduction

Since many years brake squeal is a challenging issue. Papinniemi et al. [1] referred in his paper to two research reports from the 1930s. In the meantime there are several reviews published. For example Cantoni et al. [2] provided an extensive review including 407 published papers. Among others warranty costs drive this effort to prevent brake noise issues.

The methods to reduce brake noise can be roughly divided into three main fields of research. The first approach uses minimal models of the brake system. Von Wagner et al. [3] describe some of them. These models are mostly abstract and simplified and can be analyzed analytically. The advantage is, that principal processes and measures can be described and analysed. But the outcomes are more or less abstract and a transfer to a complete brake system is difficult.

The second approach uses the Finite Element Method. Ouyang et al. [4] provides a review on numerical brake squeal analysis. These could be divided into two methods which are the Complex Eigenvalue Analysis (CEA) and the transient analysis. The CEA method is using a linearization of the brake squeal solution at steady sliding states and calculates the complex eigenvalues. A positive real part of this solution indicates an unstable system. A disadvantage of this method is that it usually predicts too many unstable modes. Thus means that many critical points are predicted, but only few of them are able to exhibit instability. Another restriction is that the solutions are proper for states close to steady states. In contrast to that the transient analysis method can analyse the effect of time-dependent loads. Furthermore it is possible to get appropriate solutions if nonlinear effects are present or the state is far from the steady state.

The third approach includes the experimental methods. There are many different ways to analyse a brake system. Regarding to the literature the majority of used measurement methods measure the velocity and the acceleration, respectively. With these measurement systems the operating deflection shapes (ODS) of e.g. the brake calliper can be detected. Marschner et al. [5] presented a method which consists of triaxial acceleration sensors which are mounted in the cooling ducts of the disc. Ouyang et al. [6] measured the displacement of the disc using 12 capacitive sensors arranged at a certain diameter. Other research groups used piezoelectric sensors as for instance [7], which can also be used as actuators to prevent vibrations.

2 Methods

During road testing the analysed vehicle axle showed squeal issues at higher disc temperature. Because of that the decision was to analyse the axle on a test bench. The chosen test bench is a combined brake and suspension test bench. Hence it is possible to test the performance of a brake system or the suspension of a vehicle.

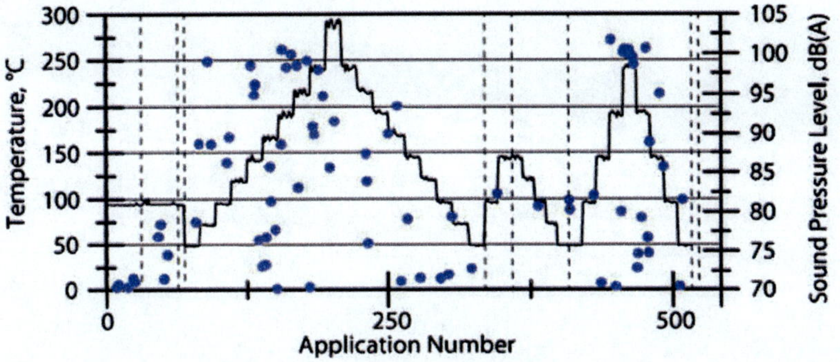

Fig. 1 Exemplary results from the SAE J2521 squeal noise matrix test original brake pads. The *solid line* denotes the temperature, the *points* denotes a squeal event the relevant sound pressure level

A special test rack has been designed for the experiments. On this rack a complete vehicle axle is installed. Using this set up a SAE J2521 test was run to evaluate the squeal noise matrix. Thereby almost 20% of 1,430 brake applications were squealing ones. See Fig. 1 for exemplary results. After these tests the aim was to analyse this brake system in detail.

For further analysis special prototype brake pads, which consist of sintered material, have been used. This has been necessary because the original brake pads mostly squeal at higher temperatures. But the used sensors cannot deal with this and limit the maximum temperature. It was not possible to find repeatable squealing operating points with the original brake pads at low temperatures.

The design of the prototype pads is optimized to obtain a maximum number of squeal events. Figure 2 shows a part of the performed SAE J2521 test run with the new prototype pads. More than 70% of the brake applications squeal. The reason for that is on the one hand the higher friction coefficient of the prototype pads and on the other hand that all known countermeasures are waived. Even an improper material composition regarding squeal behaviour is chosen. As a result it is possible to perform repeatable and because of that comparable test runs at certain operating points.

Brake squeal is a self-excited vibration generated by the contact of brake disc and brake pad. Because of that the friction force is an important parameter, which should be measured. The investigated brake system consists of a fixed calliper and the resulting force of the brake torque acts on guide pins. Thus modified guide pins are developed which measure the resulting force using strain gauges. The pins have an optimized shape; hence the cross-section looks like an I-beam. This modification is necessary to obtain an applicable signal. As a result the modified guide pins measure not only the friction force, but also the superposed vibration of this force. Figure 3 shows these modified guide pins and the mounted strain gauges.

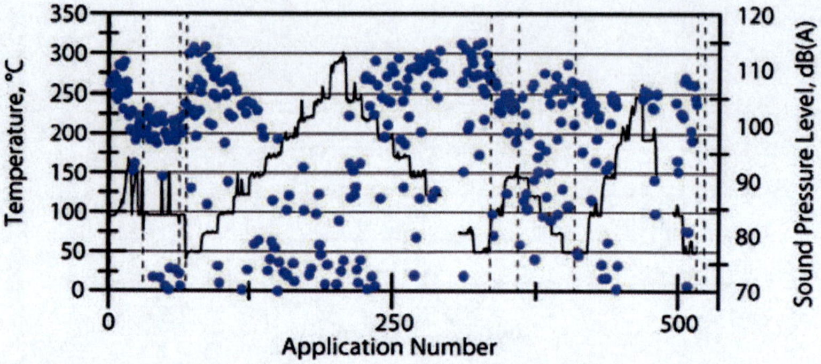

Fig. 2 Exemplary results from the SAE J2521 squeal noise matrix test prototype brake pads

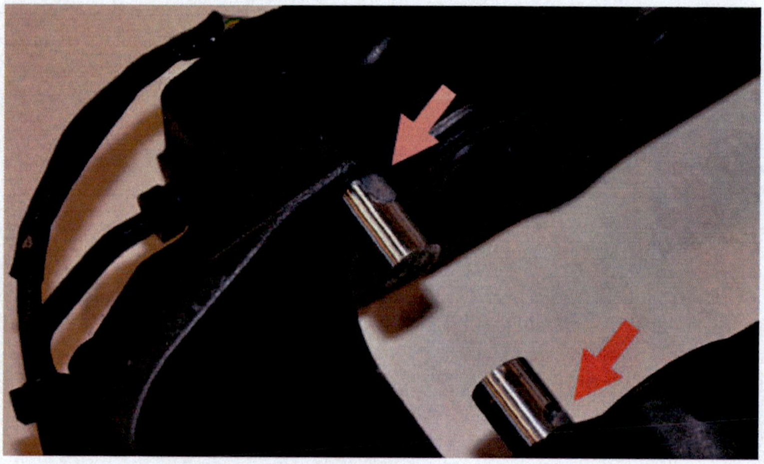

Fig. 3 Brake calliper with modified guide pins, strain gauges *marked*

3 Results

During road testing an audible squeal frequency at 1,890 Hz could be measured with a microphone. This frequency was also measured with an acceleration sensor mounted at the brake calliper. These two signals showed a good correlation.

With the invented system a more exact measurement at the test rig is possible. The measured squeal frequencies are 1,890, 2,000 Hz and the first harmonic at 4,000 Hz. Figure 4 depicts a comparison of the FFT signals from the guide pins and a reference signal obtained by a microphone. Because of the coherence of the signals it is assumed, that the measured friction force is a triggering mechanism for the vibration of the disc, hence for the noise.

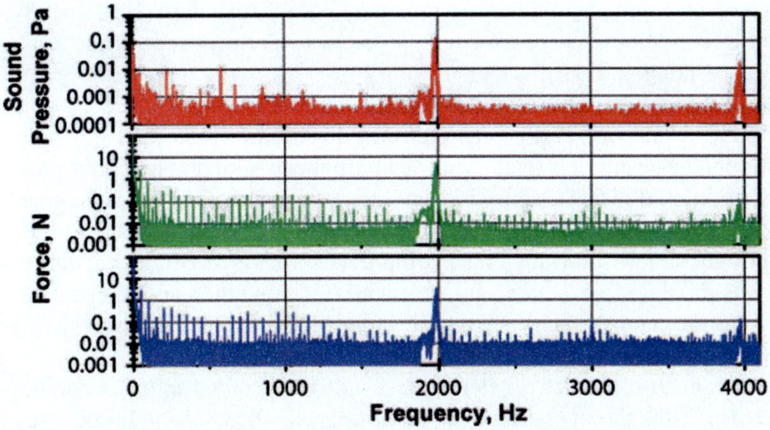

Fig. 4 FFTs of the microphone signal and the two guide pins signals

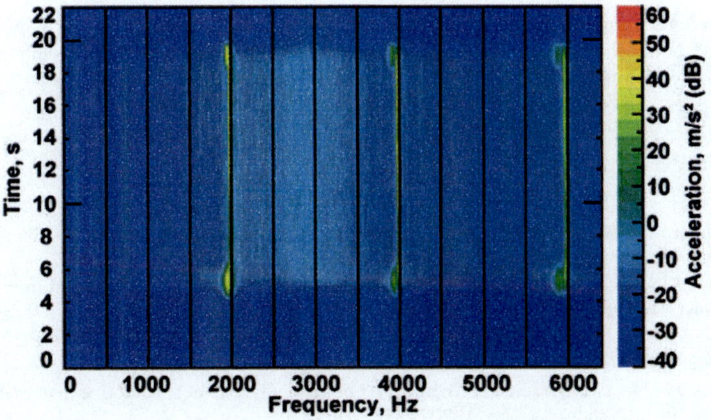

Fig. 5 Waterfall plot of the acceleration sensor mounted at the inner brake pad

Parallel performed acceleration measurements of the brake pad confirm this assumption. Figure 5 shows one waterfall plot out of the frequency analysis. The analysed squeal event lasts approximately 15 s. The shown signal is an online FFT from an acceleration sensor which is mounted on the brake pad backplate. The dominant frequency at 2 kHz and corresponding first and second harmonic are clearly visible.

4 Conclusions and Outlook

Despite decades of research brake squeal is still an elusive issue. For the presented analysis prototype brake pads which consist of sintered material are used. These pads have no countermeasures and thus they squeal very loud and at nearly every operating condition. As a result it is possible to perform repeatable test runs at certain operating points.

In order to get further insight into the triggering mechanism special guide pins have been developed. These pins measure the friction force and the superposed vibration. The results correlate with a microphone signal. Thus it is possible to obtain additional information with the presented test setup. But of course, basic analyses such as frequency analysis or modal analysis are still necessary. The special pins provide a new innovative measuring device for further fundamental research. Thus it is possible to evaluate varying material combination and to estimate squeal triggering mechanisms. Hence friction induced vibrations can be analysed using these pins.

Next miniature triaxial accelerometers will be installed into the cooling ducts of the disc. Thus the operating deflections shapes (ODS) of the brake disc can be determined. A slip ring will transmit the signal. With the obtained results a friction model and simulation procedures will be verified and improved.

Acknowledgments The authors are grateful for the financial support from MAGNA STEYR Fahrzeugtechnik and for the production of the prototype pads by MIBA Frictec Gmbh.

References

1. Papinniemi, A., Lai, J.C.S., Zhao, J., Loader, L.: Brake squeal: a literature review. Appl. Acoust. **63**, 391–400 (2002)
2. Cantoni, C., Cesarini, R., Mastinub, G., Rocca, G., Sicigliano, R.: Brake comfort: a review. Veh. Syst. Dyn. **47**, 901–947 (2009)
3. von Wagner, U., Hochlenert, D., Hagedorn, P.: Minimal models for disk brake squeal. J. Sound Vib. **302**, 527–539 (2007)
4. Ouyang, H., Nack, W., Yuan, Y., Chen, F.: Numerical analysis of automotive disc brake squeal: a review. Int. J. Veh. Noise Vib. **1**, 207–231 (2005)
5. Marschner, H., Rischbieter, F.: Three-dimensional operational deflection shape analysis of squealing disc brakes. In: Proceedings of 22nd Annual Brake Colloquium & Exhibition, Anaheim (2004)
6. Ouyang, H., Cao, Q., Mottershead, J., Treyde, T.: Vibration and squeal of a disc brake: modelling and experimental results. Proc. Inst. Mech. Eng. D J. Automob. Eng. **217**, 867–875 (2003)
7. Degenstein, T., Winner, H.: Dynamic measurement of the forces in the friction area of a disc brake during a braking process. In: Proceedings of FISITA World Automotive Congress, Yokohama (2006)

Experimental Investigation of Vibratory Peg-in-Hole Insertion for Robotic Assembly

Sigitas Kilikevičius and Bronius Bakšys

Abstract The paper aims to experimentally investigate the process of a compliantly supported peg insertion into a bush with clearance by a robot, when vibrations are provided to the bush in the axial direction. The experimental setup of robotic vibratory assembly and the investigation methodology are presented. The experiments were performed by inserting the peg which is attached to a remote center compliance device by a robot into the bush mounted on an electrodynamic shaker. Durations of insertion process stages were measured under various combinations of excitation parameters of the bush. The experiments show that parameters of vibratory excitation have an influence on the duration of insertion process. By selecting suitable excitation parameters it is possible to shorten the insertion process duration and avoid jamming of the parts to be assembled.

Keywords Vibratory assembly • Peg-in-hole insertion • Vibrations

1 Introduction

One of the newest fields of the vibrations application is a vibratory assembly. To apply the method of the vibratory assembly, one of the parts in an assembly position should be provided with vibrations of predefined direction, amplitude and frequency. The process of automatic vibratory assembly can be divided into two stages – the alignment of parts connective surfaces and the insertion of parts. During the alignment stage, due to the influence of vibrations, the compliantly supported part, being in the contact with the mating part, is able to displace and turn in respect of the later. Thus, the part-to-part alignment in the assembly position is

S. Kilikevičius (✉) • B. Bakšys
Kaunas University of Technology, Kaunas, Lithuania
e-mail: sigitas.kilikevicius@ktu.lt; bronius.baksys@ktu.lt

J. Náprstek et al. (eds.), *Vibration Problems ICOVP 2011: The 10th International Conference on Vibration Problems*, Springer Proceedings in Physics 139,
DOI 10.1007/978-94-007-2069-5_83, © Springer Science+Business Media B.V. 2011

reached, ensuring both the matching of their connective surfaces and prerequisites for unhindered assembly. The part-to-part position errors, which emerge while feeding the parts into the assembly position and locating them in assembly devices, are compensated during the alignment [1, 2]. The vibratory excitation gives a positive effect also in the stage of parts insertion. Researches show that jamming can be avoided by providing vibrations to one of the parts to be assembled, thus facilitating successful assembly [3, 4].

Up to know, the robotized vibratory assembly using vibrations for the alignment and insertion of the parts are not sufficiently analyzed. The analytical and experimental analysis of the alignment and mating, when one of the parts is provided with vibratory excitation along the two perpendicular directions, are presented in the papers [5–7]. The authors analyzed the influence of vibratory excitation parameters both on the reliability and duration of the alignment. In the paper [8] the dynamic compliance device for robotic high speed and precision chamferless assembly is proposed. Vibrations are provided to the work table in the horizontal plane by two pneumatic bellow actuators, piloted by the pseudo-random binary signal. However, the processes of alignment and insertion are random and are not stable. The influence of vibrations on the peg-in-hole insertion process is not discussed.

This presented paper considers the experimental analysis of the insertion process when the compliantly supported peg is inserted by a robot into the bush, which is excited in the axial direction. When the parameters of vibratory excitation are properly selected it is possible to obtain the minimal insertion process duration and ensure the successful insertion process.

2 Experimental Setup and Methodology of the Investigation

The experimental setup of robotic vibratory assembly was designed and made (Fig. 1). Assembly operations are performed by the robot 1 (Mitsubishi RV-2AJ). The robot gripper 2 holds the experimental remote center compliance device 3, which is attached to the peg 4. The special construction steel bush 5 was designed and made, which provides the possibility to acquire the parameters of the insertion process using the contact method (Fig. 2). The bush is mounted on the platform of electrodynamic shaker 6. The electrodynamic shaker, providing excitation to the bush, receives the electric signal of the excitation from the oscillator using the amplifier. The bush comprises the electrically insulated interdependent segments, i.e. the chamfer, two sides and the bottom.. When the peg touches the different parts of the bush, the voltage jump occurs, which is acquired by the oscilloscope and displayed on the computer screen. In such a way it is possible to track all the stages of the insertion process – peg-chamfer contact, one-point and two-point contacts of the peg with bush's hole, the end of the insertion process and to define the durations of these stages. The experiments were performed using two remote center compliance devices of different designs. One of them comprises two discs,

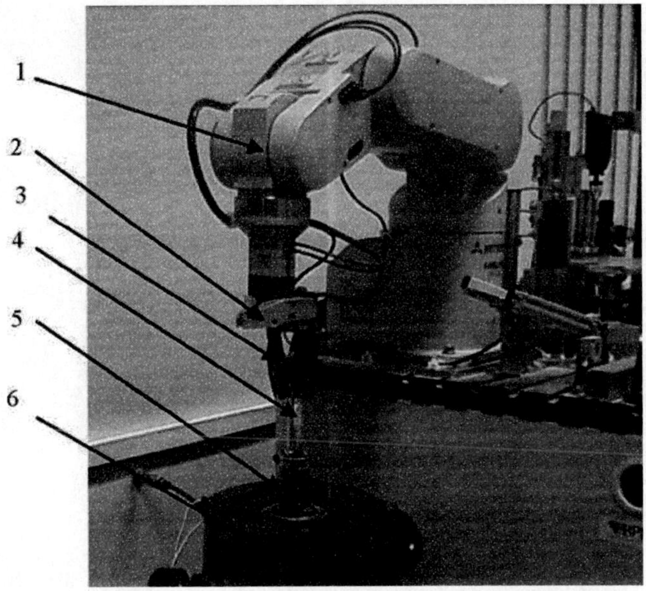

Fig. 1 Experimental setup for robotic vibratory assembly

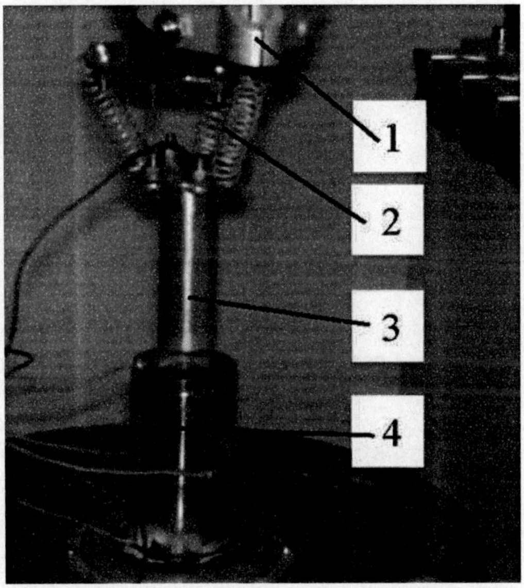

Fig. 2 Initial state of the parts in an assembly position before the insertion: 1 – robot gripper, 2 – remote center compliance device with helical springs, 3 – peg, 4 – bush

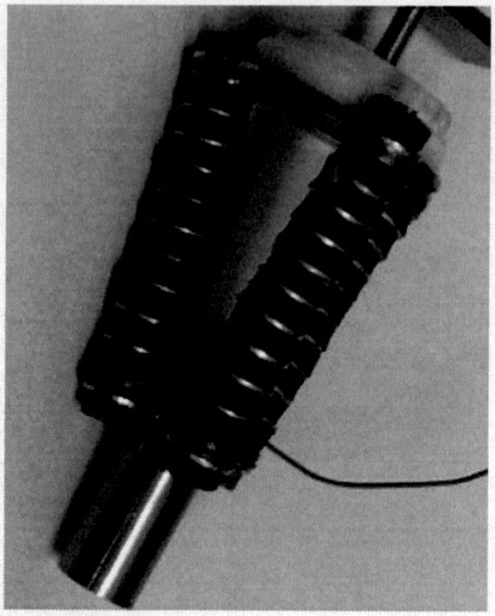

Fig. 3 Remote center compliance device with a set of metallic and rubber bushings

which are interconnected by three helical springs. The springs are allocated at a particular angle in respect of the device axis. Therefore, the compliance center of the device is located more closely to the bottom end of the peg. The other remote center compliance device (Fig. 3) is made of two discs, which are connected by means of three elastic elements, having rigidity 2.45 N/mm. They are made of a set of metallic and rubber bushings. The length of the elastic elements is 60 mm, the diameter 14 mm, the number of layers is ten.

3 Investigation of the Vibratory Insertion Process

Dependences of the durations of insertion stages on the linear positioning error of the peg ε_0 were experimentally obtained using the remote center compliance device with three helical springs (Fig. 2). The experiments were performed inserting the peg of mass $m = 0.1$ kg into the bush which diameter is $D = 20$ mm, the assembly clearance $\delta = 0.2$ mm. The other parameters of the parts arranged in the assembly position: chamfer angle of the bush $\alpha = \pi/4$ rad, initial tilt angle of the peg $\theta_0 = 0.035$ rad, initial linear positioning error $\varepsilon_0 = 2.25$ mm, remote center compliance device lateral stiffness $K_x = 2000$ N/m, axial stiffness $K_z = 5000$ N/m, angular stiffness $K_\theta = 20$ Nm/rad, distance from the lower end surface of the peg to the centre of compliance $L_C = 10$ mm, insertion speed $v = 0.18$ m/s.

When the initial linear positioning error ε_0 is increasing, the chamfer crossing duration t_1 significantly increases (Fig. 4). The duration t_2, which takes time from the initial instant of insertion to the two point contact stage, is increasing as ε_0 increases, mainly due to the increased chamfer crossing duration (Fig. 4). When ε_0

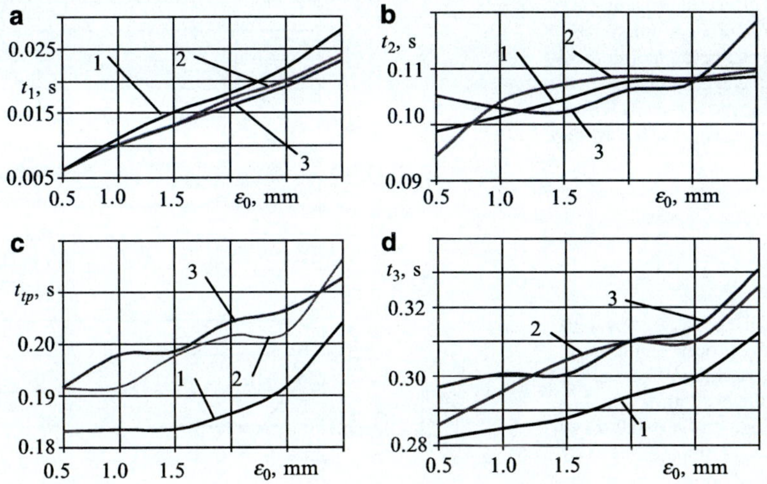

Fig. 4 Dependences of: (**a**) the chamfer crossing duration t_1; (**b**) the duration t_2; (**c**) the two-point contact stage duration t_{tp}; (**d**) the insertion process duration t_3, on the linear positioning error of the peg ε_0, when $\theta_0 = 0.0175$ rad, 1 – without the vibratory excitation, 2 – exciting vibrations of the bush with the frequency $f = 70$ Hz and amplitude $A = 1.0$ mm, 3 – $f = 70$ Hz, $A = 1.5$ mm

is increasing, the two-point contact stage duration t_{tp} increases (Fig. 4). This shows that the two point contact appears at the lower depth of the bush's hole. It is noticed, that wedging or jamming usually occurs when the two point contact appears in a small depth. Besides, the probability that the peg will jump out of the hole increases when the two point contact appears in a small depth, due to its uneven movement. Thus, the probability of successful insertion process decreases when the initial linear positioning error is increasing. The total insertion process duration t_3 increases also when ε_0 is increasing (Fig. 4). Under higher excitation amplitudes, the insertion process duration increases more significantly.

The experiments on insertion process durations were performed using the other remote center compliance device with a set of metallic and rubber bushings (Fig. 3) and bending stiffness 1,800 N/m. The experiments showed that the excitation parameters have an influence on the insertion process duration t_3. When the excitation amplitude is increasing, the insertion process duration t_3 decreases (Fig. 5). Figure 6 shows that the insertion process duration decreases also when the excitation frequency is increasing.

When the excitation frequency f is increasing, the duration t_2 slightly increases (Fig. 7), meanwhile the two point contact stage duration t_{tp} decreases (Fig. 7). According to this, the conclusion could be made that when the excitation frequency f is increasing, the two point contact appears as the peg is in a higher depth. Consequently, the increase of excitation frequency facilitates the successful insertion process.

Fig. 5 Dependences of the insertion process duration t_3 on the excitation amplitude A, when the excitation frequency: 1 – $f = 40$ Hz, 2 – $f = 60$ Hz, 3 – $f = 100$ Hz

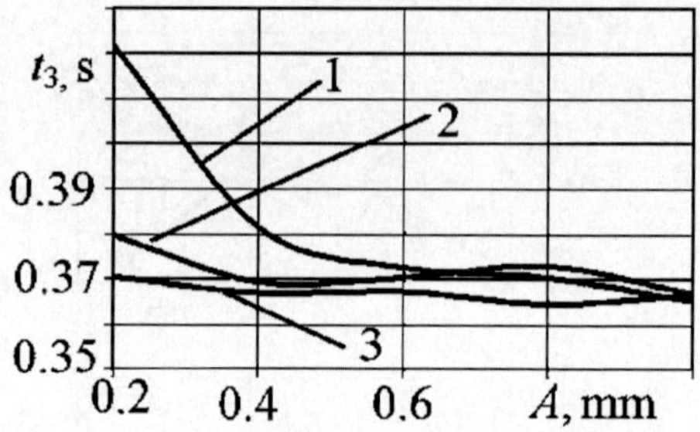

Fig. 6 Dependences of the insertion process duration t_3 on the excitation frequency f, when the excitation amplitude: 1 – $A = 0.2$ mm, 2 – $A = 0.6$ mm, 3 – $A = 0.8$ mm

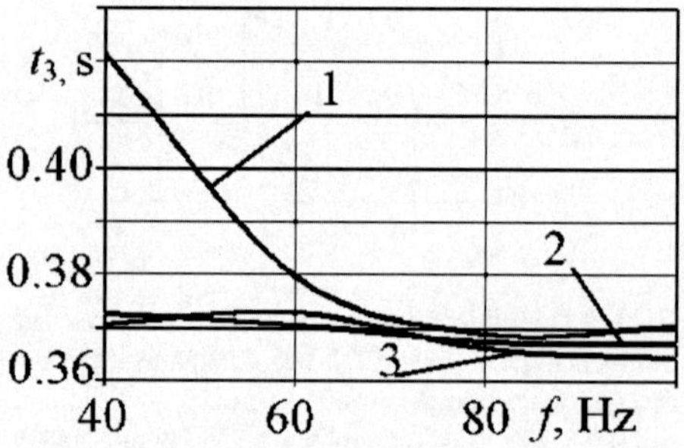

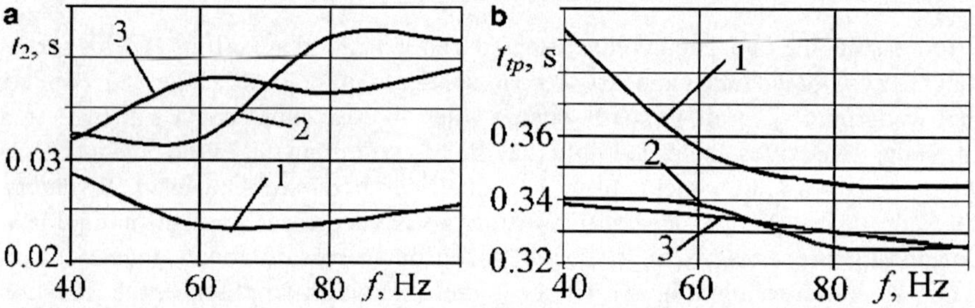

Fig. 7 Dependences of: (a) the duration t_2; (b) the two-point contact stage duration t_{tp} on the excitation frequency, 1 – $A = 0.2$ mm, 2 – $A = 0.6$ mm, 3 – $A = 0.8$ mm

4 Conclusions

The experimental analysis of vibratory insertion process is performed when the compliantly supported peg is inserted by a robot into the bush, which is excited in the axial direction. The dependencies of the durations of insertion stages on the excitation amplitude, frequency and linear positioning error are determined. The experiments showed that the excitation parameters have an influence on the insertion process duration. When the excitation amplitude and frequency are increasing, the insertion process duration decreases. As the excitation frequency increases, the two point contact appears as the peg is in a higher depth, thus jamming can be avoided. Consequently, the increase of excitation frequency facilitates the successful insertion process.

Acknowledgments This research was funded by a grant (No. MIP-44/2010) from the Research Council of Lithuania.

References

1. Bakšys, B., Puodžiūnienė, N.: Alignment of parts in automatic assembly using vibrations. Assemb. Autom. **1**(27), 38–43 (2004)
2. Bakšys, B., Baskutienė, J.: Vibratory alignment of parts during robotized assembly. J. Vibroeng. **1**(10), 69–73 (2008)
3. Bakšys, B., Fedaravičius, A., Povilionis, A.B.: Connection conditions of mobile based parts. Mechanika **5**(37), 19–25 (2002)
4. Kilikevičius, S., Bakšys, B.: Analysis of insertion process for robotic assembly. J. Vibroeng. **9**(4), 35–40 (2007)
5. Leu, M.C., Liao, H.: Parts mating aided by constant amplitude sinusoidal motions. J. Manuf. Sci. Eng. **121**(4), 802–810 (1999)
6. Leu, M.C., Liu, Y.: Modeling and analysis of parts mating in vibration assisted compliant assembly. In: 3-rd ASME Conference on Flexible Assembly Systems, Miami Beach, 33, pp. 9–20 (1991)
7. Leu, M.C., Jia, Y.: Mating of rigid parts by compliant manipulators. ASME J. Eng. Ind. **117**, 240–247 (1995)
8. Trong, D.N., Betemps M., Jutard, A.: Analysis of dynamic assembly using passive compliance. In: Proceedings of the IEEE International Conference on Robotics and Automation, Nagoya, Japan pp. 1997–2002 (1995)

Innovative Multilevel Energy Processor in Condition Monitoring

C. Cempel

Abstract Working machine can be considered as an equivalent energy processor (EP) with internally limited energy dissipation. This means, we have limited machine lifetime and breakdown time, and the observed vibration symptoms of machine condition can be incorporated into EP model. Up to now one level of EP model have been in use, but using singular value decomposition (SVD) we need two levels EP, and due to this it is possible to trace several independent faults of machine during its life. This possibility allow us to formulate concept of machine as a multilevel energy processor, and using SVD we determine the second sublevel of this multilevel EP. The present paper illustrates this way of thinking, and allows us to deepen our understanding of machine life and the wear on two levels; overall machine damage and particular faults as well.

Keywords Condition monitoring • Energy processor (EP) • Symptom observation matrix • Singular value decomposition

1 Introduction

Working machine can be modeled as an energy processor (EP), which processes input equivalent energy (*according to some control program*), into some product or another form of useful energy. As working life of the machine is in progress, the wear of its parts is starting, what can be acknowledged as internal dissipation of the input energy, it means – internal damage. The growing internal damage expresses into some evolving faults, which if not interrupted by renewal, leads to the total damage of the machine. This wear mechanism of a working machine was already

C. Cempel (✉)
Department of Mechanical and Management Engineering, Poznań University of Technology, Poznań, Poland
e-mail: czeslaw.cempel@put.poznan.pl

J. Náprstek et al. (eds.), *Vibration Problems ICOVP 2011: The 10th International Conference on Vibration Problems*, Springer Proceedings in Physics 139,
DOI 10.1007/978-94-007-2069-5_84,

modeled as an energy processor and described by the present author in some papers [1] and books [2]. It was shown there that EP has great diagnostic inference power, giving possibility to model the life of the machine, its reliability, and even to foresee system breakdown time and its residual life.

After some years of EP application and use, see for example [3], it seems that not all inference capability of EP has been uncovered and used. This concerns the possible hierarchical structure of EP, which if used properly, can lead to better understanding of some decomposition methods of Symptom[1] Observation Matrix (SOM). So, the aim of this paper is to find some good metaphor, and analogy between the multilevel EP, and symptom observation matrix processed by singular value decomposition (SVD). This decomposition is mainly used in contemporary multidimensional diagnostics of machines and other technical systems.

2 The Concept of Energy Processor (EP)

The first thinking concerning the link between the wear model of machine and emitted vibroacoustic processes during the life of the machine were published almost 20 years ago [1, 4, 5]. As a next, well clarified and more generalized model called Energy Processor (EP) was published in the book [5], see Fig. 1. By the use

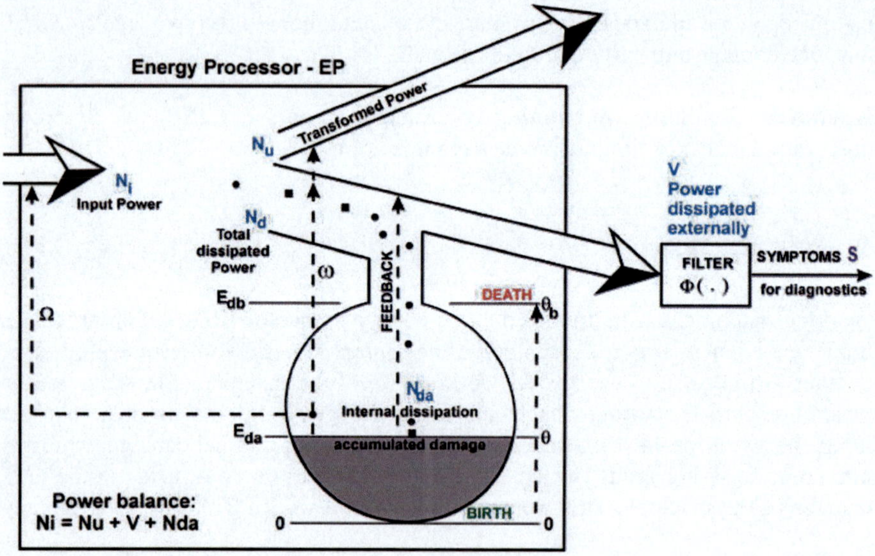

Fig. 1 One level energy processor [4]

[1]Symptom, measurable quantity covariable (*or assumed to be*) with the system condition.

of such EP model and some number of observations of one symptom measured at a real object, it was possible to assess the object symptom reliability $\boldsymbol{R(S)}$ [2], its breakdown time $\boldsymbol{\theta_b}$ and residual life time $\boldsymbol{\Delta\theta}$. This approach was used where the condition of the machine has been described by one symptom only. We may call it now the **total damage** symptom, like rms velocity used for simple rotating machinery according to ISO standards. For such simple machines it can be found that the observed symptom is of Pareto, or Weibull type, and in these cases the EP model works well, as one can see for example in [4]. This means simply, that histogram of observed symptoms is similar to Pareto or Weibull distribution, what enables to use the assessed distribution exponent γ to calculate the model of symptom life curve $\boldsymbol{S(\theta)}$, and the symptom reliability $\boldsymbol{R(S)}$.

Not repeating the whole calculation for energy processor concept, we will show only the essential relations, already mentioned above. Denoting the system lifetime as $\boldsymbol{\theta}$, externally dissipated power as $\boldsymbol{V(\theta)}$ and following [5, 6] (*see* Fig. 1) we will have the evolution of externally dissipated power in the form of breakdown time $\boldsymbol{\theta_b}$ asymptotic curve as below;

$$V(\theta) = V_o(1 - \theta/\theta_b)^{-1} \tag{1}$$

In dependence to functional and structural design of machine, the same amount of dissipated power $\boldsymbol{V(\theta)}$ may be transformed differently into the observed symptom of condition $\boldsymbol{S(\theta)}$. We denote this difference in transformation by introducing special symptom operator $\boldsymbol{S(\theta) = \Phi(V(\theta))}$, which transforms dissipated power into the observed quantity, covariable with the condition of the machine. The simplest forms of symptom operators are of exponential type, like Pareto or Weibull. They are applicable for several types of machines [4]. For Pareto type of symptom operator we will have observed symptom of system condition $\boldsymbol{S(\theta)}$ in the form as below;

$$S(\theta) = \Phi(V(\theta)) = S_o(1 - \theta/\theta_b)^{-1/\gamma} \tag{2}$$

And if the symptom reliability [2] is of Pareto type for observed symptom we have at the same time

$$R(S) = (S_o/S)^{-\gamma} \tag{3}$$

where $\boldsymbol{S_o}$ is the initial value of observed symptom of condition.

These are the essential relations describing the behavior and the possibility of diagnostic use of **one level** EP, when **total damage** symptom $\boldsymbol{S(\theta)}$ is observed, and we do know Pareto exponent $\boldsymbol{\gamma}$ from population of symptom observations. Then we can calculate all needed parameters concerning the machine safety and running, such as symptom limit value $\boldsymbol{S_l}$ and the machine residual lifetime $\boldsymbol{\Delta\theta}$.

3 Two Levels Energy Processor and Decomposition of Symptom Observation Matrix (SOM)

In some cases however, more sophisticated diagnosis of machine is needed, we need for example specification of fault type, its life evolution or strength, the symptom limit value for a given generalized fault type S_{li}, and specific residual lifetime for respective fault $\Delta\boldsymbol{\theta}_{li}$. These needs evoke a question, is that possible to use again the EP model for each generalized fault separately? It seems to be positive answer for this question, particularly when looking at the concept of multilevel energy processor given in [5], and shown below (see Fig. 2) in the two level version only. This self limitation follows from the frequent application and use of singular value decomposition (SVD) [8], or principal component analysis (PCA) in multidimensional condition monitoring. These methods, SVD in particular, give us clear identification of total damage symptom for the whole machine, and specific generalized faults for the first sublevel of decomposition. This will be seen later on, from some real examples of multidimensional condition monitoring of machines.

Looking deeply at Fig. 2 and comparing with Fig. 1 one can notice that the best condition observation for each particular fault is, if can one observe particular fault symptom originating at given energy processor $(\mathrm{EP})_i$. Following (2) this can be expressed mathematically

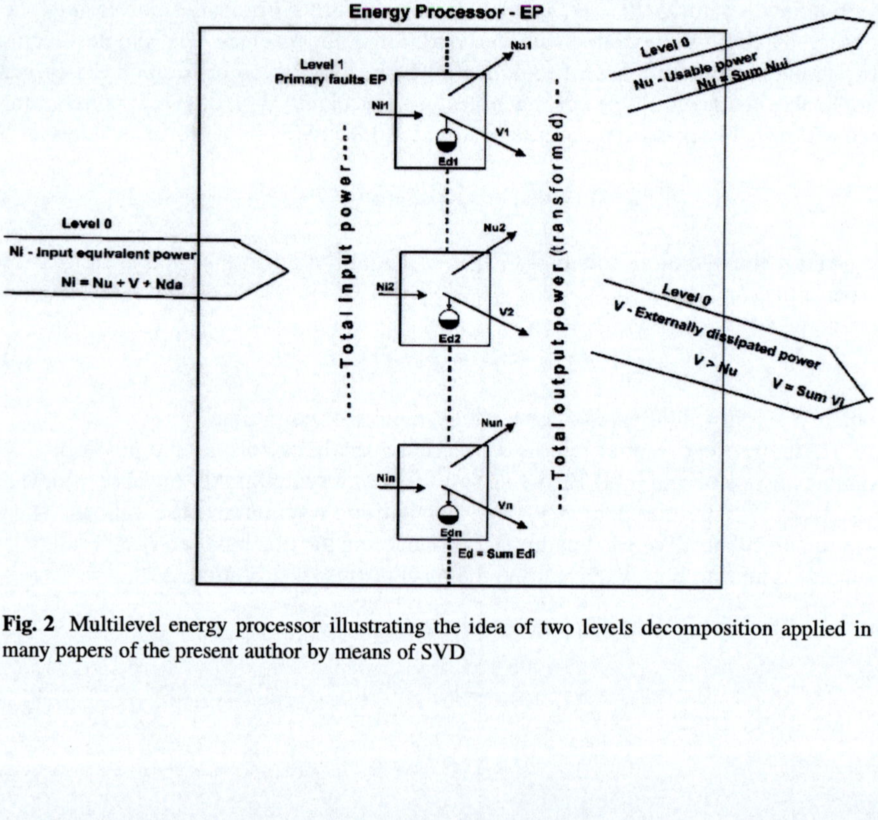

Fig. 2 Multilevel energy processor illustrating the idea of two levels decomposition applied in many papers of the present author by means of SVD

$$S_i(\theta) \equiv \Phi_i\left[V_i(\theta)\right], \quad i = 1, \ldots n. \tag{4}$$

But symptoms of condition of real machine can be observed at the machine casing, i.e. only from the outside measuring position, hence another relation including outside dissipative power ***V(θ)*** must be written here

$$S_i(\theta) \approx \Phi_i\left[V(\theta)\right] = \Phi_i\left[\sum V_i(\theta)\right], \quad i = 1, \ldots n. \tag{5}$$

Looking deeply at this new definition we can rewrite it expressing in an explicit way the relation (4) seen inside the EP, and depicting the way of symptom creation, specific to a given fault seen as *non italic term* in the first line of formula below;

$$\begin{aligned} S_i(\theta) &= \Phi_i\left[\sum V_i(\theta)\right] \approx \Phi_i\left[V_1(\theta)\right] + \Phi_i\left[V_2(\theta)\right] + \ldots \\ &+ \Phi_i\left[V_1(\theta)\right] + \ldots + \Phi_i\left[V_n(\theta)\right] \\ &= \Phi_i\left[V_i(\theta)\right] + \varepsilon\Phi_{ir}\left[V(\theta)\right], \quad i = 1, \ldots n. \end{aligned} \tag{6}$$

where $\boldsymbol{\Phi}_{\mathbf{ir}}[\boldsymbol{V}(\boldsymbol{\theta})]$ in the second line is the **residue** of our symptom operator, and it help us to express each particular symptom according to fundamental relation (4).

Repeating this decomposition ***n*** times, according to Fig. 2, we will have **n** components symptom observation vector ***S (θ)*** for our two level EP, as below.

$$\mathbf{S}(\boldsymbol{\theta}) = \begin{bmatrix} S_1(\theta) \\ S_2(\theta) \\ \ldots \\ S_n(\theta) \end{bmatrix} = \begin{bmatrix} \Phi_1\left[V_1(\theta)\right] + \varepsilon\Phi_{1r}\left[V(\theta)\right] \\ \Phi_2\left[V_2(\theta)\right] + \varepsilon\Phi_{2r}\left[V(\theta)\right] \\ \ldots \\ \Phi_n\left[V_n(\theta)\right] + \varepsilon\Phi_{nr}\left[V(\theta)\right] \end{bmatrix} \tag{7}$$

It will be good now to invent some transformation of above vector, that each residue of given component of observation vector will be close to zero, it means we need;

$$\Phi_{ir}\left[V(\theta)\right] \approx 0, \quad i = 1, \ldots n, \tag{8}$$

After such transformation, in case of some redundancy of observation, some of the components of transformed observation vector may be close to zero too, and the remaining nonzero components will be orthogonal each other, describing independent faults evolving in a machine.

Now, the above symptom observation vector (7), is observed in a discrete way along the machine lifetime ***θ*** produces symptom observation matrix (SOM), which describe the evolution of faults during the life of a machine. Applying singular value decomposition to our SOM we can obtain exactly what was described above; the total damage symptom, and particular **generalized symptoms** as the evidence of evolving faults in a machine. Due to a limitation of space we will not illustrate the SVD in detail, but one can look at this in some author papers cited in the literature.

4 Examples of Two-Level Energy Processor Use in Machinery Diagnostic Practice

As a first example of application of our idea we will take a hard diagnostic case – a huge fan for coal milling working at one of Polish thermo power station. Here the root mean square vibration velocity (V_{rms}) has been used as a symptom of condition, and initially altogether 11 symptoms at different places of fan mill aggregate structure (observation vector) were constantly monitored, over 60 weeks of a lifetime $\boldsymbol{\theta}$. We will process this case with inclusion of life time symptom to ***SOM***. How unstable and noisy the fan running environment is, one can notice from the left top picture of the Fig. 3. It is seen further (*middle left picture*), that the symptom normalization to the initial value and centering do not change much the noisy behavior of primary and generalized symptoms after ***SVD*** (*bottom left picture*).

Looking at the middle right picture of Fig. 3, one can notice that symptoms No 7,8,9,10,11 do not give substantial contribution to the three dominating generalized

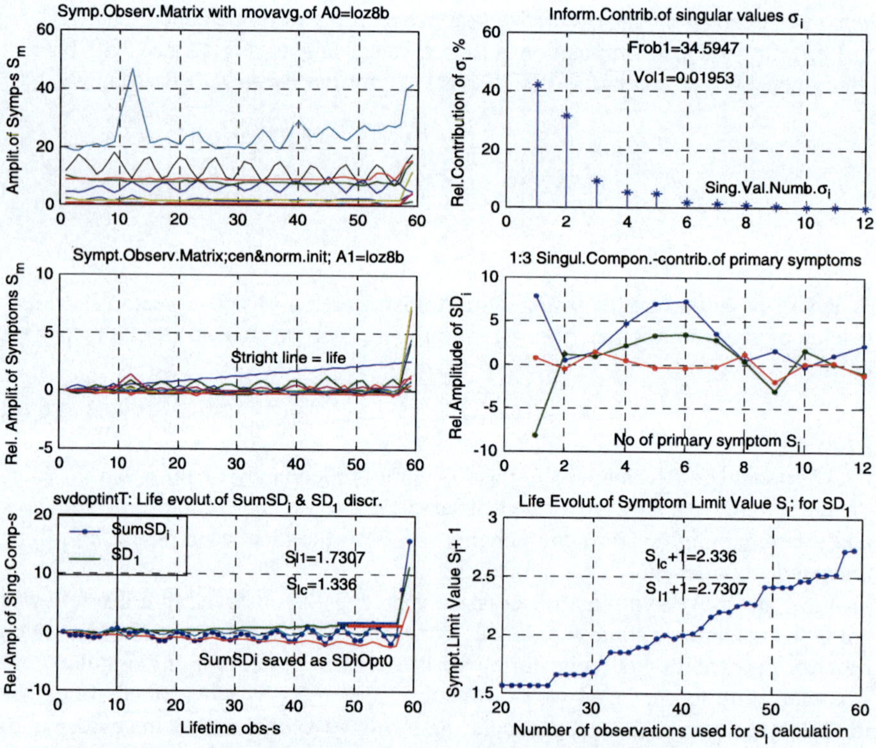

Fig. 3 Vibration condition monitoring (V_{rms}) of the coal mill fan observed at three bearings of fan and electric motor, as processed with life time symptom (*LS*) added to SOM

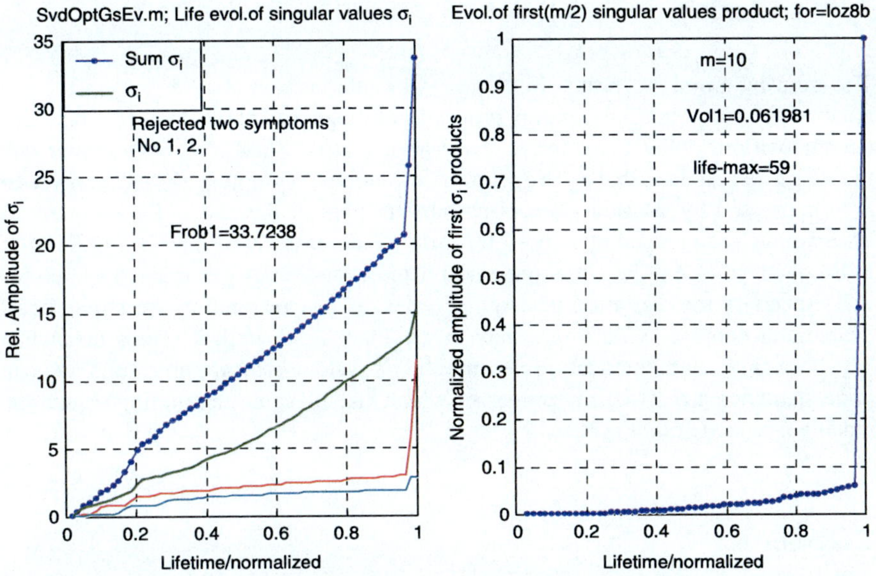

Fig. 4 The evolution of the first few singular values of the coal fan mill of Fig. 3, as the validation of separate faults evolution according to concept of two level EP

symptoms, and probably can be rejected as redundant at the first approach. With this respect please note the value of Frobenius modified measure $Frob1 = 34.59$ and the volume of the fault space $Vol1 = 0.019$, at upper right picture. One can also note here, that there are two generalized symptoms with high information contents (*picture top right*), and due to that two symptom limit values are assessed: namely S_{lc} for the total damage symptom, and S_{l1} for the first generalized symptom (*bottom pictures*).

Following the confirmation of the underlying thesis of this paper, please take a look at the picture top right of Fig. 3, where one can notice two dominating singular values, what may stand for the **two evolving** faults in the machine. Validation of this supposition one can see at the next figure left picture, where the evolution of first few singular values calculated by another program **SvdOptGsEv.m** is shown. This picture was taken from the last paper of present author [7], where the hypothesis concerning the proportionality between the singular value and system damage was fully exploited. One can see from the left picture of Fig. 4, that the concept of second level of EP has some confirmation here, at least two life curves of singular values looks like the life symptom of some advanced evolving damage, approaching the breakdown. This means that inside machine system two damages has grown constantly up to the final system breakdown.

5 Conclusion

The starting point to write this paper was supposition that there exist some analogy between multilevel energy processor of diagnosed object and singular value decomposition applied to symptom observation matrix (SOM). We have shown with the experimental data, that by proper transformation of symptom observation vector $S(\theta)$ expressed by relation (7) it is possible to neglect the residual component of observation vector decomposition. It works much easier when we apply SVD to SOM of diagnosed object, looking next for the evolution of generalized symptoms and especially the evolution of singular vales, as proportional to the strengths of developing damage. What more may mean this result? Well, it seems that when we chose right mathematical transformation of system observation results, we can understand better the wearing processes which take place in the running object; the machine, or some other system.

References

1. Cempel, C.: Generalized Tribo-vibroacoustical (TVA) model of machine in plant diagnostic environment. Bull. Pol. Acad. Sci. Tech. Sci. **38**(1–12), 39–49 (1990)
2. Cempel, C., Natke, H.G., Yao, J.P.T.: Symptom reliability and hazard for systems condition monitoring. Mech. Syst. Signal Process. **14**(3), 495–505 (2000)
3. Cempel, C., Tabaszewski, M.: Multidimensional condition monitoring of the machines in non-stationary operation. Mech. Syst. Signal Process. **21**, 1233–1247 (2007)
4. Cempel, C.: Theory of energy transforming systems and their application in diagnostics of operating systems. Appl. Math. Comput. Sci. **3**(3), 533–548 (1993)
5. Natke, H.G., Cempel, C.: Model-Aided Diagnosis of Mechanical Systems, p. 248. Springer, Berlin (1997)
6. Tumer, I.Y., Huff, E.M.: Principal component analysis of tri-axial vibration data from helicopter transmission. In: 56th Meeting of the Society of Machine Failure Prevention Technology, Virginia Beach (2002)
7. Cempel C.: The evolution of generalized fault symptoms and fault intensities as indicators of observation redundancy and coming system breakdown. Mech. Syst. Signal Process. MSSP 24 (2010) 1129–1137
8. Kielbasinski, A., Schwetlik, H.: Numeric Linear Algebra, p. 502. WNT Press, Warsaw (1992)

Effectiveness of MED for Fault Diagnosis in Roller Bearings

P. Pennacchi, Roberto Ricci, S. Chatterton, and P. Borghesani

Abstract Diagnostics of rolling element bearings is usually performed by means of vibration signals measured by accelerometers placed in the proximity of the bearing under investigation. The aim is to monitor the integrity of the bearing components, in order to avoid catastrophic failures, or to implement condition based maintenance strategies. In particular, the trend in this field is to combine in a single algorithm different signal-enhancement and signal-analysis techniques. Among the first ones, Minimum Entropy Deconvolution (MED) has been pointed out as a key tool able to highlight the effect of a possible damage in one of the bearing components within the vibration signal. This paper presents the application of this technique to signals collected on a simple test-rig, able to test damaged industrial roller bearings in different working conditions. The effectiveness of the technique has been tested, comparing the results of one undamaged bearing with three bearings artificially damaged in different locations, namely on the inner race, outer race and rollers. Since MED performances are dependent on the filter length, the most suitable value of this parameter is defined on the basis of both the application and measured signals. This represents an original contribution of the paper.

Keywords Minimum Entropy Deconvolution (MED) • Bearing diagnostics • Bearing failure

1 Introduction

Bearings are mechanical components largely used in rotating machines and employed in many industrial fields. Unfortunately, due to both the load applied and the harsh operating environment, bearings are prone to failure and may even lead to

P. Pennacchi • R. Ricci (✉) • S. Chatterton • P. Borghesani
Department Mechanical Engineering, Politecnico di Milano, Via La Masa 1, 20156 Milan, Italy
e-mail: paolo.pennacchi@polimi.it; roberto1.ricci@mail.polimi.it; steven.chatterton@polimi.it; pietro.borghesani@mail.polimi.it

J. Náprstek et al. (eds.), *Vibration Problems ICOVP 2011: The 10th International Conference on Vibration Problems*, Springer Proceedings in Physics 139,
DOI 10.1007/978-94-007-2069-5_85,

catastrophic failure of the machinery. Typical failures of rolling bearings are due to cracks or spalls on roller elements, inner ring, outer ring and seldom on the cage. For this reason, a proper diagnostics of bearings during their functioning must be performed.

Bearings diagnostics is usually performed by processing the vibration signals measured by means of transducers placed near the bearings. A lot of signal processing techniques have been implemented for this purpose. Envelope analysis is probably the most established technique: proposed for the first time in [1], it allows detecting frequencies related to bearings faults. Similar information are those provided by the spectral kurtosis [2, 3]: also in this case a distribution in frequency domain is obtained. The reliability of the cyclostationary analysis for bearings diagnostics has been tested in several applications [4]. Frequency components strictly related to the fault presence could be traced in the frequency-cyclic frequency plane. The application of these techniques is necessary for bearings diagnostics, but their effectiveness is often reduced by both environmental noise and other vibration sources. For this reason, often it is useful to enhance the measured vibration signals in order to highlight components, like transients and bursts, due to faults.

An example of signal-enhancing algorithm is represented by minimum entropy deconvolution (MED). The MED algorithm, proposed for the first time by Wiggins [5] for the study of seismic signals, has been recently considered in mechanical field.

In this paper, MED is applied to vibration signals measured on a test-rig able to test industrial roller bearings. Signals were acquired on undamaged and damaged bearings at different working conditions. The aim is to investigate the effectiveness of the MED on experimental signals.

2 Minimum Entropy Deconvolution (MED) Algorithm

The entropy of a signal is related to the amount of information included in it. This quantity depends on its randomness degree [6]. The MED is based on this assumption: it tries to minimize the randomness of a signal by minimizing its entropy. A generic signal $x(n)$ can be expressed as convolution of two terms:

$$x(n) = z(n) * w(n) + \eta(n). \tag{1}$$

where $z(n)$ is the component related to the system behavior, $w(n)$ represents the excitation and $\eta(n)$ is the random signal noise. Processing $x(n)$ with a filter $f(n)$ [7], the output signal $y(n)$ can be obtained:

$$y(n) = x(n) * f(n) = z(n) * w(n) * f(n) + \eta(n) * f(n). \tag{2}$$

The convolution between the filter and the first signal component gives functions similar to the Dirac delta function [7] whereas the convolution with $\eta(n)$ allows reducing the signal noise. The $y(n)$ in (2) is a *simple solution* since it shows the maximum possible order and it is characterized by *parsimony*. In other words, the output is composed of near zero values apart some terms assuming high values. The parsimony of a signal can be measured by means of the *Varimax norm* [6]:

$$V(y) = \sum_{j=0}^{N} y^4(j) \Bigg/ \left(\sum_{j=0}^{N} y^2(j) \right)^2 . \tag{3}$$

The signal entropy is minimized by maximizing $V(y)$ with respect to the coefficients of the filter $f(n)$. In this way, the output $y(n)$ that best fits the component $z(n)$ is calculated. The Varimax norm maximization is performed by means of an iterative procedure, by solving at each step a non-linear system. The description of the iterative process is not the aim of this paper: details can be found in [6]. However, the iterative maximization of the Varimax norm allows optimizing the filter to improve the $y(n)$ parsimony. Since the computational efforts for the iterative maximization of the norm are dependent on the length of the signal considered and, obviously, on the number of iterations required, only a part of the complete acquired signals will be analyzed and a maximum number of iterations has been set in the algorithm. Moreover, in order to improve MED performances a suitable length of the filter must be chosen. Many tests carried out on the measured signals have highlighted that best results are obtained for filter length greater than the 35% of the fault period. In this paper, filter length is equal to the 75% of the fault period.

3 Experimental Results

As stated in the introduction, the aim of the paper is the application of MED to vibration signals acquired on both undamaged and damaged industrial bearings using the test-rig shown in Fig. 1. The main shaft is driven by an electrical motor, controlled by means of an inverter and coupled to a spiral bevel gearbox with orthogonal axis. The gearbox output shaft is connected to a rotating shaft supported by two roller bearings: the bearing under test is located in the middle of the rotating shaft and it is preloaded with a simple mechanism. In order to introduce noise in measurements, mainly generated by the gearbox, another electrical motor, similar to the first one, but functioning as a brake, is connected to the rotating shaft.

Several bearings of the same type (SKF ECP 206 NJ), but in different health conditions, were tested with the test-rig of Fig. 1. Different rotational speeds were also considered. During experimental tests, vibrations in vertical and horizontal directions were measured by means of single-axis piezo-accelerometers placed on the bearing housing. Acquisitions were performed with a sample frequency of 10 kHz.

The left plot of Fig. 2 shows the raw signal of horizontal vibration acquired for a damaged bearing with a fault on the inner ring and rotating at 750 rpm. As it can be

Fig. 1 Test-rig layout. From *right* to *left*: active motor, rotating shaft with bearing under test placed in the middle, gearbox and brake motor

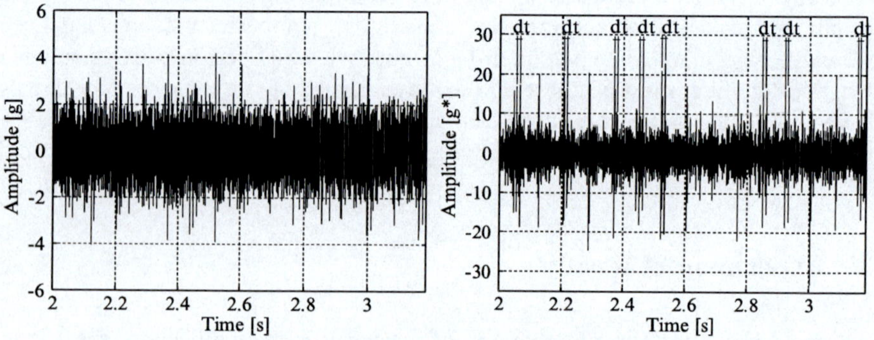

Fig. 2 Raw signal (*left*) and MED output (*right*) for a bearing damaged on inner ring

noted, only an arbitrary portion of the complete signal, lasting 15 shaft revolutions is considered. The raw signal is characterized by peak-peak amplitude of about 8 g but, no kind of structure or pattern indicates the presence of a fault.

The right plot of Fig. 2, presents the result obtained after the application of the MED algorithm to the raw signal. The amplitude of the output is obviously changed (of one order of magnitude) with respect to the original signal, due to the introduction of the deconvolution filter: for this reason also the unit of measure is distinguished by a star (g*). The shape of the output signal is very different from the original one: MED output clearly highlights peaks, which indicate transitory variations of vibration. These variations are not related to the shaft rotation, because multiple peaks are detected for each shaft revolution. By considering consecutive peaks, it is possible to calculate a time interval *dt* equal to 0.0102 s corresponding to a frequency of 98 Hz, which is very close to the theoretical BPFI (Ball Passing

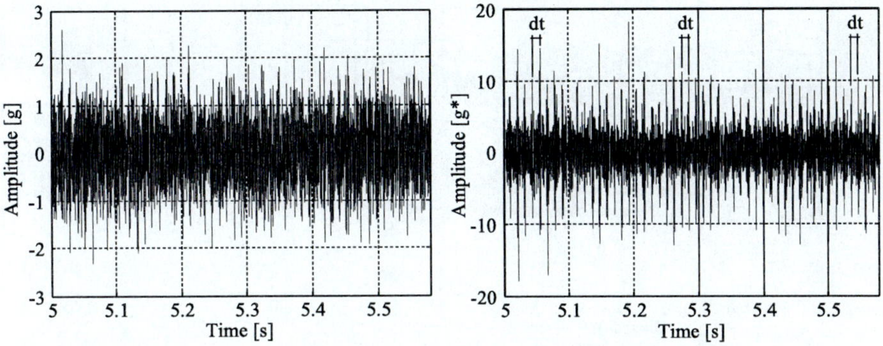

Fig. 3 Raw signal (*left*) and MED output (*right*) for a bearing damaged on outer ring

Frequency Inner) for the rotational speed considered (97.8 Hz). Therefore, the application of MED to the raw signal allows detecting the presence of a fault in the inner ring of the bearing. This detection was impossible on the basis of the original signal.

The left plot of Fig. 3 shows a portion of the vertical vibration, acquired on a bearing rotating at 1,000 rpm, with a fault on outer ring. The amplitude of vibration is similar to that of Fig. 2. The raw signal seems regular and no malfunctioning can be identified.

The result provided by the application of the MED algorithm on the original signal is shown in the right plot of Fig. 3. The deconvolution highlights a pattern of peaks not visible in the raw signal. These peaks are equally spaced in time and, also in this case, they are not strictly related to the shaft rotation period. Considering the time interval between peaks, they are separated by a *dt* of about 0.0115 s: this time gap corresponds to a frequency of 87 Hz which is very close to the theoretical BPFO (Ball Passing Frequency Outer) of the test bearing for a rotational speed of 1,000 rpm (87.4 Hz). The MED allows identifying a fault on the bearing outer ring that was impossible to be detected on the basis of the raw signal.

The left plot of Fig. 4 displays the horizontal vibration signal measured on a bearing affected by a fault on a roller element. The acquisition was performed with a shaft rotational speed of 1,250 rpm: 40 shaft revolutions are considered. The amplitude of vibration is similar to that of the previous signals considered in Figs. 2 and 3 and the trend does not indicate deviations with respect to the normal functioning.

The application of the MED algorithm to the original signal allows obtaining the result shown in the right plot of Fig. 4. MED output highlights some peaks whose pattern is similar to that obtained for a fault in the bearing inner ring. In particular, high peaks are anticipated or followed by lower peaks (a kind of "sidebands" in time). By calculating the time interval between these correlated peaks, a *dt* of about 0.01 s, corresponding to a frequency of 100 Hz, can be obtained. This frequency is very close to the theoretical $2 \times$ BSF (Ball Spin Frequency) of the bearing for the

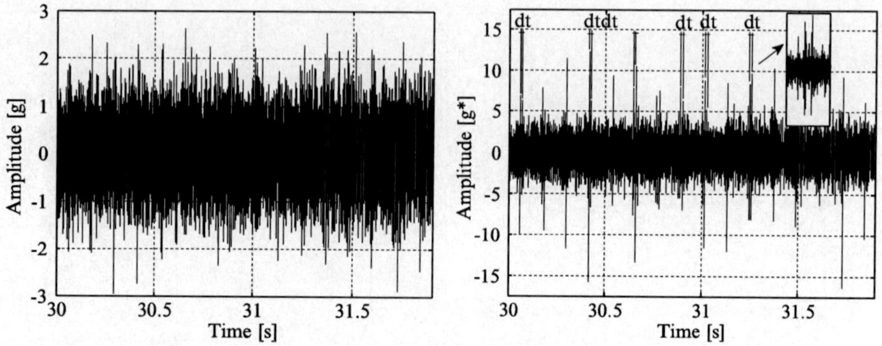

Fig. 4 Raw signal (*left*) and MED output (*right*) for a bearing damaged on rolling element

considered rotational speed (101 Hz). The occurrence of the second harmonics of the BSF is correct considering that a fault on a rolling element is excited two times for each rotation (on both inner ring and outer ring). The fault on the rolling element, which detection was impossible from the raw signal, is then identified by means of the MED.

4 Conclusions

The aim of the paper was the application of the MED technique for bearing diagnostics. MED has been applied to vibration signals measured on industrial bearings by means of a test-rig designed and realized for this purposes. Results provided by MED show that the algorithm is suitable for the detection of bearings faults and it can be used for enhancing the fault recognition in more complex applications characterized by higher environmental noise.

References

1. Darlow, M.S., Badgley, R.H., Hogg, GW.: Application of high frequency resonance techniques for bearings diagnostics in helicopter gearboxes. Technical Report, US Army Air Mobility Research and Development Laboratory, pp. 74–77 (1974)
2. Dwyer, R.F.: Detection of non-Gaussian signals by frequency domain kurtosis estimation. In: International Conference on Acoustic, Speech, and Signal Processing, Boston, pp. 607–610 (1983)
3. Antoni, J., Randall, R.B.: The spectral kurtosis: application to the vibratory surveillance and diagnostics of rotating machines. Mech. Syst. Signal Process. **20**, 308–331 (2006)
4. Antoni, J.: Cyclostationarity by examples. Mech. Syst. Signal Process. **23**, 987–1036 (2009)
5. Wiggins, R.A.: Minimum entropy deconvolution. Geoexploration **16**(1–2), 21–35 (1978)
6. Gonzàlez, G., Badra, R.E., Medina, R., Regidor, J.: Period estimation using minimum entropy deconvolution (MED). Signal Process. **41**, 91–100 (1995)
7. Sacchi, M.D., Velis, D.R., Cominguez, A.H.: Minimuym entropy deconvolution with frequency domain constraints. Geophysics **59**, 938–945 (2004)

Bearing Fault Diagnostics Using the Spectral Pattern Recognition

P. Pennacchi, P. Borghesani, S. Chatterton, and R. Ricci

Abstract In the field of diagnostics of rolling element bearings, the development of sophisticated techniques, such as Spectral Kurtosis and 2nd Order Cyclostationarity, extended the capability of expert users to identify not only the presence, but also the location of the damage in the bearing. Most of the signal-analysis methods, as the ones previously mentioned, result in a *spectrum-like* diagram that presents line frequencies or *peaks* in the neighbourhood of some theoretical characteristic frequencies, in case of damage. These frequencies depend only on damage position, bearing geometry and rotational speed. The major improvement in this field would be the development of algorithms with high degree of automation. This paper aims at this important objective, by discussing for the first time how these peaks can draw away from the theoretical expected frequencies as a function of different working conditions, i.e. speed, torque and lubrication. After providing a brief description of the peak-patterns associated with each type of damage, this paper shows the typical magnitudes of the deviations from the theoretical expected frequencies. The last part of the study presents some remarks about increasing the reliability of the automatic algorithm. The research is based on experimental data obtained by using artificially damaged bearings installed in a gearbox.

Keywords Bearing diagnostics • 2nd order cyclostationarity • Envelope analysis • Bearing characteristic frequencies

---

P. Pennacchi • P. Borghesani (✉) • S. Chatterton • R. Ricci
Dept. of Mechanical Engineering, Politecnico di Milano, Via La Masa 1, 20156 Milan, Italy
e-mail: paolo.pennacchi@polimi.it; pietro.borghesani@mail.polimi.it; steven.chatterton@polimi.it; roberto1.ricci@mail.polimi.it

J. Náprstek et al. (eds.), *Vibration Problems ICOVP 2011: The 10th International Conference on Vibration Problems*, Springer Proceedings in Physics 139,
DOI 10.1007/978-94-007-2069-5_86, © Springer Science+Business Media B.V. 2011

1 Introduction

It is widely recognized [1] that localized damages on the bearing components cover approximately 80–90% of the failures in rolling element bearings. These defects on the surface of a bearing part result, during operation, in a train of impulses which excites vibrations of the system. Theoretically, the impulse train frequency is kinematically linked to the geometry of the bearing components, but slippage in the contact rollers-races introduces small deviations. These usually do not compromise significantly the clarity of the output of the diagnostic techniques for an expert user, but assume great importance in the development of an automatic algorithm. This paper, after briefly presenting the typical patterns of damage peaks related to each type of damage, investigates their deviations from theoretical frequencies and their dependence on operating conditions. All the considerations made hereafter are based on an extensive experimental campaign performed on a specifically designed test-rig, composed of a high-power electric motor and an industrial gearbox, equipped with a complete and redundant set of accelerometers, in order to measure vibrations in proximity of each bearing.

2 Typical Shapes of Peak Patterns for Different Damage Locations

To detect localized damages, for the reasons described in [2], simple Fourier Transform of the vibration signal is not effective and more sophisticated techniques such as Envelope Analysis [3], Spectral Kurtosis [4] or 2nd order Cyclostationarity [5] have to be employed. Despite the complexity of these techniques, most of them result in a *spectrum-like* diagram with a range of frequencies on the horizontal axis and an amplitude index on the vertical axis depending on the technique used. The presence of a damage is diagnosed by the appearance of a characteristic pattern of peaks in the neighborhood of a characteristic frequency (all characteristic frequencies are described in [3]), depending on the location of the damage (inner ring, outer ring, rollers or cage). A detailed description of the typical spectral pattern characterizing each damage location is provided below, in the case of rotating inner ring and fixed outer ring.

Inner ring damage pattern is the richest in term of peaks and therefore clearly identifiable. Its main peak is located in the proximity of the theoretical Ball Pass Frequency Inner (BPFI). At least the first harmonics of the main peak are usually noticeable in the spectral indicator. Together with these BPFI harmonics, the inner ring damage is characterized by the presence of high peaks at harmonics of shaft rotational speed and at frequencies given by linear combinations between BPFI harmonics and shaft ones, called BPFI sidebands. These are due to modulation of the pulse train by the angular position of the inner ring damage relative to the bearing

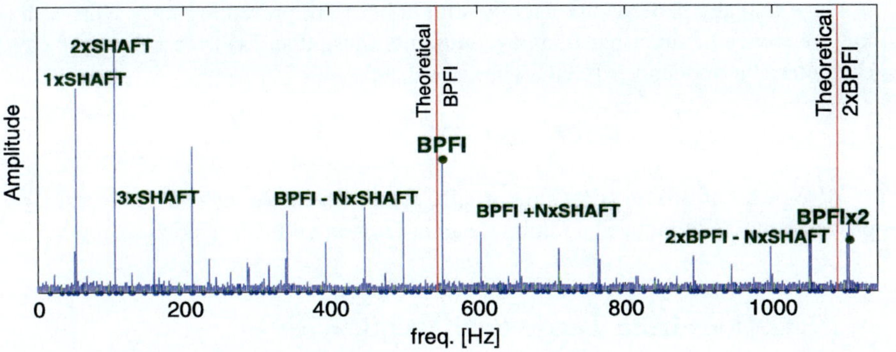

Fig. 1 Typical pattern for inner ring damage

load direction. The pattern of the inner ring characteristic frequencies (IRCF) is therefore:

$$IRCF = m \cdot BPFI \pm n \cdot f \quad m = 0, 1, 2, 3, \ldots \; n = 0, 1, 2, 3, \ldots \tag{1}$$

where f is the inner ring rotational speed.

Figure 1 shows an example of the experimental campaign, regarding an industrial cylindrical roller bearing (103.5 mm pitch diameter, 18 rollers with 15 mm diameter) mounted in the industrial gearbox.

Outer ring damage has a main peak located in the proximity of the theoretical Ball Pass Frequency Outer (BPFO). This frequency is always lower than the BPFI. The bigger the roller diameter relatively to the pitch one, the bigger the difference between BPFI and BPFO. Another remarkable difference from the inner ring pattern is the absence of sidebands, due to the constant load that is applied to the damage contact, since the damage does not change its position in time. This really simplifies and "cleans" the pattern of the outer ring characteristic frequency (ORCF), which presents only the BPFO with its super-harmonics (in most cases at least the first one):

$$ORCF = m \cdot BPFO \quad m = 1, 2, 3, \ldots \tag{2}$$

Roller damage pattern is based on two characteristic frequencies combined with super-harmonics and sidebands: the so called Ball Spin Frequency (BSF) and the Fundamental Train Frequency (FTF). The first is related to the cyclic contact of the spall on the roller with the two races, while the second represents the rotational speed of the cage that modulates the impacts governing the angular position of the damaged roller relative to bearing load. The two frequencies are found with their super-harmonics and in linear combination, in a sequence of roller characteristic frequencies (RRCF) defined as:

$$RRCF = m \cdot BSF \pm n \cdot FTF \quad m = 0, 1, 2, 3, \ldots \quad n = 0, 1, 2, 3, \ldots \tag{3}$$

Cage damage patterns are usually simple but rich, presenting only FTF with a long sequence of super-harmonics (commonly more than five). Therefore the cage characteristic frequencies (CGCF) are:

$$CGCF = n \cdot FTF \quad n = 1, 2, 3, \ldots \tag{4}$$

Also peaks at BSF or BPFO are usually present in case of cage or roller damage, probably owing to induced additional damages on other bearing components.

3 Deviations from Theoretical Frequencies

The slippage between rollers and races has two main effects: (i) it makes the impulse train cyclic rather than strictly periodic, introducing a random anticipation/delay with respect to the average period between two subsequent impacts; and (ii) it leads to an average deviation of the actual impulse train frequency from the theoretical one. The first phenomenon has been widely described [2], while the second has been neglect, since it is not so significant in the case of diagnostics performed by an expert user. On the contrary, these deviations have to be taken into account while developing an automated algorithm. Therefore, thanks to the experimental activity carried out by the authors and the variety of the examples reported in literature, it is possible to list three fundamental and very useful deviation rules:

1. The *"direction"* of the deviation (higher or lower frequency) depends strongly on the location of the damage;
2. Sidebands *"shift rigidly"* together with the main damage peak (e.g., BPFI);
3. Super-harmonics move coherently with the main peak.

The last two rules, quite obvious, are very useful for pattern recognition strategies, addressed in the next paragraph, while the first one derives from empirical observation: actual FTF and BSF are always lower than or equal to the theoretical ones. In absence of other experimental evidence, this could be explained by means of different lubrication conditions in the contacts between rollers-inner ring and rollers-outer ring, with a higher slippage on the inner ring side.

This affects not only actual FTF and BSF, but also BPFI and BPFO, since they are strictly dependent on FTF:

$$BPFI = NR\,(f - FTF)\,;\; BPFO = NR \cdot FTF \tag{5}$$

It is possible to infer that actual BPFI is always greater than or equal to theoretical one from Eq. 5, while BPFO is always lower than or equal to the theoretical one. Experimental results confirm that the deviations are higher at low load on the bearing, when slippage is allowed by low friction. The actual FTF tends towards shaft rotational speed at low loads, with consequent shift of BPFI upwards and BPFO downwards. Increasing the load on the bearing, these three frequencies tend

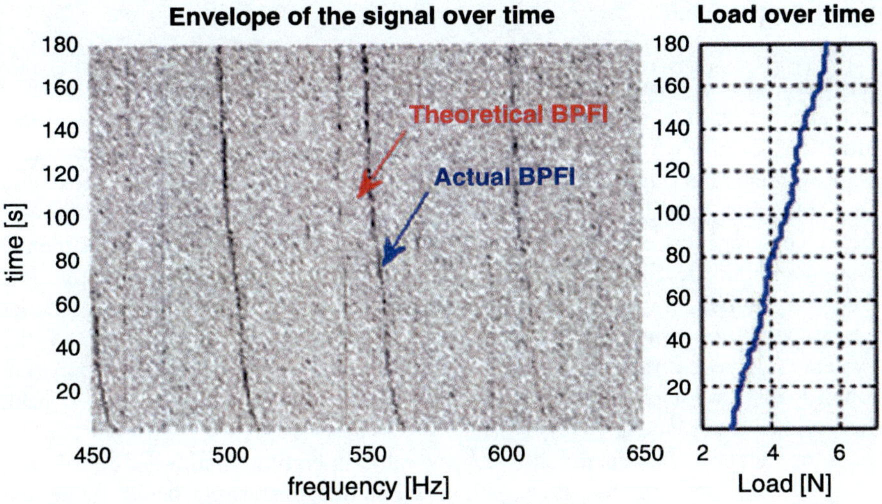

Fig. 2 Load influence on the BPFI deviation

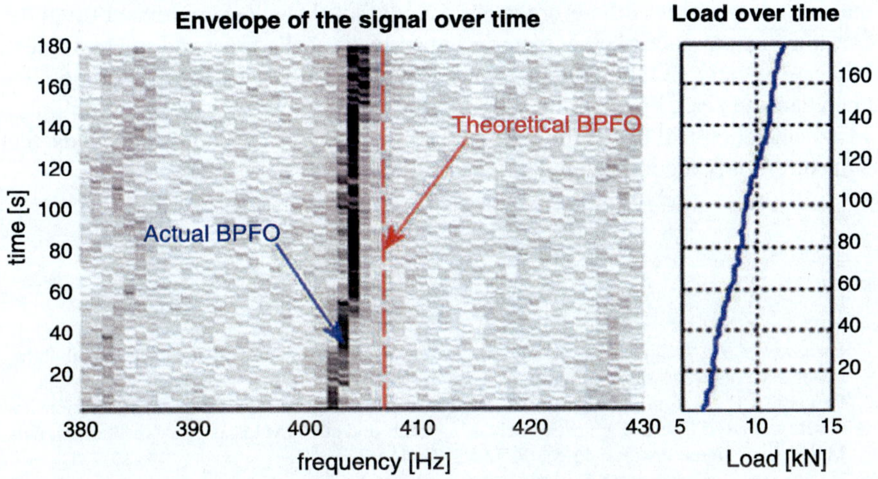

Fig. 3 Load influence on the BPFO deviation

asymptotically to the correspondent theoretical ones, decreasing the effect of the slippage. Also the actual BSF follows this behavior, by reaching asymptotically the theoretical BSF owing to increasing loads. No evidence of strong influence of speed on the deviation has been detected during the experimental campaign. An example of the effect of load on the same cylindrical roller bearing of Fig. 1, damaged on the inner ring, is given in Fig. 2, while the effect on the same type of bearing, damaged on the outer ring is displayed in Fig. 3 .

4 Conclusions

The variability of the frequency of the main damage peaks could lead an automated damage recognition algorithm to underestimate or totally neglect some critical situations, even when the damage is severe in rolling element bearings. A necessary improvement is to introduce a tolerance parameter in the pattern search, consisting in a frequency band in which the highest peak is identified as a possible damage peak. The empirical observations made by the authors during a long experimental campaign indicate that the frequency band should range from the theoretical frequency either upward or downward, depending on the type of damage under investigation (e.g. upward only for BPFI). In addition to this, it is convenient, if the system undergoes different operating conditions, to set a trigger for the acquisition when load is sufficiently high, so that the average deviations of the actual peak frequencies are reduced.

In applications characterized by unfavorable operating conditions (low load), coupled with high external noise, it is possible that misleading peaks, generated by external sources, appear in the frequency band of interest. This could happen if the band is set wide in order to cope with the high deviation/variability of the actual damage frequencies, and could lead to false alarms. In order to identify correctly the damage peaks it is therefore useful to confirm their damage-related origin by checking the presence of other pattern characteristic peaks (sideband and multiples). In fact, due to the general aforementioned deviation rules, once identified a possible main damage peak, it is possible to determine exactly the frequencies of multiples, whose deviation will be the multiple of the main peak's one, and of sidebands, that will shift rigidly together with the main peak.

References

1. Ferreira, J.L.A., Balthazar, J.C., Araujo, A.P.N.: An investigation of rail bearing reliability under real conditions of use. Eng. Fail. Anal. **10**, 745–758 (2003)
2. Randall, R.B., Antoni, J., Chobsaard, S.: The relationship between spectral correlation and envelope analysis in the diagnostics of bearing faults and other cyclostationary machine signals. Mech. Syst. Signal Process. **15**(5), 945–962 (2001)
3. Hochmann, D., Bechhoefer, E.: Envelope bearing analysis: theory and practice. In: Aerospace Conference IEEE, Big Sky, 5–12 Mar 2005, pp. 3658–3660
4. Antoni, J.: The spectral kurtosis of nonstationary signals: formalisation, some properties, and application. In: 12th European Signal Processing Conference, Vienna, 6–10 Sept 2004
5. Antoni, J.: Cyclic spectral analysis in practice. Mech. Syst. Signal Process. **21**, 597–630 (2007)

A New Conception of Bucket Wheel Excavator Cab and Seat Mounting

J. Blekta, J. Mevald, I. Petríková, J. Petříček, and A. Lufinka

Abstract One of the most important problem of coal mining by bucket wheel excavators is how to minimize influence of vibrations on driver's body. Project of Ministry of Trade and Industry of the Czech Republic solved in Technical University of Liberec was engaged in this problem. The aim of this project was to design a new cab and seat mounting of wheel excavator Schrs 1320. A suitable possibilities of cab suspensions were investigated to minimize driver body vibrations and to improve influence of mining process to drivers health. After optimization one concept was chosen. A new methodic of measured data evaluation was created for using in excavators as well as in laboratory. The new concept of dynamic seat absorber were designed also.

Keywords Vibrations • Multibody • Simulations • Measurement • Human body

1 Introduction

The bucket wheel excavators are used in the open-cast mines for mining of the uncovering soil or directly to coal mining. The interaction of bucket wheel with coal or soil is a source of strong vibrations. These vibrations are transmitted through the jib to the machine. The cab of the machine is placed near the bucket wheel and therefore vibration level is very high and deteriorating of working conditions of operators (see [1]). Vibrations transmitted to the driver sitting seat in dependence on the frequency can evokevarious illnesses, change perceptions and overall discom-

J. Blekta (✉) • J. Mevald • I. Petríková • J. Petříček • A. Lufinka
Technical University of Liberec, Liberec, Czech Republic
e-mail: jiri.blekta@tul.cz; josef.mevald@tul.cz; iva.petrikova@tul.cz; jiri.petricek@tul.cz; ales.lufinka@tul.cz

J. Náprstek et al. (eds.), *Vibration Problems ICOVP 2011: The 10th International Conference on Vibration Problems*, Springer Proceedings in Physics 139,
DOI 10.1007/978-94-007-2069-5_87,

Fig. 1 Bucket wheel excavator Schrs 1320 in Doly Nástup Tušimice

fort. Measurement and design modification were performed at the mining machine Schrs 1320 (Fig. 1) in North Bohemia mines (Doly Nástup Tušimice).

2 Analysis of the Cab Movement

The design of the optimal cabin suspension and effective vibroisolation operator inside cabin was composed from several steps. The first step was the total analysis of cab movement. Experimental measurements aimed for investigation of the cab excitation and movement.

In original conception on Fig. 2, the operator cab (a) is mounted rigidly in the cab frame (b). This frame is hung on the console (c) by two revolute joints (d) and one thread rod (e). The accelerometers for vibration measurements were placed on console, cab frame, cab deck and seat. On each part six signals of six degrees of freedom was measured. Dewetron DEWE-5000 was used in combination with MEMS accelerometers (positions 1, 2 and 3) and gyroscopic sensor (position G).

A lot of types of potential cab suspension were analyzed as 2D (Working Model) and 3D (MSC.ADAMS) models using the simple harmonic and real signals. The best reduction of vibrations was achieved by parallelogram design (Fig. 3) – cab is suspended on the frame by three wire rope fittings or alternatively by three bars with spherical bearings. For this concept the cab acceleration achieve c. one tenth of the originally values (see Fig. 4). The resonance frequency depends on the length of the bars. Cab movement is damped by friction damper placed under the cab.

Fig. 2 Example of accelerometers placing on console and cab frame

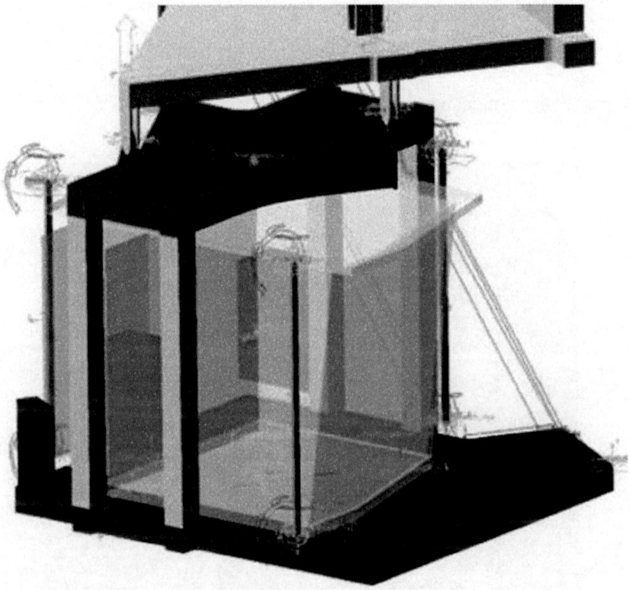

Fig. 3 Platform model in MSC.ADAMS

3 Analysis of Seat Movement in Laboratory

To investigate the spatial vibrations, an experimental device has been developed – a platform with six degrees of freedom which enables measured signal realization in laboratory (see Fig. 5).

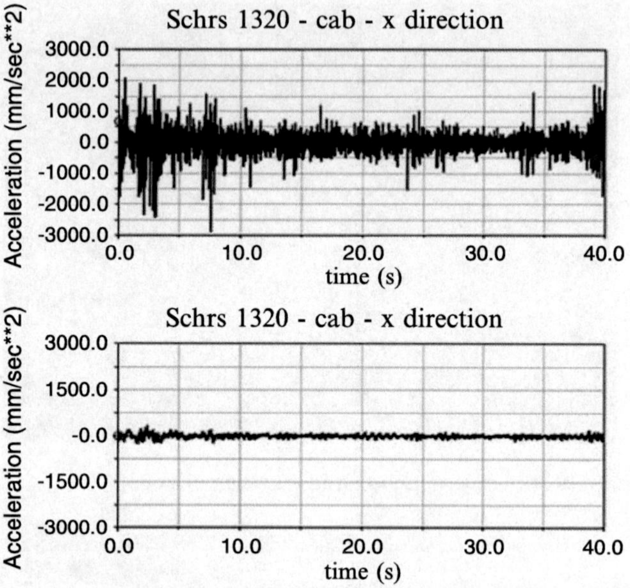

Fig. 4 Cab in parallelogram design

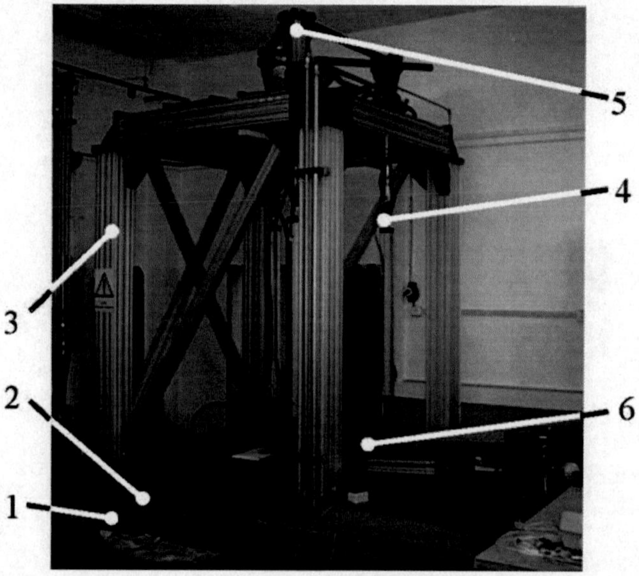

Fig. 5 Platform with six degrees of freedom (*1* concrete foundation block, *2* iron anchoring plate, *3* platform frame, *4* ball joints, *5* hydraulic motors, *6* test board)

Fig. 6 Platform model in MSC.ADAMS software

The real movement of cab deck is recorded as three-axis acceleration in three draft points. The control of six DOF of the platform are realized by six signals for six hydraulic engines. Therefore the measured data were converted into control signals in two steps.

The first step is a conversion of accelerations into translations, see [2]. These data are used for a next computation step. At the second step the multibody model of the platform was created in MSC.ADAMS/View software (see Fig. 6). Three points on platform board were determined in defined positions from measurement in real cabin. The required number of components of the translational motion converted from the measured data was applied in this points. It means three components (x, y, z) in the first point, two components (x, z) in the second point and one component (z) in the third point. The movement of the platform board was uniquely determined.

4 Design of Multi-degrees Cab Vibroisolation

The functionality of seat mechanism was tested on platform with six degrees of freedom with measured data obtained in mines during mining. Because seat mechanism is specified for vertical vibration minimizing, the seat was exposed to vibration in this direction only.

Fig. 7 MSC.ADAMS model

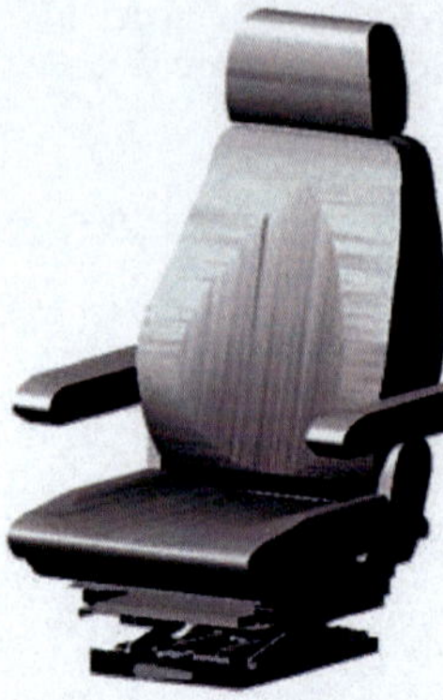

Fig. 8 Analytical seat model

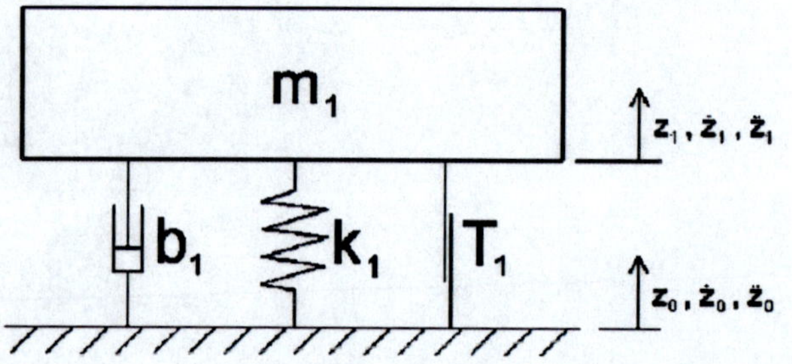

The seat acceleration and displacement transmission ratios were computed from signal measured on platform board and on seat in vertical direction. Multibody model was created in MSC.ADAMS/View software from subsystems with mass properties of real seat parts (see Fig. 7),for detail model description see [3].

Simultaneously a one mass analytical model was created in Matlab software (Fig. 8). This system was modeled as one mass analytical model and mathematically was described by the second order differential Eq. (1).

$$m_1 \cdot \ddot{z}_1 + b_1 \cdot (\dot{z}_1 - \dot{z}_0) + k_1 \cdot (z_1 - z_0) + T_1 \cdot \frac{2}{\pi} arctg\,(500 \cdot (\dot{z}_1 - \dot{z}_0)) = 0 \quad (1)$$

where k_1 is stiffness of spring,
b_1 is damping coefficient of seat damper,
T_1 is a constant setting value of dry friction representing the friction between individual parts of seat,
m_1 is mass of seat squab with driver
z_0 is vertical displacement of cab deck,
z_1 is vertical displacement of seat squab.

This equation was transferred to system of two equations of the first order. This system was solved by Runge-Kutta method with zero initial conditions.

Fig. 9 Model of dynamic absorber

Fig. 10 Seat with dynamic absorber

Parameters of models were set by comparing simulation results with measure results of real seat. Seat mounted on platform with six degrees of freedom was excited by step signals of various amplitude for this purposes. For details see [2].

It has been proposed several construction variants of dynamic absorber. In all cases overall mass of absorber has been divided to four particular masses, which each had its own suspension. This allows set a wider frequency range. Differences between variants were especially in type of used spring. Like the best it has been shown variant with masses placed on arms mounted on torsion rod. The main advantages of this solution is the possibility of changing absorber frequency by moving masses on the arms. Another advantage is low susceptibility to vibrations in the horizontal plane.

Construction of absorber is composed of four basic parts: frame, torsion rods, arms of absorber and absorber masses. The frame of absorber is welded from a rectangular sections. Torsion rods are mounted inside the frame cross sections. Preview of the absorber model created in ProEngineer software is on Fig. 9. On Fig. 10 seat with absorber is shown. Dynamic absorber is placed under the seat squab. Comparison of frequency spectrums of seat squab acceleration for seat with and without dynamic absorber is shown on Fig. 11. Absorber was set to frequency 2.3 Hz.

5 Conclusion

The significant contribution of this work is not only a theoretical analysis of the vibroisolation problem, but also practical implementation of designed modifications with good results. The methodology for evaluation of suitable types of cab suspensions was created and new cab suspension was designed like parallelogram using wire ropes. The horizontal vibrations of the cab were significantly diminished

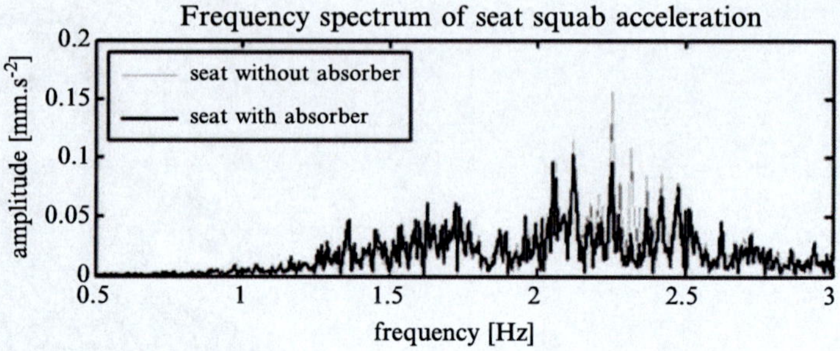

Fig. 11 Frequency spectrum of seat squab acceleration for seat with and without absorber

by this new conception. On the basis of results of experiments in laboratory with different types of seats was designed the dynamic absorber diminishing of the vibration in the vertical direction. Unique vibroisolation also had a positive response from the operators of the excavator.

Acknowledgments This paper was supported by the subvention from Ministry of Trade and Industry of the Czech Republic under Contract Code MPO FT-TA5/102 and from Ministry of Education of the Czech Republic under Contract Code MSM 4674788501.

References

1. Blekta, J., Mevald, J., Petríková, I.: Evaluation of spatial vibrations using a platform with 6 degrees of freedom. In: Proceedings of the EUCOMES 08, pp. 577–583 (2008)
2. Blekta, J., Mevald, J., Lufinka, A.: Reduction of the vibration effects actuating in general direction on colliery machine driver. In: Proceedings of the Experimental Stress Analysis 2009, pp. 43–47 (2009)
3. Petříček, J., Blekta, J., Mevald, J., Petríková, I.: Simulation model of car seat and its verification using platform with six degrees of freedom. In: Proceedings of the 48th International Scientific Conference Experimental Stress Analysis 2010, pp. 333–338 (2010)
4. Porter, R.: Mechanism design for online real-time scheduling. In: Proceedings of the ACM Conference on Electronic Commerce (EC'04). Available via CiteSeer. http://citeseer.ist.psu.edu/article/porter04mechanism.html (2004)

Investigation of Dynamics of Laser Shutter System

A. Bubulis, E. Dragasius, Vytautas Jurenas, V. Maciukiene, and S. Navickaite

Abstract In this paper, the investigation of piezoelectric unimorph laser shutting system transient vibrations damping is presented. A magnetic force technique and magneto-rheological fluid (MRF) is used for the damping of transient vibrations of the piezoelectric unimorph shutter. A methodology is presented for the analysis and design optimization of piezoelectric unimorph actuator used for laser beam shutting systems. The magnetically coupled piezoelectric unimorph cantilever and disc type piezoelectric unimorph motions are analyzed using a one-dimensional forced harmonic oscillator with a non-linear magnetic spring and MRF damper setting. Using a magneto-rheological effect for damping of transient vibrations the piezoelectric unimorph shutter has been experimentally determined.

Keywords Transient vibrations • Piezoelectric unimorph • Magneto-rheological fluid (MRF) • Damping • Laser shutter

1 Introduction

Laser technologies are used in practically every major industry, from medicine and computers, to entertainment and construction. Seeking to control laser beam in modern equipment, there are broadly used laser beam shutting systems that partially or completely blocks laser beam. Shutters are not limited to a simple periodic on–off cycle but will follow an arbitrary, varying pattern of openings and closings.

In pursuance of better dynamic characteristics of the laser beam shutting systems there is a clear tendency in using "smart" materials or materials with controllable

A. Bubulis • E. Dragasius • V. Jurenas (✉) • V. Maciukiene • S. Navickaite
Kaunas University of Technology, Kaunas, Lithuania
e-mail: algimantas.bubulis@ktu.lt; egidijus.dragasius@ktu.lt; vytautas.jurenas@ktu.lt; viktorija.maciukiene@ktu.lt; sigita.navickaite@ktu.lt

J. Náprstek et al. (eds.), *Vibration Problems ICOVP 2011: The 10th International Conference on Vibration Problems*, Springer Proceedings in Physics 139,
DOI 10.1007/978-94-007-2069-5_88,

properties application (piezo materials, shape-memory materials, suspensions with controllable rheology – electro- or magneto-rheological fluids, artificial muscles, etc.).

This paper describes improved dynamic characteristics of cantilever and disc type piezoelectric unimorph shutters. A magnetic force technique and MRF are used for the damping of transient vibrations of two types piezoelectric unimorph shutters. A methodology is presented for the analysis of piezoelectric unimorph actuators used for laser beam shutting systems. The magnetically coupled piezoelectric unimorph cantilever motion is analyzed using a one-dimensional forced harmonic oscillator model with a non-linear magnetic spring and fluid damper [1].

Transient vibrations damping in boundary piezoelectric unimorph shutters positions is the main task of the experiment. During the experiment transient vibrations damping has been created using two different working modes of magneto-rheological fluid.

2 Schemes of the Investigated Laser Beam Shutters and Working Principles

In the present approach a magnetic force technique and MRF are used for the damping of transient vibrations of the piezoelectric unimorph shutter. Transient vibrations negative affects dynamic characteristics of piezoelectric laser beam shutting system.

The bending vibrations of the piezoelectric unimorph are excited by applying an AC electric signal to the electrode of the piezo unimorph.

The magnetic field has been created using Neodymium magnet with flux density B = 1,2T.

Geometrical parameters of piezoelectric actuators used in experiment of laser beam shutting systems are shown in Tables 1 and 2.

Cantilever type piezoelectric laser beam shutter system consists of the piezoelectric unimorph actuator 1 and special ferromagnetic plate 2 for the laser beam shutting (Fig. 1). Piezoelectric effect actuates bending deformations of cantilever type unimorph 1. Laser beam 5 is blocked if electricity doesn't influence the piezo

Table 1 Geometric parameters of piezoelectric cantilever type unimorph actuator

L, mm	L_p, mm	B_m, mm	C_b, mm	d, mm
30	25	5	3.5	1...7 (variable)

Table 2 Geometric parameters of piezoelectric disc type unimorph actuator

D, mm	D_p, mm	h, mm	h_p, mm	d, mm
31	20	0.5	0.7	1...7 (variable)

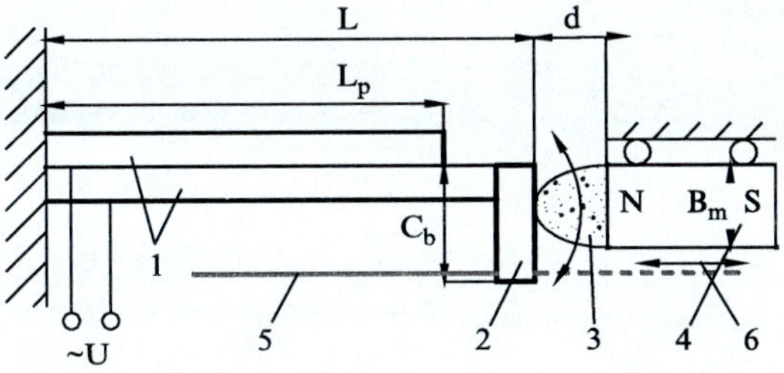

Fig. 1 The scheme of cantilever type piezoelectric unimorph laser shutting system: L_p length of piezoelectric plate, *L* length of piezoelectric unimorph 1, C_b length of special plate 2 for the laser beam 5 shutting, *d* distance between the end of unimorph 1 and permanent magnet 4, B_m thickness of permanent magnet 4, 3 MRF, 6 movement direction of permanent magnet

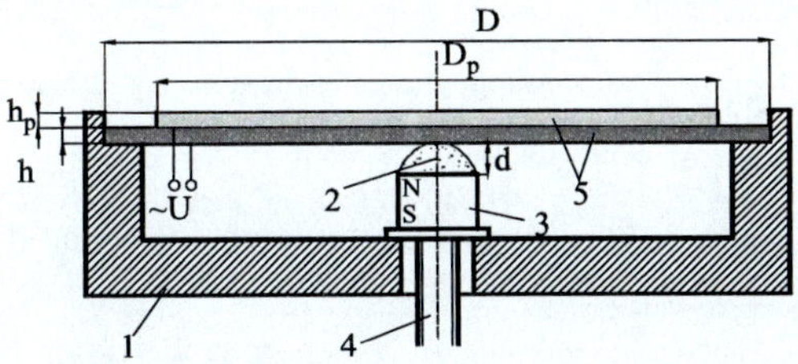

Fig. 2 The scheme of disc type piezoelectric unimorph laser shutting system with transient vibrations damping: *D* diameter of steel plate, D_p diameter of piezoelectric material plate, h_p height of piezoelectric material plate, *h* height of steel plate, *d* variable gap $d = d_0 \pm \Delta d$ between disc type piezoelectric actuator *5* and permanent magnet *3*, *1* housing, *2* magneto-rheological fluid, *4* screw for the gap d control

actuator 1, but in that case if actuator is acted by supply voltage, cantilever type unimorph actuator is bended and laser beam 5 goes without disturbing through laser shutter system [2–4]. The transient vibrations of the piezoelectric unimorph 1 are damped using MRF 3. Seeking to create magnetic field that changes viscosity of MRF 3 is used permanent magnet 4. Transient vibrations damping efficiency depends on distance d between the tip of unimorph and permanent magnet 4. The gap d is controlled by the movement of the permanent magnet 4.

Disc type piezoelectric laser beam shutter system (Fig. 2) consists of the disc type piezoelectric unimorph actuator 5, which consists of two layers: piezoelectric disc diameter is D_p and steel disc is D. Piezoelectric effect actuates bending deformations with maximum amplitude in the centre of disc. The gap d is controlled by the screw 4. Magnetic field is also directly proportional to the gap $d = d_0 \pm \Delta d$.

To design and use unimorphs rationally, it is crucial to understand their coupled electromechanical behaviours through effective modelling [5, 6].

Lumped model of laser beam shutting system with the piezoelectric actuator is presented in Fig. 3. The deflection z(t) of the piezoelectric unimorph can

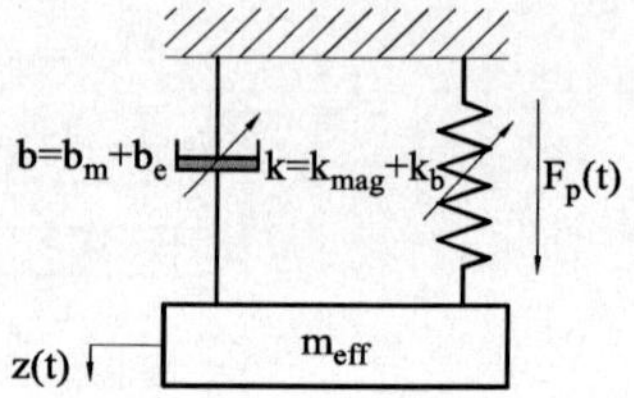

Fig. 3 Lumped model of the piezoelectric laser shutting system

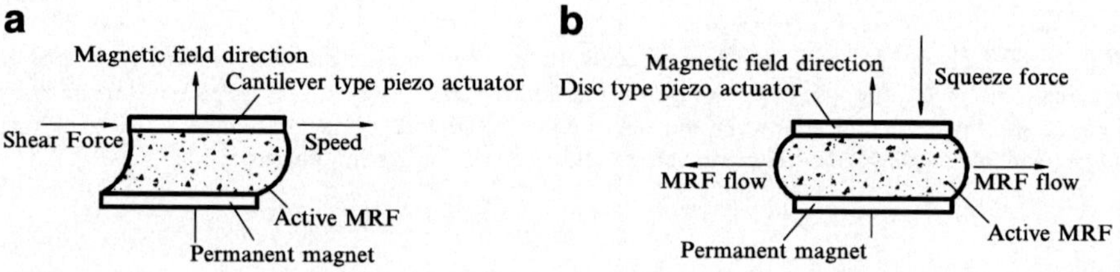

Fig. 4 Magneto-rheological fluid working principles in transient vibrations damping system: (**a**) direct shear mode. (**b**) squeeze mode

be determined by solving the differential equation for a one-dimensional forced harmonic oscillator.

$$\ddot{z}(t) + \frac{b_{eff}}{m}\dot{z}(t) + \frac{k_{eff}}{m}z(t) = \frac{F_p(t)}{m}$$

where $b_{eff} = b_e + b_m$ – total effective damping of the system; $k_{eff} = k_b + k_{mag}$ – total effective stiffness of the system, F_p – piezoelectric force, m – system mass.

3 Materials Used for Transient Vibrations Damping

Cantilever and disc type piezoelectric laser beam shutter system are illustrated in Figs. 1 and 2. Magneto-rheological fluids were used for transient vibrations damping. In cantilever type piezoelectric laser beam shutter system magneto-rheological effect is working in *direct shear mode* (for cantilever type piezo actuator) and in *squeeze mode* (for disc type piezo actuator). It is shown in Fig. 4. In experiment was used magneto-rheological fluid of type MRF – 140CG (LORD Corp, USA).

4 Experiments of Piezoelectric Unimorphs for Laser Beam Shutter

Experimental research was performed in order to evaluate the influence of the damper based on MRF to the transient resonance vibrations of the laser beam shutter. The change of the separation distance d in range from 1 mm to 7 mm between a

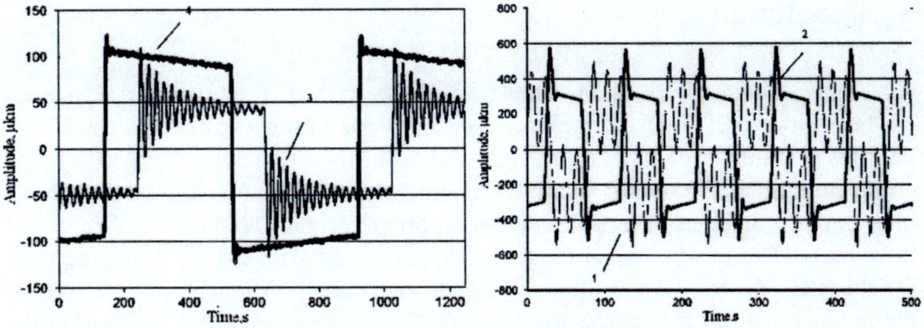

Fig. 5 Measured motion responses of the piezoelectric unimorphs: (**a**) disc type, and (**b**) cantilever type piezoelectric actuators, with the square shaped voltage (frequency – 30 Hz) applied to the piezo actuator: *1*, *3* without MRF damping; *2*, *4* with MRF damping

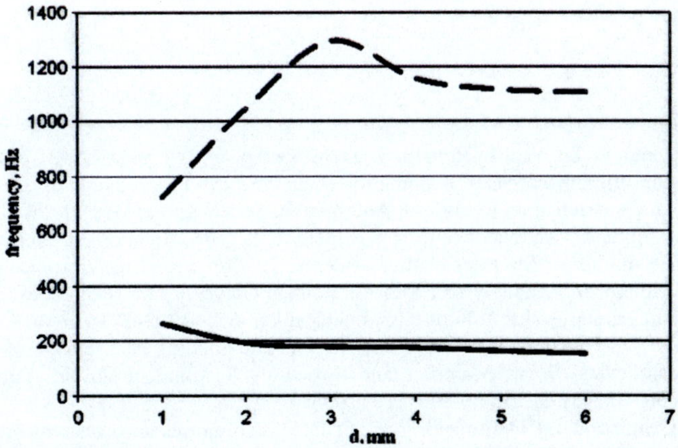

Fig. 6 Resonant frequencies of the laser beam shutter versus separation distance d between a permanent magnet and piezoelectric unimorph: ______ piezoelectric unimorph cantilever, - - - - - piezoelectric unimorph disc

permanent magnet and piezoelectric unimorph was performed by means of the high-resolution micrometer. Magneto-rheological fluid of type MRF-140CG was placed in separation gap between magnet and piezoelectric unimorph as a damper.

The measurement results of the motion response of shutter due to the square-shaped voltage supplied to piezoelectric actuator is presented in Fig. 5. The resonant frequencies of the laser beam shutter versus separation distance d between a permanent magnet and piezoelectric unimorph is presented in Fig. 6.

5 Conclusions

By applying square – shaped signals to the piezo actuator unwanted dynamics of the shutter are often a problem, especially for motion with high frequency, but also if trajectories have to be followed with high velocities. A possibility to eliminate vibrations consists in applying a damper with the MRF. With a properly dimensioned damper, it is possible to suppress residual resonance vibration of the shutter. The results reveal that magnetic coupling with the MRF can be used for the piezoelectric unimorph cantilever to increase its resonance up to 80% of the untuned natural frequency, for the piezoelectric unimorph disc increase of the resonant frequency is only 10%.

Acknowledgments This research was funded by grants (No. AUT-09/2010, TAP-32/2011) from the Research Council of Lithuania.

References

1. Sitti, M., Campolo, D., Yan, J., Ronald, S.: Development of PZT and PZN-PT based unimorph actuators for micromechanical flapping mechanisms. In: Proceedings of the 2001 IEEE International Conference on Robotics 8 Automation, Seoul, Korea, May 21–26 2001, pp. 32
2. Bubulis, A. Jūrėnas, V., Navickaitė, S., Rugaitytė, V.: Investigation of motion control of piezoelectric unimorph for laser shutter systems. J. Vibroengeering/Vibromechanika **12**(4), 533–540, Lithuanian Academy of Sciences, Kaunas University of Technology, Vilnius Gediminas Technical University. Vilnius, Vibromechanika. ISSN 1392–8716 (2010)
3. Tao, W., Paul, I.: Dynamic peak amplitude analysis and bonding layer effects of piezoelectric bimorph cantilevers. Smart Materials and Structures, Institute of Physics Publishing. PII: S0964-1726(04)72431-3, 13, pp. 203–210 (2004)
4. Donoso, A. Sigmund, O.: Optimization of piezoelectric bimorph actuators with active damping for static and dynamic loads, E.T.S. Structural and Multidisciplinary Optimization. Ingenieros Industriales Universidad de Castilla – La Mancha 13071 Ciudad Real, March 2007, pp. 54
5. Riesch, C., Reichel, E., Keplinger, F., Jakoby, B.: Characterizing vibrating cantilevers for liquid viscosity and density sensing. Hindawi Publishing Corporation J. Sensors. **2008** (Article ID 697062), 36–45 (2008), 9 pages. doi:10.1155/2008/69706
6. Smits, G., Ballato, A.: Dynamic admittance matrix of piezoelectric cantilever bimorphs. J. Microelectromech. Syst. **3**(3), 105 (1994)

Modelling of Damping Force of Polyurethane Foam in Dependence on Its Temperature

David Cirkl

Abstract The article deals with measurement and evaluation of damping properties of polyurethane (PU) foam. This problem is considered from the point of view of vibroinsulation and dynamical comfort of person sitting in a moving car. The force response to dynamical compression of PU foam specimen by harmonical course of displacement is evaluated. Frequency of exciting signal and specimen temperature is varied in predefined range. Course of damping force is simulated by explicitly defined function and also its prediction is exemplified.

Keywords Polyurethane foam • Seating comfort • Damping force

1 Introduction

The significant factors of comfort of a person sitting in a moving car are also given by statical and dynamical properties of seats. Their cushioning is predominantly made from polyurethane (PU) foam. Damping properties of this material are influenced also by loading frequency and material temperature. Mechanical properties of PU foam are standardly investigated for ambient temperature (20 °C). In real situations the material temperature can differ. Articles [1, 2] deals with evaluation of thermal comfort of sitting passengers in road–trial study and presents the temperature in contact area between seat and human varying in interval [28; 37] °C. Also other articles deals with thermal comfort of airplane seats [3] or with design of equipment capable influence thermal conditions of sitting [4].

D. Cirkl (✉)
Technical University of Liberec, Studentska 2, Liberec, Czech Republic
e-mail: david.cirkl@tul.cz

J. Náprstek et al. (eds.), *Vibration Problems ICOVP 2011: The 10th International Conference on Vibration Problems*, Springer Proceedings in Physics 139,
DOI 10.1007/978-94-007-2069-5_89,

2 Experimental Arrangement

For investigation of PU foam properties the specimen of cuboidal shape was used. Its square basement has dimensions (100 ×100) mm and height is 50 mm. The specimen was made from opened cell TDI material with density 40,99 kg/m^3.

This specimen was dynamically compressed by displacement (1) with mean $A_0 = 25$ mm, amplitude $A = 5$ mm and frequency $f \in \{0.1, 0.5, 0.7, 1, 1.5, 2, 3, 4, 5\}$ Hz. Measurement was carried out in thermal chamber assigned with loading machine Instron E3000. Environment in the chamber was controlled so that steady state temperature of specimen was $T \in \{10, 20, 30, 40, 50, 60\}$ °C.

$$x(t) = A_0 + A\sin(\omega t), \quad \omega = 2\pi f. \tag{1}$$

3 Methodology of Measurement Evaluation

In case of dynamical compression of the PU foam specimen by harmonical displacement $x(t)$ for one period we get the hysteresis course of the measured total force F as shown in Fig. 1a at the top. Let us assume that the damping force F_d is distributed around the skeleton curve of hysteresis loop symmetrically. Let us also

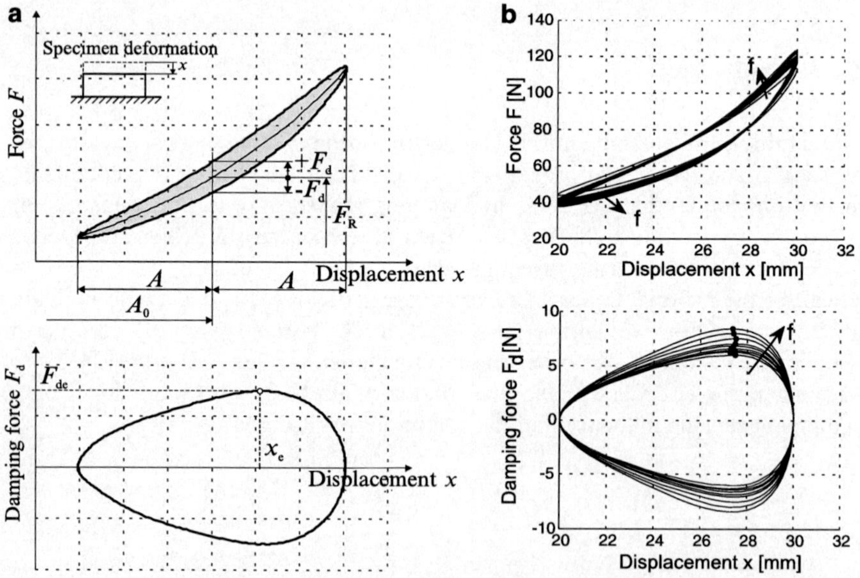

Fig. 1 Force decomposition (**a**) General scheme, (**b**) Measured total and damping force in dependence on exciting frequency, T = 20°C, • denotes maximum of damping force

assume that this skeleton curve represents just the restoring force F_R. In general it has progressive nonlinear character and shows hardening behavior with increasing exciting frequency. In Fig. 1a at the bottom there is the course of damping force in dependence on displacement x. This dependency shows a typical pear–like character with significant extreme value F_{de} at position x_e.

Work of damping force (dissipated energy) is given as a curve integral of damping force F_d with respect to x:

$$W_d = \oint F_d \, dx. \tag{2}$$

4 Measured Quantities

In Fig. 1b there are measured courses of total force response F and damping force F_d in dependence on displacement x. In this case the frequency is varied and temperature is kept constant T = 20 °C. Qualitatively the same courses are also shown with other temperatures.

From the statistical evaluation of x_e follows that there is a slight dependency on temperature. Furthermore in accordance with previously published results in [5] or [6] it is possible to say that position of extreme of damping force x_e is independent on exciting frequency. Dependence of position of extreme of damping force is then expressed by (3).

$$\hat{x}_e(\mathrm{T}) = 27.37 + 0.003075\,\mathrm{T}. \tag{3}$$

Although the foam material is strongly nonlinear in space (x, F), as mentioned by many authors [7, 8], it is not difficult to prove that the dependency of W_d on frequency and on temperature can be considered linear (see Fig. 2a,b). In all cases the coefficient of determination R^2 is greater than 0.95. Such models is possible to consider linear [9]. Thus it is sufficient to consider polynomial of the first degree to be approximating function (4).

$$\hat{W}_d(f, \mathrm{T}) = 102.9 + 5.858f - 0.6553\,\mathrm{T}. \tag{4}$$

5 Damping Force Simulation

In technical handbooks as [10] or [11] it is possible to find that damping force of elastic elements, at which damping depends on displacement x and is realized by energy dissipation inside the material, can be defined by (5). This formulation

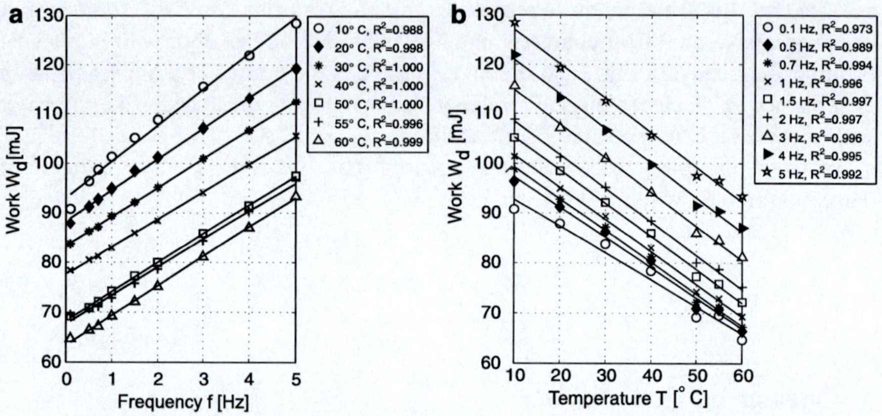

Fig. 2 Dependence of work of damping force on exciting frequency or material temperature (*marks* – measurement, *lines* – linear regression models). **(a)** $W_d = W_d(f)$, **(b)** $W_d = W_d(\mathrm{T})$,

appears to be applicable in case of description of damping force of polyurethane foam being compressed. The task then lies in finding the coefficient b_α and exponent α.

$$F_d(x, \dot{x}) = b_\alpha x^\alpha \dot{x}. \tag{5}$$

To search for the exponent of (5) we will find its extreme by putting its derivative with respect to x equal to zero. If we have unequivocally assigned x and $\dot{x}$ – e.g. by choosing certain time course – it is possible to assume the (5) as the function of one variable. Position of extreme is denoted as x_e.

$$\frac{\mathrm{d}F_d}{\mathrm{d}x} = b_\alpha \left[\alpha\, x_e^{(\alpha-1)}\, \dot{x} + x_e^\alpha \frac{\mathrm{d}\dot{x}}{\mathrm{d}x} \right] = 0, \quad \text{where} \quad \frac{\mathrm{d}\dot{x}}{\mathrm{d}x} = \frac{\ddot{x}}{\dot{x}}.$$

Specially for harmonical function of displacement x in form (1) and its time derivatives we get

$$\frac{\alpha}{x_e} - \frac{x_e - A_0}{A^2 - (x_e - A_0)^2} = 0 \quad \Rightarrow \quad \alpha = \frac{\hat{x}_e - A_0}{A^2 - (\hat{x}_e - A_0)^2}\,\hat{x}_e. \tag{6}$$

Exciting frequency f has been canceled during calculations what proves that damping force formulation (5) does not show the dependency of extreme position on this parameter, which is in accordance with observations.

Because the position of extreme x_e is defined by approximating value $\hat{x}_e$ it is possible to use (6) for exponent calculation. Another way of exponent determination is described in [12].

As has already been written above the work of damping force W_d for one loading period is given by line integral of damping force F_d with respect to displacement x. In (7) we substitute (5) for F_d and we get:

$$W_d = \oint F_d \, dx = b_\alpha \oint x^\alpha \dot{x} \, dx = b_\alpha \int_0^{2\pi/\omega} x^\alpha(t)\dot{x}(t) \frac{dx}{dt} \, dt. \tag{7}$$

Using time courses $x(t)$ from (1) and its first derivative with respect to time and differential $dx = A\omega\cos(\omega t)\,dt$, and substituting $\omega t = \varphi$ we transform this integration into angular displacement domain φ. Under consideration of constant angular velocity ω during one loading period we can write:

$$W_d = b_\alpha 2\pi f I_\alpha \quad \Rightarrow \quad b_\alpha = \frac{\hat{W}_d}{2\pi f I_\alpha}. \tag{8}$$

where

$$I_\alpha = I_\alpha(A_0, A, \alpha) = \int_0^{2\pi} [A_0 + A\sin(\varphi)]^\alpha [A\cos(\varphi)]^2 \, d\varphi \tag{9}$$

is an integral independent on frequency of harmonical exciting signal. As (8) shows the work of damping force W_d depends on frequency linearly which is in accordance with experimental observation (Fig. 2a). From (8) we express the coefficient b_α where we substitute work W_d for our purpose by approximation (4).

6 Conclusion

Equation (5) has been calculated for all combinations of exciting frequency and temperature. Selected instances are pictured in Figs. 3a,b and 4a. Error of simulation R_2 based on Euclidean norm calculation lies in interval $[0.047; 0.123]$ and is considered acceptable.

Up to this point the method of damping force simulation has been used only in cases really measured. But we can use it in entire domain of definition of quantities W_d and x_e, which means for $f \in [0.1; 5]$ Hz and for $T \in [10; 60]\,^\circ$C. Thus we can predict course of damping force for frequencies and temperatures for which was not measured. In Fig. 4b there is a simulation of damping force for example for $f = 2$ Hz and $T \in \{15, 25, 35, 45, 55\}\,^\circ$C.

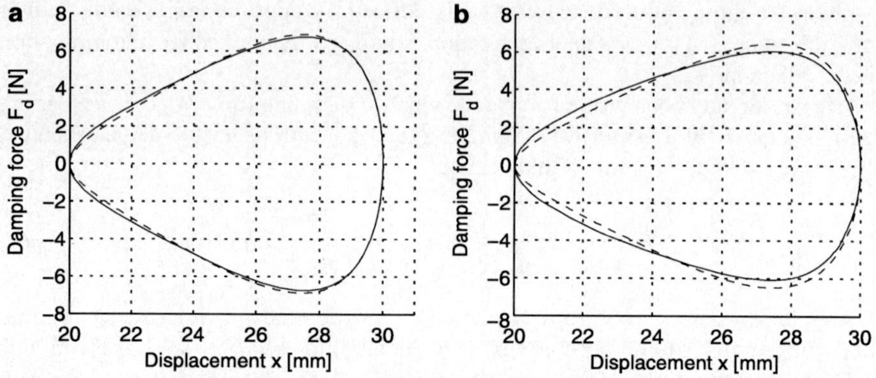

Fig. 3 Results of simulation (Measurement – *solid line*, Simulation – *dashed line*). (**a**) f = 5 Hz, T = 60°C, $R_2 = 0.047$, (**b**) f = 2 Hz, T = 30°C, $R_2 = 0.062$,

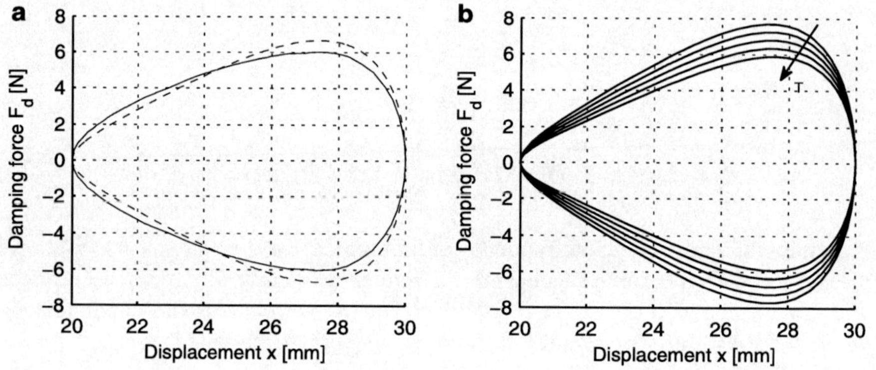

Fig. 4 Results of simulation (Measurement – *solid line*, Simulation – *dashed line*). (**a**) f = 0.1 Hz, T = 10°C, R_2 = 0.106, (**b**) Prediction of course of damping force in dependence on temperature T, f=2 Hz

Acknowledgments This work was supported by the subvention from Ministry of Industry and Trade of Czech Republic under Contract Code MPO FT–TA5/102 and from Ministry of Education of Czech Republic under Contract Code MSM 4674788501.

References

1. Cengiz, T.G., Babalik, F.C.: An on–the–road experiment into the thermal comfort of car seats. Appl. Ergon. **4**, 337–347 (2007)
2. Cengiz, T.G., Babalik, F.C.: The effects of ramie blended car seat covers on thermal comfort during road trials. Int. J. Ind. Ergon. **39**, 287–294 (2009)
3. Bartels, W.T.: Thermal comfort of aeroplane seats: Influence of different seat materials and the use of laboratorytest methods. Appl. Ergon. **34**, 393 (2003)

4. Choi, H., Yun, S., Whang, K.: Development of a temperature-controlled car-seat system utilizing thermoelectric device. Appl. Therm. Eng. **27**, 2841–2849 (2007)
5. Cirkl, D.: Mechanical properties of polyurethane foam. Ph.D. thesis (In Czech), Technical University of Liberec (2005)
6. Cirkl, D.: Modelling of damping force of polyurethane foam in dependence on its density. Shock Vib. **17**, 373–381 (2010)
7. Singh, R., Davies, P., Bajaj, A.K.: Identification of nonliear and viscoelastic properties of flexible polyurethane foam. Nonlinear Dyn. **34**, 319–346 (2003)
8. Deng, R., Davies, P., Bajaj, A.K.: Flexible polyurethane foam modelling and identification of viscoelastic parameters for automotive seating applications. J. Sound Vib. **262**, 391–417 (2003)
9. Rektorys, K., et al.: Review on Mathematics I. Prometheus Publishing, Prague (1995)
10. Julis, K., Brepta, R.: Mechanics: Dynamics, Technical Handbook (In Czech), 2nd edn. SNTL publishing, Prague (1984)
11. Brepta, R., Pust, L., Turek, F.: Mechanical Vibrations, Technical Handbook (In Czech). Sobotales Publishing, Prague (1998)
12. Cirkl, D.: Application of explicit formulation of damping force of polyurethane foam in case of harmonic kinematic excitation. Engineering mechanics. In: Proceeding of National Conference with International Participation, Svratka, Czech Republic (2008)

Part IX
Wave Problems in Solid Mechanics

Nonlinear Vibrational Behaviour of an Elasto-Pneumatic Training Tool

Claudia Körner, Hartmut Hetzler, and Wolfgang Seemann

Abstract This paper deals with the nonlinear vibrational response of a stepping-board with nonlinear elasto-pneumatic force elements. Experimental investigations often show too high vertical ground reaction forces (VGRF) between the test persons and the training tool during exercises. The goal of this contribution is to identify the main factors of the dynamical behaviour and thus the biomechanical impact on humans of this training tool. Therefore this paper presents a mechanical model in order to investigate the interaction between the nonlinear behaviour of the board and the athlete. The multiphysical modelling consists of the linear-elastic structural stiffness part and the nonlinear part due to the pneumatic components. This leads to a nonlinear ordinary differential equation.

In a first step, the vibrational behaviour is analysed with a linearised equation. In the second step, the nonlinear vibrational behaviour is investigated using an harmonic balance method. With this, the influence of various system parameters is discussed with respect to the VGRF as well as the biomechanical compatibility.

Keywords Nonlinear force element • Linearisation • Harmonic balance method • Vertical ground reaction force

1 Introduction

From a biomechanical point of view, both effectiveness of workout and the risk of injuries of an athlete are among the main interest during sport activities [1]. This paper wants to analyse these two facts for the interaction between the athlete and a gym training tool. Therefore,the vertical ground reaction force (VGRF) is

C. Körner (✉)
Karlsruhe Institute of Technology (KIT), Institute of Engineering Mechanics,
76131 Karlsruhe, Germany
e-mail: claudia.koerner@kit.edu

J. Náprstek et al. (eds.), *Vibration Problems ICOVP 2011: The 10th International Conference on Vibration Problems*, Springer Proceedings in Physics 139,
DOI 10.1007/978-94-007-2069-5_90, © Springer Science+Business Media B.V. 2011

investigated to characterise the risk of injuries [1, 2] and the mechanical power of the athlete is investigated to analyse a steady state workout situation.

2 Modelling

The investigated training tool consists of a stepping board, which is mounted on four viscoelastic-pneumatic elements (Fig. 1). It is assumed that the behaviour is predominantly influenced by these force elements. Therefore the following investigations are focused on these components.

The viscoelastic-pneumatic element consists of a compressible hollow pod, which contains a certain amount of air. Thus, it may exert forces due to structural stiffness as well as pressure of the captured air (Fig. 2).

The structural stiffness part is assumed to be linear-elastic and the structural damping part is assumed to be negligible. The equivalent spring stiffness has been determined to be $c \approx 13{,}600\,\mathrm{N/m}$ by a static force-displacement experiment.

As the elastic pod's volume changes, the pressure of the enclosed air changes. Moreover, the air chamber can freely exchange air with its surrounding through two holes on the side. Hence, a model of the pneumatic force element has to account for the air's behaviour as well as process of streaming air out and in. The time behaviour of the pneumatic force element is derived by solving the force equilibrium equation for a given displacement $u(t)$ of the top area A_1 (Fig. 3). The pneumatic force F_p reads:

$$F_p(t) = A_1(p_1(t) - p_2) = A_1\,\Delta p(t), \tag{1}$$

where p_2 means the constant external pressure, $p_1(t)$ the internal pressure and $\Delta p(t) = p_1(t) - p_2$ the pressure difference.

Inserting the change rate of volume $\dot{V} = A_1\dot{y}(t)$, with $y(t) = h_0 - u(t)$, into the time derivative of the ideal gas equation, the pressure difference Δp can be expressed by the following equation:

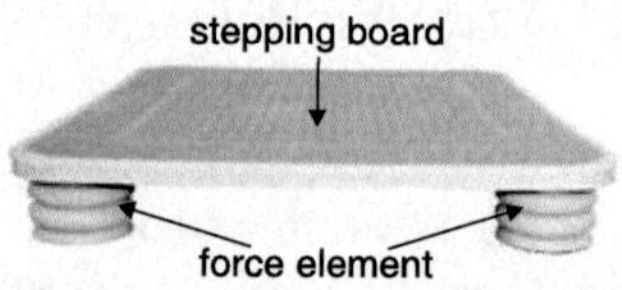

Fig. 1 Training tool

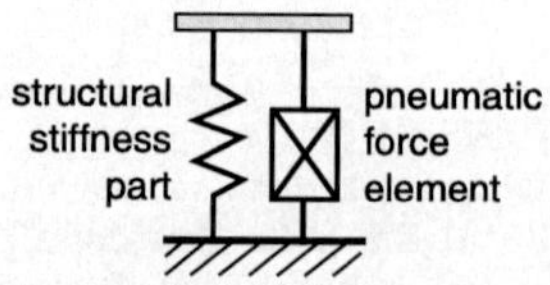

Fig. 2 Mechanical model

Fig. 3 Pneumatic component

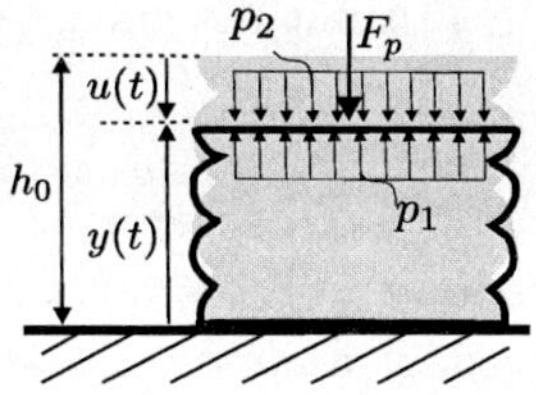

$$\Delta\dot{p} = \frac{\dot{m}_1 RT}{A_1(h_0 - u(t))} + \frac{(\Delta p + p_2)\dot{u}(t)}{(h_0 - u(t))}, \tag{2}$$

where $R = 287.059$ J/(K kg) is the specific gas constant, T the ambient temperature, $u(t)$ a given displacement and $h_0 = 0.07$ m the initial height of the pod remained constant in the following. The mass flow $\dot{m}_1$ is related to the pressure difference Δp by the incompressible Bernoulli equation. With this, Eq. 2 yields:

$$\Delta\dot{p} = \frac{-\alpha^* A_2 RT\ \sqrt{2\rho|\Delta p|}sgn(\Delta p)}{A_1[h_0 - u(t)]} + \frac{(\Delta p + p_2)\dot{u}(t)}{[h_0 - u(t)]}, \tag{3}$$

where $\rho = const.$ is the air density, A_2 the cross section of the hole and the loss factor $\alpha*$ accounts for the pressure loss through the holes.

Equation 3 is a nonlinear first order ordinary differential equation, which can be solved numerically by using e.g. Matlab. However, two analytical approaches will be presented in order to reveal the influence of the system parameters.

3 Linear Approach

In this paper, we assume a harmonic displacement

$$u(t) = \hat{u}\,\sin(\Omega t), \tag{4}$$

where $\hat{u}$ is the amplitude and Ω the angular frequency.

In a first approach a linearisation of Eq. 3 is investigated. To this end the square root in (3) is approximated by a linear function

$$\sqrt{|\Delta p(t)|}\ \mathrm{sgn}\,(\Delta p) \approx n\Delta p(t), \tag{5}$$

where n is determined by minimizing the area between the linear approximation and the original curve via $e = \int_0^{\Delta p_{max}} \left(\sqrt{\Delta p} - n\,\Delta p\right)^2 \mathrm{d}\Delta p \rightarrow$ min. This yields the factor $n = 6/\,(5\,\Delta p_{max})$, where Δp_{max} is the maximum pressure difference, which either must be guessed or be taken from numerical results.

Inserting this into a taylor linearisation for $u(t) = u_0 + \Delta u$, Eq. 3 yields

$$\Delta \dot{p} + \lambda \Delta p = \frac{p_2}{h_0 - u_0} \Delta \dot{u}, \tag{6}$$

where $\lambda = n\,\alpha^*\, A_2 R T\, \sqrt{2\rho}/\,[A_1(h_0 - u_0)]$ and $\Delta \dot{u} = \hat{u}\Omega\cos(\Omega t)$ under the condition $\hat{u}\Omega < 1$. The steady stade response of the linearised Eq. 6 yields

$$\Delta p = \frac{p_2 \hat{u}}{h_0 - u_0} \left[\frac{\lambda\, \Omega\, \cos(\Omega\, t) + \lambda^2\, \sin(\Omega\, t)}{\lambda^2 + \Omega^2} \right]. \tag{7}$$

For small values ($\hat{u}\Omega$) a good correspondance between the nonlinear and the linear solutions of $\Delta p(t)$ is given.

4 Harmonic Balance Method

Using the HBM (harmonic balance method), we assume that the harmonic displacement $u(t)$ (4) leads to a harmonic response

$$\Delta p(t) = \Delta \hat{p}\, \sin(\Omega t - \varepsilon_1)\,. \tag{8}$$

Introducing Eqs. 4 and 8 into Eq. 3 and integrating over one cycle $\Omega t = 0 \ldots 2\pi$, one finds:

$$\Delta \hat{p} = \frac{\hat{u}}{h_0} p_2 \cos(\varepsilon_1) \quad \text{and} \tag{9}$$

$$\varepsilon_1 = -\arccos\left[\frac{C^2}{2} \left(-1 + \sqrt{1 + \frac{4}{C^4}} \right) \right], \quad \text{for} \quad \varepsilon_1 = -\frac{\pi}{2} \ldots 0, \tag{10}$$

where

$$C = \frac{4.944137\, R\, T\, \sqrt{\varrho}}{\pi\, \sqrt{p_2}} \, \frac{A_2 \alpha^*}{A_1 \Omega \sqrt{h_0 \hat{u}}}, \quad \hat{u}\Omega > 0. \tag{11}$$

With this, the force F_p reads

$$F_p(t) = A_1 \Delta p(t) = \frac{\hat{u}\, A_1}{h_0} p_2\, \cos(\varepsilon_1) \sin(\Omega t - \varepsilon_1)\,. \tag{12}$$

Figure 4 shows the time behaviour of the force $F_p(t)$ in comparison with a given displacement $u(t)$. It is found that the global results of numerical and analytical solutions fit very well. Only the constant part of $F_p(t)$ is not reflected by the HBM solution, since (8) does not account for a constant part.

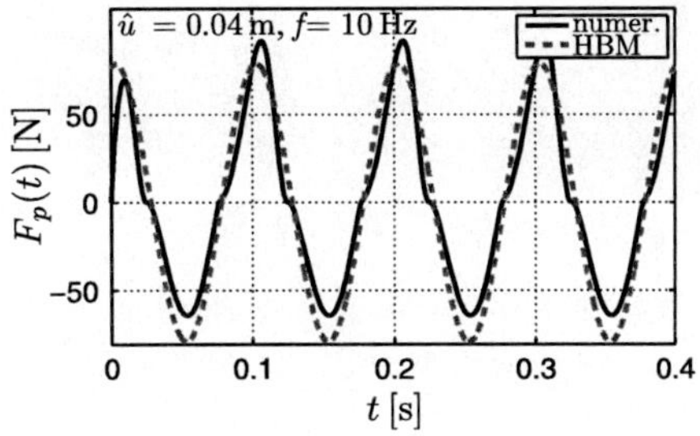

Fig. 4 $F_p(t)$ – numerical vs. HBM solution

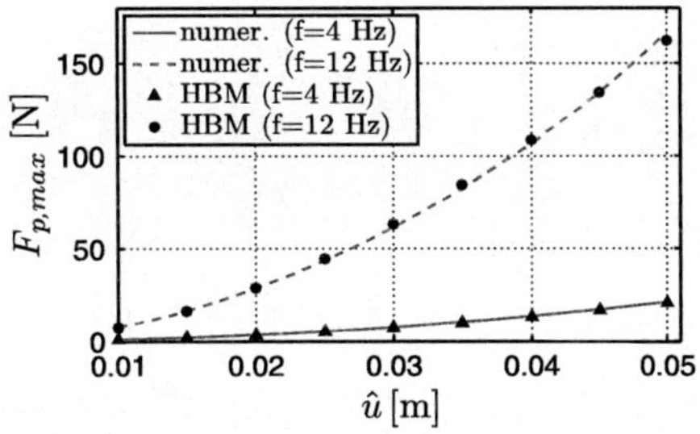

Fig. 5 $F_{p,max}$ – numerical vs. HBM solution

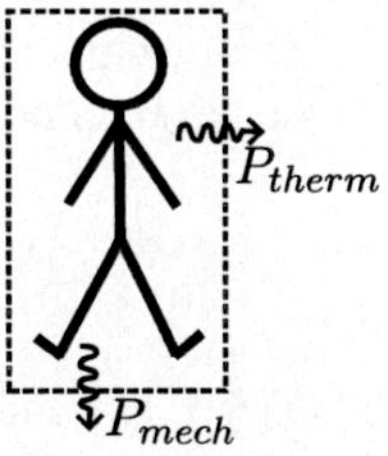

Fig. 6 Power loss

Figure 5 shows that the analytical approach is suitable for the considered range of frequencies f and amplitudes $\hat{u}$; the maximum values $F_{p,max}$ of both solutions fit very well.

5 Results and Discussion for a Steady State Workout Situation

In the following, the effectiveness of the workout is quantified by determining the power loss. During training, we assume that a human body loses energy due to two main factors: thermical power and mechanical power (Fig. 6).

Thermical power is released to the surrounding by the process of metabolism consumption, myoactivation, transpiration, etc. This fact is not represented in the paper. The second main factor is the mechanical power loss $P_{mech} = F_{VGRF}\dot{u}$. While steady state workout on a rigid ground, P_{mech} vanishes since $\dot{u} \equiv 0$. However, on a compliant ground the athlete will transmit mechanical power to the surrounding.

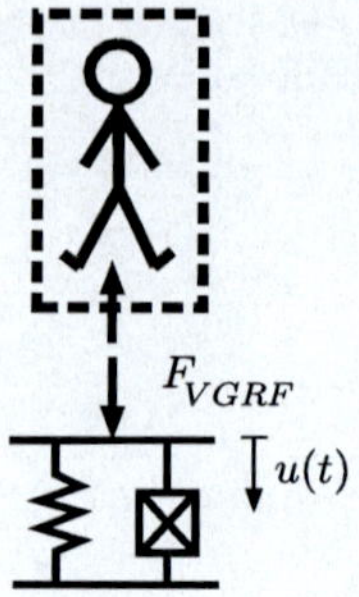

Fig. 7 P_{mech}

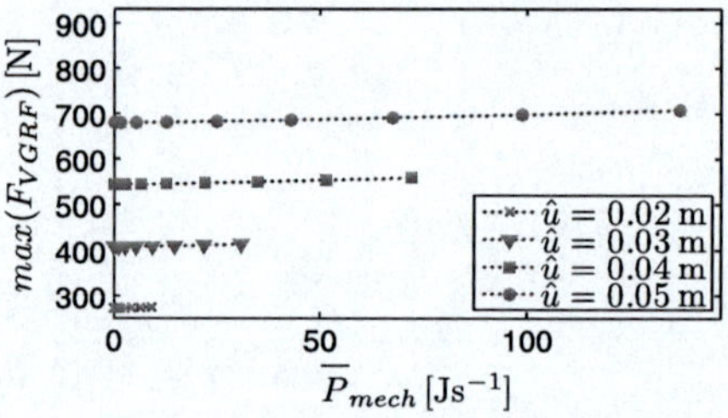

Fig. 8 Nonlinear – $max(F_{VGRF})$ vs. $\overline{P}_{mech}$, $\alpha^* = 0.7$

The mechanical power $P_{mech}(t)$ is determined by the force $F_{VGRF}(t)$, which the human exerts on the ground and the associated velocity $\dot{u}(t)$ (Fig. 7):

$$P_{mech}(t) = F_{VGRF}\,\dot{u}(t) = \overline{P}_{mech} + P_{mech,s}(t) \quad \text{and} \quad F_{VGRF} = c\,u(t) + F_p(t), \tag{13}$$

with $F_p(t)$ the analytical HBM solution from Eq. 12. The displacement is assumed harmonic as $u(t) = \hat{u}\sin\Omega t$ with $\Omega = 2\pi f$. The possible range of frequencies is $f = 1\,\text{Hz}\ldots 9\,\text{Hz}$ [3, 4]. The power $P_{mech}(t)$ is divided into a mean value $\overline{P}_{mech}$ and an oscillating part $P_{mech,s}(t)$. In the long term only the mean value

$$\overline{P}_{mech} = -\frac{\hat{u}^2\,\Omega\,A_1\,p_2\,\sin(2\,\varepsilon_1)}{4\,h_0} \tag{14}$$

contributes to the dissipated energy therefore the additional mechanical power loss due to workout on the investigated training tool.

Figures 8 and 9 show the maximum value of VGRF versus $\overline{P}_{mech}$ for the nonlinear model. Each vertical line stands for a constant amplitude $\hat{u}$ with increasing frequencies for the defined range. The loss factor is assumed to be between $\alpha^* \approx 0.3\ldots 0.7$. With the higher loss factor $\alpha^* = 0.7$ increasing frequencies give a slight increase of $\overline{P}_{mech}$, but the VGRF remains constant. Assuming a smaller loss factor $\alpha^* = 0.3$, the athlete can reach a higher effectiveness of workout by increasing his frequency but then he has to tolerate higher forces. Both figures show that an athlete should rather jump with a higher frequency and smaller amplitudes in order to have an effective workout but limited ground forces.

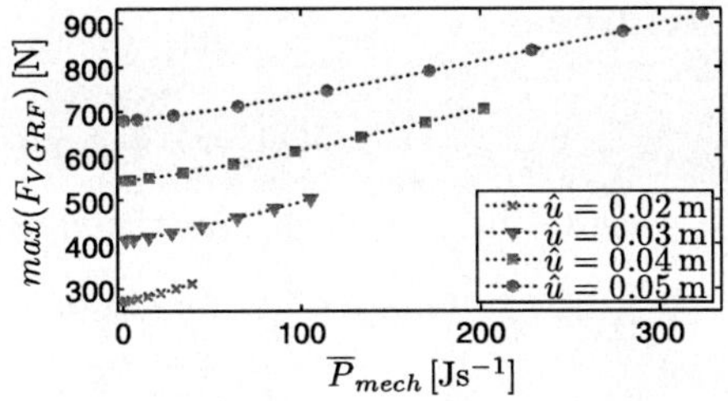

Fig. 9 Nonlinear – $max(F_{VGRF})$ vs. $\overline{P}_{mech}$, $\alpha^* = 0.3$

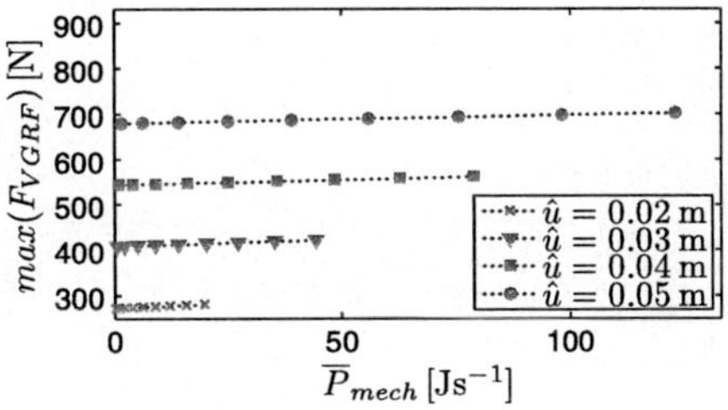

Fig. 10 Linear – $max(F_{VGRF})$ vs. $\overline{P}_{mech}$, $\alpha^* = 0.7$, $n = 0.01$

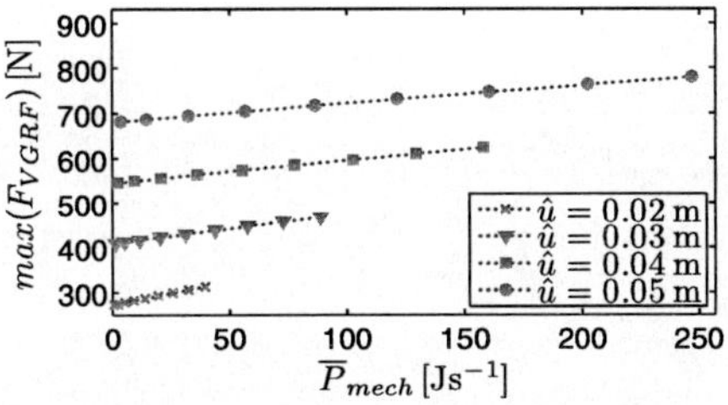

Fig. 11 Linear – $max(F_{VGRF})$ vs. $\overline{P}_{mech}$, $\alpha^* = 0.3$, $n = 0.01$

Figures 10 and 11 show a comparison of $max(F_{VGRF})$ versus $\overline{P}_{mech}$ for the linearised approach. With $\alpha^* = 0.7$, even for higher values a good correspondance of the linear and the nonlinear model is found. For stronger dissipation ($\alpha^* = 0.3$) the discrepancy between the linearised and the nonlinear model becomes significant.

6 Conclusion

This paper presents the modelling of a nonlinear force element and shows its linearisation as well as its harmonic balance approach. The investigations about interaction between the training tool and the athlete reveal the influence of the displacement amplitude and the angular frequency on VGRF. Finally the comparison of the linear and the nonlinear results shows that using nonlinear modelling is necessary.

References

1. Nigg, B., Denoth, J., Neukomm, P., Segesser, B.: Biomechanische Aspekte zu Sportplatzbelägen. Juris-Verlag, Zürich (1978)
2. Özgüven, H.N., Berme, N.: An experimental and analytical study of impact forces during human jumping. J. Biomech. **21**, 1061–1066 (1988)
3. Blickhan, R.: The spring-mass model for running and hopping. J. Biomech. **22**, 1217–1227 (1989)
4. Farley, C.T., Morgenroth, D.C.: Leg stiffness primarily depends on ankle stiffness during human hopping. J. Biomech. **32**, 267–273 (1999)

Torsional Wave Propagation in a Pre-Strained Hollow Three-Layered Circular Cylinder

S.D. Akbarov, T. Kepceler, and M. Mert Egilmez

Abstract This paper studies torsional wave dispersion in a pre-strained three-layered (sandwich) hollow cylinder made from high elastic materials the mechanical relations of which are described through the harmonic potential. The investigations are carried out within the scope of the piecewise homogeneous body model with the use of the three-dimensional linearized theory of elastic waves in initially stressed bodies. The analytical expression is obtained for the low wave number limit values of the torsional wave propagation velocity. The numerical results on the influence of the initial strains of the cylinders along the torsional wave propagation direction are presented and discussed.

Keywords Initial strains • Wave dispersion • Three-layered hollow cylinder • High elastic material • Torsional wave

1 Introduction

Torsional wave propagation in pre-strained bi-material compound cylinder has been studied in [1–4]. In the present paper these investigations are extended for three-layered (sandwich) hollow cylinders made from hyper elastic materials. It is assumed that the initial strains in the components of the sandwich hollow cylinder

S.D. Akbarov (✉)
Institute of Mathematics and Mechanics of the National Academy of Sciences of Azerbaijan, Baku, Azerbaijan

Yildiz Technical University, Istanbul, Turkey
e-mail: Akbarov@yildiz.edu.tr

T. Kepceler • M.M. Egilmez
Yildiz Technical University, Istanbul, Turkey
e-mail: Kepceler@yildiz.edu.tr; Megilmez@yildiz.edu.tr

J. Náprstek et al. (eds.), *Vibration Problems ICOVP 2011: The 10th International Conference on Vibration Problems*, Springer Proceedings in Physics 139,
DOI 10.1007/978-94-007-2069-5_91, © Springer Science+Business Media B.V. 2011

are finite and the magnitudes of these are not restricted. The mechanical relations of the materials of the cylinders are described through the harmonic potential.

2 Formulation of the Problem

We consider the sandwich hollow circular cylinder and assume that in the natural state the radius of the internal circle of the inner hollow cylinder is R and the thickness of the inner, middle and outer cylinders are $h^{(1)}$, $h^{(2)}$ and $h^{(3)}$, respectively. In the natural state we determine the position of the points of the cylinders by the Lagrangian coordinates in the cylindrical system of coordinates $Or\theta z$. It is assumed that the cylinders have infinite length in the direction of the Oz axis and the initial stress state in each component of the considered body is axisymmetric with respect to this axis and homogeneous.

For the initial state of the cylinders, we associate the Lagrangian cylindrical system of coordinates $O'r'\theta'z'$. The values related to the inner, middle and external hollow cylinders will be denoted by the upper indices (1), (2) and (3), respectively. Furthermore, we denote the values related to the initial state by an additional upper index 0. Thus, the initial strain state in the inner, middle and external hollow cylinders can be determined as follows:

$$u_r^{(k),0} = (\lambda_1^{(k)} - 1)r, \quad u_z^{(k),0} = (\lambda_3^{(k)} - 1)z \quad \lambda_1^{(k)} \neq \lambda_3^{(k)}, \quad k = 1,2,3. \tag{1}$$

We introduce the following notation:

$$r' = \lambda_1^{(k)} r, \quad z' = \lambda_3^{(k)} z, \quad R' = \lambda_1^{(1)} R \tag{2}$$

The values related to the system of coordinates associated with the initial state below, with $O'r'\theta'z'$ will be denoted by an upper prime.

Within this framework, let us investigate the axisymmetric torsional wave propagation along the $O'y_3'$ axis in the considered body. We do this investigation by using the coordinates r' and z' in the framework of the Three-dimensional Theory of Elastic Waves in Initially Stressed Bodies (TLTEWISB). We will follow the style and notation used in the paper [5].

Thus, we write the basic relations of the TLTEWISB for the case considered. The equations of motion are:

$$\frac{\partial}{\partial r'} Q'^{(k)}_{r'\theta} + \frac{\partial}{\partial z'} Q'^{(k)}_{\theta z} + \frac{1}{r'}\left(Q'^{(k)}_{r'\theta} + Q'^{(k)}_{\theta r'}\right) = \rho'^{(k)} \frac{\partial^2}{\partial t^2} u'^{(k)}_{\theta}. \tag{3}$$

The elasticity relations are:

$$Q'^{(k)}_{r'\theta} = \omega'^{(k)}_{1221} \frac{\partial u'^{(k)}_{\theta}}{\partial r'} - \omega'^{(k)}_{1212} \frac{u'^{(k)}_{\theta}}{r'}, \quad Q'^{(k)}_{\theta' z'} = \omega'^{(k)}_{1331} \frac{\partial u'^{(k)}_{\theta}}{\partial z'} \tag{4}$$

In (3) and (4) through $Q'^{(k)}_{r'\theta}$, and $Q'^{(k)}_{\theta z'}$, perturbation of the components of the Kirchhoff stress tensor are denoted. The notation $u'^{(k)}_{\theta}$ shows the perturbations of the components of the displacement vector. The constants $\omega'^{(k)}_{1221}, \omega'^{(k)}_{1212}$ and $\omega'^{(k)}_{3113}$ in (3) and (4) are determined through the mechanical constants of the inner and outer cylinders' materials and through the initial stress state. $\rho'^{(k)}$ is the density of the k-th material. By direct calculation it is obtained that for the harmonic potential there exist the following relations.

$$\omega'^{(k)}_{1221} = \omega'^{(k)}_{1212} = \frac{\mu^{(k)}}{\lambda_3^{(k)}}, \quad \omega'^{(k)}_{1331} = \frac{\lambda_1^{(k)} 2\mu^{(k)} \lambda_1^{(k)}}{\lambda_1^{(k)} + \lambda_3^{(k)}} + \frac{1}{\lambda_3^{(k)}} S_{33}^{(k),0} \tag{5}$$

where $\lambda^{(k)}$and $\mu^{(k)}$ $(k = 1, 2, 3)$ are the mechanical constants of the material k-th cylinder.

Thus, torsional wave propagation in the sandwich hollow cylinder will be investigated by the use of the Eqs. 3–5. In this case we will assume that the following complete contact and boundary conditions are satisfied:

$$Q'^{(1)}_{r\theta} = 0 \text{ at } r' = \lambda_1^{(1)} R, \quad \left\{ \begin{matrix} u'^{(1)}_{\theta} \\ Q'^{(1)}_{r\theta} \end{matrix} \right\} = \left\{ \begin{matrix} u'^{(2)}_{\theta} \\ Q'^{(2)}_{r\theta} \end{matrix} \right\} \text{ at } r' = \lambda_1^{(1)} R \left(1 + h^{(1)}/R\right),$$

$$\left\{ \begin{matrix} u'^{(2)}_{\theta} \\ Q'^{(2)}_{r\theta} \end{matrix} \right\} = \left\{ \begin{matrix} u'^{(3)}_{\theta} \\ Q'^{(3)}_{r\theta} \end{matrix} \right\} \text{ at } r' = \lambda_1^{(1)} R \left(1 + h^{(1)}/R\right) + \lambda_1^{(2)} h^{(2)},$$

$$Q'^{(3)}_{r\theta} = 0 \text{ at } r' = \lambda_1^{(1)} R \left(1 + h^{(1)}/R\right) + \lambda_1^{(2)} h^{(2)} + \lambda_1^{(3)} h^{(3)}. \tag{6}$$

In this way, investigation of the considered wave dispersion problem is reduced to the study of the eigen-value problem formulated through the Eqs. 3–5 and condition (6). Note that in the case where the initial strains are absent in the constituents in the cylinder, i.e. in the case where $\lambda_1^{(k)} = \lambda_2^{(k)} = \lambda_3^{(k)} = 1.0$, the foregoing formulation coincides with the corresponding one proposed within the scope of the classical linear theory of elastodynamics.

3 Solution Procedure and Obtaining the Dispersion Equation

The solution to the Eqs. 3 and 4 is found as follows:

$$u'^{(m)}(r') = e^{i(kz' - \omega t)} \begin{cases} A^{(m)} J_1\left(s^{(m)} k r'\right) + B^{(m)} Y_1\left(s^{(m)} k r'\right) if \left(s^{(m)}\right)^2 > 0 \\ A^{(m)} I_1\left(\left|s^{(m)}\right| k r'\right) + B^{(m)} K_1\left(\left|s^{(m)}\right| k r'\right) if \left(s^{(m)}\right)^2 < 0 \end{cases} \tag{7}$$

In (7) $J_1(x)$ and $Y_1(x)$ are Bessel functions of the first and second kind of the first order; $I_1(x)$ and $K_1(x)$ are Bessel functions of a purely imaginary argument of the first order and Macdonald function of the first order respectively.

Using the Eqs. 7, 5 and 4 we obtain the following dispersion equation for the condition (6):

$$\det \left\| \alpha_{ij} \right\| = 0, \quad i; j = 1, 2, 3, 4, 5, 6, \tag{8}$$

In order not to take up too much space here, we will not present here the explicit expressions of a α_{ij}.

4 Numerical Results and Discussions

Consider the dispersion relation $c = c(kR)$ where $c\,(= \omega/k)$ is the wave propagation velocity, k is the wave number, ω is the frequency and R is the inner radius of the inner cylinder obtained from a numerical solution of the dispersion Eq. 14. For simplification of the discussions below we introduce the notation $c_{20}^{(m)} = \sqrt{\mu^{(m)}/\rho^{(m)}}$ and assume that

$$\lambda^{(1)}/\mu^{(1)} = \lambda^{(2)}/\mu^{(2)} = \lambda^{(3)}/\mu^{(3)} = 1.5.$$

Low wave number limit values as $kR \to 0$. As usual, to obtain this limit value, each term of the corresponding dispersion equation is expanded into series form for small values of $kR \to 0$ and only the limit values of the wave propagation velocity are taken into account. For the considered problem, after doing some cumbersome mathematical transformations we obtain the following limit value for the torsional wave propagation velocity for the low wave number limit:

$$\frac{c}{c_{20}^{(1)}} = \left[\frac{\frac{\mu^{(1)}}{\lambda_1^{(1)}}\left(\xi_1'^{(1)}\right)^2 + \frac{\mu^{(2)}}{\lambda_1^{(2)}}\alpha\left(\xi_1'^{(2)}\right)^2 + \frac{\mu^{(3)}}{\lambda_1^{(3)}}\beta\left(\xi_1'^{(3)}\right)^2}{\mu^{(1)} + \mu^{(2)}\alpha\left(\frac{c_2^{(1)}}{c_2^{(2)}}\right)^2 + \mu^{(3)}\beta\left(\frac{c_2^{(1)}}{c_2^{(3)}}\right)^2} \right]^{\frac{1}{2}}, \tag{9}$$

$$\alpha = \frac{\left(\lambda_1^{(2)}\right)^4\left(\eta^{(2)}\right)^4 - \left(\lambda_1^{(1)}\right)^4\left(\eta^{(1)}\right)^4}{\left(\lambda_1^{(1)}\right)^4\left(\left(\eta^{(1)}\right)^4 - 1\right)}, \quad \beta = \frac{\left(\lambda_1^{(3)}\right)^4\left(\eta^{(3)}\right)^4 - \left(\lambda_1^{(2)}\right)^4\left(\eta^{(2)}\right)^4}{\left(\lambda_1^{(1)}\right)^4\left(\left(\eta^{(1)}\right)^4 - 1\right)}.$$

$$\eta^{(1)} = 1 + \frac{h^{(1)}}{R}, \quad \eta^{(2)} = 1 + \frac{h^{(1)}}{R} + \frac{\lambda_1^{(2)}}{\lambda_1^{(1)}}\frac{h^{(2)}}{R},$$

$$\eta^{(3)} = 1 + \frac{h^{(1)}}{R} + \frac{\lambda_1^{(2)}}{\lambda_1^{(1)}}\frac{h^{(2)}}{R} + \frac{\lambda_1^{(3)}}{\lambda_1^{(1)}}\frac{h^{(3)}}{R} \tag{10}$$

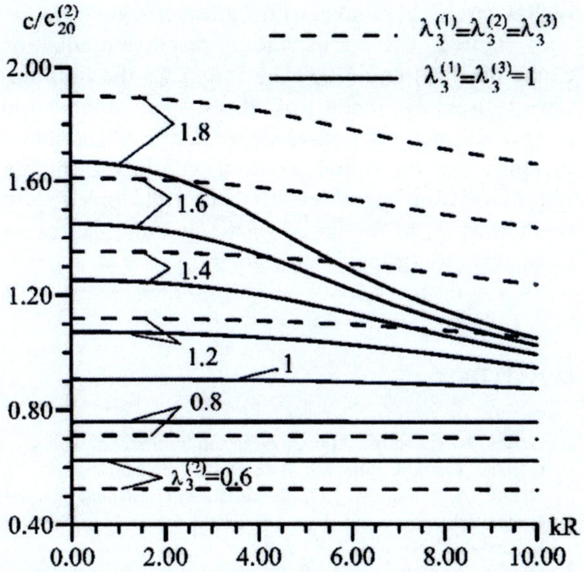

Fig. 1 Dispersion curves of the lowest first mode constructed under $\mu^{(2)}/\mu^{(1)} = \mu^{(2)}/\mu^{(3)} = 2, h^{(1)}/R = h^{(3)}/R = 0.1, h^{(2)}/R = 0.4$

Note that in the case where $\mu^{(3)} = 0$, $\lambda_3^{(2)} = 1.0$, $\lambda_3^{(1)} = 1.0$ the expression (9) coincides with the corresponding one obtained in the paper [6].

Numerical results obtained by the numerical solution to the dispersion Eq. 8. Figure 1 shows the dispersion curves obtained for the first lowest mode in the cases where $\mu^{(2)}/\mu^{(1)} = \mu^{(2)}/\mu^{(3)} = 2, h^{(1)}/R = h^{(3)}/R = 0.1, h^{(2)}/R = 0.4$.

Two cases with respect to the initial stretching in the layers are selected. In case 1 it is assumed that the initial strains exist in all layers of the cylinder and $\lambda_3^{(1)} = \lambda_3^{(2)} = \lambda_3^{(3)}$, but in case 2 it is assumed that the initial strains exist in the middle layer only. It follows from these results that the initial stretching (compression) of the layers of the cylinder causes an increase (a decrease) in the torsional wave propagation velocity. By direct comparison it can be proven that the results shown in Fig. 1 approach the corresponding values of $c/c_{20}^{(2)}$ calculated by the expression (15) as $kR \to 0$. At the same time, these results show that the values of $c/c_{20}^{(2)}$ approach to

$$c_2^{(1)}(\lambda_3)\left(= c_2^{(3)}(\lambda_3)\right) = c_{20}^{(1)}\left(2\lambda_3^{(1)}\right)^{\frac{1}{2}}\left[\left(\lambda_1^{(1)}\right)^3\lambda_1^{(1)}\left(\lambda_3^{(1)} + \lambda_1^{(1)}\right)\right]^{-\frac{1}{2}} \quad \text{as } kR \to \infty.$$

5 Conclusions

In the present paper, within the scope of the piecewise homogeneous body model with the use of the Three-dimensional Linearized Theory of Elastic Waves in Initially Stressed Bodies, torsional wave dispersion in a three-layered (sandwich) hollow circular cylinder with finite homogeneous axisymmetric initial strains has been studied. The corresponding dispersion equation was derived and analytical expressions (9) and (10) were found for the limiting values of the velocity at the lowest dispersive mode from this dispersion equation. Concrete numerical results are presented for the case where the initial strains in occur in all layers of the cylinder, and the initial strains occur in the middle layer of the cylinder only. It follows from these results that in the first lowest mode the initial stretching (compression) of the layers of the cylinder causes an increase (a decrease) in the torsional wave propagation velocity.

References

1. Ozturk, A., Akbarov, S.D.: Propagation of torsional waves in a pre-stretched compound circular cylinder. Mech. Composite Mater. **44**(1), 77–86 (2008)
2. Ozturk, A., Akbarov, S.D.: Torsional wave dispersion relations in a pre-stressed bi-material compounded cylinder. J. Appl. Math. Mech./Zeitschrift für Angewandte Mathematik und Mechanik **89**(9), 754–766 (2009)
3. Ozturk, A., Akbarov, S.D.: Torsional wave propagation in a pre-stressed circular cylinder embedded in a pre-stressed elastic medium. Appl. Math. Model. **33**, 3636–3649 (2009)
4. Kepceler, T.: Torsional wave dispersion relations in a pre-stresses bi-material compound cylinder with an imperfect interface. Appl. Math. Model. **34**, 4058–4073 (2010)
5. Akbarov, S.D., Guliev, M.S.: Axisymmetric longitudinal wave propagation in a finite pre-strained compound circular cylinder made from compressible materials. Comput. Model. Eng. Sci. **39**(2), 155–177 (2009)
6. Armenakas, A.E.: Propagation of harmonic waves in composite circular cylindrical shells. II Numerical investigation. Am. Inst. Aeronaut. Astronaut. J. **9**(4), 599–605 (1971)

Localized Nonlinear Strain Waves in Media with Internal Structure

A.V. Porubov and B.R. Andrievsky

Abstract Propagation and interaction of localized strain waves is studied using an essentially nonlinear continuum model accounting for complex bi-atomic lattice. The model describes coupling between macro-strains and internal strains giving rise to defects of an internal structure of the material. Interaction of bell-shaped localized macro-strain and kink-shaped micro-strain wave solutions is studied numerically. It is shown how predictions of the known exact traveling wave solutions may help in understanding and explanation of the processes of variations in the behavior of both localized waves. The most interesting result concerns variation in the wave velocity caused by initial conditions.

Keywords Nonlinear • Strain • Wave • Internal structure • Numerical solution

1 Introduction

Recently an essentially nonlinear model of a crystalline lattice with deep variations in the structure has been developed in Refs. [1, 2]. It allows description of the cardinal, qualitative variations of the cell properties, lowering of potential barriers, switching of interatomicconnections, arising of singular defects and other damages, phase transitions. Similar model has been suggested in Refs. [3, 4] to account for a deformable monoatomic chain endowed with rotatory molecular group. The variations in the internal structure in both systems are described by coupled nonlinear equations.

A.V. Porubov (✉) • B.R. Andrievsky
Institute of Problems in Mechanical Engineering, Bolshoy av. 61, V.O., Saint-Petersburg 199178, Russia
e-mail: Alexey.Porubov@gmail.com; bandri@yandex.ru

J. Náprstek et al. (eds.), *Vibration Problems ICOVP 2011: The 10th International Conference on Vibration Problems*, Springer Proceedings in Physics 139,
DOI 10.1007/978-94-007-2069-5_92, © Springer Science+Business Media B.V. 2011

In the notations of [1, 2] these equations are derived for the vectors of macro-displacement U and relative micro-displacement u for the pair of atoms with masses m_1, m_2,

$$\mathbf{U} = \frac{m_1\mathbf{U}_1 + m_2\mathbf{U}_2}{m_1 + m_2}, \; \mathbf{u} = \frac{\mathbf{U}_1 - \mathbf{U}_2}{a}$$

where a is a period of sub-lattice. The first variable allows us to describe macro-strains, while the second variable accounts for the reference displacement of the internal or the lattice structure. The following coupled governing equations are obtained in Refs. [1, 2] in the 1D case,

$$\rho U_{tt} - EU_{xx} = S(\cos(u) - 1)_x, \tag{1}$$

$$\mu u_{tt} - \kappa u_{xx} = (SU_x - p)\sin(u). \tag{2}$$

Choice of the trigonometric function allows us to describe translational symmetry of the crystal lattice. In Refs. [3, 4] this function accounts for finite angle rotation of the molecular group.

There exist exact traveling wave solutions to Eqs. (1) and (2) [5] that describe propagating localized wave structures. However, they cannot account for arising of such structures provided that initial and boundary conditions do not coincide with those required for the exact traveling wave solutions. In this case only numerical simulations may help to describe dynamical variations in an internal structure. Previously we considered the domain of the wave velocities where both macro- and micro- strain waves in Eqs. (1) and (2) have the bell-shaped form [6–8]. Now we consider the kink-shaped waves for u. Like in Refs. [6–8] the use of particular exact solution to design and describe numerical results is studied.

2 Exact Localized Wave Solutions

Before performing numerical results, we have to remaind previously obtained exact localized traveling wave solutions to Eqs. (1) and (2) [5]. These solutions depend only on the phase variable $\theta = x - V\,t$. Then Eq. 1 is resolved for the function u by

$$\cos(u) = 1 - \frac{(E - \rho V^2)U_\theta - \sigma}{S}, \tag{3}$$

where σ is a constant of integration. The Eq. 2 is integrated once, multiplied by u and integrated again. Then Eq. 3 is substituted in this equation finally giving an ordinary differential equation for the macro-strain $v = U_\theta$,

$$v_\theta^2 = a_0 + a_1\, v + a_2\, v^2 + a_3\, v^3 + a_4\, v^4, \tag{4}$$

whose coefficients may be found in Ref. [5]. When $a_0 = 0$, $a_1 = 0$, the ODE (4) possesses known exact localized bell-shaped traveling wave solutions of two kinds, one of them reads

$$v_1 = \frac{A}{Q \cosh(k\theta) + 1}, \tag{5}$$

whose parameters are defined for two values of σ, $\sigma = 0$ and $\sigma = -2S$ [5]. Thus, for $\sigma = 0$ we obtain

$$A = \frac{4S}{\rho(c_0^2 + c_L^2 - V^2)}, \quad Q_\pm = \pm\frac{c_L^2 - V^2 - c_0^2}{c_L^2 - V^2 + c_0^2}, \quad k = 2\sqrt{\frac{p}{\mu(c_l^2 - V^2)}} \tag{6}$$

where $c_L^2 = E/\rho$, $c_l^2 = \kappa/\mu$, $c_0^2 = S^2/(p\,\rho)$. Another kind of solution may be found in [5].

The shape of u depends upon the value of the first derivative at $\theta = 0$ in the r.h.s. of Eq. (3) [5]. One possibility reads

$$u = \pm \arccos\left(\frac{(\rho V^2 - E)U_x}{S} + 1\right) \quad \text{for } \theta \geq 0, \tag{7}$$

$$u = \pm 2\pi \mp \arccos\left(\frac{(\rho V^2 - E)U_x}{S} + 1\right) \quad \text{for } \theta < 0, \tag{8}$$

while for another one we have

$$u = \pm \arccos\left(\frac{(\rho V^2 - E)U_x}{S} + 1\right). \tag{9}$$

The solution (7) and (8) accounts for the kink-shaped profile of the wave, while the solution (9) describes the bell-shaped localized wave. The velocity intervals when one or another profile exists as well as similar analysis for $\sigma = -2S$ may be found in Refs. [5–7].

Previously the interval $(c_L^2 - c_0^2; c_L^2)$ for $\sigma = 0$ has been studied in Refs. [6–8]. Now the interval $(0; c_L^2 - c_0^2)$ is considered where the kink-shaped traveling wave solution u (7) and (8) exists. In what follows we shall study whether it may exist under modified initial conditions, also an influence of the initial velocity and the amplitude of the initial conditions on the kink existence will be investigated.

3 Evolution of Kink-Shaped Internal Structure Deviations

To solve Eqs. (1) and (2) numerically the standard MATLAB routine *ode45* is used [9]. The parameters chosen are $S = 1, p = 1, \rho = 1, c_0 = 1, c_L = 1.6, c_l = 2$, then the suitable for kinks values of V lie in the interval $(0, 1.25)$. The initial condition for v is chosen in the form (5) with $Q = Q_+$ and with initial velocity V. The condition for u is used in the form slightly differing form that of described by Eqs. 7 and 8, namely, $u = \pi(1 - \tanh(kx))$ where $k = 0.25$ is chosen to be as close as possible to the shape of the solution (7) and (8).

In the process shown in Fig. 1 the initial velocity V have been chosen equal to 1.4 that is within the interval $(c_L^2 - c_0^2;\ c_L^2)$ when no kink-shaped wave propagates according to the exact solution. However, initial velocity is not kept, it decreases by the same velocity V for u and v equal to 1.15 lying within the interval $(0; {c_L}^2 - {c_0}^2)$. As a result stable propagation of the kink u and the bell-shaped wave v with permanent shape and velocity is observed. Variation in the initial velocity does not affect the final velocity of the stable propagation. It is the amplitude of the initial condition for v that affects this velocity. We multiply previously used condition by some constant amplitude factor and found that the factor less than unity gives rise to a decrease in the final stable velocity, while the factor higher than unity yields its increase. Shown in Fig. 2 is variation in the propagating velocity caused by a decrease in the initial amplitude of v. Comparison of the last profiles in Figs. 1 and 2 demonstrates the difference in velocities.

The initial position of the inputs should coincide according to the exact solution. Difference in the relative position fails ability of the numerical solution to change the

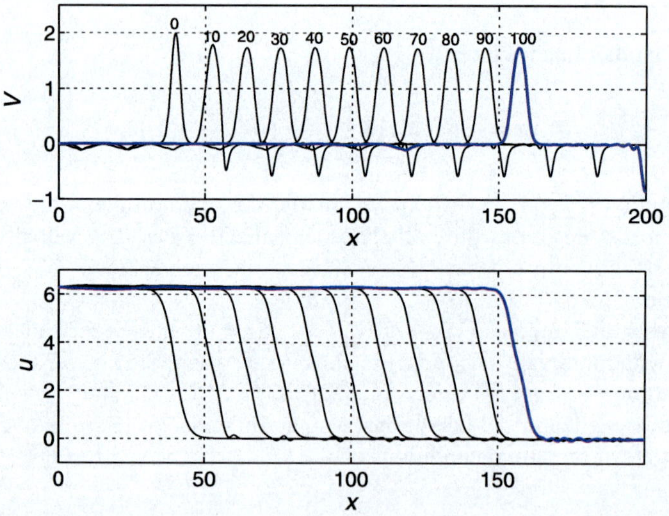

Fig. 1 Simultaneous propagation of *macro- bell-shaped* and *micro- kink-shaped* strain waves with the velocity within the interval prescribed by the exact solution. Points of time correspond to the neighboring peaks. Final wave profiles are allocated in *bold*

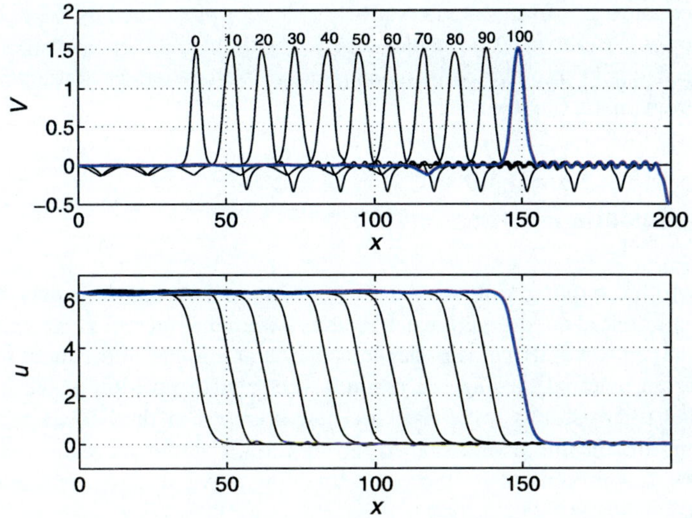

Fig. 2 Decrease in the waves velocity due to a decrease in the initial amplitude of v. Points of time correspond to the neighboring peaks. Final wave profiles are allocated in *bold*

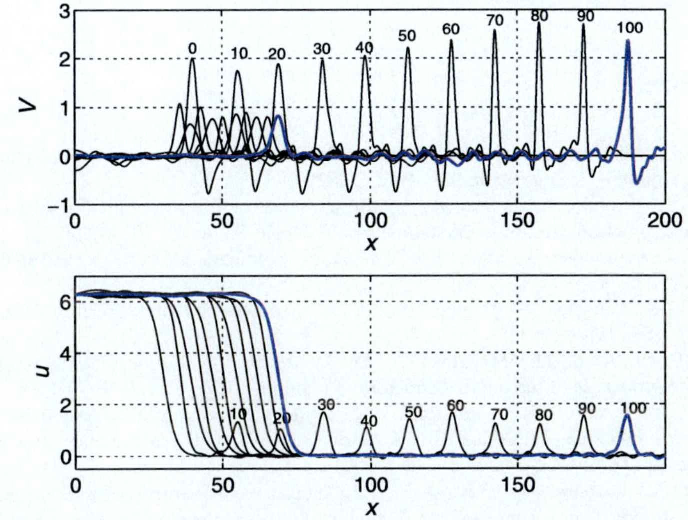

Fig. 3 Generation of moving localized *bell-shaped* micro-strain wave in a lattice from the *kink-shaped* input for u whose initial position is shifted behind that of v. Points of time correspond to the neighboring peaks. Final wave profiles are allocated in *bold*

velocity so as to shift its value within the needed interval of velocities $(0; c_L^2 - c_0^2)$. In this case the initial condition for u splits into two parts as shown in Fig. 3. The kink-shaped part propagates slowly while the new bell-shaped part moves fast with the velocity equal to 1.5 from the interval $(c_L^2 - c_0^2; c_L^2)$ prescribed by the bell-shaped exact solutions. Both waves propagate together with the bell-shaped waves v, hence the last is also split into two bell-shaped parts. It is a new effect of trapping of two corresponding parts of u by v.

4 Conclusions

It is shown that moving kink-shaped wave of internal variations u may arise in a lattice even if initial conditions have velocities within the interval of the existence of the bell-shaped wave. Its velocity depends on the amplitude of the input of macro-strain v or an external loading. Variation in the relative position of the inputs for macro- and micro-strains gives rise to a new scenario of the bell-shaped defects generation from the initial kink-shaped perturbation. The interval of the kink-shaped wave velocities prescribed by the exact traveling wave solution, is confirmed in numerics.

Acknowledgements The work has been supported by the Russian Foundation for Basic Researches, grants No 09-01-00469a and 10-01-00243-a.

References

1. Aero, E.L.: Micromechanics of a double continuum in a model of a medium with variable periodic structure. J. Eng. Math. **55**, 81–95 (2002)
2. Aero, E.L., Bulygin, A.N.: Strongly nonlinear theory of nanostructure formation owing to elastic and nonelastic strains in crystalline solids. Mech. Solids **42**, 807–822 (2007)
3. Maugin, G.A., Pouget, J., Drouot, R., Collet, B.: Nonlinear Electromechanical Couplings. Wiley, Chichester (1992)
4. Sayadi, M.K., Pouget, J.: Soliton dynamics in a microstructured lattice model. J. Phys. A: Math. Gen. **24**, 2151–2172 (1991)
5. Porubov, A.V., Aero, E.L., Maugin, G.A.: Two approaches to study essentially nonlinear and dispersive properties of the internal structure of materials. Phys. Rev. E. **79**, 046608 (2009)
6. Porubov, A.V., Aero, E.L., Andrievsky, B.R.: Dynamical properties of essentially nonlinear generalized continua. In: Maugin, G.A. Metrikin, A.V. (eds.) Mechanics of Generalized Continua, pp. 161–168. Springer, Berlin (2009)
7. Porubov, A.V., Andrievsky, B.R., Aero, E.L: Nonlinear dynamic processes in media with internal structure. In: Altenbach, H., Maugin, G.A., Erofeyev, V.I. (eds.) Mechanics of Generalized Continua, pp. 245–254. Springer, New York (2011)
8. Porubov, A.V., Andrievsky, B.R.: Influence of coupling on nonlinear waves localization. Commun. Nonlinear Sci. Numer. Simul. (2011). doi: 10.1016/j/cnsns.2011.02.028
9. Shampine, L.F., Reichelt, M.W.: The MATLAB ODE Suite, SIAM Journal on Scientific Computing, 18 (1997) 1–22

The Dispersion of the Axisymmetric Longitudinal Waves in the Pre-Strained Bi-Material Hollow Cylinder with the Imperfect Interface Conditions

S.D. Akbarov and C. Ipek

Abstract This work studies the influence of the imperfectness of the interface conditions on the dispersion of the axisymmetric longitudinal waves in the pre-strained bi-material hollow cylinder. The investigations are made within the 3D linearized theory of elastic waves in elastic bodies with initial stresses. It is assumed that the materials of the layers of the hollow cylinder are made from hyper elastic compressible materials and the elasticity relations of those are given through the harmonic potential. The shear spring type imperfectness of the interface conditions is considered and the degree of this imperfectness is estimated by the shear-spring parameter. Numerical results on the influence of this parameter on the behavior of the dispersion curves are presented and discussed.

Keywords Bi-material hollow cylinder • Imperfect interface conditions • Initial strains • Wave dispersion

1 Introduction

The review of the recent investigations related the wave propagation in the pre-strained elastic bodies was made in a paper [1]. It follows from this review and from the other analyses of the relevant references that the wave propagation in the pre-strained compound or layered cylinders has not been investigated before the last 7 years. The first attempt in this field was made by the first author and

S.D. Akbarov (✉)
Yildiz Technical University, Istanbul, Turkey
e-mail: akbarov@yildiz.edu.tr

C. Ipek
Istanbul Technical University, Istanbul, Turkey
e-mail: ipekceng@gmail.com

J. Náprstek et al. (eds.), *Vibration Problems ICOVP 2011: The 10th International Conference on Vibration Problems*, Springer Proceedings in Physics 139,
DOI 10.1007/978-94-007-2069-5_93,

the list of the corresponding references is given in the papers [2, 3] in which the axisymmetric wave propagation in the pre-strained compound cylinder made from the hyper elastic materials was studied. In these papers it was assumed that the contact condition on the interface between the inner and outer cylinders is a perfect one; i.e., it is assumed that the forces and displacements are continuous functions across the interface surface. However, in many cases (for an example, in the case where the reinforced cables are modeled as a bi-material compounded cylinders), it is unrealistic to assume a perfectly bounded interface. Consequently, in order to apply the results of the theoretical investigations to the indicated cases, it is necessary to take the imperfectness of the contact conditions into account under the study of the wave propagation in the bi-material compounded circular cylinders. Taking this statement into account, in the present work the investigation [2] is extended for the bi-material hollow cylinder with imperfect contact conditions. It is assumed that the materials of the constituents are hyper elastic compressible ones and the elasticity relations of those are described by the harmonic potential. The numerical results are presented and discussed.

2 Formulation of the Problem

We consider the bi-material compounded cylinder consisting of a hollow inner and hollow surrounding cylinders. In the natural state we determine the position of the points of the cylinders by the Lagrangian coordinates in the cylindrical system of coordinates $Or\theta z$. Assume that the cylinders have infinite length in the direction of the Oz axis and the initial stress state in each component of the considered body is axisymmetric with respect to this axis and homogeneous. Suppose that the radius of the external circle of the inner hollow cylinder is $R$and the thickness of the inner and outer cylinders are $h^{(2)}$ and $h^{(1)}$, respectively.

With the initial state of the cylinders we associate the Lagrangian cylindrical system of coordinates $O'r'\theta'z'$. The values related to the inner and external hollow cylinders will be denoted by the upper indices (2) and (1), respectively. Furthermore, we denote the values related to the initial state by an additional upper index, 0. Thus, the initial strain state in the inner solid cylinder and external hollow cylinder can be determined as follows.

$$u_r^{(k),0} = \left(\lambda_1^{(k)} - 1\right) r, \quad u_z^{(k),0} = \left(\lambda_3^{(k)} - 1\right) z, \quad \lambda_1^{(k)} \neq \lambda_3^{(k)}, \quad k = 1, 2. \tag{1}$$

Within this framework, let us investigate the axisymmetric wave propagation along the $O'z'$ axis in the considered body with the use of the 3D linearized theory elastic wave propagation in bodies with initial stresses. Thus, we write the basic equations and relations.

The equations of motion are

$$L_1\left(u_r'^{(k)}, u_z'^{(k)}\right) = \rho'^{(k)} \frac{\partial^2 u_{r'}'^{(k)}}{\partial t^2}, \quad L_2\left(u_r'^{(k)}, u_{z'}'^{(k)}\right) = \rho'^{(k)} \frac{\partial^2 u_{z'}'^{(k)}}{\partial t^2}. \tag{2}$$

The explicit expression of the operators $L_k(x, y)$ $(k = 1, 2)$ are given in [2]. The notation $u_{r'}'^{(k)}$, $u_3'^{(k)}$ shows the perturbations of the components of the displacement vector. $\rho'^{(k)}$ is a density of the k-th material.

Thus, the propagation of the longitudinal axisymmetric wave in the considered systems will be investigated by the use of Eq. 2. This equation must be supplied with the following boundary conditions

$$Q'^{(1)}_{r'r'}\Big|_{r'=R'+h'^{(1)}} = 0, \quad Q'^{(1)}_{r'z'}\Big|_{r'=R'+h'^{(1)}} = 0,$$

$$Q'^{(2)}_{r'r'}\Big|_{r'=R'-h'^{(2)}} = 0, \quad Q'^{(2)}_{r'z'}\Big|_{r'=R'-h'^{(2)}} = 0. \tag{3}$$

For the problems under consideration we select the following type imperfectness for the contact conditions

$$Q'^{(1)}_{r'r'}\Big|_{r'=R'} = Q'^{(2)}_{r'r'}\Big|_{r'=R'}, \quad Q'^{(1)}_{r'z'}\Big|_{r'=R'} = Q'^{(2)}_{r'z'}\Big|_{r'=R'},$$

$$u'^{(1)}_{r'}\Big|_{r'=R'} = u'^{(2)}_{r'}\Big|_{r'=R'}, \quad u'^{(1)}_{z'}\Big|_{r'=R'} - u'^{(2)}_{z'}\Big|_{r'=R'} = F\frac{R}{\mu^{(2)}} Q'^{(2)}_{r'z'}. \tag{4}$$

where $F$is the non-dimensional shear-spring parameter, $Q'^{(k)}_{(ij)}$ is a perturbation of components of the Kirchhoff stress tensor. The expression for determination these stresses through the displacement are given, for example, in a paper [2]. The case where $F = 0$ corresponds the perfect contact condition, but the case where $F = \infty$ corresponds the fully slipping imperfectness of the contact condition.

3 Solution Procedure and Obtaining the Dispersion Equation

Using the representation for the displacement given in [4]:

$$u'^{(k)}_{r'} = -\frac{\partial^2}{\partial r' \partial z'} X^{(k)},$$

$$u'^{(k)}_{3} = \frac{1}{\omega'^{(k)}_{1133} + \omega'^{(k)}_{1313}} \left(\omega'^{(k)}_{1111} \Delta_1' + \omega'^{(k)}_{3113} \frac{\partial^2}{\partial z'^2} - \rho'^{(k)} \frac{\partial^2}{\partial t^2} \right) X^{(k)}, \tag{5}$$

where $X^{(k)}$ is presented as follows.

$$\mathrm{X}^{(m)} = \mathrm{X}_1^{(m)}\left(r'\right)\cos\left(kz' - \omega t\right), \quad m = 1, 2. \tag{6}$$

The expression for the function $X_1^{(m)}(r')$ is determined through the Bessel functions [2].

Thus using (5) and (6) we obtain the following dispersion equation from (3) and (4).

4 Numerical Results and Discussions

Assume that $\rho^{(2)}/\rho^{(1)} = 1.0$, $\lambda^{(2)}/\mu^{(2)} = \lambda^{(1)}/\mu^{(1)} = 1.5$ ($\lambda^{(k)}$and $\mu^{(k)}$ (k = 1,2) are mechanical constants of the k-th material; these constants enter the expression of the harmonic potential [2]) and consider the dispersion curves $c = c(kR)$ and analyze the influence of the non-dimensional shear-spring parameter F on these curves for various values of elongation parameters $\lambda_3^{(2)}$ and $\lambda_3^{(1)}$. To simplify the following discussions we introduce the notation $c_{20}^{(k)} = \sqrt{\mu^{(k)}/\rho^{(k)}}$. Consider the case where $\mu^{(2)}/\mu^{(1)}$ =2, $h^{(2)}/R$ =0.5 and $h^{(1)}/R$ =1.0. Investigate the dispersion curves related the fundamental mode. First, we analyze graphs given in Fig. 1 which show how the non-dimensional shear-spring parameter F which characterizes the imperfectness of the contact conditions acts on a behavior of the mentioned dispersion curves in the case where $\lambda_3^{(1)} = \lambda_3^{(2)} = 1.0$, i.e. in the case where there is not any initial strains in the constituents of the cylinder.

It follows from this analyses that as a result of the shear-spring type imperfectness of the contact conditions, instead of the dispersion curve corresponding to the fundamental dispersive mode constructed under satisfaction of perfect contact conditions, i.e. for $F = 0$, two types of mode arise. The first (the second) appears

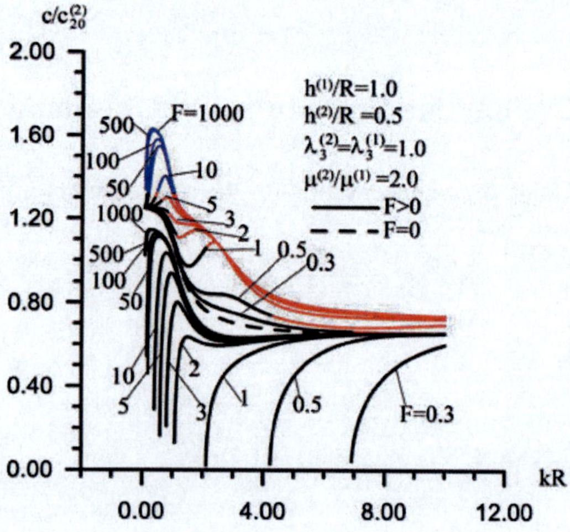

Fig. 1 The influence of the shear-spring parameter F on the dispersion curves under absent of the initial strains

below (over) the dispersion curve corresponding to the first fundamental mode constricted by $F = 0$. We denote the velocity of the wave propagation for $F = 0$ with c, but the wave propagation velocity of the first (second) branch for $F > 0$ we denote with $c_{IF}(c_{IIF})$. It follows from the numerical results given in Fig. 1 that the following relation $c_{IF} < c < c_{IIF}$ takes place.

5 Conclusions

According to the foregoing numerical results the following conclusions are reached:

- the shear-spring type imperfectness of the interface contact conditions causes two branches of the dispersion curve related to the fundamental mode to appear, the first of which disappears, but with the second approach the dispersion curve obtained for the perfectinterface case as $F \to 0$ (Fig. 2);
- the dispersion curves of the foregoing two branches of the fundamental mode approach the corresponding limit dispersion curves related to the fully slipping interface conditions as F increases;
- the shear-spring type imperfectness of the contact conditions does not change the low and high wave number limits;
- the initial strains of the layers of the compound cylinder qualitatively change only the influence of the considered imperfectness of the interface conditions on the behavior of the dispersion curves (Fig. 3).

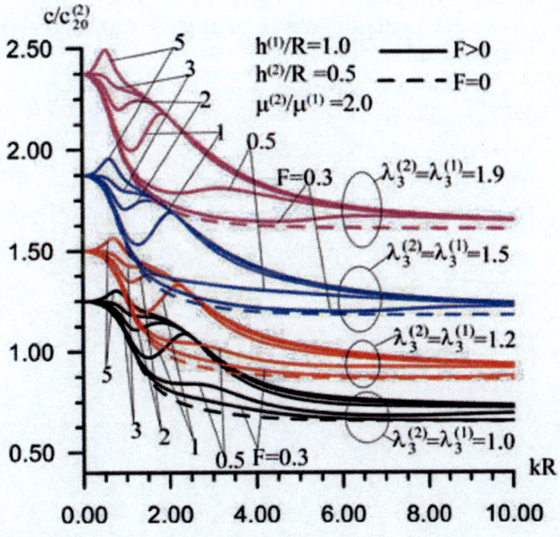

Fig. 2 The influence of the initial strains on the second branch of the dispersion curves

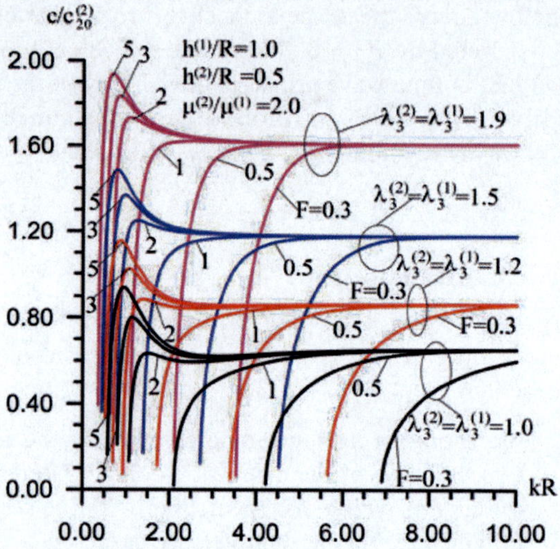

Fig. 3 The influence of the initial strains on the first branch of the dispersion curves

References

1. Akbarov, S.D.: Resent investigation on dynamic problems for an elastic body with initial (residual) stresses (review). Int. Appl. Mech. **43**(12), 1305 (2007)
2. Akbarov, S.D., Guliev, M.S.: Axisymmetric longitudinal wave propagation in a finite prestrained compound circular cylinder made from compressible materials. Comput. Model. Eng. Sci. **39**(2), 155 (2009)
3. Akbarov, S.D., Guliev, M.S.: The influence of the finite initial strains on the axisymmetric wave dispersion in a cylinder embedded in a compressible elastic medium. Int. J. Mech. Sci. **52**, 89 (2010)
4. Guz, A.N.: Elastic Waves in Bodies with Initial (Residual) Stresses. A.C.K, Kiev (2004) (in Russian)

Dispersive Wave Equations for Solids with Microstructure

Arkadi Berezovski, Jüri Engelbrecht, and Mihhail Berezovski

Abstract The dispersive wave motion in solids with microstructure is considered in the one-dimensional setting in order to understand better the mechanism of dispersion. It is shown that the variety of dispersive wave propagation models derived by homogenization, continualisation, and generalization of continuum mechanics can be unified in the framework of dual internal variables theory.

Keywords Dispersive wave • Internal variables • Microstructured solids

1 Introduction

Several modifications of wave equation are proposed to describe wave propagation in heterogeneous materials reflecting dispersion effects, such as the linear version of the Boussinesq equation for elastic crystals [1–5], the Love-Rayleigh equation for rods accounting for lateral inertia [6–10], the Maxwell-Rayleigh equation of anomalous dispersion [1], the "causal" model for the dispersive wave propagation [11], and the Mindlin-type model [12].

All the equations listed above are based either on homogenization [2, 3, 9], or on continualisation [4, 8, 11], or on generalized continuum theories [5, 12]. There is a clear need in understanding their structure from a unified viewpoint. In what follows, the description of the non-dissipative dispersive wave propagation is unified by the dual internal variable approach [13].

A. Berezovski (✉) • J. Engelbrecht • M. Berezovski
Tallinn University of Technology, Estonia
e-mail: Arkadi.Berezovski@cs.ioc.ee

J. Náprstek et al. (eds.), *Vibration Problems ICOVP 2011: The 10th International Conference on Vibration Problems*, Springer Proceedings in Physics 139,
DOI 10.1007/978-94-007-2069-5_94,

2 Thermomechanics in One Dimension

In the case of thermoelastic conductors of heat, the one-dimensional motion is governed by local balance laws for linear momentum and energy (no body forces)

$$(\rho v)_t - \sigma_x = 0, \tag{1}$$

$$\left(\frac{\rho v^2}{2} + E\right)_t - (\sigma v - Q)_x = 0, \tag{2}$$

and by the second law of thermodynamics

$$S_t + \left(\frac{Q}{\theta} + K\right)_x \geq 0. \tag{3}$$

Here t is time, subscripts denote derivatives with respect time and space, respectively, ρ is the matter density, $v = u_t$ is the physical velocity, σ is the Cauchy stress, E is the internal energy per unit volume, S is the entropy per unit volume, θ is temperature, Q is the material heat flux, the "extra entropy flux" K vanishes in most cases, but this is not a basic requirement.

3 Internal Variables

In the framework of the phenomenological continuum theory it is assumed that the influence of the microstructure on the overall macroscopic behavior can be taken into account by the introduction of an internal variable φ which is associated with the distributed effect of the microstructure. In the dual internal variable theory [13], an auxiliary internal variable ψ is used additionally. Then the free energy W is specified as the general sufficiently regular function of the strain, temperature, internal variables φ, ψ and their space derivatives

$$W = \overline{W}(u_x, \theta, \varphi, \varphi_x, \psi, \psi_x). \tag{4}$$

The corresponding equations of state are given by

$$\sigma := \frac{\partial \overline{W}}{\partial u_x}, \quad S := -\frac{\partial \overline{W}}{\partial \theta}, \quad \tau := -\frac{\partial \overline{W}}{\partial \varphi}, \quad \eta := -\frac{\partial \overline{W}}{\partial \varphi_x},$$

$$\xi := -\frac{\partial \overline{W}}{\partial \psi}, \quad \zeta := -\frac{\partial \overline{W}}{\partial \psi_x}. \tag{5}$$

The dissipation inequality (3) can be rewritten as

$$(\tau - \eta_x)\varphi_t + (\xi - \zeta_x)\psi_t - (Q/\theta + K)\theta_x + (\eta\varphi_t + \zeta\psi_t + \theta K)_x \geq 0. \tag{6}$$

Following [14], we chose the non-zero extra entropy flux K in the form

$$K = -\theta^{-1}\eta\varphi_t - \theta^{-1}\zeta\psi_t. \tag{7}$$

Such a choice allows us to reduce the dissipation inequality (6) to

$$(\tau - \eta_x)\varphi_t + (\xi - \zeta_x)\psi_t - \left(\frac{Q - \eta\varphi_t - \zeta\psi_t}{\theta}\right)\theta_x \geq 0. \tag{8}$$

In this case, the dissipation is clearly decomposed into intrinsic and thermal parts. The latter means that the dissipation inequality in the isothermal case reduces to

$$(\tau - \eta_x)\varphi_t + (\xi - \zeta_x)\psi_t \geq 0. \tag{9}$$

In the case of zero dissipation, Eq. 9 yields that the evolution equations for internal variables can be represented in the form [13]

$$\varphi_t = R(\xi - \zeta_x), \qquad \psi_t = -R(\tau - \eta_x), \tag{10}$$

where R is an arbitrary coefficient.

4 Constitutive Model

Having the evolution equations for internal variables in the non-dissipative case, we can derive a microstructure model. We start with the free energy dependence in the form

$$\overline{W} = \frac{\rho c^2}{2}u_x^2 + Au_x\varphi + \tilde{A}u_x\varphi_x + au_x\left(\frac{dF(u)}{du}\right)_x + \frac{1}{2}B\varphi^2 + \frac{1}{2}C\varphi_x^2 + \frac{1}{2}D\psi^2, \tag{11}$$

where c is the elastic wave speed, $A, \tilde{A}, B, C$, and D are material parameters, $F(u)$ is the nonlinear contribution at macroscale, a is a scaling coefficient. For simplicity, we include only the contribution of the second internal variable itself. In this case, stresses are calculated as follows:

$$\sigma = \frac{\partial\overline{W}}{\partial u_x} = \rho c^2 u_x + A\varphi + \tilde{A}\varphi_x + a\left(\frac{dF(u)}{du}\right)_x,$$

$$\eta = -\frac{\partial\overline{W}}{\partial\varphi_x} = -\tilde{A}u_x - C\varphi_x, \quad \zeta = -\frac{\partial\overline{W}}{\partial\psi_x} = 0. \tag{12}$$

The interactive internal forces τ and ξ are, respectively,

$$\tau = -\frac{\partial \overline{W}}{\partial \varphi} = -Au_x - B\varphi, \qquad \xi = -\frac{\partial \overline{W}}{\partial \psi} = -D\psi. \tag{13}$$

The evolution Eq. 10 in the case of zero dissipation take the form

$$\varphi_t = R(\xi - \zeta_x) = -RD\psi, \tag{14}$$

$$\psi_t = -R(\tau - \eta_x) = R(Au_x + B\varphi - \tilde{A}u_{xx} - C\varphi_{xx}). \tag{15}$$

It follows immediately from Eqs. 14 and 15, that the evolution equation for the primary internal variable (14) can be rewritten as the hyperbolic equation

$$\varphi_{tt} = R^2 D(\tau - \eta_x). \tag{16}$$

Accordingly, the balance of linear momentum (1) results in

$$\rho u_{tt} = \rho c^2 u_{xx} + A\varphi_x + \tilde{A}\varphi_{xx} + a[F'(u)]_{xx}, \tag{17}$$

and the evolution equation for the primary internal variable (16) gives

$$I\varphi_{tt} = C\varphi_{xx} + \tilde{A}u_{xx} - Au_x - B\varphi, \tag{18}$$

where $I = 1/(R^2 D)$ is an internal inertia measure.

4.1 *Single Dispersive Wave Equation*

To derive the single equation we make following steps. We determine the first derivative of the internal variable from Eq. 18

$$B\varphi_x = -I\varphi_{ttx} + C\varphi_{xxx} + \tilde{A}u_{xxx} - Au_{xx}. \tag{19}$$

The third mixed derivative φ_{ttx} follows from Eq. 17

$$A\varphi_{ttx} = \left(\rho u_{tt} - \rho c^2 u_{xx} - a[F'(u)]_{xx}\right)_{tt} - \tilde{A}\varphi_{ttxx}. \tag{20}$$

The appeared fourth-order mixed derivative the internal variable is calculated by means Eq. 18

$$I\varphi_{ttxx} = C\varphi_{xxxx} + \tilde{A}u_{xxxx} - Au_{xxx} - B\varphi_{xx}, \tag{21}$$

and, in its turn, the fourth-order space derivative is determined again from Eq. 17

$$\tilde{A}\varphi_{xxxx} = \left(\rho u_{tt} - \rho c^2 u_{xx} - a[F'(u)]_{xx}\right)_{xx} - A\varphi_{xxx}. \tag{22}$$

Collecting all the results (19)–(22) and substituting them into the balance of linear momentum (17) we arrive at the dispersive wave equation

$$\rho u_{tt} - \rho c^2 u_{xx} - a[F'(u)]_{xx} = \frac{C}{B}\left(\rho u_{tt} - \rho c^2 u_{xx} - a[F'(u)]_{xx}\right)_{xx}$$
$$- \frac{I}{B}\left(\rho u_{tt} - \rho c^2 u_{xx} - a[F'(u)]_{xx}\right)_{tt} + \frac{\tilde{A}^2}{B} u_{xxxx} - \frac{A^2}{B} u_{xx}. \tag{23}$$

5 Examples of Dispersive Wave Equations

5.1 Linear Dispersive Wave Equations

5.1.1 Mindlin-Type Model

The Mindlin-type model [12] corresponds to $a = 0$ (no nonlinearity) and $\tilde{A} = 0$ (no coupling between gradients) in Eq. 23:

$$u_{tt} = c^2 u_{xx} + \frac{C}{B}\left(u_{tt} - c^2 u_{xx}\right)_{xx} - \frac{I}{B}\left(u_{tt} - c^2 u_{xx}\right)_{tt} - \frac{A^2}{\rho B} u_{xx}. \tag{24}$$

The Maxwell-Rayleigh model of anomalous dispersion [1] corresponds to a special case of the latter equation with $C = 0$.

5.1.2 Causal model

Keeping the absence of nonlinearity in Eq. 23 and assuming $A = 0$ (no coupling between strain and internal variable; only gradients are coupled), we arrive at the causal model [11]:

$$u_{tt} = c^2 u_{xx} + \frac{C}{B}\left(u_{tt} - c^2 u_{xx}\right)_{xx} - \frac{I}{B}\left(u_{tt} - c^2 u_{xx}\right)_{tt} + \frac{A'^2}{\rho B} u_{xxxx}. \tag{25}$$

The higher-order dispersive wave Eqs. 24 and 25 differ from each other only by the last term in the right hand side. However, this difference is essential, because the second-order space derivative in Eq. 24 exhibits the slowing down the velocity of propagation, whereas the fourth-order derivative in Eq. 25 does not. At the same time, derivatives of the wave operator in Eq. 24 cannot be rearranged, whereas it is possible in Eq. 25 due to the additional fourth-order space derivative.

5.1.3 Unified Model

The unified model includes both couplings mentioned above

$$u_{tt} = c^2 u_{xx} + \frac{C}{B}\left(u_{tt} - c^2 u_{xx}\right)_{xx} - \frac{I}{B}\left(u_{tt} - c^2 u_{xx}\right)_{tt} + \frac{A'^2}{\rho B} u_{xxxx} - \frac{A^2}{\rho B} u_{xx}, \tag{26}$$

which generalizes both approaches [15].

5.2 *Nonlinear Dispersive Wave: Boussinesq Equation*

The obtained dispersive wave Eq. 23 can be reduced to the Boussinesq equation under following assumptions:

1. $I = 0$, which means zero microinertia.
2. $G = 0$ that corresponds to the absence of nonlinearity in microstructure.
3. $A = 0$ (no coupling between strain and internal variable; only gradients are coupled).

As a result, Eq. 23 reduces to

$$\rho u_{tt} - \rho c^2 u_{xx} - a[F'(u)]_{xx} = \frac{C}{B}\left(\rho u_{tt} - a[F'(u)]_{xx}\right)_{xx} + \frac{\tilde{A}^2}{B} u_{xxxx}. \tag{27}$$

This equation belongs to the class of the dispersive wave equations which are characterized by the so-called "Boussinesq paradigm" [16], which means: (a) bidirectionality of waves; (b) nonlinearity (of any order); (c) dispersion (of any order modelled by space and time derivatives of the fourth order at least).

This paradigm has its roots in the classical Boussinesq equation for waves in shallow water, to which Eq. 27 can be reduced by the choice of the nonlinearity function $F(u) = u^3$ [16] and $C = 0$

$$u_{tt} - c^2 u_{xx} = \left(\frac{3au^2}{\rho} + \frac{\tilde{A}^2}{\rho B} u_{xx}\right)_{xx}. \tag{28}$$

6 Conclusions

As it was shown on the example of one-dimensional wave propagation, nonlinear terms can be easily introduced in the framework of the dual internal variables approach resulting in a generalized nonlinear dispersive wave equation. A cubic macroscopic nonlinearity leads to the Boussinesq equation.

Acknowledgements Support of the Estonian Science Foundation is gratefully acknowledged.

References

1. Maugin, G.A.: On some generalizations of Boussinesq and KdV systems. Proc. Estonian Acad. Sci. Phys. Math. **44**, 40–55 (1995)
2. Santosa, F., Symes, W.W.: A dispersive effective medium for wave propagation in periodic composites. SIAM J. Appl. Math. **51**, 984–1005 (1991)
3. Fish, J., Chen, W., Nagai, G.: Non-local dispersive model for wave propagation in heterogeneous media: one-dimensional case. Int. J. Numer. Methods Eng. **54**, 331–346 (2002)
4. Askes, H., Suiker, A.S.J., Sluys, L.J.: A classification of higher-order strain-gradient models – linear analysis. Arch. Appl. Mech. **72**, 171–188 (2002)
5. Erofeyev, V.I.: Wave Processes in Solids with Microstructure. World Scientific, Singapore (2003)
6. Love, A.E.H.: Mathematical Theory of Elasticity. Dover, New York (1944)
7. Graff, K.F.: Wave Motion in Elastic Solids. Clarendon Press, Oxford (1975)
8. Maugin, G.A.: Nonlinear Waves in Elastic Crystals. Oxford University Press, Oxford (1999)
9. Wang, Z.-P., Sun, C.T.: Modeling micro-inertia in heterogeneous materials under dynamic loading. Wave Motion **36**, 473–485 (2002)
10. Wang, L.-L.: Foundations of Stress Waves. Elsevier, Amsterdam (2007)
11. Metrikine, A.V.: On causality of the gradient elasticity models. J. Sound Vib. **297**, 727–742 (2006)
12. Engelbrecht, J., Berezovski, A., Pastrone, F., Braun, M.: Waves in microstructured materials and dispersion. Phil. Mag. **85**, 4127–4141 (2005)
13. Ván, P., Berezovski, A., Engelbrecht, J.: Internal variables and dynamic degrees of freedom. J. Non-Equilib. Thermodyn. **33**, 235–254 (2008)
14. Maugin, G.A.: Internal variables and dissipative structures. J. Non-Equilib. Thermodyn. **15**, 173–192 (1990)
15. Berezovski, A., Engelbrecht, J., Berezovski, M.: Waves in microstructured solids: a unified viewpoint of modelling. Acta Mech. **220**, 349–363 (2011)
16. Christov, C.I., Maugin, G.A, Porubov, A.V.: On Boussinesq's paradigm in nonlinear wave propagation. C. R. Mecanique **335**, 521–535 (2007)

Elastic Waves and Defect Modes in Micropolar Lattices

D.J. Colquitt, A.B. Movchan, N.V. Movchan, and I.S. Jones

Abstract We analyse elastic Bloch-Floquet waves in periodic planar lattices, with micropolar rotations taken into account. This is followed by the study of standing waves corresponding to eigenmodes of certain types and defect modes. Examples include time-harmonic truss and frame structures and systems with rotational springs. The inertial properties of the system are controlled by distributing the mass throughout the junction regions in addition to the elastic links. In the analysis of dispersion properties of Bloch-Floquet waves, we derive analytical asymptotic estimates for the effective group velocities in the acoustic modes to investigate possible morphological corrections. Periodic arrays of defects are introduced by breaking several bonds leading to standing defect modes. Explicit analytical estimates are compared with the results of numerical simulations.

Keywords Asymptotics • Bloch-Floquet waves • Micropolar rotations

1 Introduction

The paper considers the micro-polar dynamic interaction between inertial elements in a periodic elastic lattice. The lattice structure maybe heterogeneous and anisotropic, and the elastic waves are dispersive. This paper is motivated by the earlier work [1–7]. In particular, analysis of the dispersion properties of waves in two-dimensional elastic discrete systems was included in [4]. The same paper

D.J. Colquitt (✉) • A.B. Movchan • N.V. Movchan
Department of Mathematical Sciences, University of Liverpool,
Liverpool, L69 7ZL, UK
e-mail: d.j.colquitt@liv.ac.uk

I.S. Jones
School of Engineering, John Moores University, Liverpool, L3 3AF, UK

J. Náprstek et al. (eds.), *Vibration Problems ICOVP 2011: The 10th International Conference on Vibration Problems*, Springer Proceedings in Physics 139,
DOI 10.1007/978-94-007-2069-5_95,

includes details of the spectral analysis for two-dimensional lattice systems. The paper [2] includes a model of dynamic defects within an elastic system induced by thermal pre-stress. Analysis of the dispersion properties of elastic waves in periodic continua with pre-stress was also presented in [1].

The analysis of two-dimensional static lattices has been presented in many texts such as [5–7]. In particular, [5–7] introduced a potential of torsional interaction between the ligaments at the junction points. After reducing the problem to a system of finite difference equations, a connection was made with the homogenised isotropic continuum, for a special class of lattices. The homogenised Lamé constants were evaluated and the static micro-polar interaction was identified.

The structure of the present paper is as follows. Section 2 deals with the main notations, geometry and governing equations of the discrete system. Low frequency approximations for micro-polar heterogenous lattices and asymptotic formulae for the effective group velocities are also derived in Sect. 2. This section also includes comparision with the earlier work [5, 6]. Section 3 discusses the dispersion properties, control of stop band width and localization in a lattice with defects. Special attention is devoted to standing waves related to rotational motion.

2 Geometry, Governing Equations and Effective Group Velocities

We consider an infinite diatomic triangular lattice in R^2, which consists of an array of lattice nodes labeled by the multi-index $\mathbf{n} = (n_1, n_2) \in \mathsf{Z}^2$. The lattice nodes are connected by thin elastic beams of length l. The interaction between nodes $\mathbf{n}$ and $\mathbf{p}$ is described by $\mathsf{A}^{(\mathbf{n},\mathbf{p})}$, that is, $\mathsf{A}^{(\mathbf{n},\mathbf{p})}$ describes the force at $\mathbf{n}$ as a result of motion at $\mathbf{p}$. For a particular node $\mathbf{n}$, we denote the set of neighboring nodes connected to $\mathbf{n}$ by elastic beams as $\mathsf{N}^{(\mathbf{n})}$ and the generalized planar displacement amplitude vector as $\mathbf{u}^{(\mathbf{n})} = (u_1^{(\mathbf{n})}, u_2^{(\mathbf{n})}, \theta^{(\mathbf{n})})^{\mathsf{T}}$. The inertial properties of the **n**th node are described by the matrix $M^{(\mathbf{n})} = \mathrm{diag}(m^{(\mathbf{n})}, m^{(\mathbf{n})}, J^{(\mathbf{n})})$, where m is the mass and J is the polar mass moment of inertia. For Bloch-Floquet waves of radian frequency ω, the equations of motion for node $\mathbf{n}$ take the form

$$\omega^2 M^{(\mathbf{n})}\mathbf{u}^{(\mathbf{n})} + \mathsf{A}^{(\mathbf{n},\mathbf{n})}\mathbf{u}^{(\mathbf{n})} + \sum_{\mathbf{p}\in\mathsf{N}^{(\mathbf{n})}} \mathsf{A}^{(\mathbf{n},\mathbf{p})}\mathbf{u}^{(\mathbf{p})} = \mathbf{0}. \tag{1}$$

The Bloch-Floquet quasi-periodicity conditions have the form $\mathbf{u}^{(\mathbf{n}+\mathbf{q})} = \mathbf{u}^{(\mathbf{n})}e^{i\mathbf{k}\cdot\mathbf{q}}$, where $\mathbf{k} = (k_1, k_2)$ is the Bloch vector and $\mathbf{q} = q_1\mathbf{t}_1 + q_2\mathbf{t}_2$. The vectors $\mathbf{t}_i$ are the lattice vectors and $(q_1, q_2) \in \mathsf{Z}^2$. The equations of motion (Eq. 1) then form a system of the generalised eigenvalue type which has the solvability condition

$$0 = \det\left[M\omega^2 - \sigma(\mathbf{k}, \omega)\right] := g(\mathbf{k}, \omega). \tag{2}$$

The matrix $\sigma(\mathbf{k},\omega)$ is the stiffness matrix of the irreducible cell and $M = \mathsf{diag}(M^{(1)}, M^{(2)})$ is the mass matrix. We note that if the inertia of elastic links within the lattice is neglected, then the matrix function σ is ω-independent. Equation 2 represents the dispersion equation of the discrete system.

2.1 The Effective Group Velocities in the Low Frequency Regime

We consider two classes of lattices: (a) *the lattice with massless beams*, and (b) *the distributed mass lattice*. In the former case the beams connecting the lattice nodes are assumed to be massless with the entire mass of the lattice being localised at the nodes. For the latter class, the mass is distributed along the lattice beams in addition to the nodes. For each type of lattice, we consider three distinct classes of interactions: (1) *the truss interaction*, where the lattice nodes are modelled as pin joints, and we consider only central interactions; (2) *the truss and torsional spring interaction*, where there is an additional interaction based on the angle between the springs; (3) *the Euler-Bernoulli beam interaction*, where the lattice nodes are connected by Euler-Bernoulli beams and the angle between the beams is fixed. In each case, the nodes of the diatomic lattice have masses m_1 and m_2 and polar moments of inertia J_1 and J_2, and each beam has density ρ.

The effective group velocity in the low frequency regime is understood as $\partial\omega/\partial\mathbf{k}$ as ω and $\mathbf{k}$ tend to zero. For each of the three interactions described above and small ω and $\mathbf{k}$, the dispersion Eq. 2 may formally be expanded in the form

$$0 = \sum_{|\alpha|>0} (\partial^\alpha g)\,(\mathbf{0},0)\,\frac{(\mathbf{k},\omega)^\alpha}{\alpha!}, \tag{3}$$

where we sum over the multi-index $\alpha = \mathsf{N}_0^3$. Keeping only terms up to $|\alpha| = 4$ in Eq. 3 and solving for non-negative ω yields the estimate for the dispersion equation in the neighborhood of the origin. Taking the gradient then yields the effective group velocities of the lattice. For the three classes of interactions described above, the effective group velocities in the low frequency limit are described below (full details may be found in [8]). It is convenient to work with non-dimensional parameters. The non-dimensional velocity is defined as $\mathbf{v} = \bar{\mathbf{v}}/\sqrt{Esl/m_1}$, where E is the Young's modulus of the lattice beams, s is their cross-sectional area and l is their length. The usual dimensional group velocity is denoted as $\bar{\mathbf{v}}$. Similarly, the masses, polar moments of inertia and density are normalised thus: $m_2 = \bar{m_2}/m_1$, $J_i = \bar{J}_i/(m_1 l^2)$ and $\rho = \bar{\rho} sl/m_1$. Here the bars denote the corresponding dimensional parameters and s is the cross-section of the lattice beams.

2.1.1 The Truss Interaction

For the case of truss interactions, the two non-dimensional effective group velocities are

$$\mathbf{v}^{(1)} \sim \frac{3}{2} \frac{\hat{\mathbf{k}}}{\sqrt{1+m_2+3\rho}}, \quad \mathbf{v}^{(2)} \sim \frac{\sqrt{3}}{2} \frac{\hat{\mathbf{k}}}{\sqrt{1+m_2+3\rho}}, \tag{4}$$

where $\hat{\mathbf{k}}$ is the normalised Bloch vector. We note that the effective group velocities of the first class of lattice (massless links) can be obtained by choosing $\rho = 0$. It is also observed that the total mass in the irreducible cell is $1 + m_2 + 6\rho$, whereas the inertia term appears as $1 + m_2 + 3\rho$ in Eq. 4. That is, there is a morphological change in the effective group velocities when the inertia is distributed across the lattice beams. It is also observed that the lattice is isotropic in the low-frequency regime (the group velocity is not a function of the magnitude of the Bloch vector).

2.1.2 The Truss and Torsional Spring Interaction

We introduce the non-dimensional parameter $\xi = \tau/cl^2$, where τ is the stiffness of the torsional spring and c is the longitudinal stiffness of the beams. The effective group velocities are then

$$\mathbf{v}^{(1)} \sim \frac{\sqrt{3}}{2} \sqrt{\frac{1+3\xi}{1+m_2+3\rho}}\hat{\mathbf{k}}, \quad \mathbf{v}^{(2)} \sim \frac{\sqrt{3}}{2} \sqrt{\frac{3+\xi}{1+m_2+3\rho}}\hat{\mathbf{k}}. \tag{5}$$

Once again, distributing the mass along the lattice beams (i.e. choosing a non-zero ρ) results in a morphological change in the effective group velocities. Based on the earlier static work [5, 6], we can derive similar expressions for the shear and pressure speeds in a special class of homogeneous triangular lattice with equivalent interactions:

$$v_s = \frac{\sqrt{3}}{2\sqrt{2}} \sqrt{\frac{3+\xi}{\mathsf{M}}}, \quad v_p = \frac{\sqrt{3}}{2\sqrt{2}} \sqrt{\frac{1+3\xi}{\mathsf{M}}}, \tag{6}$$

where M is the mass within the [monatomic] elementary cell. It is observed that Eq. 6 is consistent with Eq. 5 for the case of a monatomic lattice with massless links ($\rho = 0$, $m_2 = 1$ and $\mathsf{M} = 1$).

2.1.3 The Euler-Bernoulli Beam Interaction

The relative size of the flexural d and longitudinal c rigidities is measured by the non-dimensional parameter $\beta = 2d/cl^3$.

The effective group velocities are then

$$\mathbf{v}^{(1)} \sim \frac{3}{2}\sqrt{\frac{1+2\beta}{1+m_2+6\rho}}\hat{\mathbf{k}}\,,\ \mathbf{v}^{(2)} \sim \frac{\sqrt{3}}{2}\sqrt{\frac{1+6\beta}{1+m_2+6\rho}}\hat{\mathbf{k}}. \tag{7}$$

In contrast to the previous two interactions, distributing the mass along the lattice beams results in no morphological correction to the effective group velocities.

3 Dispersion Properties, Defects and Standing Waves

For the case of the Euler-Bernoulli beam interaction we present two dispersion diagrams in Fig. 1, for the following two cases: (1) a lattice with massless beams and (2) a lattice with inertial beams. Both dispersion surfaces show evidence of standing waves – indicated by flat regions where $\nabla\omega(\mathbf{k}) = \mathbf{0}$. In particular, we note that the first two dispersion surfaces are almost flat. Moreover, numerical simulations indicate that these surfaces correspond to standing rotational modes. Specifically, the rotational displacement is large compared with the translational components. Therefore, a simple estimate of the frequencies of these modes may be obtained by neglecting the translational motion. For the case of massless links, the estimates for the frequencies of the standing modes are

$$\omega_{\pm}^{(\mathrm{rot})} = \left(\frac{\beta}{J_1 J_2}\left(7J_1 + 7J_2 \pm \sqrt{49\left(J_1^2 + J_2^2\right) - 82 J_1 J_2}\right)\right)^{1/2}. \tag{8}$$

Taking $J_1 = 2$, $J_2 = 6$, $\beta = 0.001$ yields estimates of $\omega_{+}^{(\mathrm{rot})} = 0.0853$ and $\omega_{-}^{(\mathrm{rot})} = 0.0454$, which are in excellent agreement with the numerical simulations. A standing wave, involving relative translational motion of nodes within an irreducible cell, exists at the origin of the higher eigensurface which bounds the band gap in Fig. 1a. The presence of this standing waves allows for an efficient estimate of the upper boundary of the band gap. The estimate is found to be $\omega_T = \sqrt{(11 + 198\beta)/m_2}$. Using the same parameter values as above, we obtain $\omega_T = 1.058$, which is approximately where the upper boundary of the band gap appears in Fig. 1a. Contrasting Fig. 1a and b, it is observed that the primary difference is the presence of a number of additional dispersion surfaces in Fig. 1b in the region of the band gap in Fig. 1a. Finite element simulations in COMSOL suggest that these additional modes are related to the fundamental frequencies of the beams. For beam eigenmodes, the nodal displacements vanish and the problem reduces to that of the free vibrations of a clamped beam. Such systems have been treated extensively in the literature, see for example Graff [9]. For a Euler-Bernoulli beam with clamped ends, the first fundamental frequency is $\omega^{(\mathrm{beam})} \approx (4.964)^2\sqrt{\beta/2\rho}$. For the numerical values used to produce Fig. 1b, $\omega^{(\mathrm{beam})} \approx 0.4927$, gives the approximate location of the additional dispersion surfaces.

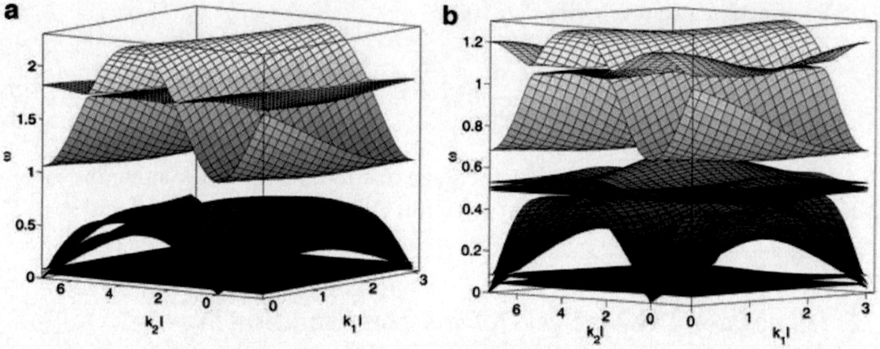

Fig. 1 The dispersion surfaces for (**a**) the lattice with massless members; and (**b**) the distributed mass lattice, for the case of the Euler-Bernoulli beam interaction. The nominal numerical values chosen were $m_2 = 10$, $J_1 = 2$, $J_2 = 6$, $\rho = 1$ and $\beta = 0.001$

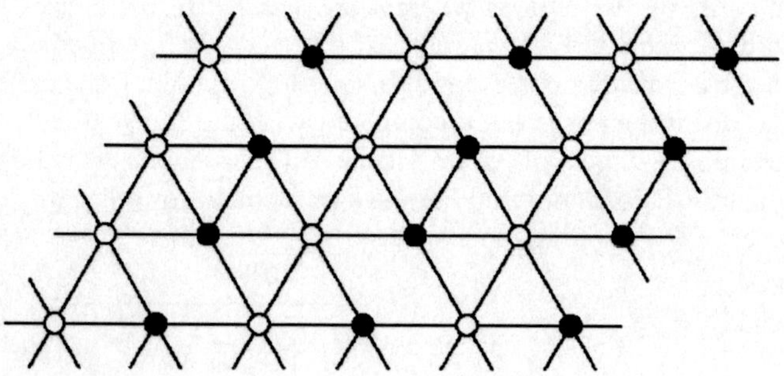

Fig. 2 The triangular diatomic lattice, with a periodic array of defects. The defects are in the form of buckled beams, where the stiffness of the buckled beams is neglected entirely

3.1 Defect Modes

We now introduce a periodic array of defects into the lattice (Fig. 2), in the form of buckled beams. Such beam buckling could occur after the application of some form of pre-stress. For example, selected beams could have a higher thermal expansion coefficient and buckle under thermal load. The beams are considered to have buckled to the extent to which they may be neglected entirely. Numerical simulations reveal a number of standing modes. In particular, we focus on two rotational modes where only similar nodes rotate. At the first mode, the two lighter nodes (indicated in white in Fig. 2) rotate, whereas at the second mode, the two heavier nodes rotate (black in Fig. 2). The estimates for these modes are found to be $\omega_1 = \sqrt{10\beta/J_1}$ and $\omega_2 = \sqrt{9\beta/J_2}$. Using the same parametric values as earlier, these estimates are in excellent agreement with the numerical solutions.

4 Concluding Remarks

Bloch-Floquet waves in infinite planar diatomic lattices with micro-polar rotations were considered. Several types of interactions were discussed and explicit estimates for the effective group velocities in the low frequency regime were derived and compared with previous results. Estimates for standing waves and defect modes were derived and compared with the results from numerical computations. The concepts and algorithms presented here have also been extended to the design of lattice systems, which allow elastic waves to be focused for a certain frequency range [8].

Acknowledgements D.J.C. gratefully acknowledges the support of an EPSRC research studentship (grant EP/H018514/1).

References

1. Gei, M., Movchan, A.B., Bigoni, D.: Band-gap shift and defect-induced annihilation in prestressed elastic structures. J. Appl. Phys. **105**(6), 63507 (2009)
2. Jones, I.S., Movchan, A.B.: Bloch-Floquet waves and controlled stop bands in periodic thermo-elastic structures. Waves Random Complex Media **17**(4), 429–438 (2007)
3. Kunin, I.A.: Elastic Media with Micro-Structure, vol. 1. Springer, Berlin (1982)
4. Martinsson, P.G., Movchan, A.B.: Vibrations of lattice structures and phononic band gaps. Q. J. Mech. Appl. Math. **56**(Part 1), 45–64 (2003)
5. Maz'ya, V.G., Nazarov, S.A., Plamenevskij, B.A.: Asymptotic Theory of Elliptic Boundary Value Problems in Singularly Perturbed Domains, vol. 2, Chap. 20. Birkhäuser, Boston (2000)
6. Morozov, N.F.: Mathematical Problems of Crack Theory. Nauka, Moscow (1984)
7. Nazarov, S.A., Paukshto, M.V.: Discrete Models and Homogenization in Problems of the Elastic Theory. Leningrad University, Leningrad (1984)
8. Colquitt, D.J., Jones, I.S., Movchan, A.B., Movchan, N.V.: Dispersion and localization of elastic waves in materials with microstructure. Proc. R. Soc. A (2011) DOI: 10.1098/rspa.2011.0126
9. Graff, K.F.: Wave Motion in Elastic Solids. Oxford University Press, London (1975)

Analytical Solution of In-Plane Response of a Thin Viscoelastic Disc Under Impact Load

Vítězslav Adámek and František Valeš

Abstract This paper concerns the analytical solution of the in-plane response of a thin viscoelastic disc to a dynamic load applied to its rim. The exact analytical relations for the Laplace transforms of radial and circumferential displacements are derived in terms of Bessel functions for the case of radial and torsional loads defined by even and odd functions of angular variable, respectively. The numerical evaluation of the analytical solution is then made for the case of an impulse radial load and transient wave phenomena are studied in the disc. With respect to the complexity of presented formulae, the multi-precision implementation of FFT based numerical algorithm for the inverse Laplace transform is used. The obtained analytical results are then compared to the results of numerical simulation performed in the finite element system MSC.Marc. The presented analytical solution can be used as a benchmark solution for the testing of numerical methods.

Keywords Analytical solution • In-plane vibration • Viscoelastic disc

1 Introduction

The problems of disc-like solids under various types of a dynamic load play an important role in many fields and practical applications. While the problems of transverse vibration of circular plates have been studied in detail already in the past, the in-plane disc response is the aim of works of the past few years. This

V. Adámek (✉)
University of West Bohemia, Pilsen, Univerzitní 22
e-mail: vadamek@kme.zcu.cz

F. Valeš
Institute of Thermomechanics AS CR, Pilsen, Veleslavínova 11
e-mail: vales@it.cas.cz

J. Náprstek et al. (eds.), *Vibration Problems ICOVP 2011: The 10th International Conference on Vibration Problems*, Springer Proceedings in Physics 139,
DOI 10.1007/978-94-007-2069-5_96,

increasing attention is mainly caused by the fact that the in-plane vibration involves higher frequencies compared with the transverse response and so it plays a role in the transmission of high frequency components from single parts to whole equipments. The non-negligible sound radiation from the in-plane vibrating discs is the second reason of mentioned interest. The authors in [1] have shown that the in-plane vibration can generate substantial sound, especially when the discs are relatively thick. This is also related with the railway transport where the impact noise generated by the wheel running over rail discontinuities is a considerable part of wheel/rail noise [2].

There exists number of works concerning the steady state in-plane vibration of stationary or rotating discs from analytical and numerical point of view. But the authors usually focus on elastic problems of solid or annular discs and mainly on the determination of natural frequencies and mode shapes under different types of boundary conditions. Such results are presented for example in [3, 4]. Problems of rotating discs under various types of edge loads are solved for instance in [5, 6].

All previous papers deal with stationary problems, but studies focused on transient disc response are relatively scarce and mainly report on numerical methods. As an example of work based on an analytical approach, we can mention the paper [7] in which the in-plane radial impact on a moving thin elastic disc is solved.

The investigation of in-plane response of a thin viscoelastic disc to dynamic edge load is the aim of this paper. The emphasis is placed on the study of transient wave phenomena and this work is based on [7, 8]. In the second mentioned paper, the relations for the Laplace transforms of radial and circumferential displacement components are derived for the case of an uniform radial pressure. With respect to the fact that we meet more general types of excitations in practical applications, the effort is to generalize the solution for more frequent types of loading. In particular, the disc response to radial and torsional excitations which are described by even and odd functions of angular coordinate, respectively, is investigated using an analytical approach.

2 Problem Formulation

Let us consider a thin disc of a constant thickness and of the finite radius r_1. Its material of mass density ρ is assumed to be isotropic and linear viscoelastic and it is modeled by the standard viscoelastic solid. This means that the elastic spring is arranged in parallel with a spring and a dashpot in series. Material parameters E_i and μ_i then correspond to Young's modulus and Poisson's ratio of the alone standing spring ($i=1$) and of the spring in series with dashpot ($i=2$). The dashpot properties are then described by normal viscosity λ, shear viscosity η and viscous Poisson's ratio ν. Hereinafter, the relation $\mu_1 = \mu_2 = \nu$ will be considered for simplicity.

It is convenient to solve the plane stress problem of the disc in the polar coordinate system $r-\varphi$ and to choose the disc centre as its origin. Consequently, the radial displacement $u_r(r,\varphi,t)$ and the circumferential displacement $u_\varphi(r,\varphi,t)$ describe the disc response. The boundary conditions are considered in the form $\sigma_r(r_1,\varphi,t)=\sigma_0(\varphi,t)$ and $\tau_{r\varphi}(r_1,\varphi,t)=\tau_0(\varphi,t)$ where $\sigma_0(\varphi,t)$ and $\tau_0(\varphi,t)$ are arbitrary even and odd functions of φ, respectively, and they can be defined on two disjoint parts of the disc rim. Finally, zero initial conditions are assumed.

3 Analytical Solution

The mathematical model describing the response of the disc consists of two coupled partial integro-differential equations of motion [8]. This system is solved using the classical method of integral transforms in combination with Fourier method of separation of variables.

3.1 Laplace Transforms of Displacement Components

The resulting relations for the Laplace transforms $\bar{u}_r(r,\varphi,p)$ and $\bar{u}_\varphi(r,\varphi,p)$ of the displacement components $u_r(r,\varphi,t)$ and $u_\varphi(r,\varphi,t)$ can be expressed formally in the same form as in [8]. These relations can be then converted to

$$\bar{u}_r = r\left\{\frac{J_1(z_1)}{z_1}P_0+\sum_{n=1}^{\infty}\left[\frac{2nJ_n(z_2)}{{z_2}^2}Q_n+\left(\frac{nJ_n(z_1)}{{z_1}^2}-\frac{J_{n-1}(z_1)}{z_1}\right)P_n\right]\cos(n\varphi)\right\},$$

$$\bar{u}_\varphi = r\left\{\sum_{n=1}^{\infty}\left[2\left(\frac{nJ_n(z_2)}{{z_2}^2}-\frac{J_{n-1}(z_2)}{z_2}\right)Q_n+\frac{nJ_n(z_1)}{{z_1}^2}P_n\right]\sin(n\varphi)\right\}, \tag{1}$$

where J_n denotes the Bessel function of the first kind ($n=0,1,2,\ldots$). The functions $z_1(r,p)$, $z_2(r,p)$ and the other quantities in (1) are defined for $p\in C$ by

$$z_1=\frac{ipr}{C_3},\quad z_2=\frac{ipr}{C_2},\quad C_k=C_k(p)=\sqrt{\left(1-\frac{\alpha}{p+\alpha}\right){c_{k2}}^2+{c_{k1}}^2}\quad \text{for } k=2,3,$$

$$\alpha=\frac{E_2}{\lambda}=\beta=\frac{G_2}{\eta},\quad c_{2k}=\sqrt{\frac{G_k}{\rho}}\quad\text{and}\quad c_{3k}=\sqrt{\frac{E_k}{\rho\left(1-\mu_k^2\right)}}\quad \text{for } k=1,2. \tag{2}$$

In the next section, we present relations for complex functions $P_n(p)$ ($n=0,1,2,\ldots$) and $Q_n(p)$ ($n=1,2,\ldots$) involved in (1) which can be derived using specified boundary conditions.

3.2 Formulae for Unknown Functions $P_n(p)$ and $Q_n(p)$

Since the boundary conditions are prescribed for the stress components σ_r and $\tau_{r\varphi}$, we need to derive the integral transforms of these stresses using constitutive and kinematic equations and using relations (1). Doing so and expanding the Laplace transforms of $\sigma_0(\varphi, t)$ and $\tau_0(\varphi, t)$ into appropriate Fourier series, i.e.,

$$\bar{\sigma}_0(\varphi, p) = \frac{A_0}{2} + \sum_{n=1}^{\infty} A_n(p) \cos(n\varphi) \quad \text{and} \quad \bar{\tau}_0(\varphi, p) = \sum_{n=1}^{\infty} B_n(p) \sin(n\varphi), \tag{3}$$

we obtain a system of equations for $P_n(p)$ and $Q_n(p)$. Its solution can be written as

$$P_0(p) = \frac{A_0(p)}{2C_{P2}(0, p)}, \quad P_n(p) = -\left(C_{Q1}(n, p)\, A_n(p) - C_{Q2}(n, p)\, B_n(p)\right) D_n(p),$$

$$Q_n(p) = (C_{P1}(n, p)\, A_n(p) - C_{P2}(n, p)\, B_n(p))\, D_n(p), \tag{4}$$

where

$$D_n(p) = \left(C_{P1}(n, p)\, C_{Q2}(n, p) - C_{P2}(n, p)\, C_{Q1}(n, p)\right)^{-1},$$

$$C_{P1}(n, p) = 2\rho C_2{}^2 n \left(\frac{J_{n-1}(z_1(r_1, p))}{z_1(r_1, p)} - \frac{(n+1)\, J_n(z_1(r_1, p))}{z_1(r_1, p)^2} \right),$$

$$C_{P2}(n, p) = 2\rho C_2{}^2 \left(\left(\frac{1}{2} K_{32} - \frac{n(n+1)}{z_1(r_1, p)^2} \right) J_n(z_1(r_1, p)) + \frac{J_{n-1}(z_1(r_1, p))}{z_1(r_1, p)} \right),$$

$$C_{Q1}(n, p) = 4\rho C_2{}^2 \left(\left(\frac{1}{2} - \frac{n(n+1)}{z_2(r_1, p)^2} \right) J_n(z_2(r_1, p)) + \frac{J_{n-1}(z_2(r_1, p))}{z_2(r_1, p)} \right),$$

$$C_{Q2}(n, p) = 4\rho C_2{}^2 n \left(\frac{J_{n-1}(z_2(r_1, p))}{z_2(r_1, p)} - \frac{(n+1)\, J_n(z_2(r_1, p))}{z_2(r_1, p)^2} \right) \tag{5}$$

and $K_{32}(p) = (C_3(p)/C_2(p))^2$.

3.3 Numerical Evaluation of the Analytical Solution

It is evident that the substitution of (2), (4) and (5) into (1) results in quite complicated formulae for the Laplace transforms of desired displacement components the exact inverse transform of which is practically impossible. The inversion is hindered not only by the complexity of derived relations but mainly by the presence of branch points. Hence, the FFT based numerical algorithm in combination with ε–algorithm

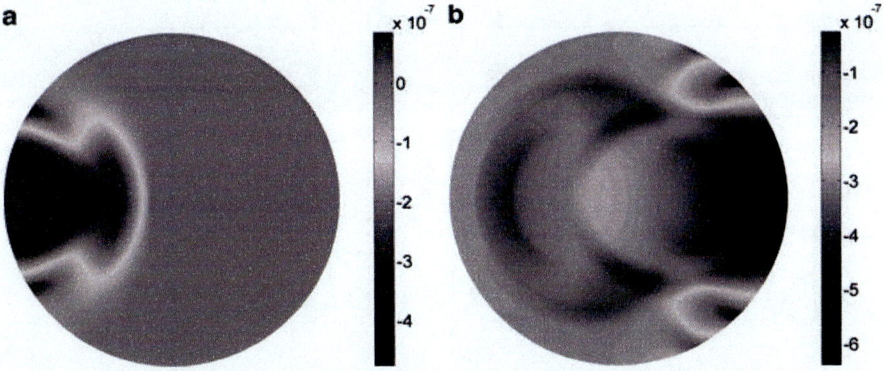

Fig. 1 Analytical results of spatial distribution of displacement u_x at (**a**) $t = 8\,\mu$s and (**b**) $t = 40\,\mu$s

as the sequence accelerator was used for this purpose. To avoid the loss of the analytical solution accuracy, the multi-precision implementation in Maple language was used so results of arbitrary precision could be obtained [9].

The numerical evaluation of derived analytical solution was made for the time interval $t \in \langle 0, 40\rangle\ \mu$s and for the disc of radius $r_1 = 20\,$mm and of the following material properties: $\rho = 1100\,\mathrm{kg\,m^{-3}}$, $\nu = \mu_1 = \mu_2 = 0.4$, $\eta = 10432.9\,\mathrm{Pa\,s^{-1}}$, $\lambda = 29212.0\,\mathrm{Pa\,s^{-1}}$, $E_1 = 3.603 \times 10^9$ Pa and $E_2 = 7.918 \times 10^8$ Pa. The boundary conditions were represented by functions $\sigma_0(\varphi, t) = -\sigma_0\,(H(t) - H(t - t_0))$ for $\varphi \in -\langle\alpha_0, \alpha_0\rangle$ and by $\tau_0(\varphi, t) = 0$, where $H(t)$ denotes the Heaviside function in time. This means that only the pressure impulse of amplitude σ_0 and of time duration t_0 was assumed in the analysis. The calculations were made for $\sigma_0 = 10^6$ Pa, $t_0 = 2\,\mu$s and $\alpha_0 = \pi/20$.

Figure 1 presents the distribution of the horizontal displacement u_x which was calculated on the basis of u_r and u_φ evaluation. The primary waves propagating from the loaded area can be clearly seen from Fig. 1a which depicts the situation at $t = 8\,\mu$s. The final distribution of u_x at the end of analysis, which is the result of primary and reflected waves interaction, is then shown in Fig. 1b.

4 Numerical Simulation and Results Comparison

To verify the correctness of the analytical solution derivation and its evaluation, the numerical simulation in the finite element software MSC.Marc was carried out. This approach also helped us to find out the capabilities of used numerical model.

The disc mesh consisted of about 16,000 four-noded isoparametric elements of basic size 0.2×0.2 mm. The Newmark integration method with time step 2.4×10^{-8} s was used. The obtained numerical results were then compared with the analytical ones. The comparison of analytical and numerical results for $r = r_1$

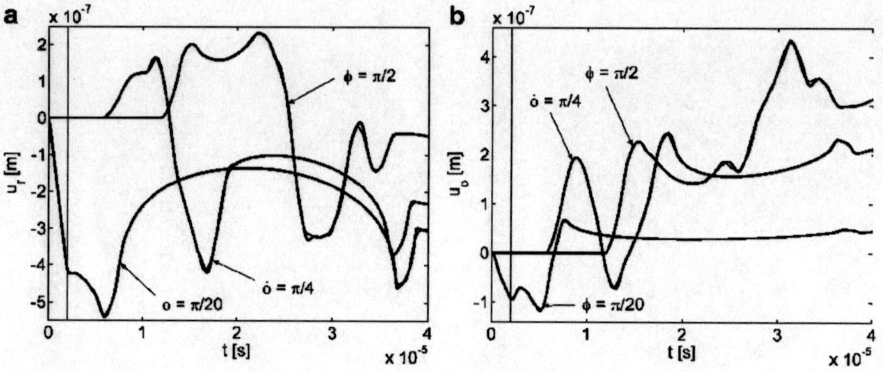

Fig. 2 Comparison of analytical (*gray curves*) and numerical (*black curves*) results for $r = r_1$ and different values of φ – displacement components (**a**) u_r and (**b**) u_φ

is presented in Fig. 2 where time paths of u_r and u_φ for three different values of φ are depicted. It should be emphasized that the angle $\varphi = \pi/20$ corresponds right to the end of loaded rim area. It is obvious that curves representing numerical and analytical solution are nearly identical. The gray vertical lines in Fig. 2a and b mark the duration of applied loading. The comparison of results for other values of r has shown that the smaller radius r is, the better consistency is achieved.

5 Conclusion

This paper presents the analytical solution of the in-plane response of a thin viscoelastic disc to radial and torsional edge load described by even and odd functions of angular coordinate, respectively. The solution is evaluated for the case of an impulse radial load and the results obtained are then compared with those of numerical simulation. Presented solution can be used as a benchmark solution in the testing of numerical methods in future. Moreover, the numerical evaluation of the analytical solution is in many cases more effective than the numerical simulation based on FEM and hence it can be used in optimization problems requiring multiple evaluations of desired quantities, e.g., problems of material properties identification etc. This is the principal direction of future work we want to proceed.

Acknowledgements This work has been supported by the grant GA CR 101/09/P082 and by the project MSM 4977751303.

References

1. Lee, H., Singh, R.: Self and mutual radiation from flexural and radial modes of a thick annular disk. J. Sound Vib. **286**(4–5), 1032–1040 (2005)
2. Thompson, D.J., Jones, C.J.: A review of the modelling of wheel/rail noise generation. J. Sound Vib. **231**(3), 519–536 (2000)
3. Fu, Z., Lin, S., Xian, X.: Vibration of annular plate concentrators with conical cross-section. J. Sound Vib. **321**(3–5), 1026–1035 (2009)
4. Bashmal, S., Bhat, R., Rakheja, S.: In-plane free vibration of circular annular disks. J. Sound Vib. **322**(1–2), 216–226 (2009)
5. Chen, J.-S., Jhu, J.-L.: In-plane response of a rotating annular disk under fixed concentrated edge loads. Int. J. Mech. Sci. **38**(12), 1285–1293 (1996)
6. Koh, C.G., Sze, P.P., Deng, T.T.: Numerical and analytical methods for in-plane dynamic response of annular disk. Int. J. Solids Struct. **43**(1), 112–131 (2006)
7. Červ, J., Slavíková, J.: Motion and stress-state of a thin disc under radial impact load. Acta Tech. ČSAV **32**(2), 113–133 (1987)
8. Adámek, V., Valeš, F.: Thin viscoelastic disc subjected to radial non-stationary loading. Appl. Comput. Mech. **4**(1), 5–14 (2010)
9. Abate, J., Valkó, P.P.: Multi-precision Laplace transform inversion. Int. J. Numer. Meth. Eng. **60**(5), 979–993 (2004)

Travelling Waves in Non Linear Elastic Solids with Multiple Microstructures

Franco Pastrone

Abstract In non classical mechanics naturally arises the problem of the propagation of nonlinear waves in solids with different internal structural scales. Here we make use of a suitable model of one dimensional microstructured solids to describe the behavior of internal structures with two different scales. Hence we have an elastic material composed by a macrostructure, a first microstructure (say a mesostructure) and a second microstructure at some smaller scale. The choice of suitable microstrains functions φ and ψ at the two levels respectively, of the microdisplacement u, of their time derivatives as strain velocities, allows us to obtain the field equations via a variational principle. In a particular case a sixth order PDE is obtained, with characteristic hierarchical structure, where the three levels hierarchy and the various coefficients may reflect the dominance of one structural level over the other ones in wave propagation. This equation is integrated in terms of elliptic functions. Using the same basic model, the case of two concurrent microstructures is studied.

Keywords Hierarchies of waves • Microstructures • Nonlinear waves

1 One Dimensional Solid with Hierarchical Microstructure

We consider a one-dimensional (1 + 1D) microstructured model with two different scale levels applied for the microstructure. Instead of the two-scale elastic system, containing both macro- and microstructures, we introduce a material, which is

F. Pastrone (✉)
Department of Mathematics, University of Torino, Torino 10123, Italy
e-mail: franco.pastrone@unito.it

J. Náprstek et al. (eds.), *Vibration Problems ICOVP 2011: The 10th International Conference on Vibration Problems*, Springer Proceedings in Physics 139,
DOI 10.1007/978-94-007-2069-5_97,

supposed to be a compound of a macrostructure, a first level microstructure and a second level microstructure at much smaller scale. The last may be interpreted as a nanostructure, to some extent (see [2, 5, 6]).

Therefore, following the model, we deal with three different scalar functions: the one for the macrostructure and two for the microstructures, one for each scale level. The model of a material is the one-dimensional manifold, and we consider the material coordinates in space x and in time t; and the functions $v = v(x,t)$ for the macrostructure, $\varphi = \varphi(x,t)$ and $\psi = \psi(x,t)$ respectively for the first and the second scale level in microscale. The macro body is supposed to be purely elastic, and both the first and second level microstructures satisfy the same generalized elasticity hypothesis as well, therefore the existence of an internal strain energy is assumed.

A particular choice of the strain energy function W defines different nonlinear models, see [2]; in this paper we consider it in the following form:

$$W = \frac{1}{2}\alpha v_x^2 + \frac{1}{3}\beta v_x^3 - A_1\varphi v_x + \frac{1}{2}B_1\varphi^2 + \frac{1}{2}C_1\varphi_x^2 - A_2\varphi_x\psi + \frac{1}{2}B_2\psi^2 + \frac{1}{2}C_2\psi_x^2 \tag{1}$$

This function is the generalization of the strain energy function for nonlinear elastic solids with one microstructure level to our case, where the introduction of the cubic term v_x^3 represents the nonlinear behavior of the matrix.

The field equations can be derived as in [2] via a variational principle:

$$\begin{cases} \rho v_{tt} = \alpha v_{xx} + \left(\beta v_x^2\right)_x - A_1\varphi_x \\ I_1\varphi_{tt} = C_1\varphi_{xx} + A_1 v_x - B_1\varphi - A_2\psi_x \\ I_2\psi_{tt} = C_2\psi_{xx} + A_2\varphi_x - B_2\psi \end{cases} \tag{2}$$

where α, β and A_i, B_i, C_i $(i = 1,2)$ denote material constants.

To obtain the governing equation in dimensionless form, it is necessary to introduce some suitable parameters and constants (see [3]) and two different parameters δ_i, $i = 1,2$, characterizing the ratio between the microstructure and the wave length L, and ε, accounting for small but finite elastic strain magnitude:

$$\delta_1 = (l_1/L)^2, \qquad \delta_2 = (l_2/L)^2, \qquad \varepsilon = v_0 \ll 1 \tag{3}$$

where v_0 is the intensity of the initial excitation and the values l_1 and l_2 represent the size of the microstructural elements. Introducing the macrostrain $v = v_x$ (the term "strain" is used for brevity only; in fact, it is the longitudinal displacement gradient component, while expressions for genuine strains are nonlinear with respect to v) and the dimensionless variables

$$u = v/v_0, \quad X = x/L, \quad T = (c_0/L)t$$

and substituting them into the previous system, we obtain the following coupled dimensionless equations:

$$\begin{cases} u_{TT} = \dfrac{\alpha}{\rho c_0^2} u_{XX} + \dfrac{\beta\varepsilon}{\rho c_0^2}\left(u^2\right)_{XX} - \dfrac{A_1^* l_1}{\varepsilon\rho c_0^2}\varphi_{XX} \\ \varphi = \dfrac{A_1^* l_1 v_0}{B_1} u - \dfrac{A_2^*\sqrt{\delta_2}}{B_1}\psi_X + \dfrac{\delta_1}{B_1}\left[C_1^*\varphi_{XX} - \rho I_1^* c_0^2 \varphi_{TT}\right] \\ \psi = \dfrac{A_2^*\sqrt{\delta_2}}{B_2}\varphi_X + \dfrac{\delta_2}{B_2}\left[C_2^*\psi_{XX} - \rho I_2^* c_0^2 \psi_{TT}\right] \end{cases} \tag{4}$$

The slaving principle [5] can now be used for further transformations. This procedure allows us to write one function in terms of the other; on this way we can obtain the governing equation for the function $u(x,t)$ only. To this end, we determine the variable ψ in terms of φ and its derivatives from $(4)_3$. Then the Eq. 4_3 can be used to express φ in terms of derivatives of u. This expression will eventually be substituted into Eq. 4_1 to obtain the one differential equation for u.

The resulting equation can be written as:

$$u_{TT} + \alpha_1 u_{XX} + \alpha_2 (u^2)_{XX} + (\alpha_3 u_{XX} + \alpha_4 u_{TT})_{XX} + (\alpha_5 u_{4X} + \alpha_6 u_{TTXX} + \alpha_7 u_{4T})_{XX} = 0 \tag{5}$$

where the α_i are constant coefficients explicitly defined in [3].

The Eq. 5 above may be considered as the hierarchical equation in terms of u, where two different levels of microstructure are expressed in five different dispersive terms, and the higher order terms contain the parameters of the second level of microstructure.

We have obtained a sixth order PDE that is hardly to be solved explicitly in general case. However, we will find some exact travelling wave solutions of the PDE (5), when the equation can be reformulated in terms of the phase variable $z = x \pm Vt$ in the corresponding ODE, as follows:

$$(V^2 + \alpha_1)u^{(\mathrm{II})} + \alpha_2 (u^2)^{(\mathrm{II})} + \left(\alpha_3 + V^2\alpha_4\right) u^{(\mathrm{IV})} + \left(\alpha_5 + V^2\alpha_6 + V^4\alpha_7\right) u^{(\mathrm{VI})} = 0 \tag{6}$$

where V is the velocity of wave propagation.

Following the method introduced by Samsonov in [7], upon the introduction of z and integration twice with corresponding conditions at infinity $|z| \to \infty \Rightarrow u, u' \to 0$ the Eq. 6 may be rewritten as the nonlinear ODE of the fourth order:

$$u^{(\mathrm{IV})} + a u^{(\mathrm{II})} + b\, u^2 + c u = 0 \tag{7}$$

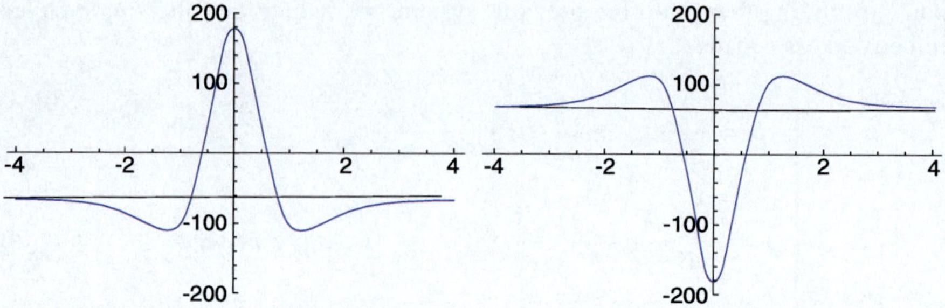

Fig. 1 $a = -8$, $b = -1.5$ for the figure on the *left*, $a = -8$, $b = +1.5$ for the figure on the *right*

where obviously:

$$a = (\alpha_3 + V^2\alpha_4)/\chi; \quad b = \alpha_2/\chi; \quad c = (\alpha_1 + V^2)/\chi; \quad \chi = \alpha_5 + V^2\alpha_6 + V^4\alpha_7 \tag{8}$$

Following the method described in [7], the exact solution to the ODE (7) in terms of elliptic functions, containing only poles as the critical singularities, can be found in the following form:

$$u = M\wp^2(x; g_2, g_3) + S\wp(x; g_2, g_3) + K \tag{9}$$

where the coefficients, M, S, K and invariants g_i of the Weierstrass elliptic function $\wp$ are defined in [3].

In the appropriate limit the Weierstrass elliptic function $\wp$ may be further reduced to the elliptic Jacobi $cn-$function and, in due course, to the bounded solution u_0 in terms of $\cosh^{-2}$ function, i.e., to *the solitary wave solution*, as follows:

$$u_0 = s\cosh^{-4}(x) + q\cosh^{-2}(x) + p;$$

$$p = -c/b = -\frac{-18928 + 3640a - 31a^2}{507b}; \; q = \frac{140(52 + a)}{13b}; \; s = -840/b, \tag{10}$$

which has a form of the so called "mexican hat" (Fig. 1):

The approach used to obtain these solutions is similar to that introduced and grounded in [7], and can be applied to explicitly solve different higher order ODE, e.g., the fifth order KdV and the fifth order mKdV equations.

In the first step from $(4)_3$ the expansion

$$\psi = \frac{A_2^*\sqrt{\delta_2}}{B_2}\varphi_x + \frac{A_2^*\sqrt{\delta_2}\delta_2}{B_2^2}\left[C_2^*\varphi_{xxx} - \rho I_2^* c_0^4 \varphi_{xTT}\right]$$

is obtained. Upon substitution into (4)$_2$, which also is to be expanded, this yields

$$\varphi = \frac{A_1 v_0}{B_1} u + \frac{\delta_1 A_1 v_0}{B_1^2} \left[C_1^* u_{XX} - \rho I_1^* c_0^4 u_{TT} \right] + \frac{\delta_2^2 A_1 (A_2^*)^2 v_0}{B_1^2 B_2^2} \left[C_2^* u_{XXXX} - \rho I_2^* c_0^4 u_{XXTT} \right]$$

Finally this expression is inserted in (4)$_1$ resulting in the partial differential equation

$$u_{TT} = \left(\frac{\alpha \varepsilon L B_1 + A_1^2 v_0}{\varepsilon L \rho c_0^4 B_1} \right) u_{XX} + \frac{\beta \varepsilon L}{\rho c_0^4} (u^2)_{XX} - \frac{\delta_1 A_1^2 v_0}{\varepsilon L \rho c_0^4 B_1^2} \left[C_1^* u_{XX} - \rho I_1^* c_0^4 u_{TT} \right]_{XX}$$
$$+ \frac{\delta_2^2 A_1^2 (A_2^*)^2 v_0}{\varepsilon L \rho c_0^4 B_1^2 B_2^2} \left[C_2^* u_{XX} - \rho I_2^* c_0^4 u_{TT} \right]_{XXXX}$$

2 Concurrent Microstructures

Instead of a hierarchy of microstructures, one can be interested in concurrent microstructures, as introduced in [1], namely in two, or more, microstructures which act at the same scale level and interact with the macrostructure as well. We can obtain the fiels equations as done before just introducing a different expression of the strain energy function, for instance:

$$W = \frac{1}{2} \alpha v_x^2 + \frac{1}{3} \beta v_x^3 + A_1 \varphi v_x + A_2 \psi v_x + \frac{1}{2} B_1 \varphi^2 + \frac{1}{2} C_1 \varphi_x^2 + A_{12} \varphi \psi$$
$$+ \frac{1}{2} B_2 \psi^2 + \frac{1}{2} C_2 \psi_x^2 \tag{11}$$

where φ and ψ denote the microstrains of the two concurrent microstructures.

Hence the field equations read:

$$\begin{cases} \rho v_{tt} = \alpha v_{xx} + \beta (v_x^2)_x + A_1 \varphi_x + A_2 \psi_x \\ I_1 \varphi_{tt} = C_1 \varphi_{xx} - A_1 v_x - B_1 \varphi - A_{12} \psi \\ I_2 \psi_{tt} = C_2 \psi_{xx-} - A_2 v_x - B_2 \psi - A_{12} \varphi \end{cases} \tag{12}$$

where $\alpha, A_i, B_i, C_i, A_{12}$ $(i = 1, 2)$ denote material constants.

With the substitution $v_0\, u = v_x$, from (12)$_1$ we derive:

$$\rho v_0 u_{tt} = \alpha v_0 u_{xx} + \beta v_0^2 (u^2)_{xx} + A_1 \varphi_{xx} + A_2 \psi_{xx}. \tag{13}$$

Using the change of parameters already recalled in the previous Section we can obtain the dimensionless equations

$$\begin{cases} \dfrac{\rho v_0 c_0^2}{L} u_{TT} = \alpha \dfrac{v_0}{L^2} u_{XX} + \dfrac{\beta v_0^2}{L^2} \left(u^2\right)_{XX} + \dfrac{A_1^* l_1}{L^2} \varphi_{XX} + \dfrac{A_2^* l_2}{L^2} \psi_{XX} \\ B_1 \varphi + A_{12} \psi = -A_1^* l_1 v_0 u + \dfrac{l_1^2}{L^2} \left(C_1^* \varphi_{XX} - \rho I_1^* c_0^2 \varphi_{TT}\right) \\ A_{12} \varphi + B_2 \psi = -A_2^* l_2 v_0 u + \dfrac{l_2^2}{L^2} \left(C_2^* \psi_{XX} - \rho I_2^* c_0^2 \psi_{TT}\right) \end{cases} \tag{14}$$

This system can be re-written in the simpler formal way:

$$\begin{cases} u_{TT} = \alpha_1 u_{XX} + \beta_1 \left(u^2\right)_{XX} + \alpha_2 \varphi_{XX} + \alpha_3 \psi_{XX} \\ B_1 \varphi + A_{12} \psi = \Phi \\ A_{12} \varphi + B_2 \psi = \Psi \end{cases} \tag{15}$$

with the obvious meaning of the coefficients α_1 and β_1 and of the right-hand terms in $(15)_{2,3}$ Φ and Ψ. Iff $B_1 B_2 - A_{12}^2 \neq 0$, the algebraic system $(15)_{2,3}$ admits a unique solution and φ and ψ are linear combinations of Φ and Ψ.

Now we have two possibilities:

1. $A_{12} = 0$, the concurrent microstructures are fully independent. This case has already been briefly studied in [1], where the equations for the macro and micro structures have been obtained neglecting the cubic term in W. We can add here that, using the slaving principle, one can reach an approximate equation in u only:

$$\frac{\rho v_0 c_0^2}{L} u_{TT} = \left[\alpha \frac{v_0}{L^2} - \frac{\left(A_1^* l_1\right)^2}{L^2} v_0 - \frac{\left(A_2^* l_2\right)^2}{L^2} \right] u_{XX} + \frac{\beta v_0^2}{L^2} \left(u^2\right)_{XX} \tag{16}$$

   which is a well known nonlinear PDE widely studied elsewhere.
2. $A_{12} \neq 0$, the microstructures are coupled. In a very similar way, we obtain a leading equation of the type:

$$u_{TT} - \mathcal{A} u_{XX} = \frac{\beta v_0^2}{L^2} \left(u^2\right)_{XX} \tag{17}$$

   where $\mathcal{A}$ briefly denotes the set of coefficients of u_{XX} analogous to the square brackets in (16), but it contains A_{12}, namely the coupling constant.

Another possibility is to write the system of three second order equations (15) as an equivalent sixth order equation in u and generalize the procedure followed in [4, 5], for example.

Indeed one can imagine higher order coupling terms introducing in W products of derivatives of φ and ψ, namely terms containing $\varphi_x\psi$, $\psi_x\varphi$, $\varphi_x\psi_x$, but for need of brevity we do not go further in this direction.

3 Conclusions

The problem of the propagation of nonlinear waves in solids with different internal structural scales is studied. The general model developed in [2] and [6] has been used. In the case of one microstructure a sixth order PDE is obtained and the hierarchy of waves is clearly obtained. Using the same basic model, the case of two concurrent microstructures is studied and by means of the slaving principle one can reach meaningful approximate equations.

References

1. Berezovski, A., Engelbrecht, J., Peets, T.: Multiscale modeling of microstrucutured solids. Mech. Res. Commun. **37**(6), 531 (2010)
2. Casasso, A., Pastrone, F.: Wave propagation in solids with vectorial microstructure. Wave Motion **47**, 358 (2010)
3. Casasso, A., Pastrone, F., Samsonov, A.M.: Travelling waves in microstructure as the exact solutions to the 6th order nonlinear equation. Acoust. Phys. **56**(6), 871 (2010)
4. Engelbrecht, J.: Complexity in mechanics. Rend. Sem. Matme. Univ. Pol. Torino **67**(3), 293 (2009)
5. Engelbrecht, J., Pastrone, F., Braun, M., Berezovski, A.: Hierarchy of waves in nonclassical materials. In: Delsanto, P.P. (ed.) Universality of Nonclassical Nonlinearity, pp. 29–48. Springer, New York (2007)
6. Pastrone, F.: Wave propagation in microstructured solids. Math. Mech. Solids **10**, 349 (2005)
7. Samsonov, A.M.: Strain Solitons in Solids and How to Construct Them. Chapman and Hall/CRC, Boca Raton (2001)

Parallel Implementation of Triangular Cellular Automata for Computing Two-Dimensional Elastodynamic Response on Arbitrary Domains

Michael J. Leamy and Adam C. Springer

Abstract In this research we report parallel implementation of a Cellular Automata-based simulation tool for computing elastodynamic response on complex, two-dimensional domains. Elastodynamic simulation using Cellular Automata (CA) has recently been presented as an alternative, inherently object-oriented technique for accurately and efficiently computing linear and nonlinear wave propagation in arbitrarily-shaped geometries. The local, autonomous nature of the method should lead to straight-forward and efficient parallelization. We address this notion on symmetric multiprocessor (SMP) hardware using a Java-based object-oriented CA code implementing triangular state machines (i.e., automata) and the MPI bindings written in Java (MPJ Express). We use MPJ Express to reconfigure our existing CA code to distribute a domain's automata to cores present on a dual quad-core shared-memory system (eight total processors). We note that this message passing parallelization strategy is directly applicable to computer clustered computing, which will be the focus of follow-on research. Results on the shared memory platform indicate nearly-ideal, linear speed-up. We conclude that the CA-based elastodynamic simulator is easily configured to run in parallel, and yields excellent speed-up on SMP hardware.

Keywords Wave propagation • Cellular automata • Computational mechanics • Distributed computing • MPI

M.J. Leamy (✉) • A.C. Springer
Georgia Institute of Technology, Atlanta, GA, USA
e-mail: Michael.Leamy@me.gatech.edu; acspringer@me.com

J. Náprstek et al. (eds.), *Vibration Problems ICOVP 2011: The 10th International Conference on Vibration Problems*, Springer Proceedings in Physics 139,
DOI 10.1007/978-94-007-2069-5_98,

1 Introduction

The cellular automata paradigm [1, 2] has recently been adopted to simulate wave propagation in two-dimensional linear and nonlinear elastic domains of arbitrary shape [3, 4]. The approach shares an idea central to all cellular automata modeling, which is domain discretization using autonomous cells (usually rectangular or hexagonal) whose state is updated via simple rules. In the cited elastodynamic work, a rule set for non-uniform triangular cells has been developed which allows domains of multiply-connected, arbitrary shape to be simulated efficiently and accurately. In fact, the method has been shown to effectively avoid spurious oscillations at the front of sharp wave fronts without the need for specialized treatment (unlike other methods – e.g., the finite element method). The method is briefly reviewed next – full details can be found in the cited work, and Java source code can be found on the first author's research web page.

Figure 1 provides a graphical overview of the method. Non-uniform triangles are employed to discretize a domain into multiple state machines termed automata. The rule set governing the temporal update of each cell is first arrived at using a balance of momentum applied to a target cell which sums forces present on each face. In doing so, the computation of strains is necessary, which is done by categorizing the strains as either Type I (derivatives in the normal direction) or Type II (tangential derivatives). Numerical evaluation of the spatial derivatives then follows from simple finite difference expressions using the appropriate states of neighboring automata. By choice, we use only von Neumann and what we term secondary von Neumann neighbors. Finally, a forward-Euler time integration of the momentum

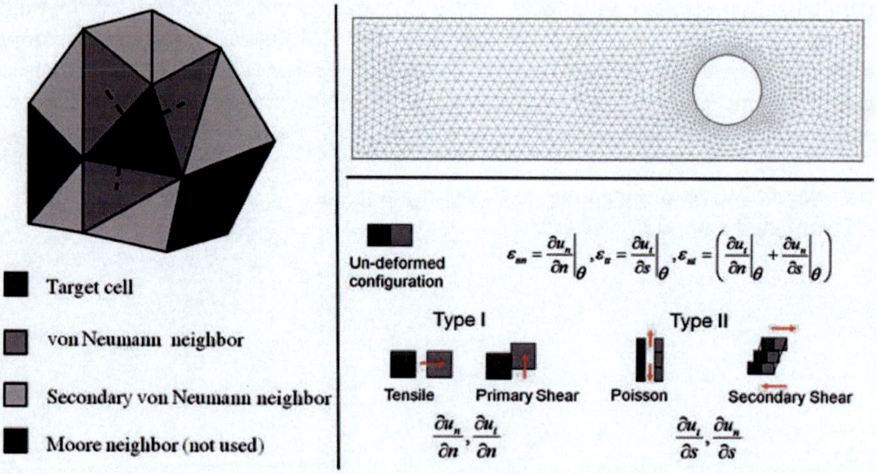

Fig. 1 *Left*: target cell (or target automaton) geometry with neighbors identified. *Top Right*: multiply-connected domain with automata mesh. *Bottom Right*: rectangular automata illustrating strains needed for rule set development

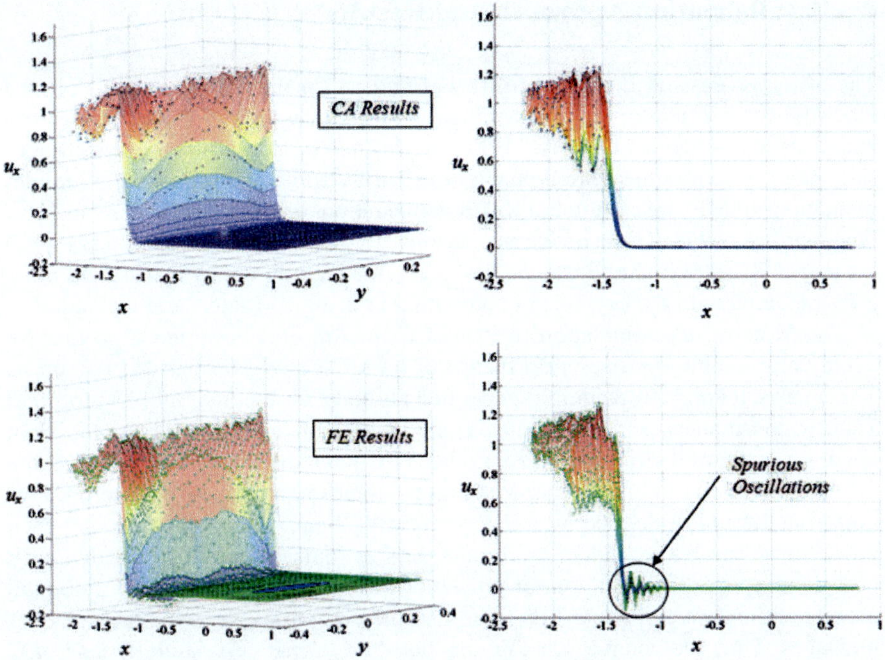

Fig. 2 Comparison of x-displacement results at a snapshot in time (finite element [*bottom*] vs. cellular automata [*top*]) for a sharply loaded domain with an interior hole – see Fig. 1 for the meshed domain. The *left subfigures* employ an isoperimetric perspective, while the *right subfigures* employ a perspective from the y-axis. The material simulated is aluminum, and the loading occurs on the left boundary in the form of an imposed Dirichlet boundary condition at the start of the simulation (Figure reproduced from [3])

equation yields the target cell's explicit rule set. Simulation proceeds by requesting that each automata update their state (displacement and velocity components) at each time step. Note that the method relies solely on local interactions, avoids partial differential equations and their complexity, and is fully object-oriented. In fact, the traditional process of assembling and solving a matrix set of equations is traded for requests to the automata objects to update their state.

Simulation results from the elastodynamic CA approach have been compared to those from other methods, including commercially-available finite element simulators, and excellent agreement has been documented [3, 4]. For smooth loading histories, the two methods yield nearly identical results. For sharply discontinuous loading, such as experienced during impact, the CA approach avoids spurious oscillations (see Fig. 2), which are a well-known artifact in interpolation-based approaches such as the finite element method. For this reason, it appears the CA method is particularly attractive for studying elastodynamic problems where loading results in sharp wave fronts.

2 Parallelization Approach and Results

The Message Passing Interface (MPI) was chosen for parallelizing the CA-based elastodynamic simulator on both symmetric multiprocessor (SMP) shared-memory and distributed-memory clusters. MPI is the *de facto* standard for passing messages (i.e., data), which is the central enabler of distributed computing. While C implementations of MPI are common, Java implementations are not; furthermore, the Java language has received little notice as a serious High Performance Computing (HPC) language. To justify using Java, we first performed a preliminary assessment of its HPC potential with the example of computing Pi using a Monte Carlo technique.

The Monte Carlo simulation of Pi was written in both Java and C so that we could compare the runtime performance of both C-based MPI and MPJ Express. The approach employs randomly-generated numbers on the interval [0,1] for both x and y coordinates, and then tallies a one if the pair lies inside of the top right quadrant of the unit circle, and a zero otherwise. Note that the probability of being inside the circle is ¼ of Pi. After n iterations, the estimate for Pi is therefore four times the total tally divided by n. By performing 'inner' iterations in which each processor generates n random (x,y) pairs, and m 'outer' iterations in which each processor reports back to a master processor their Pi estimate after performing an inner iteration, we can control the ratio of time computing Pi to the time passing messages. Thus the comparison captures both the inherit calculation speeds of C and Java, and the message passing overheads.

The Monte Carlo estimates of Pi were run on a 64-bit Windows 7 machine with two Intel Xeon E5506 quad-core processors and 24 GB of available RAM. The C simulation employed the MPICH2 implementation of MPI as well as Cygwin, a Linux emulator, to facilitate message passing [5]. The Java version utilizes MPJ Express [6] and runs natively through the NetBeans IDE. The coded algorithms are nearly identical in terms of the number of lines of code needed, variable types employed, and overall program flow. Figure 3 illustrates scaling results from the simulation. The graph on left shows the weak scaling comparison (fixed problem

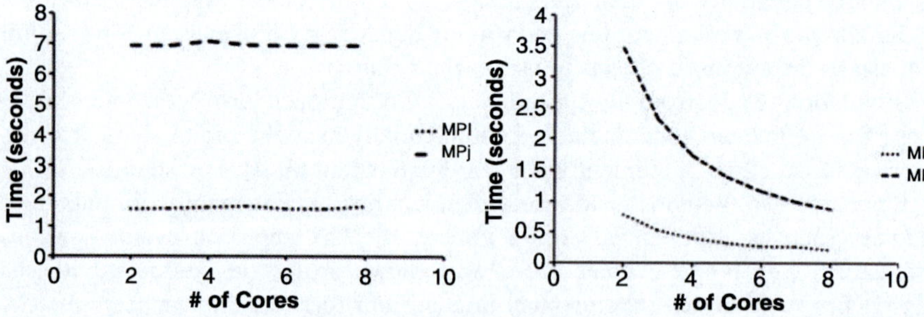

Fig. 3 Weak and strong scaling, C vs. Java, using a parallel Monte Carlo algorithm for computing Pi

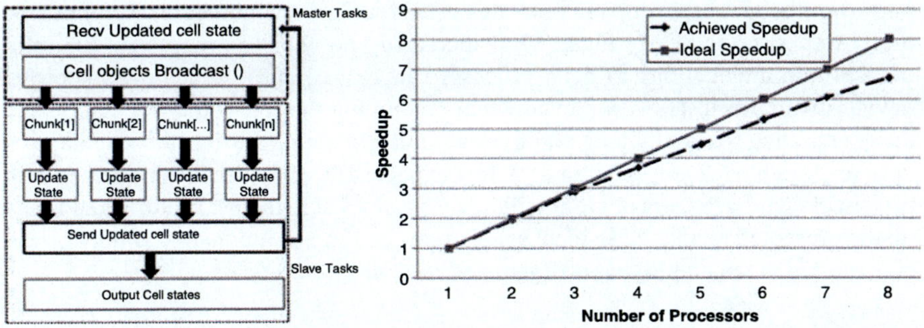

Fig. 4 *Left*: Graphical depiction of the parallel calculation flow. *Right*: Speedup as a function of cores employed on a multi-core shared memory system

size per core) using ten million iterations per core. Strong scaling is also provided and is used to illustrate the manner in which the solution time varies with the number of cores employed for a fixed total problem size. We again used ten million iterations as our problem size. The results indicate that, as implemented, Java performs on the same order, or better, than C. This conclusion has also been reached in studies conducted by the MPJ Express developers [6].

Development of a distributed CA algorithm amounts to splitting an array of automata objects into 'chunks' which can then be computed on each processor. Since each cell (or automaton) is a Java object encapsulating data and methods (e.g., pointers to neighbors and a step method), the entire object can be passed to a processor as a single entity. After completing a state update (i.e., a step), the object requires only that neighbor states residing on other processors be passed to it before completing another state update. We accomplish neighbor state sharing using blocking send/receive MPJ Express communications. Communications are often the bottleneck in a parallelization strategy due to their non-negligible latency. In order to decrease the number of communications per processor, we populate cells on a processor by exploiting the neighbor information already stored by the CA cells in such a way that when a cell is added to a processor, its neighbor cells are next added to the processor. These neighbor cells are also placed in a first-in, first-out queue such that their neighbors can be added to the same processor. This process repeats until the processor has a predetermined number of cells. The straight-forward CA parallelization approach described can be contrasted with traditional elastodynamic simulation methods, such as most finite element approaches, which require a sparse linear system solver and complex decomposition schemes prior to HPC deployment [7]. Figure 4 provides a graphical depiction of one version of this parallel algorithm where the Master processor coordinates all sends and receives.

On a symmetric multi-core, shared memory system, we analyzed the performance of our algorithm using a ratio of the best sequential running time to the parallel run time. Figure 4 illustrates the achieved speedup as a function of the number of employed cores while simulating just over 20,000 cells. For comparison purposes, the ideal speedup curve is also provided. For low numbers of processors, the CA speedup is nearly ideal. With increasing number of processors, the ratio of MPI communications to cells processed increases, and the achieved speedup deviates from ideal. However, even when employing the full eight processors, and thus competing for computing resources with the operating system, the speedup is a very respectable 6.7 (ideal being 8.0). It is expected that inexpensive machines with processors well in excess of eight will be on the market in the near future, and based on the presented results, near-ideal speedup of the CA elastodynamic simulator can be expected when employing a large proportion of those processors.

3 Concluding Remarks

This paper describes a new simulation technique in solid mechanics for computing elastodynamic response in arbitrary two-dimensional domains using multiple processors. The paper documents near-linear speedup with respect to the number of available processors on shared memory systems. The developed method is notable for its straight-forward formulation based on local interactions, its ability to accurately simulate sharp wave fronts, and its compatibility with both modern object-oriented software paradigms and parallel processing techniques. Follow-on work will address performance on distributed-memory computer clusters, and may also consider distributed computing using GPU-based systems.

References

1. Chopard, B., Droz, M.: Cellular Automata Modeling of Physical Systems. Collection Aléa-Saclay 6, xii, 341 p. Cambridge University Press, Cambridge, England/New York (1998)
2. von Neumann, J.: Theory of Self-Reproducing Automata. University of Illinois Press, Urbana (1966)
3. Hopman, R.K., Leamy, M.J.: Triangular cellular automata for computing two-dimensional elastodynamic response on arbitrary domains. J. Appl. Mech. Trans. Asme. **78**(2), 021020 (2011)
4. Leamy, M.J.: Application of cellular automata modeling to seismic elastodynamics. Int. J. Solids Struct. **45**(17), 4835–4849 (2008)
5. Noer, G.: Cygwin: a free win32 porting layer for UNIX applications. In: Proceedings of the 2nd USENIX Windows NT Symposium, Seattle, WA, USA, Usenix Association (1998)
6. Baker, M., Carpenter, B., Shafi, A.: MPJ express: towards thread safe Java HPC. In: IEEE International Conference on Cluster Computing, Barcelona, Spain (2007)
7. Paszynski, M., Kurtz, J., Demkowicz, L.: Parallel, fully automatic hp-adaptive 2d finite element package. Comput. Methods Appl. Mech. Eng. **195**(7–8), 711–741 (2006)

Wave Localisation in Structured Elastic Plates

S.G. Haslinger, N.V. Movchan, A.B. Movchan, and R.C. McPhedran

Abstract The paper presents the results of recent work on the modelling of flexural waves in elastic plates constrained periodically by rigid pins. It includes an outline of the analysis of the transmission problem for a stack of rigid pin gratings incorporating lateral shifts. We use a recurrence algorithm to determine reflection and transmission matrices which characterise the filtering of plane waves by the structured interface. The representations of scattered fields use the quasi-periodic Green's function for a single grating. Both propagating and evanescent fields are taken into account. A special attention is given to the analysis of trapped modes which may exist within the system of rigid pin gratings. Analytical findings are accompanied by numerical examples.

Keywords Biharmonic operator • Flexural waves • Trapped waves

1 Introduction: Geometry and Governing Equations

The paper extends the results of the earlier work [1, 2] on transmission of time-harmonic waves by periodic gratings. The emphasis here is on the case of flexural vibrations of thin elastic plates constrained periodically by several rows of rigid pins acting as a structured interface. The phenomenon we are interested in is the enhancement in transmission by the system, and its dependence on the internal

S.G. Haslinger (✉) • N.V. Movchan • A.B. Movchan
Department of Mathematical Sciences, University of Liverpool, L69 7ZL, UK
e-mail: sgh@liv.ac.uk

R.C. McPhedran
School of Physics, University of Sydney, Sydney, NSW 2006, Australia

J. Náprstek et al. (eds.), *Vibration Problems ICOVP 2011: The 10th International Conference on Vibration Problems*, Springer Proceedings in Physics 139,
DOI 10.1007/978-94-007-2069-5_99,

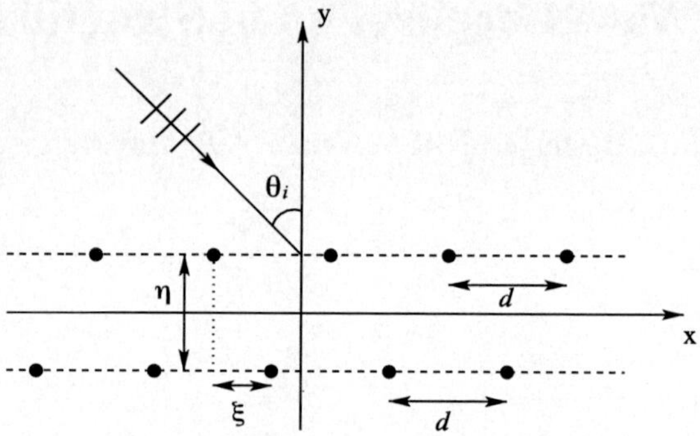

Fig. 1 Interface consisting of a stack of two gratings of rigid pins with relative horizontal shift ξ. The periodicity of each grating is d and the separation of the gratings is denoted by η. Plane waves are incident on the upper grating with angle of incidence θ_i

structure of the interface. In this analysis we also make a connection with the work [3], which contains approximations of "trapped modes" within a pair of gratings in an elastic plate.

We consider a thin elastic plate containing periodic gratings. Each grating consists of an infinite number of rigid pins arranged periodically with spacing d apart. A finite number of these gratings are arranged parallel to one another, with separation η. In this paper we study vertical (varying η) and horizontal shifts (varying ξ) of these gratings and compare results with those derived in [2] for stacks in which all gratings are directly below one another. A pair of shifted gratings representing a structured interface is shown in Fig. 1.

We model the propagation of flexural waves through the structured interface. We consider plane waves that propagate freely through the homogeneous material until they reach the stack of gratings whereupon they are reflected and transmitted. We derive expressions for the reflected and transmitted fields, in which we allow for both evanescent and propagating waves. Special attention is given to vibration modes, which enhance transmittance across the stack of gratings.

1.1 Governing Equations

Let W be a solution of the scattering problem for the biharmonic operator (see [2]):

$$\Delta^2 W(\mathbf{x}) - \beta^4 W(\mathbf{x}) = 0, \tag{1}$$

where $\beta^2 = \omega\sqrt{\rho h/D}$ with $D = Eh^3/(12(1-\nu^2))$ (the flexural rigidity of the plate), ρ is the mass density, ω is the angular frequency, and h is the plate thickness. In addition, E is the Young modulus and ν is the Poisson ratio.

The boundary conditions on the circular boundaries of the inclusions are Dirichlet clamping conditions:

$$W\Big|_{r=a} = 0, \quad \frac{\partial W}{\partial r}\Big|_{r=a} = 0. \tag{2}$$

In what follows, we consider the limit as $a \to 0$, corresponding to an array of rigid pins constraining the plate.

For the grating stack problem, the displacement field W is the sum of the incident wave and the scattered field (to be determined):

$$W = W^{(i)} + W^{(s)}. \tag{3}$$

The field W satisfies the Bloch quasi-periodicity condition along the horizontal x-axis:

$$W(\mathbf{x} + md\mathbf{e}^{(1)}) = W(\mathbf{x})e^{i\alpha_0 md}, \tag{4}$$

where $m \in \mathbb{Z}$, d is the period, and α_0 is the Bloch parameter, $\alpha_0 = \beta \sin \theta_i$, where θ_i is the angle of incidence (see Fig. 1).

The solution of (1) is conveniently expressed in the form

$$W(r,\theta) = W_H(r,\theta) + W_M(r,\theta), \tag{5}$$

where W_H and W_M satisfy the Helmholtz and modified Helmholtz equations respectively,

$$(\Delta + \beta^2)W_H = 0, \qquad (\Delta - \beta^2)W_M = 0. \tag{6}$$

The incident field is represented by plane waves of two types:

1. Propagating (and evanescent) solution of the Helmholtz equation

$$W_{i,\,H}(\mathbf{x}) = \frac{1}{\sqrt{|\chi_0|}} \exp\{i(\alpha_0 x - \chi_0 y)\}, \tag{7}$$

where $\alpha_0^2 + \chi_0^2 = \beta^2$. Here χ_0 is real and positive for a propagating solution. For the evanescent solution, χ_0 is pure imaginary, with positive imaginary part.

2. Evanescent solution of the modified Helmholtz equation

$$W_{i,\,M}(\mathbf{x}) = \frac{1}{\sqrt{|\hat{\chi}_0|}} \exp\{i(\alpha_0 x - \hat{\chi}_0 y)\}, \tag{8}$$

where $\alpha_0^2 + \hat{\chi}_0^2 = -\beta^2, \ \ \hat{\chi}_0 = i\tau_0, \ \ \tau_0 > 0.$

As in [2], we also use similar plane wave series expansions to describe the reflected and transmitted waves.

2 Recurrence Algorithm for Shifted Gratings

We follow a similar procedure to that outlined in [2]. For a single grating, we define the matrices $\mathbf{R}^{\pm}$ and $\mathbf{T}^{\pm}$ to characterise the reflection and transmission of an incident wave meeting the grating from above (+) or below (−).

$$\mathbf{R}^{\pm} = \begin{pmatrix} \mathbf{R}^{\pm}_{HH} & \mathbf{R}^{\pm}_{HM} \\ \mathbf{R}^{\pm}_{MH} & \mathbf{R}^{\pm}_{MM} \end{pmatrix}, \quad \mathbf{T}^{\pm} = \begin{pmatrix} \mathbf{T}^{\pm}_{HH} & \mathbf{T}^{\pm}_{HM} \\ \mathbf{T}^{\pm}_{MH} & \mathbf{T}^{\pm}_{MM} \end{pmatrix}. \tag{9}$$

We then build a stack of gratings by placing an additional grating on top of the preceding layers.

2.1 *Propagation Matrices for Shifted Gratings*

When a grating is a component of a stack, it is convenient to adjust the phase origin to halfway between successive grating elements using propagation matrices. We define two propagation matrices, one for the vertical direction and the other to account for the horizontal shifts. With the phase origins for reflected and transmitted fields placed at $y = \pm\eta/2$ and $x = \xi/2$ where ξ represents the relative horizontal shift, the propagation matrices are defined as follows:

$$\mathscr{P} = \begin{pmatrix} P & 0 \\ 0 & P \end{pmatrix}, \quad \text{where} \quad P = [\delta_{tp} e^{i\tilde{\chi}_p \eta/2}], \tag{10}$$

with $\tilde{\chi}_p = \chi_p$ if p corresponds to a Helmholtz-type wave, and $\tilde{\chi}_p = \hat{\chi}_p$ if p corresponds to a plane wave of modified Helmholtz type, and

$$\mathscr{Q} = \begin{pmatrix} Q & 0 \\ 0 & Q \end{pmatrix}, \quad \text{with} \quad Q = [\delta_{tp} e^{-i\alpha_p \xi/2}], \tag{11}$$

where $\alpha_p = \alpha_0 + \frac{2\pi p}{d}$ varies with the order p of the wave, with p being an integer.

2.2 *Recurrence Relations for a Stack of Gratings*

For our recurrence procedure, we have similar relations for the reflection and transmission coefficients for a stack of $s+1$ gratings (see [2]), but the introduction of $\mathscr{Q}$ and its conjugate affect the forms of the matrices $\mathscr{R}^{\pm}$ and $\mathscr{T}^{\pm}$ with phase origins at $x = \xi/2$ and $y = \pm\eta/2$. We use three pairs of matrices in the recurrence procedure and subsequent numerical calculations:

$$\begin{aligned}
\mathscr{R}_1^+ &= \overline{\mathscr{Q}}\mathscr{P}\mathbf{R}_1^+\mathscr{P}\overline{\mathscr{Q}} \quad \text{and} \quad \mathscr{T}_1^+ = \overline{\mathscr{Q}}\mathscr{P}\mathbf{T}_1^+\mathscr{P}\overline{\mathscr{Q}},\\
\mathscr{R}_1^- &= \overline{\mathscr{Q}}\mathscr{P}\mathbf{R}_1^-\mathscr{P}\mathscr{Q} \quad \text{and} \quad \mathscr{T}_1^- = \overline{\mathscr{Q}}\mathscr{P}\mathbf{T}_1^-\mathscr{P}\mathscr{Q},\\
\mathscr{R}_s^+ &= \mathscr{Q}\mathscr{P}\mathbf{R}_s^+\mathscr{P}\overline{\mathscr{Q}} \quad \text{and} \quad \mathscr{T}_s^+ = \mathscr{Q}\mathscr{P}\mathbf{T}_s^+\mathscr{P}\overline{\mathscr{Q}}.
\end{aligned} \tag{12}$$

Here the subscript s corresponds to a stack consisting of s gratings. The above matrices are used to derive the following recurrence relations:

$$\mathscr{R}_{s+1}^+ = \mathscr{R}_1^+ + \mathscr{T}_1^- \mathscr{R}_s^+ (\mathbf{I} - \mathscr{R}_1^- \mathscr{R}_s^+)^{-1} \mathscr{T}_1^+, \tag{13}$$

$$\mathscr{T}_{s+1}^+ = \mathscr{T}_s^+ (\mathbf{I} - \mathscr{R}_1^- \mathscr{R}_s^+)^{-1} \mathscr{T}_1^+, \tag{14}$$

where $\mathscr{R}_i^\pm$, $\mathscr{T}_i^\pm$, $\mathscr{P}$, $\mathscr{Q}$, P and Q are defined by (10)–(12).

The recurrence procedure without the horizontal shifts is described in detail in [2]. To evaluate the matrices $\mathbf{R}_{s+1}^+$, $\mathbf{T}_{s+1}^+$, we re-phase with $\mathscr{Q}$ accordingly.

3 Conclusion: Trapped Modes for a Pair of Gratings

In [2], attention was given to the existence of trapped modes between a pair of gratings, characterised by very sharp transmittance at a certain frequency. The model computation was carried out for an angle of incidence of 30° for a symmetrical pair of gratings where the rigid pins were aligned directly above one another. Here we investigate how shifting the gratings relative to one another affects the frequency at which we observe enhanced transmittance across the stack. We also note that in [3], analysis was conducted to determine explicit representations for trapped modes between a pair of unshifted gratings, for propagating waves only.

In Fig. 2, we show the transmitted energy as a function of β for a pair of gratings of rigid pins with unit periodicity, for various horizontal shifts ξ of one of the layers. For all of the shifts (dotted line represents no shift, solid line is a shift of $0.2d$ and thick line is a shift of $0.4d$), a very sharp transmittance is observed. This enhanced transmittance for a pair of gratings corresponds to a trapped mode. The value β_* of β, for which the trapped mode occurs, increases with increasing shift up to half of the period ($\xi = 0.5d$), where it attains its maximum. Its minimum (given in [2]) corresponds to the case of zero shift. The frequency β_* as a function of the horizontal shift ξ is shown in Fig. 3. The curve is symmetric about $\xi = 0.5$, as expected for a pair of identical gratings incorporating a horizontal shift.

Further analysis will include investigating the quality factors of the resonant transmittances for various shifts. This will determine the internal structure of

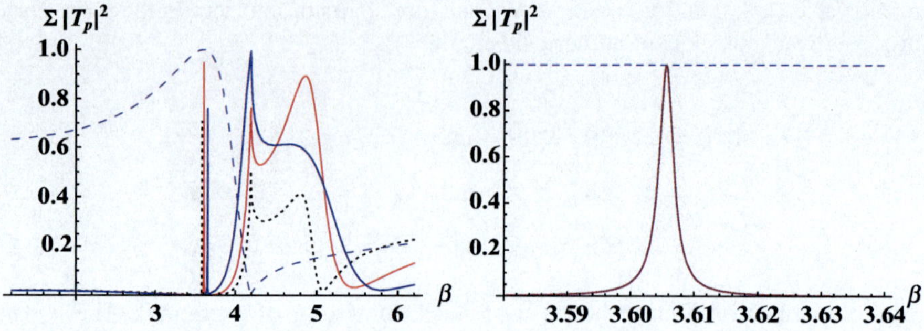

Fig. 2 (*Left*) Normalised transmitted energy $\sum|T_p|^2$ as a function of β for pairs of shifted gratings of rigid pins with $\xi = 0,\ 0.2d,\ 0.4d$, and normalised reflected energy for a single grating (*dashed line*). For all gratings, $d = 1$, $\eta = 1$ and $\theta_i = 30°$. The unshifted case (*dotted line*) has the lowest frequency corresponding to enhanced transmittance. As the shift ξ increases from $0.2d$ (*solid line*) to $0.4d$ (*thick line*), this frequency β_* increases. (*Right*) Detail in the neighborhood of the enhanced transmittance $\beta_* = 3.6056$, for $\xi = 0.2d$

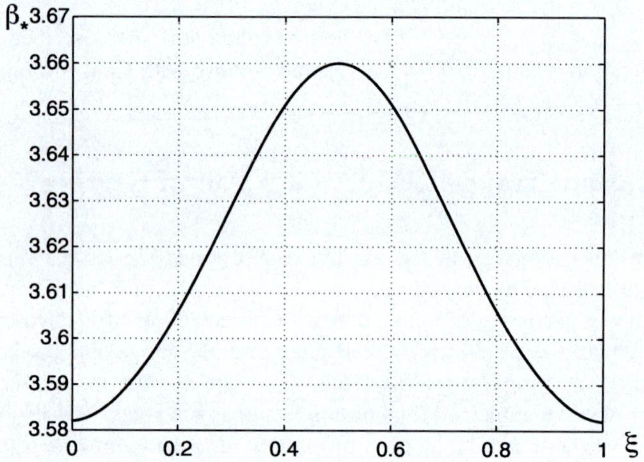

Fig. 3 The frequency β_* of enhanced transmittance versus horizontal shift ξ

the interface required to produce maximum transmittance. We shall study stacks incorporating more gratings and various alternative arrangements of the gratings which may provide interesting applications in wave-guiding and filtering of flexural waves in plates.

Acknowledgements SGH gratefully acknowledges the financial support of the Duncan Norman Charitable Trust through the Duncan Norman Research Scholarship. The participation of R.C.M. in this project was facilitated by a grant from the Research Centre in Mathematics and Modelling of the University of Liverpool.

References

1. Botten, L.C., Nicorovici, N.A., McPhedran, R.C., Martijn de Sterke, C., Asatryan, A.A.: Photonic band structure calculations using scattering matrices. Phys. Rev. E **64**, 046603 (2001)
2. Movchan, N.V., McPhedran, R.C., Movchan, A.B., Poulton, C.G.: Wave scattering by platonic grating stacks. Proc. R. Soc. A **465**, 3383–3400 (2009)
3. Evans, D.V., Porter, R.: Penetration of flexural waves through a periodically constrained thin elastic plate in *vacuo* and floating on water. J. Eng. Math. **58**, 317–337 (2007)

Part X
Flow-Induced Vibrations, Fluid-Structure Interaction, Vibroacoustics, Aeroelasticity

Time-Delayed Feedback Control for Flutter of Supersonic Aircraft Wing

Shu Zhang, Yu Huang, and Jian Xu

Abstract An active control technique called servo delayed feedback control is proposed to control the flutter of supersonic aircraft wing. It's motivated to increase the critical flow velocity. Firstly, the servo delayed feedback control is designed based on a two-dimensional airfoil so that delayed differential equations are modelled for the controlled system under consideration. Then, the stability of the system without time delay and with time delayed feedback control are considered analytically and flutter boundary of the parameters in the delayed feedback control system is predicted when time delay varies. Finally, numerical simulation for time domain with MATLAB/SIMULINK software is made to demonstrate the effectiveness of the theoretical result. The results show that, critical flow velocity can be increased by regulating the quantity of time delay and the provided strategy of delayed feedback to control the flutter in supersonic aircraft wing system is not only valid but also easily applied to engineering structures.

Keywords Flutter • Stability • Supersonic • Time delayed feedback control

1 Introduction

Flutter instability, which takes place when hypersonic vehicle achieves supersonic long-range flights, may decrease aircraft performance [1]. Therefore, studying the flutter properties and control is in great significance. Aeroelasticity is the field that describes the response and stability properties of physical systems under the interaction of structural, inertial and aerodynamic forces. Various aerodynamic loads have

S. Zhang • Y. Huang • J. Xu (✉)
Tongji University, Shanghai, China
e-mail: zhsh886@yahoo.com.cn; huangyu@tongji.edu.cn; xujian@tongji.edu.cn

J. Náprstek et al. (eds.), *Vibration Problems ICOVP 2011: The 10th International Conference on Vibration Problems*, Springer Proceedings in Physics 139,
DOI 10.1007/978-94-007-2069-5_100,

been developed over the past decades [2–4]. For example, the Theodorsen function or the Wagner function is used by many authors to model the aerodynamic loads [5–7].

If one considers time delay in a control loop, it is well known that such delay has an important effect on dynamical behaviour [8–10]. Librescu et al. [11] considered the open/closed-loop aeroelasticity of 2-D lifting surfaces in an incompressible flow field and the time delay in a control loop. They emphasized the implications of the time delay on the feedback control and its complex role. Their result represented that for small time delays the system remained stable, but with increasing delay the response became unbounded, indicated that aeroelastic instability occurred. Following Librescu's work, Zhao [12] investigated also the traditional models in aero-elastic control for a 2-dof airfoil in low speed flow field. Both Librescu's and Zhao's works indicated that the delay in the control loop could induce the instability of the airfoil which resulted in the flutter. However, in our past research, it was found that delay could be taken as a variable parameter or an active control parameter. Thus, time delay may be used as a simple but efficient measure to control motions of a system for different applications [8, 13]. Motivated by such idea, we plan to show that the delay in the control loop may also benefit for flutter suppression of the airfoil and provide the method to determine the range of the delay to realize the control.

2 Aero-Elastic Model with Delayed Feedback Control

The structure in which airfoil oscillates in pitch and plunge is shown in Fig. 1.

In this paper, we assume that the flow is incompressible to simplify the theoretical analysis so that we can focus on the possibility of suppressing the flutter by utilizing the delay in supersonic flights. Assume that the actuator is located in the place

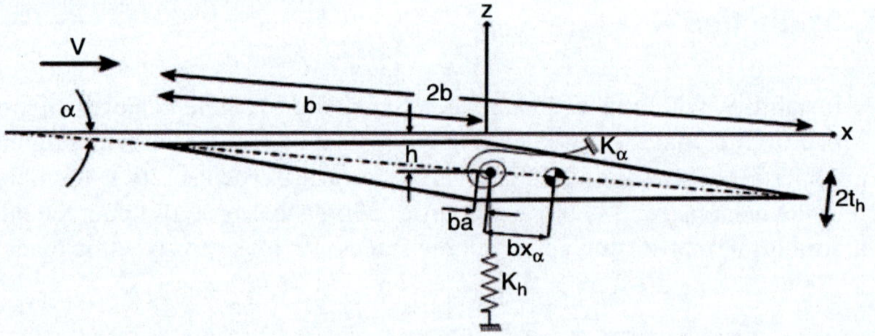

Fig. 1 Scheme for a 2-dof airfoil, where V is the flow velocity, b the semi-chord length, ba the distance from the elastic axis Z to mid-chord, bx_α the distance of the elastic axis from center of mass, the plunging deflection is denoted by h, positive down ward direction, the pitching angle is α, positive nose up and k_h and k_α are plunging and torsional stiffnesses, respectively

which has a distance from mid-chord, denoted by x_1 and the control strategy is implemented by the plunging velocity as a time-delayed feedback signal. Then the aero-elastic model with delayed feedback control for the airfoil in supersonic flow field is given by

$$\begin{cases} m\ddot{h} + mx_\alpha b\ddot{\alpha} + k_h h = -4c_\infty \rho b \left(\dot{h} - ba\dot{\alpha} + V\alpha\right) + g_v \dot{h}(t-\tau), \\ mx_\alpha b\ddot{h} + mr_\alpha^2 \ddot{\alpha} + k_\alpha \alpha = -4c_\infty \rho b^2 \left[\frac{b}{3}\dot{\alpha} - a\left(\dot{h} - ba\dot{\alpha} + V\alpha\right)\right] \\ \qquad + g_v \left(x_1 - ba\right) \dot{h}\left(t - \tau\right), \end{cases} \tag{1}$$

where m is airfoil mass, c_∞ the speed of sound, ρ the air density, $V = M_a c_\infty$, r_α the radius of inertia, g_v the velocity feedback gain and $\tau \geq 0$ the time delay. (1) can be rewritten as

$$\dot{\mathbf{X}} = \mathbf{A}\left(Ma\right)\mathbf{X} + \mathbf{B}\left(g_v\right)\mathbf{X}(t-\tau), \tag{2}$$

where $\mathbf{X} = \left[h \ \alpha \ \dot{h} \ \dot{\alpha}\right]^{\mathrm{T}}$, $\mathbf{A}$ and $\mathbf{B}$ are given by the coefficients in (1).

3 Stability Analysis

It follows from (1) that $\mathbf{X}^* = (0,0,0,0)$ is an equilibrium. To investigate the stability of the trivial equilibrium, substituting $\mathbf{X} = \bar{\mathbf{X}}e^{\lambda t}$ into (2) yields

$$D\left(\lambda,\tau\right) = \det\left(\lambda\mathbf{I} - \mathbf{A}\left(Ma\right) - \mathbf{B}\left(g_v\right)e^{-\lambda\tau}\right) = H\left(\lambda\right) + L\left(\lambda\right)e^{-\lambda\tau} = 0 \tag{3}$$

where $H(\lambda)$ and $L(\lambda)$ are polynomial functions of λ. For the case $\tau = 0$, one has $D\left(\lambda,\tau\right) = H\left(\lambda\right) + L\left(\lambda\right)$. Then, one can employ the Routh-Hurwitz criterion to analyze the stability of the equilibrium and determine the critical value of M_a.

Notice that we are exploring the possibility that whether there exist some values of the delay and the feedback gain such that the critical velocity for the airfoil flutter can be increased. Thus, we assume that for $\tau = 0$, there is a pair of complex eigenvalues on the right half plane and other eigenvalues locate in the left half plane. We hope that the complex eigenvalue in the right half plane crosses imaginary axis back to the left half plane and the other eigenvalues still locate in the left half plane for $\tau \neq 0$. To this end, assuming that one of the eigenvalues is given by $\lambda = \mathrm{i}\omega$ and substituting it into (3), one may solve that $\sin\left(\omega\tau\right) = \Delta_1(\omega)$ and $\cos\left(\omega\tau\right) = \Delta_2(\omega)$. Using $\sin^2\left(\omega\tau\right) + \cos^2\left(\omega\tau\right) = 1$ yields $F\left(\omega\right) = 0$ where $F\left(\omega\right)$ is a polynomial function of ω. It should be noticed that there exist positive real roots of ω for $D\left(\mathrm{i}\omega,\tau\right) = 0$ and $\tau \geq 0$ if and only if $F\left(\omega\right) = 0$. Substituting such ω into $\sin\left(\omega\tau\right) = \Delta_1(\omega)$, one can obtain the critical value of the delay, denoted by τ_c, which is expressed as a function of the critical velocity. For $\tau = \tau_c$, the complex eigenvalues mentioned above cross the imaginary axis and enter the left half plane when $\mathrm{Re}\left(\frac{\mathrm{d}\lambda}{\mathrm{d}\tau}\,|_{\tau_c}\right) < 0$.

4 Numerical Examples

To illustrate that the provided method mentioned above is feasible, we give some numerical examples in this section. The parameters in use for the numerical simulation are taken from [6] as follows: $m = 209.95kg, c_\infty = 340m/s, \alpha = 0.15$, $r_\alpha = 0.5m$, $b = 0.5m$, $ba = 0.075m$, $\rho = 1.2256kg/m^3$, $k_h = 1552788N/m$, $k_\alpha = 956584Nm/rad$, $bx_\alpha = 0.125m$, $x_1 = -0.5m$.

For the case without the delayed feedback ($\tau = 0$), using Routh-Hurwitz criterion, one can obtain that the critical velocity for the flutter of the airfoil is $3.1M_a$. The eigenvalues of (1) change with the variation of the flight speed, as shown in Fig. 2a. Using SIMULINK package in software MATLAB, we plot the time history of the plunging deflection and the pitching angle for $M_a = 2.8$, 3.1 and 3.2, respectively, as shown in Fig. 2b, c.

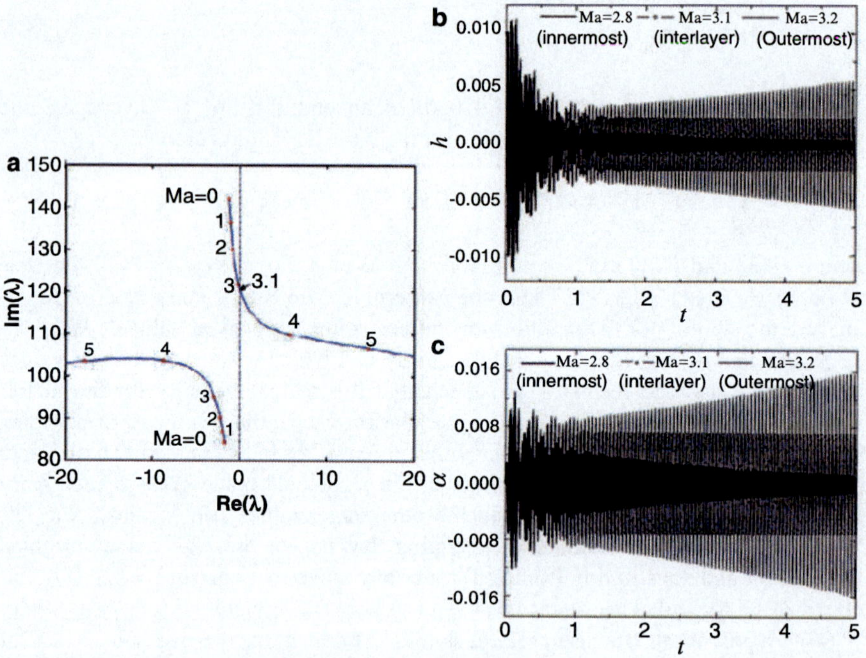

Fig. 2 Distributions of eigenvalues of Eq. 1 in **(a)** and time history of Eq. 1 for **(b)** plunging deflection and **(c)** the pitching angle, all figures for $\tau = 0$

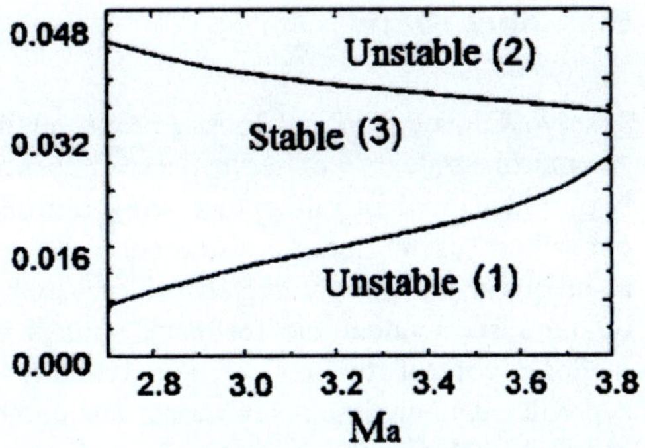

Fig. 3 Critical boundary in (Ma, τ), where stable and unstable regions are shown for $g_v = -270$

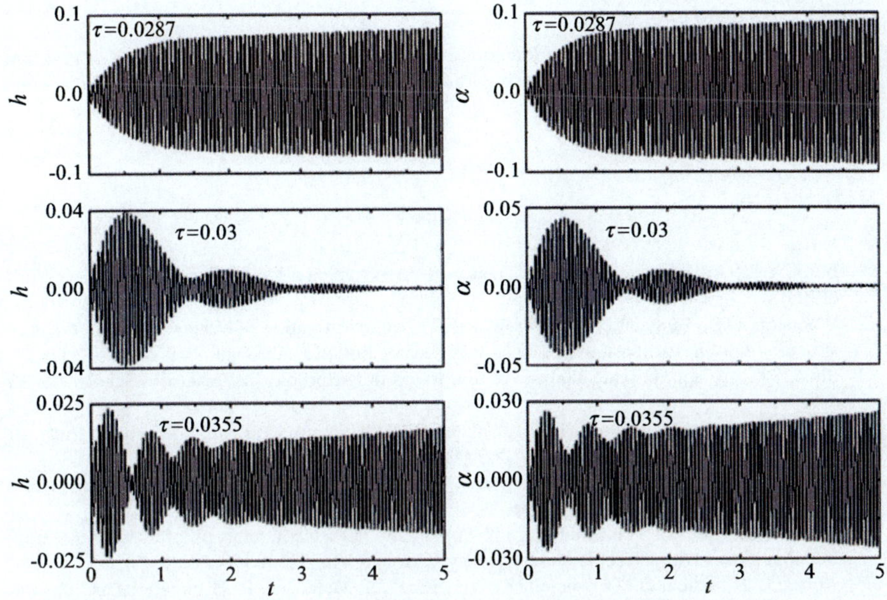

Fig. 4 Time history plots of (1) by SIMULINK for the system with delayed feedback control when $g_v = -270$, $Ma = 3.8$ for plunging deflection (*left*) and pitching angle (*right*)

For the case $\tau \neq 0$, one can establish the relation between the critical delay and the critical flight velocity by the procedure in section "Stability Analysis" for any given feedback gain, as shown in Fig. 3.

It can be seen from Fig. 3 that a suitable choice for values of the delay can increase the flight critical velocity efficiently. The trivial equilibrium is unstable in regions (1) and (2), and stable in region (3). To verify the results in Fig. 3, we plot the time history of the plunging deflection and the pitching angle for different values of the delay which locate in three regions of Fig. 3, using SIMULINK package in software MATLAB, as shown in Fig. 4. It follows from Fig. 4 that the analytical result is in good agreement with the numerical simulation.

5 Conclusions

A servo delayed feedback loop is proposed to control the flutter of the 2-dof airfoil. A control strategy, i.e. servo delayed feedback control, is designed based on a two-dimensional airfoil, so that delayed differential equations are modelled for the controlled system under consideration. We show that the delay can be considered as an active control parameter to increase the critical flight velocity. The stability of the system under the feedback control without time-delay and time delayed feedback control are considered analytically and flutter boundary of the parameters is predicted when time delay varies. The numerical simulation for time domain with MATLAB/SIMULINK software is performed to demonstrate the effectiveness of the theoretical result.

Acknowledgments The research work reported here was made possible by the State Key Program of National Natural Science of China under Grant No. 11032009 and Shanghai Leading Academic Discipline Project in No. B302.

References

1. Fung, Y.C.: An Introduction to the Theory of Aeroelasticity. Dover Publications, New York (1993)
2. Woolston, D.S., Runyan, H.L., Andrew, R.E.Y.: An investigation of effects of certain types of structural nonlinearities on wing and control surface flutter. J. Aeronaut. Sci. **24**, 57–63 (1957)
3. Shen, S.F.: An approximate analysis of non-linear flutter problems. J. Aerosp. Sci. **26**, 25–32 (1959)
4. Jr McIntosh, S.C., Jr Reed, R.E., Rodden, W.P.: Experimental and theoretical study of nonlinear flutter. J. Aircr. **18**, 1057–1063 (1981)
5. Yang, Z.C., Zhao, L.C.: Analysis of limit cycle flutter of an wing section in incompressible flow. J. Sound Vib. **123**, 1–13 (1998)
6. Hyun, D.H., Lee, I.: Transonic and low-supersonic aeroelastic analysis of a two-degree-of-freedom airfoil with a freeplay non-linearity. J. Sound Vib. **234**, 859–880 (2000)
7. Abbas, L. K., Chen, Q., O'Donnell, K., Valentine, D., Marzocca, P.: Bifurcations and chaotic behavior of hypersonic lifting surfaces with freeplay including the stiffness and damping nonlinearities. In: Proceedings of the 48th AIAA/ASME/ASCE/AHS/ASC Structures, Structural Dynamics, and Materials Conference, Honolulu, HI, pp. 23–26 (2007)
8. Xu, J., Chung, K.W.: Effects of time delayed position feedback on a van der Pol–Duffing oscillator. Phys. D **180**, 17–39 (2003)
9. Xu, J., Yu, P.: Delay-induce bifurcations in a nonautonomours system with delayed velocity feedbacks. Int. J. Bifurc. Chaos **14**, 2777–2798 (2004)
10. Xu, J., Chung, K.W., Chan, C.L.: An efficient method for studying weak resonant double Hopf bifurcation in nonlinear systems with delayed feedbacks. SIAM J. Appl. Dyn. Syst. **6**, 29–60 (2007)
11. Librescu, L., Marzocca, P., Silva, W.A.: Aeroelasticity of 2-D lifting surfaces with time-delayed feedback control. J. Fluids Struct. **20**, 197–215 (2005)
12. Zhao, Y.H.: Stability of a two-dimensional airfoil with time-delayed feedback control. J. Fluids Struct. **25**, 1–25 (2009)
13. Zhao, Y.Y., Xu, J.: Effects of delayed feedback control on nonlinear vibration absorber system. J. Sound Vib. **308**, 212–230 (2007)

Investigation of the Flutter Suppression by Fuzzy Logic Control for Hypersonic Wing

Dongxu Li, Qing Luo, and Rui Xu

Abstract This paper presents a fundamental study of flutter characteristics and control performance of an aeroelastic system based on a two-dimensional double wedge wing in the hypersonic regime. Dynamic equations were established based on the modified third order nonlinear piston theory and some nonlinear structural effects are also included. A set of important parameters are observed. And then aeroelastic control law is designed to suppress the amplitude of the LCOs for the system in the sub/supercritical speed range by applying fuzzy logic control on the input of the deflection of the flap. The overall effects of the parameters on the aeroelastic system were outlined. Nonlinear aeroelastic responses in the open- and closed-loop system are obtained through numerical methods. The simulations show fuzzy logic control methods are effective in suppressing flutter and provide a smart approach for this complicated system.

Keywords Aeroelasticity • LCO • Hypersonic • Nonlinear • Fuzzy logic control

1 Introduction

Nowadays, hypersonic aeroelasticity are becoming a hot topic since the great demands of the high speed flight. Flutter, which belongs to dynamic aeroelastic area, is the most concerned issue as it can cause catastrophic problems to the whole vehicle. The system can exhibit different dynamic responses, such as Limit Cycle Oscillations (LCO) or chaotic vibration, owning to the existence of the

---

D. Li • Q. Luo (✉) • R. Xu
National University of Defense Technology, Changsha, China
e-mail: dongxuli@263.net; qingluo321@gmail.com; alpc_xurui@139.com

J. Náprstek et al. (eds.), *Vibration Problems ICOVP 2011: The 10th International Conference on Vibration Problems*, Springer Proceedings in Physics 139,
DOI 10.1007/978-94-007-2069-5_101, © Springer Science+Business Media B.V. 2011

nonlinearities [1]. The contribution of the nonlinearities can be either detrimental or beneficial [2]. Determination of the flutter boundary and forecast of the dynamic behavior around the critical states [3] are the main interests of the aeroelasticity scholars in the past [4]. Librescu [5] studied the dynamic behavior of a 2-D wing model in the vicinity of the flutter boundary and derived the expression of computing the linear flutter speed through Hopf-bifurcation theory. Abbas [6] had researched systematically the bifurcation of the LCOs considering both the structural and aerodynamic nonlinearities and presented the evolution of the complex dynamic response under variable flight speed. Active flutter control has been studied widely [4] in the recent days because of the limitations of the passive control methods, such as penalties of mass or structure [2]. As a kind of smart control methodology, fuzzy logic control usually has advantages over other control methods when the precise statements of the system can't be given out [7]. G. Karpouzian [8] implemented a non-adaptive fuzzy logic controller on a two degree airfoil, and revealed its power of solving inherently complex problems.

In this paper, general equations of aeroelastic model of 2-D wing with a flap in hypersonic regime are derived using the piston theory. Parametric studies are achieved by comparing amplitude of the LCOs of different system parameters that characterize the whole system. Finally, a kind of fuzzy logic control strategy is implemented on the aeroelastic system, and the simulation demonstrated the effectiveness and robustness of this smart control methodology.

2 Aeroelastic Model

Figure 1 shows typical 2-D section of a hypersonic wing. Using the Lagrange Equations, the aeroelastic dynamic equations can be found as follows [9]. The inertial effects of the control surface are ignored. Thus the flap is not involved in the dynamics of the wing but simply as an excitation device. The closed-loop aeroelastic

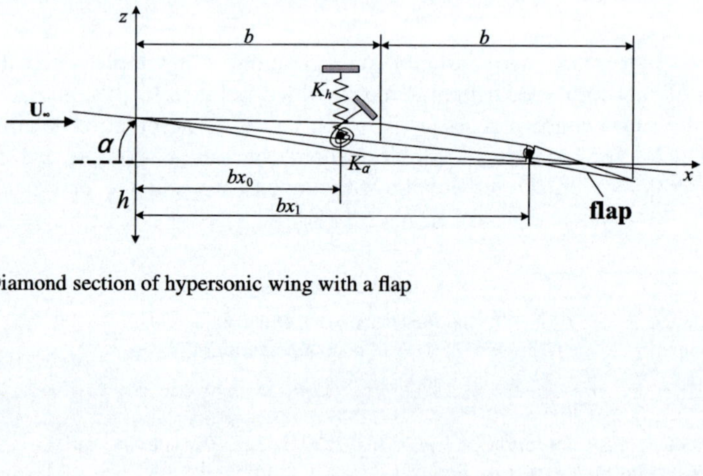

Fig. 1 Diamond section of hypersonic wing with a flap

governing equations of a 2-D wing featuring plunging-pitching -flapping motion in dimensionless form are,

$$\ddot{\xi}(\tau) + \chi_\alpha \ddot{\alpha}(\tau) + 2\zeta_h \left(\frac{\omega'}{V}\right) \dot{\xi}(\tau) + \left(\frac{\omega'}{V}\right)^2 \xi(\tau) = l_a(\tau) + l_c(\tau)$$

$$\left(\frac{\chi_\alpha}{r_\alpha{}^2}\right) \ddot{\xi}(\tau) + \ddot{\alpha}(\tau) + \frac{2\zeta_\alpha}{V} \dot{\alpha}(\tau) + \frac{1}{V^2} \alpha(\tau) + \frac{\eta}{V^2} \alpha^3(\tau) = m_a(\tau) + m_c(\tau) \quad (1)$$

where $\xi(\tau) = h(t)/b$ is the plunging displacement (positive downward), α is the pitching angle (positive nose up), $\tau = U_\infty t/b$, $V = U_\infty/b\omega_\alpha$ is the flight speed (or flow velocity); $\omega' = \omega_h/\omega_\alpha$ is the ratio of the plunging frequency and pitching frequency, μ is reduced mass ratio; χ_α static unbalance about the elastic axis, r_α is the gyration radius of the wing section; $\zeta_h = c_h/2m\omega_h$ and $\zeta_\alpha = c_\alpha/2I_\alpha\omega_\alpha$ are the linear plunging and pitching damping coefficients; l_a and l_c are aerodynamic and control force, m_a and m_c aerodynamic and control moment, respectively.

Considering there are nonlinearities of cubic stiffness in the pitching degree while the plunging stiffness is linear. Pitching moment can be expressed as [5],

$$\overline{M}_\alpha = K_\alpha \alpha(t) + \eta K_\alpha \alpha^3(t) \quad (2)$$

η is the nonlinear stiffness factor. For hypersonic regimes, piston theory is an efficient tool for the aeroelastic analysis in an engineering view [10]. Local pressure can be obtained through Taylor's series in the third-order approximation [11]. The aerodynamic force and moment can be obtained through integration of the pressure on the upper and lower surface of the wing [2, 5].

Then rewrite the Eq. 1 in the state space form.

$$\begin{aligned} \dot{\mathbf{X}} &= \bar{\mathbf{A}}\mathbf{X} + \mathbf{P}\psi(y) + \mathbf{G}\beta + \mathbf{G}_d\dot{\beta} \\ y &= \mathbf{C}^{\mathrm{T}}\mathbf{X} \end{aligned} \quad (3)$$

Where β is flapping displacement, and all the auxiliary matrix can be found in reference [5, 11].

3 Fuzzy Logic Control for the Aeroelastic System

Fuzzy logic control is very effective for the situation in which the controlled systems are very complex. In this study, Mamdani-type fuzzy inference system, which consists of two inputs and one output, was constructed by using the Fuzzy Toolbox in Matlab. The controller receives the pitching displacement and the pitching velocity, while gives the control flapping displacement as a control variable. Triangular and trapezoidal shape membership functions are chosen both for inputs and output. The membership function plots are shown in Fig. 2. All rules have

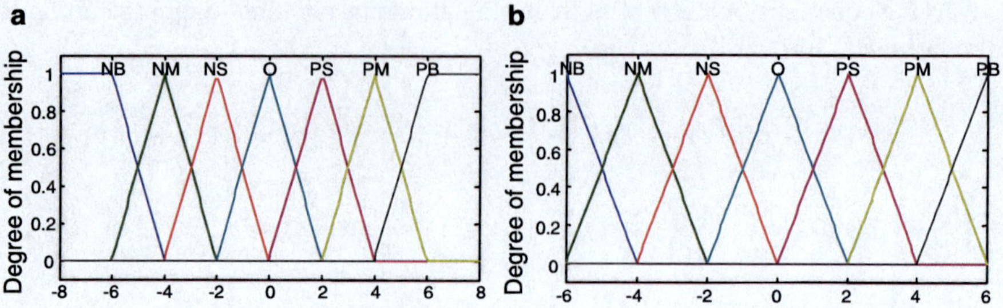

Fig. 2 Membership function plots. (**a**) Pitching displacement and velocity, (**b**) Flapping displacement

Table 1 Fuzzy IF-THEN rule base

$\dot{\alpha}\|\beta\|\alpha$	NB	NM	NS	O	PS	PM	PB
NB	NB	NB	NM	NM	NS	O	O
NM	NB	NB	NM	NM	NS	O	O
NS	NB	NB	NM	NS	O	OM	PM
O	NB	NM	NS	O	PS	PM	PB
PS	NM	NM	O	PS	PM	PB	PB
PM	O	O	PS	PM	PM	PB	PB
PB	O	O	PS	PM	PM	PB	PB

Table 2 Structural and aerodynamic parameters

Parameters	Value	Parameters	Value	Parameters	Value
b	1.5 (m)	r_α	0.5	**H** (altitude)	10 (km)
x_0	0.65	χ_α	0.25	ρ_∞	0.4135 (kg/m^3)
x_1	0.75	η	20	γ	1.4
ω'	1.0	ζ_h, ζ_α	0.0,0.0	μ	100
ω_α	80 (rad/s)				

weights to 1 and use the AND-type logical operator, shown in Table 1. These rules are constructed by employing the knowledge about the oscillating wing.

4 Results and Discussion

4.1 *Characteristic of LCOs Around the Critical Points*

The baseline parameters of the system are shown in Table 2. The simulations are separated into two parts: first, fundamental studies of the effects of the system parameters on the amplitude of the LCOs; second, a simple example is considered through the aeroelastic response of the closed loop system under different flight speed to estimate the effectiveness the fuzzy logic control. The initial pitching placement $\alpha_0 = 0.08722$ *rad*, and the others are all zero.

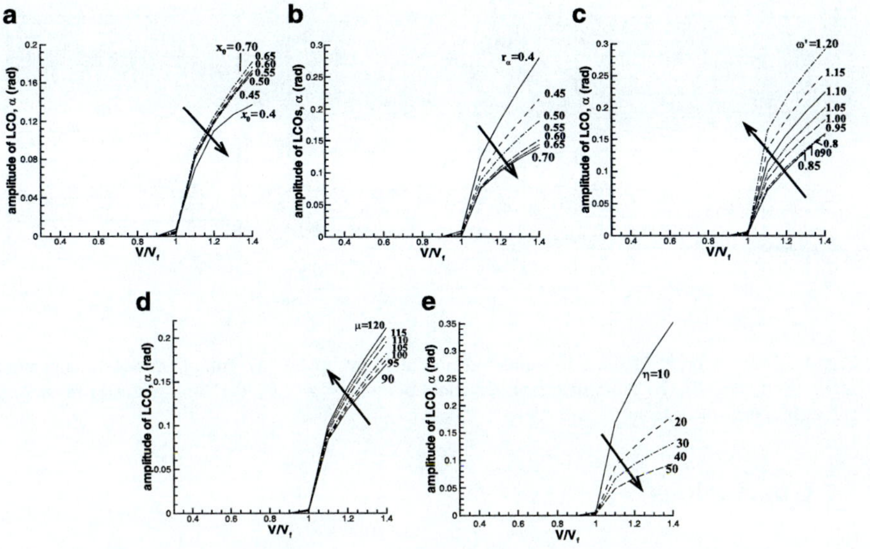

Fig. 3 Effects of system parameters on the amplitude of the LCOs. (**a**) Elastic axis, (**b**) gyration radius, (**c**) frequency ratio, (**d**) reduced mass ratio, (**e**) nonlinear stiffness factor

Figure 3 shows the effects of system parameters on the amplitude of the LCOs in the vicinity of the critical states. Generally, the amplitude of LCOs increases as the flight speed increases and the existence of the nonlinearities can both delay and precede the LCOs. In the Fig. 3a, c, d, as the parameters (x_0, ω' and μ) increases, the amplitude increases. While in Fig. 3b, e, as the parameters (r_α and η) increases, the amplitude decreases. For different parameters the amplitude of the LCOs exhibit differently in the subcritical and supercritical states.

4.2 *Control Effects*

Aeroelastic responses are obtained by solving Eq. 3 under different flight speed conditions in Fig. 4. $V/V_f < 1$ and $V/V_f > 1$ denotes the subcritical and supercritical states, respectively. Apparently, the control effects are excellent. In Fig. 4a, b, the control flapping displacement becomes zero when the system is under control. While in Fig. 4c, we can see that, for increasing the flight speed, the flap (control surface) can't 'calm down' even when the system is stable, which is not favorable for the real system during the flight.

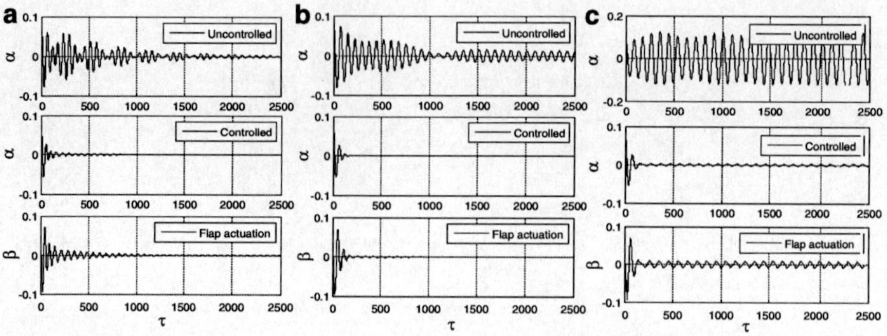

Fig. 4 Time-histories of the wing under different flight speed (**a**) Time-histories in subcritical regime ($v/v_f = 0.8$), (**b**) Time-histories in critical point ($v/v_f = 1.0$), (**c**) Time-histories in subcritical regime ($v/v_f = 1.2$)

5 Conclusions

An aeroelastic model of 2-D wing with a flap was developed. Both fundamental parametric studies and active control design were conducted. Generally, the parameters investigated above contribute greatly to the whole system. Through the numerical simulation, the fuzzy logic control reveals its effectiveness of suppressing the flutter, while disadvantages also exist which can be seen in section "Control Effects". But it can be a powerful tool to expand the flight envelope if we do have a 'feel' of the behavior of the system.

References

1. Dowell, E.H., Edwards, J.W., et al.: Nonlinear aeroelasticity. J. Aircr. **40**(5), 857–74 (2003)
2. Rao, V.M., Behal, A., Marocca, P., et al.: Adaptive aeroelastic vibration suppression of a supersonic airfoil with flap. Aerosp. Sci. Technol. **10**, 309–15 (2006)
3. Marzocca, P., Librescu, L., et al.: Flutter, postflutter, and control of a supersonic wing section. J. Guid. Control. Dyn. **25**(5), 962–970 (2002)
4. Mukhopadhyay, V.: Historical perspective on analysis and control of aeroelastic responses. J. Guid. Control. Dyn. **26**(5), 673–684 (2003)
5. Librescu, L., Chiocchia, G., et al.: Implications of cubic physical/aerodynamic nonlinearities on the character of the flutter instability boundary. Int. J. NonLinear Mech. **38**, 173–199 (2003)
6. Abbas, L.K., Chena, Q., et al.: Numerical studies of a non-linear aeroelastic system with plunging and pitching freeplays in supersonic/hypersonic regimes. Aerosp. Sci. Technol. **11**, 405–418 (2007)
7. Ross, J.R.: Fuzzy Logic with Engineering Applications. McGraw-Hill, New York (1995)
8. Karpouzian, G.: Fuzzy control of an aeroelastic wing section. AIAA-98-1797 (1998)
9. Wright, J.R., Cooper, J.E.: Introduction to Aircraft Aeroelasticity and Loads. Wiley, Chichester (2007)
10. Ashley, H., Zartarian, G.: Piston theory – a new aerodynamic tool for the aeroelastician. J. Aerosp. Sci. **23**(10), 1109–1118 (1956)
11. Friedmann, P.P., McNamara, J.J., et al.: Aeroelastic analysis of hypersonic vehicles. J. Fluids Struct. **19**, 681–712 (2004)

On the Interaction of Compressible Flow and Human Vocal Folds

J. Prokopová, M. Feistauer, J. Horáček, A. Kosík, and V. Kučera

Abstract The paper is devoted to the numerical solution of interaction of compressible flow with elastic structure with application to the simulation of flow in vocal folds. It is described by the coupling of compressible Navier-Stokes equations and dynamical elasticity system, solved by the combination of the discontinuous Galerkin method for the solution of the flow field and conforming finite elements for the elasticity problem.

Keywords Biomechanics of voice • Discontinuous Galerkin finite element method • Fluid-structure interaction • ALE method

1 Introduction

The simulation of flow-induced vibrations of elastic body plays an important role in many areas, for example development of aircrafts and turbines, in civilengineering, car industry or medicine.

The paper is concerned with the numerical simulation of the biomechanical problem of compressible flow in interaction with human vocal folds. The flow field is described by the compressible Navier-Stokes equations considered in a time dependent domain of the form of a channel representing the human vocal tract. The governing equations written in the ALE form are discretized by the discontinuous Galerkin finite element method. It is well-known that this technique is more suitable

J. Prokopová (✉) • M. Feistauer • A. Kosík • V. Kučera
Charles University, Faculty of Mathematics and Physics, Sokolovská 83, 18675 Praha 8,

J. Prokopová • J. Horáček • A. Kosík
Institute of Thermomechanics, Academy of Sciences of the Czech Republic, v.v.i., Dolejškova 1402/5, 182 00 Praha 8,
e-mail: jarkaprokop@post.cz

J. Náprstek et al. (eds.), *Vibration Problems ICOVP 2011: The 10th International Conference on Vibration Problems*, Springer Proceedings in Physics 139,
DOI 10.1007/978-94-007-2069-5_102,

for the solution of the singularly perturbed problems of fluid dynamics than standard finite element methods [1]. The deformation of the structure is described by the dynamical linear elasticity equations. They are discretized by the conforming finite element method. Flow and elasticity problems are coupled via interface conditions. Some numerical results are presented.

2 Governing Equations for Flow in a Time Dependent Domain

We deal with compressible flow in a bounded domain $\Omega_t^f \subset I\!R^2$ depending on time $t \in [0, T]$. We assume that the boundary of Ω_t^f consists of four disjoint parts $\partial\Omega_t^f = \Gamma_I \cup \Gamma_O \cup \Gamma_{W_t} \cup \Gamma_F$, where Γ_I and Γ_O represent the inlet and outlet, respectively, Γ_{W_t} represents moving impermeable walls and Γ_F represents fixed impermeable walls.

Compressible viscous flow is described by the continuity equation, the Navier-Stokes equations and the energy equation, written in the conservative form [1]

$$\frac{\partial w}{\partial t} + \sum_{s=1}^{2} \frac{\partial f_s(w)}{\partial x_s} = \sum_{s=1}^{2} \frac{\partial R_s(w, \nabla w)}{\partial x_s} \text{ in } \Omega_t^f,\ t \in [0, T], \tag{1}$$

where

$$w = (\rho, \rho v_1, \rho v_2, E)^T \in I\!R^4,$$

$$f_s(w) = (\rho v_s, \rho v_1 v_s + \delta_{1s} p, \rho v_2 v_s + \delta_{2s} p, (E + p) v_s)^T,\ s = 1, 2,$$

$$R_s(w, \nabla w) = \left(0, \tau_{s1}^V, \tau_{s2}^V, \tau_{s1}^V v_1 + \tau_{s2}^V v_2 + k \frac{\partial \theta}{\partial x_s}\right)^T,\ s = 1, 2,$$

$$\tau_{ij}^V = \lambda \delta_{ij} \operatorname{div} v + 2\mu d_{ij}(v),\ d_{ij}(v) = \frac{1}{2}\left(\frac{\partial v_i}{\partial x_j} + \frac{\partial v_j}{\partial x_i}\right),\ i, j = 1,$$

We use the following notation: ρ – density, p – pressure, E – total energy, $\mathbf{v} = (v_1, v_2)$ – velocity vector, θ – absolute temperature, $c_v > 0$ – specific heat at constant volume, $\gamma > 1$ – Poisson adiabatic constant, $\mu > 0, \lambda$ – viscosity coefficients, $k > 0$ – heat conduction coefficient, τ_{ij}^V – components of the viscous part of the stress tensor. We set $\lambda = -2\mu/3$. The vector functions $\boldsymbol{f}_s$ are inviscid fluxes of the quantity $\mathbf{w}$ in the directions x_s and $\boldsymbol{R}_s$ represent viscous terms.

System (1) is completed by the thermodynamical relations

$$p = (\gamma - 1)\left(E - \rho\frac{|v|^2}{2}\right),\quad \theta = \frac{1}{c_v}\left(\frac{E}{\rho} - \frac{1}{2}|v|^2\right). \tag{2}$$

In order to simulate flow in a time-dependent domain, we employ the Arbitrary Lagrangian-Eulerian (ALE) method [2]. This method is based on the use of a regular, one-to-one ALE mapping $\mathscr{A}_t$ of the reference configuration $\Omega^f_{ref} = \Omega^f_0$ onto the computational domain Ω_t^f at time t (the so-called current configuration). In virtue of this we rewrite system (1) in the ALE form

$$\frac{D^{\mathscr{A}} w}{Dt} + \sum_{s=1}^{2} \frac{\partial g_s(w)}{\partial x_s} + w \mathrm{div} z = \sum_{s=1}^{2} \frac{\partial R_s(w, \nabla w)}{\partial x_s}, \tag{3}$$

where g_s, $s = 1,2$, are modified inviscid fluxes $g_s(w) = f_s(w) - z_s w$, $s = 1,2$, $D^{\mathscr{A}}/Dt$ denotes the so-called ALE-derivative and $z = \partial \mathscr{A}_t / \partial t$ is the domain velocity [3].

System (3) is equipped with the initial condition $\boldsymbol{w}(\boldsymbol{x},0) = \boldsymbol{w}^0(\boldsymbol{x})$, $\boldsymbol{x} \in \Omega_t{}^f$, and the boundary conditions:

$$\text{Inlet } \Gamma_I\text{:} \quad \rho|_{\Gamma_I \times (0,T)} = \rho_D, \quad v|_{\Gamma_I \times (0,T)} = v_D = (v_{D1}, v_{D2}),$$

$$\sum_{j=1}^{2} \left(\sum_{i=1}^{2} \tau_{ij}^V n_i \right) v_j + k \frac{\partial \theta}{\partial n} = 0 \quad \text{on } \Gamma_I \times (0,T);$$

$$\text{Fixed impermeable wall } \Gamma_F\text{:} \quad v = 0, \quad \frac{\partial \theta}{\partial n} = 0;$$

$$\text{Moving wall } \Gamma_{W_t}\text{:} \quad v = z(t), \quad \frac{\partial \theta}{\partial n} = 0;$$

$$\text{Outlet } \Gamma_O\text{:} \quad \sum_{i=1}^{2} \tau_{ij}^V n_i = 0, \quad \frac{\partial \theta}{\partial n} = 0, \quad j = 1,2. \tag{4}$$

Here $\boldsymbol{n}$ is the unit outer normal to $\partial \Omega_t{}^f$. We have $\Gamma_{W_t} = \mathscr{A}_t(\Gamma_W)$, where we denote $\Gamma_W = \Gamma_{W_0}$.

The space discretization of the problem is carried out by the discontinuous Galerkin finite element method. For the time discretization we use a semi-implicit scheme, where the ALE-derivative is approximated by the first-order backward difference and the remaining terms are treated with the aid of a linearization and time extrapolation [3, 4].

3 The Dynamical Problem of Elasticity

In the sequel, $\Omega^b \subset IR^2$ will denote a bounded domain representing an elastic body. Elastic deformations of this body are described by the dynamic equations

$$\rho^b \frac{\partial^2 u_i}{\partial t^2} + C\rho^b \frac{\partial u_i}{\partial t} - \sum_{j=1}^{2} \frac{\partial \tau_{ij}^b}{\partial x_j} = 0, \quad \text{in } \Omega^b \times (0,T), \quad i = 1,2, \tag{5}$$

where $u(x,t) = (u_1(x,t), u_2(x,t))$, $x \in \Omega^b$, $t \in (0,T)$, is the displacement, $(\tau_{ij}^b)_{i,j=1}^2$ represents the stress tenzor fulfilling the generalized Hooke law for isotropic material

$$\tau_{ij}^b = \tilde{\lambda} \operatorname{div} u \delta_{ij} + 2\tilde{\mu} e_{ij}, \quad i,j = 1,2, \tag{6}$$

with the Lamé coefficients $\tilde{\lambda}$, $\tilde{\mu}$. Often the Young modulus E^b and the Poisson ratio σ^b are used. They are defined by $E^b = \frac{\tilde{\mu}(3\tilde{\lambda}+2\tilde{\mu})}{\tilde{\lambda}+\tilde{\mu}}$, $\sigma^b = \frac{\tilde{\lambda}}{2(\tilde{\lambda}+\tilde{\mu})}$. Further, $e_{ij}(u) = \frac{1}{2}\left(\frac{\partial u_i}{\partial x_j} + \frac{\partial u_j}{\partial x_i}\right)$, $i,j = 1,2$, are the components of the strain tensor and ρ^b is the density of the solid material.

In reality, mechanical systems dissipate mechanical energy. The expression $C\rho^b \frac{\partial u_i}{\partial t}$, where $C \geq 0$, is a dissipative damping of the system. The formulation of the dynamic elasticity problem (5) is completed by the initial conditions $u(x,0) = 0$ and $\frac{\partial u}{\partial t}(x,0) = 0$, $x \in \Omega^b$, and boundary conditions on the boundary $\partial\Omega^b = \Gamma_W \cup {\Gamma_D}^b$, where $\Gamma_W \cap {\Gamma_D}^b = \emptyset$, $\Gamma_W = \Gamma_{W_0}$ and ${\Gamma_D}^b$ is a fixed part of the boundary:

$$\sum_{j=1}^{2} \tau_{ij}^b n_j = T_i^n \quad \text{on } \Gamma_W \times (0,T), \qquad i = 1,2, \tag{7}$$

$$u = 0 \quad \text{on } \Gamma_D^b \times (0,T). \tag{8}$$

By T_i^n, $i = 1,2$, we denote the components of the normal stress. In order to derive the space-time discretization, we formulate the problem in a weak sense and apply the finite element method using continuous piecewise linear elements. This leads to a second order system of ordinary differential equations. For the time discretization we use the Newmark scheme.

4 Coupled Problem

Up to now the fluid flow and the deformation of the elastic body have been considered as two separate problems. Now we need to take into account the mutual interaction of the fluid and the body on the common boundary

$$\Gamma_{W_t} = \left\{x \in I\!R^2;\ x = X + u(X,t),\ X \in \Gamma_W\right\}.$$

The ALE mapping $\mathcal{A}_t$ is determined with the aid of a special stationary linear elasticity problem. Then the domain velocity z is computer at time t, the flow problem is solved and the aerodynamical surface force

$$T_i^n = -\sum_{j=1}^{2} \tau_{ij}^f n_j, \quad i = 1,2, \tag{9}$$

acting on the body Ω_b on Γ_W is determined. Here $\boldsymbol{n}$ is the unit outer normal to the domain $\partial\Omega^b$ and $\tau_{ij}^f = -p\delta_{ij} + \tau_{ij}^V$, $i, j = 1, 2$, are the components of the stress tensor of the fluid.

The coupling procedure follows this outline: Let the domain $\Omega_{t_n}^f$, the ALE mapping $\mathscr{A}_{t_n}$ and the domain velocity z are known at time t_n. Then we solve the flow problem in the domain $\Omega_{t_n}^f$ and obtain the approximate solution $\boldsymbol{w}_h$. This allows us to compute the components $\tau_{ij}{}^f$ of the stress tensor and the surface force acting on the domain Ω^b on the part of the boundary Γ_W can be determined. Solving the problem of elasticity we get the approximate displacement $\boldsymbol{u}_h$ at time t_n. In this way we get the domain $\Omega_{t_{n+1}}^f$, compute the ALE mapping $\mathscr{A}_{t_{n+1}}$ and the approximation of domain velocity z_h at time t_{n+1}. Then we come to the next time step t_{n+1}. (This algorithm represents the so-called loose (weak) coupling. More complex and accurate is the strong coupling.)

5 Numerical Experiments

Here we present numerical results obtained with the aid of the described coupling technique and applied to the interaction of airflow in the domain Ω_t^f, representing the human vocal tract, with human vocal folds represented by the domain Ω^b. The length of the domain Ω_t^f is 0.065 m and the inlet width of this domain is 0.0087 m.

We use the same time step $\tau = 4.35 \cdot 10^{-5}$ s for the flow problem and the structural problem. The computational process starts by the solution of the flow problem in the domain $\Omega_{t_0}^f$ at the initial time $t_0 = 0$ s. We use the data $\mu = 1.5 \cdot 10^{-5}\,\mathrm{kg\,m^{-1}\,s^{-1}}$, $k = 2.428 \cdot 10^{-2}\,\mathrm{kg\,m\,s^{-3}\,K^{-1}}$, $c_v = 721.428\,\mathrm{m^2\,s^{-2}\,K^{-1}}$, $Re = 10453$. At the inlet we prescribe the velocity vector $v_{in} = (4, 0)\mathrm{m\,s^{-1}}$ and the density $\rho_{in}^f = 1.225\,\mathrm{kg\,m^{-3}}$. At the outlet the pressure $p_{out} = 97611$ Pa is prescribed.

We assume that the vocal folds are isotropic bodies with the constant material density $\rho^b = 1040\,\mathrm{kg\,m^{-3}}$ and the damping coefficient $C = 0.1\ \mathrm{s^{-1}}$. The values of the Young modulus E^b and the Poisson ratio σ^b are different in four subdomains of Ω^b. See Fig. 1.

Figure 2 shows horizontal and vertical displacement of the elastic body.

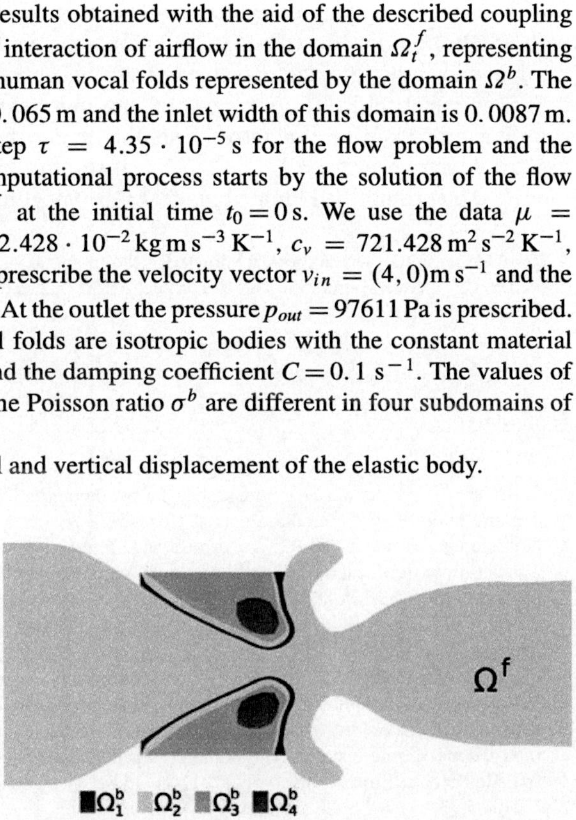

Fig. 1 The scheme of the vocal tract: The material parameters are $E^b = 100$ kPa, $\sigma^b = 0.4$ in Ω_1^b, $E^b = 1$kPa, $\sigma^b = 0.495$ in Ω_2^b, $E^b = 8$ kPa, $\sigma^b = 0.4$ in Ω_3^b, $E^b = 12$ kPa, $\sigma^b = 0.4$ in Ω_4^b

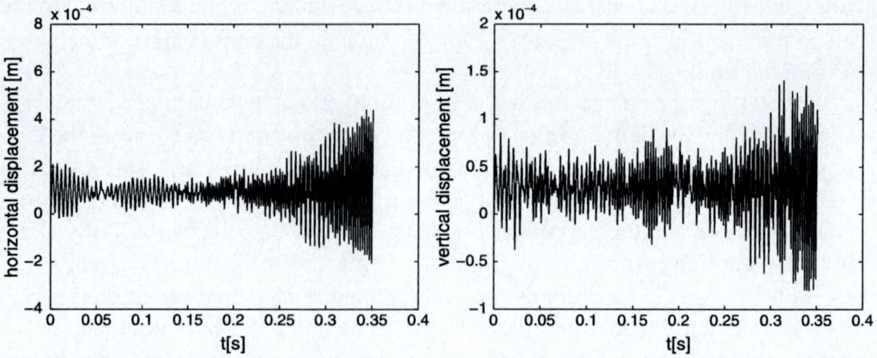

Fig. 2 Horizontal and vertical displacement of a point lying inside the area $\Omega_3{}^b$ of the upper vocal fold

6 Conclusion

The developed method for the numerical simulation of the interaction of fluid flow with an elastic body, described by the compressible Navier-Stokes equations and the dynamic elasticity system, was successfully applied to the simulation of air flow in human vocal folds. The structure of the flow field computed by the described coupling procedure is qualitatively similar to the flow patterns obtained in the case of driven vocal folds [3] and to a realistic threshold of aeroelastic instability modelled by a simplified lumped model of the vocal folds [5].

Acknowledgements The authors acknowledge the financial support of the project P101/11/0207 of the Czech Science Foundation and the project MSM 0021620839 of the Ministry of Education of the Czech Republic.

References

1. Feistauer, M., Felcman, J., Straškraba, I.: Mathematical and Computational Methods for Compressible Flow. Clarendon Press, Oxford (2003)
2. Nomura, T., Hughes, T.J.R.: An arbitrary Lagrangian-Eulerian finite element method for interaction of fluid and a rigid body. Comput. Methods Appl. Mech. Eng. **95**, 115–138 (1992)
3. Feistauer, M., Česenek, J., Horáček, J., Kučera, V., Prokopová, J.: DGFEM for the numerical solution of compressible flow in time dependent domains and applications to fluid-structure interaction. ECCOMAS CFD 2010, Lisbon (2010)
4. Feistauer, M., Kučera, V., Prokopová, J.: Discontinuous Galerkin solution of compressible flow in time-dependent domains. Math. Comput. Simulat. **80**, 1612–1623 (2010)
5. Sváček, P., Horáček, J.: Numerical approximations of flow induced vibrations of vocal folds. In: International Conference on Numerical Analysis and Applied Mathematics (ICNAAM 2010), AIP Conference Proceedings, vol. 1281, pp. 131–134. American Institute of Physics, Melville (2010)

Numerical Simulation of Interaction of Human Vocal Folds and Fluid Flow

A. Kosík, M. Feistauer, J. Horáček, and P. Sváček

Abstract Our goal is to simulate airflow in human vocal folds and their flow-induced vibrations. We consider two-dimensional viscous incompressible flow in a time-dependent domain. The fluid flow is described by the Navier-Stokes equations in the arbitrary Lagrangian-Eulerian formulation. The flow problem is coupled with the elastic behaviour of the solid bodies. The developed solution of the coupled problem based on the finite element method is demonstrated by numerical experiments.

Keywords Fluid-structure interaction • Incompressible viscous flow • ALE method • Biomechanics of voice

1 Introduction

The phonation onset in biomechanics of voice is an important characteristic of human voice production. From the point of view of the flow-induced vibrations the phonation onset can be characterized as a state of the system when it is loosing the aeroelastic stability, i.e. when the airflow parameters like subglottal pressure or airflow rate cross a limit value and the system becomesunstable by flutter.

A. Kosík (✉) • M. Feistauer
Charles University in Prague, Faculty of Mathematics and Physics, Prague, Czech Republic
e-mail: adam.kosik@atlas.cz; feist@karlin.mff.cuni.cz

J. Horáček
Academy of Sciences of the Czech Republic, v. v. i., Institute of Thermomechanics,
e-mail: jaromirh@it.cas.cz

P. Sváček
Czech Technical University in Prague, Department of Technical Mathematics, Prague, Czech Republic
e-mail: Petr.Svacek@fs.cvut.cz

J. Náprstek et al. (eds.), *Vibration Problems ICOVP 2011: The 10th International Conference on Vibration Problems*, Springer Proceedings in Physics 139,
DOI 10.1007/978-94-007-2069-5_103, © Springer Science+Business Media B.V. 2011

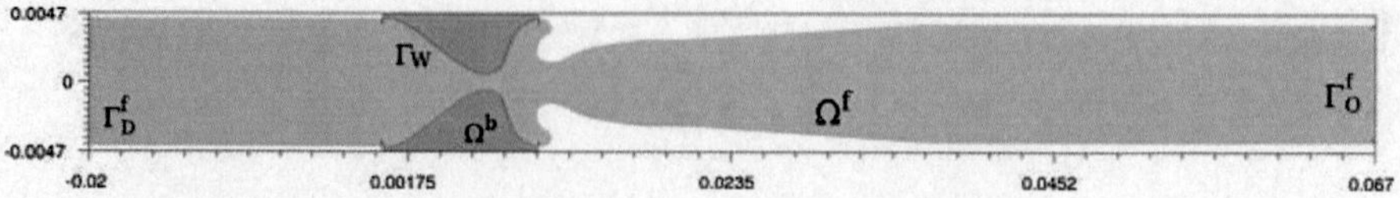

Fig. 1 The domain occupied by the fluid and the parts of the boundary

Frequency-modal analysis of a simplified three-mass model of the vocal fold in interaction with a potential flow separated at the superior edge of the vocal fold showed that the two eigenfrequencies are coupled when the instability occurs [1]. A similar linear stability analysis was performed on a 2-D continuum model of vocal folds in potential flow by Zhang [2]. Here the instability threshold is studied in the time domain using 2-D FE model of the vocal folds coupled with 2-D FE model of the viscous incompressible flow (Fig. 1).

2 The Equations Describing Fluid Flow in a Moving Domain

We deal with incompressible viscous flow in a bounded domain $\Omega_t^f \subset \mathbb{R}^2$ depending on time $t \in [0, T]$. By $\boldsymbol{v} = \boldsymbol{v}(\boldsymbol{x}, t)$ we denote the velocity and by $p = p(\boldsymbol{x}, t)$ the kinematic pressure (i.e., pressure divided by the density of the fluid ρ^f), $\boldsymbol{x} \in \Omega_t^f$, $t \in (0, T)$ and η denotes the kinematic viscosity. The incompressible viscous flow is described by the system of the Navier-Stokes equations equipped with initial and boundary conditions [3].

In order to simulate flow in a moving domain, we employ the Arbitrary Lagrangian-Eulerian (ALE) method. This method is based on a special mapping of the reference configuration Ω_0^f onto the deformed, actual configuration Ω_t^f. We reformulate the Navier-Stokes equations in the ALE form [4]:

$$\frac{D^{\mathscr{A}}}{Dt}\boldsymbol{v} + ((\boldsymbol{v} - \boldsymbol{w}) \cdot \nabla)\,\boldsymbol{v} + \nabla p - \eta \Delta \boldsymbol{v} = 0 \quad \text{in } \Omega_t^f, \tag{1}$$

$$\nabla \cdot \boldsymbol{v} = 0 \quad \text{in } \Omega_t^f. \tag{2}$$

Here $\frac{D^{\mathscr{A}}}{Dt}$ is so-called ALE derivative and $\boldsymbol{w}$ denotes the domain velocity.

System (1)–(2) is equipped with the initial condition

$$\boldsymbol{v}(\boldsymbol{x}, 0) = \boldsymbol{v}_0, \quad \boldsymbol{x} \in \Omega_0^f, \tag{3}$$

and boundary conditions.

We assume that $\partial\Omega_t^f = \Gamma_D^f \cup \Gamma_O^f \cup \Gamma_{W_t}$, where Γ_D^f, Γ_O^f and Γ_{W_t} are mutually disjoint. On Γ_D^f, representing the inlet and impermeable fixed walls, we prescribe the Dirichlet boundary condition, on the impermeable moving walls Γ_{Wt} we assume that the fluid velocity $\boldsymbol{v}$ is equal to the domain velocity of the elastic body and on the outlet Γ_O^f we prescribe the so-called "do-nothing" boundary condition:

$$\boldsymbol{v}|_{\Gamma_D^f} = \boldsymbol{v}_D, \quad \text{on } \Gamma_D^f, \tag{4}$$

$$\boldsymbol{v}|_{\Gamma_{W_t}} = \boldsymbol{w}|_{\Gamma_{W_t}}, \quad \text{on } \Gamma_{W_t}, \tag{5}$$

$$-(p - p_{ref})\boldsymbol{n} + \eta \frac{\partial \boldsymbol{v}}{\partial \boldsymbol{n}} = 0, \quad \text{on } \Gamma_O^f. \tag{6}$$

Here $\boldsymbol{n}$ is the unit outer normal to $\partial\Omega_t^f$ and p_{ref} is a prescribed reference outlet kinematic pressure.

There are several possibilities how to carry out the space-time discretization. In order to develop a stable, accurate scheme, which can easily treat complicated boundaries, we apply the finite element method (FEM). Here, the Taylor-Hood P_2/P_1 finite element pair satisfying the Babuška-Brezzi condition is used. In order to avoid spurious oscillations in approximate solutions in the case of high Reynolds numbers we apply the *streamline diffusion method* together with *div-div* stabilization of pressure. For the time discretization we use a second-order two-step backward difference formula using the computed approximate solution $\boldsymbol{v}^{n-1}$ in $\Omega_{t_{n-1}}$ and $\boldsymbol{v}^n$ in Ω_{t_n} for the calculation of $\boldsymbol{v}^{n+1}$ in the domain $\Omega_{t_{n+1}}$. For the numerical solution of incompressible flow we use the program FEMFLUID developed by P. Sváček.

3 The Linear Dynamical Problem of Elasticity

In what follows, $\Omega^b \subset \mathbb{R}^2$ will be a bounded domain representing the elastic body of the form of vocal folds. We denote by $\boldsymbol{u}(\boldsymbol{x}, t)$, $\boldsymbol{x} \in \Omega^b$, $t \in (0, T)$, the displacement of the body and the strain tensor as

$$e_{ij}(\boldsymbol{u}) = \frac{1}{2}\left(\frac{\partial u_i}{\partial x_j} + \frac{\partial u_j}{\partial x_i}\right), \quad i, j = 1, 2. \tag{7}$$

The deformation of the vocal folds is modelled by the generalized Hooke law for isotropic bodies [5]:

$$\tau_{ij}^b = \lambda \operatorname{div} \boldsymbol{u} \delta_{ij} + 2\mu e_{ij}, \quad i, j = 1, 2, \tag{8}$$

where $(\tau_{ij}^b)_{i,j}^2 = 1$ denotes the stress tensor and λ and μ are the Lamé coefficients related to the Young modulus E and to the Poisson ratio σ as

$$E = \frac{\mu(3\lambda + 2\mu)}{\lambda + \mu}, \quad \sigma = \frac{\lambda}{2(\lambda + \mu)}. \tag{9}$$

The dynamic equations of an elastic body have the form

$$\rho^b \frac{\partial^2 u_i}{\partial t^2} + C\rho^b \frac{\partial u_i}{\partial t} - \sum_{j=1}^{2} \frac{\partial \tau_{ij}^b}{\partial x_j} = 0, \quad \text{on } (0,T) \times \Omega^b, \quad i = 1, 2. \tag{10}$$

The expression $C\rho^b \frac{\partial u_i}{\partial t}$, where $C \geq 0$, is a dissipative damping of the system and ρ^b denotes the density of the solid material. We complete the elasticity problem with initial and boundary conditions. The initial conditions read

$$\boldsymbol{u}(\cdot, 0) = \boldsymbol{0}, \quad \frac{\partial \boldsymbol{u}}{\partial t}(\cdot, 0) = \boldsymbol{0}, \quad \text{in } \Omega^b. \tag{11}$$

Further, let $\partial\Omega^b = \Gamma_W \cup \Gamma_D^b$, where Γ_W and Γ_D^b are two disjoints parts of $\partial\Omega^b$. Let surface force $\boldsymbol{T}$ be prescribed on the boundary Γ_W and let the part of the boundary Γ_D^b be fixed:

$$\sum_{j=1}^{2} \tau_{ij}^b n_j = T_i^{\boldsymbol{n}} \quad \text{on } \Gamma_W \times (0,T), \qquad i = 1, 2, \tag{12}$$

$$\boldsymbol{u} = 0 \quad \text{on } \Gamma_D^b \times (0,T). \tag{13}$$

We are looking for the displacement $\boldsymbol{u}$ satisfying Eq. (10) and initial and boundary conditions (11)–(13).

Further, we carry out the space-time discretization. We reformulate the problem in a weak sense and apply the finite element method using continuous piecewise linear elements. The semi-discretized problem can be written as a second order system of ordinary differential equations. For the time discretization we apply the Newmark scheme. In each time-step we get a linear algebraic system with symmetric positive definite matrix. The solution of this system was realized by the solver, which is based on the conjugate gradient method.

4 Solution of the Coupled Problem

Up to now we considered that the fluid flow and the deformation of the elastic body are separated processes. On the common boundary of both domains we need to take into account mutual action of the fluid and the body. We denote as the common boundary set Γ_{W_t}, defined as

$$\Gamma_{W_t} = \left\{ \boldsymbol{x} \in \mathbb{R}^2;\ \boldsymbol{x} = \boldsymbol{X} + \boldsymbol{u}(\boldsymbol{X}, t),\ \boldsymbol{X} \in \Gamma_W \right\}. \tag{14}$$

The domain Ω_t^f is defined by the displacement $\boldsymbol{u}$ of the part Γ_W at time t. The ALE mapping $\mathscr{A}_t$ is defined with the aid of a special stationary linear elasticity problem. If we know the computational domain Ω_t^f obtained by the fluid at time t, we can solve the problem describing the flow and we can assign the surface force T^n acting on the body Ω_b on the part of the boundary Γ_W. This surface force is defined by the equation

$$T_i^n = -\sum_{j=1}^{2} \tau_{ij}^f n_j, \quad i = 1, 2, \tag{15}$$

where $\boldsymbol{n}$ is the unit outer normal to the domain $\partial\Omega^b$ and τ_{ij}^f are components of the stress tensor of the fluid:

$$\tau_{ij}^f = \rho^f \left(-p\delta_{ij} + \eta \left(\frac{\partial v_i}{\partial x_j} + \frac{\partial v_j}{\partial x_i} \right) \right), \quad i, j = 1, 2. \tag{16}$$

The procedure of implementation is following. At first the computation of the flow problem is carried out in the domain $\Omega_{t_n}^f$. We get the approximate velocity v_h and pressure p_h and can compute the components of stress tensor τ_{ij}^f. The stress tensor gives us the surface force acting on the domain Ω^b on the part of boundary Γ_W. Further, we solve the elasticity problem and get the approximate displacement u_h at time t_n. In this way we get the domain $\Omega_{t_{n+1}}^f$, compute the ALE mapping $\mathscr{A}_{t_{n+1}}$, and the approximation of domain velocity w_h at time t_{n+1}. Thus we obtain the datum $w|_{\Gamma_{W_{t_{n+1}}}}$ in the boundary condition (5). Now the flow problem can be solved at time t_{n+1}.

5 Numerical Experiments

We consider model of human vocal folds and vocal tract [6]. The vocal folds represented by the domain Ω^b have different material characteristics in different subdomains. We use the same time step $\tau = 5 \cdot 10^{-5}$ s for the solution of the coupled flow and elasticity problem and the input data $\eta = 1.5 \cdot 10^{-5}$ m^2.s^{-1}, $\rho^f = 1.17$ kg.m^{-3}, $C = 0.1$ s^{-1}, the initial velocity $\boldsymbol{v}^0 = 0$ and the inlet and outlet pressure $p_{in} = 600$ Pa, $p_{out} = 0$ Pa.

As for the domain Ω^b we distinguish subdomains with different material characteristics. We consider the material density is the same for all subdomains $\rho^b = 1,040$ kg.m^{-3}, but the values of Young modulus E and Poissone's ratio σ are different, see Fig. 2.

On the other part of boundary Γ_W we prescribe the conditions for the movable boundary. We retrieve the surface forces $\boldsymbol{T}^n$ by solving the problem of fluid flow. The computational process starts by the solution of the flow problem in the domain $\Omega_{t_\alpha}^f$ at the initial time $t_\alpha = -10^{-4}$ s. At time $t = 0$ the structure is released and solution of the complete fluid-structure interaction started.

Figure 3 shows velocity streamlines and displacement of the computational domain at several time instants.

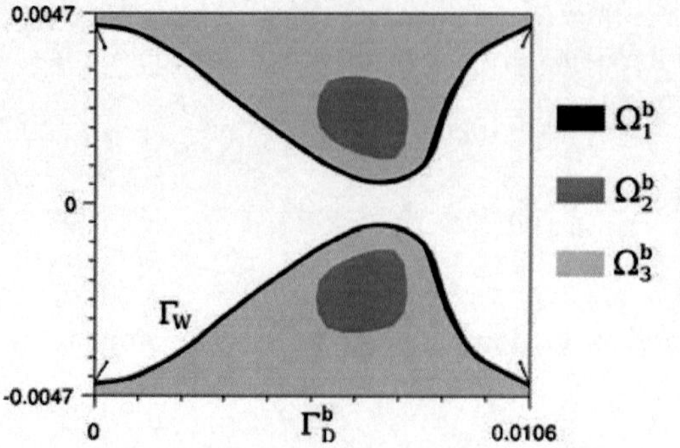

subdomain	E	σ
Ω_1^b	25000	0.49
Ω_2^b	65000	0.4
Ω_3^b	8000	0.49

Fig. 2 The model of vocal fold. Material characteristics

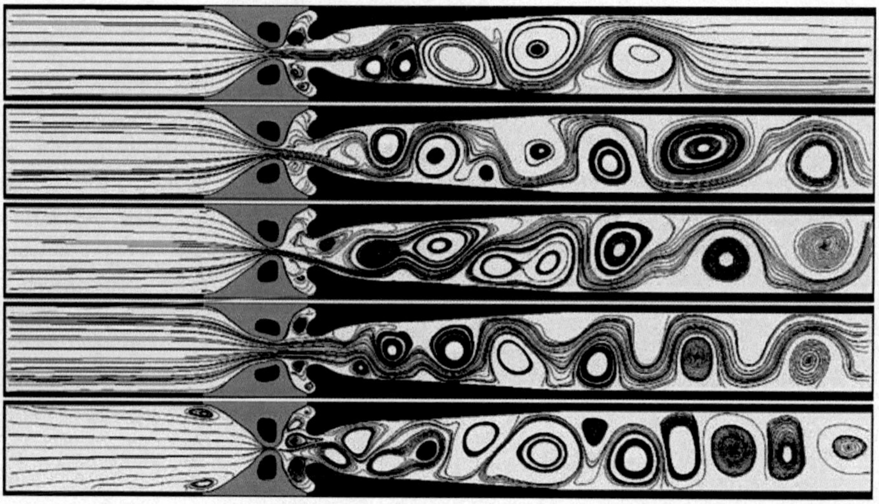

Fig. 3 Streamlines at time instants t = 0.005, 0.0425, 0.04615, 0.0597 and 0.0605 s

6 Conclusion

The numerical method for solving the flow-induced vibration of an elastic body in incompressible viscous flow has been developed and applied to the simulation of airflow in interaction with human vocal folds. The results are in good agreement with previous simulations using simplified models and with known physiological data (see, e.g. [1]).

Future work will be focused on the simulation of the complete closure of the channel and on a non-linear elasticity model in the coupled problem.

Acknowledgements The authors acknowledge the financial support of the projects P101/11/0207 and 201/08/0012 of the Czech Science Foundation.

References

1. Horáček, J., Švec, J.G.: Aeroelastic model of vocal-fold-shaped vibrating element for studying the phonation. J. Fluids Struct. **16**(7), 931–955 (2002). doi:10.1006/jfls.454
2. Zhang, Z.: Characteristics of phonation onset in a two-layer vocal fold model. J. Acoust. Soc. Am. **125**(2), 1091–1102 (2009)
3. Feistauer, M.: Mathematical Methods in Fluid Dynamics. Longman Scientific & Technical, Harlow (1993)
4. Nomura, T., Hughes, T.J.R.: An arbitrary Lagrangian-Eulerian finite element method for interaction of fluid and a rigid body. Comput. Methods Appl. Mech. Eng. **95**, 115–138 (1992)
5. Nečas, J., Hlaváček, I.: Mathematical Theory of Elastic and Elasto-Plastic Bodies, An Introduction. Elsevier, Amsterdam (1981)
6. Feistauer, M., Česenek, J., Horáček, J., Kučera, V., Prokopová, J.: DGFEM for the numerical solution of compressible flow in time dependent domains and applications to fluid-structure interaction. In: Pereira, J.C.F., Sequeira, A. (eds.) ECCOMAS CFD 2010. Lisbon, 14–17 June 2010, CDROM, ISBN 978-989-96778-1-4

Dynamic Stability of a Steady Flow in a Planar Channel

Karolina Bach, Hartmut Hetzler, and Wolfgang Seemann

Abstract In flexible channels with internal flow the steady state may loose stability by divergence or flutter, thus leading to undesirable dynamic behaviour. A lot of investigations have already been done on this subject. The aim of this contribution is to investigate and compare systematically the influence of various parameters on the stability behaviour of the coupled problem. Therefore, a simple, yet general model is built. The fluid is assumed to be inviscid and irrotational and both compressible and incompressible fluid are considered. The steady flow is bounded by a planar channel with a rigid and a thin, flexible wall. The latter is modelled as a continuum on a viscoelastic foundation, which exhibits bending and extensional stiffness. The influence of the various characteristics on the complex eigenvalues and therefore on the stability of the steady state is discussed.

Keywords Flow-induced vibrations • Fluid-structure-interaction • Stability analysis

1 Introduction

Fluid-structure interaction is an important subject in many fields of engineering and therefore many investigations in this field have already been done. Several problems were investigated (see [1–4] for example), varying in the assumptions regarding compressibility of the fluid, external damping of the structure, various types of stiffness and magnitude of oscillations. The aim of this contribution is to investigate and compare systematicallythe influence of various parameters and assumptions on

K. Bach (✉)
Karlsruhe Institute of Technology (KIT), Institute of Engineering Mechanics, Germany
e-mail: karolina.bach@kit.edu

J. Náprstek et al. (eds.), *Vibration Problems ICOVP 2011: The 10th International Conference on Vibration Problems*, Springer Proceedings in Physics 139,
DOI 10.1007/978-94-007-2069-5_104,

the stability of the steady state. Therefore, a simple and thus analytically accessible model is built, which includes various characteristics. Non-dimensional variables and parameters are introduced for convenience.

2 Problem Statement

As a model for fluid-structure interaction a planar and infinitely long channel is studied (cf. Fig. 1). The problem is described in Cartesian coordinates $\hat{x}$, $\hat{y}$. The fluid (density $\hat{\rho}_F$) is assumed to be inviscid and irrotational, allowing the introduction of a velocity potential $\hat{\phi}_F$, so that $\mathbf{v}_F = (\hat{u}_F, \hat{v}_F)^T = \nabla\hat{\phi}_F$ holds. Furthermore, an isentropic state change is assumed and both incompressible and compressible flow are considered. Therefore, the two-dimensional flow can be described by the continuity and Bernoulli equation and, for a compressible fluid, by the ideal gas law.

As the problem is assumed to be independent of the $\hat{z}$-axis, the flexible wall is modelled as a one-parametric continuum (constant thickness B, density ρ_S) on a viscoelastic foundation (stiffness and damping constants per unit area c and d). This structure exhibits bending stiffness D, is preloaded by a constant axial tension S and is exposed to a surface pressure $\hat{p}_F$ due to the fluid. Furthermore, the non-permeability condition at the rigid and flexible walls has to be fulfilled [1].

The fluid field variables and the structure displacement are split into steady quantities and fluctuations:

$$\hat{p}_F = \hat{p}_0 + \hat{p},\ \hat{\rho}_F = \hat{\rho}_0 + \hat{\rho},\ \hat{\phi}_F = \hat{\phi}_0 + \hat{\phi},\ \hat{u}_F = \hat{u}_0 + \hat{u},\ \hat{q}_S = \hat{q}_0 + \hat{q}. \tag{1}$$

Furthermore, non-dimensional variables and parameters are introduced using height H, density $\hat{\rho}_0$ and velocity $a_0 = \sqrt{\kappa \frac{\hat{p}_0}{\hat{\rho}_0}}$ as reference parameters [5]:

$$x = \frac{\hat{x}}{H},\ y = \frac{\hat{y}}{H},\ q = \frac{\hat{q}}{H},\ t = \frac{a_0\hat{t}}{H},\ \rho = \frac{\hat{\rho}}{\hat{\rho}_0},\ p = \frac{\hat{p}}{a_0^2\hat{\rho}_0},\ \phi = \frac{\hat{\phi}}{a_0 H}, \tag{2}$$

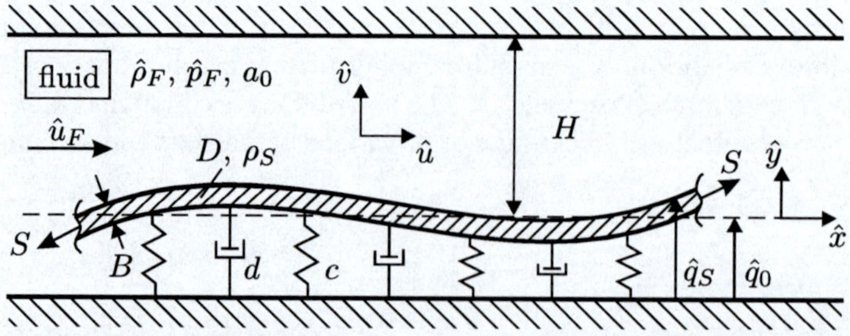

Fig. 1 Model of the fluid-filled channel

$$\lambda^2 = \frac{cH^2}{B\rho_S a_0^2},\ \mu = \frac{dH}{B\rho_S a_0},\ \beta = \frac{S}{B\rho_S a_0^2},\ \delta = \frac{D}{B\rho_S a_0^2 H^2},\ U = \frac{\hat{u}_0}{a_0},\ \alpha = \frac{\hat{\rho}_0 H}{\rho_S B}. \tag{3}$$

The structural deflections $q(x,t)$ about the steady state are assumed to be small and of the order of a small parameter ε. Based on this assumption a calculation of perturbations is carried out [1]. To analyse the stability of the steady state the perturbation problems of the zeroth- and first-orders are considered and the coupled first-order problem for a compressible fluid under consideration of the steady-state solutions is described by the set of differential non-dimensional equations.

3 Results

To analyse the influence of the various parameters, the eigenvalues are displayed versus the investigated parameter. All remaining parameters are fixed, for the dimensionless wavenumber k a constant positive value is used.

It has been shown in [5], that in absence of damping and axial flow the eigenvalues of the coupled problem come in pairs of real values with opposite signs. Increasing λ, δ, β, and k cause an increase of the absolute values. In regions were the characteristics of different eigenvalues of the uncoupled system ($\alpha = 0$) cross, strong coupling effects for $\alpha > 0$ are observed and for an incompressible fluid an added mass effect is shown.

In the following the effects of foundation damping μ, axial flow-velocity U and geometry ratio α_0 are studied. In Fig. 2 the complex eigenvalues for the problem without axial flow are shown versus the damping parameter μ for a compressible and an incompressibe fluid. The first observation is that for the latter only one pair of corresponding eigenfrequencies exists while for the former an infinite number of pairs can be found (here only a detail is outlined) [5]. By looking at the corresponding eigenmodes of the uncoupled system ($\alpha = 0$) this turns obvious. The horizontal lines in Fig. 2a belong to the so called acoustic modes [1] of the fluid; in Fig. 2b they are not present since only a compressible fluid can oscillate itself [6].

While for $\mu = U = 0$ the eigenvalues are always real and come in pairs with opposite signs, Fig. 2 shows the complex eigenvalues for $\mu \neq 0$. Since the imaginary parts are always positive, the steady state remains stable.

Figure 3 shows the eigenvalues of the coupled system ($\alpha > 0$) for varying U for both $\mu = 0$ and $\mu = 1$. It can be seen that a growth of U also causes complex eigenfrequencies. Due to the transport velocity the eigenvalues are not symmetric to the abscissa. In contrast to the former example the imaginary parts may become negative. In case of an incompressible fluid (Fig. 3b) for $U = 0$ the real eigenvalues have opposite signs. Without damping ($\mu = 0$) the real eigenvalues converge with

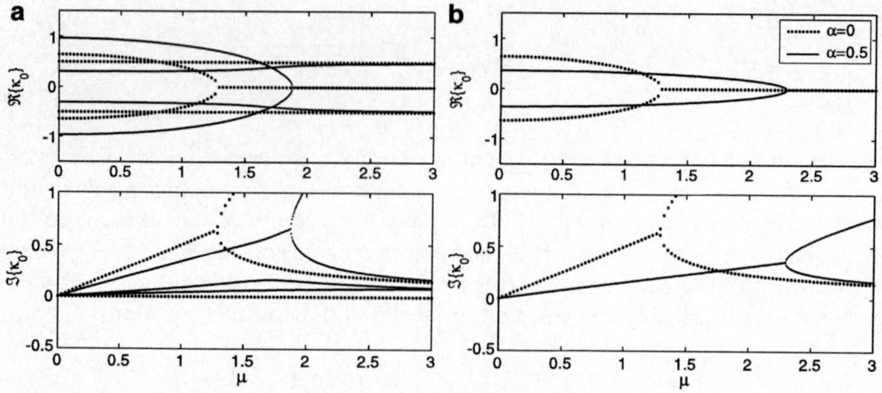

Fig. 2 Eigenvalues vs. damping parameter ($U = 0$) **(a)** compressible fluid **(b)** incompressible fluid

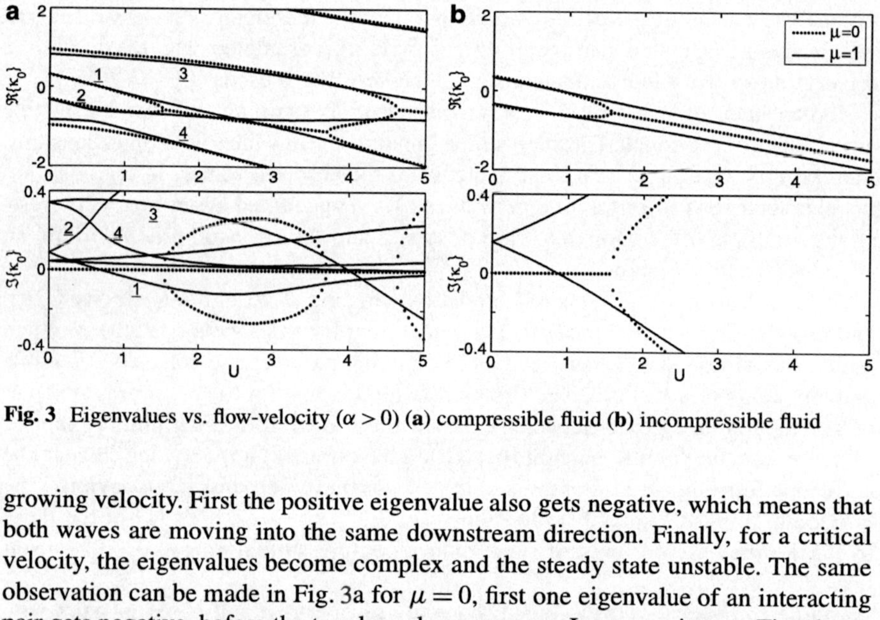

Fig. 3 Eigenvalues vs. flow-velocity ($\alpha > 0$) **(a)** compressible fluid **(b)** incompressible fluid

growing velocity. First the positive eigenvalue also gets negative, which means that both waves are moving into the same downstream direction. Finally, for a critical velocity, the eigenvalues become complex and the steady state unstable. The same observation can be made in Fig. 3a for $\mu = 0$, first one eigenvalue of an interacting pair gets negative, before the two branches converge. In comparison to Fig. 3b not only two, but an infinite number of eigenvalues are interacting with each other. This interaction of different modes is the reason, why for the compressible fluid a restabilisation is observed before getting unstable again.

Additionally for $\mu > 0$ the influence of external damping is investigated. In Fig. 3b it can be seen that the steady state gets unstable for smaller values of U than for the undamped system. Both eigenvalues become negative again, but in contrast to $\mu = 0$, from the moment when both waves move downstream the steady state looses stability by divergence and then by flutter. For clarity in Fig. 3a

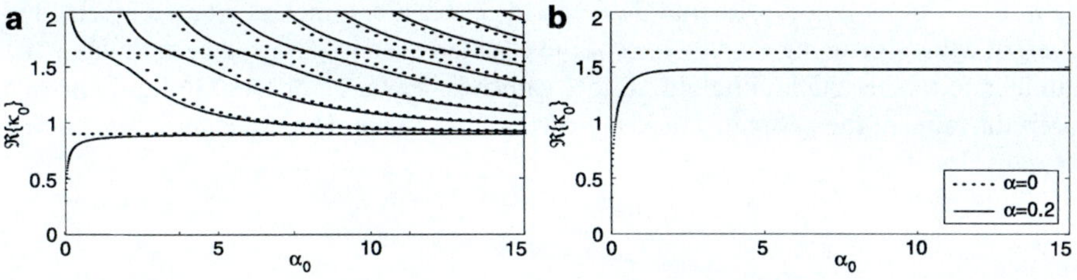

Fig. 4 Eigenvalues vs. geometry ratio ($U = \mu = 0$) (**a**) compressible fluid (**b**) incompressible fluid

the eigenvalues for $\mu = 1$ are numbered. The imaginary part of eigenvalue 1 also gets negative for a smaller flow-velocity than for $\mu = 0$ and divergence is directly followed by flutter-instability. The difference to $\mu = 0$ is, that no restabilisation is observed. Before the imaginary part of 1 becomes positive again, the imaginary part of eigenvalue 3 gets negative. Obviously damping prevents the steady state from getting stable again.

Finally the influence of H is investigated. Therefore the dimensionless geometry ratio α_0 is varied while reference length B is fixed. For $U = \mu = 0$ the eigenvalues are real and symmetric to the abscissa, hence in Fig. 4 only the positive real parts are displayed. Since $\coth(\pi) = 1.004$ holds, for an incompressible fluid and the parameters used for Fig. 4b the influence of the channel height almost vanishes for $\alpha_0 > \frac{\pi}{k} = 3.49$. The Horizontal for $\alpha = 0$ in this figure belongs to the structural mode, for $\alpha > 0$ the eigenvalues of the coupled system are increasing with α_0 up to a limit value. In case of a compressible fluid (Fig. 4a) for $\alpha = 0$ the characteristics belonging to the structural and acoustical modes are displayed and for $\alpha > 0$ above a certain value of α_0 the eigenvalues of the coupled system are decreasing and also seem to aspire to limit value.

4 Conclusion

For a given wavenumber a channel conveying compressible fluid has an infinite number of eigenfrequencies and the uncoupled system exhibits acoustical and structural modes. In contrast for an incompressible fluid only one pair of eigenvalues exists. In the absence of axial flow the steady state is stable. Growing axial velocity causes complex eigenvalues and the steady state is only stable up to a critical flow velocity. For a compressible fluid in the absence of external damping a restabilisation is observed before getting unstable again. External damping causes complex eigenvalues. The critical flow velocities in case of external damping are smaller than without damping and therefore the damping has a destabilising effect on the steady state of the coupled problem with axial flow. Without damping and axial flow the eigenvalues for varying geometry ratio are real and symmetric to the abscissa. For an incompressible fluid and the chosen parameters increasing channel height causes growing eigenvalues of the coupled system, above a certain value the influence of the channel height almost vanishes. For a compressible fluid above a certain value of the geometry ratio a growth of the channel height causes descending eigenvalues.

References

1. Wauer, J.: Nonlinear waves in a fluid-filled planar duct with a flexible wall. In: Proceedings of the IUTAM Symposium on Recent Developments in Non-linear Oscillations of Mechanical Systems, pp. 321–332. Kluwer Academic Publishers, Dordrecht/Boston (2000)
2. Kornecki, A.: Aeroelastic and hydroelastic instabilities of infinitely long plates I. Solid Mech. Arch. **3**, 381–440 (1978)
3. Kornecki, A.: Aeroelastic and hydroelastic instabilities of infinitely long plates II. Solid Mech. Arch. **4**, 241–345 (1979)
4. Horáček, J., Zolotarev, I.: Acoustic-structural coupling of vibrating cylindrical shells with flowing fluid. J. Fluid. Struct. **5**, 487–501 (1991)
5. Bach, K.: Stability of a steady flow guided by flexible walls. Proc. Appl. Math. Mech. (PAMM). **10**, 353–354 (2010)
6. Wauer, J.: Kontinuumsschwingungen. Vieweg+Teubner, Wiesbaden (2008)

Friction-Maintained Dynamic Stability

Keiko Anami, Noriaki Ishii, Charles W. Knisely, Takuma Tsuji, Tatsuya Oku, and Shigeki Sato

Abstract Mechanical systems relying on Coulomb friction to maintain dynamic stability may suffer a dynamic instability if exposed to an initial displacement exceeding a system-specific threshold. In fluid systems, even small values of negative damping are sufficient to drive the dynamic instability with sufficiently large initial displacement. The Tainter gate failures at the Folsom dam in 1995 and at the Wachi dam in 1967 are two well-known failures. To aid in preventing a recurrence, the authors engaged in a decade long research program that provided evidence that both gates failed due to an essential dynamic instability mechanism that all Tainter-gates may possess. This paper presents measurements suggesting "friction-maintained dynamic stability" of a full-scale 50-ton Tainter gate. Accompanying gate model studies showed that the gate can fail when exposed to an initial displacement exceeding a threshold value. The present study should serve to alert gate designers, owners and operators that many Tainter gates which have not yet failed may, nonetheless, have a high susceptibility to failure if and when they are exposed to a sudden input of energy resulting in an initial displacement exceeding the gate-specific thresholddisplacement.

K. Anami (✉)
Department of Mechanical Engineering, Ashikaga Institute of Technology, 268-1 Omae-cho, Ashikaga, Tochigi 326-8558, Japan
e-mail: anami@ashitech.ac.jp

N. Ishii • T. Tsuji • S. Sato
Osaka Electro-Communication University, Neyagawa, Japan
e-mail: ishii@isc.osakac.ac.jp

C.W. Knisely
Bucknell University, Lewisburg, PA, USA
e-mail: knisely@bucknell.edu

T. Oku
Mayekawa MFG. Co., Ltd., Tokyo, Japan
e-mail: tatsuya-oku@mayekawa.co.jp

J. Náprstek et al. (eds.), *Vibration Problems ICOVP 2011: The 10th International Conference on Vibration Problems*, Springer Proceedings in Physics 139, DOI 10.1007/978-94-007-2069-5_105, © Springer Science+Business Media B.V. 2011

Keywords Hydraulic gate • Flow-induced vibration • Dynamic instability • Friction-maintained dynamic stability • Initial displacement

1 Introduction

Tainter gates (also known as radial gates) are frequently used for water-level regulation in impoundment dams. An example of a Tainter gate is shown in Fig. 1. This particular gate was installed in the Folsom dam in California. During operation on July 17, 1995, one of the Folsom dam gates failed (see Ishii [1], for details). Vibration tests of a remaining gate [2], undertaken as part of the failure investigation, identified the two significant natural vibration modes illustrated in Fig. 1. One mode was the rigid-body rotational lifting vibration of the whole gate about the trunnion pin, as denoted by Θ in Fig. 1. The second significant mode was a relatively low frequency streamwise bending of the skinplate (see the dotted lines), as denoted by Ψ in Fig. 1. Under specific conditions, these two vibrations can easily couple through the inertial and the hydraulic load, and induce intense self-excited vibration [3]. This mechanism may have caused, or contributed to, the Folsom dam gate failure, as well as other gate incidents at other dams.

As part of the ongoing investigation of the dynamic instability of the Tainter gates, field vibration tests were conducted on a full-scale operational Tainter gate in Japan. During the in-water vibration tests with small gate openings, intense self-excited vibrations were clearly observed. The measured data suggest that the observed vibration was a "friction-maintained steady vibration".

This paper aims to explain why some Tainter gates fail and why many others do not. Initially, studies are presented of a 50-ton Tainter gate vibrating at small amplitudes identified as "friction-maintained steady vibration." Damping due to Coulomb friction on the side seals and on the trunnion pin, maintains the gate stability at small vibration amplitudes. Subsequently, laboratory studies of a model gate show that a gate with "friction-maintained steady vibration" can fail readily

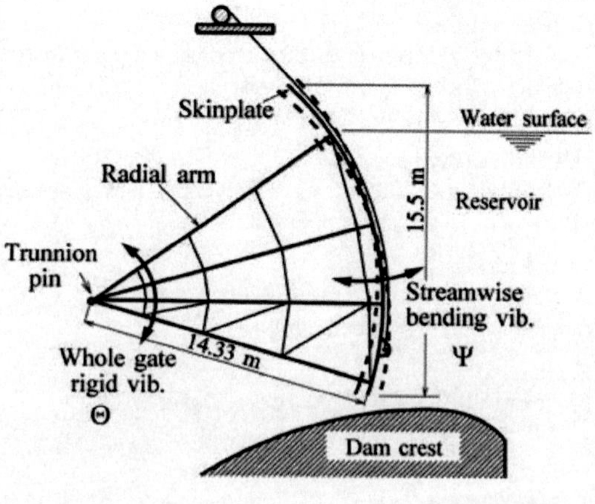

Fig. 1 *Side view* of an 87-ton Tainter gate from the Folsom Dam in California, showing two predominant natural vibration modes

when exposed to an initial displacement exceeding a threshold value. A small initial displacement, exceeding the system threshold, overcomes the damping effects and triggers a violent self-excited vibration.

2 Friction-Maintained Steady Vibration of Tainter Gate "C"

Field vibration tests were undertaken on a 50-ton Tainter gate, denoted as gate "C" [4]. When the gate submergence depth was 9.27 m, intense self-excited vibrations occurred at small discharge openings. Representative vibration traces measured at the bottom center of the skinplate are shown in Fig. 2. The upper trace shows the up-and-down vibration and the lower trace shows the streamwise vibration. Both waveforms show a clear exponential growth of the vibration amplitude. The vibration frequencies were 6.79 Hz.

Detailed observations of the gate during the vibrations revealed that the soft rubber seals, attached to each side of the skinplate, were pressed against the side seal plate embedded in the concrete pier by the large hydraulic pressure. The rubber seals themselves experienced repeated elastic expansion and compression in the up-and-down direction. However, the rubber seals were not sliding over the side seal plate, but rather were tightly pressed to it.

The rubber seals were soft, and thus acted as a spring, allowing spontaneous vibration. With increasing vibration amplitude, the elasticity of rubber seal is significantly reduced, and Coulomb friction between the rubber seal and the side seal plate resists any further amplification of the vibration, thus maintaining a steady, moderate amplitude vibration.

If, during this steady vibration, a sufficiently large trigger displacement had been introduced to overcome the amplitude-dependent damping effect of the Coulomb friction (equivalent damping is inversely proportional to amplitude), the observed

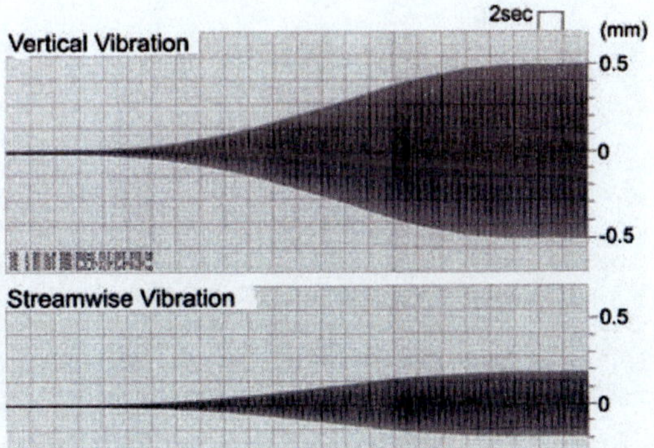

Fig. 2 Self-excited vibration waveforms at the *bottom center* of the skinplate, with a gate submergence d_0 of 9.27 m

friction-maintained stability at a steady vibration amplitude would have been lost and Tainter gate "C" would have failed completely.

Our theoretical calculation predicts that an intense, self-excited vibration will overcome the friction-maintained steady vibration, when triggered by a small initial displacement of 2.9 mm in the vertical direction. Gate failure is predicted when the streamwise vibration amplitude at the bottom spanwise center of the skinplate reaches a value of 15.8 mm. As a result, the 50-ton Tainter gate "C" would fail essentially instantaneously.

3 Model Experiments of Friction-Maintained Steady Vibration

In order to confirm both the existence of friction-maintained steady vibration and the loss of stability predicted in the previous section, tests were undertaken on a gate model. A three-dimensional 1/13-scaled model gate of the Folsom dam Tainter gate is shown in Fig. 3. The fundamental natural vibration modes were reproduced in this model gate. This equivalent model gate has a height of 1,110 mm, width of 900 mm, skinplate radius of 1,000 mm and the total mass of 42.5 kg.

The in-air natural streamwise vibration frequency of the skinplate, $\Omega_{na\psi}$, was 35.0 Hz. The in-water natural streamwise vibration frequency $\Omega_{nw\psi}$ was lowered to 5.92 Hz, due to the water added mass with a gate submergence depth of 1.1 m. The inherent in-water damping ratio was $\zeta_{nw\theta} = 0.0267$ for the whole gate rotational vibration and $\zeta_{nw\psi} = 0.0414$ for the skinplate streamwise vibration.

Based on a previous study [5], violent coupled-mode self-excited vibrations are expected when the in-water natural vibration frequency of the skinplate streamwise vibration, $\Omega_{nw\psi}$, is slightly smaller than the natural vibration frequency of the whole gate rotational vibration around the trunnion pin, $\Omega_{nw\theta}$. Based on this expectation, the whole gate rotational vibration frequency $\Omega_{nw\theta}$ was adjusted to be 5.99 Hz, slightly larger than the skinplate streamwise bending vibration frequency of $\Omega_{nw\psi} =$

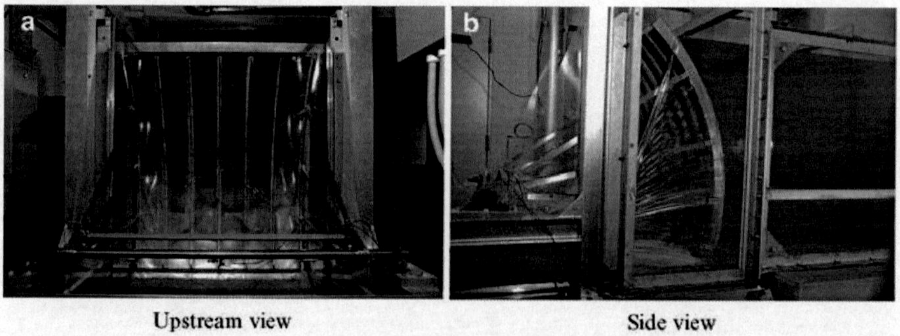

Fig. 3 Three-dimensional 1/13-scaled equivalent model gate of the Folsom dam Tainter gate. (**a**) Upstream view, (**b**) side view

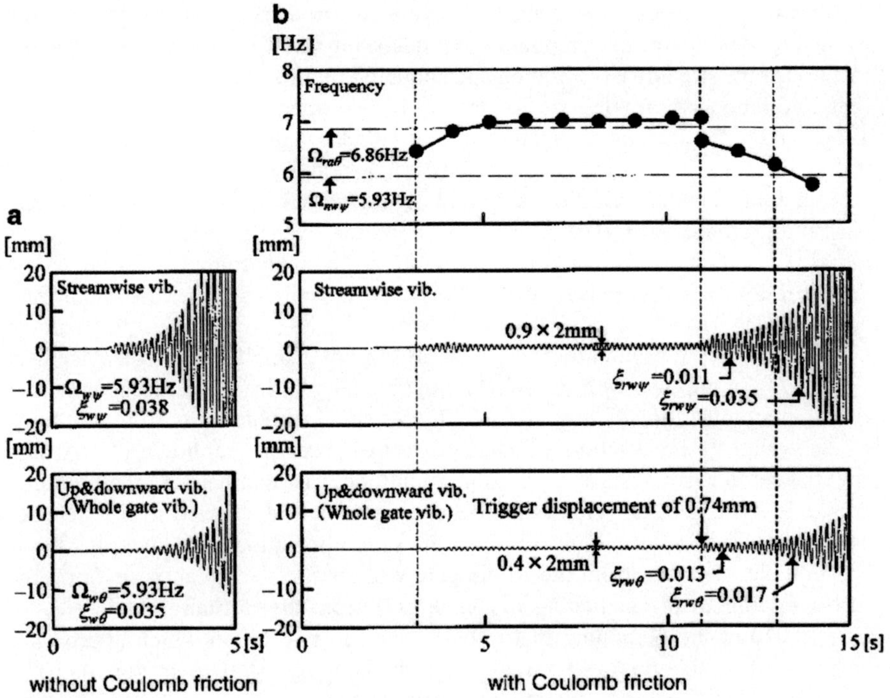

Fig. 4 Model experiment of friction-maintained steady vibration with frequency ratio of 0.86. (**a**) Without Coulomb friction, (**b**) with Coulomb friction

5.92 Hz, yielding a frequency ratio of skinplate streamwise natural vibration to whole gate natural vibration, $\gamma_{nw}(\equiv\Omega_{nw\psi}/\Omega_{a\theta})$ of 0.99. With this frequency ratio and with no additional Coulomb friction on the sides of the skinplate, a very intense dynamic instability with the excitation ratio $\xi_{w\psi}$ and $\xi_{w\theta}$ of 0.035 was observed, as shown in Fig. 4a, where the upper trace shows the skinplate radial vibration and the lower trace shows the skinplate tangential vibration, representing the skinplate streamwise vibration and the whole gate vibration around the trunnion pin, respectively.

Under the same conditions for which this intense dynamic instability was observed, small rubber seals with a length of 180 mm and the width of 60 mm were attached to the sides of the skinplate, to add Coulomb friction damping (that all operational gates possess) to the model gate. With these rubber seals, the in-water skinplate streamwise natural vibration frequency was 5.93 Hz, which is nearly same as in the non-friction study, while the whole gate rotational vibration frequency $\Omega_{ra\theta}$ was increased to 6.86 Hz due to spring effect of the rubber seals. The damping ratio $\zeta_{nw\theta}$ increased to 0.0326 due to the added Coulomb friction damping. The frequency ratio of skinplate streamwise natural vibration to the whole gate natural vibration, $\gamma_{rnw}(\equiv \Omega_{rnw\psi}/\Omega_{ra\theta})$ was 0.86. With these conditions, the gate was raised to a small gate opening, and spontaneous weak vibration occurred, as shown starting at about 3 s in Fig. 4b, where the amplitude approaches a small constant value of about 0.9 mm for the streamwise vibration and about 0.4 mm for the whole gate vibration. These vibration amplitudes are limited by the dissipative effects of friction and correspond to the friction-maintained steady vibrations observed during the testing of the full-scale Tainter gate "C", previously shown in Fig. 2.

With these steady vibration amplitudes, the rubber seals are expanding and contracting due to their elasticity, but do not slide over the side walls. They are tightly pressed against the side walls. The vibration synchronizes with the whole gate vibration with the frequency of 6.86 Hz, as shown in the upper frequency plot in Fig. 4b.

Subsequently, at around the 11 s mark in Fig. 4b a small vertical displacement trigger of 0.74 mm was introduced while the model gate was undergoing the weak friction-maintained steady vibrations. With this small displacement trigger exceeding only slightly the friction-maintained steady vibration amplitude of 0.4 mm, an exponential increase in the vibration amplitude was recorded. As the vibration amplitude increased, the vibration frequency gradually decreased to 5.93 Hz for the streamwise vibration, indicating that the synchronization of the coupled-mode vibration has switched from the whole gate vibration to the streamwise vibration. Corresponding to the switching of synchronization, the excitation ratio $\xi_{rw\theta}$ of 0.011–0.013 at the beginning of excitation increases to 0.035 which approaches the level of excitation ratio for no friction damping, shown in Fig. 4a. The effectiveness of Coulomb friction damping decreases with increasing vibration amplitude, yielding ultimately an instability that grows nearly as rapidly as it did with no additional Coulomb friction.

4 Conclusions

Friction maintained stability is clearly better than instability, but it also provides a false sense of security. Any event (including a seiche, an earthquake or landslide triggered surge, or even just corrosion of the trunnion pin or hoisting chains) that inputs a sudden step change in the gate opening has the potential to shatter the false sense of security provided by friction maintained stability. The combination of field testing on a full-scale gate and laboratory testing of a model gate fitted with side seals to simulate the full-scale Coulomb friction has demonstrated the suddenness of the transition from friction-maintained stability to complete instability and gate failure. Gate designers, owners and operators are urged to take note of mechanism presented here.

Acknowledgments The authors wish to acknowledge the financial support provided by the Grants-in-Aid for Scientific Research from Japan Society for the Promotion of Science.

References

1. Ishii, N.: Folsom Dam gate failure evaluation and suggestions, 1st report submitted to U.S. Bureau of Reclamation (Aug 1995)
2. Ishii, N.: Folsom Dam gate failure evaluation based on modal analysis and suggestion, submitted to U.S.B.R (Nov 1995)
3. Anami, K., Ishii, N.: Flow-induced coupled-mode vibration of Folsom Dam Radial Gates in California. In: Proceedings of 1999 ASME PVP Conference, vol. 369, Boston, MA, pp. 343–350 (1999)
4. Anami, K., Ishii, N., Knisely, C.W., Oku, T.: Field measurement of dynamic instability of A 50-Ton Tainter-Gate. In: Proceedings of 2007 ASME PVP Conference, San Antonio, TX, PVP2007-26543 (CD-ROM) (2007)
5. Anami, K., Ishii, N.: Flow-induced dynamic instability closely related to Folsom Dam Tainter-Gate failure in California. In: Ziada, S., Staubli, T. (eds.) Flow Induced Vibration, pp. 205–212. Balkema, Rotterdam (2000)

Sedimentation of a Spherical Solid Particle in a Compressible Viscous Fluid Under Vibration

A. Petrov and M. Shunderyuk

Abstract The differential nonlinear equation of movement of a spherical heavy solid particle in a fluid in the field of a standing acoustic wave which is created by the applied vibration is studied. Isentropic dependence of pressure on density in the form of the cubic polynom is accepted, allowing to obtain the analytic solution for a standing wave. Speed and density in a wave during the initial moment of time look like a harmonious standing wave. For time, inversely proportional to amplitude of a wave, there comes a gradient catastrophe. Further wave characteristics are described by means of discontinuous functions. Movement of a spherical heavy solid particle in the field of such wave before the gradient catastrophe is investigated. The results are compared to the solution in case of linear equation of movement. The implication of the results to the separation of particles in a fluid is studied.

Keywords Sedimentation • Viscous compressible fluid • Gradient catastrophe • Separation of particles in a fluid

1 Introduction

The problemin consideration arises when studying a novel method of particles' separation by means of ultrasonic vibration. In Fig. 1 the design of the experimental separator used in [1] is shown. In the channel standing waves are obtained by means of a PZT mounted on the back side of a silicon chip. The PZT provides standing waves both orthogonal to the liquid flow as well as in the plane of the silicon chip.

When particles are exposed to the acoustic force they will be collected in the nodes and anti-nodes of the standing wave. Because of the laminar conditions in

A. Petrov (✉) • M. Shunderyuk
Institute for Problems in Mechanics RAS, Moscow, Russia
e-mail: shunderyuk@gmail.com; petrovipmech@gmail.com

J. Náprstek et al. (eds.), *Vibration Problems ICOVP 2011: The 10th International Conference on Vibration Problems*, Springer Proceedings in Physics 139,
DOI 10.1007/978-94-007-2069-5_106, © Springer Science+Business Media B.V. 2011

Fig. 1 Experimental particle separator

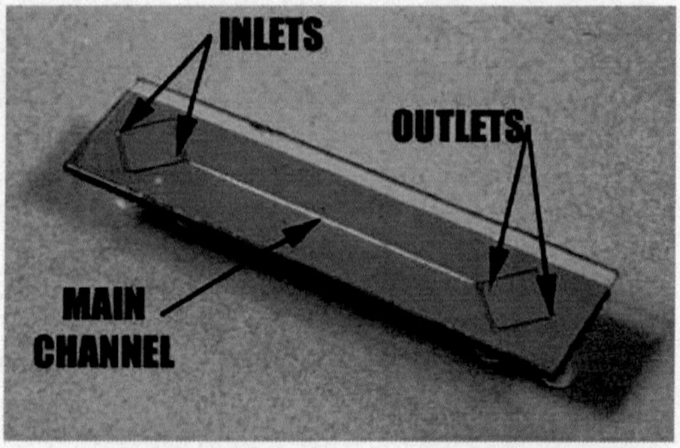

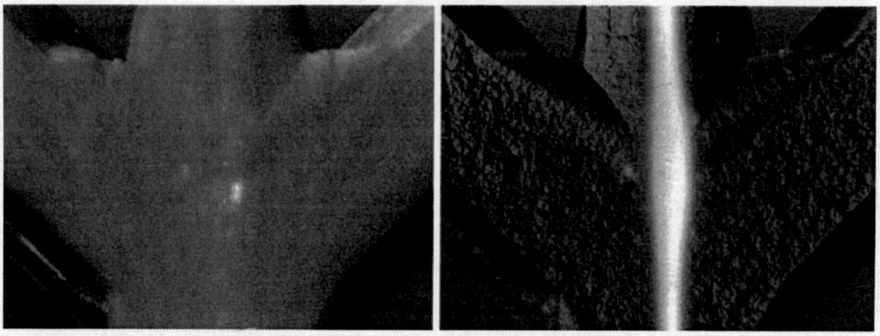

Fig. 2 The end of the channel without (*left*) and with (*right*) ultrasonic vibration. Suspension with 5 μm particles flowing in water through the 350 μm separation chip. Particle flow 0.01 ml/min

a narrow flow channel, particles once collected in a lateral position will stay there even after leaving the acoustic force field, forming long bands of particles in the channel. When operating the separation channel in its fundamental resonance mode ($\lambda/2$) a single band formation will be observed in the middle of the channel (Fig. 2). This fact can be used, for example, to separate particles with different densities. The figures are taken from [1].

The aim of this article is to study how non-linearity in the standing wave can affect on the band formation.

2 Equation of Movement of a Spherical Particle

A solid spherical particle with radius a and density ρ_s in an inviscid compressible fluid with density and velocity fields $\rho(x,t)$ and $v(x,t)$ is subjected to the action of the force equal to the sum of the added mass force and the Archimedes force [2]:

$$F_1 = -\frac{1}{2}\rho(x,t)\,V\left(-\frac{1}{2}\ddot{x} + \frac{3}{2}w(x,t) + g\right), V = \frac{4}{3}\pi a^3, w = \frac{\partial v}{\partial t} + v\frac{\partial v}{\partial x} \quad (1)$$

where g is the gravity force acceleration and V is the particle's volume.

In a viscous fluid at small Reynolds numbers, the particle is additionally subjected to the action of the Stokes $F_{St} = -6\pi\mu a\dot{x}$ [2], where μ is the dynamic viscosity of the fluid. With account of these forces, we obtain the following Cauchy problem:

$$\left(\rho_s+\frac{1}{2}\rho(x,t)\right)\frac{d^2x}{dt^2}=\frac{3}{2}\rho(x,t)w(x,t) - (\rho_s-\rho(x,t))\,g-\frac{6\pi\mu a}{V}\left(\frac{dx}{dt} - v(x,t)\right) \quad (2)$$

3 Fields of Velocity and Density

Standing ultrasonic wave with frequency ω and amplitude A satisfies wave equation:

$$\frac{\partial \rho}{\partial t} + \frac{\partial \rho v}{\partial x} = 0, \qquad \rho\left(\frac{\partial v}{\partial t} + v\frac{\partial v}{\partial x}\right) + \frac{\partial p}{\partial x} = 0 \quad (3)$$

We will accept following dependence between pressure and density:

$$p = (p_0 + B)\left(\frac{\rho}{\rho_0}\right)^{\gamma_s} - B, \quad B = \frac{c^2\rho_0}{\gamma_s} - p_0 \quad (4)$$

where c is velocity of sound in fluid and $\gamma_s = 3$. Also we will consider small fluid compressibility regarding applied vibration, which means

$$b = \frac{A\omega}{c} \to 0$$

It is proved that the Eq. 3 has following analytical solution when $\gamma_s = 3$:

$$v = bc\cos\frac{(x - vt)\omega}{c}\sin\frac{\rho\omega t}{\rho_0}, \rho = \rho_0\left(1 - b\sin\frac{(x - vt)\omega}{c}\cos\frac{\rho\omega t}{\rho_0}\right) \quad (5)$$

At the moment $t_c = 1/(\omega b)$ the Jacobian of system (5) becomes zero and here comes the gradient catastrophe.

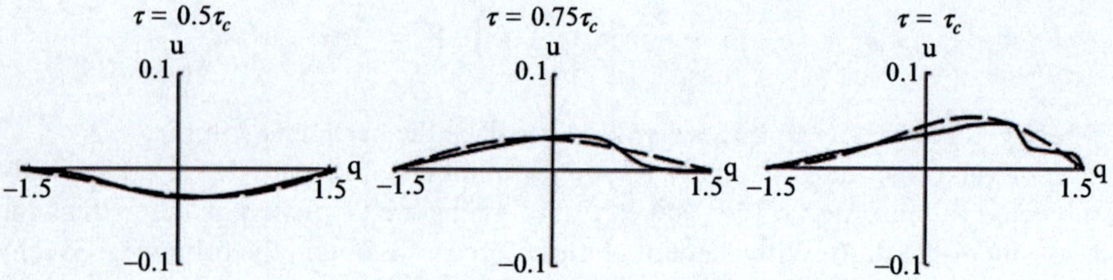

Fig. 3 Precisevelocity wave (*solid line*) and asymptotic (*dashed*) one at different times

Then we introduce the nondimensional variables:

$$x = \frac{c}{\omega}q,\ t = \frac{\tau}{\omega},\ v = cu,\ \rho = \rho_0 r,\ w = c\omega W,\ \mu = \frac{a^2\omega(\rho + 2\rho_s)}{9}\delta,\ \rho_s = s\rho_0,$$

$$g = \frac{9}{2}\frac{\omega c b^2 \gamma}{(s-1)(2s+1)},\ b = \frac{2s+1}{3}\varepsilon^2,\quad \frac{4s+1}{12} = \sigma \tag{6}$$

We can also obtain explicit asympotics for the fields of velocity and density by expanding system (5) to the Taylor series with the small parameter b (Fig. 3):

$$u = b\cos q \sin\tau - \frac{1}{2}b^2\tau \sin 2q \cos 2\tau + O(b^3),$$

$$r = 1 - b\sin q\cos\tau + \frac{1}{2}b^2\tau\cos 2q \sin 2\tau + O(b^3) \tag{7}$$

4 Averaged Equation of Movement

By substituting changes (6) to the Eq. 2 we receive following dimensionless equation of movement:

$$\ddot{q} + \frac{2s+1}{2s+r}\delta(\dot{q} - u) = \frac{3r}{r+2s}W - \frac{(s-r)\,(2s+1)}{(s-1)\,(2s+r)}\gamma\varepsilon^4 \tag{8}$$

Then using asympotics (7) we expand the equation to the Taylor series with small parameter ε ($b \sim \varepsilon^2$):

$$\dot{q} = \varepsilon p,\ \dot{p} = -\varepsilon\delta p + \varepsilon\cos q\cos\tau - \varepsilon^3(\gamma + \sigma\sin 2q)$$
$$- \frac{2s+1}{6}\varepsilon^2 \sin 2q(2t\sin 2t - \cos 2t) \tag{9}$$

Notice that here $\gamma = 0$ in case $g = 0$.

Time can be eliminated from the Eq. 9 by applying to it Krylov-Bogoliubov-Mitropolsky averaging method on the interval $[T; T + 2\pi]$ with arbitrary $T < \tau_c$, which gives following substitution for the orders up to ε^3:

$$q = \tilde{q} - \varepsilon^2 \cos\tilde{q}\cos\tau - 2\varepsilon^3 \tilde{p}\sin\tilde{q}\sin\tau + \varepsilon^2\delta\cos\tilde{q}\sin\tau$$

$$\begin{aligned} p = \tilde{p} &+ \varepsilon\cos\tilde{q}\sin\tau + \varepsilon\delta\cos\tilde{q}\cos\tau - \varepsilon^2\tilde{p}\sin\tilde{q}\cos\tau + \\ &+ \varepsilon^2\delta\tilde{p}\sin\tilde{q}\sin\tau - \varepsilon\delta^2\cos\tilde{q}\sin\tau + \\ &+ \frac{1}{3}\cos\tilde{q}\left(3\tilde{p}^2\sin\tau + \sin\tilde{q}\left[(2s+1)(\cos 2T - \cos 2\tau)\tau + \sin 2\tau\right]\right) \end{aligned} \quad (10)$$

For the averaged equation we then have

$$\dot{q} = \varepsilon p, \dot{p} = -\varepsilon\delta p - \varepsilon^3\gamma - \frac{1}{4}\varepsilon^3(4\sigma - 1)\sin 2q + \frac{1}{12}\varepsilon^3(12\sigma + 1)\cos 2T\sin 2q \quad (11)$$

By choosing in this equation $T = \pi/4 + 2\pi n$ we obtain a single averaged equation for the initial one on the interval $[\pi/4; \tau_c]$:

$$\dot{q} = \varepsilon p, \dot{p} = -\varepsilon\delta p - \varepsilon^3\gamma - \frac{1}{4}\varepsilon^3(4\sigma - 1)\sin 2q \quad (12)$$

The solution of similar equation was obtained in [3]. It is proved that in case of $\gamma = 0$ (no gravity) points $q = \pi n$ are the stability points of the system (12).

5 Numeric Results

The following values of the system's parameters were chosen for numerical calculations:

$$\rho = 1\frac{\mathrm{kg}}{\mathrm{m}^3}, \quad \rho_s = 1.5\frac{\mathrm{kg}}{\mathrm{m}^3}, \quad a = 5\mathrm{mkm}, \quad \mu = 10^{-3}\frac{\mathrm{g}}{\mathrm{cm}\cdot\mathrm{s}}, \quad g = 0\frac{\mathrm{m}}{\mathrm{s}^2}$$

$$A = 2\cdot 10^{-6}\mathrm{cm}, \omega = 2\pi\cdot 2\cdot 10^6 s^{-1} \quad (13)$$

In order to check the quality of averaged equation in the Fig. 4 presented are the solutions of the initial precise equation and the averaged one. As it predicted in theory considerable difference can be seen only at the end of the interval $[\pi/4; \tau_c = 6\cdot 10^3]$.

In the Fig. 5. several solutions of the averaged equation with different initial coordinates and velocities are shown. The resulting band formation consists with experimental picture in the Fig. 2.

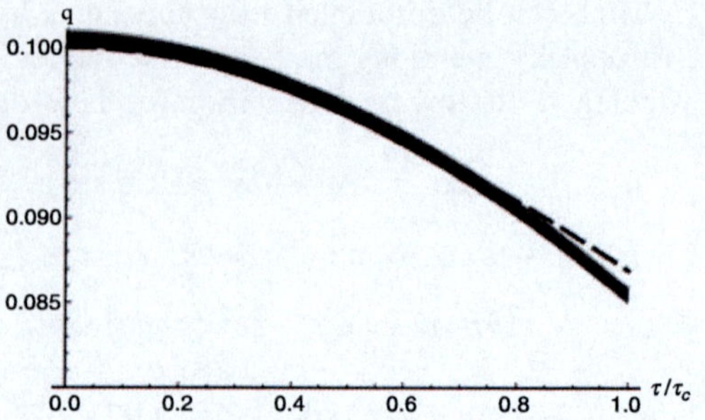

Fig. 4 The solutions of the precise equation (*solid line*) and the averaged one (*dashed line*)

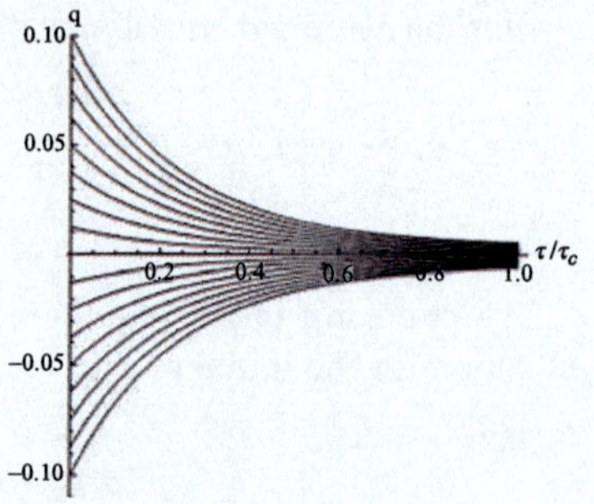

Fig. 5 Several solutions of the averaged equation with different initial conditions

6 Conclusions

The differential nonlinear equation of movement of a spherical heavy solid particle in a fluid in the field of a standing acoustic wave which is created by the applied vibration was studied. By using Krylov-Bogoliubov-Mitropolsky averaging method we proved that gradient catastrophe that comes at time, inversely proportional to amplitude of a wave, does not affect the stability of the velocity field's antinodes. This consists with the experimental results.

Acknowledgments The research work reported here received financial support from RFBR (No. 08-01-00251) and the Departmental analytical program "Development of Scientific Potential of High School (2009–2010)" (2.1.1/3604).

References

1. Fahlander, M.: Ultrasonic particle separator. Project report, Experimental Biophysics, LTH, Lund University. URL: http://nanobio.ftf.lth.se/~biokurs/projects/projects2005/10C-Ultrasonic.pdf (2005)
2. Lamb, H.: Hydromechanics. Dover, New York (1945)
3. Petrov, A.G.: Analiticheskaya gidrodinamica (Analytical Hydrodynamics). Fizmatlit, Moscow (2009)

On the Suppression of Flow-Generated Self-Excited Vibrations of a Valve

Horst Ecker and Aleš Tondl

Abstract Self-excited vibrations of valves due to the flowing medium represent an unfavorable effect and danger for some systems, e.g., pipe systems. The dynamics of a basic valve system is modeled by using an additional differential equation expressing the effect that the resulting dynamic force acting on the valve is delayed with respect to the static force component. The static force is created by the inertial force of the flowing medium at a steady state position of the valve. To enhance the dynamic stability of the basic system, the valve seat body is designed as a moveable subsystem that is connected to the inertia reference frame by a visco-elastic element. The additional damping of the subsystem in combination with the additional degree of freedom represents the passive means of vibration suppressing. The conditions for stabilization of the equilibrium position are presented using the transformation of the differential equations of the disturbed motion into the quasi-normal form. Stability charts show within which range of parameters stabilization of the system can be achieved.

Keywords Flow-induced vibrations • Valve oscillations • Vibration suppression

H. Ecker (✉)
Institute of Mechanics and Mechatronics, Vienna University of Technology,
A-1040 Vienna, Austria
e-mail: horst.ecker@tuwien.ac.at

A. Tondl
Zborovská 41, CZ-150 00 Praha, Czech Republic
e-mail: tondl@seznam.cz

J. Náprstek et al. (eds.), *Vibration Problems ICOVP 2011: The 10th International Conference on Vibration Problems*, Springer Proceedings in Physics 139,
DOI 10.1007/978-94-007-2069-5_107,

1 Introduction

Self-excited systems represent an important class of oscillatory systems. Self-excited vibrations can have a negative effect on the safe operation of various machines and devices like valves. They occur especially if the valve is only partly open. Therefore, vibration suppression can be an important problem. In previous works [1] and [2] passive and active means for suppressing self-excited vibration have been investigated. A damped mass-spring subsystem was used to control the motion of the valve, see Fig. 1 (left). Active suppression was realized by parametric excitation using a periodic variation of the spring stiffness of the additional subsystem.

In this contribution a different design is used, which shall give the designer more freedom to choose a suitable system, resp. its parameters. The basic idea of this passive system is to consider a movable valve seat body with visco-elastic mounting and elastic connection to the valve. A sketch of that system is shown in Fig. 1 (right). The valve and the seat body are denoted m_1 and m_2, the connecting stiffness elements are k_1, k_2 and k_3, viscous damping for the valve and the seat are denoted b_1, b_2, respectively.

The model for the self-excitation mechanism is based on the idea that the dynamic force due to the pressure by the flowing medium acting on the valve and the seat is delayed with respect to the static force due to the inertial force of the flowing medium for the steady state of the non-moving valve. This time delay can be modelled by an additional first-order differential equation, which reads

$$\dot{P} - \alpha[F(y) - P] = 0, \tag{1}$$

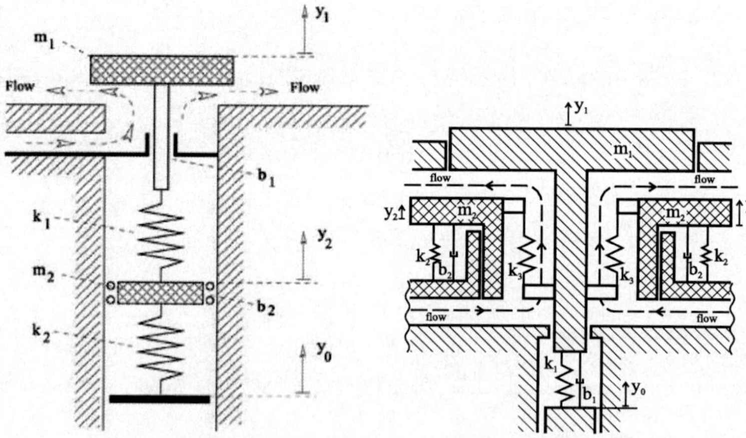

Fig. 1 Schematics of different valve systems. (*Left*) as used in [1], [2], (*Right*) System as used in this study with moveable valve seat body. The valve is operated by displacement y_0

where P represents the dynamic force, $F(y)$ the static force, y the relative deflection of the valve, and α is a positive constant (see [3], Chap. 12).

2 System Description and Equations of Motion

The valve having mass m_1 is connected by a spring of stiffness k_1 to the operating device. Operation of the valve (opening and closing) is realized by changing the value of y_0 accordingly. The valve seat body has a mass of m_2 and is mounted by spring and dampers k_2, b_2 to the reference system. Additionally, a spring k_3 connects the valve and the seat body. The deflections of the masses, i.e., the degrees of freedom, are denoted y_1, y_2. For the sake of simplicity let us consider linear visco-elastic elements. The static pressure force acting on the valve can be expressed by a function $F(y_1 - y_2)$, which is a monotonically decreasing function. Considering the above presented description of the system, the following differential equations of motion can be derived

$$\begin{aligned} m_1\ddot{y}_1 + b_1\dot{y}_1 + k_1(y_1 - y_0) + k_3(y_1 - y_2) &= P, \\ m_2\ddot{y}_2 + b_2\dot{y}_2 + k_2 y_2 - k_3(y_1 - y_2) &= -\gamma P, \\ \dot{P} - \alpha_0[F(y_1 - y_2) - P] &= 0. \end{aligned} \tag{2}$$

By introducing dimensionless deflections

$$y_i / y_0 = x_i, \qquad (j = 1, 2) \tag{3}$$

and using a time transformation

$$\omega_1 t = \tau, \qquad \omega_1 = \sqrt{k_1/m_1}, \qquad \mathrm{d}/\,\mathrm{d}t = (\dot{}), \qquad \mathrm{d}/\,\mathrm{d}\tau = ()'. \tag{4}$$

Equations 2 can be transformed into the dimensionless form

$$\begin{aligned} x_1'' + \kappa_1 x_1' + x_1 - 1 + q_3^2(x_1 - x_2) &= f, \\ x_2'' + \kappa_2 x_2' + M q_2^2 x_2 - M q_3^2(x_1 - x_2) &= -\gamma M f, \\ f' - \alpha[\Phi(x_1 - x_2) - f] &= 0, \end{aligned} \tag{5}$$

with abbreviations

$$\kappa_1 = \frac{b_1}{m_1\omega_1}, \quad \kappa_2 = \frac{b_2}{m_2\omega_1}, \quad f = \frac{P}{k_1 y_0}, \quad q_2^2 = \frac{k_2}{k_1},$$
$$M = \frac{m_1}{m_2}, \quad \alpha = \frac{\alpha_0}{\omega_1}, \quad \Phi(x_1) = \frac{F(x_1)}{k_1 y_0}, \quad q_3^2 = \frac{k_3}{k_1}$$

3 Stability of the Equilibrium Position

The equilibrium positions of the valve masses are characterized by constant values x_{10}, x_{20} which can be determined from the following algebraic equations:

$$x_{10} - 1 + q_3^2(x_{10} - x_{20}) = \Phi(x_{10} - x_{20}),$$
$$q_2^2 x_{20} - q_3^2(x_{10} - x_{20}) = -\gamma\Phi(x_{10} - x_{20}). \tag{6}$$

By inserting for the states x_1, x_2, f the expressions $x_1 = x_{10} + u_1$, $x_2 = x_{20} + u_2$ and $f = \Phi(x_{10} - x_{20})$ which are based on the disturbances u_1, u_2, v of the equilibrium solutions, the following differential equations for the disturbed motion of the system in a first-order approximation are obtained:

$$u_1'' + \kappa_1 u_1' + u_1 + q_3^2(u_1 - u_2) = v,$$
$$u_2'' + \kappa_2 u_2' + M q_2^2 u_2 - M q_3^2(u_1 - u_2) = -\gamma M v,$$
$$v' - \alpha[\Phi'(u_1 - u_2) - v] = 0, \tag{7}$$

where $\Phi' = \partial\Phi(x_{10} - x_{20})/\partial(x_{10} - x_{20})$ with $\Phi' \leq 0$ because $\Phi(x_{10} - x_{20})$ is a decreasing function.

Assuming that the damping terms are relatively small, these equations can be transformed into the quasi-normal form by using the transformation

$$u_1 = z_1 + z_2, \qquad u_2 = a_1 z_1 + a_2 z_2 \tag{8}$$

which leads to the following relations (see [4], Appendix I, for more details):

$$z_1'' + \Omega_1^2 z_1 = \frac{1}{a_1 - a_2}\left\{-\kappa_2(a_1 z_1' + a_2 z_2') + a_2[\kappa_1(z_1' + z_2') - v]\right\},$$
$$z_2'' + \Omega_2^2 z_2 = \frac{1}{a_1 - a_2}\left\{\kappa_2(a_1 z_1' + a_2 z_2') - a_1[\kappa_1(z_1' + z_2') - v]\right\},$$
$$0 = v' - \alpha[\Phi'(z_1 - z_2) - v]. \tag{9}$$

with coefficients

$$a_1 = 2Mq_3^2 \Big/ \left[M(q_2^2 + q_3^2) - 1 - q_3^2 + \sqrt{(1 + q_3^2 - M(q_2^2 + q_3^2))^2 + 4Mq_3^4}\right], \tag{10}$$

$$a_2 = 2Mq_3^2 \Big/ \left[M(q_2^2 + q_3^2) - 1 - q_3^2 - \sqrt{(1 + q_3^2 - M(q_2^2 + q_3^2))^2 + 4Mq_3^4}\right], \tag{11}$$

$$\Omega_{1,2}^2 = \frac{1}{2}\left[(1+q_3^2+M(q_2^2+q_3^2)) \mp \sqrt{(1+q_3^2-M(q_2^2+q_3^2))^2+4Mq_3^4}\right]. \tag{12}$$

Let us consider the stability of single frequency vibrations only, i.e., we will separate the first mode and the second mode vibration stability. Of course, system stability requires that both modes are stable. When investigating the first mode we shall suppose that $z_2 = 0$, which means that the following equations will be analyzed:

$$z_1'' + \Omega_1^2 z_1 = \frac{1}{a_1 - a_2}\{-\kappa_2 a_1 z_1' + a_2 \kappa_1 z_1' - v\},$$
$$0 = v' - \alpha[\Phi'(z_1) - v], \tag{13}$$

with $\Phi' = \frac{\partial \Phi}{\partial z_1}$. For this reduced system the characteristic equation reads

$$\lambda^3 + \left[\alpha + \frac{1}{a_1 - a_2}(a_1\kappa_2 - a_2\kappa_1)\right]\lambda^2$$
$$+\left[\Omega_1^2 + \alpha\frac{1}{a_1 - a_2}(a_1\kappa_2 - a_2\kappa_1)\right]\lambda + \alpha\Omega_1^2 + \alpha\Phi'\frac{a_2}{a_1 - a_2} = 0. \tag{14}$$

By using the Routh-Hurwitz criterion the following conditions are obtained:

$$\alpha\left(\Omega_1^2 + \Phi'\frac{a_2}{a_1 - a_2}\right) > 0, \tag{15}$$

$$\left(\alpha + \frac{\bar{\kappa}}{a_1 - a_2}\right)\left(\Omega_1^2 + \alpha\frac{\bar{\kappa}}{a_1 - a_2}\right) - \alpha\Omega_1^2 - \alpha\Phi'\frac{a_2}{a_1 - a_2} > 0, \tag{16}$$

where the symbol $\bar{\kappa} = a_1\kappa_2 - a_2\kappa_1$ is used.

Considering the supposed shape of function Φ, then Φ' is negative and the condition Eq. 15 is always met because $a_2 < 0$. It can be supposed that $\bar{\kappa}$ is of relatively small value and therefore the term with $\bar{\kappa}^2$ may be neglected. Consequently, condition Eq. 16 can be expressed in an abbreviated form:

$$\left(\kappa_1 - \frac{a_1}{a_2}\kappa_2\right)(\alpha^2 + \Omega_1^2) + \alpha\Phi' > 0. \tag{17}$$

In the same manner we can obtain the condition for the stability of the vibration for the second mode ($z_1 = 0$):

$$\Omega_2^2 - \frac{a_1}{a_1 - a_2} > 0, \tag{18}$$

$$(a_1\kappa_1 - a_2)(\alpha^2 + \Omega_2^2) + \alpha a_1 \Phi' > 0. \tag{19}$$

Again the first condition Eq. 18 is always satisfied. Comparing the conditions for the first and second vibration modes it can be expected that by increasing $-\Phi'$ the condition for the first mode will not be met first.

4 Numerical Example

To verify the stability conditions as obtained from the analysis in the previous section, a numerical example is shown. Based on the differential equations (7) for the perturbed system, an eigenvalue analysis was carried out numerically by using the software package MATLAB.

For the numerical approach it is necessary to define function $\Phi' = \partial\Phi/\partial\Delta u$, which appears in the first-order differential equation of Eq. 7. For this example it is sufficient to capture the behavior of the valve system by a linear function $\Phi' = \frac{\partial\Phi}{\partial\Delta u} = k_\Phi < 0$.

For an exemplary study the parameters of the perturbed system Eq. 7 are chosen as listed in the caption of Fig. 2. Both diagrams show the real part of the two largest eigenvalues of the system. The figure to the left is a parameter study of the mass ratio $M = \frac{m_1}{m_2}$. Due to a positive real part of the first eigenvalue, the system is unstable for approx. $M \leq 1$. For $M \geq 5$ the second mode becomes unstable. The diagram to the right explains the influence of the connecting stiffness ratio q_3^2. For this assumed system, at least $q_3^2 > 7$ must hold to ensure stability. These results are in agreement with the conditions derived in the previous section.

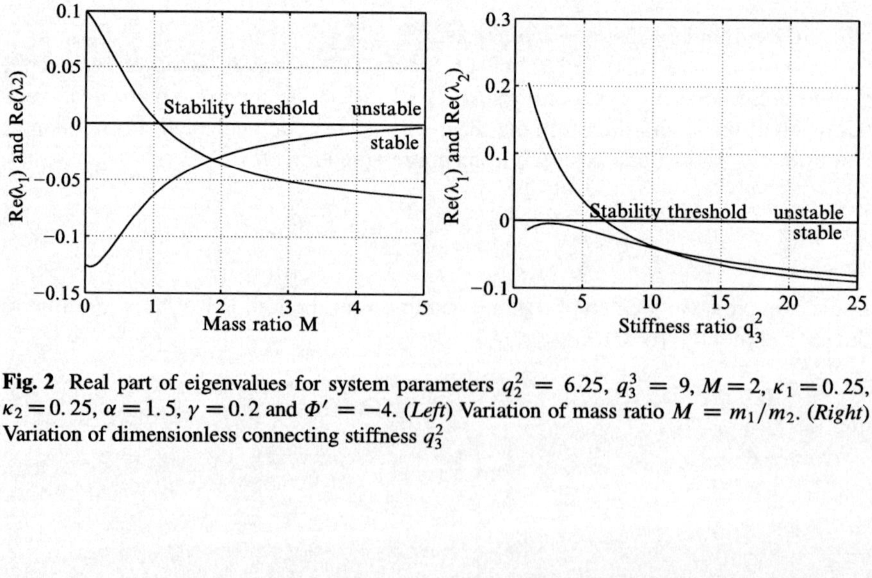

Fig. 2 Real part of eigenvalues for system parameters $q_2^2 = 6.25$, $q_3^3 = 9$, $M = 2$, $\kappa_1 = 0.25$, $\kappa_2 = 0.25$, $\alpha = 1.5$, $\gamma = 0.2$ and $\Phi' = -4$. (*Left*) Variation of mass ratio $M = m_1/m_2$. (*Right*) Variation of dimensionless connecting stiffness q_3^2

5 Conclusion

There are various possibilities for a sophisticated valve design to minimize the danger of flow-induced self-excited vibrations. The new design presented here is analyzed analytically and useful stability criteria are derived to get insight into the influence of the relevant system parameters on the stability behavior. The numerical stability analysis can be used to optimize the design and maximize the safety margin with respect to the stability threshold.

References

1. Tondl, A., Ecker, H.: Suppression of self-excited vibrations in valve systems: passive means. In: Pešek, L. (ed.) Proceedings Dynamics of Machines 2007 (DOM 07). Czech Committee of the EUROMECH Society, Praha (2007)
2. Ecker, H., Tondl, A.: Suppression of Self-Excited Vibrations in Valve Systems: active Means. In: Pešek, L. (ed.) Proceedings Dynamics of Machines 2007 (DOM 07). Czech Committee of the EUROMECH Society, Praha (2007)
3. Tondl, A.: Quenching of Self-Excited Vibrations. Academia, Prague and Elsevier Science Publisher, Amsterdam (1991)
4. Tondl, A.: On the Interaction Between Self-Excited and Forced Vibrations. Monographs and Memoranda, vol. 20. National Research Institute for Machine Design, Prague (1976)

FE Modelling of the Fluid-Structure-Acoustic Interaction for the Vocal Folds Self-Oscillation

Pavel Švancara, J. Horáček, and V. Hrůza

Abstract The flow induced self-oscillation of the human vocal folds in interaction with acoustic processes in the simplified vocal tract model was explored by three-dimensional (3D) finite element (FE) model. Developed FE model includes vocal folds pretension before phonation, large deformations of the vocal fold tissue, vocal folds contact, fluid-structure interaction, morphing the fluid mesh according the vocal folds motion (Arbitrary Lagrangian-Eulerian approach), unsteady viscous compressible airflow described by the Navier-Stokes equations and airflow separation during the glottis closure. Iterative partitioned approach is used for modelling the fluid-structure interaction. Computed results prove that the developed model can be used for simulation of the vocal folds self-oscillation and resulting acoustic waves. The developed model enables to numerically simulate an influence of some pathological changes in the vocal fold tissue on the voice production.

Keywords Biomechanics of human voice • Numerical simulation of voice production • Fluid-structure-acoustic interaction • Self-oscillation

1 Introduction

Human voice production is a direct consequence of the flow-induced vibration of the vocal folds interacting with acoustic spaces of the vocal tract. For better understanding, the mechanism of phonation for healthy as well as for pathological

P. Švancara (✉) • V. Hrůza
Institute of Solid Mechanics Mechatronics and Biomechanics, Brno University of Technology, Brno, Czech Republic
e-mail: svancara@fme.vutbr.cz

J. Horáček
Institute of Thermomechanics, Academy of Sciences of the Czech Republic, Brno, Czech Republic
e-mail: jaromirh@it.cas.cz

J. Náprstek et al. (eds.), *Vibration Problems ICOVP 2011: The 10th International Conference on Vibration Problems*, Springer Proceedings in Physics 139,
DOI 10.1007/978-94-007-2069-5_108,

voices it is important to study this problem as the fluid-structure-acoustic interaction problem. Several experimental and computational studies on this problem have been published. Main limitations of experimental techniques [1, 2] are in difficulties to obtain detailed spatial and time information on quantities such as the intraglottal pressure, velocity distribution, impact stress etc. Literature describing the computational models includes reduced-order structural models [3], models of fluid flow [4] and FE models [5, 6]. The main advantage of the FE models is their ability to deal with complex vocal folds and acoustic spaces geometry and ability to solve fluid-structure-acoustic interaction combining tissue elasticity with aerodynamics.

Recently, a 3D FE model of flow induced oscillations of the vocal folds in interaction with acoustic spaces of a simplified vocal tract modelled as a straight prismatic rectangular channel was developed by the authors [7, 8]. The constant inflow velocity of the air supplied by the lungs is the driving parameter of this model. In the present paper new FE model with a 3D shaped vocal tract was developed to study the effect of real vocal tract shape on acoustics.

2 Complex 3D Finite Element Model of the Voice Production

FE model of the vocal folds together with the FE models of the trachea and the simplified acoustic spaces of the human vocal tract shaped for simulation of phonation of vowel [a:] are shown in Fig. 1. The vocal tract model of the acoustic spaces for the Czech vowel [a:] was created by converting the data from the magnetic resonance images [9].

The FE model was developed within the program system ANSYS 13.0. The vocal folds are modelled by a three layered tissue – epithelium, lamina propria and muscle. Homogenous and isotropic material of each layer was assumed with Young modulus 20 kPa for epithelium, 3 kPa for lamina propria and 65 kPa for muscle. Poisson ratio 0.49 and density 1,040 $\text{kg}\cdot\text{m}^{-3}$ was used for all three layers. Proportional structural damping was used with constants $\alpha = 150$ and $\beta = 3 \times 10^{-4}$. Collisions of the vocal folds during phonation are modelled by the symmetric surface to surface contact pair elements on faces of the vocal folds. Motion of the vocal folds is computed by transient analysis with time step $\Delta t = 1.5 \times 10^{-4}$ s within the ANSYS/Structure environment taking into account the large deformations and the vocal folds contact. The flow of air is modelled as unsteady viscous compressible and laminar using ANSYS/Flotran code using the material properties: speed of sound 343 m/s, fluid viscosity 1.8135×10^{-5} $\text{kg}\cdot\text{m}^{-1}\cdot\text{s}^{-1}$ and density 1.205 $\text{kg}\cdot\text{m}^{-3}$. The compressible Navier-Stokes equations capture acoustic wave propagation phenomena in fluid. Constant airflow velocity at the entrance to the subglottal space is prescribed as only one driving parameter (0.4 m/s used here). Zero acoustic pressure is prescribed on the upper end of the vocal tract.

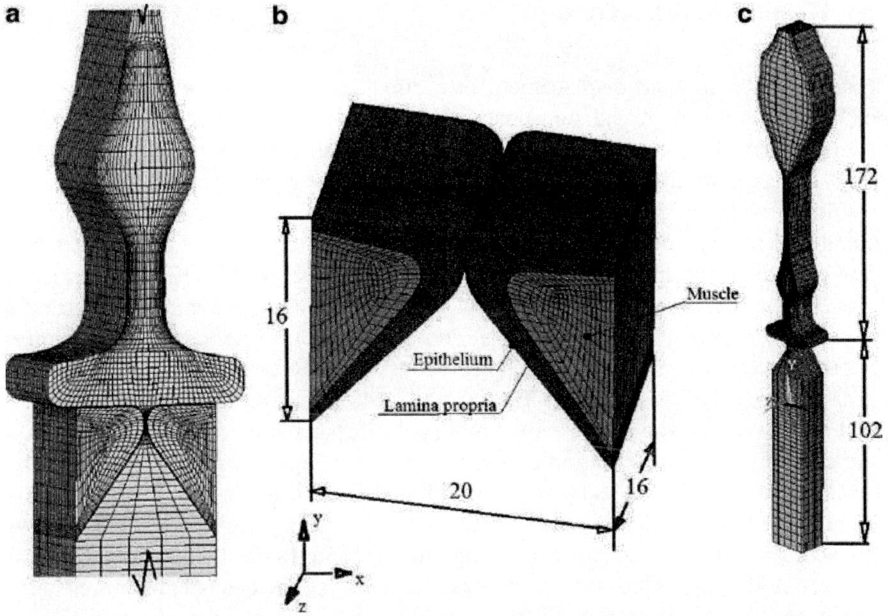

Fig. 1 (**a**) FE model of the vocal folds together with a part of the model of the vocal tract, (**b**) detail of the FE model of the three layered tissue of the vocal folds, (**c**) complete FE model of the acoustic spaces of the trachea and the vocal tract for vowel/a/(all dimensions are in mm)

FE model consists of 9,744 linear 8-node structural elements and 24,084 linear 8-node fluid elements. Before the fluid-structure interaction simulation starts the vocal folds are prolonged in the longitudinal direction by ¼ of original length and pushed slightly into the contact. This is performed as static analyses in several steps taking into account the large deformations and contact. Fluid-structure interaction is solved by iterative partitioned solution procedure in ANSYS. Where the results of the flow solution are transferred as loads on the vocal folds surface, then the vocal folds motion is computed and then again the fluid flow is solved. This solution is iteratively repeated until a needed convergence limits for stress and displacement residuals on interface are reached. The fluid domain mesh is morphed in each time step according to the vocal folds movement. In order to represent moving boundaries in fluid domains the Arbitrary Lagrangian-Eulerian (ALE) approach implemented in ANSYS is utilized. For purpose of modelling of the vocal folds closure the distance between the faces of the vocal folds is monitored and if exceeds a defined minimal value, the mesh is not more deformed at these nodes and the flow velocity is set to zero.

3 Results and Discussion

Example of computed displacement in x direction of selected nodes on the face of the left and right vocal fold positioned in the middle of vocal folds length is shown in Fig. 2 together with fluid velocity and acoustic pressure in corresponding fluid node on the axis of symmetry of the simplified vocal tract model. From the results we can see that prescribed constant airflow velocity at the entrance to the subglottal space produce vocal folds self-oscillations that are stabilized after first few periods of the transient regime. Fundamental vocal fold oscillation frequency was 98 Hz, maximum intraglottal fluid velocity was about 30 m/s and maximum intraglottal pressure was about 700 Pa. The computing time per oscillation period is approximately 6 h on two Intel Xeon E5520 (8 cores/16 threads) and enabled GPU accelerator capability on Nvidia Tesla C1060 (240 thread processors).

Figure 3 shows the numerically simulated displacement of the vocal folds in x direction during one oscillation period at eight time steps from 0.0297 to 0.0360 s.

Computed airflow velocity in the middle cross-section along the vocal tract at eight time instances during one oscillation period is shown in Fig. 4. Figure 5 shows the acoustic pressure and corresponding power spectral densities computed at two selected nodes above the vocal folds and near the lips. The acoustic pressure is also nearly periodic after first four oscillation periods. In the spectra we can observe peak of fundamental frequency 98 Hz and many higher harmonics as reported for human voice [1]. From the spectrum of the acoustic pressure near the lips we can see that some harmonics are amplified by the resonant frequencies of the vocal tract and producing so called "formants" typical for the modelled vowel [a:].

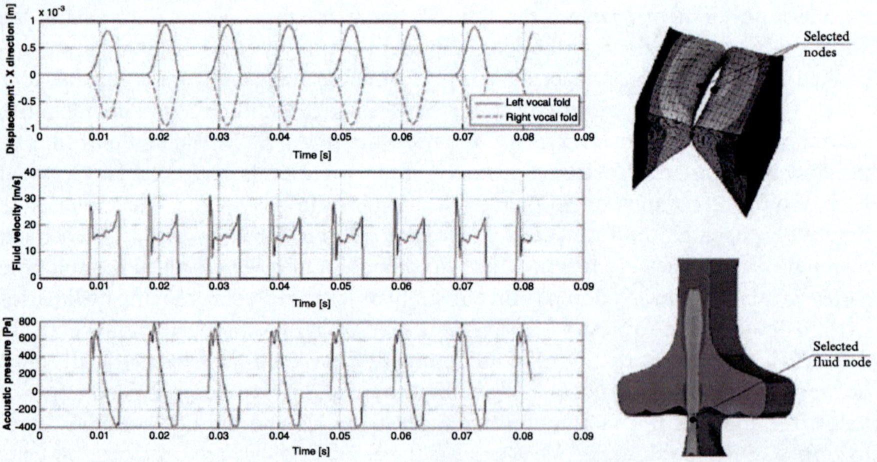

Fig. 2 Numerically simulated displacement in x direction of selected nodes on the face of the *left* and *right* vocal fold (*top graph*), and the fluid velocity (*middle graph*) and acoustic pressure (*bottom graph*) in a fluid node on the axis of symmetry of the channel modelling the vocal tract

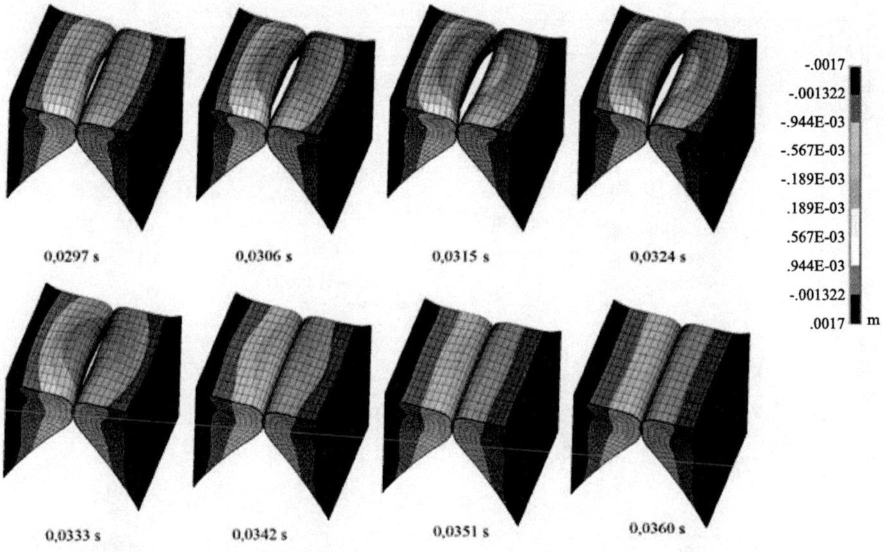

Fig. 3 Computed displacement of the vocal folds in x direction at eight time steps during one oscillation period (0.0297–0.0360 s)

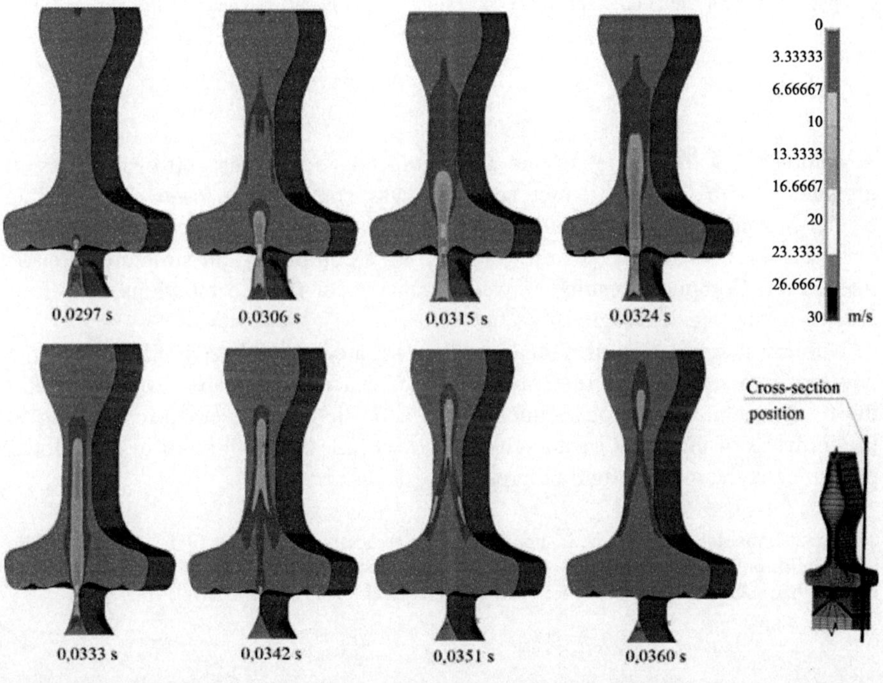

Fig. 4 Numerically simulated airflow velocity on cross-section through the vocal tract during one oscillation period at eight time steps from 0.0297 to 0.0360 s

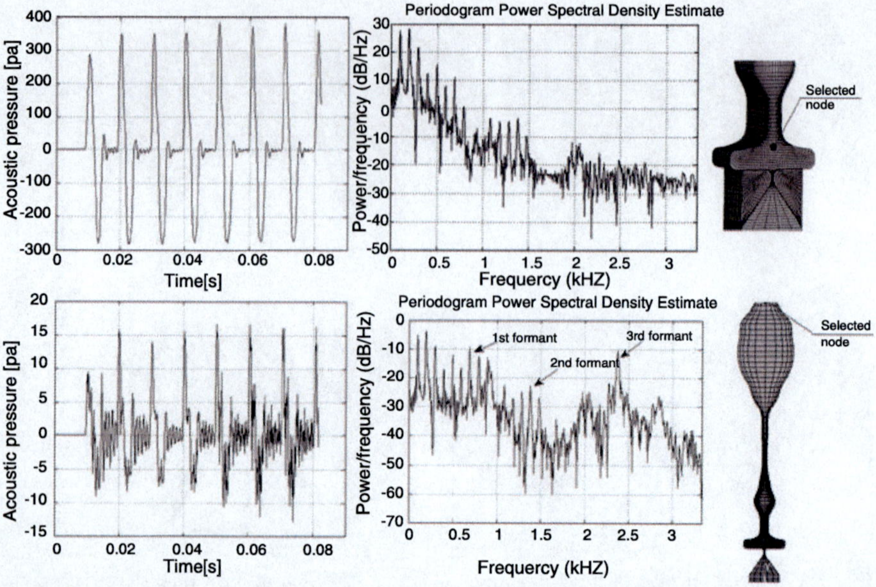

Fig. 5 Acoustic pressure (*left graphs*) and corresponding power spectral densities (*right graphs*) computed in the selected nodes above the vocal folds (*upper panel*) and near the lips (*lower panel*)

4 Conclusions

A complex 3D FE model of the flow induced vocal folds self-oscillations in interaction with the vocal tract acoustics was created. The used compressible Navier-Stokes equations enable to model propagation of the acoustic waves in the vocal tract model and therefore modelling a complete fluid-structure-acoustic interaction. Computed results showed regular vocal folds oscillations after four periods of the transient regime.

Numerical simulation of phonation demonstrated very close similarities of the developed model with a real production of voice in humans [10], especially the fundamental oscillation frequency, the acoustic resonances corresponding to the formants of the vocal tract cavity for vowel [a:], the amplitudes of intraglottal pressure and the mean subglottal pressure.

Acknowledgments The research is supported by the Academy of Sciences of the Czech Republic by the pilot project "Biomechanics of human voice – acoustic properties of vocal cavities and creation of 3D FEM phonation model" within the research plan No AV0Z20760514.

References

1. Titze, I.R.: The Myoelastic Aerodynamic Theory of Phonation. National Centre for Voice and Speech, Denver/Iowa City (2006)
2. Šidlof, P., Švec, J.G., Horáček, J., Veselý, J., Klepáček, I., Havlík, R.: Geometry of human vocal folds and glottal channel for mathematical and biomechanical modelling of voice production. J. Biomech. **41**, 985–995 (2009)
3. Horáček, J., Šidlof, P., Švec, J.G.: Numerical simulation of self-oscillations of human vocal folds with Hertz model of impact forces. J. Fluids Struct. **20**, 853–869 (2005)
4. Zhao, W., Zhang, C., Frankel, S.H., Mongeau, L.: Computational aeroacoustic of phonation, part I: numerical methods, acoustic analogy validation, and effects of glottal geometry. J. Acoust. Soc. Am. **112**, 2134–2146 (2002)
5. Alipour, F., Berry, D.A., Titze, I.R.: A finite–element model of vocal-folds vibration. J. Acoust. Soc. Am. **108**, 3003–3012 (2000)
6. Thomson, S.L., Mongeau, L., Frankel, S.H.: Aerodynamic transfer of energy to the vocal folds. J. Acoust. Soc. Am. **118**, 1689–1700 (2005)
7. Švancara, P., Horáček, J., Hrůza, V.: Development of FE model of interaction between oscillating vocal folds and acoustic space of the vocal tract. In: Proceedings of the 6th International Conference on Voice Physiology and Biomechanics, University of Tampere, Tampere, Finland, pp. 163–167 (2008)
8. Švancara, P., Hrůza, V., Horáček, J.: FE model of flow induced vocal folds oscillations in interaction with acoustic spaces of the vocal tract. In: Proceedings of the 3 rd Advanced Voice Function Assessment International Workshop, Technical University of Madrid, Madrid, Spain, pp. 17–20 (2009)
9. Radolf, V.: Direct and inverse task in acoustics of the human vocal tract. Ph.D. thesis, Faculty of Mechanical Engineering, Czech Technical University, Prague, 95 p (2010)
10. Titze, I.R.: Principles of Voice Production. Prentice Hall, London (2000)

Mathematical Modeling of Flow in Human Vocal Tract

Petra Pořízková, Karel Kozel, and Jaromír Horáček

Abstract This study deals with the numerical solution of a 2D unsteady flow of a compressible viscous fluid in a channel for low inlet airflow velocity. The unsteadiness is caused by a prescribed periodic motion of the channel wall. Unsteady flow fields for inlet Mach number $M_\infty = 0.012$ and frequency 100 Hz are presented.

Keywords Bio-mechanics of human voice • CFD • Finite volume method • Low mach number • Unsteady flow • Viscous compressible fluid

1 Introduction

Fluid-structure interaction problems can be treated in numerous engineering and other applications.

A current challenging question is a mathematical and physical description of the mechanism for transforming the airflow energy in the glottis into the acoustic energy representing the voice source in humans. The voice source signal travels from the glottis to the mouth, exciting the acoustic supraglottal spaces, and becomes modified by acoustic resonance properties of the vocal tract.

P. Pořízková (✉)
Czech Technical University in Prague, Faculty of Mechanical Engineering, Department of Technical Mathematics, Karlovo náměstí 13, Prague 2, 121 35, Czech Republic
e-mail: puncocha@marian.fsik.cvut.cz

K. Kozel • J. Horáček
Institute of Thermomechanics AS CR, Dolejškova 5, Prague 8, Czech Republic
e-mail: Karel.Kozel@fs.cvut.cz; jaromirh@it.cas.cz

J. Náprstek et al. (eds.), *Vibration Problems ICOVP 2011: The 10th International Conference on Vibration Problems*, Springer Proceedings in Physics 139,
DOI 10.1007/978-94-007-2069-5_109,

Acoustic wave propagation in the vocal tract is usually modeled from incompressible flow models separately using linear acoustic perturbation theory, the wave equation for the potential flow [1] or the Light-hill approach on sound generated aerodynamically [2]. In reality, the airflow coming from the lungs causes self-oscillations of the vocal folds, and the glottis completely closes in normal phonation regimes, generating acoustic pressure fluctuations. In this study, the channel is harmonically opening and nearly closing in the narrowest cross-section of the channel, making the investigation of the airflow field in the glottal region possible. For phonation of vowels, the frequencies of the vocal folds oscillations are in the region from ca 82 Hz for bass up to ca 1,170 Hz for soprano in singing voice, the airflow velocity in the trachea is approximately in the range of 0.3–5.2 ms^{-1} taking into account the tracheal diameter in humans in the range 14.5–17.6 mm [1].

2 Mathematical Model

To describe the unsteady laminar flow of a compressible viscous fluid in a channel, the 2D system of Navier-Stokes equations was considered as a mathematical model. The Navier-Stokes equations were transformed to non-dimensional form. The reference variables are inflow variables (marked with the infinity subscript): the speed of sound $\hat{c}_\infty = 343\,\mathrm{ms}^{-1}$, density $\hat{\rho}_\infty = 1.225\,\mathrm{kg\,m}^{-3}$, temperature $\hat{T}_\infty = 293.15\,\mathrm{K}$, dynamic viscosity $\hat{\eta}_\infty = 18 \cdot 10^{-6}\,\mathrm{Pa{\cdot}s}$ and a reference length $\hat{L}_r = 0.02\,\mathrm{m}$. The system of Navier-Stokes equations is expressed in non-dimensional conservative form [3] as:

$$\frac{\partial \mathbf{W}}{\partial t} + \frac{\partial \mathbf{F}}{\partial x} + \frac{\partial \mathbf{G}}{\partial y} = \frac{1}{\mathrm{Re}} \left(\frac{\partial \mathbf{R}}{\partial x} + \frac{\partial \mathbf{S}}{\partial y} \right). \tag{1}$$

$\mathbf{W}$ is the vector of conservative variables $\mathbf{W} = [\rho, \rho u, \rho v, e]^T$ where ρ denotes density, u and v are the components of the velocity vector and e is the total energy per unit volume. $\mathbf{F}$ and $\mathbf{G}$ are the vectors of inviscid fluxes and $\mathbf{R}$, $\mathbf{S}$ are the vectors of viscous fluxes. The static pressure p in $\mathbf{F}$ and $\mathbf{G}$ is expressed by the state equation in the form

$$p = (\kappa - 1) \left[e - \frac{1}{2} \rho \left(u^2 + v^2 \right) \right], \tag{2}$$

where $\kappa = 1.4$ is the ratio of specific heats.

General Reynolds number in (1) is computed from reference variables $\mathrm{Re} = \hat{\rho}_\infty \hat{c}_\infty \hat{L}_r / \hat{\eta}_\infty$. The non-dimensional dynamic viscosity in the dissipative terms is a function of temperature in the form $\eta = (T/T_\infty)^{3/4}$.

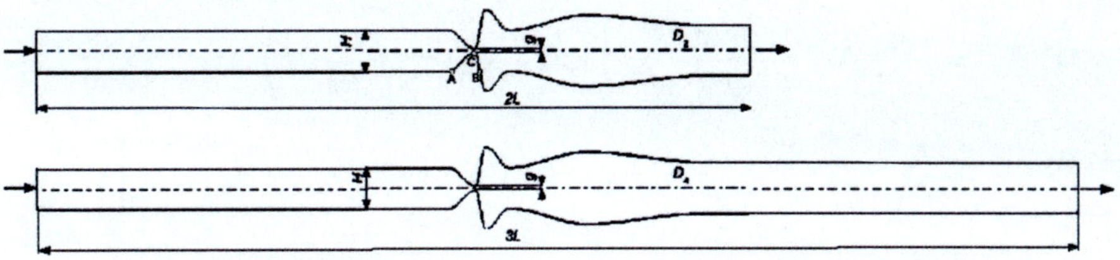

Fig. 1 The computational domains $D_{2,4}$. *Top*: D_2, $2L = 16$ (320 mm). *Bottom*: D_4, $3L = 24$ (480 mm). The width $H = 0.8$ (16 mm)

3 Computational Domain and Boundary Conditions

The bounded computational domains $D_{2,4}$ used for the numerical solution of flow field in the channel are shown in Fig. 1. The domains are identical in the inlet and middle part of the channel. The outlet part of the D_4 is prolonged. The upper and the lower boundaries are the channel walls. A part of the walls changes its shape between the points A and B according to a given function of time and axial coordinate. The gap width (in point C) was oscillating between the minimum $g_{min} = 0.4$ mm and maximum $g_{max} = 2.8$ mm.

The boundary conditions are considered in the following formulation:

1. Upstream conditions: $u_\infty = M_\infty$, $v_\infty = 0$, $\rho_\infty = 1$, p_∞ is extrapolated from D.
2. Downstream conditions: $p_2 = 1/\kappa$ and $(\rho, \rho u, \rho v)$ are extrapolated from D.
3. Flow on the wall: $(u, v) = (u_{wall}, v_{wall})$ and $\frac{\partial T}{\partial \mathbf{n}} = 0$ $(T = \kappa p/\rho)$.

The general Reynolds number in (1) is multiply with non-dimensional value M_∞ H represents kinematic viscosity scale and for computation of the real problem inlet Reynolds number $\mathrm{Re}_\infty = \hat{\rho}_\infty \hat{c}_\infty M_\infty H \hat{L}_r / \hat{\eta}_\infty$ is used.

4 Numerical Solution

The numerical solution uses finite volume method (FVM) in cell centered form on the grid of quadrilateral cells. In the time-changing domain, the integral form of FVM is derived using the ALE formulation. The ALE method defines homomorphic mapping of the reference domain $D_{t=0}$ at initial time $t = 0$ to a domain D_t at $t > 0$ [4].

The explicit predictor-corrector MacCormack (MC) scheme in the domain with a moving grid of quadrilateral cells is used. The scheme is the second order accurate in time and space [3]:

$$\mathbf{W}_{i,j}^{n+1/2} = \frac{\mu_{i,j}^{n}}{\mu_{i,j}^{n+1}}\mathbf{W}_{i,j}^{n} - \frac{\Delta t}{\mu_{i,j}^{n+1}}\sum_{k=1}^{4}\left[\left(\tilde{\mathbf{F}}_k^n - s_{1k}\mathbf{W}_k^n - \frac{1}{Re}\tilde{\mathbf{R}}_k^n\right)\Delta y_k - \left(\tilde{\mathbf{G}}_k^n - s_{2k}\mathbf{W}_k^n - \frac{1}{Re}\tilde{\mathbf{S}}_k^n\right)\Delta x_k\right],$$

$$\overline{\mathbf{W}}_{i,j}^{n+1} = \frac{\mu_{i,j}^{n}}{\mu_{i,j}^{n+1}}\frac{1}{2}\left(\mathbf{W}_{i,j}^{n} + \mathbf{W}_{i,j}^{n+1/2}\right) - \frac{\Delta t}{2\mu_{i,j}^{n+1}}\sum_{k=1}^{4}\left[\left(\tilde{\mathbf{F}}_k^{n+1/2} - s_{1k}\mathbf{W}_k^{n+1/2} - \frac{1}{Re}\tilde{\mathbf{R}}_k^{n+1/2}\right)\Delta y_k - \left(\tilde{\mathbf{G}}_k^{n+1/2} - s_{2k}\mathbf{W}_k^{n+1/2} - \frac{1}{Re}\tilde{\mathbf{S}}_k^{n+1/2}\right)\Delta x_k\right], \tag{3}$$

where $\Delta t = t^{n+1} - t^n$ is the time step, $\mu_{i,j} = \int\int_{D_{i,j}} dx dy$ is the volume of cell $D_{i,j}$, Δx and Δy are the steps of the grid in directions x and y, vector $\mathbf{s}_k = (s_1, s_2)_k$ represents the speed of edge k.

The physical fluxes $\mathbf{F}$, $\mathbf{G}$, $\mathbf{R}$, $\mathbf{S}$ on the edge k of the cell $D_{i,j}$ are replaced by numerical fluxes (marked with tilde) $\tilde{\mathbf{F}}$, $\tilde{\mathbf{G}}$, $\tilde{\mathbf{R}}$, $\tilde{\mathbf{S}}$ as approximations of the physical fluxes. The higher partial derivatives of velocity and temperature are approximated using dual volumes V'_k (see [3]).

To stabilize computation the Jameson artificial dissipation $AD(W_{i,j})^n$ is added to the MC scheme [5]. Since the artificial dissipation term is of third order, the overall accuracy of the scheme is of second order. The vector of conservative variables $\mathbf{W}$ can be computed at a new time level $\mathbf{W}_{i,j}^{n+1} = \overline{\mathbf{W}}_{i,j}^{n+1} + AD(W_{i,j})^n$.

The grid used in the channel has successive refinement cells near the wall (see [6]). The minimum cell size in y-direction is $\Delta y_{min} \approx 1/\sqrt{Re}$ to capture the boundary layer effects.

5 Numerical Results

The numerical results were obtained (using a specifically developed program) for the following input data: uniform inflow Mach number $M_\infty = 0.012$ ($\hat{u}_\infty = 4.116\,\mathrm{ms}^{-1}$), Reynolds number $\mathrm{Re}_\infty = 4,481$, atmospheric pressure $p_2 = 1/\kappa$ (102,942 Pa) at the outlet, and wall oscillation frequency $\hat{f} = 100$ Hz. The computational domain contained 750 $\times$ 100 cells in D_2 and 1,050 $\times$ 100 cells in D_4.

The computation has been carried out in two stages. First, a numerical solution is obtained, when the channel between points A and B has a rigid wall fixed in the middle position of the gap width. Then this solution is used as the initial condition for the unsteady simulation (see [6]).

Figure 2 shows the steady numerical solutions which are the initial conditions for unsteady computation of the flow field in $D_{2,4}$. The maximum Mach number computed in the domain D_2 was $M_{max} = 0.190$ (65 ms^{-1}) and in D_4 was $M_{max} = 0.226$ (77.5 ms^{-1}).

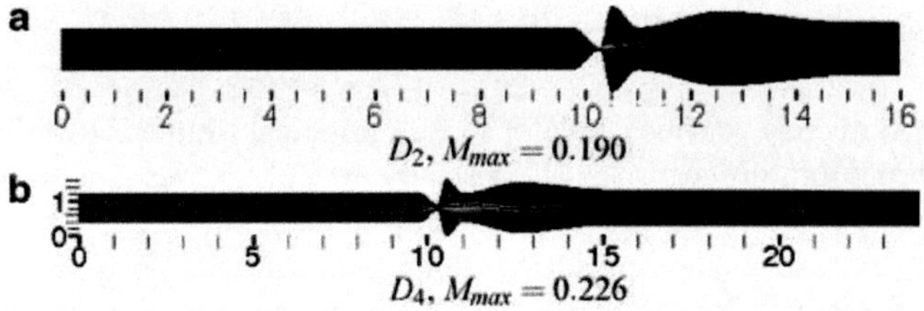

Fig. 2 The steady numerical solution computed in D_2 (**a**) and in D_4 (**b**) $-M_\infty = 0.012$, Re = 4,481, $p_2 = 1/\kappa$, D_4 – 1,050 × 100, D_2 – 750 × 100 cells, $g = 1.6$ mm

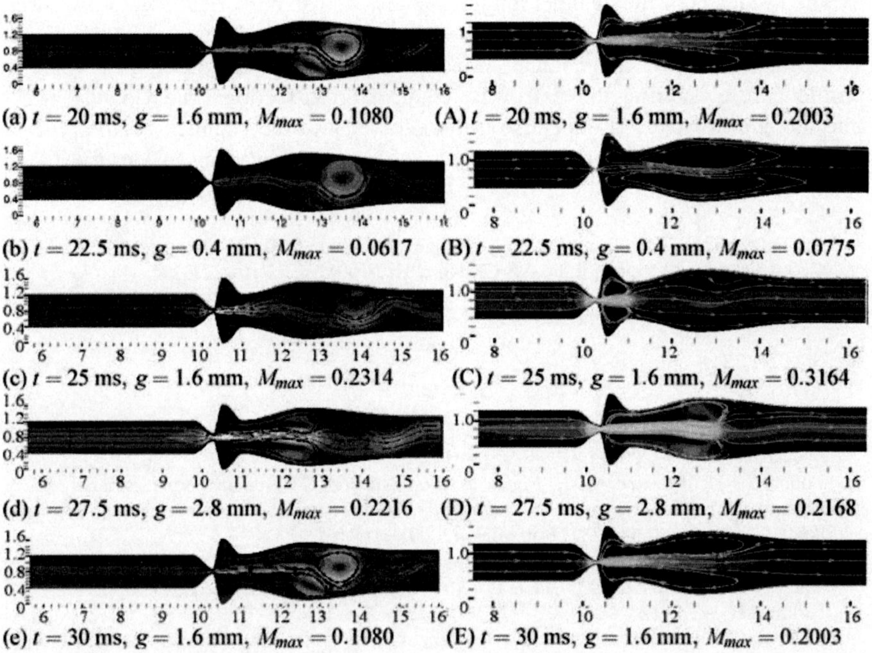

Fig. 3 The numerical simulation of the airflow during the third oscillation cycle computed in D_2 (*left column*, the inlet part of the channel is abridged) and in D_4 (*right column*, the inlet and outlet part of the channel is abridged) – $\hat{f}$ = 100 Hz, $M_\infty = 0.012$, Re = 4,481, $p_2 = 1/\kappa$, D_2 – 750 × 100, D_4 – 1,050 × 100 cells. Results are mapped by iso-lines of Mach number and by streamlines

The convergence to the steady state solution is supervised using the L_2 norm of momentum residuals (ρu). The residuals indicates the non-stationary solution which is caused by eddies separated in the un-movable glottal orifice and floating away.

The unsteady solution in domains D_2 and D_4 during one cycle of oscillation is shown in Fig. 3. The absolute maximum of Mach number in period is $M = 0.270$ ($92.6\,\mathrm{m \cdot s^{-1}}$) in D_2 (left column) at time $t = 26.12$ ms and $M = 0.328$ ($112\,\mathrm{m \cdot s^{-1}}$) in D_4 (right column) at $t = 25.625$ ms.

6 Discussion and Conclusions

Special program code has been developed for numerical simulation of the airflow in a channel with vibrating walls for 2D unsteady flow of viscous compressible fluid.

The numerical solution in the channel showed large vortex structures developed in the supraglottal space moving slowly downstream and decaying gradually. It was possible to detect a "Coandă phenomenon" in the computed flow field patterns. A similar generation of large-scale vortices, vortex convection and diffusion, jet flapping, and general flow patterns were experimentally obtained in physical models of the vocal folds by using PIV method in [7–9].

The influence of the channel length was tested. The numerical solution showed more streamlined flow pattern due to the prolonged outlet part of the channel. The prolongation of the channel inlet part resulted in the developed velocity profile before entering the narrowest channel cross-section. The Coandă effect (the direction of the jet) depends on the geometry of the channel, on the type of domain meshing and on the computational scheme, but is was not so sensitive to the coarseness of mesh.

Acknowledgements This contribution was partially supported by Research Plan MSM 6840770010 and by the grants of the GA CR No. 201/08/0012 and P101/11/0207.

References

1. Titze, I.R., The Myoelastic Aerodynamic Theory of Phonation. National Center for Voice and Speech, Iowa City (2006). ISBN 0-87414-122-2
2. Zöner, S., Kalteenbacher, M., Mattheus, W., Brücker, C.: Human phonation analysis by 3d aero-acoustic computation. In: Proceedings of the International Conference on Acoustic NAG/DAGA, pp. 1730–1732. Rotterdam (2009)
3. Fürst, J., Janda, M., Kozel, K.: Finite volume solution of 2D and 3D Euler and Navier-Stokes equations. In: Neustupa, J., Penel, P. (eds.) Mathematical Fluid Mechanics, pp. 173–194. Birkhaüser, Berlin (2001)
4. Honzátko, R., Horáček, J., Kozel, K.: Solution of inviscid incompressible flow over a vibrating profile. In: Beneš, M., Kimura, M., Nataki, T (eds.) COE Lecture Notes, vol. 3, pp. 26–32. Kyushu University, Fukuoka (2006). ISSN 1881-4042
5. Jameson, A., Schmidt, W.,Turkel, E.: Numerical solution of the Euler equations by the finite volume methods using Runge-Kutta time-stepping schemes. AIAA 1981–1259 (1981)
6. Punčochářová-Pořízková, P., Horáček, J., Kozel, K., Fürst, J.: Numerical simulation of unsteady compressible low Mach number flow in a channel. Eng. Mech. **17**(2), 83–97 (2010)
7. Šidlof, P.: Fluid-structure interaction in human vocal folds. PhD thesis, 88, Charles University in Prague, Faculty of Mathematics and Physics, Prague (2007)
8. Neubauer, J., Zhang, Z., Miraghaie, R., Berry, D.: Coherent structures of the near field flow in self-oscillating physical model of the vocal folds. J. Acoust. Soc. Am. **121**(2), 1102–1118 (2007)
9. Horáček, J., Šidlof, P., Uruba, V., Veselý, J., Radolf, V., Bula, V.: PIV measurement of flow-patterns in human vocal tract model. In: Proceedings of the International Conference on Acoustic NAG/DAGA 2009, pp. 1737–1740. Rotterdam (2009)
10. Punčochářová, P., Kozel, K., Horáček, J., Fürst, J.: An unsteady numerical solution of viscous compressible flows in a channel for low mach numbers. J. Comput. Appl. Math. Mech. **8**(2), 175–191 (2007). ISSN 1586-2070

Optimal Design of a Torsional Tuned Damper for Marine Diesel Engines Using Fluid-Structure Interaction Analysis

Young-Cheol Kim, D.H. Lee, T.Y. Chung, D.Y. Ham, and Y.B. Kim

Abstract A torsional tuned damper is usually used in order to reduce the torsional vibration of the crank shaft system in marine diesel engines. The damper consists of leaf springs, fluid chambers, fluid channels, and intermediate masses. The leaf springs provide the stiffening force to the shaft system, and the fluid chambers and channels give the damping force. In this paper, FSI (fluid-structure interaction) analysis by using FEM is carried out for the calculation of the stiffness and damping coefficients of the designed damper. The numerical calculation result about the equivalent damping coefficients is compared to the value obtained from a simple damping simulation model.

Keywords Torsional tuned damper • Viscous damping • Optimal design • FSI analysis • Diesel engine

1 Introduction

Torsional vibration of a crank shaft due to torque variation is inevitable in marine diesel engines. Such excessive torsional vibration can be usually reduced by installation of a torsional tuned damper. There are several types of torsional damper, which can be selected by engine speed. The low-speed large engines use usually a viscous fluid damper with leaf type springs. However, a viscous fluid damper with sleeve ring is used in the high-speed small engines. Torsional vibration analysis of

Y.-C. Kim (✉) • D.H. Lee • T.Y. Chung • D.Y. Ham
Korea Institute of Machinery and Materials, Daejeon, Korea
e-mail: kyc@kimm.re.kr; ldh@kimm.re.kr; tychung@kimm.re.kr; dyham@kimm.re.kr

Y.B. Kim
Se Young Industry Co., LTD, Seoul, Korea
e-mail: seyoung2897@naver.com

J. Náprstek et al. (eds.), *Vibration Problems ICOVP 2011: The 10th International Conference on Vibration Problems*, Springer Proceedings in Physics 139,
DOI 10.1007/978-94-007-2069-5_110,

the crank shaft system should be given considerations in the early design stage, thereby identifying the significant resonance condition. The tuned torsional damper acts as a dynamic absorber with mass and spring.

Lee [1–3] compared the dynamic characteristics of a diesel engine propulsion shafting system with different dampers. Kim [4, 5] analyzed the static and dynamic characteristics of a damping flexible coupling with leaf type springs. Lee [6] proposed the procedure to get the optimal parameters of a spring-viscous damper in a diesel engine using the effective modal value.

This paper presents the static and dynamic characteristics of a viscous fluid damper with leaf type springs. The 1-way FSI (fluid-structure interaction) analysis by using FEM (finite element method) is carried out for the calculation of the stiffness and damping coefficients of the damper. The spring deformation by the external torsional vibration is firstly calculated in transient state, and then the fluid flow analysis is carried out. The numerical calculation result about the equivalent damping coefficient is compared to the value obtained from a simple damping simulation model.

2 Torsional Tuned Damper

2.1 Structure of the Damper

Figure 1 shows the structure of the torsional tuned damper which is attached to the end of the propulsion shaft for a diesel engine system. This damper behaves as a dynamic absorber with mass, spring and damping elements. The damper consists of the 24 pairs of leaf springs, the fluid chambers, the fluid channels, and the intermediate masses. The material of the springs is a kind of stainless steel. The leaf springs provide the tuned stiffening force to the shaft system, and the fluid chambers

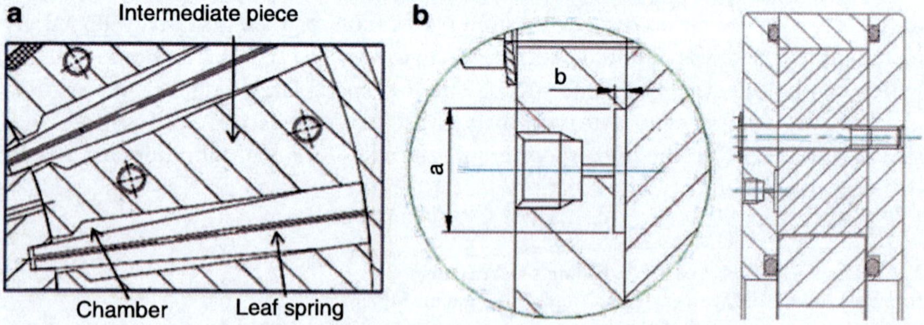

Fig. 1 A torsional tuned damper; (**a**) the cross section of damper and (**b**) a fluid channel on a cover for damping force

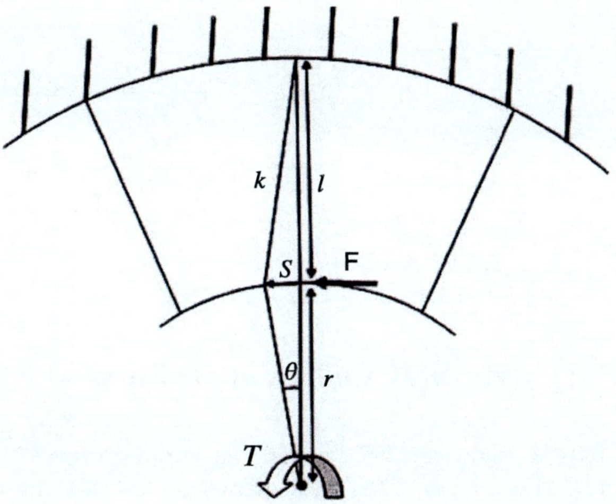

Fig. 2 The relationship between a torque and a displacement in the torsional damper

and channels give the damping force to the system. If the external forces are exerted to the damper, the springs are deformed, and the fluid chambers between the springs become narrow. Therefore the fluid flow is occurred through the channels on the cover, and the vibratory engine system can get the damping forces from the damper.

2.2 *Specification of the Optimal Designed Damper*

The design specification of a damper can be get from the natural frequency analysis for engine propulsion for a diesel engine system. The design target of the damper obtained by the optimization method of [6] is proposed that the moment of inertia is 13 kg-m^2, the damping coefficient is 2,300 Nms/rad, and the stiffness coefficient is 2 MN/rad. The amplitude of torsional harmonic vibratory torque obtained from engine vibration analysis is 10 kNm. Then, the force exerted on the end point of a leaf spring can be calculated asthe following (Fig. 2).

$$F = \frac{T}{nr} = \frac{10^4}{24 \times 0.22} = 1893\,\mathrm{N} \tag{1}$$

Therefore, the linear stiffness coefficient of a pair of leaf spring, the deflection at the end point of a spring, and the linear damping coefficient through a fluid channel on the cover can be calculated as the followings.

$$k_s = \frac{k_\theta}{nr^2} = \frac{2 \times 10^6}{24 \times 0.22^2} = 1.72\,\mathrm{MN/m} \tag{2}$$

$$\delta = \frac{F}{k_s} = \frac{1893}{1.72 \times 10^6} = 1.1\,\text{mm} \tag{3}$$

$$c_s = \frac{c_\theta}{nr^2} = \frac{2300}{24 \times 0.245^2} = 1600\,\text{Ns/m} \tag{4}$$

3 FSI Analysis of the Damper

3.1 *Structural Analysis of the Damper*

Figure 3 shows the FEM model of a pair of springs of damper using ANSYS module [7]. The number of the mesh elements is 34,085, the vibratory frequency is 15 Hz, and the number of the transient calculation time step during one vibration cycle (0.067 s) is 60. As a calculation result, it is verified that the maximum deformation at the end point of the spring is 1.04 mm, which is close to the design target. Figure 4a represents the SN curve obtained from material fatigue test for the same material, stainless steel. From the FEM calculation, it is confirmed that the maximum stress is 528 MPa, and the damper has a sufficient safety margin and service life. Figure 4b shows the rate of gap change at the end point between a pair of springs, which is made by deflection of only one spring. It is observed that the frequency of the rate of gap change is twice of the external vibration frequency, and the two sudden rate of gap change is occurred during one vibration cycle. This fact gives the nonlinearity

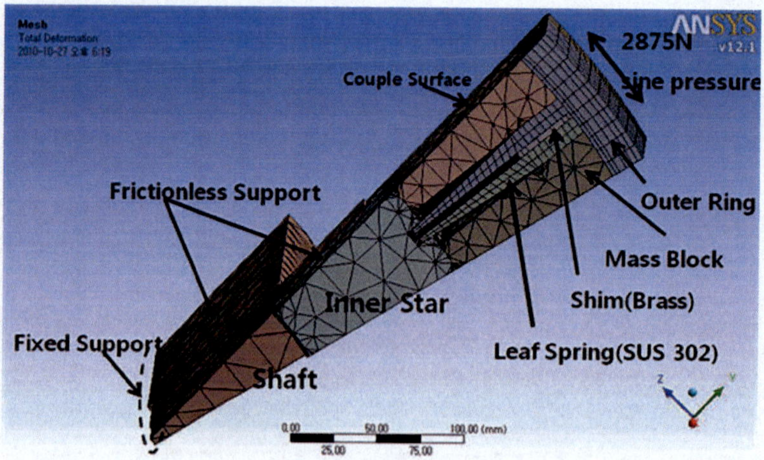

Fig. 3 The FEM modeling for the stiffness calculation of a pair of springs

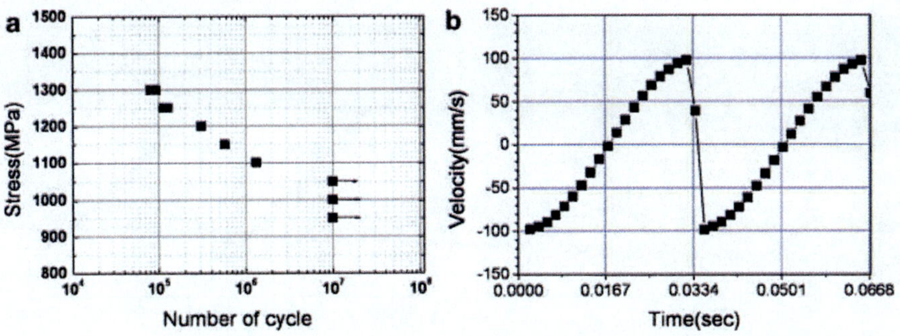

Fig. 4 (**a**) SN curve of the spring and (**b**) the rate of gap change between a pair of springs

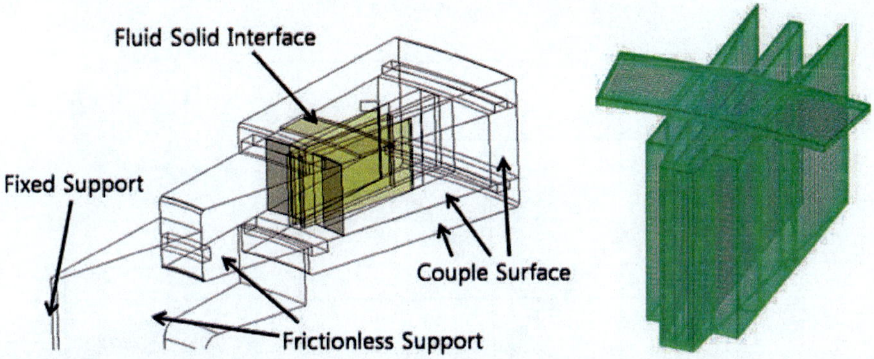

Fig. 5 (**a**) The FSI modeling and (**b**) the detailed moving mesh for the fluid

to the damper system. The vibration velocity at the end of a pair of springs is represented as the following.

$$v[\mathrm{mm/s}] = 82.0 \times \cos(30\pi t) \tag{5}$$

3.2 *Fluid Flow Analysis of the Damper*

Figure 5a shows FSI modeling using the CFX module [7]. If the structural deformation information is linked to the CFX module, then the fluid velocity and pressure fields can be calculated during one vibration cycle (0.067 s). Figure 5b shows the moving mesh generated by ANSYS ICEM-CFD. The number of the mesh elements is 610,000, and the ISO grade of the oil viscosity is 220. As a calculation result, Fig. 6a represents the flow velocity vector in the rectangular channel on the cover. It can be observed that the fluid flows to the largest neighboring gap through the channel. Figure 6b represents the flow resistance damping force due to

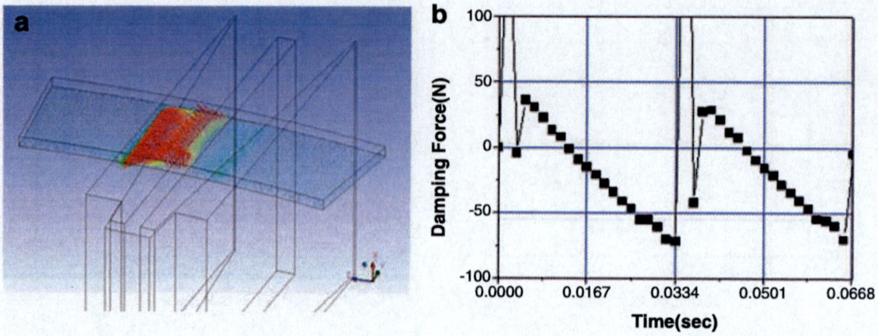

Fig. 6 (**a**) Flow velocity vector in an oil channel and (**b**) the calculated damping force

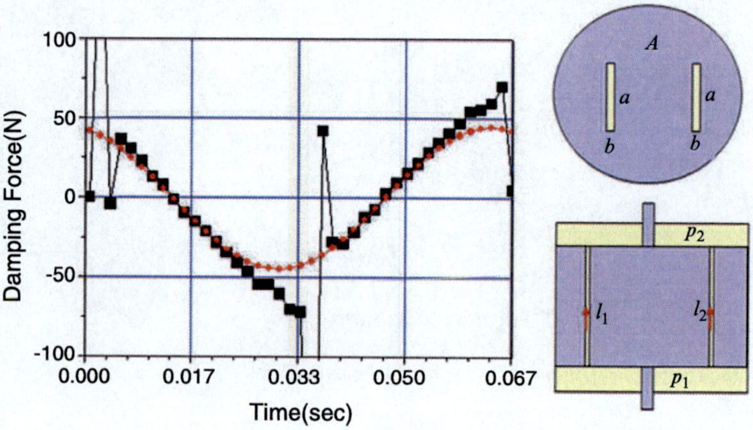

Fig. 7 (**a**) The approximated damping force obtained from FSI analysis and (**b**) a equivalent simple dashpot model

the external vibration. It is observed that the frequency of the damping force is twice of the external vibration frequency, and the sudden change in the damping force is occurred by the periods.

3.3 Damping Coefficient

The damping force is approximated to the cosine curve such as Fig. 7a to get the equivalent damping coefficient as the following.

$$F_d[\mathrm{N}] = 44.5 \times \cos(30\pi t + 0.28) \tag{6}$$

So the approximated damping coefficient can be obtained from the ratio between the external vibration velocity and damping force as the following.

$$c_{eq} = \frac{F_d}{v} = \frac{44.5}{0.082} = 542\,\mathrm{Ns/m} \tag{7}$$

The equivalent simple damping dashpot model is proposed to confirm the availability of the damping coefficient obtained using 1-way FSI technique. Force equilibrium and continuity equation in the dashpot [8] can be derived as the followings.

$$q = \frac{ab^3}{12Al\eta}F = AU \tag{8}$$

where U is velocity of the dashpot. The simple model equation for the damping coefficient in a rectangular channel is as the followings.

$$c = \frac{12A^2l\eta}{ab^3} \tag{9}$$

Therefore, the damping coefficient of the equivalent dashpot model with two different rectangular channels is as the followings.

$$c_{eq} = \frac{c_1c_2}{c_1 + c_2} = 448\,\mathrm{Ns/m} \tag{10}$$

The result for the equivalent dashpot model has a good agreement with 17% difference from FSI result. If the original height of the rectangular channel is changed from 2 to 1.3 mm, then the design target can be satisfied.

4 Conclusions

The FSI simulation of the torsional damper is successfully carried out to get the stiffness and damping coefficients. The validity of the method is confirmed by comparing the results obtained from the equivalent simple dashpot model.

References

1. Lee, D.C., Lee, B.W., Park, Y.N., Park, B.H.: A study on the dynamic characteristics and performance of Geislinger type torsional vibration damper for two stroke, low-speed diesel engine. J. Korean Soc. Mar. Eng. **16**(5), 329–340 (1992)
2. Lee, D.C.: Improvement of dynamic characteristics of torsion on the marine propulsion shafting system with elastic rubber coupling. J. Korean Soc. Noise Vib. Eng. **13**(12), 923–929 (2003)

3. Lee, D.C., Barro, R.D., Kim, S.H.: A study on the torsional vibration of super large two stroke low speed engines with tuning damper. J. Korean Soc. Noise Vib. Eng. **19**(1), 64–75 (2009)
4. Kim, J.S., Jei, Y.G., Jeong, J.H., Jeon, H.J.: A theoretical study on the dynamic characteristics of damping flexible coupling(I). J. Korean Soc. Mar. Eng. **18**(1), 11–22 (1994)
5. Kim, J.S., Jei, Y.G., Jeong, J.H., Jeon, H.J.: A experimental study on the dynamic characteristics of damping flexible coupling(II). J. Korean Soc. Mar. Eng. **18**(1), 23–31 (1994)
6. Lee, D.H., Chung, T.Y., Kim, Y.C., Shin, Y.H.: Study on the parameter decision of spring-viscous dampers for torsional vibration reduction of diesel engine shafting system. J. Korean Soc. Noise Vib. Eng. **20**(12), 1168–1175 (2010)
7. http://ansys.com
8. Kim, Y.C., Lee, D.H., Chung, T.Y., Kim, Y.B.: Analysis of the stiffness and damping coefficients of a torsional vibration damper. In: Proceeding of the Korean Society of Noise and Vibration Engineering Twentieth Anniversary Spring Conference, pp. 215–216 (2010)

Author Index

J. Náprstek et al. (eds.), , Springer Proceedings in Physics 139,
DOI 10.1007/978-94-007-2069-5, © Springer Science+Business Media B.V. 2011

L